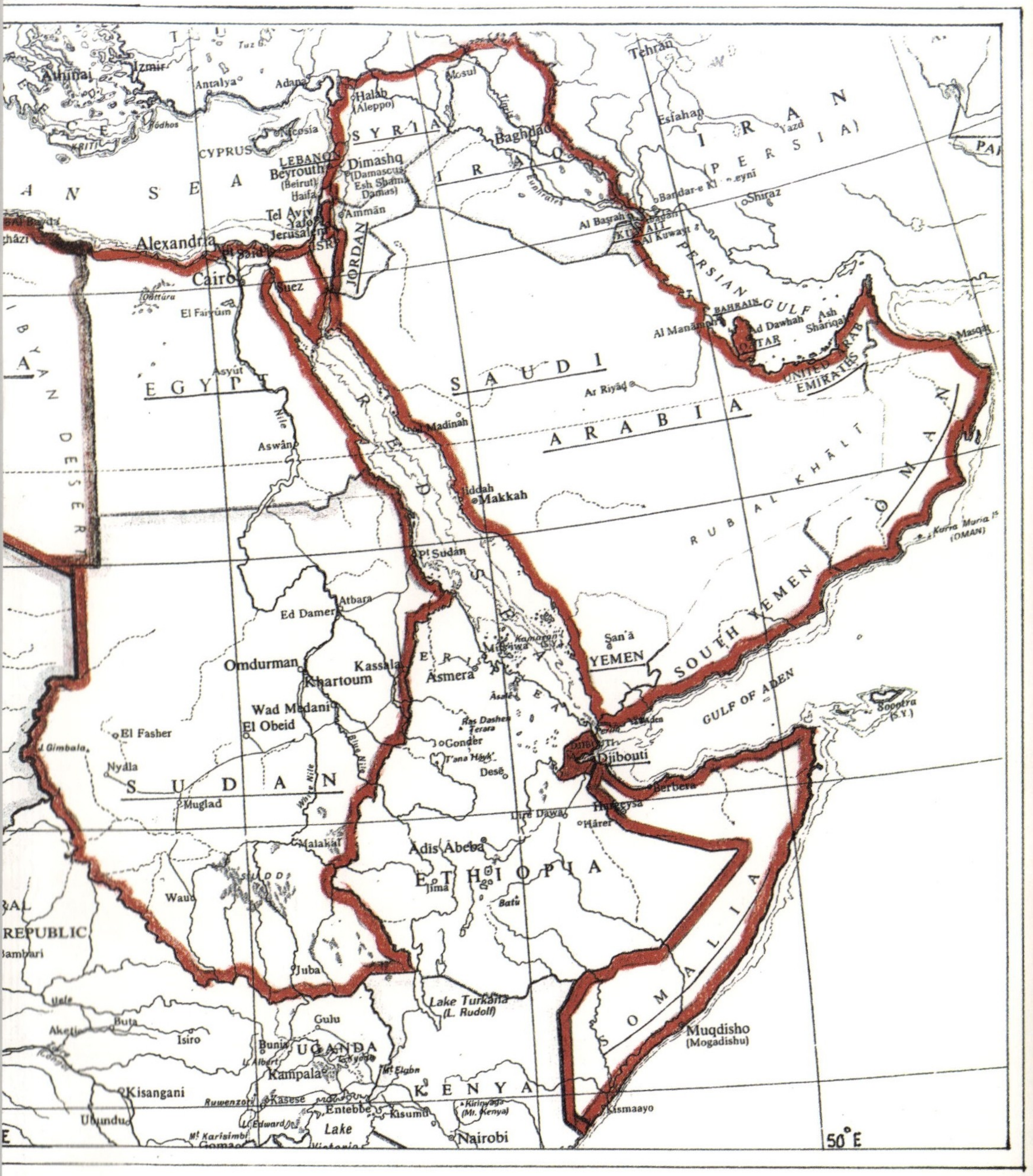
Athinai
Izmir
Antalya
Adana
Rodhos
CYPRUS
Nicosia
Halab (Aleppo)
Mosul
SYRIA
LEBANON
Beyrouth (Beirut)
Haifa
Dimashq (Damascus Esh Sham Damas)
Tel Aviv-Yafo
Jerusalem
Ammān
JORDAN
Baghdad
IRAQ
Tehran
IRAN (PERSIA)
Esfahan
Yazd
Shiraz
Al Basrah
KUWAIT
Al Kuwayt
PERSIAN GULF
BAHRAIN
Al Manamah
QATAR
Ad Dawhah
Ash Shariqah
UNITED ARAB EMIRATES
Masqat
Alexandria
Pt Said
Cairo
Suez
El Faiyûm
Asyût
EGYPT
Aswân
Nile
SAUDI ARABIA
Ar Riyād
Al Madinah
Jiddah
Makkah
RED SEA
RUB AL KHĀLĪ
OMAN
Kuria Muria Is. (OMAN)
SOUTH YEMEN
YEMEN
San'ā
GULF OF ADEN
Socotra (S.Y.)
Pt Sudan
Atbara
Ed Damer
Omdurman
Khartoum
Kassala
Asmera
ERITREA
Wad Medani
El Obeid
El Fasher
Nyala
SUDAN
Muglad
White Nile
Blue Nile
Malakal
Wau
Juba
Ras Dashen Terara
Gonder
T'ana Hāyk'
Desē
DJIBOUTI
Djibouti
Berbera
Hargeysa
Hārer
Dire Dawa
Ādis Ābeba
ETHIOPIA
Jīma
Batu
Lake Turkana (L. Rudolf)
SOMALIA
Muqdisho (Mogadishu)
Kismaayo
Gulu
UGANDA
Bunia
Kampala
Kasese
Entebbe
Kisumu
KENYA
Kirinyaga (Mt. Kenya)
Nairobi
Lake Victoria
Isiro
Buta
Kisangani
REPUBLIC
Bambari
LIBYAN DESERT
50° E

WATER RESOURCES AND HYDROMETEOROLOGY OF THE ARAB REGION

by

MAMDOUH SHAHIN

Water Resources Engineering Consultant
Formerly Professor Cairo University, Giza, Egypt
and
IHE-Delft, The Netherlands

Springer

A C.I.P. Catalogue record for this book is available from the Library of Congress.

ISBN-10 1-4020-4577-8 (HB)
ISBN-10 1-4020-5414-9 (e-book)
ISBN-13 978-1-4020-4577-6 (HB)
ISBN-13 978-1-4020-5414-3 (e-book)

Published by Springer,
P.O. Box 17, 3300 AA Dordrecht, The Netherlands.

www.springer.com

Printed on acid-free paper

Inner cover image

Map of the Arab Region. Source: The Times Concise Atlas of the World pp. 82–83.

To

THE PEOPLE OF THE ARAB REGION

The author dedicates this book

Mamdouh Shahin
Voorburg, The Netherlands June 2006

FOREWORD

The great importance of detailed data and their use in generating interdisciplinary scientific knowledge on the development and utilisation of the water systems of the Arab region, with its widespread water stress and desert areas, is well established. With increasing consumption of water in the region backed up by the growing population, the per capita availability of freshwater in the coming years is expected to cause severe scarcity, unless some fundamental and anticipatory changes are introduced in the management of their water systems. Towards that end, such a rich book will prove to be a great help. The book is the last one written by the author and the depth of data presentation and analysis clearly indicates the professional maturity with which it has been written. Accordingly, the book is the final expression of the lifelong work and scholarship of Professor Mamdouh Shahin.

The Arab Region has been the home of some of the earliest human civilisations and their history is also a continuum of human endeavours towards ensuring the access to water over an expanding space and time. The presentation of the important elements of history of the region as related to water will prove to be a rich reference. At present this region is faced with some critical challenges in the development and utilisation of water systems. There are several books on the water situation in the region. The present publication adds a lot of value to the store of that knowledge on the water resources of the region. However, unlike many technical presentations this book does not directly start with the stream flow data and hydrological features of the Arab Region. Not only have chapters with detailed information on the precipitation and physiographic features of the region been included in the beginning, the rich history of the human civilisation in the Arab Region and its links with water have been identified as the starting point for the book. Keeping this interdisciplinary commitment, in the last chapter the author has addressed the critical approaches alleviating water scarcity in the region, in the absence of which, sensitive issue of sharing common water resources in the region is surely expected to affect the peace and socio-political stability both at the country and regional levels. In between the book, there are very useful chapters which describe the water system of the region with painstaking details. The information on the groundwater resources of the region is of particular value.

Socio-economic stability at the national and regional levels is closely related to water. For example, increasing the pumping head of groundwater or surface water for irrigation leads to the rise of the price of water, and hence, of crop production. The situation becomes very difficult when the pumped aquifer is located along a seacoast. At the regional level conflicts between neighbouring countries could result and water sharing is governed then by diplomatic or even military strength of the concerned states. Reduction in the availability of water for irrigation would lead to malnutrition and increased dependence on imported food. Similarly, when water is in premium, the quality of water used may drop, thus affecting the health condition in general and hygienic situation in particular. Over-drafting of groundwater from coastal aquifers might result in groundwater pollution, which can be too difficult to repair in a short period of time. This means that the situation requires a more efficient use of the available water.

The author has added a new dimension to the picture of water systems in the Arab Region by not limiting to the traditional ideas of water availability. By drawing attention to several non-traditional sources of freshwater, like by desalinisation, and by stressing on more efficient and multiple use of water before being drained out, the book opens important areas in water systems management. As the author points out, research and development in these two crucial directions would find encouragement from the book and such changes are most welcome in the region. I am sure that not only scholars from the Arab region working on water systems, but also a wide number of students of water systems all over the world will benefit from this rich and worthy publication. The book will prove to be really useful in addressing the water stress and reduce the possibility of the scarcity of this vital natural resource to continue to threaten the much-deserved peace in the Arab Region.

Jayanta Bandyopadhyay
Professor and head
Centre for Development and Environment Policy
Indian Institute of Management, Kolkata (India)

February 2006

PREFACE

Back in The Netherlands from a trip to Oman, the United Arab Emirates and Al-Bahrain in the spring of 1995, the author became fully convinced that the time was ripe to draft the manuscript of a technical publication addressing the issue of aridity and scarcity of water in the Arab Region. That thought was strongly motivated by the author's accumulated experience during many years of full-time and part-time working in Egypt, frequent consultancy missions to Iraq, Yemen, the Sudan, Mauritania, Morocco and Tunisia, added to his consultative activities to UNESCO/ROSTAS (Regional Office for Science and Technology in the Arab States) in relation to water resources in the Arab Region.

By mid 1995 the time that was still left to the author, before his retirement from his post with the then International Institute for Hydraulic, Infra-structural and Environmental Engineering (IHE), Delft, The Netherlands, was limited. Accordingly, it was not possible to embark on the scarcity of water resources in the Arab Region in the length and depth they deserve. Shortly after the publication of "Hydrology and Scarcity of Water Resources in the Arab Region" in 1996 as IHE Monograph I, by Balkema Publishers, 1996, the author sought advice of some of the scientists and water resources engineers who were busy with the same issue. The agreement was unanimous that an extended and more elaborate version of the book covering a wider range of items would be more useful than its predecessor.

The former Kluwer Academic Publishers (KAP), The Netherlands, before as well as after its merger with Springer Verlag have shown great interest in having the new manuscript to become one of the publications belonging to their water resources library.

As the title implies, the book is prepared with the aim of presenting to the user a comprehensive, fairly detailed text on Water Resources of the Arab Region based on the relevant meteorological and hydrological data, which are available at the national and international agencies involved in these matters.

Before dwelling upon the preface properly, the subject matter of this publication is confined solely to the Arab Region (AR). As such, it does not deal specifically with any of the regions: Middle East (ME), Arab Middle East (AME) or the Middle East and North Africa (MENA) Regions. This does not eliminate the fact that there is a strong geographical overlapping between all these regions. The Arab Region as referred to in this text can be defined as a region comprising twenty-two states, all members of the Arab League and their official language is Arabic. Despite the fact that Turkey and Ethiopia are two major contributors to streamflow in certain areas in the region, and Israel is sharing with its neighbouring Arab States the available surface and groundwater resources, they are, by virtue of this definition, not included in this treatise.

The book is composed of twelve chapters starting with an historical introduction, which reviews the ancient civilisations in the Arab Region, and the extent and type of hydraulic works the ancient settlers used to secure accessibility to water, mitigate the harmful effects of floods and to become accustomed to sailing and fishing in the seas and rivers. The region is the birthplace of three holy religions believed by its inhabitants. The chapter further highlights some of the water-related issues which the holy books contain, such as water formation and occurrence, distribution, use and management. Equally, there is mention of meteorological matters of interest to mankind. Last, but not least, they deal with instructions, restrictions and pieces of advice, which constitute the foundation of what is currently called water law.

Chapter 2 presents some of the physiographic features and geologic settings of the three sub-regions: western (Mauritania, Morocco, Algeria, Tunisia and Libya), central (Egypt, Sudan, Djibouti, Somalia and Comoros) and the eastern (Syria, Lebanon, Palestine, Jordan, Iraq, Kuwait, Bahrain, Qatar, Saudi Arabia, Oman, United Arab Emirates and Yemen). The physiography comprises land level and form, and slope and topography including plains, plateaus, mountains and deserts. It also includes sabkhas, shallow lakes (chotts), other lakes, wadis and rivers. The geological sequence and properties of the water-bearing formations are highlighted. Special attention is given to those formations of importance to groundwater abstraction.

The climate of the region is presented in Chapter 3. Climatic factors are: air circulation and wind, temperature, air humidity, daytime and sunshine hours, radiation, cloudiness, precipitation and evaporation and evapotranspiration. Using these factors, the climate of the various parts of the region has been classified as hyper-arid, arid, semi-arid, etc. In view of their immediate influence on the yields of water resources, Chapters 4 and 5 discuss the results of the analyses of precipitation data, evaporation from water surfaces and evapotranspiration from irrigated lands in the region. Readers interested in these data and their sources can refer to Appendix I installed on the CD Rom at the end of the book.

Chapters 6 and 7 discuss the principles underlying the occurrence of surface runoff and streamflow. The former deals with measurement, analysis and modelling of river flow. The rivers considered are: the Nile in the Sudan and Egypt, the Shebelle and Juba in Somalia, the Tigris and Euphrates in Iraq and Syria, the Jordan River in Jordan, and the Sebou and Zeroud in Morocco and Tunisia, respectively. Chapter 7 is devoted to wadis and wadi flow in several parts of the region, especially in countries of the Arabian Peninsula such as Saudi Arabia, Oman and United Arab Emirates, and in the western sub-region such as Mauritania, Morocco, Algeria and

Tunisia. In this chapter too, some of the principles underlying the occurrence of wadi flow, extreme floods and losses are discussed and a number of modelling studies summarised. Appendix II contains hydrologic data related to perennial rivers and wadis while Appendix III contains water quality data. Chapter 8 is closely related to Chapters 6 and 7. It reviews two major items; namely, erosion and sedimentation in drainage basins and storage reservoirs. Sample results obtained from experimental basins and measurements taken over large areas are discussed. Likewise, sediment yield of drainage basins, and sediment discharge of flowing streams and their concentration and transport characteristics are presented using case studies from some countries of the region. Certain aspects of reservoir siltation such as measurement and monitoring, and application of conceptual and mathematical models are discussed in Chapter 8 and case studies from Jordan, Yemen and Algeria as well as other countries are presented.

Chapter 9 provides the reader with a fairly detailed account of the groundwater resources in the Arab Region. After a brief description of the groundwater basins, the resources so far known are dealt with country-wise. Similar to surface water resources, quantitative as well as qualitative parameters characterising some of the groundwater resources are listed in tabular forms in Appendixes II and III, respectively. It is customary to call streamflow (Chapters 6 and 7) and groundwater flow (Chapter 9) by conventional water resources. Chapter 10 reviews the non-conventional (new) water resources currently available in the Arab Region. The non-conventional resources principally comprise the reuse of municipal and agricultural drainage waters and the desalination of brackish and seawater. Rain harvesting, water transfer from one location to another or from country to country, virtual water embedded in imported articles such as grains and meat, and abstraction of deep or fossil groundwater (hundreds of meters below ground surface) are less widely used in comparison with wastewater reuse and desalination of sea or brackish water. The chapter briefly reviews some of the technologies used to make these waters accessible and safe for drinking, domestic and agricultural purposes. After having presented a fairly large amount of data, analysis, discussion and conclusions concerning the conventional and non-conventional water resources, surface and subsurface, Chapter 11 limits itself solely to water storage. Storage dams are built on rivers and wadis and in wadi basins in the region with the aim of reducing the amounts of water lost by preventing the streams from flowing to a neighbouring sea or gulf, flow regulation and protection against severe floods. The use of retention dams built on the main wadi channels or in wadi basins have been dealt with at a fair length in this chapter. Reservoirs formed by retention dams are a tradition typifying the Arabian Peninsula and North Africa in order to artificially recharge the groundwater formations. This is true, especially where yield and/or quality of water is (are) deteriorating because of the excessive abstraction and encroachment of saline water. Some of the new technologies including management strategies of reservoir construction and operation are given coupled with case studies from some Arab countries.

Chapter 12, the last chapter of the book, lists estimates for the near future concerning the population growth, the demands of the various sectors and the availabilities of water to meet those demands. It also discusses the measure of water scarcity or water poverty. Other than the imbalance caused by the steadily growing difference between the demand and available water, issues such as conflicts between the states in the region, and between states in the region and others outside the region aggravate the adverse impacts of scarcity. Such impacts include increasing lack of adequate quantities of water of suitable quality to cope with drinking, hygienic and domestic requirements. The same, though to different standards, applies to water needed for agricultural activities. Besides, there is the fear that the water challenge might seriously threaten the fragile stability of the region. Effective bilateral and multilateral agreements between the riparian countries of the water resources are discussed.

The preparation of the present work became possible only with the support of many individuals and institutions, to all of whom the author is deeply indebted. Despite the fact that it is not possible to list all their names, some cannot be missed.

The author, while gathering the meteorological and hydrological data and other indispensable material concerning the region, has been sincerely assisted by the staff members of the libraries of the Royal Dutch Meteorological Institute (KNMI), Agricultural University of Wageningen, IHE and Technical University, Delft, The Netherlands, Dr T. Maurer of the Global Runoff Data Centre (GRDC) and the Deutscher Wetterdienst Germany; Unesco, Paris; Unesco Office for the Arab States (ROSTAS), Egypt; Dr B. Jovkov, Frankfurt, Germany; and Eng. Ms. M. Boukharouba, Algeria.

Dr R. Herschy, Reading, and Dr J. Sutcliffe, Oxfordshire, both of international Hydrology Consultants, U.K., have kindly undertaken the review of the draft manuscript and provided the author with valuable comments and views. In addition to reviewing the draft manuscript, Professor J. Bandyopadhyay, Head of the Centre for Development and Environment Policy, Kolkata, India, has gladly written a foreword to the book. The author wishes to seize the opportunity of drafting this preface to express his deepest sense of gratitude to these outstanding colleagues for the assistance they kindly provided.

Last, but certainly not in the least, the author would like to express his thanks to Mrs. P. van Steenbergen, Senior Publishing Editor, Geosciences Division Springer, The Netherlands and her assistants Ms M. Jonckheere and Mrs. R. Balk both of Springer Publishers, The Netherlands, for their genuine help and endless patience during the years of book preparation and production of the book. Thanks are also due to Mr. E. el-Ghazaly for his patience and skill in solving computer problems, which appeared every now and then, and to Mrs. S. Jones for the editorial corrections of the script.

The present book "Water Resources and Hydrometeorology of the Arab region" is the third in a series of books written by the author concerning water resources in certain parts of the world. Its predecessors are "Hydrology and Water Resources of Africa" published by the formerly Kluwer Academic Publishers (KAP), 2002, and

"Hydrology of the Nile Basin" published by Elsevier Science Publishers, 1985. As the author's professional life is nearing its end, this book will be the last one to be written. It is sincerely hoped that the series, which covers a total geographical area of about 35 million square kilometers will remain a useful reference to all individuals and agencies involved in water resources science and technology in the regions they cover. As life has to go on, it is equally hoped that the new generations will be able to expand the author's works in more depth and extent.

Mamdouh Shahin
June 2006

Voorburg (Z.H.)
The Netherlands

TABLE OF CONTENTS

CHAPTER 1

HISTORICAL INTRODUCTION

1.1. ANCIENT CIVILISATIONS

The historical introduction presented in this chapter applies to the Arab Region. The region as dealt with in this text is an almost continuous stretch of land extending from northwest Africa in the west to the Gulf in the east. The surface area of the region is about 14 million square kilometres; bounded by the North Atlantic Ocean in the west; Southern and Eastern Mediterranean in the north and east respectively; and the Gulf of Aden and the Indian Ocean in the far southeast. The Red Sea is the water body separating the central part of the region from its eastern flank. The population of the region, which currently surpasses 300 million inhabitants, is expected to reach or exceed 600 million by the year 2030.

At present, the Arab Region (AR) comprises 21 sovereign states, all members of the United Nations Organisation (UN), plus one authority (Palestinian Authority), observant member of the UN, still awaiting the declaration of its statehood. The 22 countries are member states of the League of Arab Nations, and all of them have Arabic as their official language. By virtue of this definition, the Arab Region (AR) is somewhat different from the Middle East (ME), Arab Middle East (AME), or the Middle East and North Africa (MENA) Region.

It is an established fact that most of the ancient civilisations developed in the Middle East along the shores of the Euphrates, Tigris and Nile Rivers as well as along the southern and eastern coasts of the Mediterranean Sea. Civilisations have also flourished in some areas mostly lacking perennial flowing streams such as Yemen and certain parts of Al-Maghreb Al-Arabi (the western flank of the Arabic Region).

Water played three simple roles in primitive, evolving civilisations: it was needed for drinking and domestic purposes; it had to be avoided as a natural hazard such as flooding and inundation; and it served as a fishing, sailing and hunting environment. Water resources technology, according to Yevjevich (1992), played an important role in ancient civilisations of the Babylonian, Egyptian, Hittite, Greek, Roman, Chinese and other empires. Over the past 9,000 years, communities in AR have developed sophisticated water management systems. Throughout the millennia precipitation has been the primary source of water for a large area; therefore, collecting and storing rainwater for drinking, washing, livestock, irrigation and agricultural purposes was of utmost importance. Some of the masonry aqueducts used for water supply to large ancient cities around the Mediterranean Sea stand as historical monuments and witnesses to the advanced technology of those times. The same holds for drainage of groundwater by galleries (e.g. qanāts) with access wells to supply water to human agglomerations and settlements.

1.2. HIGHLIGHTS ON THE OLD CIVILISATIONS IN THE ARAB REGION IN RELATION TO WATER RESOURCES PLANNING, UTILISATION AND MANAGEMENT

Two major civilisations began to flourish more than 5,000 years ago in the Middle East region. These were the civilisations of the Egyptians in the Nile Valley and the Sumerians along the Tigris and Euphrates in Mesopotamia (the land between two rivers). Other civilisations, though not as powerful or of dominant influence as those of the ancient Egyptians and the Sumerians, penetrated their roots in southern Saudi Arabia, Yemen and in North Africa.

1.2.1 The Egyptian Civilisation

Archaeological findings have shown that primitive man lived along the Nile River long before the dynastic history of Egypt's Pharaohs (rulers of Egypt) began. By the year 6000 B.C., the Nile Valley was the scene of a Neolithic Age, marked by such characteristic elements of settled life as agriculture, domestic animals and pottery. By 3500 B.C. many tribes living in the Nile Valley had coalesced into the kingdoms of Upper and Lower Egypt, ruled by pre-dynastic kings.

In about 3100 B.C. Upper and Lower Egypt were united under one ruler known as King Mena or Menes. Political union of the Nile Valley north of the First Cataract (just south of Aswân) was of particular significance as it brought use and management of the river under one authority (Nyrop et al., 1976). It has been claimed that Menes was the first to build banks along the Nile to confine its water. According to Biswas (1972) it is likely that King Scorpion around 3200 B.C. was the first Egyptian ruler to occupy himself with water resources management works.

Although the ancient Egyptians traded down the Red Sea and up the Nile beyond Khartoum, the present capital of the Sudan, the sources of the Nile were apparently unknown to them, and no records of the Nile Basin beyond Egypt can be found in the early history. Several theorems had been developed about the source of the Nile. As an example, Juba II, the King of Mauritania, had suggested one of those theorems around 20 A.D. It says that the source of the Nile lies in Western Mauritania-not far from the North Atlantic Ocean. From there it moves underground for some distance up to a certain lake in Mauritania. From that lake the Nile continues its underground lateral travel for a twenty days' journey to the source point Nigris-on the Sudanese-Ethiopian border-and further through Ethiopia in a northerly direction.

This matter remained largely ambiguous till the middle of the nineteenth century when exploratory missions searching for the Nile and its tributaries began. Shahin (2002) gave a brief account of most of those missions.

The unification of Upper and Lower Egypt by King Menes had served well for centralising and improving the control on the river. By 3000 B.C. the Egyptians had a developed agriculture as well as a complex town and city life. King Menes is also accredited with building Memphis as capital of his kingdom, damming the Nile above it and diverting the river water to a man-made channel between two hills. Later, he dug a lake and a canal to connect it to the river. By this way he created a moat around the capital Memphis to protect the kingdom from his enemies.

The Sadd el Kafarah (dam of the Pagans) was built between 2950 and 2750 B.C. near Ma'âdi[1], some 30 kilometres (km) south of what is presently known as Cairo-the capital of Egypt. According to Murray (1955), the Kafarah Dam was constructed of masonry without mortar. The dimensions of the dam were: 105 metres (m) long at the top, 80 m base width and 11.5 m height of crest above the lowest bed level of Wadi (valley) el-Garawi across which the dam was built. In view of the lack of sound engineering design of the dam, particularly the absence of a spillway, the life age of the dam was quite short.

The reign of Menes was succeeded by 30 dynasties. Even in the troubled times of the 9th, 10th and 11th dynasties, when the central power became weak, the independent princes continued the efforts to improve irrigation. Prince Kheti (2125 B.C.) praised an irrigation canal he built at the present location of the town of Asyût in Middle Egypt saying; "I brought a gift (of water) for the city, I substituted a channel of ten cubits (1 cubit is ~ 0.5 m) and excavated for it upon the arable land." He added "I equipped a gate for its head and supplied water in the highlands which had never seen water before. I caused the water of the Nile to flood over the ancient landmarks. Every neighbour was supplied with water and every citizen had Nile water up to his heart's desire."

Man-dug lakes were used during the Middle Kingdom of Egypt (2160–1790 BC) for controlling high flood-flows of the Nile by storing the excess water in them. Pharaoh Amenemhet III (1850–1800 B.C.), the most prestigious ruler of that period, is known to have ordered the excavation of Lake Moeris (presently known as Birket Qārûn (birket means lake or pool of water) at the northern edge of the Faiyûm Depression in Middle Egypt. During Nile floods, water was diverted to Lake Moeris by means of a canal. At the end of the high flow season, the stored water of the lake was returned to the Nile. By doing so, the lake, with an estimated storage capacity of 5×10^9 m^3, was made available for the next flooding season. The locations of the lake with respect to the Nile Valley, and the Faiyûm Depression are shown on Figure 1.

---

[1] Names of places and objects of interest are written here as they appear in the Times Concise Atlas of the World (80 edition). Names appearing in the sections dealing with religions are kept as given in accredited English translations of the Holy Books.

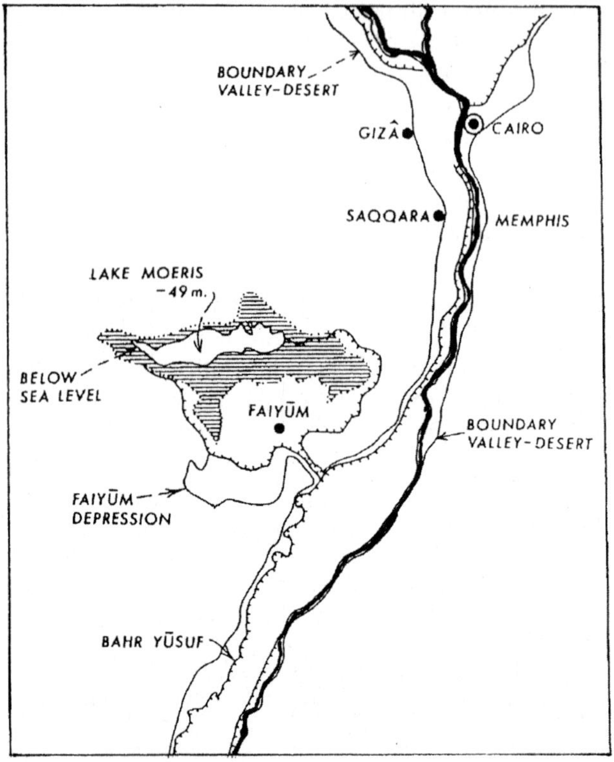

Figure 1. Part of the Nile Valley showing the Faiyûm Depression (from Biswas, 1972)

Almost 1,750 years after the kingdom of Pharaoh Menes had come to an end, the Egyptians built another dam, the Quatinah dam, during the reign of King Sethi I (1319–1304 B.C.) on the Orontes River (Nahr el-'Āsi) near Homs in Syria. The dam, which is still existing, is a rock fill dam 2 km long and more than 6 m high.

From the earliest historical period one finds a well organised 'River Hydrology and Irrigation Department' in the ancient Egyptian administration, which was probably independent of the then Ministry of Public Works. One often encounters the title 'Chief of Irrigation' among the notable officials in that time. An important task of this department was the observation of the rise and fall of the Nile. For this purpose gauges known as 'Nilometers' were built. Records of the Nile levels can be traced to earlier than 3000 B.C. The fragments that remained of the so-called Palermo stone show the highest marks of the Nile for every year. The kings, in their anxiety over the rise of the Nile, had a Nilometer built at Memphis, the then capital of Egypt. The second Nilometer was built at the Elephantine Island near the Cataract at Aswân, where the rise should be 12 ells (the el; zirā' in Arabic, is an old measure of length) above that at Memphis.

Each province in Egypt had its own central office known as the 'Water House'. The task of the chief of this office was to direct all the 'Dyke Masters', 'Chief of Canal Workmen', 'Inspectors of forced labour or Corvée', 'Observers of the Nilometers', 'Inundation Inspectors' and 'Operatives of Dykes and Dams' under his command. There were also special courts known as 'Water Tribunals' to deal with conflict on water and water-related matters. The forced labour (corvée) was employed in the period from April to June each year. It consisted of the digging and clearing of canals and ditches so as not to obstruct the flow of water.

The number and types of river gauges or Nilometers increased in the course of time. In the earliest type, high water levels were marked on cliffs on the banks of the river. The Nilometer at Semna Cataract in the northern part of the Sudan and close to the southern frontier of Egypt, discovered by Lepsius between 1842 and 1843 is just an example. The second type of Nilometers utilised flights of steps, which led down to the river. The third and most advanced type used conduits to lead the water from the river to a well or cistern where the levels were marked on the walls of the well or on a central pillar (Biswas, 1972).

It may be of interest to mention here that the skill of the ancient Egyptians to build irrigation canals had enabled them to build navigable canals too. To avoid the danger that might arise in the south of the country, King Mernere (2400 B.C.) excavated the first canal at Elephantine. This canal was later widened and deepened by Sesostris-III (1875 B.C.). Protected by fortresses and other protection works; the canal commanded the trade with Nubia. One should not forget that the Egyptian Kings had attempted at several periods to construct a canal connecting the Nile with the Red sea. Due to some engineering difficulties that canal was never able to serve for a long time.

The Hyksos (Persians), who invaded the Middle Kingdom of Egypt, seem to have neglected the irrigation canals. However, the empire builders of the New Kingdom (after 1580 B.C.) took an interest in irrigation matters.

Murray (1955) reported that by the time of the IV Dynasty, the Old Egyptians had become skilful at sinking shafts through soft rocks like limestone or sandstone. Two wells were sunk in sandstone on the road to quarries in the Nubian Desert at distances of 63 and 77 km from the Nile River. The estimated depth of these wells is over 70 m. Well sinking was also practised during the 11th Dynasty (2160–2000 B.C.) on the road to other quarries where 12 wells were claimed to have been dug. The same reference adds that the well, which still exists, is 32 m deep and is of great antiquity. One of the objects that still exists and demonstrates the skill of the ancient Egyptians in exploiting groundwater is the well-known Joseph's well (in Arabic Bīr Yūsuf) near Cairo. This well is nearly 100 m in depth and dates back to say 1700 B.C.

There are evidences that King Seti I and his successor Ramsis II, and later Ramsis IV, all in the period from 1313 B.C. to 1160 B.C., had dug several deep wells on the roads from the vicinity of the River Nile in the southern part of Upper Egypt to the locations of gold mines in the Eastern Desert. Additionally, Murray (1955) mentioned that the earliest inscription indicating to the installation of water well in an Egyptian Oasis (Dakhla Oasis) in the Western Desert dates back to around 905 B.C. There, the clay deposits overlying the upper face of the groundwater aquifer are 30–100 m thick.

One of the greatest achievements in the utilisation of groundwater in ancient times was the building of tunnels, commonly referred to as 'qanāts', which spread in the Middle and Near East around 700 BC. Shortly after 512 B.C., Darius I, King of Persia, sent Admiral Scylox to El-Khârga Oasis in Egypt. According to Butler (1933), Scylox introduced the Persian method of irrigation by means of underground conduits (qanāts) fed by water from the sandstone formations where it used to be collected in faults. A longitudinal profile and cross section of an unlined qanat are shown in Figure 2.

With the rise of the Hellenism around 600 B.C., it can be stated that the first Ptolemaic began his energetic work on the canals, and undertook extension of the arable land in Faiyûm by irrigation. Ptolemy I sponsored viticulture both in the Faiyûm and Delta, which meant that higher lands had to be irrigated throughout the whole year. The low lands south of Lake Moeris in the Faiyûm (Figure 1) suffered from defective drainage; hence they were preserved for the production of papyrus and fowling. Different types and forms of water lifting devices such as waterwheels had been developed and tried out. Generally, the improvement of irrigation, drainage of marshes and careful irrigation of sandy soil and stony borders of the desert contributed to the Ptolemaic extension of the Pharaonic system. There was extensive tree planting on the river embankments. The extension of Hellenistic irrigation was due to the introduction of goof iron tools in irrigation and agriculture, which was almost a revolution.

The Greco-Roman period did not witness except a meagre effort to collect quantitative data about the level of the Nile. The ancient gauging sites, however, remained in operation until the Arab conquest of Egypt in 641 A.D. Additionally, the latest Ptolemies seemingly neglected irrigation. The pressure of taxes led to the scarcity of

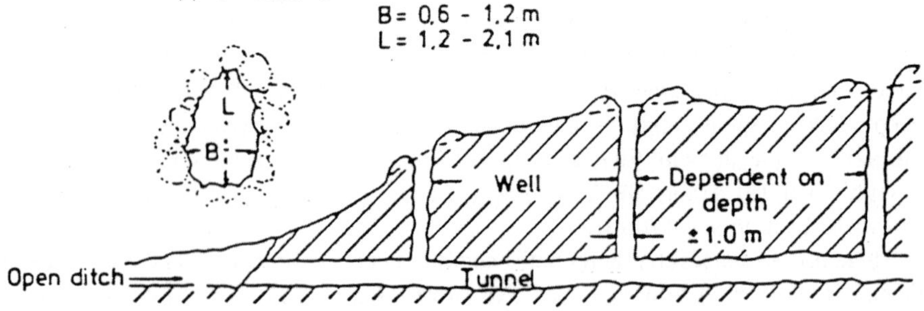

Figure 2. Longitudinal and cross sections of a qanāt

labour, gradual depopulation and neglect of the dykes and canals. Under their reign there was a shortage of available land as the limits of irrigation had been reached.

The interest of the Romans in Egyptian irrigation methods increased, as Egypt became the granary of Rome and the supply of corn floated down the Nile to be shipped to Rome at Alexandria. The Roman Emperor August not only reorganised the 'Irrigation Service', but also ordered his soldiers to clear out the canals and construct new ones. There were overseers for the flooding and draining of basins, and overseers of weirs and sluices, which were maintained by a certain tax for costs of repairs. The rank of 'river workman' was assigned to those working in dredging the canals. The final downfall of the Egyptian irrigation system took place in the late Roman and Byzantine times (around 200 A.D.). The decline of central power and above all the oppressive taxes led to depopulation of villages, shortage of labour and maintenance funds to clean the canals and repair their structures.

Other than land irrigation in the Nile Valley and Delta, the development of irrigation in El-Khârga Oasis, in the Western Desert, continued at a fair speed during the Ptolemaic period (323–30 B.C.). Artesian water was fully exploited to meet the requirements of cultivation. Thompson and Gardner (1932) claimed that mining of groundwater increased largely during the Byzantine period (395–638 A.D.) with the spread of Christianity in Egypt and the subsequent establishment of the desert settlements near the oases. Later, those remarkable works were left to deteriorate.

The most notable work of the Arab rulers upon their conquest of Egypt (641 A.D.) was the building of certain gauges along the course of the Nile to indicate the water level at different river stages. According to Popper (1951), the Egyptian Orthodox Christians (Copts of Egypt) began recording the Nile water level some 20 years earlier to the Arab conquest. The Copts guarded the Nilometers in that time and remained guarding them upon the request of the Arab rulers until 861 A.D., after that the job was made public to all Egyptians irrespective of their faith, whether being Copt or Moslem.

The first Arab Nilometer was constructed in 711 A.D. (Tousson, 1925)/715 A.D. (Popper, 1951) at the southern end of the Rodha Island opposite to Old Cairo, and was reconstructed or renovated in 861 A.D. The two important sources of information regarding this monumental work are Prince Omar Tousson (1925) and Professor William Popper 1951. These two sources provide the reader with slightly different pieces of information about the construction and repairs of the Nilometer (Reusing, 1994). Figure 3 shows the scale on the central pillar and the crown of the structure.

1.2.2 The Sumerian, Babylonian Assyrian and Nabatean Civilisations

Brief accounts of achievements and developments in water resources technology during these civilisations are presented in the next paragraphs according to their historical order of occurrence in each of Iraq, Jordan, Syria and Palestine.

1.2.2.1 Iraq

Mesopotamia or 'the land between two rivers' presently known as Iraq is well known with its magnificent ancient civilisations. By the two rivers here is meant the Tigris and the Euphrates. This land was crowned with its great architecture, art and scientific work. Mathematics and geometry, principles of building construction and waterworks developed there long before Jesus Christ. The cities of Ur (sometimes referred to as Ur of the Chaldeans), hometown of Abraham at least for quite sometime, Nimrod with its fabulous sculptures and Babel with its gardens are all landmarks of civilisations that began some 7000 years ago. Lamentably, several manuscripts and objects of unparalleled historical significance, all distinguished sources of information of the Iraqi civilisation, have been lost in the last few years.

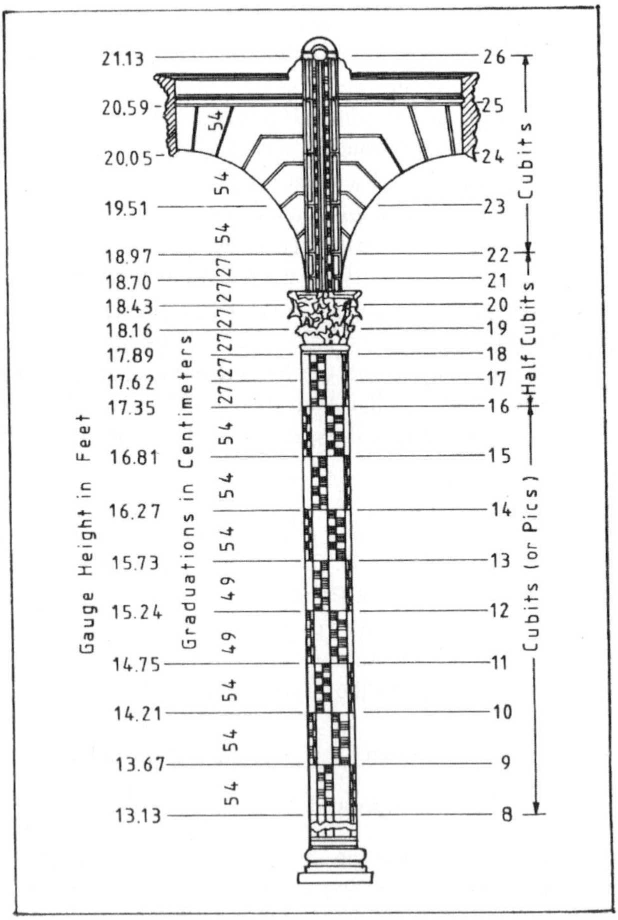

Figure 3. The Nile gauge at Rodha, Cairo, Egypt

Euphrates and Tigris in Mesopotamia were in the deep past two rivers, unlike the Nile in Egypt, with highly unpredictable floods. Simultaneous flooding of these two rivers used to produce excessive calamities. The Sumerians had to develop the art of building flood protection works such as earthen embankments. King Hammurabi was reputed by his profound interest in the construction of hydraulic works as well as water law. The most complete codification of the Sumarian and Babylonian laws appears clearly in the well-known Code of Hammurabi (1750 B.C.). The laws concerning irrigation were thoroughly written and primarily aimed at avoiding any misuse of water or carelessness that might lead to flood disastrous consequences. The Code of Hamurabi has largely influenced many of the modern water laws, especially in Europe.

The 'Ur Babylonian Tablet' is one of the great achievements of the Babylonian rulers around the eighteenth century B.C. It is a tablet full of drawings and writings in Cuneiform (characters used in the Old Babylonian and Assyrian languages; known in Arabic as Mismāriyah). On each side of the tablet there is a number of problems related to dam designs, water wells, canals, etc and their respective solutions.

King Sanherib (sometimes written Sennacherib), almost seven centuries B.C., kept himself busy developing the water resources of his kingdom, Assyria, and introducing new crops to Assyria. He had a special affinity for inventing water-lifting devices. It is quite possible that he had invented the cylinder and screw to lift water for irrigating the hanging gardens of Babel.

Garbrecht (1987) summarised the achievements established by Sennacherib as diversion of river water by means of weirs placed across the Khosr River; excavation of a 16-m long irrigation canal; canalisation of 18 water sources (springs and streams) at the northwest of Nineveh (Ninawa in Arabic); transfer of water from the catchment area of the Gomel River by means of a conveyance canal discharging water into an aqueduct across a water course near the village of Jerwan; and diversion of a mountainous stream into the seasonal water course, Wadi el-Milh. Figure 4 is a map showing the water supply system of the city of Nineveh, the then Assyrian capital. According to Gerbracht (1987), the cuniform clay tablets from the Nineveh Library reveal that fruits, sesame, cotton and grain crops were grown on the intensively cultivated land around the city. Beside this, water had become available all the year round for drinking and domestic purposes.

1.2.2.2 Jordan

The earliest permanent settlements in Jordan, e.g. Jericho-Tell es-Sultan, Ain Ghazal, Es-Sifiya and Al-Baseet, which date to the Pre-Pottery Neolithic B (PPNB, 7600–6400 B.C.), were situated next to perennial water resources. As such, the inhabitants of those settlements had easy access to water for drinking, agriculture and drinking of their domesticated animals all the year round. The so-called 'mega villages', which belonged to the Late Pre-Pottery Neolithic B (LPPNB, 6500–6000 B.C.), were up to 14 ha in size and had certain underground chambers for temporary

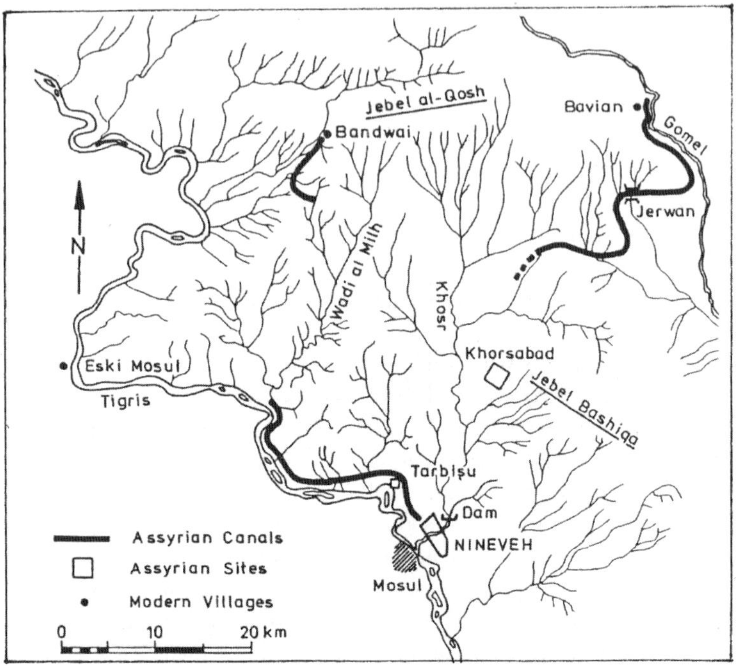

Figure 4. Water supply system of Nineveh (Assyria, about 700 B.C.)

storage of water. Smaller villages of PPNB, situated north of Nabatean City of Petra, southern Jordan, may have had some kind of simple water collecting and/or storage system as no natural springs existed in their close vicinity. It is likely that water may have been transported in containers made of animal skin or wood treated with bitumen (Bienert, 2000).

There is secure archaeological evidence as early as in the Chalcolith period of a proper water management system by a system of runoff control and water storage. As early as the 4th millennium B.C., water storage facilities existed in Tell Jawa (tell in Arabic means hill) on the Wadi Räjil in northeastern Jordan close to the Jordanian-Syrian border. The said wadi drains the southern part of Jebel ed-Drūz. As the geology of the region does not allow recharging groundwater and no springs or permanent streams were available to secure a steady and year-round water supply, a storage dam was built to help store floodwater. So, the only runoff produced by winter rains was retained by the dam and diverted and stored by canals, in pools and cisterns was available to secure the survival of the population (Bienert, 2000). Figure 5 is a sketch of Tell Jawa's water systems cited in Bienert (2000).

Building a covered staircase to reach a deep spring somewhere in the northern Jordan valley, digging wells and construction of bell-shaped cisterns for collecting and storing water for public use became well known and practiced in the successive

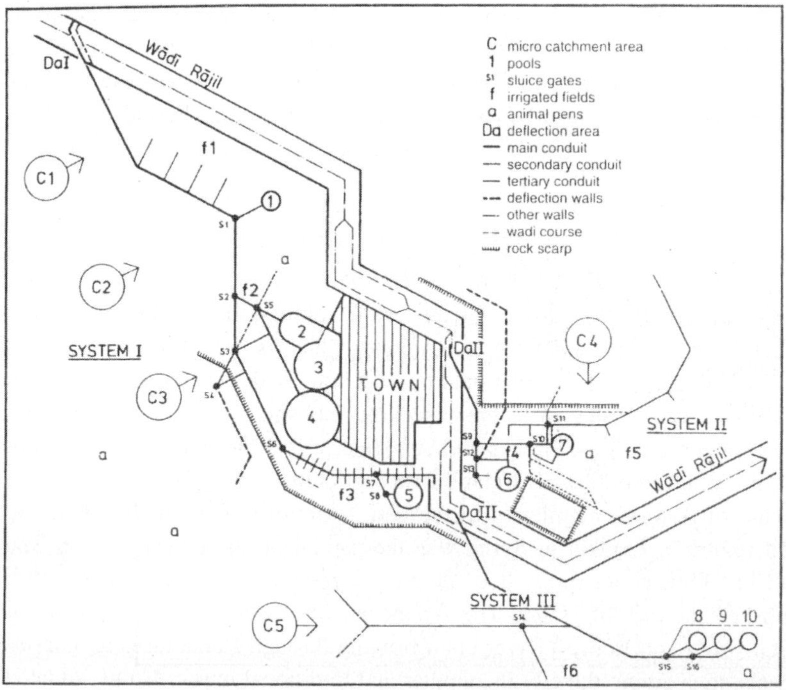

Figure 5. Sketch of Jawa's water systems (from Bienert, 2000, source: Helms, 1981)

stages of the Bronze Age (3200–1500 B.C.). Cisterns were first constructed in large numbers of the Iron Age. The inscription on the stele that was set up about 850 B.C. by King Mesha of Moab in his capital Dhibn, included the King's recommendation to his people.

"Make you every man a cistern in his house."

The Hellenistic period (323–63 B.C.) witnessed a steady progress in the use of private cisterns. The peak was reached in the Roman Period (63 B.C.–324 A.D.) and later in the Byzantine Period (324–630 A.D.) where large numbers of cisterns (30–50 m^3 average capacity) and reservoirs were built all over the country. An example of the reservoirs built by the Romans is the basin in Madaba; 117 m long, 100 m wide and 10 m deep, having a capacity of 117×10^3 m^3.

It is claimed that the Nabateans moved during the 5th Century B.C., probably from northwestern Arabia, to Edom in southern Jordan. The water that sustained them was mostly surface runoff produced by the scanty rainfall and stored in cisterns. The Nabateans in the course of time developed their sophisticated hydraulic systems. Firstly, they developed the available water harvesting methods, which had been in use since the Iron Age, such as cisterns. The terracing of hillsides is perhaps the most spectacular and visible of those early methods. Secondly, they seem to have adopted a technique to cover large reservoirs that had appeared in the Hellinistic world by at least the third century B.C. The method of roofing became typical for almost every Nabatean settlement and was used not only for cisterns but also for houses throughout the Nabatean Kingdom until the early Islamic Period. The Petra Region of Jordan is crowded with many hydraulic structures installed during the Nabatean Civilisation.

1.2.2.3 Syria

Many Syrian cities had witnessed remarkable growth of their trade in the period from the year 2000 B.C. till the middle Ages. Aleppo, which is one of the oldest cities of the world, had become an international trade centre. With the continuous growth of trade, Syria became a haven to those seeking business from the neighbouring peoples: Egyptians, Persians, Greeks and Assyrians. Next to trade, one should not lose sight of the development of irrigation and agriculture in Syria.

The Island of Arward, mentioned in the Old Testament, and which belongs to the coastal region of Latakia in Syria, was the capital of an ancient empire. The island still has Phoenician ruins. It is claimed that Arados, the King of Arward, had established the city of Hama. The Arameans who occupied the city made it their capital. Al-'Āsi (Orontes) River upon flowing through Hama turns its famous water wheels, thus giving the city its popular and traditional music. Homs, situated between Aleppo and Damascus, and Palmera, known as the bride of the desert, are both ancient Syrian cities. They have been frequently mentioned in the Assyrian manuscripts.

1.2.2.4 Palestine

As early as 1500 B.C., conduits or tunnels (sinnōrs) were built in Palestine and Syria to transport water from the downhill streams to townships uphill. Each sinnōr had a secret intake on the stream downhill and the other end was located within the city's boundary. Entrance to the sinnōr was possible by means of a shaft provided with a flight of stairs. According to Biswas (1972) two springs were joined by a sinnōr in the city of Tell Ta'annek in Palestine around 1500 B.C. The same reference claims that King Hezekiah constructed the Siloam sinnōr between the Kidron Valley and Mount Zion in Palestine around 700 B.C.

Toethmosis III, Pharaoh of Egypt (1479–1425 B.C.), occupied the land of Canaan: presently Palestine, Israel, Syria and Jordan. The Sumerians (1300 B.C.) used the 'Canal of Hamri' to supply water to the township of Hamri near Nippur, the capital of Mesopotamia, during the high flow season. A network of channels was also used for land irrigation and drainage. The low-lying lands were preserved as marshland for growing cane-reed, which had several functions. Reed mats were used for reinforcing the dykes protecting the arable land. A new era began around 1230 B.C. in the land of Canaan after Moses had led the Jewish people to the 'Promised Land'.

Three independent states were established in Jordan; Edom in the south, Moab in the middle and Ammon in the north. The rise of those states can be attributed to the growth of trade and business with the Arabian Peninsula. According to Meijer (1997) the prosperity of those states became an attractive target to invading powers; the Assyrians (732 B.C.), Babylonians (612 B.C.) and the Persians (539 B.C.).

It might be of interest to mention here that quantitative measurements of rainfall were made in Palestine around the first century A.D. This appears in the Mishnah, a Jewish book that records cultural and religious activities over nearly four centuries.

Measurements of rainfall were associated with agriculture, especially the phases of growth of grains and other plants. The unit depth of rainfall used to express the crop water needs was known as 'Tefah', which is equivalent to 90 mm.

1.2.3 The Yemenite Civilisation

In the eighth century B.C. there were rich kingdoms in southern Saudi Arabia. Those kingdoms reached a high level of civilisation according to the standards of that time. The flourishing and prosperity of those states was strongly supported by their location along the caravan routes extending from Iraq, Syria and Egypt in the north down to Saudi Arabia and Yemen in the south. The most important kingdom of that time was that of Queen Bilqis (Sabā in the Qu'rān and Sheba in the Bible) and King Solomon (Sulaymān) in the 10th century B.C. Sabā was a city in Yemen, said to have been three days' journey from the city of San'ā, the present capital of Yemen. It is claimed that King Solomon's interest in studying the flow of water led him to develop the hypothesis "All the rivers run into the sea, yet the sea is not full; unto the place from whence the rivers come thither they return again."

Around 800 B.C. plenty of water works were already located along the borders of the Arabian Desert, and for some centuries later Sabā was still a happy and prosperous country, amply irrigated from Maārib dam. Its roads and perhaps its canals were skirted by gardens on both sides, right and left: at any given point, one could always see two gardens. It produced fruits, spices, and frankincense, and got the name of Arabia the Blest (*Arabia Felicia* as often appears in old literature) for that part of the country.

The Maārib Sadd (dam), which was constructed somewhere between 1000 and 700 B.C. in Yemen, has acquired such a wide reputation to qualify it to be one of the Seven Wonders of the World. The dam was 600 m long and 10 m high, with a complex inlet pipe system to allow the dam to receive the runoff running in Wadi Adana, some 70 km from the present location of the township of Maārib. The dam thus was able to irrigate a surface area of not less than 1,000 ha. The large dam by 600 years B.C. had an ingenious system of dams, sluices and irrigation canals. The existing rock outcrops and lava tracks were skillfully used in the building of these structures. The torrential rains were thus collected and thereafter shared as wished. The construction and management of the dam were evidence of good planning and coordination, and for that time a high standard of technology. The water used for irrigating the land made the Maārib area flourish and become rich. The prosperity resulting from the dam and it's ancillary works called for constructing certain works to protect the city of Maārib.

The dam did not last long; instead it showed signs of bursting some decades after its construction. The agriculture was badly neglected and the Ethiopians invaded the country. After a number of minor breaches the dam finally collapsed in 570 B.C. (Hoff, 1995). According to the Qur-ān (Sura 34, verse 16) "...the people of Sabā had beautiful gardens with good fruit. Then the people turned away from God, and to punish them, he burst the dam, turning the good gardens into gardens bearing bitter fruit."

The Romans had a bitter experience in 26 B.C. when they sent a force of 130 ships and more than 11,000 troops with Gallius, Eparch of Egypt, in command. After six months, during which battles were won but no treasures gained, the Romans retreated, and the Himyar dynasty of Sabā knew how to get the administration of their land anew in their hands. This situation, however, did not last long as Yemen became a fertile ground for intensive struggle, mainly for commercial reasons, between the Romans-Egyptians, the Christian invaders coming from Ethiopia, the Jewish and later the Moslems.

Many Yemenis played a major role in spreading the Islam in the Middle East and North Africa. This move brought many Yemenis to several Arab countries. Tareq, who reached and occupied Gibraltar (Jebel al Tarik) in 700 A.D. is recognised as one of the top Moslem heroes.

1.2.4 The North-Africa Civilisation

There is hardly any well-documented information about water civilisation in North Africa in the ancient times. Probably one of the few exceptions is the water meter developed and used some 3000 years B.P. at the Ghadames Oasis, situated in what is presently known as Libya. This meter is still being used in the same location without modification. According to Biswas (1967,1972), the discharge, around 50 l s^{-1}, of a small spring called Ain el Faras, is collected in a basin and distributed through a system consisting of a main canal, two secondary canals and a number of ditches for irrigating the neighbouring land. A container, consisting of a jar with a perforation in its bottom, is lowered into the water well and pulled up when filled. Water is allowed to flow into the irrigation system through the hole in the jar, and the operation is repeated over and over again. Each cycle needs about three minutes to be completed. The water commissioner decides the length of time, as measured continuously by filling and emptying of the jar, that a given landowner may divert the entire flow of the nearest ditch onto his field. Upon the arrival of the last cycle of water to which a participant is entitled, he gets a signal to shut off his channel irrigation inlet so as to allow the water to flow down and reach the inlet of the next land owner.

The process is repeated incessantly until the last participant along the ditch has received all of the water released to him while his quota of cycles is being counted off at the location of the source well. After that last field has been supplied with its share of water the entire process is repeated, beginning with the field nearest the supply reservoir. The complete turn of the rotation lasts about 12 days.

1.3. HIGHLIGHTS ON THE HOLY RELIGIONS IN RELATION TO WATER AND WATER RESOURCES

1.3.1 From magic to religion

Sec. 1.2 of this chapter summarised the civilisations developed by the different communities which inhabited the land area currently known as the Arab Region. It reviews the interest of people for having water to meet their needs and sustain their lives. It also highlights the methods and technologies they used to develop, utilise and manage the then available water resources. They all go back in history to several centuries before the birth of Christ.

Alatas (1963), upon reflections on the theory of religions, has reaffirmed the conclusion that religion came into being as a result of the need to explain the phenomena of nature and the processes that make such phenomena be utilised for man's welfare. The imperious need of tracing the causes of events has driven man to discover or invent a deity. Magic in many lands and amongst many races has been practiced with the intention of controlling the forces of nature for the good of man. The institutionalisation of magic in man's life gradually led to the emergence of some of the elites to become magicians. Some of those magicians became chiefs and kings.

The transition from magic to religion is evidenced by the fact that instances have been witnessed where gods were also believed to be magicians. Magic was met with successes and failures. However, the understanding of religion as a propitiation or conciliation of powers superior to man, which are believed to direct and control the course of nature and human life, kept steadily growing with time.

The belief of man in God, the Lord Almighty, and consequently in the Holy Religions increased substantially after the birth of Jesus Christ, a little over 2,000 years B.P. The Holy Religions; Judaism, Christianity and Islam have since become the beacon that guides all believers to do righteousness.

1.3.2 Water as described in the Holy Books

The reader of any of the Holy Books will find many a phrase linked to water such as precipitation, runoff and flood, clouds and storms, wind movement and air circulation, rivers and oceans, surface and subsurface water, freshwater and saline water, etc. All these elements compose what is referred to collectively as 'Hydrologic Cycle.' Matters related to water resources and their repercussions on mankind and other living creatures, and the environment they live in are all included in these books as well. They provide guidelines concerning the use, conservation and management of water resources as well as water rights.

During the Middle Ages, all philosophers and interpreters of the Holy Scripture, thought that water springs have their origin in the sea. Certain passages in the Bible tell us that water escapes through holes in the bottom of the sea, flows through subterranean channels, and thence is elevated to the springs. Meinzer (1942) mentioned that it is not these passages but its medieval interpretation that is in conflict with the modern concept of the hydrologic cycle. Besides, Herschel (1899) reported that Hero of Alexandria had a clearer understanding of what is involved in measuring flowing water than any of his compatriots or predecessors had and that he recommended a much better practice than was anywhere shown in the Roman water law.

The Bible states that the creation of heavens and earth (dry land) necessitated having a sky to divide the vapour above from the water below. A river from Aden, the garden where Adam was placed, flowed to through the garden to water it. That river later divided into four branches; the Pishon ran along the border of Babylonia, the Gihon[2] traversing the full length of the Land of Cush (Ethiopia), the Tigris and Euphrates traversing the plain of Mesopotamia. The Bible also gives a fascinating account of the first catastrophic hydrologic event, i.e., the flood of Noah. In the context of analysis of time series; when to a maximum value of a certain process there is a maximum, and when to a minimum value other event there is a minimum, one can speak of 'Noah' effect. The dreams of Egypt's Pharaoh regarding seven fat cows eaten up by other seven skinny cows coming out of the Nile, and the seven plump and full heads of grain swallowed by seven withered heads coming out of the same stalk, were interpreted by Joseph as examples of what is currently known as persistence or Joseph effect. Both Noah and Joseph effects can be observed while analysing geophysical, meteorological and other data regarding these effects. Mandelbrot and Wallis (1968) were among the pioneers who have emphasised the role of these two effects in the field of hydrology.

The reader of the Qur-ān will find in many verses (Āyats) mention of water and water-related matters. For example one can read in Sũra 23, Āyats 18 and 19: Allah (God) sends down water from the sky according to due measure. He causes this water to soak in the soil; and can let it easily drain off. With the moisture retained in the soil, gardens of date palms, vineyards can grow bearing abundant fruits; from which people can eat and have enjoyment. Also, Āyat 53 of Sũra 25 mentions that Allah tells us that he is who let the two bodies of flowing water, one palatable and sweet and the other salt and bitter, yet he has made a barrier between them, a barrier that is not to be passed. Similar to the Bible, the Qur-ān too gives an account of the Noah flood. Besides it contains many statements related to occurrence and use of rainfall, river flow, waves and circulation of air masses, and quality of water.

---

[2] "And the name of the second river [is] Gihon: the same [is] it compasseth the whole land of Ethiopia." *(Genesis 2:13, King James version of the Bible).*

1.3.3 Establishment of the water law

In the previous sub-sections accounts of the main achievements in the areas of water resources utilisation and management as part of the various civilisations, which began some 7,000 years B.C. and covered most of the Arab Region, have been highlighted. Most of those achievements were essentially based on the developments in water resources technology. The art of water level observation and recording, rainfall measurement, building of dykes and dams and other hydraulic structures, digging canals and tunnels and well drilling were the major items comprised in that technology.

In sub-section 1.2.2.1 we mentioned that the most complete codification of the Sumarian and Babylonian laws appears clearly in the well-known Code of Hammurabi (1750 BC). The laws concerning irrigation were thoroughly written and primarily aimed at avoiding any misuse of water or carelessness that might lead to flood disastrous consequences. Despite the fact that the Code of Hammurabi has largely influenced many of the modern water laws, it was practically ignored by the nomads (Bedouins) who were crossing the Arabian Desert searching for water and less harsh living conditions.

Before Muhammad assumed his prophecy, in the period of Jahilyya (ignorance, lack of knowledge) some fourteen centuries ago, there were no established water regulations. In certain districts of the Arabian Peninsula, e.g. south of Arabia, where water is relatively plentiful, ownership of water was private and selling water was a common trade. When the nomads came to live next the settled populations, possession of water wells and springs led sometimes to bloody struggles. This situation lasted for sometime till the teachings of Prophet Muhammad spread among the people of Arabia and the necessary laws with regard to sharing the available water resources were laid down. He declared that free access to water was the right of the Moslem community. The general principle laid by the Prophet is that water is indispensable and priceless element of purification (Caponera/FAO, 1973). The Holy Qur-ān expressed this general formula by stating:

"We made from water every living thing."

Sometime later, several precepts and schools of interpretation (Mazhabs) were initiated regarding the free access to water for human beings and for animals. The water rights comprise the rights for drinking, irrigation, fishing and sailing. The ensuing doctrines include articles concerning maintenance of rivers, navigable and non-navigable waterways. Specifications regarding the forbidden area surrounding a well field or along a river or canal are also included. If it surrounds a well, it is forbidden to dig another well next to it so as to avoid depleting the groundwater aquifer. Establishing special funds to finance community activities related to water resources such as clearing irrigation canals, maintenance of navigable waterways, etc. are discussed in the various doctrines.

The water legislations currently applied in the Arab Region are founded on the principles laid down by Islam. The Arabs brought those principles along with them to other parts of the world some centuries ago. Presently there is a growing tendency to regard water as a natural resource that constitutes an essential component of the national wealth of each state. Such a tendency is based on the principle of public interest to which private water ownership rights, whether they exist or happen to be established for private water use rights, constitute a basic constraint. It therefore appears that modern tendencies in water legislation aim at institutionalising the concept of community of interest, which concept constitutes the traditional basis of Moslem Customary Water Law (Caponera/FAO, 1973).

1.4. FROM A SHINING PAST TO A GLOOMY PRESENT

Many of the aspects of ancient civilisations and Holy Religions concerning water have been briefly reviewed in the previous sections. It is hoped that such a review will provide an essential background to all potential users of the book, not only from within the Arab region, but to those from outside the region in the first place.

After reaching the peak of its glorious past, some areas of the Arab Region have been suffering for ages, and some are still suffering till present from very slow development. This situation is the inevitable consequence of internal disputes, fights and wars, and most definitely foreign occupation. The region, in the course of the last twenty centuries or so till a few decades ago, was under alternating or collective occupation by the Romans, Greek, Persians, Ethiopians, Byzantines, Ottomans, British, French and Italians.

In compliance with the charter of the United Nations Organisation, many Arab States started to become independent and sovereign. This took place between the end of the 1950s and beginning of the 1960s. As the sovereignty and independence they obtained are not always up to the extent they should be, many of the Arab nationals regard this state of affairs as unacceptable and needs immediate and radical change.

Four major facts are dominating the Arab Region since the beginning of the twentieth century. These are:

– The discovery, production and export of oil and natural gases on a very large scale have led the industrial nations to depend heavily on the Arab oil producing countries for a great part of the energy they need.

  It is true that the oil producing and exporting countries are currently enjoying a high state of welfare in this age of 'oil civilisation.' The average income per capita in certain countries can be ranked among the world's highest incomes. However, one fails to observe that the wide majority of the population are really enjoying their basic human rights such as freedom of speech and expression, democratic representation in the government and parliament, social justice, etc. One should not forget that oil money has brought to the oil producing countries foreign experts, technicians, and skilled and unskilled labourers. However, the proportion of workmanship originating from the poorer Arab countries is limited.

  A considerable proportion of the oil revenues go as savings and/or investments in the highly industrialised countries, also for purchasing luxurious articles mostly made in these countries. The investments within the oil producing countries themselves are certainly growing but at a slower pace than actually needed. Moreover, only limited sums of money are spent on industries and the rest is exhausted in building luxurious infrastructures. Probably, with the exception of the Sultanate of Oman, 60 to 75% of the workmanship in the other oil states comprising the Arabian Peninsula is employed in the services sector whereas the remaining 25 to 40% works in the production sector such as agriculture and industry. Remarkable enough is that in case of any serious disputes between the oil producing countries and the industrial countries, the latter do not hesitate to freeze the assets of the oil countries!

- The establishment of the State of Israel in 1948 within the Arab Region and the subsequent conflicts with the neighbouring Arab States have resulted in several problems that so far are left unresolved.

  The creation of a new State in the region, filling it with people imported from many parts of the world and stretching its mighty arms to surface as well as groundwater resources in the surrounding areas has diminished the share of water per capita in the neighbouring countries. The present consumption per capita is at 139 and 172 m^3 for the Palestinians in the Gaza strip and the West Bank respectively. The average annual consumption of the Israelis is 411 m^3 (Eckstein & Eckstein, 2003). In his article "A Dwindling Natural Resource" in the section of World News of the Washington Post, Rendel (1992) stated that there is no binding international law regarding conflicts on water issues. Rather, the claim of Israel's Arab neighbours-the Palestinians, Jordanians, Lebanese and Syrians-is based on evolving theories of international water law, which have been moving away from the 'historic use' argument invoked by Israel to justify its firm grip on regional water.
- Most of the Arab States are administered by undemocratic ruling systems. The freedom of the individual in general and the human rights do not rank high. This is creating a sort of impermeable barrier between the society and the administration, a situation that limits considerably the community participation in several matters. Water resources management and development projects is just an example.
- The region, since the end of World War-II till present, is witnessing a long chain of armed conflicts in many of its parts. One can mention here the 1948, 1956, 1967 and 1973 wars between some of the Arab States including Palestine and Israel; the war between Iraq and Iran, 1980–88; the Iraqi invasion of Kuwait in 1990; the first Gulf war in 1991 and the invasion of Iraq by the United States forces and their allies, which began in March 2003 and is still going on. The inadequate functioning of the Iraqi's infrastructures including the water supply and sanitation facilities, which were mainly caused by the sanctions that began in 1991 has deteriorated further since then.

Definitely, there are many more facts and events, though relatively minor, which together with the major ones, are hampering any profound development of the region. The above-mentioned facts and events and many more have reacted in the near past and are still reacting at present on the situation of water resources development in the Arab Region. One can list many examples of the adverse reactions, some of which are as follows:

- Despite its limited water resources, the region is experiencing an explosive growth of population. The high birth rate, the decreasing rate of mortality, and the large number of foreign workers who together with their families are working and living in the oil producing countries without proportionate increase in the available water resources are causing an accelerated decline of the share per capita of the renewable water resources as well as the deterioration of water quality in certain areas.
- The policy adopted by loan granting agencies seems to be under strong pressure imposed by the super powers. This became evident when the World Bank, after its preliminary approval to finance the construction of the Aswân High Dam, Egypt, had to withdraw its offer. This change of attitude has forced Egypt to nationalise the Suez Canal, direct its request to the then Soviet Union, and to go for a war in 1956.
- The conflicting situation between the Sudanese Government in the north and the southern Sudan, which lasted more than two decades, has only recently come to an end. Despite the reconciliation of the two parties, the implementation of water resources conservation schemes in the Upper Nile Basin, aiming at saving water from being lost in the swamps of the southern Sudan, is beyond thinking, at least for the time being. This is true especially that the Government of the Sudan is presently facing a strong resistance to administer the Darfur Province.
- Till the present time many parts of the Arab Region are deprived from clean, freshwater supplies and from adequate sanitary drainage. The existing high level of illiteracy, and the lack of financial means and technological know-how are the main reasons behind this situation. One should not forget the low-level of safety and security, which earmarks the living condition in the hot spots like Iraq, Palestine Somalia and southern Sudan. It goes without saying that the unfit condition of the existing water supply and sanitation systems underlies the main cause behind the water-related and water-transmitted diseases in those parts of the region.
- During World War-I, 1914–1919, the British engineers undertook the task of remodelling the cross sections of irrigation canals in Egypt, thereby converting lift irrigation to free flow irrigation. The objective was to save the fuel used for running the pumps that lift water for land irrigation and thus save it for the military operations only. The conversion of the irrigation system from pumped flow to gravity flow has contributed to the perpetual rise of water table levels in the irrigated fields with all its adverse consequences on soil properties and crop yields. Excavation of extensive networks of drainage canals and construction of costly field drainage systems is still underway so as to alleviate the detrimental effects of high water table levels.
- The Iraqis who for centuries long used to enjoy adequate water supply and irrigation systems are presently without any fair access to either freshwater supplies or to healthy sanitation facilities. The existing sewage systems have

become almost destroyed or too old and no longer able to function satisfactorily. This situation is a grave menace to the majority of the Iraqis. The local water engineers have estimated the time they need to renew the existing facilities and systems, assuming that the necessary funds and technical know-how are made available, to be in the order of 20 years.

- The breaks in the record of the Nile water levels at the Rodha gauge near Cairo can easily be observed from the readings of the Nilometer placed there and already described in sub-sec. 1.2.1. The breaks, which combined together lasted more than 350 years, practically coincide with the period of Ottoman occupation of Egypt. These particular series of river water levels are the longest hydrologic time series available all over the world (Shahin, 1985). They have been used by many a scientist for detecting, studying and verifying some aspects of river hydrology, e.g. persistence and extreme-value events in stochastic hydrology, long-term storage in reservoir hydrology, etc. Besides, the uninterrupted lengths of the maxima and minima series have been employed in detecting the climatic changes in the Middle East. The scientific as well as monumental values of these series, beyond any doubt, would have been much better if there were no interruptions or missing records.

  Similar, though short-term, breaks in the series of water levels and discharges of the Nile River system are taking place in the southern Sudan since 1983 up till now. Likewise, meteorological as well as hydrological measurements in Algeria have suffered from the gaps and interruptions during and after the war of liberation. One can also observe the breaks in the meteorological and hydrological measurements in other parts of the region. All these breaks and interruptions, which mean missing data, have several demerits. It is a known fact that long and uninterrupted time series in geophysical and atmospheric sciences reveal more information regarding the performance of the processes underlying the variable in question compared to short series. The availability of reasonably long time series is an essential requirement before examining the issue of atmospheric warming and evaluating the subsequent changes in precipitation and runoff. The determination of the design storage capacities of reservoirs and planning the strategies needed for their operation are other examples worth to be mentioned.

It is hoped that the less shining present of the Arab Region, compared to those periods of more illuminating past, will not last long nor will deteriorate any further. As long as human life is dominated by unseen holistic, existential and interdisciplinary inertias, things keep changing. We say holistic because man is moved by an ultimate purpose in life without which his existence seems meaningless and incoherent. We say existential because the pressing problems of existence follow him at every step, challenging or yielding the strength of his ultimate commitment. We say interdisciplinary because human life is a unified, integrated and dynamic activity, which cannot be understood not to say explained, by any other method than the unified approach of man's learning and skill. These three major motivative powers are profoundly issuing from man's belief in knowledge, faith and religion.

To move forward with the development of water resources of the Arab Region, which is part of the overall development of the region, one has to remember the good and bad experiences from the lessons of the past as well as the present. When and how the Arabs are expected to seriously consider improving their gloomy present and develop their region is definitely beyond the scope of this book.

CHAPTER 2

PHYSIOGRAPHY AND GEOLOGY OF THE ARAB REGION

2.1. BRIEF DESCRIPTION OF THE PHYSICAL SETTING OF THE ARAB REGION

The Arab Region is almost a continuous surface of land situated between 01°35′ south (S) and 37°14′ north (N) latitudes, and between about 17° west (W) and 60° east (E) longitudes. The region extends from the Atlantic Ocean in the west to the Persian Gulf in the east, and from the southern shore of the Mediterranean Sea in the north to southern Sudan and the Gulf of Aden in the south. The Red Sea penetrates the Arab Region from the southeast to the northwest, thereby separating the eastern part (Al-Mashreq Al Arabi) from both the central and western parts (Al Wasat and Al Maghreb Al Arabi respectively). One can also divide the Arab Region into the Arab Middle East subregion and the North African subregion. The Arab States located east of the Red Sea belong to West Asia. In this text we shall adopt the division of the Arab Region into three subregions: the western, middle and eastern. Table 1 includes the division of the Arab countries between these subregions together with some relevant data.

The Arab Region covers a surface area of about 14×10^6 km^2, partly in Africa and partly in west Asia. By virtue of its geographical location between the Tropics of Cancer and Capricorn, at latitudes 23°30′ N and 23°30′ S respectively, a substantial part of the region has a tropical climate.

Dry, warm deserts cover the greater part of the surface area of the region. The vast desert, which extends from eastern Mauritania in the far west to Morocco and Algeria, and further to southern Tunisia, is known as the Great Sahara (actually the Arabic word sahara means desert). The name changes to Libyan Desert in Libya, and Western Desert in the western part of Egypt. The Sahara covers almost the entire northern part of the African continent, being approximately 7×10^6 km^2 in area. A map showing the major parts of the Sahara in North Africa with its ergs (basins filled with dune sands) and sand streams from 1925 to 1960 has been composed by Wilson (1971) and shown in Figure 1. The Eastern and Nubian Deserts cover the areas situated east of the Nile River in Egypt and the Sudan to the western shore of the Red Sea. The landscape of the Nubian Desert is varied; it is mainly occupied

Table 1. Subregions, official names of states, capital cities, surface areas and lengths of coastal lines (World Guide 1999–2000, 1999)

Subregion	Offical name of the country	Capital	Surface area, km^2	Coastal line, km
Western	Al-Jumhuriyah al-Islamiyah al-Muritaniyah (Mauritania)	Nouakchott	1,025,520	600
	Al-Mamlaka al-Maghribiyah (Morocco)	Rabat	446,550	2,170
	Al-Jumhurriya al Jazairia ash-Shaabiya (Algeria)	Algiers	2,381,741	1,200
	Al-Jumhuriyah at-Tunisiyah (Tunisia)	Tunis	163,610	1,000
	Al-Jamahiriyah al-'Arabiyah al-Libiyah ash-Sha'biyah al-Ishtrakiyah (Libya)	Tripoli	1,759,540	980
Central	Jumhuriyat Misr al'Arabiyah (Egypt)	Cairo	1,001,449	2, 990
	Jumhuriyat as Sudan (Sudan, the)	Khartoum	2,505,813	720
	Republique de Djibouti (Djibouti)	Djibouti	23,200	314
	Jumhuriaydda Soomaliya (Somalia)	Mogadiscio	637,657	3,000
	Republique Federal et Islamique des Comores (Comoros)	Moroni	2,235	
Eastern	Al-Jumhouriya al Arabiya as-Suriya (Syria)	Damascus	185,180	300
	Al-Jumhuriyah al-Lubnaniyah (Lebanon)	Beirut	10,400	210
	Palestine State (Palestine)	Al-Quds	6,220	75
	Al-Mamlakah al-Urdunniya al-Hashimiyah (Jordan)	Amman	97,740	25
	Al-Jumhuriyah al-'Iraqiyah (Iraq)	Baghdad	438,317	40
	Dawlat al-Kuwait (Kuwait)	Al-Kuwait	17,818	499
	Al Mamlaka al' Arabiyah as Sa'udiyah (Saudi Arabia)	Riyadh	2,149,690	2,255
	Mamlakat al-Bahrayn (Bahrain)	Al-Manamah	622	161
	Dawlat Qatar (Qatar)	Doha	11,000	563
	Dawlat al Imarat al' Arabiyah al-Muttahidah (United Arab Emirates/UAR)	Abu Dhabi	83,600	1,318
	Sultanat Uman (Oman)	Muscat	212,457	2,092
	Al-Jumhouuriya al-Jamaniya	San'a	527,968	1,800

by a structural plateau with elevations of 350–1,000 m. It descends stepwise from east to west and is crossed by many wadis. In the west, the plateau is covered by deposits of eolian sands with a barchan relief. Sparse small shrubs are encountered only at times when water flows in the wadis.

A number of deserts occupy most of the area between the Red Sea in the west and the Persian Gulf (in some texts the Arabian Gulf) in the east. These are; the desert of Ar'Rub al Khali (meaning the empty quarter), which is situated in the south of the Arabian Peninsula between Yemen and Saudi Arabia; Great Nefud Desert in the middle of Saudi Arabia; Little Nafud (Dahna); Al Hasa; Al Hamada or Badeyat Esh'Sham in the north (Esh'Sham is an old name of the land surface currently occupied by Syria and Lebanon) and others. The Badeyat Esh'Sham Desert forms a large part of the surface of Syria and Iraq. In the west, the Arabian Peninsula is

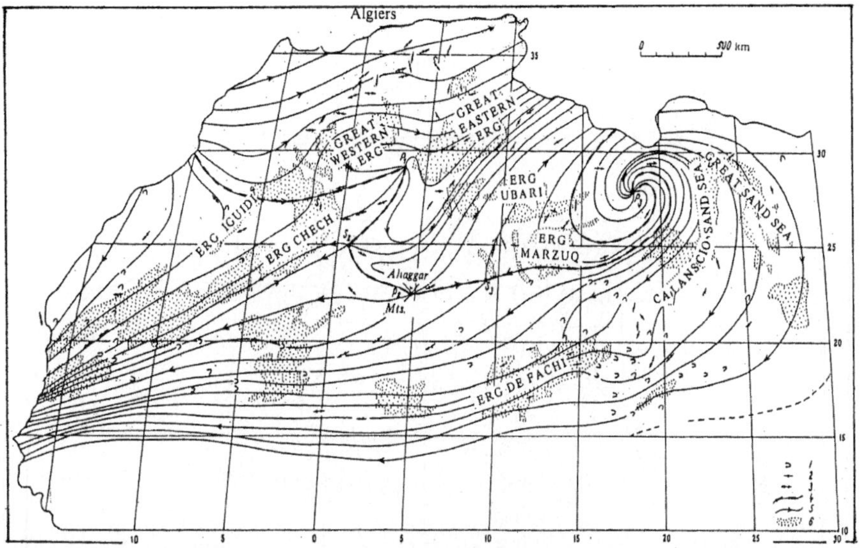

1- barchan sands; 2- seifs and yardangs; 3- resultant direction of winds (after Dubief, 1952); 4- borders of sand streams; 5- lines of sand streams; 6- ergs. High–pressure peaks: P_1) Tademait; P_2) Hajjar; P_3) Libyan. Low-pressure regions ('saddles'): S_1) Erg Chech; S_2) Tanzerouft; S_3) Ténéré.

Figure 1. Prevailing winds and sand streams from 1925 to 1960 over North Africa (after Wilson, 1971)

fringed by the chain of mountains running along the length of the Red Sea coast – the Hejaz and Asir.

The distribution of sand deserts in the Arabian Peninsula is closely linked to its geological structure and wind regimes in the Quaternary and Recent history. The map in Figure 2 shows the sand deserts, the gravel belts, plains of weathered granites and predominant sand movement in the Arabian Peninsula (Brawn, 1960).

Some mountain heights formed of volcanic lava or crystalline basic rocks protrude here and there at different locations in the above-mentioned deserts. Examples of these protuberances are Al-Hoggar (2,900 m) in Algeria; Uweinat (1,850 m) between Libya, Egypt and the Sudan; and Darfur and Kurdufan in the Sudan. The peak of the Tibesti Mountain in Libya exceeds 3,000 m a.m.s.l. (meters above mean sea level).

In addition to the inland heights, there are also chains of mountains running parallel to the shores of the Mediterranean Sea and the Red Sea. The Atlas Mountain Range, which occupies the northwest extremity of the Arab Region, reaches an elevation of 4,165 m a.m.s.l. at Toubkal in Morocco. The mountain ranges in the eastern part of the region extend along the Arabian-African graben. The Eastern and Western Lebanon Mountain Ranges extend along both ends of this graben. Elevations of the peaks of the western and eastern ranges are 3,083 and 2,814 m a.m.s.l. respectively.

The Arabian Shield, which begins in the west at the coast of the Atlantic Ocean, terminates in the east by a tectonically unstable zone. This zone is represented in

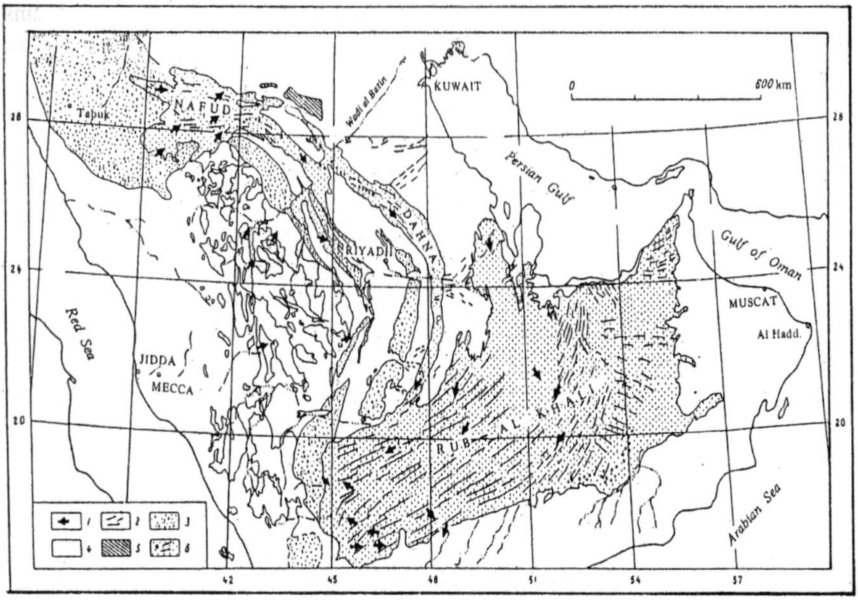

1- predominant direction of sand movement; 2- gravel belts; 3- sandstone; 4- plains of weathered granites; 5- deflation basins; 6- sands of linear and star (pyramidal) dunes.

Figure 2. Sand deserts of the Arabian Peninsula (after Brawn, 1960)

the east by the Omani Mountain Range, with an elevation of the peak of Jebel Lakhdar of more than 3,000 m a.m.s.l. In the northeast there is the Zagros Range with a peak elevation of 3,700 m a.m.s.l., and in the north the Taurus Mountain Range. Two parallel ranges of Mountains separated by Al Ghab Valley are to be found in Syria. The Western Range runs parallel to the Mediterranean Coast, with two peaks: Amman, 1,800 m and Al Alawyya, 1,500 m a.m.s.l. It continues along the coast across Lebanon to reach a peak elevation of more than 3,000 m. The eastern range extends parallel to the western range and has a number of peaks, most important of which is Jebel el Sheikh at an elevation of 2,814 m a.m.s.l.

Mountainous heights penetrate into the Arabian Peninsula along the eastern coast of the Red Sea. In the northern stretch of the seacoast they are referred to as Al Hijāz Mountains and rise up to 2,000 m a.m.s.l. In the middle stretch the mountain range is called Al Asīr Mountains Range, which rises up in some locations to an elevation of 2,200 m a.m.s.l. The Yemen Range of Mountains extends itself along the southern stretch of the Red Sea and rises up to 3,700 m a.m.s.l. This range bends and changes its direction at Bab el Mandab in the southern extremity of the Red Sea. From there it runs parallel to the Gulf of Aden under the name of Hadhramout Mountains Range with elevations up to 2,400 m a.m.s.l. The mountain range continues in a west-east direction along the coast of the Arabian Sea under the name of Dhufar Mountains reaching an elevation up to 2,000 m, and Oman Mountains along the Gulf of Oman. The Oman Mountains are split into the eastern range (Al Hajar Ash'Sharqui) with peak heights up to 2,100 m and the western range (Al Hajar Al Gharbi) with peak heights up to 2,670 m a.m.s.l. All these mountain ranges are referred to collectively as the Arabian Shield. This shield has a strong effect on the precipitation in the Arabian Peninsula, especially the coastal areas.

Another range of mountains extends along the western shore of the Red Sea in Egypt and the Sudan. Similar to the mountain ranges in the east, the ground elevation increases in the direction of the south to reach an elevation of more than 2,200 m a.m.s.l., e.g. the heights of Mounts Erba and Odah in the Sudan. The elevation increases further to more than 2,700 m a.m.s.l. near the Sudanese-Ethiopian frontiers.

The Somali Mountains can be viewed as a natural extension of the Ethiopian Plateau. They slope towards the east and terminate by narrow coastal plains along the Gulf of Aden.

The Arabian Deserts contain not only high plateaus and mountain ranges, but also low-lying areas. Depending on their natural characteristics, the low-lying areas can be in the form of oases, flat depressions (playas), sabkhas (sometimes written sebkhas) or swamps. The Western Desert in Egypt contains Siwa, Farafra, Al Baharîya, El Faiyūm, Dunqul, Kharga and Dakhla Oases. In Libya there is a number of oases such as Al Jaghbūb, Al Kufrah and Jalo.

The dry watercourses (wadis) should be viewed as the basic form of the ancient, erosional relief of the region. They have their origins in inselbergs and highlands, and terminate in closed intermontane depressions or at the sea or ocean coast. The Arab region is characterised by some large depressions. The bottom of Wadi Araba in Jordan is at an elevation of between 200 m and 392 m b.m.s.l. (meters below mean level of the Dead Sea). The Wadi Tharthar, a vast depression in Central Iraq between the Tigris and Euphrates Rivers, is another example. The third, the Qattara Depression, is a very large depression in northwest Egypt, with its northeastern extremity close to the Mediterranean Sea. The deepest point in the bottom of this depression is at 156 m b.m.s.l. (meters below the mean level of the Mediterranean Sea near Alexandria). Shallow salt lakes in the western states of the region are known as Chotts, e.g., Chott Jerid in Tunisia and Chott el-Hodna in Algeria. In addition, there are several spots within the low-lying areas serving as local drainage basins to the groundwater and/or torrents running off the higher grounds. The soils of these spots contain a high percentage of colloidal material. Depending on their geographic location, they are called as Sebkha, Khebrat, etc. In the south of Algeria one finds the Sebkha de Tinduf, Sebkha Mekerrhane and Sebkha Azzel Matti. As the water content in these sabkhas rises up to the level of saturation or inundation the sabkhas are then called swamps. A typical example of these wetlands are the swamps of Bahr el-Jebel and Bahr el-Ghazal in southern Sudan. The swamps get a different name in the eastern part of the Arab Region. In Iraq they are called Ahwār (singular is hawr). The area from Basrah in the south of Iraq to Kut on the Tigris

River to the northwest is crowded with many ahwars, of which Suq ash'Shuyukh and As'Sadiyah Ahwar are the most important ones.

The Coastal plains, which extend parallel to the seashores in the Arab Region, are of two distinct categories. The first category comprises the desert coastal plains such as Tihama and Aden in Yemen, and the coastal plains in Oman and Somalia. The other category comprises those plains situated between the seashores and the ranges of coastal mountains. The width of these plains narrows and widens depending on the distance between the sea and the respective coastal mountain range.

The considerable difference in elevation between the different parts of the region can be observed from the approximate topographic map shown in Figure 3.

In the subtropical and tropical deserts of Africa and Arabia, sclerophyllous shrubs and perennial grasses predominate but succulents are also present. In contrast to deserts in the temperate zone, the vegetations in these regions are more sparse. The most sparse vegetation is found in sand-pebble deserts. There is hardly any vegetation in areas of saline crusts. No plants can be encountered for several kilometres in the moving sands of extremely arid sand deserts. In the oases, where groundwater is close to the ground surface, groves of date palms, fruit trees and agriculture grow exuberantly (Petrov, 1976). This state of affairs causes the population density in and around them to be several times the overall average density in the desert areas.

A number of Rivers and other watercourses traverse the surface of the Arab Region. Rivers are, by definition, perennially flowing streams while other watercourses can be intermittent or ephemeral, known as wadis (in French oued is used for wadi). The large rivers: the Nile, Tigris and Euphrates, and Sénégal, traversing the lands of Egypt, Syria and Iraq, and Mauritania, respectively have their sources in areas lying outside the region. Some tributaries of the Nile have their sources fully or partly rising in the Sudan. Some rivers, despite their small basin areas, are shared by more than one Arab country, or by Arab and non-Arab countries. The Khabour, Belikh and Sajur Rivers are examples of small rivers, yet shared between two countries, Syria and Turkey in this case.

While some countries of the Arab Region are not traversed by any perennial stream, they are intersected by major and minor wadis. The duration of flow is basically determined by the density, depth and frequency of rainstorms. Some of the rain spells produce flash or torrential floods. Such floods can bring along excessive amounts of silt-laden water in a short period of time. It is often too difficult to control or manage these kinds of floods. It is equally difficult to measure the flood volume with high level of accuracy. Examples can be found in Algeria, Tunisia, Oman, and many more countries. One of the demerits of such floods, despite the large amounts of water they can bring along, is the excessive erosion and subsequently the sedimentation of the erosion products.

An approximate map showing the locations of the perennial and major wadis is presented in Figure 4. Tables 1 and 2, both of Appendix II (hydrometric data), give the catchment areas of both perennial rivers and major wadis in the Arab

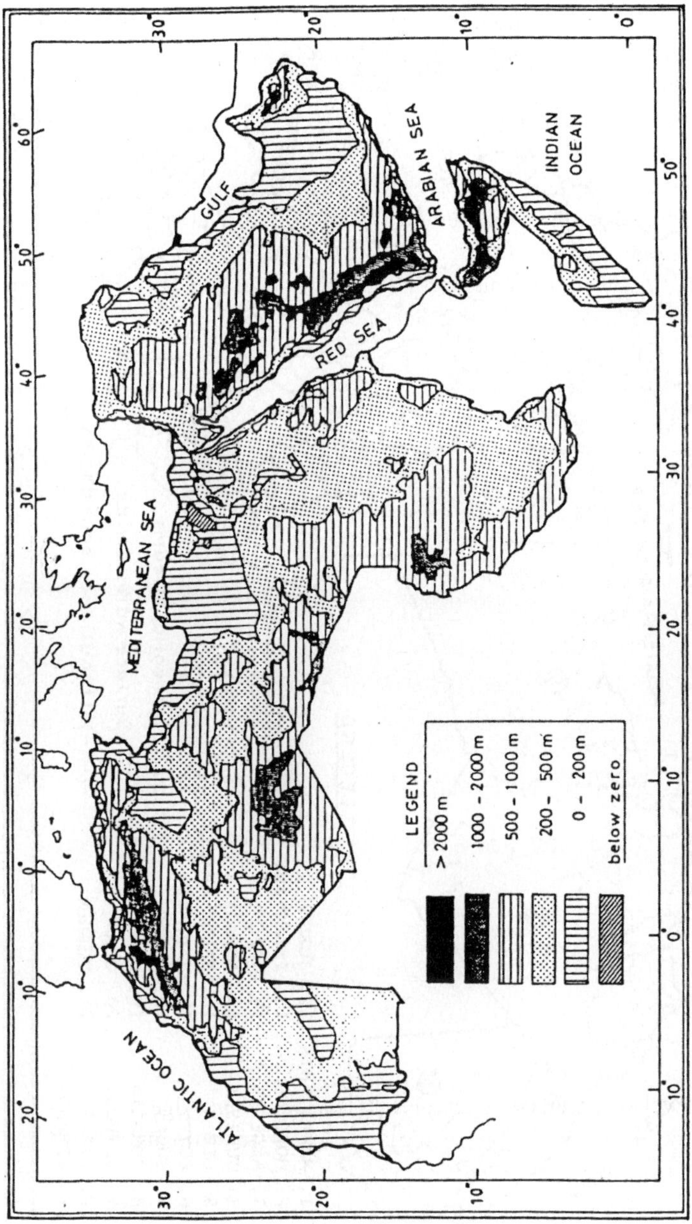

Figure 3. Topographic map of the Arab Region

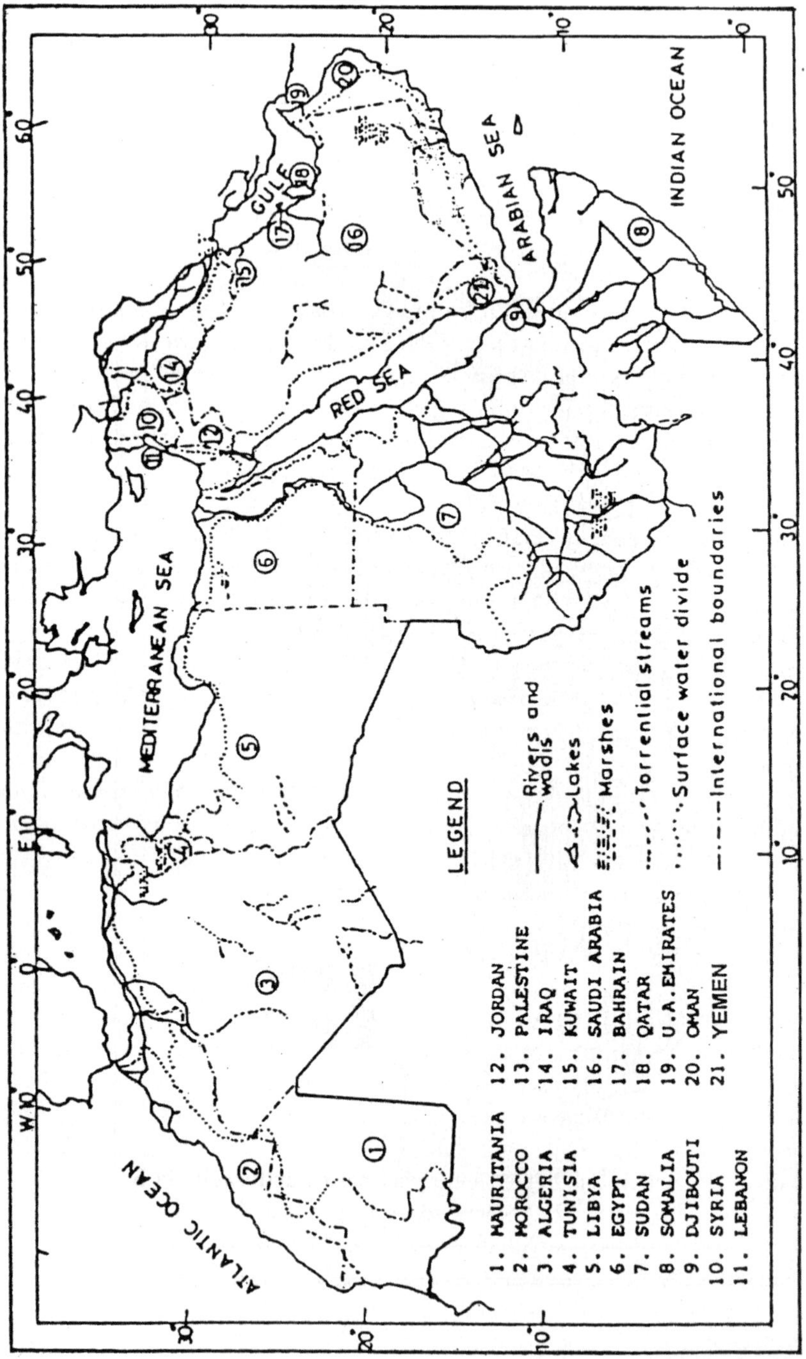

Figure 4. Map showing the perennial rivers and major wadis in the Arab Region

Region, together with the long-term mean annual rainfall on these catchments and the corresponding annual flows.

2.2. BRIEF DESCRIPTION OF THE GEOLOGICAL SETTING OF THE REGION

The geological history of the Arab Region dates back to the Pre-Cambrian Age, i.e. more than 500×10^6 years ago, as can be seen from the Geological Scale or Table shown in Figure 5.

The African Shield was formed in the Pre-Cambrian Age. It has been suggested that the Arabian Shield, which occupies the area presently known as the Arabian Peninsula, formed a part of the African Shield. The Arabian Shield has three major components. These are: the Western Arabian Shield forming the central Najd-Hijaz and Asir Highlands in Saudi Arabia, the Yemen Plateau, which extends to Aden, and the Southern Arabian Shield along the Arabian Sea coast.

The tectonic movements that took place much later than the Pre-Cambrian Age caused the Arabian Shield to split from the African Shield. The split of the Arabian part of the shield from the Nubian part, which occurred before the end of the Cretaceous period (135–65 million years BP, Figure 5), is considered to be the first indication to the geological destruction that resulted in the formation of the huge depression filled by the Red Sea. It is also believed that the Arabian Shield had been subject to a slight rotation of a few degrees, and that a tear in the earth's crust had occurred, thereupon resulting in a slow displacement of the shield starting from the Red Sea.

The southern stretch of the arch engulfing the Anatolian and Iranian Fold Mountains frames the Arabian Peninsula in the north and northeast. The Mesopotamian Foredeeps are located in a zone of marginal troughs with thick Mesozoic and Cainozoic sediments of marine geosynclinal facies.

The thickest accumulation of sediments occurs at the eastern side of the Arabian Gulf. It decreases in a northwesterly direction and the zone becomes somewhat shallower and narrower util it forms in the north of Syria a small seam along the southern margin of the folded mountains. The marginal troughs gradually pass southwards into a wide zone of the unstable shelf of the Arabian Platform. The sediments of this zone represent the transition from geosynclinal facies to the predominantly continental depositional environment in the foreland of and on the Nubian-Arabian Shield.

The unstable shelf is subdivided into local basins and swells in Syria, Jordan, the eastern province of Saudi Arabia and in the Arabian Gulf States. The basins offered locally favourable conditions for the accumulation of euxinic rocks and the swell areas were subjected to terrigeneous influence. The transition zone from the unstable to the stable shelf extends from Central Sinai to the Gulf of Aqaba and from there follows the Wadi Araba Depression in Jordan. In Syria, it trends eastwards and south eastwards whereas in Saudi Arabia the transition trends southwards and

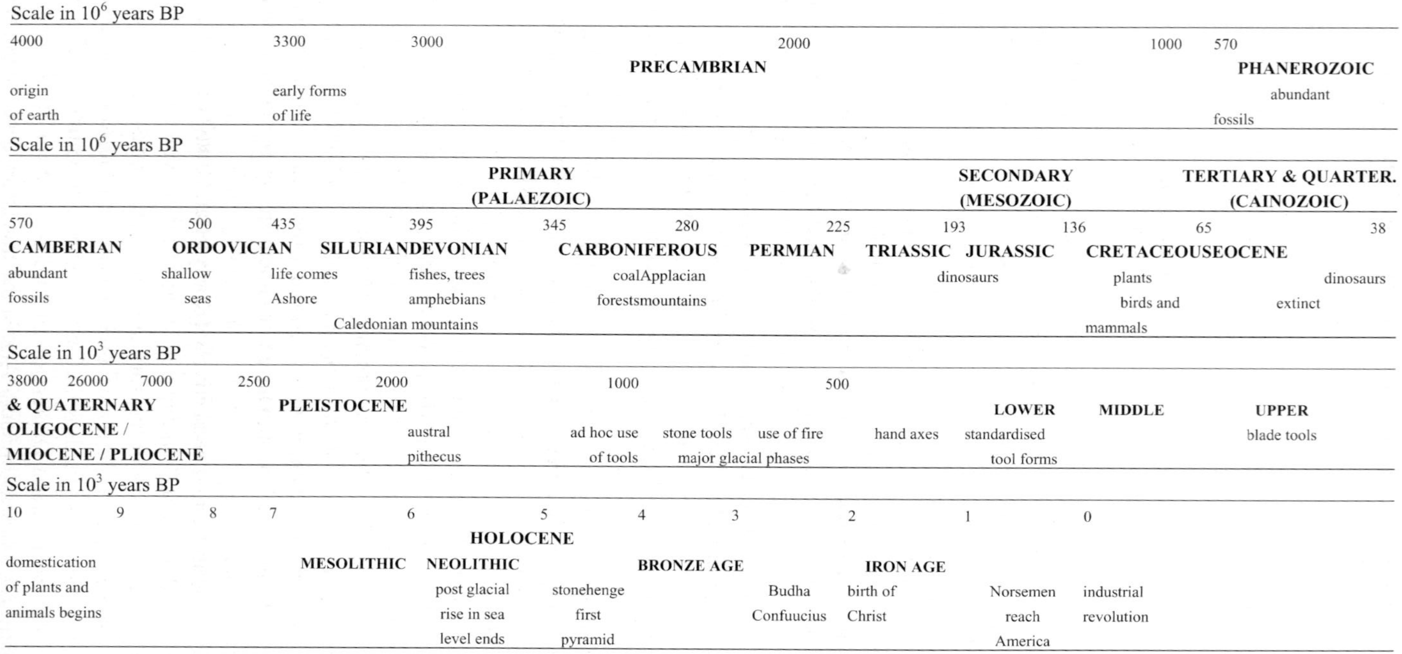

Figure 5. Schematic representation of the Geological scale (redrawn from the Times Concise Atlas of the World, 1980)

approximately parallels the Arabian Gulf. In the zone of the stable shelf, neritic, littoral and continental sediments are intertwining.

The Nubian-Arabian Shield, regionally emerging southwestwards, occupies the southern part of the Sinai Peninsula and the areas adjacent to the Gulf of Aqaba and the Red Sea as far as approximately the northern border of Yemen. Clastic sediments of continental origin and rocks of the Pre-Cambrian basement complex characterises these areas (Bender, 1974).

The outcropping of the Crystalline Shield covers vast areas in the Arab Region. The most important outcroppings can be found in Mauritania, the southern and western parts of Algeria, the southern part of Libya, east and southwest of the Sudan, and southeast of Egypt. These areas comprise certain parts of the African Shield. The sediments overlying the Crystalline Basement at some of the depressions in North Africa and the Arabian Peninsula can be up to some kilometres thick.

Certain parts of the sedimentary cover had been formed as a consequence of frequent inundation of the sea, i.e. in a marine environment. Those parts of the sedimentary cover that escaped the inundation by seawater, mostly in the southern part of the Arab Region, which is located in Africa, were deposited in a continental environment. They are basically deposits from the Nubian sandstone, covering the area located between the northern Sudan, the Western Desert in Egypt and Libya. Similar deposits, known as Continental Intercalary were formed in the southern part of Algeria. The Tertiary Period witnessed alternating deposition of sand and sandstones, and calcareous lacustrine sediments. These deposits have formed the area known as Al-Hamada, which can be found in the western part of the Arab region in North Africa, and known as the Continental Terminal. Additionally, several basins made up of sandstones belonging to the Maestrichtian-Eocene period or from the Quaternary Age are spread here and there in the area covered by the Atlas Mountains. The rock outcrops in the northern parts of Morocco and Algeria are mostly made up of impermeable marls.

The depression of the Dead Sea, which extends across Al Beqā and Al Ghab Plains in Syria, can be regarded as one of the most important tectonic movements in the eastern part of the Arab Region. This has caused a shift of the block situated east of the depression by about 150 km in a northerly direction. Despite the fact that the region is still active, the movement is slowing down with time as a result of the collision of the Arabian Plate with the Euro-Asian Plate. As a consequence of the mentioned structural changes, certain land areas are developed along the Euphrates in Iraq and Syria, thus delineating the course of the river channel. The maps in Figures 6 and 7 show the approximate geological build up and the tectonic movement in the Arab Region respectively (Khory et al., 1986).

The Tadmur (Palmyra) sedimentary basin was formed under the huge pressure forces acting in a northerly direction resulting from the increasing distance between the two extremities of the Red Sea. A similar sedimentary basin was formed during the Triassic and Jurassic Ages in the area where Sangār Mountain is located. In both basins Carbonic-evaporite sediments were deposited and accumulated. The thickness of the sediments at certain locations reaches up to 2,500 m. After the

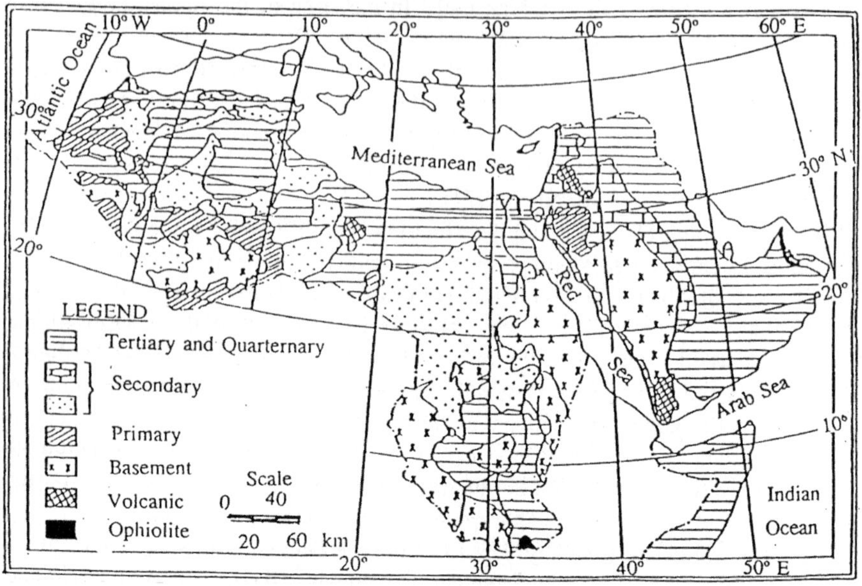

Figure 6. Geological map of the Arab Region (redrawn from Khory et al., 1986)

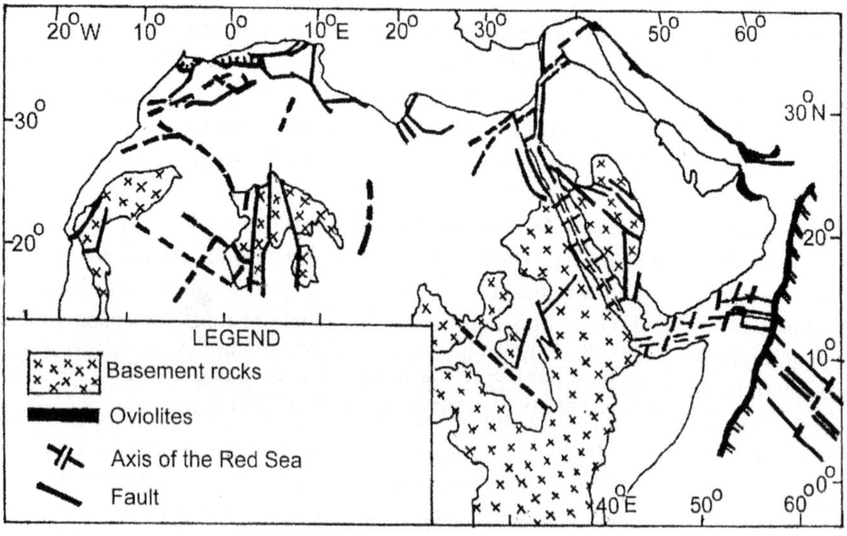

Figure 7. Tectonic map of the Arab Region (Redrawn from Khory et al., 1986)

sedimentation process came to a halt it continued in the Tertiary and Quaternary Ages. The sediments filling the Tadmur and Sangār basins responded much more than other basins to the pressure forces acting on them causing a severe warping and fissuring of their strata. The huge uplift, which occurred to these sedimentary basins in the Tertiary Period, stopped the transgression of the sea and led to the formation of a chain of mountains.

The volcanic activity in the eastern wing of the Arab Region is relatively recent as it began only in the Palaeo-Pleistocene Period. The most important areas that have been strongly influenced by this volcanic activity are the Horan Plateau, ed Drūz Mountain (Jebel ed-Drūz) in Syria, northern Jordan and Saudi Arabia. Likewise, mountain ranges parallel to the coasts of the Mediterranean Sea and the Red Sea were formed in the areas adjacent to the grabens in Syria, Lebanon, Palestine, Egypt and Saudi Arabia. These chains form the recharge areas of the groundwater basins in the eastern subregion, including the Arab Peninsula.

Equally important to the tectonic movements that led to the formation of mountains are the movements that led to the gradual subsidence of the Arabian Shield in the Rub al Khali, Arabian Gulf, Euphrates and Damascus regions. A group of sandy deposits that accumulated in the large sedimentary basins in earlier times outcrop in the subsurface of the western and northern parts of Saudi Arabia. While moving northwards, these thick sandy deposits lost their importance as groundwater aquifers due to the strong inclination of the strata embodying them. The sedimentation continued in the Secondary and Tertiary Ages thereby accumulating deposits mainly in the form of carbonate rocks. The Secondary Age sediments in Syria, Lebanon, Palestine and Jordan are mostly dolomite-calcareous sediments. Similar deposits were formed in the basin of the Arabian Peninsula during the Tertiary Period (Palaeocene and Eocene, Figure 5). It is worthwhile to mention that the Tertiary sediments in the East Mediterranean Basins are mostly made up of marl.

Sedimentation proceeded in the Quaternary in most of the basins mentioned above. However, the deposited sediments were formed in the same environment. As such, they are not all similar, instead, they are of sand, gravel or calcareous.

2.3. PHYSICAL AND GEOLOGICAL SETTINGS OF THE COUNTRIES MEMBERS OF THE ARAB REGION

While presenting this section, we would like to remind the reader that the countries comprising the Arab Region will be arranged in the following order:

- The western subregion comprises Mauritania, Morocco, Algeria, Tunisia and Libya. The total surface area of this subregion is 5,782,091 km^2.
- The central subregion comprises Egypt, Sudan, Djibouti, Somalia and Comoros Islands. The total surface area is 4,170,354 km^2.
- The eastern subregion comprises Lebanon, Syria, Palestine, Jordan, Iraq, Kuwait, Saudi Arabia, Bahrain, Qatar, Oman, United Arab Emirates (UAE) and Yemen. This subregion has a total surface area of 3,740,992 km^2.

2.3.1 The Western subregion states

2.3.1.1 Mauritania

The Republic of Mauritania lies in the far west of the Arab Region between 27° 17′ and 14° 40′ N latitudes and 17° 03′ and 04° 54′ E longitudes. The country covers a surface area of about 1,030,700 km^2. It is bounded on the north and northwest by the Western Sahara, on the west by the Atlantic Ocean, on the south by the Sénégal and Mali and on the east and northeast by Algeria. The coastline along the Atlantic Ocean is about 660 km long.

Mauritania is essentially a Saharan state, with nearly two-thirds of its surface situated in the Sahara Desert. This vast surface is almost flat; there is hardly any difference in elevation between one location and another to encounter. In the desertic part of Mauritania there are sandstone plateaus such as the Adrar, Tagant and Karêt rising to between 300 and 500 m a.m.s.l. Some isolated peaks protrude out of the high plateaus. Peaks, such as Kedit el Jill (915 m) can be rich in minerals. Nearly 40% of the surface of the country is covered with ergs (sand dunes). These ergs keep shifting their locations under the influence of the northeastern trade winds, which are blowing continually towards the south. The southern sector of the Atlantic littoral of the African Sahara lies in Mauritania, but without any well-defined physiographic limits inland.

The surface of the country is marked by coastal dunes and playas (sebkhas), inland from which long lines of high, reddish sand dunes extend for hundreds of kilometres inland over barren surfaces of sand and rock. The inner part of the littoral is a sparse grazing area, plagued by limited water and scanty vegetation (Meigs, 1966).

The present state of knowledge of the geology of Mauritania still has a number of uncertainties that are unsettled. However, the geological/geomorphologic chronology can be approximately set as follows.

Sedimentation, metamorphism, accumulation of magma and Pre-Cambrian folding had been smoothened by a flat polygenic surface some time before the beginning of the Primary Age. From the Infra-Cambrian to the Caboniferous times, a series of thick sedimentary layers accumulated between the two geologic units composing the Taoudeni and Tindouf Basins. The Basement rock of the African Shield outcrops between the Sedimentary Basins of Tindouf in the north, Taoudini in the east and the Sénégalo-Mauritanian in the west in the form of a dome. It assumes the shape of a crescent 400 km wide in the north and about 70 km at the level in the south.

The chain of Mauritandies (Figure 8) was heavily folded to the extent of having the products of their erosion moved. By the end of the Primary Age, the recapturing of erosion material, a process that occurred during the Cretaceous Period, must have led to uncovering the Pre-Cambrian surface in the northern and northeastern parts of the country. Since then, only inselbergs, which are composed of granite and highly metamorphosed rocks had been left as evidence of the past, intense orogenic activity.

The edges of the sedimentary basins, which receded backward, rested against the highly elevated plateaus while as sedimentation kept accumulating in the middle

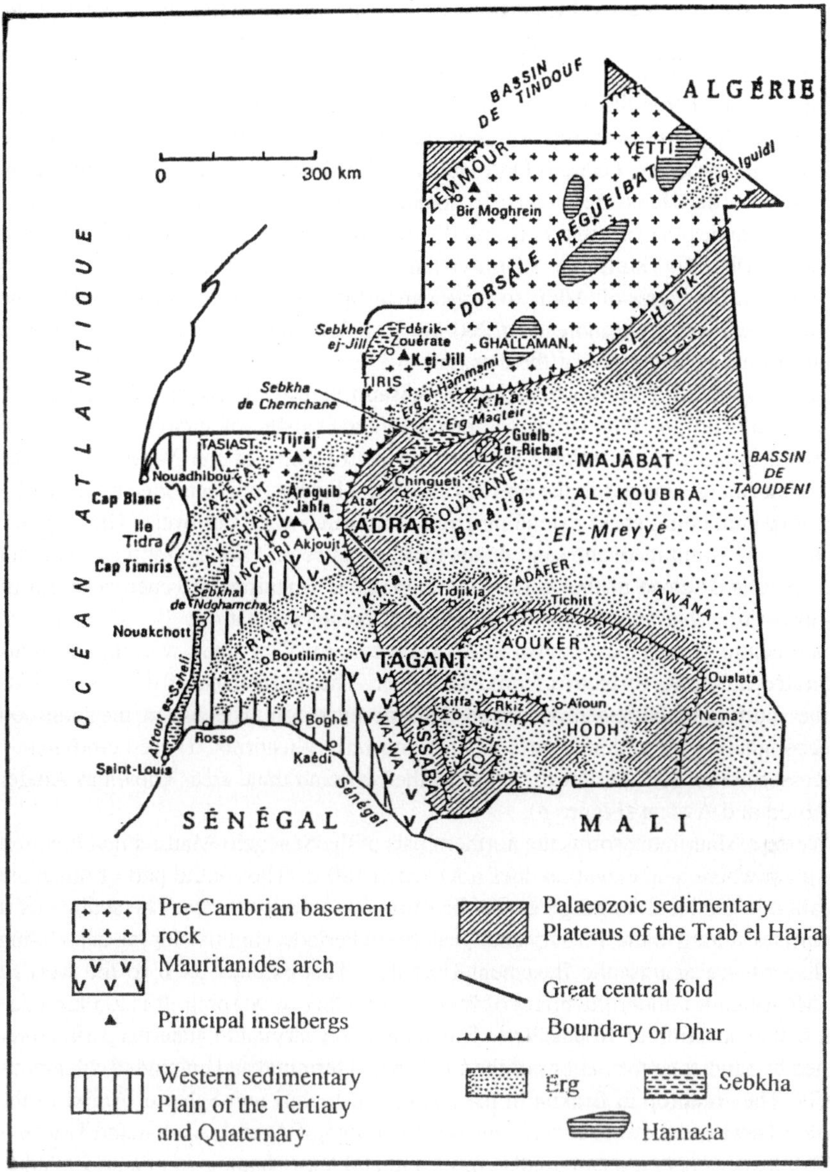

Figure 8. Relief map of Mauritania (from Pitte, 1977)

and gradually filled the hollow basins. Halfway down the Tertiary, the redeemed deposits found their way to the Hamada Basin where they became entrapped and fossilised here and there. In the west a slow subsidence washed down the formation to a new basin, known as Sénégalo-Mauritianian, where sedimentation continued their accumulation. Lastly, the Quaternary deposits are decorating the true landscape by their magnificent authenticity. The alternation of the dry and wet episodes is responsible for the build up of the ergs, cuirasses, etc.

The rim of the Taoudeni Basin is known as the land of stone dust (Trab el Hajara) contrary to the loose sand and/or lightly consolidated sediments predominant in the east and in the west. Cambro-Ordovician and Caboniferous sediments appear in the south of the Hank area, which is located in the north of the Taoudini. These sediments appear vigorously as from 10° longitude. They constitute the skeleton of the Adrar Plateau. The Cambro-Ordovician sediments disappear under the sands of Al Khatt in the south of Adrar to reappear in the Tagant Plateau (Figure 8). The Assaba, a narrow strip more than 200 km in length and not exceeding 25 km in width, forms an extension of the Tangant.

The Majâbat al Koubra situated in the eastern part of the country is, in general, of fluviatile origin. It is often claimed that the manner of endorheic spread of the sand dunes there is related to the past hydrography of the Niger Basin. The so-called Bnâig, situated in the northwest of the Majâbat, presents the structure of a gigantic, undulating sheet extending from the northeast to southwest. The distance between the summits of the sand dunes there is between 2 and 3 km; they ravel out into 'barchans' close to the Adrar Mountain. These manifest the encroachment of desertification that approaches the limits of Sahelian vegetation.

The centre of Majâbat, known as Mreyyé, is formed of small winding cushions separated hundreds of metres from each other. The dominance of the trade winds in the south is much less pronounced than the monsoon. The rise of the monsoon is accompanied by violent swirls of dust and rare rainstorms. These aerodynamic processes are responsible for the relief of the two sand dune areas known as Adâfer el Abiod and Awâna (Figure 8).

Western Mauritania forms the northern part of the Sénégalo-Mauritanian Basin. It is a plain where land elevation does not exceed 100 m. The central part of this plain is called in the local language as the direction opposite to Mecca. The sediments it comprises belong to the Tertiary and Quaternary Periods, slightly sloping and resting in discordance against the Basement Complex. This extends itself to the west of the Mauritanides under the effect of faulting, bending or warping. It is 2.5 km deep below the elevation of Nouakchott. The ancient Tertiary and Quaternary are represented by sandstone formations of the Continental terminal and entangled calcareous banks. They outcrop at Brakna in the south, and Inchiri and Sahel el Abiod in the north. Elsewhere, they are covered by the sand dunes of Azefal, Akchar and Trarza.

The western ergs are essentially fossilised, and their shaping can be attributed to the dry and wet episodes of the Oligocene Period. The trades and westerly winds blow across the coastal areas. Should these two streams have no effect on the overall relief, it would be the Sahelian vegetation that undertakes proper fixation of the western ergs. The landscape of the strip of land bordering the Atlantic as well as the Sénégal Valley (Chemama) is largely due to the late part of the Quaternary. It also has the landscape of a mangrove as can be found at Cape Timiris, less than 100 km north of Nouakchott.

The landscape changes little by little until the Nouakchottian shore becomes lagoonal. The lagoons have dried up under the effect of desertification during the course of the last 3,000–4,000 years. The Aftout es Sahili is a long, straight depression situated slightly beneath the sea level. It stretches itself from the mouth of the Sénégal River up to Nouakchott. It currently separates the reddish dunes of the off shore bar. A string of 'sebkha's' separated by sills occupies the bottom, and are used for extracting gypsum. When inundated by heavy winter rains these sebkhas came in contact with the River Sénégal (Pitte, 1977).

2.3.1.2 Morocco

The Kingdom of Morocco is situated in northwest Africa between latitudes 27° 30′ and 36° N and longitudes 01° and 13° W, thus covering a surface area of 446,550 km^2. The total length of the coastal line is about 2,170 km, of which 470 km along the Mediterranean Sea and approximately 1,600 km along the Atlantic Ocean. Morocco (Al Maghreb, meaning west in Arabic) is bordered on the east by Algeria and by the Western Sahara on the south. Of all Arab States, Morocco is the closest country to Europe. The Strait of Gibraltar, which separates Morocco from Spain, is 13 km wide.

With the exception of the low-lying coastal strip along the Atlantic Ocean, a vast surface area of Morocco is covered by the Atlas Mountains. This mountain range, along its length from north to south, comprises four massifs. The Rif Mountains in the north with their top Tidirhine (2,459 m a.m.s.l.), form a buffer along the Mediterranean Coastline. The Atlas Mountains form a barrier across the middle of the country.

They comprise three distinct ranges. The High Atlas (in French Haut Atlas) is 740 km long and its top at Toubkal is 4,165 m high. The Middle Atlas (Moyen Atlas) trends away from the High Atlas in a northerly direction rising to a peak elevation of 3,340 m at Naceur. The Anti Atlas extends southward from the High Atlas to the Atlantic, and can be as high as 1,250 m a.m.s.l. The area situated between Al Rif massif and the Middle Atlas is a low-lying plateau, often referred to as the Gap of Taza. History tells that the invaders coming from the east used this gap as a gateway to threaten the country.

Morocco lies on the rim of the African Plate. The Atlas and Rif Mountain Ranges, which extend from southwestern Morocco to northeastern Tunisia, are the result of complex mountain-building processes that occurred during the Tertiary Period (between 66.4×10^6 and 1.6×10^6 years ago). Limestone and sandstone deposited on an older surface of the Hercynian Age (roughly 374×10^6 to 286×10^6 years old) were pressured upward and folded and faulted by northward movements of the African Plate. This plate is colliding against the Eurasian Plate. The persistent structural instability in the mountain ranges can be seen in the periodic occurrence of earthquakes, such as the one that struck Agadir, Morocco in 1960, and Algiers, Algeria in 2003, just a few days before the first drafting of this section. Both events were devastating indeed. Additionally, northwest Africa is characterised by the existence of a major fault, which extends from the Atlantic Ocean in the west to Syria and Turkey in the Eastern Mediterranean. This fault separates the uplifted mountainous areas in the west from the low-lying desert areas in the south.

The primary rocks cover the central region of Morocco. They extend further from the Atlantic Ocean towards the inland. Most of these rocks are Carbonate rocks and in some locations they alternate with sandstones. In northern Morocco and along the seacoast the rocks belong to the Secondary and Tertiary Periods, most important of which are the Oligocene and Miocene Ages. The deposits in the northern half of the Coastal Plain adjacent to the ocean are mainly sandstone belonging to the Secondary period, while the rocks in the southern half of the plain down south to Mauritania are bounded on the east by marl and on the southeast by basic rocks.

Layers of sandstone alternating with marine calcareous rocks were deposited during the Tertiary and Quaternary Periods (Eocene and Pleistocene) Periods. These deposits can be frequently found in Al Hamada-Tounassine region in Morocco and Algeria. The marly facies and calcareous dolomite rocks of the Jurassic are spread in the Rif Basin. In the region of Middle Atlas and its extension towards the Atlantic one finds calcareous formations belonging to the Jurassic Age.

East of the Rif and the Atlas Ranges is the Moulouya Basin, a semi-arid land created by the eroding force of the Moulouya River. Farther east are the high Plateaus of eastern Morocco, which are extensions of landforms of neighbouring Algeria.

2.3.1.3 Algeria

The Republic of Algeria is situated in the northwestern corner of Africa between 19° and 38° N latitudes, and 9° W and 12° E longitudes. With its surface area of 2,381,741 km^2, Algeria is the second largest country of the Arab Region. The country has a long coastline of about 1,200 km overlooking the west Mediterranean Sea.

Algeria is bounded by Tunisia and Libya on the east, Western Sahara and Morocco on the west, and Niger, Mali and Mauritania on the south. The surface of Algeria can be subdivided into three physiographic units (Figure 9). These subdivisions can be briefly discussed as follows:

- The Atlas Mountainous region comprises two mountain ranges. The Mesozoic Tell Atlas runs along the narrow coastal line of the country in the north. The height of these mountains varies between 300 and 600 m in the northwest of the country. The highest peak (2,308 m) lies at the top of the Jurjura massif. A number of deep, spectacular ravines intersect this mountain range. South of the Tell Atlas lies the monotonous High Plateau of Algeria. Bordering this plateau to the south is a series of low mountains – the Ksour, Amour, and Oulad Naïl-collectively termed the Saharan Atlas, which is the second range of mountains. The two mountain ranges are separated by a group of saltwater shallow lakes or depressions known as Chotts, most important of which is the Eastern Chott (locally known as Chott ech Chargui). The two ranges merge and join the Aurès Mountain whose top is located at Chelia (2,308 m).

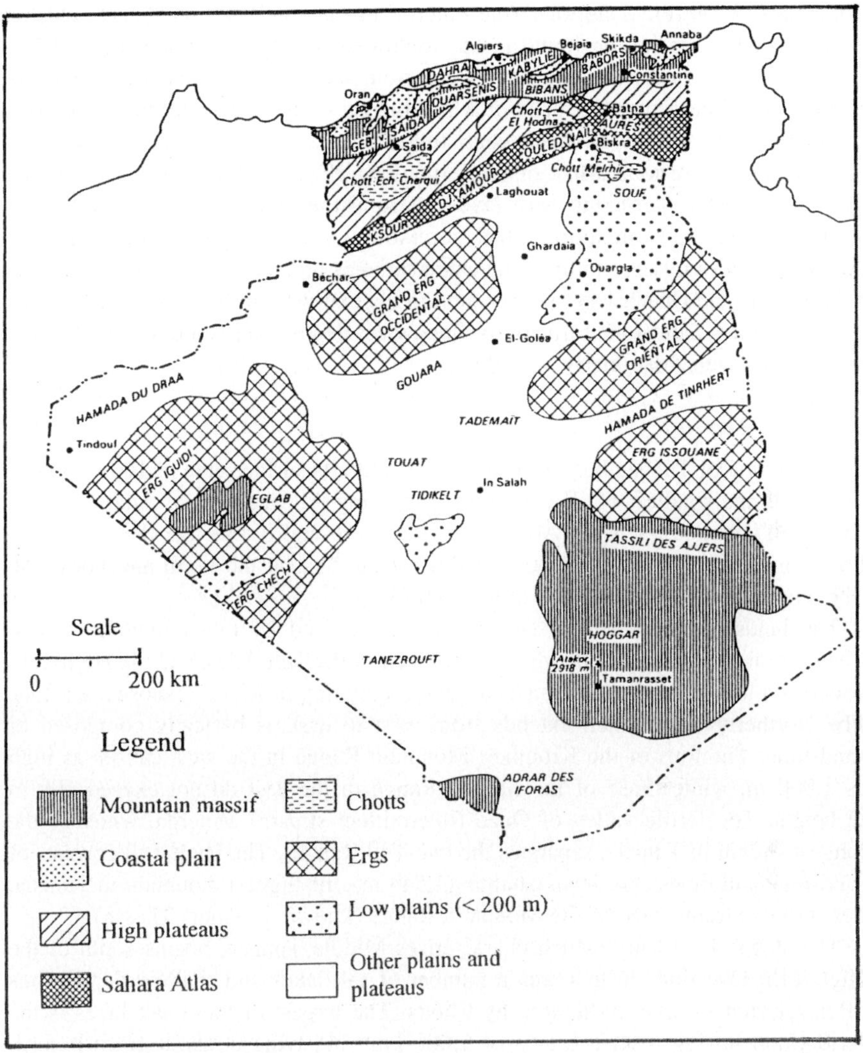

Figure 9. Physical setting of Algeria

- The Southern and Southeastern Heights are isolated heights on the surface of the northern part of the Great Desert or Sahara and known as the Hoggar Plateau. The peak elevation of this plateau is at 2,918 or 3,003 m a.m.s.l.
- Low Plains are located between the Hoggar Plateau and the Atlas Mountains. The elevation of these plains varies from 350 m in southwest to 100 m in the northeast. In some locations it can be as low as 21 m b.m.s.l. The dune sands of the Great Western Erg (Grand Erg Occidental) cover these Plains.

The rocks of the Pre-Cambrian Age outcrop in the south and southeast, and the decomposed rocks can be found in the southwest of Algeria. While the rocks of the Primary and Secondary Periods extend from south to north, some of the sandy formations ‘Continental Intercalary’, which belong to the Secondary Period, extend close enough to the coastal areas where they alternate with deposits of the Tertiary and Quaternary Periods. Most of the formations of the Tertiary consist of marine calcareous rocks alternating with layers of sandstone ‘Continental Terminal’. In view of the excessive faulting of these formations, which occupy a large surface of Algeria, one can observe in them the characteristics of karst formations. Marl rocks belonging to the Quaternary can be found in the extreme north of the country close to the seacoast. Sandstone formations, floodwater deposits and dune sands are often interbedded with these marly formations.

2.3.1.4 Tunisia

The Republic of Tunisia lies in North Africa. With its surface area of 163,610 km^2, Tunisia is the smallest of all countries comprising Al-Maghreb Al-Arabi. Tunisia shares its boundaries with Algeria on the west and Libya on the southeast. Furthermore, it is bounded by the Mediterranean Sea on the north and northeast. The length of the coastal line is about 1,000 km.

The landscape of northern Tunisia is characterised by two mountain ranges. These are the Northerly Tell, which is a branch of the Tell Atlas, and the High Tell, which is a branch of the Saharan Tell, both originating in Algeria (subsec. 2.3.1.4). The Northerly Tell, which extends from west to east, is basically composed of sandstone. The tops of the Kroumeri Mountain Range in the west can be as high as 1,000 m, while those of the Mogod Range in the east do not exceed 400 m in height. The fertile valley of Oued (intermittent stream) Mejerda, which is the longest stream in Tunisia, separates the two Tell Ranges. The High Tell consists of sandstone and limestone. Jebel Chambi (1,544 m), the highest mountain in Tunisia, lies in the extreme west of the Mountain Range (known as Monts Tbessa).

The steppe landscape, which characterises Middle Tunisia, begins south of the High Tell. One finds in this area a number of salt lakes and shallow depressions often referred to, like in Algeria, by Chotts. The largest of these salt lakes is the Jerid Chott having a surface area of 5,000 km^2 and lying at about 16 m b.m.s.l. The lowest part of Tunisia is Chott el Gharsa, 23 m b.m.s.l. The extreme south of Tunisia lies in the Sahara Desert, a large part of it consists of sand plateaus interrupted by chains of hilly ground such as the Dahra.

Tunisia generally consists of Sedimentary Basins. These can be subdivided into regions separated one from the other by the Southern Atlas Fault situated along the Saharan Atlas. The latter extends from Algeria in the west to Tunisia in the east.

- The Northern Zone is occupied by a considerable length of the Atlas Range, which extends from southwest to northeast Tunisia. The stretch of the Tell and Saharan Atlas approach one another in this zone where the most favoured agricultural lands lie north of the rain watershed, the ‘Dorsale’. The northern zone is geologically characterised by uniform distribution of its warps, which are intersected by tectonic faults in two perpendicular directions. This structural pattern has helped in dividing the said zone into small twisted units, especially in the extreme north. There, some faults accompanied with salt stones from the Triassic and other tectonic formation belonging to the Oligocene can be found.

- The Coastal Zone is less active compared to the previous unit. It encompasses two sedimentary basins belonging to the Miocene and Palaeocene Periods. As the area situated in the extreme north is characterised by its many warps and is dominated by rather impermeable formations, most of the rainfall flows as a surface runoff, especially so that it enjoys a dense drainage network.
- The North-western Zone represents an extension of the Atlas Mountains. Cretaceous formations, especially those deposits of the Cinemania Age, dominate this zone. They show up as calcareous layers overlying layers of marl. The next important layers are those calcareous deposits belonging to the Eocene Age. All these calcareous layers overlie the high grounds and form a certain component of their built up. The low-lying plains, which separate the high peaks, are deposits of the Quaternary Age.
- The Middle Zone is characterised by its abraded formations belonging to the Miocene Age. These formations usually exist beneath the sediments of the hollow warps, which have a vast extent and were deposited prior to the Quaternary Age.
- The Desert Plateau is characterised by its inactive geological formations over the time from the Triassic up to the Quaternary. They have the structure of a monoclinal dip extending toward the Eastern Erg. These formations on their eastern end overlook the Jeffara Coastal Plain, Libya, which comprises sedimentary deposits too.
- The Southern Zone is characterised by the outcropping of the oldest geological units, which belong to the Permian and Triassic Periods. These formations have abrasive nature while the calcareous rocks and marl dominate the deposits of the Jurassic Period. The deposits of the Lower Cretaceous have a vast aerial extent, and consist mainly of sand and sandstone. The Chotts are encountered in sandy Miocene formations and extend towards the Algerian Desert in the west

2.3.1.5 Libya

The Djamahiriyya (People's) al Arabbiyya al Libbyya is located between latitudes of 33° 01′ and 19° 30′ N and longitudes of 9°10′ and 25° 01′ E. With a surface area of 1,759,540 km^2, Libya ranks the fourth in area after Zaïre, the Sudan and Algeria, in Africa. It also ranks the fourth in area among the Arab countries after The Sudan, Algeria and Saudi Arabia. Libya borders Egypt on the east, the Sudan on the southeast, Chad and the Niger on the south, Algeria on the west, and Tunisia on the northwest. The northern zone of Libya is bounded by a coastline 980 km in length running along the Mediterranean Sea.

The northwestern Coastal Plain along the Mediterranean Sea is known as the Jifārah Plain, sandy steppes with stunted vegetation and occasional muddy depressions into which water seeps at certain times. The Plain is a triangular area of about 20,000 km^2, bounded on the south by the bare rock cliff known as Jabal Nafūsah or simply the Jabal. The Jifārah is a low-lying area; its topography rises slowly from sea level along the coast to 200 m at the foot of the escarpment of Jabal Nafūsah. The steep edge of the Jabal, which can be regarded as the beginning of the African Plateau, is serrated by dry watercourses. The average height of Jabal Nafūsah is 500 m. In some places, however, it can be up to 760 m in height but falls eastwards (Stamp & Morgan, 1972).

In several locations there are lava deposits and old craters that indicate past volcanic activity in this area. A high plateau begins south of Jabal Nafūsah and extends several hundreds of kilometres in a southerly direction and gradually rises in elevation to reach a height of about 1,200 m. At the northeast side of this high plateau a number of wadis runs off during the rainy season.

The high Plateau generally has a desert landscape made up of sand, gravel and rock formations. A number of depressions and oases lie south of the High Plateau, thus changing the desert landscape. Another High Plateau, 1,000 m in height, lies to the south of these depressions. This High Plateau extends itself towards the west and south and joins the mountains in Algeria and Niger respectively.

The low-lying area along the Gulf of Sidra lies relatively to the south that it almost makes part of the desert. Here a number of salt lakes exist, of which Sabhat (Sebkha or Sabkhat) Ghuzzayil is the deepest one (47 m b.m.s.l.). South of the low-lying area along the Gulf of Sidra a number of offshoots of high volcanic plateaus extend for several hundreds of kilometres. The highest elevation of these plateaus is 1,200 m. South of these volcanic plateaus lies a number of depressions, and in the far south a new high plateau including offshoots of the Tibesti Mountain Range of Chad. The highest mountaintop, Bette, within Libya is 2,286 m.

The Jabal al Akhadar, a steep coastal mountain, lies in northeast Libya along the Mediterranean Sea. This mountain rises from the seaside and reaches a maximum elevation of 600 m. Behind it stands a low-lying plateau, which does not belong to the Libyan Desert, but rather to another desert known as the Great Sand Sea. It consists of gigantic dunes of very fine sand intercepted here and there by a number of oases.

The southeastern part of Libya is made up of high desert plateaus with scattered locations of inactive volcanoes and lava formations. A number of mountains exist along the Libyan frontiers with Egypt and the Sudan. Al Kufrah Oasis, famous for its groundwater resources is situated in the middle of these mountains (Mourik et al, 1986).

The principal geologic features of Libya can be summarized in the following:

- The Basement Rock outcrops on the heights of Fezzan Mountain south of Libya. These deposits belong to the sandstones of the Cambrian-Ordovician.
- The northern part (north of Al'Azīzīyah parallel) of the Jifārah Plain is defined by the Miocene transgressive series, which thickens towards the north. The total thickness of the Miocene-Pliocene-Quaternary formations can reach 600 m along the coast. This sequence consists of thick sandy and calcareous beds with clay intercalations. The Miocene overlies the Triassic limestones and Jurassic evaporates in the west, and the Lower Calcareous sandstones in the east. The southern

part of the plain lies between Al'Azīzīyah parallel and Jabal Nafūsah. There, thin Quaternary mantel covers Upper Jurassic alternating sandstones, clay and dolomites in the southwestern part, Triassic dolomitic limestones in the central part and the Lower Cretaceous sandstones in the eastern part. Furthermore, the Basement Rocks outcrop at places on the Tāwurghā Heights and Jabal Nafūsah.

- The deposits of the Palaeozoic Age are found in Al Hammada al Hamra Plateau in the form of a syncline extending from east to west. They consist of limestone rocks and marl.
- The canyons of the Hūn region form the natural divide line between the basins of Sirte and Al Hammada al Hamra. The Tertiary deposits in the Hūn region and the Basin of Sirte seem are well developed, dominated by evaporites and carbonate rocks. The Miocene deposits at Misrātah (Misurata) in northwestern Libya do not penetrate inland except for a few kilometres.
- Al Kufrah Basin south of Tazerbo is well delineated on its periphery by Palaezoic outcrops composed mainly of continental sandstones. The central part of the basin (more than 250,000 km^2) is occupied by Mesozoic continental sandstone outcrops occasionally covered by sand dunes or alluvial deposits. North of Tazerbo, a Cretaceous and Lower Tertiary marine transgression overlies the Mesozoic and Palaeozoic sandstones. After the Eocene a continental environment was established. Post-Eocene deposits consist chiefly of sand, sandstone and clay with some sandstone. Their total thickness can be up to 900 m in the centre of As Sarīr. Further north of Jālu, north of Tazerbo, the post-Eocene continental deposits pass gradually to shallow marine carbonaceous formations with many evaporitic intercalations characteristic of lagoonal and estuarine conditions (Pallas, 1987).

Figure 10 shows a general geologic map of most of the Libyan territory. It also includes a cross section extending from Tazz Ulli in the west of the country to Sarir Tibesti through Marzuq Depression and further to the southeast through al Kufrah Depression (after Hecht et al., 1965).

2.3.2 The Middle subregion states

2.3.2.1 Egypt

The Arab Republic of Egypt occupies the northeastern corner of Africa. It lies between latitudes 22° and 32° N, and longitudes 25° and 34° E. Egypt is bounded on the north by the Mediterranean Sea, on the south by the Republic of Sudan, and on the west by Libya and the east by the Red Sea. The total surface area of Egypt is 1,001,449 km^2, with about 4% under plough. The coasts have a total length of about 2,990 km, of which 990 km fringing the Mediterranean and about 1,900 km looking over the Red Sea. Cairo, the Capital of Egypt, was founded around 1,350 years BP.

The Nile River and its branches traverse the Egyptian land from Aswân in the south to the Mediterranean Sea in the north. The Nile Valley, south of Aswân, is a desert land. It is an extension of the Sahara Desert. An outcropping of granite and basalt rocks, known as the first cataract, obstructs the river channel at Aswân. The

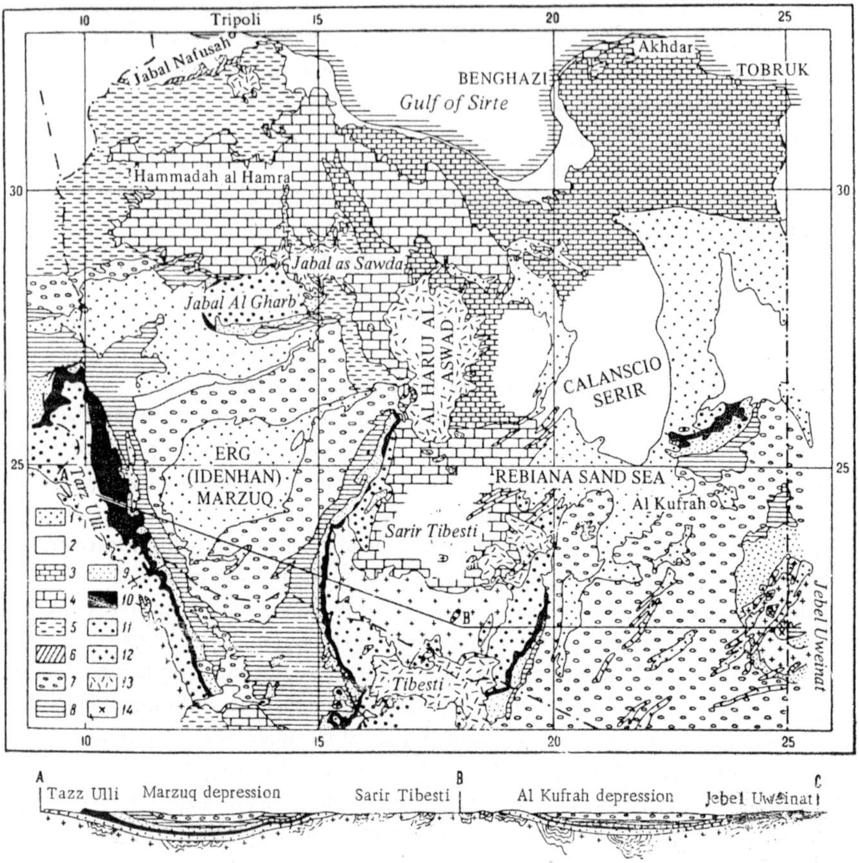

1) Recent sands; 2) Pliocene-Quaternary deposits; 3) Miocene-Oligocene; 4) Eocene and Palaeocene; 5) Upper Cretaceous; 6) Triassic and Jurassic, mainly marine; 7) Lower Cretaceous-Permian, marine and terrestrial; 8) Carboniferous; 9) Devonian; 10) Silurian; 11) Ordovician and Cambrian; 12) Pre-Cambrian; 13) Basalt, partly acid volcanics; 14) Post-Carboniferous diorite-granite.

Figure 10. General geological map of Libya and schematic section of the Murzuq and Kufrah depressions along the lines A-B-C (after Hecht et al., 1965)

surface covering Upper Egypt extends for a distance of 490 km between Aswân and Asyût. On both sides of the Valley in this stretch there are ranges of rock hills about 300 m a.m.s.l. From Asyût up to Giza in the north, close to Cairo, the Valley extends for a distance of 410 km forming the so-called Middle Egypt. About 24 km north of Cairo, the Nile bifurcates into two branches: the Rosetta and the Damietta. These two branches confine between them the Nile Delta.

The Suez Canal lies east of the Nile Delta, and extends for about 170 km from the Gulf of Suez up to the Mediterranean Sea. The reach of the southern Mediterranean Coast between Port Said and Alexandria is about 350 km long. Within this distance the Manzala, Burullus, Edku and Mariut Lakes are located.

The remaining 96% of the Egyptian land is a desert. This can be divided into the Eastern Desert, occupying 25% of the total surface of Egypt, and the Western desert, occupying the remaining 71%. The Eastern Desert comprises the Sinai Peninsula in the north, with a surface area of 61,000 km^2, and the rest, about 150,000 km^2, is located between the western Coast of the Red Sea and the Nile River. With the exception of the coastal zones in Sinai and the southern zone, the central rocky plateau in Sinai can be as high as 3,000 m a.m.s.l., and the peaks of the rocky mountains in the south as high as 3,200 m a.m.s.l. The southern part of the Eastern Desert is a sterile area characterised by a mountain range that can reach 2,000 m a.m.s.l. The Desert border running along the western coast of the Red Sea has a number of relatively small seaports such as Ra's Ghareb, Quseir, Safâga and Hurghada.

The Western Desert extends from the Nile Valley in the east to the boundary between Egypt and Libya in the west, and from the Mediterranean Coast in the north to the boundary between Egypt and the Sudan in the south. The Qattara Depression is located in the northern part of the desert at about 80 km south of the seacoast. The Siwa and Baharîya Oases, and the Wadi el Natroun are all situated in the northern part of the desert while as the oases of Farafra, Kharga and Dakhla are located in the southern part of the Western Desert.

Igneous, metamorphic and crystallised rocks covered the surface of Egypt during the Archeozoic and Proterzoic Eras (some 500×10^6 y B.P., i.e. 500 million years before present). Following that time the Mediterranean Sea covered all the surface of Egypt as well as vast areas of the Western Desert and the Sudan for a period estimated as $300–325 \times 10^6$ y, during which biological sea sediments were deposited. Most of these deposits, however, were washed by weathering factors while some other deposits were buried down under the sedimentary rocks that came later. The fossils of that time, some of which can still be found in Sinai, belong to the Carboniferous Period. At the end of the said period, the sea receded for a period of $50–75 \times 10^6$ y.

The recession of the sea was followed by a subsidence of a considerable surface in North Africa including Egypt and the Sudan. Consequently, the sea inundated the subsided areas for some $50–70 \times 10^6$ y during the Mesozoic Era, thereby depositing a layer of 1,500 m thick of sand and pebbles. The lowest one-third was deposited in the Jurassic Period and the remaining two-thirds in the Cretaceous Period. The middle third is often referred to as Nubian Sandstone. By the end of the Mesozoic Era, the sea receded gradually and a number of changes took place in the Cainozoic Era (50×10^6 y BP, Figure 5). Most of the surface of Egypt subsided during the Eocene Period, thus permitting the sea once more to flood the country and deposit layers of calcareous stones. These deposits form the chain of hills east and west of the Nile Valley. Again the sea receded to about Cairo in the Oligocene Period. A number of minor changes followed during the Miocene, Pliocene and Pleistocene Periods (Figure 11).

The plateaus that make up the greater part of the surface of Egypt consist in the extreme south west of metamorphic rocks and granites. These are covered by

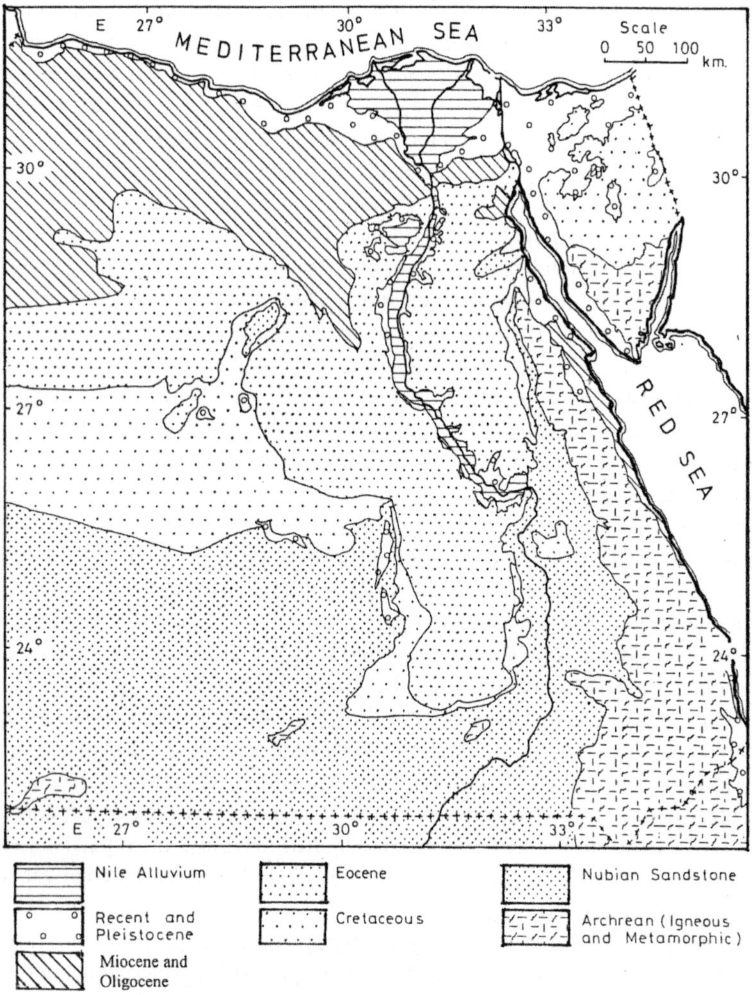

Figure 11. Geological map of Egypt (with some approximations)

sediments sloping gently towards the north, and changing, as a result of weathering, to great flat-topped, table-like hills. North of the southern outcrops of ancient rocks are found the wide stretches of Nubian sandstone, then northward wide expanses of limestone.

The principal oases in the Western Desert are El Kharga and El Dakhla. Each oasis rests on Nubian sediments consisting of alternations of clays, shales and sandstones. These sediments have been gradually exposed as a result of erosion.

It has been suggested that the Nile was made up of several distinct systems, which became joined at a much later stage in geological history. Each of these systems is related to a different structural setting and/or a different geological period. The Eonile (oldest Nile) began to form its valley in the Upper Miocene Age. It cut its gorge at a much lower level than the present sea level. The bottom of the canyon formed by the Eonile reached depths varying from 170 m at Aswân to more than 2,500 m north of Cairo and to even greater depths in the northern delta embayment (Said, 1962). As a result of faults and shifts, the Eonile changed its course from northwest to a more northerly direction. In the Lower Pliocene time the Eonile became covered with sea sediments along its course up to Aswân. Later, sediments were brought by the Paleonile during the Upper Pliocene some 3.2×10^6 years ago. The sedimentation was augmented by the contribution of the Red Sea hills via the then existing wadis, which carried the sediment-laden torrents. By the end of the Paleonile sedimentation, the Eonile canyon was completely filled up and the surface of the Delta became more or less even, with a gentle northward slope.

The time from 1.85×10^6 to 0.7×10^6 y B.P. was a period of harsh cooling and dryness in Egypt. It is possible that the Paleo-Proto-Nile (the then Nile in Egypt) ceased to flow, and Egypt itself became a desert. The abraded material was transported by the blowing wind and began to form the large depressions in the Western desert. Deposition of gravel and sands was quite active in the Protonile period, i.e. from 0.7×10^6 to 0.5×10^6 y B.P. The deposited material took the form of terraces parallel to the modern Nile Valley.

The tectonic movements that took place from the end of the Protonile period up to 125,000 y B.P. helped to establish some hydraulic connection between the Nile in Egypt (the Prenile) and the Ethiopian Plateau. Accordingly, larger floods and sediments were carried by the Prenile than by the Proto or by the Paleo-Proto Niles. As a result, a large Delta was developed, with sediments extending into the sea. From 125,000 up to 30,000 y B.P. the Prenile was followed by the Pre-Neonile. In the wet episodes of the said period heavy rainfalls resulted in coarse sand and gravel deposits. The late Acheulean pluvial, which ended 35,000 y ago, was an important one. The Prenile sediments were laid down by the river, and the final shape of the present Nile Valley was formed (Said, 1981). During the last wet episode, the Neolithic pluvial from 10,000 to 5,000 years B.P., the annual rainfall in the southern part of Egypt was in the range of 100–300 mm.

2.3.2.2 The Sudan

Around 3000 B.C., the northern part of the Sudan called Kush by the Egyptians and Nubia by the Greeks, began to be ruled by the Pharaonic Egypt. This dependence lasted until the rulers of Napata, a Sudanese State that was formed in the eighth century B.C., conquered Egypt in 730 B.C. and became independent.

The Sudan is situated in northeast Africa between 23° and 03° N latitude, and 22° and 30° E longitude. The Sudan has international borders with Egypt, Ethiopia, Kenya, Uganda, Zaïre, Central African Republic, Chad and Libya. The Sudan is the largest country in the Arab Region. It has a surface area of 2,505,813 km^2 and a coastal line 720 km along the Red Sea.

The main physiographic feature of the Sudan is its vast sedimentary basin, which is traversed by the Nile River from the south to the north for a distance not less than 2,000 km. The edges of the plains rise near the eastern and southern frontiers where they become surrounded from the east by the Ethiopian Plateau and the Red Sea Hills, and by the Equatorial Lake Plateau from the south.

The Sudan Plains cover almost 85% of the surface of the country while as hills and mountains cover 10% of the land. The remaining 5% is occupied by water bodies. The Central Plains are almost flat; they consist of heavy clay soil, which expands upon wetting and shrink with dryness. The clay plains are surrounded from the west by dune sand plains, which are suitable for raising pastures and grazing as well as seasonally rain-fed agriculture. The country surrounding the clay plains from the south is characterised by its elevation, steep slope, intense precipitation and high content of hematite. These conditions make it suitable for raising coffee and tea plantations as well as for growing equatorial forest trees. The map in Figure 12 illustrates the geology of the Sudan.

The Basement Complex in the Sudan covers about 42% of its total surface area. Its formations are basic, impervious rocks belonging to the Pre-Cambrian Era. Some of the rocks are weathered and fissured. The upper formations of the Primary Era had been subject to weathering as a consequence of the tectonic movements. During the Secondary Era (Mesozoic), the deposition of the sand and the Nubian sandstone on top of the primary deposits continued until the Lower Cretaceous Period. The thickness of the sand deposits of the Mesozoic can reach up to 3,000 m. The sedimentary deposits, which took place in the Tertiary and Quaternary, cover 48.6% of the surface area of the country while mountains and plateaus cover the remaining 9.3%.

Four major heights, separated from each other, can be found in the Sudan: Jebel Marra in the west, Anquesna Mountain in the east, the Dongotona and Imatong in the south, and the Nuba Mountains in the middle. The Nile River system in the Sudan consists basically of the Bahr el-Jebel, Bahr el-Ghazal, the Sobat, the White Nile, the Blue Nile, the Atbara and the main Nile.

From the outlet of Lake Albert down to Nimule, in Southern Sudan and close to the Uganda-Sudan border, the Albert Nile flows for a distance of 225 km as a broad, sluggish stream, fringed with swamps and lagoons. It meanders through a narrow flood plain between a hilly country on either side so that the area of the swamp is well defined. Below Nimule, the river changes its name and character. The placid, swamp-bounded stream descends in a series of rapids from the plateau to the Sudan plain lands to become the Bahr el-Jebel (literally meaning the sea mountain). A short distance below Nimule, some torrential streams join the Bahr el-Jebel and the river course twists itself through a sharp bend and its direction changes suddenly to the west then to the north and northwest up to Mongalla. The Tertiary-Quaternary Umm Ruwaba formation outcrops in the stretch between Juba and Mongalla. This formation consists of unconsolidated sands, silts and clays of variable thickness. Erosion surfaces were formed on the divides and valleys as subsidence progressed in the Umm Ruwaba Basin. The main river channel bifurcates at the Bahr el-Zeraf

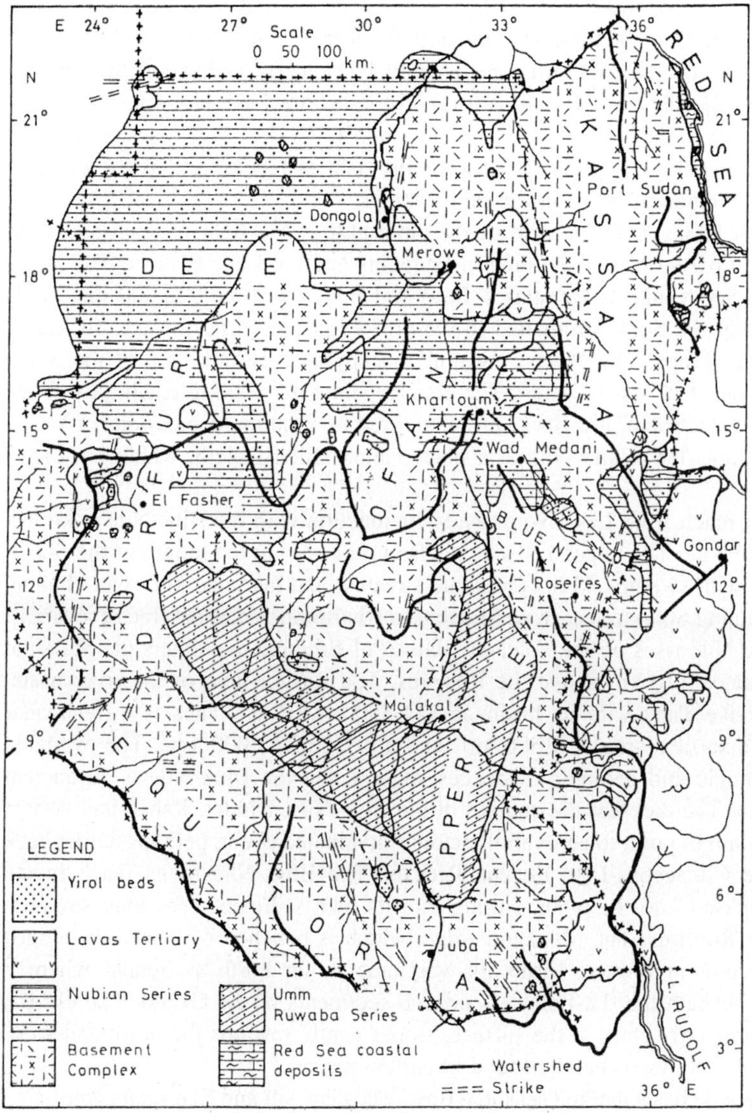

Figure 12. Geological map of the Sudan (from Andrew, 1948)

cuts and the area becomes known as the Sudd region. The swamps covering certain parts of the region temporarily and permanently are caused by over bank floods, and not by longitudinal inundation of the river.

The Bahr el-Arab joins Bahr el-Ghazal at the northern end of the swamps. Along Bahr el-Ghazal and to the south and east of it are large areas of permanent and temporary swamps, which are fed by a number of streams. The country where the upper courses of these streams flow is entirely covered by a savannah forest. On the lower reaches of these streams and along the Ghazal itself are large areas of swamps. Since the tributaries are usually blocked by vegetation they hardly contribute to the flow in the main stream. Approaching Lake No, a large shallow lagoon, the dry land nears the river on the north and the river loses its defined banks. Here the sluggish Bahr el-Ghazal joins Bahr el Jebel after having wasted a tremendous volume of water in the said swamps.

The Sobat Basin includes most of the plain east of the Bahr el-Jebel and Bahr el-Zeraf, and parts of the Abyssinian Mountains and the Lake Plateau. The Basin area of the Sobat and its tributaries–Baro and Pibor–is about 225,000 km^2. It is suggested that the Sobat originated in the post-Oligocene times on the surface of the Ethiopian Traps. The Machar swamps occupying a certain area of the Sobat Basin is a source of loss of water. The country along the 300 km reach, just above the mouth of the Sobat, is a flat plain intersected by swampy depressions or khors. The Sobat joins the White Nile at a place called Hillet-Doleib, which is near to Malakal.

The river reach from Lake No down to its junction with the Blue Nile is known as the White Nile. The upper 120 km, i.e. from Lake No to the mouth of the Sobat, is crowded with swamps, khors (seasonal streams) and lagoons. The river channel in the remaining 800 km, i.e. from Malakal to Khartoum, is almost free of swamps.

The Blue Nile rises on the volcanic plateau of Ethiopia and enters the Sudan at a point situated 170 km above the Roseires. The surface area of the river basin, including Lake Tana with its basin, is 324,350 km^2. Lake Tana is a freshwater body with a surface area of 3,500 km^2 and a basin of 14,000 km^2. The available geomorphologic and geologic evidences suggest that the Blue Nile is an ancient river system. The fact that the courses of the Blue Nile, Dinder, Rahad and Atbara are parallel might point to an overall tectonic-dominated event, perhaps dating back to the Late Cretaceous-Late Eocene Periods. The Blue Nile in its reach in the Sudan is incised into the plain, cutting into its own sediments, for long stretches north of Er-Roseires. The Managuil ridge, which is benched by a well-developed old valley-floor or terrace, forms its west side as far north as Sennar where it disappears beneath the flat-lying, superficial sediments of the Gezira. The Gezira, with predominantly clays at the surface, slopes gently towards the north and west. Sand spreads and dunes can be found in certain places.

The Blue Nile has a marked seasonal flow. The clay, silt and fine sand comprising the suspended load, up to 4,000 ppm in the floodwater, are mostly derived from Ethiopia. The Nile silt and the wadi-fill, which consists of sand and gravel, are the youngest superficial deposits in the Gezira (Whiteman, 1965). The Gezira formation occupies a large part of the triangle between the White and Blue Niles and extends up the Blue Nile Valley. It rests on the Nubian formation and is overlain by qoz (local name for sand and gravel) and wind-blown sand. Gravely sand, up to 10 m thick, occurs at the base of the formation and changes upwards into clayey sand 40 m thick. The upper part of the formation consists of dark clays known locally as 'Gezira Clay' (Berry & Whiteman, 1968).

South of Khartoum the White Nile as far as Shambe, and the Blue Nile as far as Er-Roseires, flow mainly over alluvial deposits. Downstream, the alluvial deposits are more restricted, and the Main Nile flows largely over the Basement Complex and the Nubian formations, except in the Shendi and Kerma basins where there are extensive tracts of alluvium. In the Khartoum area the thickness of alluvium is highly variable.

At Khartoum the Blue Nile joins the White Nile and the combined waters flow for 1,885 km to Aswân in Egypt through a region of Nubian sandstone overlying an old eroded land surface of crystalline rocks. These rocks offer a greater resistance to the river's action than does the softer Nubian sandstone. The river channel thus consists of a series of placid reaches of mild slope separated by rocky rapids (cataracts), where the slope is greater and the flow more turbulent.

Ninety km north of Khartoum, the 6th Cataract obstructs the channel of the Main Nile. Further northwards, the land surface is developed mainly on the Nubian formation. During the early Tertiary, a depression developed over the present Nile Valley north of Sabaloka (6th Cataract). This depression was occupied by a freshwater lake in which the so-called Hudi Chert formation was laid down (Berry & Whiteman, 1968). With the lowering of the sea level in the Oligocene Period, the Early Tertiary Nile slowly cut its course through that formation and the rest of the Nubian formation, and eventually 'locked' itself in the igneous and metamorphic rocks of the Basement Complex.

The River Atbara, the last tributary of the Nile, joins the Main Nile at 320 km below Khartoum. It is 880 km long, and the greater part of its catchment, 112,400 km^2, lies in Ethiopia. The high peaks are more than 3,500 m a.m.s.l. and most of the eastern watershed is above 2,500 high. The Takazze or Setit, with a basin area of 68,000 km^2, is the principal source of the Atbara. The river flow is highly seasonal and the concentration of the suspended sediment during the peak flood is the highest among all tributaries of the Nile River system.

From Atbara to Berber, 387 km from Khartoum, the Main Nile runs in a northerly direction. The valley is wide and flanked by flat to gently rolling country. The river valley below the 5th Cataract, northwards and north-westwards, is mainly incised in a flat to gently undulating surface cut across metamorphosed Basement Complex rocks. Gravel and silt terraces occur but in general they are not extensive. The river runs from the 5th Cataract to the 4th Cataract, a distance of 350 km, across the Basement Complex. The drop in water level between the two ends of this stretch is 88 m and the average gradient is 2.5×10^{-4}. The 4th Cataract, 80 km upstream from Karima, is composed of biotite gneiss, often seamed by pegmatite and quartz veins. Northwards to Dongola there is an extensive gravel and silt terrace. West of this terrace the Nubian formation is covered with a skin of coarse, rounded, predominantly quartz gravel.

The 3rd Cataract is located 480 km downstream from the 4th Cataract. The Cataract is composed of biotite gneiss, locally associated with marble bands to the west. The gneiss is traversed by compact granitic bodies, which form hard bands at right angles to the river, and give rise to the rapids. North of the cataract the Nile passes between two high hills, in the order of 400 m high. The Nubian formation and the Basement Complex are cut out by dykes and volcanic necks. A short distance below the 3rd Cataract the river bends east and west until it reaches the 2nd Cataract. The Semna (2nd) Cataract is formed by a bar of hard gneiss and crushed granite, thus confining the Nile and narrowing its width to just 40 m. The Main Nile flows below the 2nd Cataract with an average natural slope of 7.7×10^{-5} till it reaches the 1st Cataract at Aswân after 340 km.

An elaborate description of the physiography and topography of the Nile Basin can be found in Volumes I, V and VIII of the Nile Basin (Hurst et al., 1931, 1938 and 1950, respectively), Shahin (1985, 2002) and Sutcliffe & Parks (1999).

2.3.2.3 Djibouti

The country is situated in the Horn of Africa between 13° and 10° N latitudes and 41° and 44° E longitudes. With its 23,000 km^2 surface area, Djibouti is one of the smallest African Countries. Djibouti, formerly known as the French Somaliland or 'Territoire Français des Afars et des Issas' (The French Territory of Afars and Issas) is bordered on the north, west and south by Ethiopia and on the southeast by Somalia. The Red Sea and the Gulf of Aden form the eastern boundary of Djibouti, the length of the coastal line along the Gulfs of Aden and Tadjoura is about 314 km.

Djibouti lies mainly in the Danakil Plain, which is a part of the East African Rift. The two largest patches of lowland littoral, 20 km wide at the widest, are in the northern half of Djibouti. The northern one of these patches, the Doumeira plain at the Eritrean border, is a wind swept uninhabited flat of gravely reg. At the international border the Weima, the largest wadi of the country, reaches the Red Sea. Water can be found at shallow depths in its sandy bed. Water from other wadi beds is pumped to supply nearby towns and villages.

2.3.2.4 Somalia

The Somali Republic also lies in the Horn of Africa, the costal strip in the extreme eastern corner of Africa ending in Cape Guardafui. The country is bounded on the east by the Indian Ocean, on the north by the Gulf of Aden, on the northeast by Djibouti, and on the northwest and west by Ethiopia and Kenya respectively. The country has a surface area of 637,657 km^2 and coastal line 3,000 km long.

The northern part of Somalia consists of a narrow coastal strip along the Gulf of Aden. A steep mountain, an offshoot of the Ahmar Mountain in eastern Ethiopia, borders the coastal strip. The elevation of the Ahmar Mountain exceeds 2,000 m a.m.s.l. The highest mountain top in Somalia is the Shimber Berris (2,407 m). A high plateau with an average height of 600–900 m lies south of this mountain. The plateau is sloping gradually eastward and presenting a series of steep scarp faces toward the coastal plain and the Indian Ocean.

The eastern and southern parts of Somalia consist of a broad low-lying coastal strip changing gradually in a westerly direction to merge into high plateaus of Ethiopia and northeast Kenya. As such, the southern part of the country is almost entirely desert with the exception of a fertile area crossed by two rivers, the Juba

(Giuba) and Shebeli, originating in the highlands of Ethiopia. A short distance above its mouth at Kismayu (Chisimaio) on the Indian Ocean, the Shebeli River joins the Juba at a place called Galib and the combined waters flow down to the ocean. The rivers originating within Somalia itself, such as the River Nogal, which springs from the northern mountains, are not of much significance as far as their flow volumes are concerned.

2.3.2.5 The Comoro Islands (Comoros)

These are a group of small volcanic islands situated in the Indian Ocean off the coast of Mozambique, 300 km from both Madagascar and the African mainland. The Comoro Islands are located between 11° and 13° S (south of the Equator) and 44° E. The main islands are the Grand Comore (1,148 km^2), Mayotte (374 km^2), Anjouan (424 km^2), and Mohéli (290 km^2). The natural rain forests have been largely cleared for agriculture. They have a combined surface area of 2,235 km^2 (The World Guide, 2000). There are small lengths of fringing reef formation, totalling about 60 km.

The Grand Comoro Island possesses an active volcano and the permeability of the ground is very high. Consequently there are no permanent streamsand and the depth of the water table is quite low.

2.3.3 The Eastern subregion states

2.3.3.1 Syria

The Syrian Arab Republic has a surface area of 185,180 km^2. This area is confined between latitudes of 32° and 37° N and longitudes 36° and 42° E. Turkey, Iraq, Lebanon and Jordan bound Syria on the north, east, south and west respectively. The Eastern Coast of the Mediterranean Sea forms the northeast boundary of Syria. The length of the Syrian Coast along the Mediterranean Sea is roughly 300 km. The country is traversed by a number of rivers, notably: the Euphrates (Al Furat); Al Khabour; Al Āsi (Orontos); As'Senn; Balikh; Barada; Afrien; and Al A'waj. Some of these rivers are international having their basins shared with other countries like Turkey, Lebanon, etc.

The Syrian landscape is characterised by five major physiographic units. These are:

i The Mediterranean Coastal strip extends for 170 km long from the Border with Turkey in the north down to the mouth of Al Kebir River in the south. It is a narrow plain running parallel to the western slopes of the coastal mountain range.

ii The Coastal Mountains consist of two parallel ranges separated by el-Ghab Depression. The western range has three mountains: Amanus (1,800 m), Akr'a (1,700 m), and Al Alawyya (1,500 m). The eastern range consists of Al Akrad (1,200 m), Samān (870 m), Harim (800 m), Az'Zawiya (880 m), and the Anti-Lebanon Range, which has a maximum elevation of 2,800 at Hermon Mountain in Syria.

iii The dislocated plains form the northern segment of the Red Sea-Dead Sea-Jordan Valley-Beqa'a Rift Valley system, which is part of the African graven within Syria. The dislocated plains separate the eastern and the coastal mountain ranges, and consist of Al Ghab, Al-Rouj and Al-Omq plains.
iv The Inland Mountains in eastern and southern Syria are formed up of volcanic highlands at the Golan (1,200 m) and Jebel Al-Arab (1,800 m) and the high lands of Tadmur Range (1,400 m) and many more ranges.
v The Inland Plains and Plateau, which comprise Al-Jezira, Badiat (meaning desert) Al-Sham (old name of the area presently occupied by Syria and Lebanon), and Damascus, Homs, Hama, Aleppo (Halab in Arabic) and Huran Plains.

The structural pattern of the maritime zone of Syria is generally arched block uplifts dipping towards the west. The other major structure is the Ghab Graben, which forms the northern segment of the Rift Valley System. It forms a low-lying area parallel to and east of the Alaouite range of the coastal mountain. The geological structure of the inland areas is a reflection of the general topography of eastern Syria. The only complex structure occurs in the folded and step faulted Tadmur Range and the Daw Basin (ECWA, 1981).

The Sedimentary rocks cover a large part of the country's surface. The sediments comprise the Jurassic and Middle Cretaceous dolomite and limestone, the Upper Cretaceous and Palaeocene marl and chert, the Paleogene and marine Miocene limestone, and the lagoonal Miocene evaporites. Additionally, the Pliocene conglomerates include sand, marl and lacustrine formations, and the Quaternary conglomerates such as alluvium and lacustrine formations. The thickness of each of these formations is variable. The average thickness of the Paleocene formation (marl, marl limestone and chert) can be up to 200 m while the thickness of the same formation deposited in the Upper Cretaceous can be up to 800 m.

The Volcanic rocks cover the surface of the country in the northwest and the southeast as can be seen from the geological map of Syria, Figure 13. The rocks are composed mainly of basic and basaltic flows, tuff and volcanic ash.

The so-called Green Rocks occur mainly north of Latakia along the northwestern coastal zone of Syria. These rocks belong to the Triassic period and are composed mainly of oviolite series, clastic rocks and other igneous and sedimentary rocks.

2.3.3.2 Lebanon

The Lebanese Republic is situated along the eastern shore of the Mediterranean Sea between latitudes 33°03′ and 34°41′ N and longitudes 35°06′ and 36° 05′ E. The surface area of the country is 10,400 km^2 and the length of the coastal line is roughly 210 km.

Lebanon has generally a mountainous topography. It is dominated by two parallel mountain ranges extending from north to south. These are the Coastal Mountain Range and the inland or eastern Anti-Lebanon Range. The two ranges are separated by a structural depression. The coastal plain is about 210 km long, varying in width from 2 to 20 km, and characterised by the occurrence of several heads and bays. The Coastal mountain range extends parallel to the seacoast. The peak elevation of

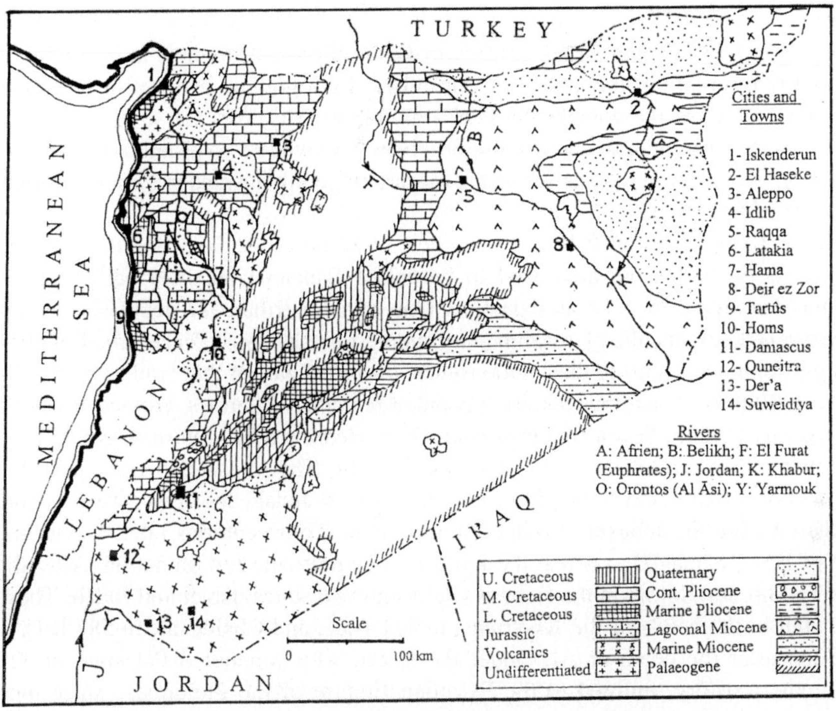

Figure 13. Geological map of Syria (reproduced from the ECWA Report, 1981)

this range is 3,083 m a.m.s.l., and average height of 2,200 m. Several wadis and rivers drain westwards to the sea, the Litani River is the most important one. To the south of Beirut the elevation of the mountain range drops to just 400–600 m a.m.s.l. The Anti-Lebanon Range forms the boundary between Lebanon and Syria. There are two peaks; the northern peak is 2,629 m a.m.s.l. and the southern 2,814 m a.m.s.l. Mount Hermon, which comprises the southern portion of the eastern range slopes steeply towards the Hasabani River in the south.

The Beqa'a Valley is a depression separating the eastern from the western mountain ranges. The width of the valley varies, as already mentioned, from 2 up to 20 km or more. The average land level is 900 m a.m.s.l. It ends in Lake Homs in Syria in the North and Jebel Al-Arabi (1,508 m) in the south.

The rock units exposed in Lebanon are predominantly sedimentary successions ranging from Jurassic to recent. They consist of carbonate rocks essentially limestone and dolomites, chalky marl and marl. Coarse clastics, lacustrine gypsum and marl up to 800 m thick were deposited in the Upper Miocene period. Limited occurrences of Pliocene sandstone and shale and Quaternary alluvium can be found in the northern coastal areas. Quaternary deposits also occur in patches along the coast and within the Beqa'a Valley.

2.3.3.3 Palestine

The ongoing Palestinian-Israeli conflict makes it impossible to draw at this moment a definitive picture of the landscape and other physiographic features of Palestine. The geography of the country, surface area, capital city, etc. are so far unsettled. As such, the presentation and discussion given here will be confined to the present situation. However, it may be advisable to start by giving the reader a brief historical account of Palestine.

Some 6,000 years B.P. the Canaanites, a Semitic people from the Arabian Peninsula, settled in an area used to be called Canaan land, later called by the Romans around 135 A.D. as Palestine. One of their tribes, the Jebusites, built a fortified settlement called Urusalim (Jerusalem), i.e. the city of Peace! The Egyptian Kings, aiming at securing their trade routes, occupied part of that land around 3200 B.C. In 2000 B.C., another Semitic people, Abraham's Hebrews, crossed Palestine on their way south. Seven centuries later, some Hebrew tribes returned from Egypt, following Prophet Moses to the land of Canaan. There, a fierce fight over the possession of the land took place, but the sons of Judah, the Jews, were unable to exterminate the Jebusite dwellers of Jerusalem. Four centuries later, David, the son of Issac, managed to defeat the Jebusites and unite the Jewish nation. After the death of his son Solomon the Hebrews split into two states, Israel and Judah. These later fell in the hands of the Assyrians, in 721 B.C., and Chaldeans, in 587 B.C.

Soon after the death of Alexander the Great, who conquered Palestine in 322 B.C., the territory returned to the Egyptian Empire of the Ptolomies. Since then the Syrians and the Romans kept invading the country until the Jews were finally expelled from Jerusalem in 135 A.D. The Byzantines followed the Romans until 611 A.D. when the Persians invaded it. The Arabs conquered Palestine in 634 A.D., and with the exception of short periods in the 11th, 12th and 13th centuries, Palestine had Arab rulers for 1,000 y and Islamic governments for 15 centuries (a View from the South, 1999/2000).

The Ottoman Empire conquered Jerusalem in 1516 and maintained power there until the end of World War I (1914–1919). The defeat of the Turks in that war and the British Mandate, through the League of Nations, gave Britain the power to try to enforce its controversial promise to the Zionist Movement during the wartime concerning the establishment of a national homeland for the Jews. The Arab States rejected the plan put forward by Britain in 1936 for the partition of Palestine, and rebellion broke out, lasting until 1939 when Britain gave up that plan. Instead, in 1947 the UN General Assembly approved another partition plan. Since 1948, when Israel unilaterally proclaimed itself an independent State, a succession of wars broke out between Israel and its neighbours of the Arab States. The imbalance of weaponry and military experience, and the continuous financial and military support from the Western countries to Israel, added to lack of interest of mighty nations to enforce the relevant Security Council resolutions, brought the conflict to a different phase. Israel went on occupying Palestinian land and water resources, destroying Palestinian homes and properties, and building more Jewish settlements. Figure 14(a) shows the growth of surface area of Israel under the Israeli administration since 1947 till the present and the subsequent decline of the size of Palestine in the same time This illegal expansion is met with steadily growing attacks and fierce reprisal (Intifadha) from the Palestinians, and the whole territory has ever since become plagued with violence and unrest.

The total surface area of the West Bank and Gaza Strip at this moment is about 6,220 km^2, which is almost one-fourth the original surface area of Palestine when it was under the British Mandate. The respective surface areas of the West Bank and the Gaza Strip are about 5800 km^2 ($\sim$ 125 km long times 46 km average width) and 360 km^2 ($\sim$ 45 km long times an average width of 5–7 km).

Despite its small size, the West Bank is characterised by a widely variable landscape. Fertile plains can be found in the northwestern part. All wadis rising in the western basins drain towards the Mediterranean Sea through Israel. Moving towards the east one finds a chain of highlands and mountain ranges. The mountains in the north are called Mountains of Sumaria, and those in the south mountains of Juda. The elevation of the surroundings of Nablus and Ramallah is 940 m and 880 m a.m.s.l. respectively. Several wadis intersect the eastern high lands. Further to the east one finds the eastern hill slopes and the depression of the Jordan Valley known as El Ghor. This low-lying depression collects the runoff carried by the wadis rising in the high ground and running along the eastern slopes. The map in Figure 14(b) shows the main types of landscape in the West Bank.

The total surface area of the drainage basins of the wadis flowing towards the Jordan Valley is about 2,540 km^2. Nearly all these wadis are of seasonal nature.

An approximate idea of the geology of the West Bank can be seen from the cross section shown in Figure 15.

The landscape of the Gaza Strip is essentially a foreshore plain. A sandy beach stretches all along the coast, bound in the east by a ridge of sand dunes up to 40 m high. The area occupied by the coastal dunes varies in width from 1.5 to 4.3 km. There is a break in that ridge northwest of Gaza city leaving a space for a narrow valley to cut through the dunes for a distance of about 10 km south of Gaza. This valley has been formed by Wadi Gaza, the only source of surface water in the Gaza Strip before it reaches the Mediterranean Sea (Mogheir, 1997).

Since time immemorial the area currently called the West Bank contains a large number of water wells and springs. Inadequate maintenance and/or over drafting of water to meet the needs of the increasing population have turned the quality of water in some of these sources to be unsuitable at least for drinking and other domestic purposes.

2.3.3.4 Jordan

The Kingdom of Jordan is situated between latitudes 29°30′ and 33°30′ N and longitudes 35° and 39° E. The surface area of the country is 97,740 km^2. Jordan is practically a land-locked country, except for a short coastal line of about 25 km along the Gulf of Aqaba on the Red Sea. Three major typographical units dominate the landscape of Jordan. These are the Highlands, the Eastern Upland Plateau and the Low-lying lands.

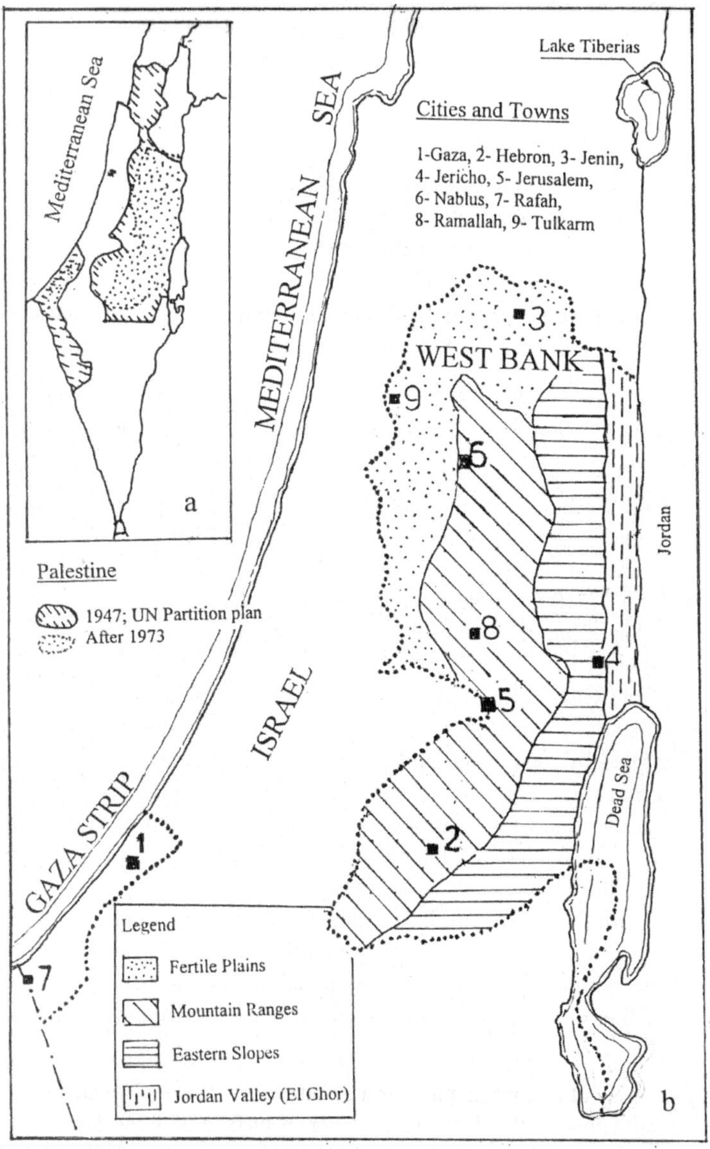

Figure 14. (a) The decline of area allocated to Palestine since the UN 1947 partition plan to that after the 1973 war; (b) Physiographic units composing the West Bank and the Gaza Strip (reproduced from the Country Report of Palestine, 1986)

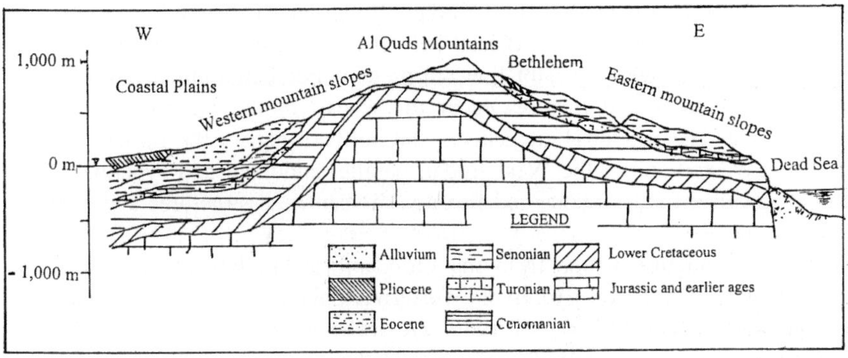

Figure 15. Cross section through the West Bank showing its geology (from The Country Report of Palestine, 1986)

The Highlands comprise the western and eastern escarpments, which flank the Jordan Valley-Dead Sea Depression on both sides. The eastern escarpment extends southwards to flank the Wadi Araba area. Small and large canyons cut the Western and Eastern Highlands bordering the Jordan Valley-Wadi Araba Depression, thus forming dissected ridges and rugged relief. The Western Highlands drain their runoff westwards to the Mediterranean Sea through several wadis, and so eastwards to the Jordan River. The Eastern Highlands are intersected by a number of valleys located in the zone between the Syrian border to the southern end of the Dead Sea. Some of these wadis are the Yarmouk River, Wadi Zarqa, Wadi Mujīb and Wadi Hāsā. They are drained eastwards to the Upland Plateau by small and shallow rivulets. The highest elevation of the Western Highland is about 1,000 m a.m.s.l. near Ramallah. The peak elevation of the Eastern Highland is 1,856 m a.m.s.l.

The Eastern Upland Plateau borders the Eastern Highland from the west and extends to the border with Saudi Arabia. It is a fairly dissected plateau formed from flat lying sediments. The elevation of the ground surface ranges from 1,000 m to less than 600 m near the Saudi Arabian border. The internal drainage basins of the plateau are characterised by the existence of mudflats or playas such as those in basin of Al Azraq.

The Low-lying areas comprise the Jordan Valley-Dead Sea-Wadi Araba depression, which extends about 360 km from Lake Tiberias to the Gulf of Aqaba. This depression is part of the Rift Valley. The land surface in Jordan valley is generally flat, with elevation ranging between 200 and 392 m b.m.s.l. The Jordan River traverses thc valley at its middle part for a distance of 105 km. The Dead Sea, which forms the middle zone of the depression, covers a surface of about 1,000 km^2 in area, 80 km in length and 15 km as average width. The lowest level of the Dead Sea bed is 792 m below the mean level of the Mediterranean Sea. The landscape in Wādi Araba is almost flat with slightly rolling terrain at its middle zone between the Dead Sea and the Gulf of Aqaba. The ground level ranges between 250 and 392 m b.m.s.l. near the Dead Sea. The width of the wbdi in Jordan is between 5 and 20 km.

The depression comprising the Jordan Valley-Dead Sea-Wbdi Araba, as mentioned above, is part of the African Rift. It is primarily a graben bordered on both sides east and west by fault line scarps of major longitudinal faults. Block foldings, undulations and flexures occur to the east of the Jordan Valley-Wādi Araba Graben. The structural elements in the northern part of the Eastern Plateau trend NE-SW, and so are the West Jordan parallel anticlinal and synclinal structures. Transversal fault systems and flexures dissect these structures and result in SE-NW trend.

Rock units ranging in age from the Pre-Cambrian to Recent outcrop at the subsurface in Jordan (Figure 16). In the Western Highlands, calcareous sediments of the Upper cretaceous and Lower Tertiary age are exposed. Limestone to dolomitic limestone with marl, shale, chalk and chert of the same age occur in the Eastern Highlands and the Eastern Plateau. The total thickness of these sediments range from 150–800 m. sandstone and sandy facies with some shale cover a relatively small area. The total thickness of these sediments is 1,900 m, of which 1,600 m belong to Early Palaeozoic and 300 m to the Lower Cretaceous. Basalt outcrops overlie the fluviatile gravels of the Middle Pleistocene. They occur mainly in the area comprising the southern extension of the Jabal ed'Drūz up to Wādi Sirhān near the Saudi Arabian border. The clastic sediments of marine to continental origin belonging to the Neogene-Quaternary ages occupy a great part of Wādi Araba-Jordan Valley Depression. They exist also in the flood plains of the main rivers and wādies as the Jafr and Azraq depressions. The Pre-Cambrian Basement Complex occupies the extreme southwestern corner of Jordan along the Gulf of Aqaba and in the eastern escarpment of Wādi Araba. Figure 16 is a geographic map of Jordan (reproduced from the Country report of Jordan, 1986).

2.3.3.5 Iraq

The Republic of Iraq lies to the northeast of the Arabian Peninsula between latitudes 28°59′ and 37°07′ N and longitudes 38°49′ and 46°40′ E. The surface area of the country is 434,924 km^2. Iraq is bounded on the north by Turkey, on the northwest by Syria, west by Jordan, southwest and south by Saudi Arabia, south by Kuwait, and on the east by Iran. The shore length along the Arabian Gulf being too short, about 40 km, makes Iraq almost land-locked.

Iraq is generally a flat country and the plain occupying the area between the Tigris and Euphrates Rivers is quite large in area, bounded on the west and south by desert plateaus rising to 200–300 m a.m.s.l. The plain is bounded on the north and east by the Zagros Mountains with maximum elevation up to 3,700 m a.m.s.l. The mountain foot hills lie to the southwest of Zagros-Torous Range. The Al-Jazira area lies to the south of these mountain foot hills area and has a desertic landscape, except the strip along the rivers. Its elevation ranges from 50 to 500 m a.m.s.l.

Iraq is situated at the northeastern rim of the unstable shelf of the Arabian shield. A major geosyncline trending northwest-southeast comprising most of Iraq is the predominating geologic structure. This structure and the resultant synclinoriums and anticlioriums have been formed when the earth's crust in Iraq was subjected to

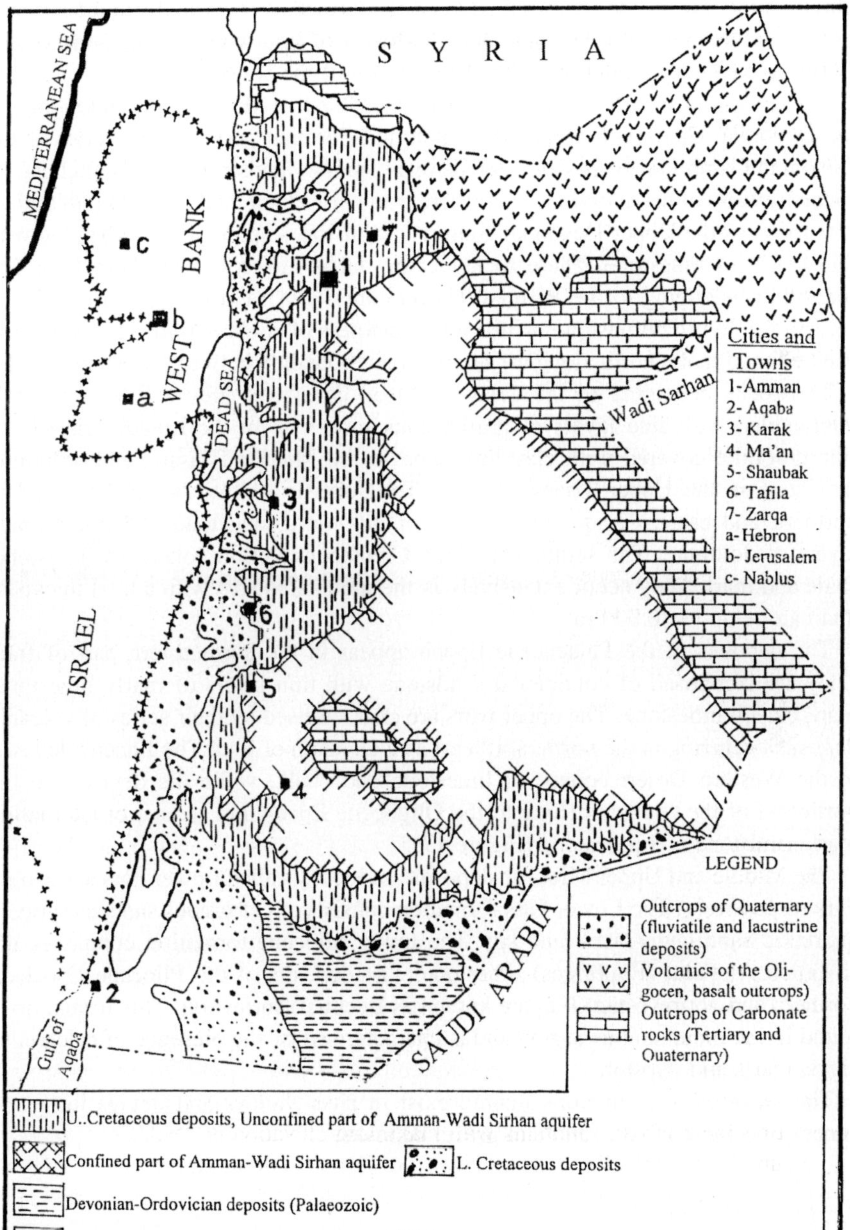

Figure 16. Geological map of Jordan (from the Country Report of Jordan, 1986)

the tangential forces originating in the Iranian Plateau and blocked by the Arabian Shield in the south. The geosynclinal belt thus formed starts away from the Arabian Shield and extends into the south and southwest of Iraq with a general thickening of the sedimentary sequences in the central part of the country.

The rocks of the Palaeozoic Era are composed of three series belonging to the Devonian Cambrian, Upper Devonian and Permian periods. The rocks of the Devonian Cambrian are basically sandstones, quartz, mudstones, marls, and can be found in many of the anticlines in northern Iraq with a total thickness of up to 600 m. The rocks of the Upper Devonian comprise sandstones and marls interbedded with thin layers of limestone. These rocks occupy large surfaces in northern Iraq. The formation belonging to the Permian period consists of sandstones, marls and thin layers of dolomitic limestone, with a total thickness of about 800 m. This formation can be found in the north along the international boundary between Iraq and Turkey.

The rocks belonging to the Triassic Period consist of grey and yellow marls interbedded with fine layers of argillaceous limestone in the Lower Triassic, of fine dolomite covered with clay limestone in the Middle Triassic, and dolomitic limestone in the Upper Triassic Period. The Jurassic sediments can be found in northern and eastern Iraq. They consist of dolomitic sandstone and fine-grained layers of dolomite. The sediments of the Cretaceous Period consist of limestone, shale and marl. They occur extensively in the mountains area with a total thickness that can be up to 10,000 m.

The deposits of the Palaeocene Epoch appear in the northeastern part of Iraq. They are composed of continental sandstone with thin layers of marly limestone, marl, clayey limestone. The upper parts are characterised by their series of volcanic deposits occurring in the north, northeast as well south of Iraq. The Eocene deposits in the Western Desert consist of limestone and chalk, while the deposits in the northwest of the country belong to the Oligocene Epoch and consist of alternating marl, sandstone and limestone (Figure 17).

The Middle and Upper Miocene deposits are locally called Lower Fars and Upper Fars respectively. The Lower Fars formation consists of calcareous sand and gravel, quartzitic sand and gravel, and silt. The lower Miocene formation comprises the so-called Al Furat (Euphrates) limestone. The deposits of the Pliocene Epoch in northern and northwestern Iraq are known as Bechtiari formation. This formation is found in the mountainous region and is characterised by the presence of calcareous clays, marls and gypsum.

The Quaternary formations mainly exist in river valleys and depressions, and generally consist of silt, sand and gravel deposits.

2.3.3.6 Kuwait

The State of Kuwait lies at the northwest corner of the Arabian Gulf between latitudes 28°30′ and 30° 05′ N and longitudes 46°30′ and 48°30′ E. The total surface area of the country is 17,818 km^2 and the length of its coastal line along the Gulf is about 300 km.

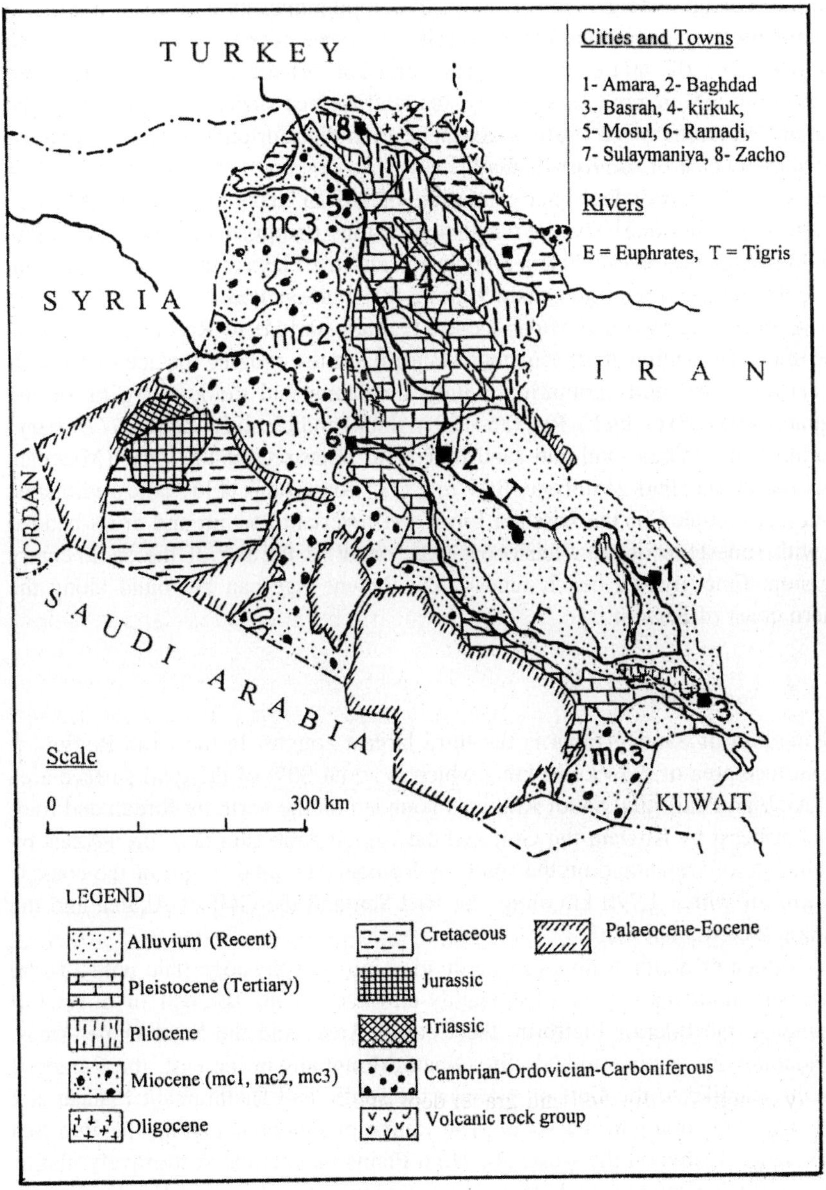

Figure 17. Geologic map of Iraq (UN Technical Cooperation for Development, Study Reports on Water, No. 9, 1982)

Kuwait is located at the edge of the northern extremity of the Arabian Peninsula, on the southwestern edge of the Gulf syncline. The regional dip towards the northeast is at an average rate of 0.2%. The topography of Kuwait is related to the regional tectonic structure of the country. It is generally flat or undulating with occasional low hills and shallow depressions. The ground level ranges from sea level to about 300 m a.m.s.l. Generally, the land surface slopes northeastward. The Jal Az'aor escarpment is a straight landscape extending for almost 60 km along the northwestern coast of the Kuwait Bay.

The Ahmadi ridge, which is a gentle rise of 137 m, separates the Burgan Plain from the Gulf Coast. The valleys and depressions comprise the Wadi Al Batin, Wadi Al Musannat, Al Rawdatein Depression and the Playas (sabkhas or sebkhas). The Aeolian sand dunes occur mainly along the southern coast of Kuwait.

Rock units ranging from Mid-Eocene (Tertiary period) to Recent (Quaternary period) form the sedimentary sequence occurring at the country surface of Kuwait. The Tertiary rock units comprise mainly limestone and dolomite rocks of the Dammam (180–220 m thick), Rus (75–120 m thick) and Umm Er-Raduma (Tertiary) formations; calcareous sand and sandstone with limestone of the Ghar (Miocene, 275 m thick) and Fars (Pliocene, 105 m thick) formations. Fluviatile sediments, sand-gravel complex of the Dibdiba formation, overlain by thin and cross-bedded sand with limestone of the Pleistocene Age occur at the top of the sedimentary succession. Fine grained beach sand of the Recent Age can be found along the southern coast of Kuwait.

2.3.3.7 Saudi Arabia

The Kingdom of Saudi Arabia is the third largest country in the Arab Region. It has a surface area of 2,149,690 km^2, which is about 80% of the total surface area of the Arabian Peninsula. Saudi Arabia is bounded on the north by Jordan and Iraq; on the northeast by Kuwait, the Gulf and the United Arab Emirates; on the east by the Sultanate of Oman and on the south by Yemen. The total length of the coast is 2,255 km, of which 1,770 km along the Red Sea and the Gulf of Aqaba, and the rest along the Arabian Gulf.

The surface of Saudi Arabia can be divided into six physiographic units. These are: the Mountainous regions, the Highly-elevated plains, the belt of down hill escarpments, the Interior Platform, the Coastal Areas and the Sand Dune Areas. The Mountainous regions include the Omani Mountains in the east, the Range of mountains parallel to the Gulf of Aden in the south, and Hadhramaut Plateau and the Yemeni Mountains in the south. The Range of mountains along the Red Sea bounds Saudi Arabia on the west. The High Plains include the vast gravely plains. The down hill escarpments include the western slopes, the slope of Jebel Al-Tuwaiq. The Central Plateau, which is basically formed by the deposits of the Tertiary Age, is located between the hill foot slopes and the Coastal Plains. The coastal areas are generally flat areas comprising shallow depressions (Sebkhas) and small pebbly plains. The sand dune areas cover almost one-half of the surface of Saudi Arabia.

The two principal desert areas are the rub el-Khali (meaning the empty quarter) in the south and An'Nafud Desert in the north of the central part.

Saudi Arabia lies in a relatively stable structural zone, characterised by regular gentle dips and absence of folds or thrusts. Two major geologic structural features are recognised: the Arabian Shield and the Arabian Shelf, where most of the sediments were deposited in the east. The Arabian Shield comprises the western wing forming Central Najd, Hijaz and Asir regions, the Aden Plateau in Yemen, and the southern wing along the Arabian Sea Coast. The Arabian Shelf includes the Interior Platform and Basins occupied by thick sediments within which gentle folding and doming occur. It includes the Interior Homocline, the Interior Platform and the Basins such as the Rub Al-Khali, northern part of the Arabian Gulf and the Dibdibba basins.

The main rock outcrops in Saudi Arabia are the crystalline Pre-Cambrian Basement rocks of the Arabian Shield, which cover nearly one-third of the surface of the country. Basalt outcrops of the Upper Cretaceous to the Tertiary can be found in the extreme southwest of the Hijaz plateau. Other basalt outcrops of the Late Tertiary to the Quaternary occur over the entire western part of the Arabian Shield. The surface covered by basalt deposits in Saudi Arabia is about 5% of the total surface of the country.

The remaining two-thirds of the surface of Saudi Arabia is underlain by Sedimentary formations (Arabian Shelf) covering the geological time scale from the Cambrian Age to the Recent Age. The total estimated thickness of the stratigraphic column of sedimentary sequences is around 5,500 m. The sequence of deposits comprising the lithological column appears as follows: Carbonates with some shale and marl (800 m) belonging to the Tertiary-Upper Cretaceous; sandstones with other clastic deposits (450 m) belonging to the Mid-Cretaceous; carbonates with rare sulphates (1,350 m) belonging to the Lower Cretaceous-Jurassic, sandstones with rare carbonates (750 m) belonging to the Triassic Age; carbonates with some shale (150 m) belonging to the Permian Age and sandstones with shale-carbonates (2,000 m) belonging to the Pre-Cambrian time.

The map in Figure 18 shows the geology of Saudi Arabia (reproduced from Report Studies No. 9, UN Technical Cooperation for Development, 1981).

2.3.3.8 Al-Bahrain

The Kingdom of Bahrian comprises thirty-three islands in the (Arabian) Gulf. Five principal islands comprise 90% of the total surface area of the country. The largest of all islands is the Bahrain Island having an oblong shape about 48 km long and 13–16 km wide. The surface area of the country is 622 km^2 and the length of the coastal line along the (Arabian) Gulf is roughly 120 km.

The topography of the country reveals a few pronounced characteristics. It is a fairly flat plain rising up to 60 m a.m.s.l. in the mid south of the island. It is also characterised by the existence of many Sebkhas (saline low flats). The northern coastal plain is narrow, extending inland up to 5–6 km where a rocky area starts. The elevation of the rocky area varies from 30 to 45 m a.m.s.l. It extends southwards for

Figure 18. Map showing the geology of Saudi Arabia (from Study Reports on Water, No. 9, 1982, UN Technical Cooperation for Development)

several kilometres east and west for a great part of the island's width. The average ground level in Al Muharraq Island is generally less than 5 m a.m.s.l., though some of the heights reach up to about 10 m.

The Bahrain Island consists of an elongated anticlinal dome of Eocene limestone dipping 1°–5°, covered at its periphery by more recent deposits. The geological structure of the country, like Qatar, belongs to the Interior Platform with strata dipping gently towards the east. The monoclinal structure is disturbed locally by almost parallel striking anticlines with N-S axis. The three main anticlines are the Qatif, Dhahran and Bahrain anticlines.

The rock units in Al-Bahrain are similar to those of Qatar and the eastern coastal belt of Saudi Arabia. These units include the Umm-Er-Raduma formation (Palaeocene), which is basically a dolomitised limestone with a thickness of up to 320 m. They also include the Rus formation (Lower Eocene) consisting of carbonate sediments, marls and clays with a thickness from 35 to 70 m, Dammam formation (Lower-Middle Eocene) consisting of carbonate sediments, clays, dolomitic limestone and shale with a total thickness of 120–150 m. The Neogene (Miocene) complex consists of clayey marl, sandy rocks, limestone and sandy limestone (Dam formation). It is generally argillaceous sediments at the bottom and calcareous at top. The thickness of this formation in Bahrain is in the order of 120 m.

2.3.3.9 Qatar

The State of Qatar is a Peninsula of a surface area of about 11,000 km^2 protruding into the Arabian Gulf for a maximum length of about 180 km and a maximum width of 85 km. Qatar is situated between latitudes of 24°30′ and 26°10′ N and longitudes of 50°45′ and 51°40′ E.

The highest point on the surface of Qatar has an elevation of 103 m a.m.s.l. Hundreds of almost circular depressions of 2 to 3 km diameter have been formed as a result of collapsed structures, which occurred in the Rus formation, are scattered on the surface of the peninsula. These depressions have produced an almost rolling terrain in northern Qatar. They are often crater-like in shape with a depth reaching 20 m in southern Qatar. Sebkhas and playas (saline flats) can be found along the coast of Qatar with a total area estimated as 700 km^2, while sand dunes are to be found in southeast Qatar.

The Qatar Peninsula forms part of the Interior Platform of the Arabian Peninsula. This platform is characterised by a thick accumulation of sedimentary rocks with low gentle undulations dipping north-northeast and a series of north-south trending folds and gentle arching. A wide anticlinal structure exists in Qatar, with its axis coinciding with the centre line of the country. There are other anticlines and structures like faults, fractures and joints (Figure 19).

The geological sequence of Qatar is composed of Tertiary limestones and dolomites with interbedded clays, marls and shales overlain by Quaternary and Recent deposits. The Mesozoic succession of deposits comprises the Upper Jurassic carbonate rocks and The Cretaceous sequence, which consists of limestones, dolomites, sandstones and shales. This sequence can be up to 800 m in thickness. The Paleogene succession includes the Umm Er-Raduma formation (Palaeocene) consisting of dolomites, chert, marl and clay up to 300 m thick, Rus formation (Lower Eocene) consisting of dolomitic limestone and limestone with thickness up to 120 m, and the Dammam formation consisting of dolomitic limestone and marl. The Upper Dammam formation outcrops over the entire surface of Qatar.

The Neogene succession comprises the marine limestone and clay of the Dam formation (Middle Miocene). This formation crops out in the southwest of Qatar with a thickness up to 80 m. This formation comprises also continental gravels, sands and conglomerates of the Upper Miocene or Mid-Pliocene. The Quaternary deposits are diverse. They vary from beach rocks to continental gravels; depression fills of silt and clay, Sabkha deposits of silts and sands, and aeolian sands.

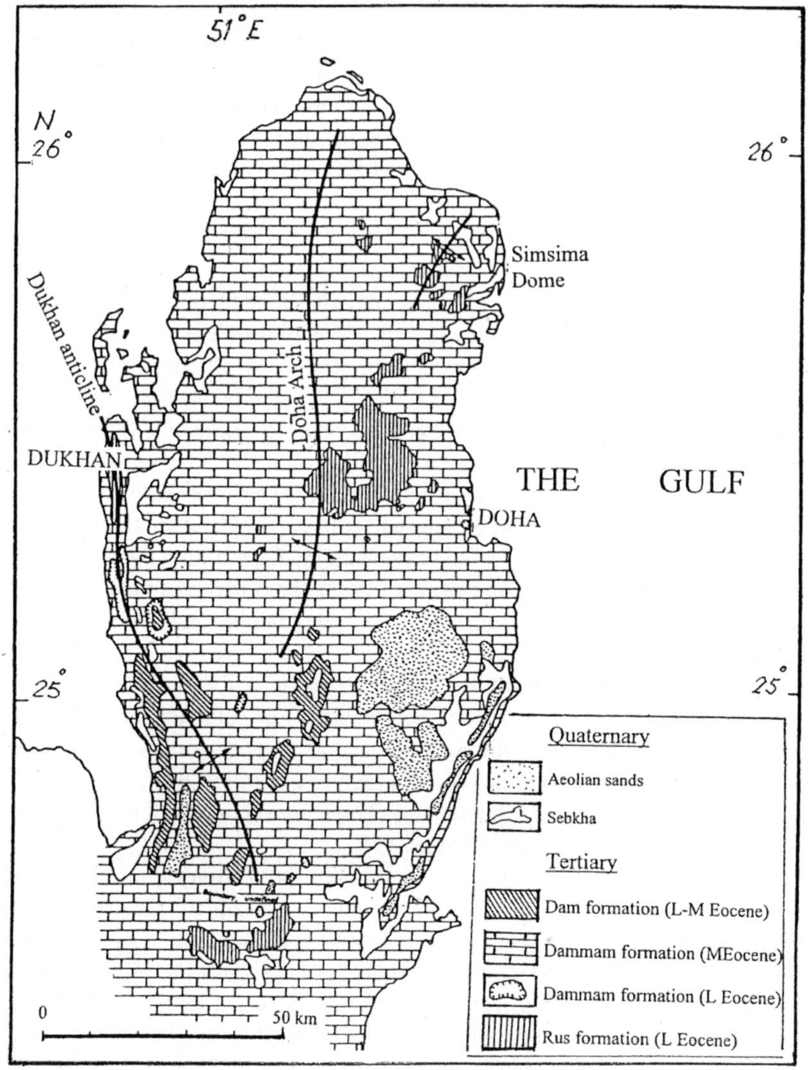

Figure 19. Simplified map of the geology of Qater (from Study Reports on Water, No. 9, 1982; UN Technical Cooperation for Development)

2.3.3.10 United Arab Emirates (UAE)

This country is composed of seven Emirates: Abu-Dhabi (67,340 km^2), Dubai (3,885 km^2), Shārjah (2,590 km^2), Ra's al-Khaymah (1,684 km^2), Umm al Qaywayn (777 km^2), Ajman (259 km^2), Al Fujayrah (116 km^2). The United Arab Emirates lies in the southeastern corner of the Arabian Peninsula to the north of

Oman and Rub al Khali Desert. The Emirates has a total surface area of 83,000 km^2, including the Gulf Islands, which count to about 200. The country is located between latitudes 22° 30′ and 22° 00′ N and longitudes 51° 00′ and 56° 30′ E. It is bounded on the west by the Arabian Gulf and on the east by the Gulf of Oman.

Physiographically speaking, the country can be subdivided into a number of regions. These are: the Mountainous region in the middle and the east, the Batinah Coastal Plain, the Sand Dune Areas in the centre (Central Desert Foreland) and south (Southern Desert), the Buraimi Oasis and the Western Coastal Plain along the Arabian Gulf (Figure 20).

The mountain ranges, which run south from the Musandam Peninsula, are basically Tertiary folded mountains with an extremely complex and unidentified structure. Rock units from Palaeozoic to Upper Cretaceous outcrop in the mountainous region of the United Arab Emirates. They show a considerable amount of similarity to the rock units exposed and existing in the subsurface of the Omani mountains (sec. 2.3.3.11).

Mentioned rock units belong to the Hawasina Group and the Sama'il Nappe Group. The former consists of a base of conglomerates, overtopped by calcareous

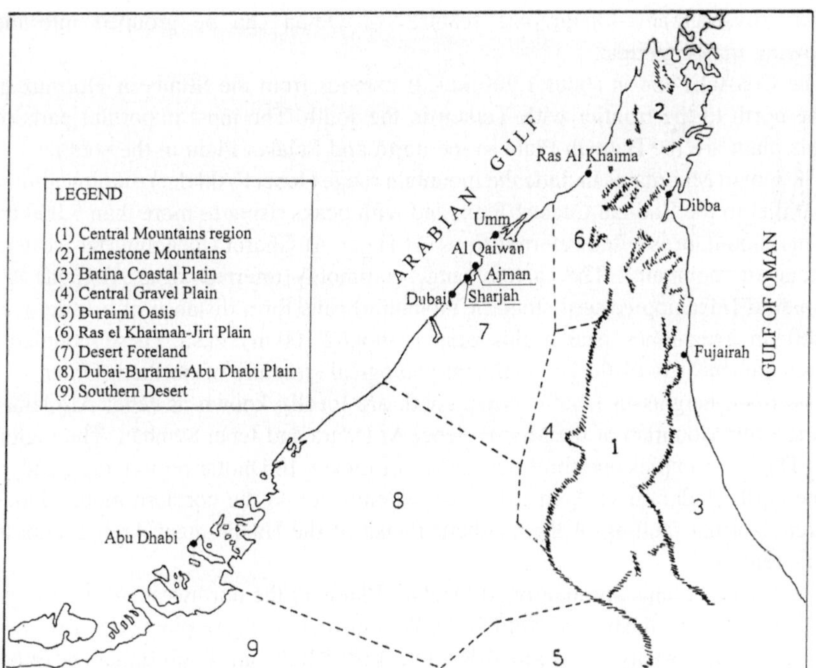

Figure 20. Map showing the physiographic units of the United Arab Emirates (from the Report of ECWA, 1981)

shales with a succession of limestone conglomerates, cherty limestone and chert of mixed ages. The Sama'il Nappe Group consists of gabbros, basic dykes and intrusives. Palaocene and Eocene rocks were deposited after these rock units. These deposits form the Umm er'Raduma, Rus and Dammam formations. Marine Oligocene deposits followed these units. At this stage, the main folding and uplifting structures took place. A repeated inthrusting took place in the north of the UAE. This brought a 3,500 m thick of Triassic-Jurassic Lower Cretaceous limestone to overlie the Hawasina/Sama'il Nappe rock units. Since then the rock exposures have been subjected to major erosion activity; this has continued over long periods. As such, the vast outwash gravel plains have been formed, as well as the thick layers of conglomerates that fill many of the drainage channels, thus forming the alluvial fans.

2.3.3.11 Oman

The Sultanate of Oman occupies most of the southeastern corner of the Arabian Peninsula. The country is bordered on the north by the Arabian Gulf, from the east by the Gulf of Oman, from the southeast by the Arabian Sea and on the west by the United Arab Emirates and Saudi Arabia, and from the southwest by Yemen. The surface area of Oman is 212,457 km^2 and the total length of its coastal line is about 1,700 km.

The physiographic-topographic features of Oman can be grouped into the following major groups:

- The Coastal Plain is about 1,700 km. It extends from the Straits of Hormuz in the north to the frontier with Yemen in the south. The most important parts of this plain are the Batinah Plain in the north and Salalah Plain in the south.
- The Oman Mountains include the mountain range (Jebel l'Akhdar) running almost parallel to the Batinah Coastal Plain and with peaks rising to more than 3,000 m. This mountain is often referred to as Al-Hajar Al-Gharbi (meaning the western stone or mountain). The eastern range commonly referred to as Al-Hajar Al-Sharqui (meaning eastern stone or mountain) runs for a distance of a little over 200 km. The highest peak of this range is about 2,100 m a.m.s.l. These mountains are characteristic of the predominant geological structure of northern Oman.
- The main heights in Dhofar down south are locally known as Jebel Al-Qamar (meaning Mountain of the Moon), Jebel Al Ddara and Jebel Samhān. The region of Dhofar occupies one-third the surface of Oman. In Dhofar region, the trend of the north Hadhramaut Arch exists and extends across the northern mountainous area. Normal faulting of the southern flanks of the Hadhramaut Arch occurs in this region.
- The Interior Plains are mainly Al Dhahira Plains in the northwest, Al-Sharquiya Plains in the northeast and Nejd in the Dhofar region. These plains are traversed by four major wadis: Al-Wadi Al-Kabier, Wadi Khalafein, Wadi Bahla and Wadi Sumeil.

In northern Oman, the Omani Mountains constitute the core around which the present geological features and the more recent strata have been formed. The main geological sequences in Oman are the Basement rocks comprising igneous and metamorphic rocks. They outcrop mainly in Jebel Al Akhdar. The Basement rocks are overlain by massive bedded limestone and dolomites with some thin argillaceous, clayey carbonates. These deposits, which are referred to as Hajar-Super Group, can be up to 3,000 m thick. They belong to the mid-Permian to the Late Cretaceous Age.

The Sumeini Group, consisting of sedimentary sequences of Permo-Triassic limestone, dolomites, sandstones and marls, succeed the sediments of the Hajar Group from above. The Hawasina Group, in turn, succeeds the Sumeini Group from above. This former, i.e. Hawasina, belongs to the Triassic-Upper Cretaceous Age. It consists of quartz sandstones, cherts, silicified limestones and shales. The Sama'il Nappe Group, which consists of basic and ultra basic rocks up to 3,000 m in thickness, overlies the Hawasina Group. The Sama'il group is believed to belong to the Mid-Late Cretaceous Period.

The marine deposits of the Maestrichtian and early Tertiary Age rocks overlie uncomfortably the old rock units. These deposits consist of calcareous sandstone, marls, reef limestone and Conglomerates. They are locally known by the names Umm er-Raduma, Rus and Dammam formation. The Alluvial and recent deposits overtop the stratigraphic column in Oman. They are well developed along the wadis and all along the Coastal Plains there. The deposits comprise gravels, sands, silt and conglomerates.

2.3.3.12 Yemen

The Republic of Yemen has a total surface area of 482,683 km^2 located in the southwestern corner and the southern portion of the Arabian Peninsula along the Gulf of Aden. It is situated between latitudes 12°00′ and 19°10′ N and longitudes 42°00′ and 53°50′ E. It is bordering Saudi Arabia on the north and west, Oman on the west, the Red Sea on the east and the Gulf of Aden on the south. The total length of the coastal line is about 1,800 km.

The country is characterised by a number of physiographic units. These are:

- The Maritime Plain of Tihama extends for about 440 km along the Red Sea Coast. It has an altitude of up to 400 m and an average width of about 45 km. This plain has a flat to rolling terrain, intersected by almost east-west running wide and shallow wadis draining their flood water towards the Red Sea.
- A Coastal Plain extends along the Gulf of Aden south east and south of the country. This plain extends inland for 50 km and rises gradually from about sea level near the Gulf to a few hundred meters inland.
- The Foothills and Middle-Height Western Mountain Slopes comprise the area encountered between contour lines of 400 and 2,000 m altitude. The terrain wherein wadis have cut deep canyons with vertical or steep slopes in some cases can be described as rugged.
- The Central Highlands comprising the upper parts of the Central Mountain Range have an altitude varying from 2,000 to 3,000 m, with peaks often exceeding 3,500 m. This terrain is less rugged than the previously described western

slopes. The eastern slopes are mild and terminate in the interior region of the high plateau.

- The Mountain Plains (Interior Plateau) are almost flat with altitude ranging between 2,200 and 2,700 m a.m.s.l. The drainage system of this area is mostly eastwards. The Coastal Mountain Belt in the south, which rises in some places to about 2,000 m a.m.s.l., joins the Interior Plateau in the north.
- The Eastern Mountain Slopes comprise a large part of the surface of Yemen. This area ranges in altitude from say 1,000 to 2,000 m a.m.s.l. The direction of land drainage is due east towards the Rub al Khali Desert.

The tectonic movements of the Yemen Plateau are generally minor compared to those occurring along the escarpments facing the Red Sea Rift Valley system on the west. The major effect of those movements is that they have opened passages for the rise of magma, which led to the formation of extensive volcanic extrusions in Yemen. The Jurassic and Cretaceous strata have been subjected to gentle folding everywhere in Yemen. The volcanic province forms an uplifted horst, and most of the faults are parallel to the Red Sea Rift. The Tihama is down-faulted some thousands of meters with respect to the rest of the country along an axis roughly coincident with the position of the present mountain front.

The southern part of Yemen can be subdivided into three geological structures. These are: The Hadhramaut Arch, which is a sedimentary plateau encompassing the south-eastern region, the uplifted basement block in the west, and the depressed area covered by volcanic rocks of the Aden Trap Series in the southwest.

The stratigraphy of Yemen is quite close to the rest of the Arab Peninsula. Rock units from Jurassic to Recent Age outcrop in the subsurface. Basement Complex of the Pre-Cambrian Age, consisting of igneous and metamorphic rock, outcrops in the south, southwest and northeastern parts of the country. Examples can be found in Radah and Sadah in the north as well as in the foothills and in the midland belt. This formation forms a general north south folding. The deposits of the Ordovician Age overlie the Pre-Cambrian rocks. These deposits consist mostly of well-sorted and cross-bedded quartizitic sandstone with conglomerates. They are commonly referred to as Wajid sandstone. The Kohlan and Amran series belong to the Lower and Upper Jurassic respectively. The former occurs at the foothills and midland areas. It overlies directly the Basement Complex. The formation consists of shale, sandstone and conglomerates. The Amran Series is widely distributed on the Highlands of Sadah-Jabal Musawar-Amran and Ma'reb regions. It consists mainly of limestone, marl and shale. The Tawilah Group and Majd-Zir Series deposits belong to the Cretaceous-Palaeocene Age. Their outcrops consist of coarse cross-bedded sandstone with conglomerates, gravel and shale. They occur mainly in the middle of the Central Highlands. The Yemen Volcanics is composed of pyroclastic rocks including basalts and tuffs. They belong to the Tertiary and Quaternary Ages and can be found in an extensive area south of San'ā'. The alluvial deposits of the Recent Time come on the top of the stratigraphic column. They are composed of silt, clay, sand and alluvial valley fills with loess deposits in the mountain plains.

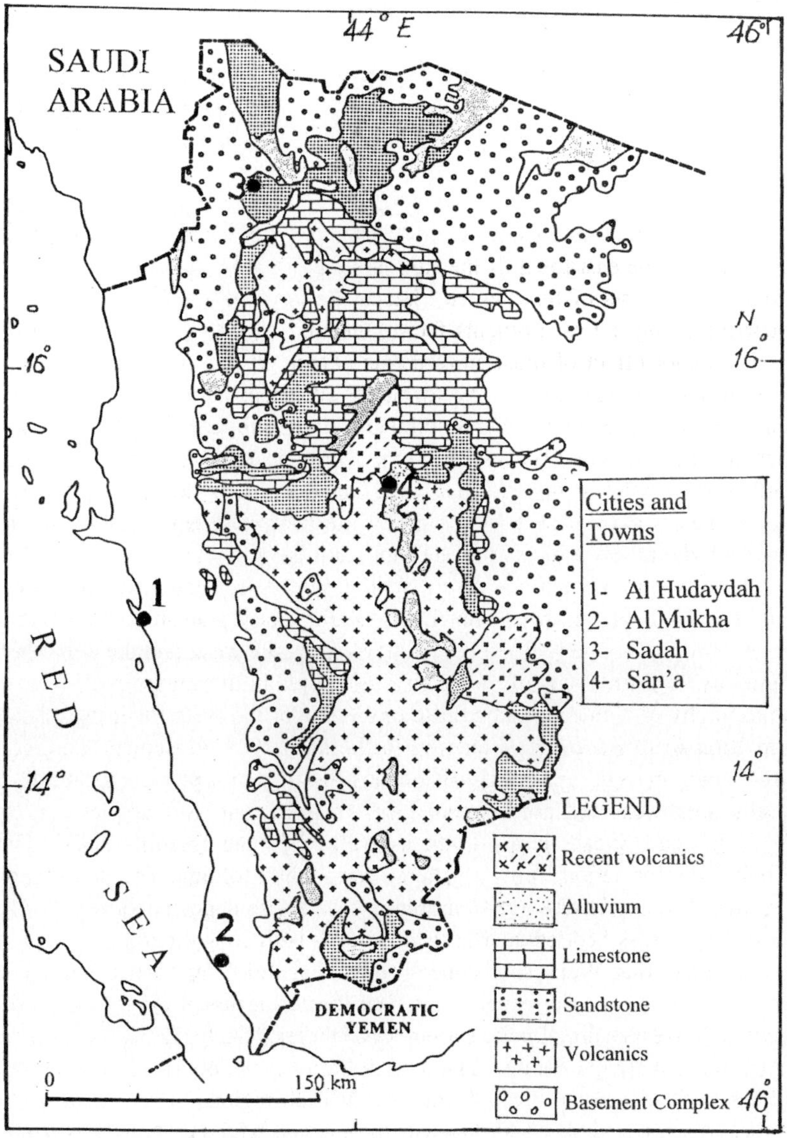

Figure 21. Map showing the geology of a certain part of northern Yemen (from Study Reports on Water, No.9, 1982; UN Technical Cooperation for Development).

An approximate idea about the locations of the abovementioned deposits in the northern part of Yemen, formerly known as Yemen Arab Republic, is shown on figure 21 (from Study Reports on Water, No.9, 1982; UN Technical Cooperation for Development).

CHAPTER 3

CLIMATE OF THE ARAB REGION

3.1. GENERAL CLIMATE OF THE ARAB REGION

Most of the great deserts lie in or close to the Tropics of Cancer and Capricorn, where for a large part of the year the midday sun stands very high. This causes the rise of the average daily temperature of the desert land. In dry deserts clouds are uncommon, so that these regions have no shield against sunlight. According to Simons (1967) "The fundamental cause of deserts is the pattern of world circulation of winds, and this is not likely to change suddenly of its own accord."

The greater part of the Arab Region lies in an area dominated by its dry climate 'desert climate'. The only exceptions are the coastal strips adjacent to the Mediterranean Sea or the Atlantic Ocean where temperature is less than the inland temperature and the air is more humid.

The Mediterranean climate is a characteristic of the Arab countries in the northwest of Africa, where mountain ranges, deep valleys, and coastal strips of varying exposure give marked and rapid local variations in temperature and rainfall. Precipitation generally increases with elevation. Though rare at sea level, frost and even snow occur at comparatively low elevations and are usual at greater heights (Stamp & Morgan, 1972). This class of climate shares its hot dry summers with the neighbouring desert lands. The only exceptions are the moderate elevations and the littoral areas. The cool moist winters derive their rainfall from the westerly air streams, which bring the rain through a succession of depressions or lows to lands nearer the poles. In the northwestern parts of Africa eastward to about mid-Algeria, the stable forms of precipitation (light, general rain) predominate. But from eastern Algeria eastward the showery, unstable rain associated with cool air predominates. The ratios of unstable (U) precipitation to stable (S) precipitation are as shown in Table 1 (Trewartha, 1962).

The Mediterranean climate characterises also the western areas in Syria, Lebanon, Jordan and Palestine, i.e. the eastern subregion outside the Arabian Peninsula and Iraq, which are adjacent or close to the eastern shore of the Mediterranean Sea. The dominating climate of the coastal strips is mild and dry in the summer, rainy and very cold in winter, mild and rather cold and rainy in the transitional periods. The western inland districts are dominated by rather mild and dry summers, light rainy

Table 1. Ratio of unstable (U) to stable (S) forms of precipitation over the Mediterranean Sea at different locations in North Africa (Trewartha, 1962)

U/S for Season and Year	Location North of West Algeria	North of East Algeria	North of Egypt
Winter	0.28	0.77	1.60
Spring	0.32	0.58	0.71
Autumn	0.67	0.76	2.25
Year	0.40	0.80	1.50

and cold in winter. The transitional periods have rather mild temperature and light rain. The Euphrates and its tributaries affect the climate of the eastern districts of Syria. The dominant climate there is hot and dry in summer, cold and slightly rainy in winter and rather hot and almost rainless in the transitions.

The dominant climate of the desert occupying vast surfaces in Jordan, Syria and Iraq is rather warm in the daytime and severely cold at night-time during winter. It is excessively hot and dry in summer, and usually hot and rainless in the transitions.

The prevailing climates in the Arabian Peninsula depend on the proximity of water resources and topography of each area in question. Three major classes of climate dominate the Arabian Peninsula. These are:

- The hot desert climate prevails in two zones. The first is the narrow coastal strip extending along the Red Sea from about 25 °C to the Gulf of Aden. Local circulations influence strongly the climate of this zone. The second zone is the vast desert plain extending parallel to the coasts of the Arabian Gulf, covering more than half the surface area of the peninsula. Both zones enjoy small amounts of rain in winter and early spring.
- The hot steppe climate prevails in two zones. The first is the elongated plateau extending parallel to the Red Sea from the extreme north to the extreme south, including Jebel Shefa Al Qusseim, Nejd, Tarique and Hadhramaut. The second zone is the Oman Plateau extending along the western coast of the Gulf of Oman. The climate of this class is a little milder than the previous class with some rains in spring and winter.
- The warm temperate rainy climate with a dry winter prevails in the triangular plateau of the southwestern corner of the peninsula. This zone includes the Asir Mountains of Saudi Arabia, most of the territory of North Yemen and the norther part of Southern Yemen. Rain falls throughout the year with maxima in spring and summer. The climate of the winter season is generally dry.

The Central subregion comprising Egypt, the Sudan, Djibouti, Somalia and the Comoros extends from north of 31° N to south of 11° S. The subregion is bisected by more than 42 parallels. Certain parts of the subregion are adjacent to the Mediterranean Sea in the north, the Red Sea in the northeast and east, the Gulf of Aden and the Indian Ocean in the east and southeast. An intensive network of tributaries of the Nile traverses the surfaces of the Sudan and Egypt whereas the surface of Somalia is traversed by the Shibeli and the Juba Rivers. Despite these water bodies, most of the surface of the Central subregion is occupied by desolate deserts and waste lands.

The coastal strip along the Mediterranean Sea in northwestern and northern Egypt enjoys Mediterranean type climate. The northeastern littoral shares a moderate desert climate of mild winters with occasional light rains, and warm, dry summers: a climate where most sub-tropical and tropical crops can be raised with irrigation. The Red Sea littoral is humid during the day. Winters are comfortable whereas nights may be rather chilly. South of the southern border of Egypt in the Sudan, certain changes in climate appear. Temperatures are truly tropical, with winters warm and summers hot and the extreme aridity of the Egyptian littoral is replaced by normal desert climate. Along the 550 km stretch, where the Sudan reaches the seacoast, the mean daily maximum temperature in June-July is around 42 °C dropping to say 27 °C by night. The entire littoral from Bab-al Mandab to the Horn of Africa can be classed as a hot desert, with high temperatures throughout the year, and with a scanty rainfall regime related to the alternating northeasterly (winter) and southwesterly (summer) monsoonal winds. The rainfall along the south coast of the Gulf of Aden, from Bab-el Mandeb to Cape Guardafui, Somalia, is scanty. It ranges from about 130 mm y^{-1} at Djibouti to 10 mm y^{-1} at Bender Cassim near Guardafui. However, there are strong evidences of heavy dew fall (Meigs, 1966).

The climate of Lower Egypt is mild with some rain showers, mainly over coastal areas. The annual rainfall, mostly in winter, declines rapidly with distance from the seacoast. Upper Egypt is practically rainless with warm sunny days but rather cool nights. Between the Nile Valley and the Red Sea lies the Eastern Desert of Egypt. This is a dry, inhospitable region where temperatures are high, with a slight relief from the heat at night.

The remaining area of the surface of Egypt is part of the desert belt extending from Africa's west coast through the Arabian Peninsula to deep in Asia in the east. This area together with vast surfaces in Libya and the Sudan have desert climate. The Saharan or desert climate covers the area from south of Cairo in Egypt to Atbara in the Sudan. There is hardly any rainfall in this area and the climate can be described as hyper arid.

Local climates in the Sudan are strongly affected by the movement of the Inter Tropical Convergence Zone (ITCZ) across the country. The ITCZ is the surface where the moist southerly winds (monsoons) blowing from the Indian Ocean meets the dry northeasterly winds blowing from central Asia. Tropical climate covers the area south of Atbara. While as the seasonal variation in air temperature in the Sudan Plains decreases from north to south, the rainfall increases in a southerly direction. It is a generally accepted concept to consider the isohyet of 400 mm y^{-1} as the lower limit of Savannah land. As such, most of Central Sudan has a Savannah climate. Less than this depth, the climate can be described as steppe. It is rather difficult to say where the steppe changes to desert. However, the divide line is likely to be located between isohyets of 200 and 250 mm y^{-1}. The swamps and marshes in southern Sudan help to maintain the air humidity in the Sudd area quite high.

The eastern part of the Sudan is characterised by its highlands. The temperature is generally modified by the elevation. It is agreeable during summer though fairly cold in winter by day and chilly by night. The topography of the ground helps to cause a heavy rainfall. The Central Plains are surrounded from the west by sand dunes. Temperature is warm in the winter and hot in summer, and most of the rain occurs in the period July-September and ceases in the period November-April.

The Comoros Islands have a tropical climate with a cool, dry season from May to October and a warm, rainy season from November to April. The temperature varies between 15 °C and 35 °C (WHO, 1996).

After highlighting the general climate of the Arab Region, it is worthwhile to substantiate it with some information about the different factors shaping the climate.

3.1.1 Air circulation and wind

The cool season in the Mediterranean zone is affected by the depressions caused by the polar front arising in and moving across this zone Several depressions (up to 60 per year) pass over Gibraltar, of which 30% or so reach northeast Tunisia. Some of these depressions move farther and reach the Persian Gulf (Figure 1). These depressions regulate the winter weather in north Morocco, Algeria and Tunisia. Of about 20 depressions arising over central Algeria, the large majority moves eastwards in the direction of Cyrenaica, Libya. The most serious atmospheric disturbances are associated with those rare depressions that cross Central Libya and the Qattara Depression, Egypt, towards Cyprus. These air depressions are accompanied by

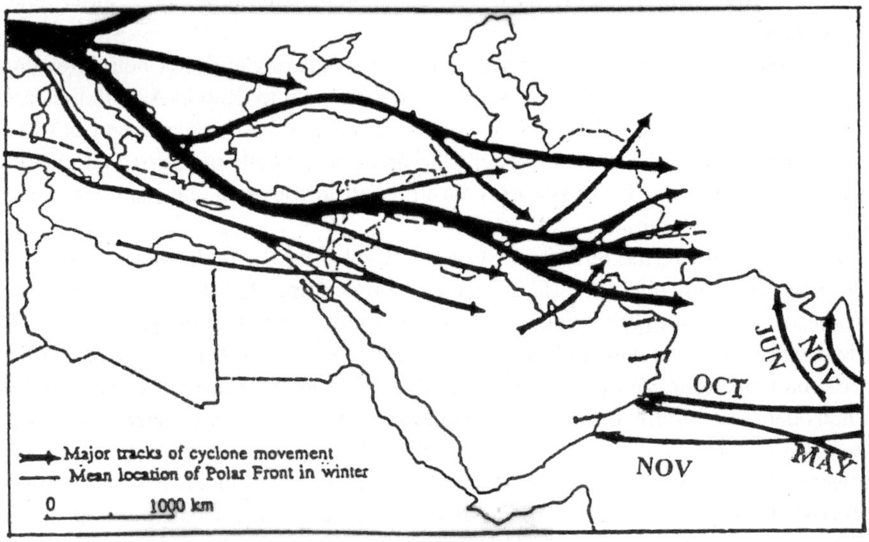

Figure 1. Cyclone tracks over the Middle East and Arabian Gulf (Beaumont et al., 1976 and Ministry of Water Resources, Oman, 1995)

violent dust storms and advection of exceptionally hot air. They are also responsible for those very strong winds, the 'Sirocco' and the Khamasīn (dust storms that last about 50 days). Nearly 50% of the depressions to which these winds belong occur in April and May. A few of these depressions move farther and reach the Persian Gulf. The disturbances they cause do not last except two or three days during which the coastal strips of the United Arab Emirates and Oman get light rain showers. What is called Khamasīn in Egypt is known as Ghibli in Libya; Shlour in Syria and Lebanon; and Shargi in Iraq.

The northerly and northeasterly winds blowing on Lower Egypt are an extension of the 'etesian' (seasonal) winds that blow across the eastern Mediterranean. The meridional mountains of Lebanon, other than blocking the effect of the westerly circulation of air in winter, exert strong influence on the climate. Similarly, the Zagros and Kurdistan Mountains cut off the large Mesopotamian Plain from the northeast. The Mediterranean air is transported across this vast plain from the Mediterranean in the west to the Persian Gulf in this east (Martyn, 1992).

The Sahara belt between 10° and 20° N is the warmest part of Africa, where continental tropical air masses arrive from the north as the trade wind 'Harmattan' and the humid equatorial masses from the south. The circulation over this part of Africa is typically monsoonal, with seasonal changes in wind direction, northward-prolonged dry seasons and correspondingly shortened rainy seasons.

The relatively warm Mediterranean area attracts cool masses of maritime polar air and sometimes maritime Arctic air from the Atlantic, and continental air from Europe. These air masses are rapidly transformed, especially from November to January, when the sea is warmer than the land, and become unstable.

The atmospheric circulation and weather in summer months are determined by the continental depressions over Mesopotamia and the Persian Gulf, over the southern Sahara and southern Sudan, and by the high pressure areas lying over central Africa and the wedges of the Azores High extending over the Mediterranean. The average atmospheric pressure over the Persian Gulf area (Kuwait, Basra, Abadan, Abu Shahr, Dhahran, Al Bahrain and Sharja) in January is 1019 millibar (mb) and falls down to 997 mb in July. In summer, locally very strong pressure gradients can be developed as the result of intense solar heating, producing a range of small dust storms characteristic of the interior desert of Saudi Arabia. At about the same time, in July, the convergence of the Harmattan and the Guinea monsoon reaches 20° N parallel up to the Khartoum region in the east. The Khartoum temperature is also affected by the Indian monsoon.

Part of the air front east of the Nile serves in separating two air masses from each other. The first is the drier and cooler air masses from the Arabian Peninsula blowing in from the east and northeast and travelling north up the Red Sea. The other is the warmer and moister air masses moving northwest and northwards off the Indian Ocean. A front forms across the Red Sea between the air mass from the Gulf of Aden and Saudi Arabia, and that from the Mediterranean in the north.

The Gulf countries, Yemen and Somalia are under the effect of the seasonal monsoon winds. The southern part of Yemen is dominated by the southeastern winds during the winter season from October to April, and the northeasterly winds in the summer months, June-August. Northwest winds blow on Oman and Qatar during the months from November to April, while the southwesterly winds predominant in the period June-September. Kuwait too is exposed to the same winds, i.e. the cold north-westerly winds during the winter and early spring seasons, while the hot and moist south-westerly winds blow during the second half of the spring season and early summer. Saudi Arabia is dominated during the longest part of the year by the northerly winds, which blow from the eastern Mediterranean toward the Persian Gulf.

3.1.2 Temperature

The average air temperature during January, the coolest month of the year, is around 5 °C or less, especially in the region of the Atlas Mountains in Morocco. Likewise, the January temperature in the highlands in north Syria, Lebanon, east of Iraq is 7–8 °C. It rises to 15 °C and further to 20 °C upon travelling in a southerly or easterly direction. The January temperature in the northern part of Egypt, along the Mediterranean coast is in the range of 12–13 °C, increasing to 16 °C or more along the Red Sea coast, especially in the south. It continues its rise to reach 23–24 °C in the Sudan at places like Port Sudan. In southern Sudan and the lowlands of Somalia, the January temperature is in the range of 28–29 °C. The coolest month in southern Sudan is July and August instead of December and January. The average July-August temperature amounts to 26 °C.

The January temperature in the inland areas in Syria, Jordan and Iraq is 7–8 °C. This temperature increases considerably to reach 23 °C and further 25 °C at Jeddah and Al Hudaydah along the Red Sea, in Saudi Arabia and Yemen respectively. The temperature at other coastal locations in the Arabian Peninsula is a few degrees lower in the north and a few degrees higher in the south. Winter temperature at inland areas in Saudi Arabia and Yemen are much lower. It is around 13–14 °C at Riyadh, Saudi Arabia, and San'a, Yemen. The elevation has a considerable effect on temperature.

The April average temperature reaches about 10 °C in the mountainous areas of the North African countries. It increases to 25 °C in the east and south. The warmest period of the year in the Arab Region can occur in any month, from March in southern Sudan to June in the Eastern Sahara, July in Western Sahara and August on the Mediterranean and Red Sea coasts. The northward movement of the Intertropical Convergence Zone (ITCZ) is responsible for this variation. The summer average temperature in July exceeds 20 °C along the sea and ocean coast while the inland average temperature is above 30 °C and even 36–38 °C in Western Sahara. As a matter of fact, it reaches in some July days 50 °C or a little higher. Examples can be found in the Sahara and certain parts of Iraq, Saudi Arabia and the Sudan. Temperature is similarly high along the Red Sea coasts, and especially in the Gulf of Aden where 36 °C is common. In the warmest part of the year, afternoon temperatures exceed 40 °C in much of the Sahara and in the Gulf of Aden, and 45 °C in Algeria.

The temperature along the northern, northwestern coasts of Africa drops in the autumn season by about 5 °C compared to the summer temperature. Similar temperature drop can be observed along the eastern coast of the Mediterranean Sea and on the high mountains. The drop in temperature inland is highly noticeable.

Several sources of information as textbooks and publications on climate include maps showing the mean isothermal lines for the different regions of the world over the whole year and/or certain parts of the year, especially for January and July. Outstanding maps for the Near East including the Arabian Peninsula can be found in Volume 9 of the World Survey of Climatology (ed. Takahashi & Arakawa, 1981).

3.1.3 Humidity

The relative humidity of air is generally low in most parts of the Arab Region. However, exceptions can easily be observed in the coastal plains and land strips adjacent to seas and oceans; intensely irrigated areas, e.g. the Nile Delta area in Egypt and the Gezirah Plain in the Sudan; land areas surrounding large lakes and reservoirs.

The mean annual relative humidity along the Mediterranean coast is about 70%, decreasing rapidly to say 30% or less a few kilometres inland. The coastal areas along the Gulf of Oman and the Persian Gulf are in the order of 85%, though varying from one month to another month. Similar variation happens along the eastern coast of the Mediterranean Sea where the average humidity is between 70% and 80% during the winter, and 55% and 60% in the summer time. The humidity figure falls considerably inland and reaches its lowest value in desert areas. This observation supports the results obtained from a study of atmospheric moisture content. The results listed in Table 2 show that the air over the great deserts is normally very dry (Lockwood, 1974). This dryness is partly caused by the lack of evapotranspiration and partly due to lack of horizontal advection. Similar to temperature, the elevation of any location above mean sea level is an important factor affecting the atmospheric humidity.

3.1.4 Length of daytime

The longest day lasts from 12.2 h in the south of the Sudan to 14.7 h in north Tunisia and Algeria; the shortest day is correspondingly 11.9 h to 9.4 h. In the south of the Arabian Peninsula the length of the daylight during the year varies from 12.8 h on the longest day to 11.4 h on the shortest; in northern Syria and Iraq the times are 14.2 h and 10.0 h respectively. The limited differences between the length of daylight during the year in the southern parts of the Arab Region makes for a more or less constant inflow of solar heat which is modified only by cloudiness (Martyn, 1992).

3.1.5 Sunshine

The January effective sunshine in the Sahara and in southern Sudan ranges from 250 to 300 hours. This amount reduces to about 200 h of direct sunlight per month

Table 2. Mean relative humidity at isobaric levels for radiosonde stations in the Sahara and Arabia (from Lockwood, 1974; after Drozdov & Grigor'eva, 1963)

Station name and coordinates	Country	Isobaric level, mm	Month January	July
Bir Moghein (Fort Trinquet)	Algeria	850	33	16
25° 14′ N, 11° 35′ W, 360 m		700	22	22
		500	15	31
C. Bechar	Algeria	850	44	17
31° 37′ N, 02° 13′ W, 780 m		700	33	22
		500	22	28
Oualef el Arab	Algeria	850	28	14
27° 04′ N, 01° 06′ E, 275 m		700	22	17
		500	15	21
Cairo	Egypt	1,000	47	38
30° 08′ N, 31° 24′ E, 68 m		850	49	36
		700	40	24
		500	37	15
Habbaniya	Iraq	1,000	62	N.A.
33° 22′ N, 43° 34′ E, 45 m		850	53	15
		700	33	18
		500	26	7

on the Mediterranean Coast. In July, the situation changes to some extent. The longest hours of sunshine are now enjoyed by areas farther north compared with areas receiving the longest sunshine duration in January. This can be attributed to the northerly movement of the maximum intertropical cloudiness zone and the polar front from the Mediterranean region. The zone north of 15° N latitude has over 250 h of sunshine, the northern Sahara has over 350 h and the regions around the Great western Erg and the Nubia Desert have as much as 400 h. In the extreme south of the Sudan there are less than 150 h and not much more than 250 h in Mauritania.

The annual sunshine amount reflects the great influence of cloudiness. There are over 3,600 h of direct sunshine in the central and eastern Sahara and over 4,000 h where the borders of Libya, Chad, Egypt and the Sudan meet. In the belt encompassing the Tibesti Mountains, the southern Libyan Desert, the Nubian and Arabian Deserts there are more than 4,000 h, which is 90% of the possible hours of sunshine.

North and south of the sunniest regions are areas with less sunshine because of the greater cloudiness in winter. Clouds blowing in off the Atlantic reduce the sunshine to 2,600–3,000 h on the coastal plain of Morocco and in the High Atlas. In other parts of Morocco and the Cyrenaica Peninsula, Libya, the total is under 3,000 h. For example, the annual sunshine hours in southern Mauritania are between 2,900 and 3,000 h and between 2,000 and 2,200 h in Equatorial Sudan.

Owing to cloudiness and other factors like atmospheric pollution, both natural and man-induced, the annual effective sunshine varies from 3,100 h in northern Iraq and Lebanon Mountains to 3,300 h in the Bab el Mandab Straits and 3,600–3,800 h in the interior of the Arabian Peninsula. The relative sunshine in that area is high, from 70 to 90%. North of 17° N parallel, the summer months are the months that have the longest duration of sunshine, and the winter months that have the shortest duration. In January, the number of sunshine hours ranges from 150 h in Lebanon, northwest Syria and northern Iraq to 300 h and more in southwest of Saudi Arabia The sunshine in June-July in the Syrian Desert and north Saudi Arabia is above 400 h. North of the Tropic of Cancer (23° 30′ N) it is 350 h in June decreasing southwards to 200–250 h in Yemen.

3.1.6 Radiation

The global radiation, like many more climatic factors, varies within the Arab Region from one location to another. The annual solar radiation is 84% of the possible maximum in Agadir, Morocco; 90% in Cairo, Egypt; 87% in Ahaggar Mountains; and 82–83%, at Wadi Halfa, El-Fasher and Khartoum, the Sudan. The effective annual amounts are very high: over 8,370 MJ m^{-2} in the Sahara, and 9,210 MJ m^{-2} in the Nile Valley between the first and fourth cataracts. From this very highly insulated Saharan region, the solar radiation declines in all directions to 7,536 MJ m^{-2} on the Atlantic, 6,280 MJ m^{-2} in the western Mediterranean and 6,700 MJ m^{-2} on th Cyrenaica coast, Libya. Similar to the western Mediterranean, southern Sudan gets 6,280 MJ m^{-2}. The annual solar radiation in West Asia between 20° and 30° N is high. It exceeds 8,370 MJ m^{-2}, but falls down northwards to 6,300–6,700 MJ m^{-2}.

The January solar radiation over North Africa rises in a southerly direction, from 335–419 MJ m^{-2} in the north to 670 MJ m^{-2} in the south. In July, the solar radiation distribution is reversed. The highest amount of energy, over 921 MJ m^{-2}, reaches the central and western Sahara, along with Tripolitania, Cyrenaica and the Egyptian coast: westwards this amount of energy decreases to 750–840 MJ m^{-2}. The southern Sudan does not receive more than 335–420 MJ m^{-2}, i.e. as much as the Mediterranean receives in January. The January solar radiation is least over western Syria (170–210 MJ m^{-2}) and rises sharply to about 380 MJ m^{-2} along the line extending from the Gulf of Aqaba in the west to central Mesopotamia in the east. The southward increase is much less rapid, and reaches 586 MJ m^{-2} in the Bab al Mandab Straits.

On the Mediterranean coasts the summer radiation amounts are 335–419 MJ m^{-2}, while Eastern Sudan receives over 586 MJ m^{-2}. The highest totals in summer are recorded in Jordan and the Syrian Desert; the July figure is about 900 MJ m^{-2}. At the same time the radiation falls to 754–796 MJ m^{-2} in the south of the Arabian Peninsula.

The global radiation, R_G, reaching the ground or water surface is not always measured. Instead, it can be estimated approximately from the extraterrestrial radiation, R_A, reaching the outer surface of the atmosphere using regression equation of the type:

$$(1) \qquad R_G = R_A(a + b(n/N))$$

where,
a = regression constant
b = regression coefficient
n/N = actual to possible sunshine duration

The available measurements in Egypt have been employed for calibrating a and b using the method of least squares, and found as 0.22 and 0.55 respectively (Mehanna, 1976). The mean values for ten observation stations in the Sudan have yielded a different pair of results; 0.350 and 0.368 for a and b respectively (El-Agib, 1995).

3.1.7 Cloudiness

Evaporation and evapotranspiration depend largely on the degree of cloudiness of the sky. Cloudiness is usually expressed in tenths or in oktas, i.e. eighths. For example, sky cloudiness of 3 oktas means three-eighths or three and three-quarter tenths, i.e. 0.375 is covered with clouds.

The cloudiness in the Arab Region is known to be relatively small, in the order of 2–3 oktas in the winter season and less than that in summer time, except in locations close to the Equator. A rather old, yet useful, map of the cloudiness in January and July is shown in Figure 2. Saddler (1968 cited in Lockwood, 1974) prepared that map to give the cloudiness, averaged over two years, based on satellite observations.

3.1.8 Precipitation

The northern countries of the Arab region are exposed in winter time to some scattered depressions. The cold fronts blow from the atmospheric heights centred over Siberia and the Atlantic. During their journey across the Mediterranean, they become saturated with moisture and cause precipitation. The amount precipitated reduces gradually upon crossing the mountain ranges scattered along the seacoasts. As such, the rainfall that reaches Jordan, northern Iraq and Saudi Arabia is scanty. It also becomes insignificant a few kilometres inland from the seacoast in Libya and Egypt.

In summer the atmospheric heights are mainly centred over the Indian Ocean. The seasonal winds (monsoon) blow dragging along with the air masses, which have become saturated with water near the coasts of Somalia and the Arabian Sea, is causing rainfall along the coastal strips and down hills. These winds, however, do not penetrate much inland because of the mountain heights along the Arabian Sea and the Somali Coast.

It is important to remember that the mountain ranges overlooking the Red Sea, the Mediterranean Sea, and the Atlantic and Indian Oceans play a major role in the distribution of the precipitation in the Arab Region. The orographic effect of these mountains is to minimise the rain from falling inland. Instead, most of the rain falls close to the mountain tops, slopes and down hills on the sea side. The tops of the

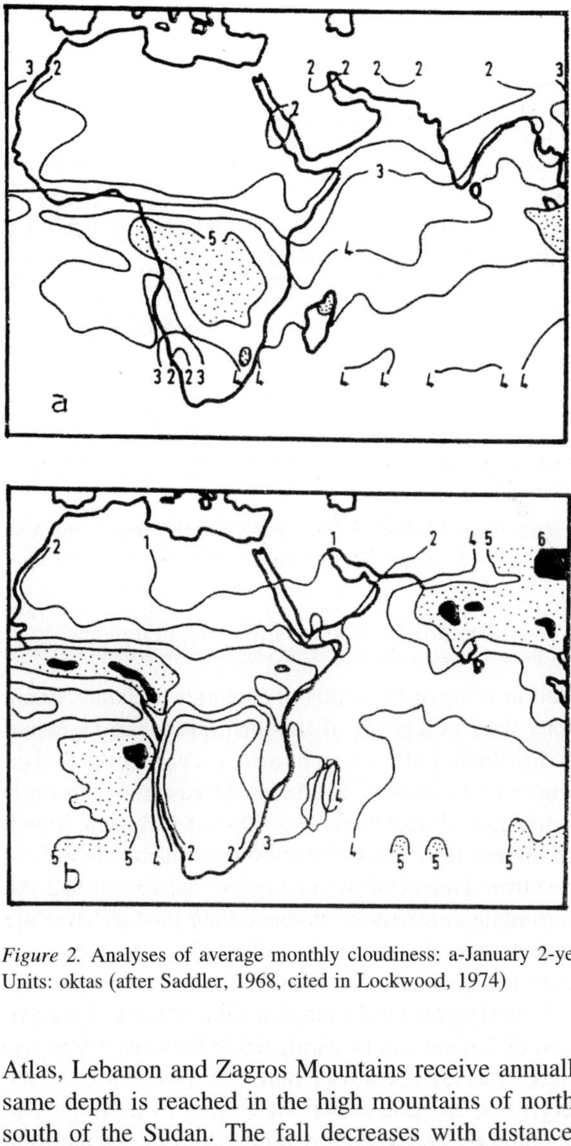

Figure 2. Analyses of average monthly cloudiness: a-January 2-year average, b: July 2-year average. Units: oktas (after Saddler, 1968, cited in Lockwood, 1974)

Atlas, Lebanon and Zagros Mountains receive annually more than 1,000 mm. The same depth is reached in the high mountains of northern Yemen and the extreme south of the Sudan. The fall decreases with distance from the sea or ocean; the topography and relief of the area in question essentially control the rate of decrease. The annual rainfall can be as scanty as 25 mm or less at the desert fringes and other locations (Figure 3).

In addition to the orographic precipitation, some of the Arab countries like Syria, Lebanon, Iraq, Jordan and northern Saudi Arabia receive convective rainstorms during the spring season. These rainstorms are usually characterised by their short duration and high intensity, which can reach 100 mm h^{-1}. With such a density the rainstorm can produce the so-called flash flood, which may have destructive

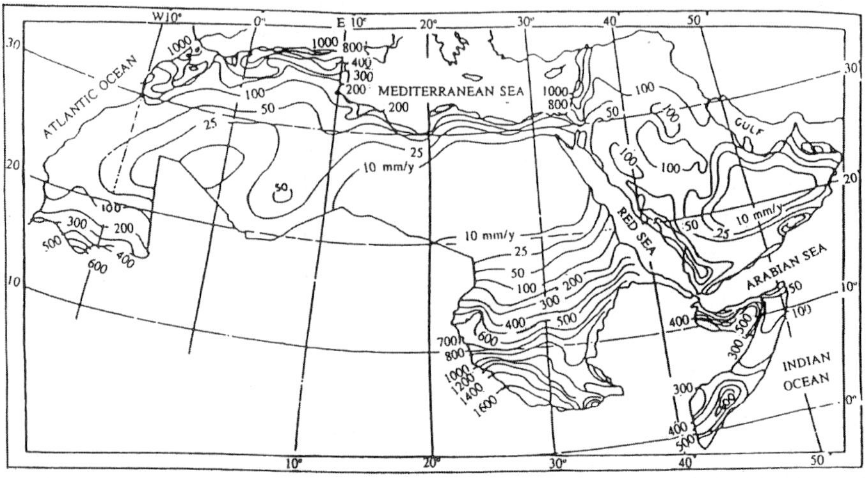

Figure 3. Mean annual precipitation totals (from Agricultural Development Organisation for the Arab Countries, 1980)

effects. However, it has to be remembered that the total rainfall depth remains small compared to winter rain.

The principal or major rainfall in some of the southern countries in Saudi Arabia and the Sudan occurs in summer time as a result of seasonal rains. The mountain ranges play a major role in the distribution of these rains also, except for the Sudan. The saturated air masses coming from the Atlantic and Indian Oceans move towards the Congo in a northeasterly direction. From there, the Ethiopian Plateau, which stands high at the southeastern corner of the Sudan cause the air masses to deflect to the north. In view of the relative flatness of the country in the Sudan and the absence of inland heights, the saturated air masses continue their passage over the Sudan up to say 1,300 km near Atbara, where rainfall practically diminishes. As such, the annual rainfall diminishes gradually from more than 1,200 mm in the extreme south of the Sudan to 25 mm or less at the Atmour Desert north of Atbara.

The average rainfall in the Arab Region can be estimated at between 1,900 and $2{,}000 \times 10^9$ m^3, with an average rate of about 140 mm y^{-1}. Two thirds of the surface area of the region receives no more than 300×10^9 m^3 y^{-1} at an average rate of less than 100 mm y^{-1}. Slightly above 2×10^6 km^2, i.e. 15% of the total surface area receives about 320×10^9 m^3 y^{-1}, with an average rate of between 100 and 300 mm y^{-1}. The remaining amount; 1300×10^9 m^3, falls annually over 20% of the total surface area of the region at a rate exceeding 300 mm y^{-1}.

3.1.9 Evaporation and evapotranspiration

The general climate of the Arab Region, as already mentioned above, is characterised by high mean air temperature, low relative humidity, long hours of sunshine and the availability of adequate heat energy at land and water surfaces. All these factors combined help augment the loss of water by evaporation and evapotranspiration from water and vegetated surfaces respectively.

Observation of free water evaporation along the southern and eastern shores of the Mediterranean has yielded figures in the order of 1,000 mm y^{-1} increasing to 2,000 mm y^{-1} or more inland eastwards. The same situation occurs upon moving from the North African coast inland towards the Sahara, and then decreases gradually with increase in rainfall and relative humidity until it becomes in the order of 1,500 mm y^{-1} in southern Sudan along its border with Uganda. Summer evaporation in the Arabian Peninsula and the Gulf area often exceeds 15 mm d^{-1}. This value is often exceeded as one moves inward to desert areas.

Similar to evaporation, observations or estimates of potential evapotranspiration have shown that it varies both temporally and spatially. In the Gulf area the potential evapotranspiration can be as low as 2.5 mm d^{-1} in winter and reaches as high as 12 mm d^{-1} in summer. As a general rule, the summer rate in the Arabian Peninsula is 3 to 4 times the potential evapotranspiration rate in wintertime. The annual potential evapotranspiration ranges from about 800 mm along the eastern shore of the Mediterranean Sea and the extreme northwest of Africa to 2,500 mm in certain parts of the Arabian Peninsula. It is worth mentioning that a potential evapotranspiration of 2,000 mm y^{-1} is quite normal in the Arab Region. This amount, however, cannot be reached unless adequate supply of water is secured. As this condition does not exist in the Arab Region, except in limited areas, the evapotranspiration that actually occurs is mostly less than the potential level. The actual evapotranspiration hardly exceeds 100 mm y^{-1} in most of the areas of the region. It increases, however, to 300 mm y^{-1} in the northwestern corner of Africa and certain parts of Lebanon and Syria. It increases further to 600 mm y^{-1} or more in southern Sudan.

Before reviewing the climate of the Arab Region in more detail, it is worthwhile to inform the reader of the raw data used in preparing certain parts of the text of this book. These data are kept available in 14 Tables included in Appendix I entitled "Meteorological Data."

Table 1- Names and coordinates of meteorological stations and the countries to which they belong,
Table 2- Average daily daytime hours for the 12 months of the year,
Table 3- Mean daily air temperature,
Table 4- Mean daily maximum temperature at selected stations,
Table 5- Mean daily minimum temperature at selected stations,
Table 6- Mean daily relative humidity; monthly and annual,
Table 7- Average daily duration of sunshine at selected stations,
Table 8- Sky cloudiness at selected stations,
Table 9- Short wave radiation flux at the outer surface of the atmosphere,
Table 10- Global radiation at selected stations,
Table 11- Mean daily wind speed at selected stations,
Table 12- Total rainfall; monthly and annual,
Table 13- Annual rainfall series for selected stations,
Table 14- Evaporation measurements at selected stations, and Sources of information

The data presented in the above listed Tables have been recorded at a total of 200 meteorological stations. The distribution of those stations is as follows:

No. of Stations	Country	No. of Stations	Country	No. of Stations	Country
6	Mauritania	10	Somalia	6	Kuwait
10	Morocco	1	Comoros	1	Bahrain
14	Algeria	10	Syria	3	Qatar
12	Tunisia	4	Lebanon	9	S. Arabia
17	Libya	5	Jordan	1	UAE
30	Egypt	6	Palestine	4	Oman
23	Sudan	11	Iraq	16	Yemen
1	Djibouti				

It has to be remembered that, with the exception of daytime hours, mean daily air temperature, and monthly and annual rainfall, not all stations possess complete records of other climatic parameters like sunshine duration, cloudiness, etc. Additionally, the length of record is not always the same for all stations. Most stations, however, have at least 30 uninterrupted years of continuous record.

The locations of those 200 stations together with their names and code numbers are plotted on three maps; Figures 4(a), 4(b) and 4(c) for the western, central and eastern subregions respectively.

3.2. CLASSIFICATION OF CLIMATE

Review and discussion of standard methods used for classifying the climate are available in the relevant literature (e.g. Gentili, 1972). Certain factors such as air temperature, precipitation, potential evapotranspiration and number of wet days per year are frequently included in those methods.

De Martonne (1926) developed an expression for the index of aridity, *IA*, a follows:

$$IA = P/(T + 10) \qquad (2)$$

where:

P = long-term average annual precipitation, mm
T = long-term average annual temperature, °C

The limit of a true desert was set by De Martonne at $IA = 5$. The dry steppe boundary, i.e. the limit of dry farming without irrigation is drawn where the *IA* has a value of 10.

A certain area with $P = 210$ mm y^{-1} and $T = 20$ °C has, according to Eq (2), an aridity index of 7. As such this area can be regarded as a true desert.

Köppen (1931) classified the climate using the annual rainfall, P, in inches, mean annual temperature, T, in ° F, the rainfall in the wettest month of the winter half compared to the rainfall in the driest month of the summer half of the year or the wettest month of the summer half compared to the driest month of the winter half of the year.

The general expression he obtained is:

$$r = (cT) - x \tag{3}$$

where:

r = critical value of rainfall to be compared to P,

c = constant = 0.22, and

x = a variable quantity, depending on the distribution of monthly rainfall and the ratio of the rain in the wettest month of one half of the year to the rain in the driest month in the other half of the year. The three values set for x are 1.5; wettest month of summer has at least 10 times as much rain as the driest month of winter, 4.25; the wettest month in the summer has less than 10 times as much rain as the driest month of winter or the wettest month in winter has less than 3 times as much

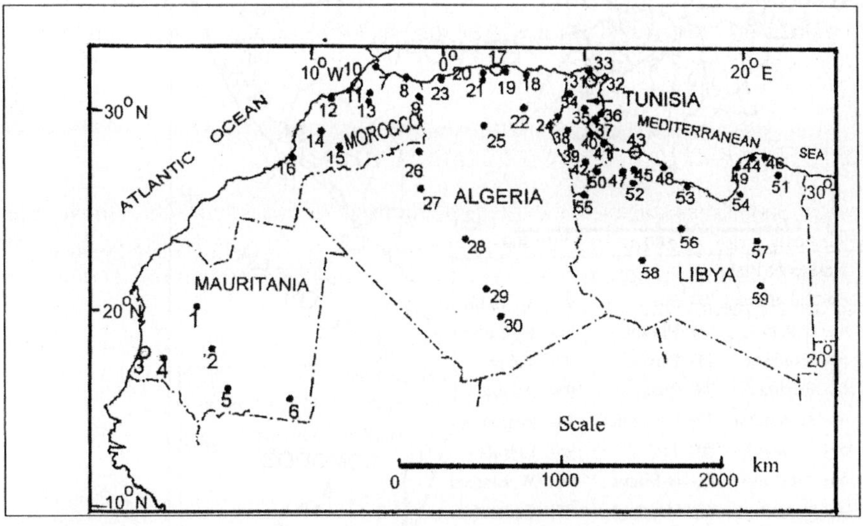

Names of Places
1- Atar, 2- Tidjika, 3- Nouakchott, 4- Boutilimit, 5- Kiffa, 6- Nema, 7- Tanger, 8- Melilla, 9- Oudja, 10- Rabat, 11- Meknes, 12- Casablanca, 13- Ifrane, 14- Marrakech, 15- Ouarzazate, 16- Agadir, 17- Alger, 18- Cap Carbon, 19- Maison Blanche, 20-Miliana, 21- Beni Abbes, 22- Biskra, 23-Oran, 24-Tebessa, 25- Laghouat, 26- C. Bechar, 27- El Golea, 28- In Salah, 29- Ouallen, 30- Tamanrasset, 31- Bizerte, 32- Kelibia, 33- Tunis, 34- Jendouba, 35- Kairaouan, 36- Monastir, 37- Sfax, 38- Gafsa, 39- Tozeur, 40- Gabes, 41- Djerba Mellita, 42- Remada, 43- Tripoli, 44- Shahat, 45- Garian, 46 -Derna, 47- Aziziyah, 48- Misurata, 49- Beni Ghazi (Benina), 50- Nalut, 51- A. Nasser A.P., 52- Mizda, 53- Sirte, 54- Ajdabiyah, 55- Gadames, 56- Hon, 57- Gilao, 58- Sebha, 59- Kufrah.

Figure 4(a). Names and locations of meteorological stations used for describing the climate of the western subregion

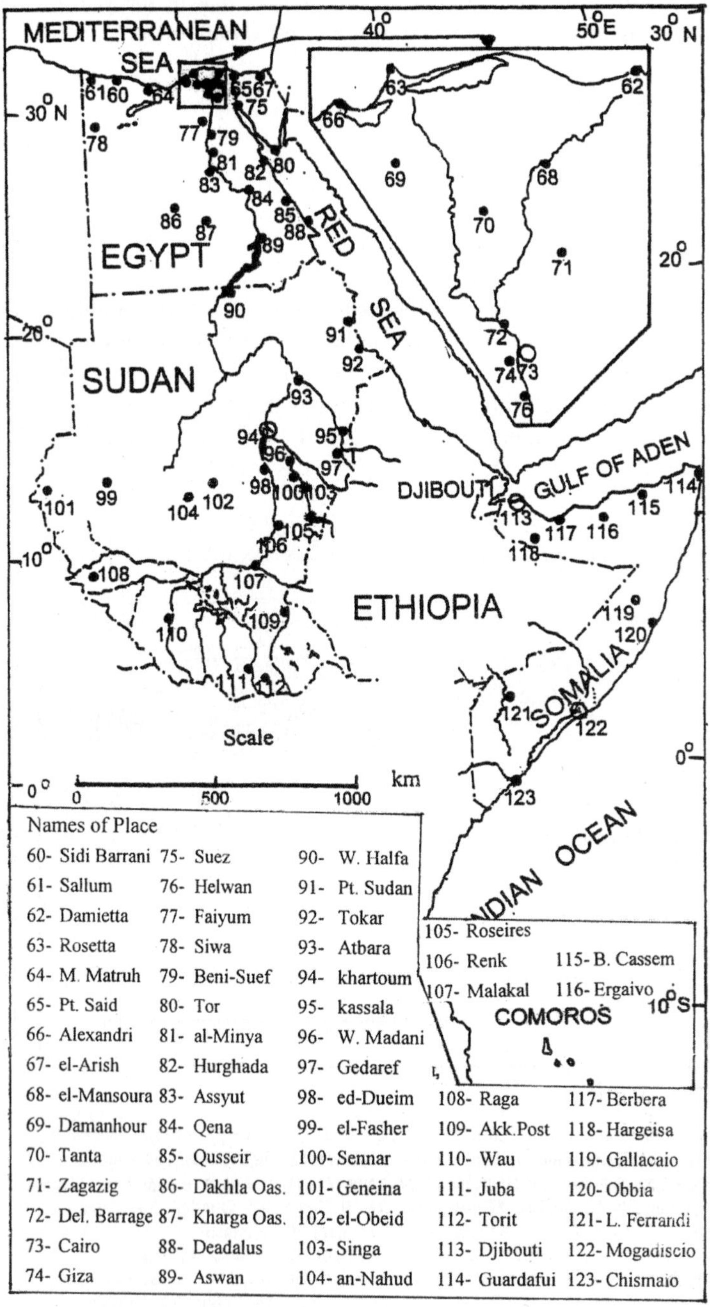

Figure 4(b). Names and locations of meteorological stations used for describing the climate of the central subregion

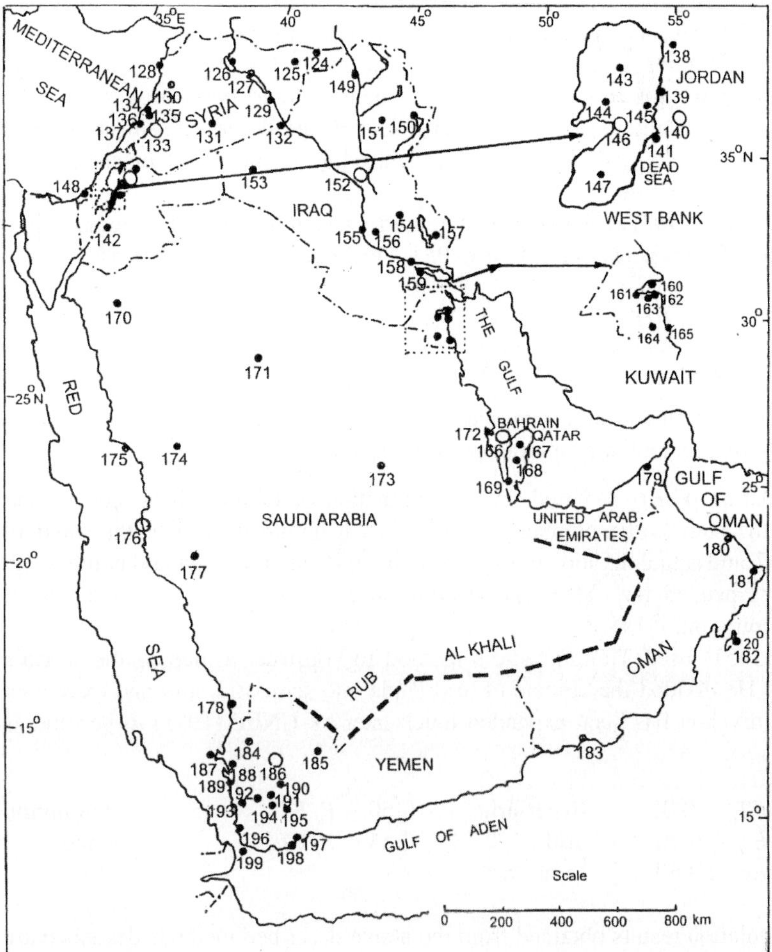

Names of Places

124- El Qamishliye, 125- Al Haseke, 126- Aleppo, 127- Raqqa, 128- Lattakia, 129- Deir ez Zor, 130- Hama, 131- Palmyra, 132- Abu Kamal, 133- Damascus, 134- Tripoli, 135- Rayack, 136- Kasara Observatory, 137- Beyrouth, 138- Irbid, 139- Salt, 140- Amman, 141- Madaba, 142- Shaubak, 143- Nablus, 144- Ramallah, 145- Jericho, 146- Jerusalem, 147- Hebron, 148- Gaza, 149- Mosul, 150- As Sulaymaniyah, 151- Kirkuk, 152- Baghdad, 153- Ar Rutba, 154- Kut al Hai, 155- Najaf, 156- Ad Dywaniyah, 157- Amarah, 158- Nasiriyah, 159- Basrah, 160- Failaka Island, 161- Ash Shuwaikh, 162- Omariyah, 163- Kuwait A.P., 164- Al Ahmadi, 165- Al Ahmadi Port, 166- Al Manamah, 167- R. al Faras, 168- Al Utoriyah, 169- Abu Samra, 170- Tabuk, 171- Hail, 172- Dahran, 173- Ar Riyadh, 174- Medina, 175- Yanbu, 176- Jidda, 177- At Taif, 178- Gizan, 179- Sharja, 180- Muskat, 181- Sur, 182- Al Masirah, 183- Salalah, 184- Hajjah, 185- Marib, 186- Sana'a A.P., 187- Kamaran Island, 188- Dahi, 189- Hodeydah A.P., 190- Riyan, 191- Yarim, 192- Rihab, 193- Zabid (Gerba), 194- Ibb, 195- Al Barh, 196- Al Mokha, 197- Al Khod, 198- Aden Khormaksar, 199- Barim Island.

Figure 4(c). Names and locations of meteorological stations used for describing the climate of the eastern subregion

rain as the driest month of summer, and 7.0; the wettest month in the winter half has at least 3 times as much rain as the driest month of summer.

For $P < r$; the climate can be described as true desert, for $2r < P < r$; the climate is semi-desert; and for $P > 2r$; the climate is neither desert nor semi-desert. If the mean annual temperature, T, is less than 64.4° F, the desert or semi-desert can be described as cold, and if it is greater than 64.4° F, the desert or semi-desert is hot.

Thornthwaite (1948) developed a different expression introducing an index called Precipitation Evapotranspiration Index, *PEI*. This index can be written as:

$$PEI = P/ET_o \tag{4}$$

where:

P = annual precipitation in mm, and
ET_o = annual potential evapotranspiration in mm.

The calculation of Et_o is based on the summation of monthly heat indexes and adjusting the sum to the number of days in each month instead of the standard number, 30, and actual duration of daytime hours instead of the standard number, 12. Expressions needed for obtaining the adjusted ET_o are given in the chapter on evapotranspiration.

Meigs (1951) used Thornthwaite's method to construct a map of the world's arid lands. He divided the climate of arid lands into semi-arid, arid and extremely arid. This division has been expanded much later by UNEP (1993) to become as follows:

$0.00 \le P/ET_o < 0.05$	hyper-arid	$0.50 \le P/ET_o < 0.65$	semi-humid
$0.05 \le P/Et_o < 0.20$	arid	$0.65 \le P/ET_o$	humid
$0.20 \le P/Et_o < 0.50$	semi-arid		

The calculation results obtained from the above-described methods described are presented in Table 3.

Table 3. Climate classes of the countries of the Arab Region

Ser. No.	Station Name	De Martonne A	DS SA	SA/ SH	H	Thornthwaite A/ HA	SA	SH	H	Koppen D	SD	NDNSD h/c
Mauritania												
1	Atar	x				A				h		
2	Tidjika	x				A				h		
3	Nouakchott	x				A				h		
4	Boutilimit	x				A				h		
5	Kiffa		x			A				h		

(continued)

6	Nema		x			A				h		
Morocco												
7	Tanger			x					x			c
8	Melilla			x			x				h	
9	Oudeja			x			x				c	
10	Rabat			x			x					c
11	Meknes			x					x			c
12	Casablanca			x			x					c
13	Ifrane				x				x			c
14	Marrakech		x			A					h	
15	Ouarzazate	x				A				h		
16	Agadir		x				x			h		
Algeria												
17	Alger			x					x			c
18	Cap Carbon			x					x			h/c
19	M. Blanche			x					x			c
20	Miliana			x					x			c
21	Benni Abbes	x				HA				h		
22	Biskra	x				A				h		
23	Oran			x			x					c
24	Tebessa			x			x					c
25	Laghouat		x			A				c		
26	C. Bechar	x								h		
27	El Golea	x				HA				h		
28	In Salah	x				HA				h		
29	Ouallen	x								h		
30	Tamanrasset	x				HA				h		
Tunisia												
31	Bizerte			x					x			c
32	Kelibia	x								c		
33	Tunis			x								h/c
34	Jendouba			x				x				c
35	Kairouan		x				x				h	
36	Monstair			x			x					h/c
37	Sfax		x				x				h	
38	Gafsa		x			A				h		
39	Tozeur	x				A				h		
40	Gabes		x			A				h		
41	Djerba Mellita		x				x				h	
42	Remada	x				A				h		
43	Tripoli City		x				x				h	
44	Shahat	x							x			c
Libya												
45	Garian			x			x				h/c	

(continued)

(Continued)

Ser. No.	Station Name	De Martonne A	DS SA	SA/ SH	H	Thornthwaite A/ HA	SA	SH	H	D	Koppen SD	NDNSD h/c
46	Derna		x				x				h	
47	Aziziyah		x				x				h	
48	Misurata		x				x				h	
49	Benina		x				x				h	
50	Nalut	x				A				h		
51	A. Nasser A.P.	x				A				h		
52	Mizdah	x				A				h		
53	Sirte		x			A				h		
54	Ajdabiyah	x				A				h		
55	Ghadames	x				HA				h		
56	Hon	x				HA				h		
57	Gilao	x				HA				h		
58	Sebha	x				HA				h		
59	Kufrah	x				HA				h		
Egypt												
60	Sidi Barrani		x			A				h		
61	Sallum	x				A				h		
62	Damietta	x				A				h		
63	Rosetta		x			A				h		
64	Mersa Matruh	x				A				h		
65	Port Said	x				A				h		
66	Alexandria		x			A				h		
67	El Arish	x				A				h		
68	El Mansurah	x				A				h		
69	Damanhour	x				A				h		
70	Tanta	x				A				h		
71	Zagazig	x				A				h		
72	Delta Barrage	x				A				h		
73	Cairo	x				HA				h		
74	Giza	x				HA				h		
75	Suez	x				HA				h		
76	Helwan	x				HA				h		
77	Faiyum	x				HA				h		
78	Siwa	x				HA				h		
79	Beni Suef	x				A				h		
80	Tor	x				A				h		
81	Minya, El	x				HA				h		
82	Hurghada	x				HA				h		
83	Asyut	x				HA				h		
84	Qena	x				HA				h		
85	Quseir	x				HA				h		
86	El Dakhla	x				HA				h		
87	El Kharga	x				HA				h		
88	Deadalus Isl.	x				A				h		
89	Aswan	x				HA				h		

The Sudan												
90	Wadi Halfa	x				HA				h		
91	Port Sudan	x				A				h		
92	Tokar	x				HA				h		
93	Atbara	x				HA				h		
94	Khartoum	x				A				h		
95	Kassala		x			A				h		
96	Wad Medani		x			A				h		
97	Gedaref			x			x				h	
98	Ed-Dueim		x			A				h		
99	El Fasher		x			A				h		
100	Sennar			x			x				h	
101	Geneina			x			x				h	
102	El Obeid		x				x			h		
103	Singa			x			x				h	
104	An-Nahud		x				x			h		
105	Er Roseires			x			x				h	
106	Renk			x			x				h	
107	Malakal			x			x					h
108	Raga			x				x				h
109	Akobo Post			x				x				h
110	Wau			x				x				h
111	Juba			x				x				h
112	Torit			x				x				h
Djibouti												
113	Djibouti Serp.	x				A				h		
Somalia												
114	C. Guardafui											
115	Ben. Cassim	x				HA				h		
116	Erigavo			x			x				c	
117	Berbera	x				HA				h		
118	Hargeisa			x			x				h	
119	Gallacaio	x				A				h		
120	Obbia	x				A				h		
121	Lugh Ferrandi		x			A				h		
122	Mogadiscio			x			x				h	
123	Chismaio			x			x				h	
Comoros Islands												
C.I. 1	Moroni				x				x			
Syria												
124	El Qamishliye			x			x					h
125	El Haseke			x			x				h	
126	Aleppo			x			x				c	
127	Raqqa		x				x				h/c	
128	Lattakia			x					x			h
129	Deir ez Zor		x			A				h		
130	Hama			x			x				h	
131	Palmyra	x				A				h		
132	Abu Kamal	x				A				h/c		
133	Damascus		x				x				c	

(continued)

Table 3. (Continued)

Ser. No.	Station Name	De Martonne A	DS SA	SA/ SH	H	Thornthwaite A/ HA	SA	SH	H	D	Koppen SD	NDNSD h/c
Lebanon												
134	Tripoli			x			x					h
135	Rayack			x					x			c
136	Kasara Observ.			x					x			c
137	Beyrouth			x					x			c
Jordan												
138	Irbid School			x			x					c
139	Salt			x					x			c
140	Amman A.P.			x			x				c	
141	Madaba			x			x					c
142	Shubak School			x			x				c	
Palestine												
143	Nablus			x					x			h
144	Ramallah			x					x			h
145	Jericho		x			A				h		
146	Jerusalem			x			x					c
147	Hebron			x					x			c
148	Gaza			x			x				h	
Iraq												
149	Mosul			x			x				h	
150	Sulymaniyah			x					x			h
151	Kirkuk			x			x				h	
152	Baghdad	x				A				h		
153	Ar Rutba	x				A				h		
154	Kut al Hai	x				A				h		
155	An Najaf	x				A				h		
156	Dywaniyah	x				A				h		
157	Al Amarah		x			A				h		
158	An Nasiriyah	x				A				h		
159	Basrah	x				A				h		
Kuwait												
160	Failaka Isl.	x				A				h		
161	Shuwaikh	x				A				h		
162	El Omariyah	x				A				h		
163	Kuwait A.P.	x				A				h		
164	Al Ahmadi	x				A				h		
165	Al Ahmadi Port	x				A				h		
Bahrain												
166	11 Manamah	x				HA				h		
Qatar												
167	R. al Faras	x				HA				h		
168	Al Utoriyah	x				HA				h		
169	Abu Samra	x				HA				h		
Saudi Arabia												
170	Tabuk	x				A				h		

171	Hail	x								h		
172	Dahran	x				A				h		
173	Er Riyadh	x								h		
174	Medina	x				A				h		
175	Yanbu	x				HA				h		
176	Jidda	x								h		
177	Al Taif		x			A				h		
178	Gizan	x				HA				h		
United Arab Emirates												
179	Sharja	x				A				h		
Oman												
180	Muscat	x				A				h		
181	Sur	x				A				h		
182	Al Masirah	x				HA				h		
183	Salalah	x				A				h		
Yemen												
184	Hajjah			x			x				h	
185	Marib	x				HA				h		
186	San'a A.P.		x				x				h	
187	Kamaran Isl.	x				HA				h		
188	Dahi	x				A				h		
189	Al Hodeydah	x				A				h		
190	Riyan	x				HA				h		
191	Yarim		x					x				c
192	Rihab		x				x				h	
193	Zabid (Gerba)		x			A					h	
194	Ibb				x				x			h
195	El Barh		x			A				h		
196	Al Mocha	x				HA				h		
197	El Khod	x				HA				h		
198	Aden/Khormaksar	x				HA				h		
199	Barim Island	x				HA				h		

Abbreviations used:
HA = hyper-arid, A = Arid, DS = dry steppe, SA = semi-arid, SH = semi-humid, H = humid, D = Desert, SD = semi-desert, NDNSD = neither desert nor semi-desert, c = cold and h = hot

3.3. CLIMATE OF THE ARAB COUNTRIES

3.3.1 Western subregion

3.3.1.1 Mauritania

The considerable amount of sunshine the country is receiving all the year round as a function of the excessive influx of solar radiation and the limited cloudiness of the sky, causes the air temperature to be generally high. The mean daily temperature varies from a minimum of 25.3 °C for Nouakchott on the Atlantic coast to a maximum of 30.7 °C for Nema, deep inland. Table 3, Appendix I, shows that December or January is the month of minimum temperature (20.0–23.5 °C). The hottest month falls in the period May-September (29.0–36.9 °C) depending on the location of the point in question. The range of variation is minimum (about 8 °C) close to the Atlantic coast in a north-southerly direction and increases up to 14 °C in a west-easterly direction.

The mean relative humidity at the coastal stations is, as one should expect, greater than the inland stations. The maximum and minimum relative humidity figures for Nouakchott (station No. 3) on the ocean are 41–42% in the winter period, December-February, and 73–74% in the summer period, July-August, respectively. The inland station of Nema (station No.6) has a minimum relative humidity of 14% in March-April, and a maximum of 68% in August. Almost the same figures can be found for another inland, station, Kiffa (station No. 5), where a minimum of 17% occurs in April and a maximum of 69% occurs in August as well.

The high temperature, the relative dryness of air and the perseverance of the tradewinds are the main reasons behind the high rates of evaporation. In the absence of measurements of evaporation from Colorado sunken pan or Class A pan, the only significant rates of evaporation are those obtained from the Piche readings: the annual total over the entire surface of the country exceeds 3.5 m (Toupet & Pitte, 1977).

The northern half of Mauritania is practically dry, with annual rainfall much less than 100 mm; a greater part of it receives even less than 50 mm y^{-1}. The southern half of the country is bounded by isohyets of 100 mm y^{-1} on the north and 500 mm y^{-1} on the far southeast (see Figure 5). The rainy season begins in May or June and terminates in October or November. The rain spells are generally moderate, though concentrated over a short duration each. The characteristics of each spell depend on the amount of rain carried by the Monsoon winds.

3.3.1.2 Morocco

The country is located between the cool, temperate region in the north, and the hot, arid region in the south. Next to this, Morocco is surrounded from the north and west by the Mediterranean Sea and the Atlantic Ocean. The average width of the country inland is quite small compared to its length along the coast. The climate of the coastal strip along the Mediterranean, and the area occupied by the Rif Mountains belongs to the Mediterranean type. The Central plains too are to some extent under the influence of the Ocean. The areas south and east of the Atlas range of mountains are strongly influenced by the harsh desert conditions. The range of Atlas Mountain stands as a barrier against the air streams blowing from the ocean. All these factors strongly affect the climate of Morocco.

The air temperature in Morocco is generally cooler than the surrounding countries. The mean daily temperature for the whole year, varies from a maximum of 19.5 °C at Marrakech (station No. 14; 466 m a.m.s.l.) to 17.0 °C at Oudja near the coast (station No. 9; 470 m a.m.s.l.) and further to 10.9 °C at Ifrane (station No. 13; 1,640 m a.m.s.l.). The winter temperature, December-January, at Ifrane is at or less than 3 °C increasing to about 22 °C in July, the hottest month. The corresponding range of temperature variation is 19.7 °C. A slightly larger range, 20.3 °C can be found in Ouarzazate (station No. 15; 1,136 m a.m.s.l.). The lowest and highest figures for temperature at Marrakech are 11.8 °C and 28.2 °C, respectively, with a range of variation of 16.4 °C. The range

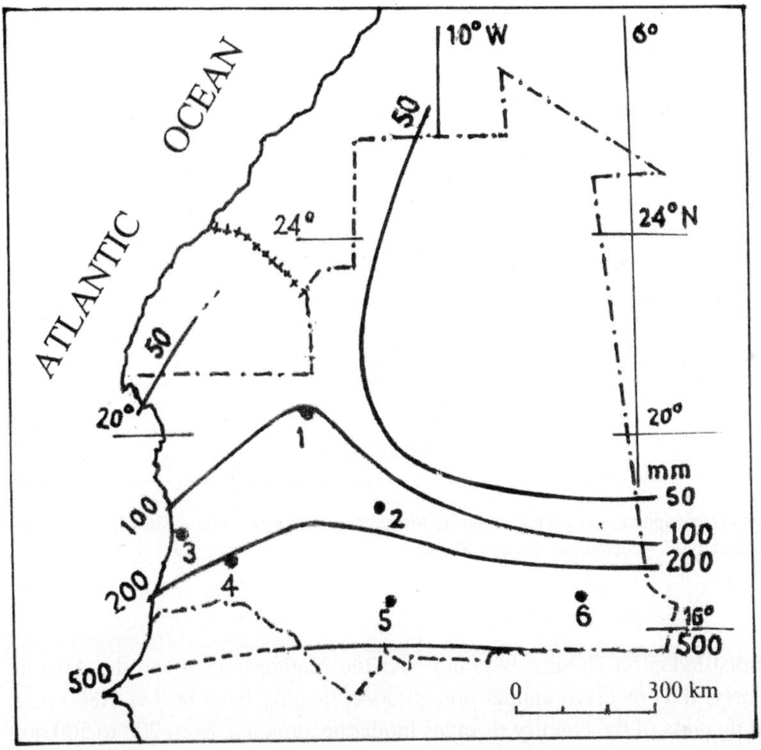

Names of places
1- Arar, 2- Tidjika, 3- Nouakchott, 4- Boutilimit, 5- Kiffa, 6- Nema

Figure 5. Isohyetal map of Mauritania (partly from Toupet & Pitte, 1977)

of variation between winter and summer for coastal places, e.g. Casablanca (9.8 °C) and Mellila (11.5 °C), is much less than those figures.

The relative humidity is generally high as a large part of the territory is strongly influenced by the ocean. The coastal areas are usually more humid than the inland desert. January and July are the months of maximum and minimum relative humidity, respectively.

Rainfall in Morocco is dominated by three main factors; winter cyclonic depressions, topography and convection. The precipitation, in general, decreases from north to south and from west to east. It also increases with altitude above mean sea level, especially in the northern and southern slopes that are exposed to the moist-laden air streams. Broadly speaking the maximum fall occurs during November, December or January. July and August are the driest months of the year. The long-term average precipitation at Ifrane, Rabat and Ouarzazate is plotted against time, in months, as shown in Figure 6. The northwestern part of Morocco benefits

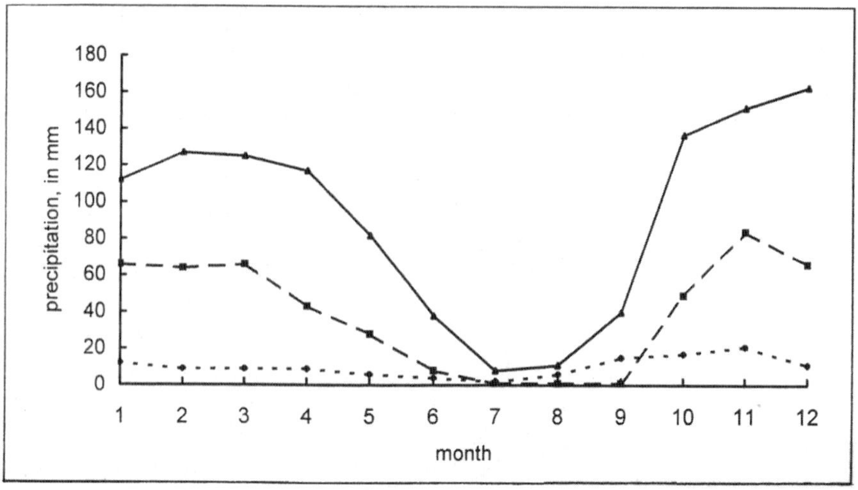

Figure 6. Monthly precipitation at three locations in Morocco: —— Ifrane (station No. 13), – – – Rabat (station No. 10) and - - - - Ouarzazate (station No. 15)

from the moist-laden air streams blowing over the Mediterranean and the Atlantic simultaneously, thus receives annual precipitation ranging from 600 to 1,800 mm. The remaining parts of the country receives moderate amounts, from 200 to 500 mm in the middle coastal areas, decreasing to smaller amounts in the range from 40 mm to 200 mm in desert areas.

On the peaks of the High sand Middle Atlas in Morocco snow may be seen during October, while at heights of 1,000–2,000 m it may lie for a week or so in the winter. In the High Atlas winter blizzards are frequent and the snow can lie a meter or more in depth, while some north or west facing slopes can retain a snow cover for 6–9 months. Snow is never seen at Rabat or Casablanca and rarely at Tangier; all are coastal towns.

3.3.1.3 Algeria

The vast surface Algeria occupies is entrenched between the temperate and Mediterranean climate zones in the north and the harsh, arid desert climate in the south. In winter, the eastward moving cyclones crossing Europe bring frequent outbreaks of cold northwest air, which penetrate into the northern part of the country and meet with warmer, moist air. The resultant vertical instability then leads to the development of active depressions, which result in relatively heavy precipitation and frequent gales in the Mediterranean. The early part of the spring season is a continuation to winter. It is a windy time around the coastline but perhaps because of the small temperature differential between land and sea, there is relatively little rain and cloud, except where orographic influences intervene.

In summer, a period of calm, sunny weather prevails. Heat lows develop over the Sahara and the resulting subsidence over the cooler, large sea area provide the reason for the northwest to northeast winds that prevail during this time of the year. The months of April-May are characterised by the cyclonic winds called Sirocco or Khamasin, already described in an earlier section. The Saharan depression and to some extent the other depressions moving southeast from the southwestern Mediterranean strongly affect the Algerian and Tunisian coast.

Reviewing the temperature data in Table 3, Appendix I, one finds that the coastal stations (Alger, M. Blanche, Miliana and Oran) have a mean daily temperature averaged for the whole year below 18 °C. The same observation can be obtained for those inland stations located at a high elevation above mean sea level. Examples of the latter are Tebessa (station No.8, 816 m a.m.s.l.) and Laghouat (station No.9, 767 m a.m.s.l.). These two stations have mean daily temperatures of 15.6 °C and 17.9 °C, respectively. Places far south in the Sahara are generally too warm, e.g. Ouallèn (station No. 29) with a mean temperature of 28.1 °C. Exception from this rule can be observed for highly elevated locations such as Tamanrasset (station No. 30) where the mean daily temperature is 21.9 °C.

All the fourteen Algerian meteorological stations examined here, despite their locations and altitudes, agree on a certain number of items. January is the coldest month while July or August is the hottest month of the year. The mean daily temperature in January, except for highly elevated places, is above 10 °C. The mean daily temperature reaches up to 24-26 °C for the coastal stations and up to 38 °C for inland, desert places. The high elevation of the desert stations, such as Tamanrasset (1,378 m a.m.s.l.) helps to modify the summer heat. The altitude of Tamanrasset station being about 1,000 m above that of In Salah (station No. 28) results in a difference of about 10 °C in the July temperature of the two stations. This amount corresponds to a lapse rate of almost 1 °C for every 100 m rise in elevation. The ratio of the July/August temperature to the January temperature is around 2.5 for the coastal areas. This ratio jumps to 3–4 for inland and desert areas. The distribution pattern of the mean daily temperature among the months of the year for three different stations; Miliana (No. 20), Biskra (No. 22) and Ouallèn (No. 29) are shown in Figure 7(a).

The relative humidity, except for the coastal areas, is generally low: C. Béchar, 36%; In Salah, 32%; Ouallèn, 21%; and Tamarasset, 27%. It is greater in the winter months and smaller in summer. The coastal areas are characterised by their high humidity, with an average of 70–75%. Because of the permanent availability of the source of moisture, i.e. the sea, there is hardly any change in the humidity level from winter to summer. The monthly distribution of relative humidity is practically uniform (Table 6, Appendix I).

The annual precipitation in Algeria varies from 1,500 mm on the mountain heights in the northeast of the country to less than 100 mm in the northern part of the desert area. Deep south in the Sahara the annual rainfall decreases further to become in the order of 10 mm. Areas along the seacoast or close to it receive rainfall in no less than 9 months, beginning in September

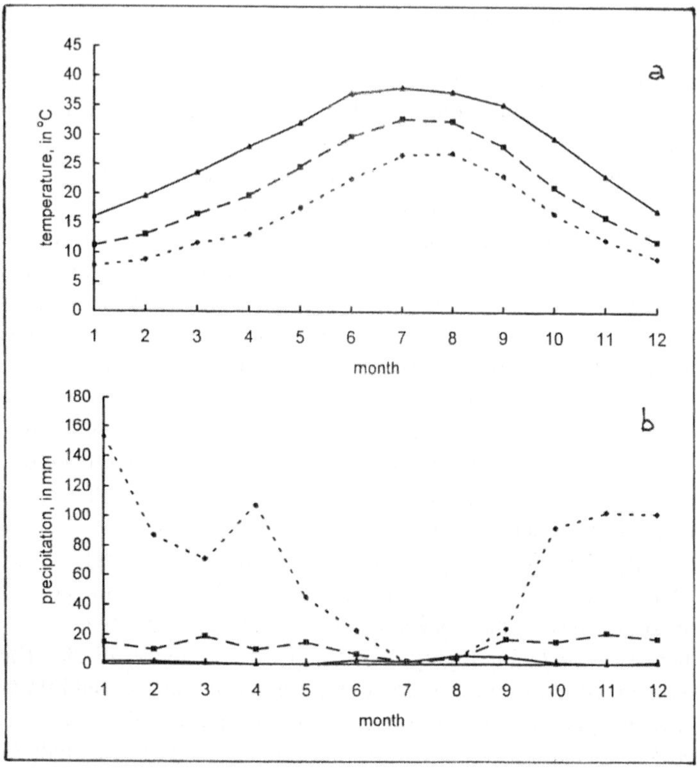

Figure 7. (a) Monthly temperature and (b) rainfall, both for stations: Miliana (No. 20) - - - -, Biskra (No. 22) – – – –, and Ouallèn 3 (No. 29) ——

and ending in May. The rainfall in the remaining period, June-August, is limited to a few millimetres per month. In the desert areas, where annual rainfall is of limited amount, there is hardly any distinction between one season and another as far as rainfall is concerned. The curves in Figure 7(b) show the distribution of monthly rainfall for the same locations already used for preparing the curves in Figure 7(a).

3.3.1.4 Tunisia

The coldest period in Tunisia is December and January; the mean temperature then is in the range of about 10–13.5 °C. The mean daily temperature of the warmest period, June-July, is about 25.5–32 °C, depending on the location with respect to the sea and the topography and relief of the land. As a rule the coastal areas enjoy warmer winters and cooler summers compared to inland areas. Bizerte (station No. 31), which is located on the seacoast in the north, has a monthly range of 14.5 °C. Not much different from this is the monthly range at Gabès (station No. 38), 15.6 °C. Despite its being in south Tunisia, about 400 km south of Bizerte, its location on the coast of the Gulf of Gabès helps to improve the climate there.

This is not the case at Tozeur (station No. 39), which is located far from the sea in the north or the east; the range of mean daily temperature between summer and winter is 21.1 °C. (see Table 3, Appendix I).

Tunisia is one of the few Arab countries that depend largely on rainfall in agriculture. As such, it is worthwhile to present here the rainfall situation in the different parts of the country. The rainy season generally begins in September and ends in May. The general pattern of precipitation is that of the Mediterranean. As Tunisia, similar to other countries of North Africa, is situated in a climatic zone of climatic discontinuity and instability, the depth of rainfall is far from being uniform. A small variation in the pattern of blowing winds and /or the relief of the country can cause an exaggerated effect on rainfall. The rainfall in September and October 1969 and end of March 1973 has been extremely exceptional, indeed.

The coastal zone in the northwestern part of Tunisia receives annually more than 600 mm and has 120 rainy days. The average annual rainfall at Bizerte (station No.31) is 685 mm. The second zone, which covers the larger part of central Tunisia, receives between 400 and 600 mm annually, with 60–100 rainy days. Table 12, Appendix I, gives the mean annual rainfall in Tunis (station No. 33) as 461 mm. The third zone, which is located south of the second zone, receives rainfall in the range from 200 to 400 mm in 40–70 rainy days. The fourth and last zone receives an annual rainfall less than 200 mm and has around 30 rainy days each year. Gafsa (station No. 38) and Gabès (station No. 40) are located in this zone. The long-term mean annual rainfall is 163 and 174 mm respectively. Similar to Algeria, the extreme south of Tunis hardly receives any rainfall of significance. The four rainfall zones are indicated on the map in Figure 8.

Analysis of long-term annual rainfall series has shown that the ratio of the largest to the smallest rainfall varies with the zone where the rain gauging station is located. The value this ratio assumes is: 2.6, 3.8, 3.7, 4.9, 5.9, 6.3, 12.6, 18.1 and 13.6 for stations No. 34, 31, 33, 32, 35, 37, 38, 39 and 40, respectively. This is solid proof that the coefficient of variation of annual rainfall decreases with increasing mean annual rainfall.

3.3.1.5 Libya

The climate of Libya, similar to Algeria and Tunisia, is influenced by the Mediterranean in the north and the Sahara in the south. However, the lack of continuous and high relief compared to its neighbouring countries of North Africa, renders the coastal area vulnerable to the flows of hot, dry air blowing from the south. The surface of Libya can be subdivided into five climatic zones. These are:

- Sub-tropical Mediterranean zone, which is small in extent and bounded by Jabal Al Akhdar.
- Semi-Mediterranean zone, with a narrow coastal strip in the west that widens in the east to include the whole of the Jiffara Plain and the eastern part of Jabal Nafusah and terminates at Misurata to reappear along the narrow strip between Benghazi (Benina) and Derna.

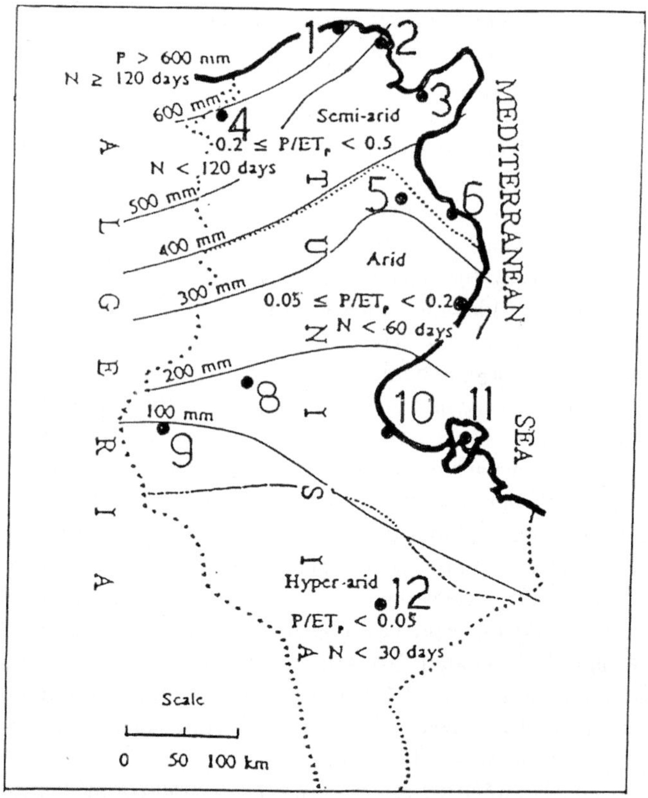

Names of places
1-Bizerte, 2- Kelibia, 3- Tunis, 4- Jendouba, 5- Kairouan, 6- Monastir,
7- Sfax, 8- Gafsa, 9- Tozeur, 10- Gabes, 11- Djerba Mellita, 12- Remada

Figure 8. Isohyetal map of Tunisia (partly from the Country Report of Tunisia on its Water resources and their utilisations, 1986)

- Steppe zone, which lies along the southern slopes of Jabal Nafusa and Jabal Al Akhdar. It also includes the area between Ajdabiyah and the eastern frontier with Egypt.
- Desert zone, which extends over the rest of the country with the exception of the mountainous areas in the south and southwest.
- Mountainous zone, which can be found in the southwest (Ghat) and south (Tibesti).

The mean daily temperature in winter at the Mediterranean coast varies from less than 10 °C to above 14 °C; rising up to 28 °C in summer. Further from the sea lies the steppe zone (Figure 9) where the temperature ranges are wider. At al-Aziziyah (station No. 47) the world's highest temperature was recorded: 58 °C in September 1922. However, the average for July in the steppe zone is 25–28 °C. The July

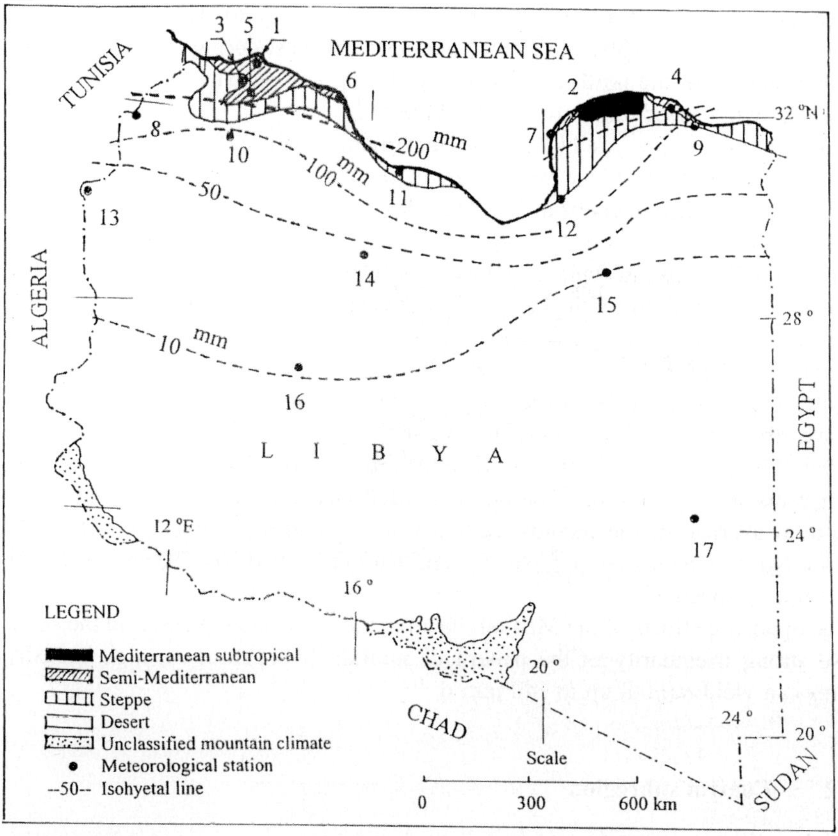

Names of places
1-Tripoli City, 2- Shahat, 3- Garian, 4- Derna, 5- Aziziyah, 6- Misurata, 7- Benina, 8- Nalut, 9- Nasser A.P., 10- Mizdah, 11- Sirte, 12- Ajdabiyah, 13- Ghadames, 14- Hon, 15- Gialo, 16- Sebha, 17- Kufrah.

Figure 9. The different climatic zones and isohyetal lines of annual rainfall in Libya

average for the same station, i.e., No. 47, is 28.8 °C. The daily variations are even marked in the desert zone. The averages range from 26 °C to 41 °C for July and from 8 °C to 16 °C for January. At Ghadames (station No. 55) the absolute minimum is −6 °C in January and −8 °C in February; the absolute maximum is 50.6 °C in June and 49.8 °C in August.

The relative humidity generally falls from north to south. The highest levels in the desert locations occur in winter and the smallest in the summer. The figures for Sabha (station No. 58) are 40% in January and 15% in July. The relative humidity figure for areas along or close to the coast is almost the same for summer as for winter.

The force and direction of winds often vary from day to day throughout the seasons. The winds usually blow from the north and east during the period May–October, and from the north and southwest during the period from November to April. The wind speed varies from 10–30 km h^{-1}.

The evaporation rates are high, with a maximum in June-July and a minimum in December-January. The Piche evaporation increases from north to south, the annual range for the coastal area is from 1,700 to 2,500 mm and for the interior from 4,000 to 5,000 mm.

The rainfall occurs in winter, mainly from October to March, and the amount decreases from north to south. The maximum annual rainfall (400–600 mm) occurs along the Jabal Al Akhdar. The rainfall at Shahat (station No. 44), 569 mm y^{-1}, is an example. The range declines to 200–400 mm towards Misurata in the west and Ajdabiyah-Derna in the east. The long-term annual rainfall at Derna (station No. 46) and Misurata (station No. 48) are 256 and 246 mm respectively. The 100 mm isohyet (Figure 9) runs from west to east almost parallel to the seacoast at an average distance of 150 km. The desert rainfall is usually less than 25 mm. The long-term average of the records available at Ghadames (station No. 55), Gialo (station No. 57), Sebha (station No. 58) and Kufrah (station No. 59) are 28, 10, 11 and 3 mm, respectively.

The opposing effects of the Mediterranean in the north and the desert in the south cause strong irregularity in the pattern of rainfall. Short-duration, high-intensity storms can yield rainfall up to 140 mm d^{-1}.

3.3.2 Central subregion

3.3.2.1 Egypt

The climate of Egypt is determined by the pressure system in each season, the moving depressions, the Mediterranean and to a lesser extent the Red Sea. Orography plays a minor in the general climate though it can have strong local effect.

The winter season usually lasts from December up to end of February. It is characterised by the passage of diversity of depressions, which are the main cause of the weather in this season. The climate of Lower Egypt in this period is usually mild and accompanied by some rainy showers, mainly on coastal areas. The mean daily range of temperature between day and night is around 12 °C, and reduces to 8 °C along the seacoast. Upper Egypt has a slightly higher temperature by day and a higher range of temperature variation between day and night in the winter season, no less than 16 °C. The summer lasts from June till September. The climate of Lower and Middle Egypt, being affected by the cool Mediterranean waters, is warm during the daytime and rather cool by night. The mean daily temperature increases in a southerly direction and with distance from the seacoast. The mean August temperature increases from 26.2 °C for Alexandria to 28.7 for Assiut and further to 33 °C at Aswan. The range of diurnal variation of temperature in the summer can be as much as 18 °C. Only those places on the seacoast like Hurghada, Quseir and Deadalus Island have much less diurnal range of temperature variation.

The spring, which lasts from March up to May is characterised by the Khamasin winds prevailing for almost 50 days bringing dust and rendering the climate warm and dry. The other transitional season, autumn, is not too different from spring. Khamasin-like depressions begin to cross Egypt during late October and cause a breakdown of the settled summer regime. The exception here is that the autumn depressions are less vigorous than the Khamasin depressions in the spring and move slower eastwards (Griffiths & Soliman, 1972).

The cloudiness is generally greater in Lower Egypt, especially in the coastal areas, than in Upper Egypt. It reaches a maximum in winter and a minimum in summer. In a tropical, arid country such as Egypt, the sky cloudiness does not exceed 4 oktas.

The rainy season in Egypt lasts from October to March. Rain falls in the form of showers and its amounts may vary significantly from year to year in the same location, and from one location to a nearby location in the same month or season. There are some instances where the daily rainfall exceeded 100 mm, and in other instances the winter rainfall at the wettest places did not exceed 5 mm. A review of the rainfall data listed in Table 12, Appendix I, shows that the annual rainfall increases from a little less than 200 mm along the Mediterranean coast and falls sharply to about 25 mm at the apex of the Nile Delta. South of Cairo the total rainfall drops to just 4 mm at Al Minya (station No. 81) in Middle Egypt, then to 1 mm at Aswan (station No. 89) in the southern part of Upper Egypt. The areal distribution of annual rainfall in Egypt is shown in Figure 10.

3.3.2.2 The Sudan

Throughout the great east-west extent of the Sudan region of tropical Africa north of the equator, the region is bordered on its northern side by the Sahara, while southward it merges into the tropical wet climates along the margins of the Gulf of Guinea and the Atlantic Ocean in the west, and the Congo Basin farther to the east. Over the Sudan, and all but the southern margins of the lands bordering the Gulf of Guinea, there occurs a reversal of wind direction during the annual march of seasons, with northerly Saharan air, or trade winds, prevailing at low sun and maritime southerly currents at high sun, a seasonal reversal designated as a monsoon (Trewartha, 1962).

The climate of the northern part of the country is hot and arid or hyper-arid. Tokar (station No. 92) is an example of this climate where the mean daily temperature varies from a minimum of 24.4 °C in January-February, the coldest period, to a maximum of 35.0 °C in the hottest period, July-August. Travelling southward to the central part of the Sudan, the climate changes to semi-desert in the northern half and savannah in the southern half. The stations at Wad-Medani (No. 96) and Malakal (station 107) represent these two types of climates respectively. Station No. 96 has its lowest mean daily in January at 24.2 °C. The highest mean daily temperature has two values, 32.5 °C in May and 30.2 °C in October. The lowest

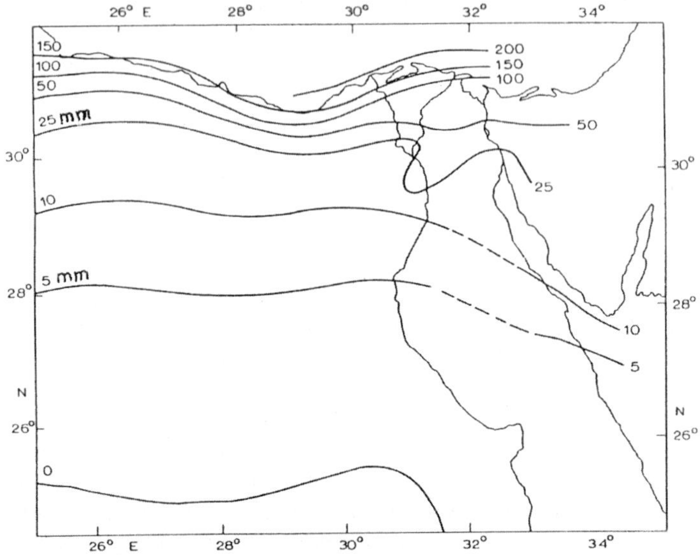

Figure 10. Annual rainfall in Egypt

mean daily temperature, 26.2 °C, for station 107, occurs in August and the highest mean temperature of 31.0 °C in April. The climate of the swamps in southern Sudan has the least range of variation between the months of lowest and highest mean daily temperature. The August and April temperatures at Wau (station No. 110) are 26.0 °C and 29.8 °C respectively, bringing the range to just 3.8 °C only. The range between the lowest and highest temperatures for stations No. 92, 96 and 107 is 10.6, 8.3 and 4.8 °C respectively. The distribution patterns of temperature for these four stations are shown in Figure 11(a).

The relative humidity level is quite low in the interior of northern Sudan and somewhat higher along the Red Sea. In either case the maximum humidity occurs in the period December-February and the minimum in the period May-August. The low humidity of air in the semi-arid and savannah zones begins in February/March and lasts till June. It increases from the end of June and reaches the maximum value in the period August-October.

The northern part of the Sudan is practically dry. Station No.92, for example, receives on average a total depth of 82 mm. This amount increases steadily and reaches 346 mm at Wad Medani (station No. 96), and further to 784 mm at Malakal (station No. 107) more to the south. The southern part including the swamps receives the greatest annual depth of rain. This is usually between 1,000 and 1,200 mm depending on the location and the year. The station at Wau (No. 110) receives on the average 1,111 mm y^{-1}. The distribution of annual rainfall between the months of the year for the four mentioned stations is presented graphically in Figure 11(b).

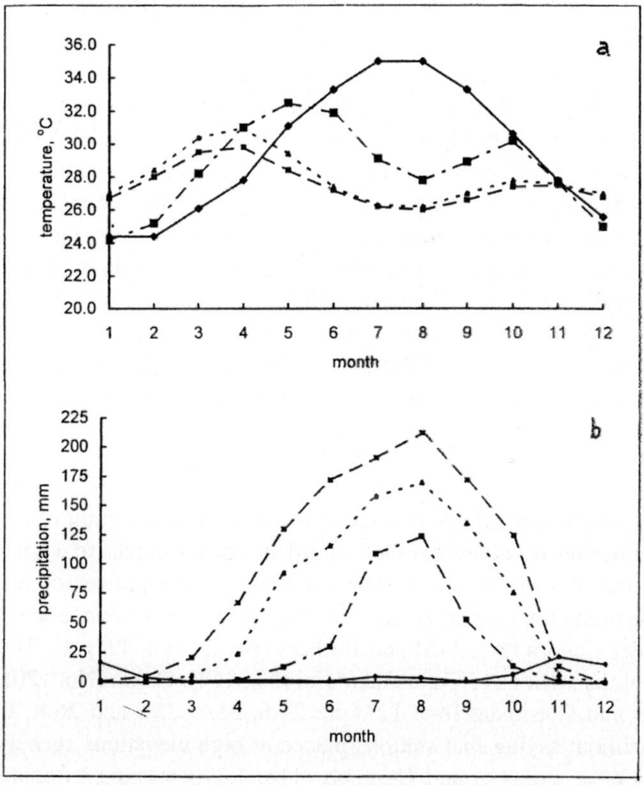

Figure 11. Temperature (a) and rainfall (b) variation with month of the year at four stations in the Sudan; .... Tokar (No. 92), - - - - Wad-Medani (No. 96), —— Malakal (No. 107) and – – – Wau (No. 110)

The general decline of temperature coupled with increase of humidity upon moving from north to south causes annual Piche evaporation to decrease from 5,370 mm at station No. 96 to 3,290 mm at station No. 107 and further to 2,780 mm at station 110. The Red Sea humidity also helps to suppress evaporation along its coast.

3.3.2.3 Djibouti

The climate of Djibouti can be classified as arid to hyper-arid. The mean daily temperature varies between 25.6 °C in January to 35.0 °C in July, with an annual mean of say 30 °C.

The relative humidity is generally high due to the influence of the Red Sea and Gulf of Aden. However, humidity figures are relatively low in the period June-August, 40–55%, and increases from September to May, 67–76%. The average relative humidity for the year as a whole is 64%.

The overall mean precipitation in Djibouti is 147 mm y^{-1}. It varies from 50 mm y^{-1} in the north east of the country to about 300 mm y^{-1} in the zone west of Tadjoura. The long-term rainfall as recorded at Djibouti Serpent (station No. 113) is 135 mm. Griffiths (1972) mentioned that at the same station (No. 113) the ten-year averages, beginning with 1901and ending 1970 are 121, 122, 133, 144, 139, 131 and 134 mm y^{-1}. He added that the variation is small and well within the limits expected from a homogeneous population of rainfall data. This remark does not, however, eliminate the fact that annual rainfall is significantly irregular, and extreme rainfalls have been observed in a number of years. The smallest annual rainfall ever recorded since 1901 was 9.5 mm in 1980 and the largest 693 mm in 1989. Table 12 shows that rainfall exceeds 10 mm mo^{-1} from October up to April. The rainfall in this period is above 80% of the annual total rainfall. The remaining months of the year receive amounts less than 10 mm each.

3.3.2.4 Somalia

Most of Somalia faces southeastward towards the Indian Ocean. The terrain of the desert littoral is facing northward along the Gulf of Aden. Owing in part to a zone of upwelling cooler water offshore the ocean, temperatures of the warm season are less extreme than those along the Gulf of Aden. The long-term average temperature at the stations of Bender Cassim (No. 115) and Berbera (No. 117) is 30.0 °C. The average temperatures at the stations of Cape Guardafui (No. 114), Obbia (No. 120), Mogadiscio (No. 122) and Chisimaio (No. 123) are 25.6, 26.7, 27.2 and 26.8 °C respectively. It goes without saying that stations placed at high elevations such as Erigavo (No. 116, 1,730 m a.m..s.l.) and Hargeisa (118, 1,370 m a.m.s.l.) enjoy cooler temperatures, the respective mean annual temperatures are 17.1 °C and 21.6 °C. The winter (cold-temperature season) in the northern part of the country covers the period from December to February/March leaving the longer part of the year to summer season warm-temperature season). The stations located in the southern half of the country have their lowest temperatures in July-September. The difference in timing is due to the characteristics of the Inter Tropical Convergence Zone (ITCZ). The temperature assumes a more uniform distribution as one travels further to the south. The difference between the highest and lowest mean daily temperatures as can be seen from Figure 12(a) is 11.1 °C for Bender Cassim, 4.9 °C for Erigavo and 2.2 °C for Chisimaio.

The mean relative humidity in Somalia, like many more countries, is high in the vicinity of the seacoast and reduces in the interior. The annual figures for Bender Cassim (station No. 115), Obbia (station No. 120) and Mogadiscio (station No. 122) are 61, 67 and 84% respectively. The air humidity at highly elevated places in the interior is less and decreases further with elevation above m.s.l. Long-term observations give annual figures of 57% and 51% for Hargeisa (station No. 118) and Erigavo (station No. 116) respectively. The fluctuation of the humidity level from month to month at locations close to the coast of the Indian Ocean is extremely limited. This is not the case with the stations in the interior where the pattern

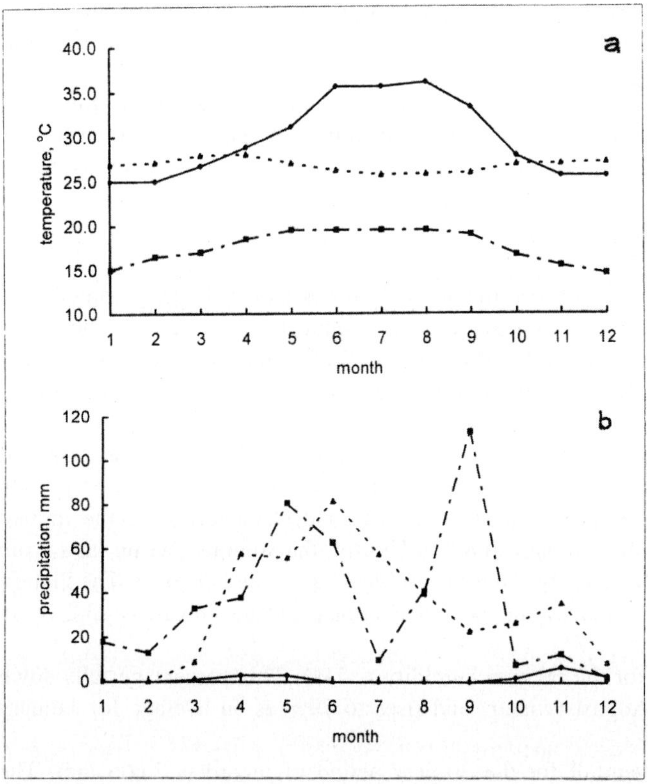

Figure 12. (a) Temperature and (b) rainfall at three stations in Somalia; Bender Cassim (station No. 115) ——, Erigavo (station No. 116) - - - and Chisimaio (station No. 123) – – –

can assume a U-shape as for station No. 118 and inverted U-shape as for stations No. 116 and 121.

The annual rainfall in Somalia is widely variable in time and space. The north-eastern half has a slight predominance of rainfall during the winter monsoon, though with occasional convectional storms at other seasons. The southern half is definitely within the tropical regime, with maximum rainfall in summer. The data in Table 12, Appendix I, shows that rainfall generally increases from north to south. The exception from this general rule is the rainfall in highly elevated places, no matter their geographic position. The rainfall at Erigavo and Hargeisa is above 400 mm y^{-1}. The annual rainfall increases from 165 mm (in some references 201 mm) at Obbia to 399 mm at Mogadiscio, and further to 468 mm at Chisimaio, all along the Indian Ocean coast in southeast Somalia. The rainy season for the last two stations begins in April and lasts till end of November. The major part of the rainy season in the mountainous region in north Somalia is a little shorter, it lasts from March to September. The rainfall has two peak values; the first occurs in April/May and the second in August/September. A fair idea about the pattern of rainfall variation from month to month can be obtained from the set of graphs shown in Figure 12(b).

Some of the data about sunshine, cloudiness and global radiation are reasonably abundant but still do not present a good coverage for the whole country. The connection between cloud and rainfall would lead one to suspect a wide variation in mean month clouds from year to year (Griffiths, 1972).

3.3.2.5 Comore Islands

The published material about the climate of Comoros Islands is quite limited. The author has to depend entirely on the statistical summary prepared by The Royal Dutch Meteorological Institute (KNMI, 2002). That summary covers the 30-year period, 1971–2000. The only exception is that the record of sunshine duration covers 10 years only.

Due to the geographic location of the islands in the Indian Ocean south of the Equator (11°42′ S), the temperature is characterised by being relatively high, 25.4 °C, averaged over the whole year. The second characteristic is the limited fluctuation from month to month, 3.8 °C. Thirdly, the summer season lasts from November to April/May and the winter from May/June to end October. The highest temperature is 27 °C occurring in January-February and the lowest is 23.2 °C in August.

The annual figure for the relative humidity is 71%. This percentage falls down to 65% in July and August (winter) and rises to 79% as an average for January (summer).

The mean annual rainfall for the 30-year period of record is 2,666 mm. The largest monthly rainfall, round 350 mm, falls steadily to about 285 mm in March to rise to about 318 mm in April. The monthly rainfall declines again from May to September after which it rises again until the highest peak is reached in January the year after.

The total duration of sunshine is about $2,612 \text{ h y}^{-1}$. The maximum duration, 237 h, occurs in October and the minimum, 177 h, in February.

3.3.3 Eastern subregion

3.3.3.1 Syria

The climate of Syria belongs to the hot temperate zone of the Mediterranean climate which is generally characterised by a rainy winter and a hot dry summer, and spring and autumn as transitional seasons. This general situation is somewhat modified by the topography and relief of the country. The climate of Syria can be subdivided into four major zones.

- The coastal zone can be represented by locations such as Safita, Baniyās, Tartūs and Lattakia (station No.128). This zone is characterised by heavy winter rain and a dry, moderate summer. The range between the mean summer temperature and mean winter temperature is about 20 °C or less.

- The second zone is the interior, which is characterised by a rainy winter and a hot dry summer. Damascus (station No. 133) is an example that represents this zone. The temperature difference between summer and winter is in the order of 25 °C.
- The third zone is the mountainous zone. This zone is characterised by its cold winters, moderate summers and a relatively small difference between summer and winter temperatures. Locations with elevation at or higher than 1,000 m a.m.s.l. receive annual rainfall that may exceed 1,000 mm.
- The fourth climatic zone is the desertic zone. It receives annual rainfall hardly exceeding 100 mm in the winter season, whereas summers are dry and hot. Palmyra (station No. 131), which represents this zone, receives 132 mm y^{-1} as a long-term average.

In general one finds January the coldest months of the year while July and August are the hottest. This observation holds for all stations listed in Table **??**. Additionally, the temperature frequently falls below 0 °C in winter but hardly below −10 °C. It may rise during the daytime to above 45 °C in the summer season. Snowfall covers all parts of the country with elevation at or higher than 1,500 m a.m.s.l. The mean relative humidity at a certain point depends strongly on the location. It varies from 20 to 50% in the interior and from 70 to 80% in the coastal zone, both in the summer season. These ranges change to 60 to 75% in the interior and 60 to 70% in the coastal areas respectively, both in winter months. Figure 13 shows an isohyetal map of annual rainfall in Syria over the 30-year period, 1955–1984.

3.3.3.2 Lebanon

Three climatic zones cover the Lebanon territory. These zones and their respective climates are as follows:

- The coastal zone comprises the narrow Mediterranean strip of land, with an average width of say 10 km. This zone has a mild, dry summer, and cold and rainy winter. The transitions have mild temperature and light rainfall. The mean daily temperature in January, the coldest month, varies between 10 °C and 15 °C, and in August, the warmest month, between 25 and 28 °C. The mean value of the relative humidity of the air varies from month to month. However, the range of variation is rather limited due to the dominant influence of the sea. The monthly precipitation is in the range of 150–200 mm in January and falls down to almost zero in the period from May to August.
- The hilly zone includes the western chain of mountain ranges running approximately in a northeasterly direction through the whole territory. The dominating climate of this zone is mild and dry in summer as in the coastal zone. The winter is very cold with heavy rainfall. The transition seasons are relatively cold and rainy. The January temperature varies between 5 °C and 10 °C while as the August temperature between 20 °C and 25 °C. The January precipitation can be as much as 300 mm and decreases to less than 25 mm in July and August.
- The inland zone represents those parts east of the western mountain ranges. It also includes thc other mountain ranges along the east border of the country.

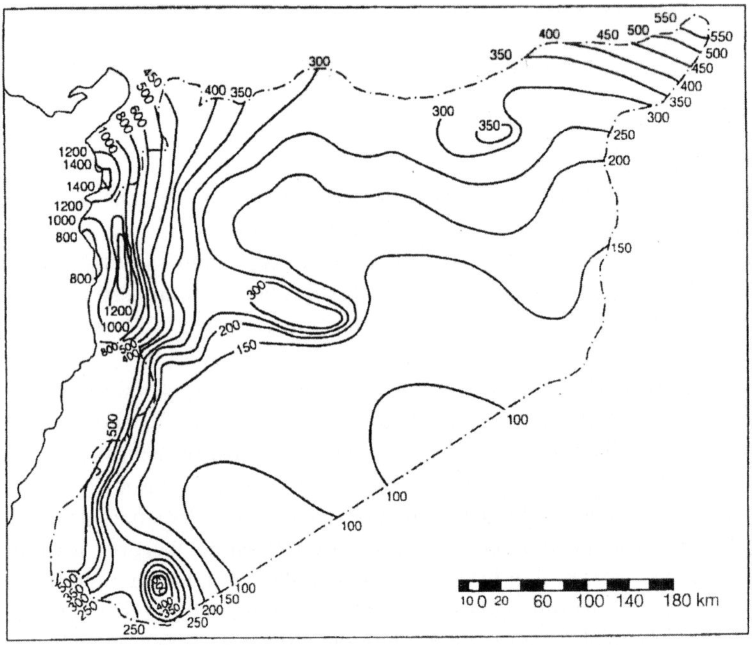

Figure 13. Average total annual precipitation (mm) for the period 1955–1984 in Syria (partly from Al Shalabei et al., in Jeftic et al., 1992(b))

The climate is generally hot, dry in summer, and cold and wet in winter. The winter temperature (January) is ranging between 0 °C and 5 °C rising to between 18 °C and 25 °C in August. (e.g. Rayack, station 135). The air humidity decreases from about 80% to 45% in these two months respectively. The precipitation varies between 50 mm and 200 mm in January and between 10 and 40 mm in August, depending on the location and elevation of the point in question.

The range of mean daily temperature between summer and winter is almost the same, 14.0 °C, for the coastal stations at Beyrouth (station No. 137) and Tripoli (station No. 134). It is also about the same, 19 °C, for the inland stations at Rayack (No. 135) and Kasara Observatory (No. 136). The spatial variation of annual precipitation in Lebanon can be observed from the isohyetal map shown in Figure 14.

3.3.3.3 Jordan

The climate of Jordan can be subdivided into three climate zones depending on the geographic location of each area. The three subdivisions are as follows:

- The low-lying (Ghor) area, which comprises the Jordan River and the Dead Sea, is characterised by a rather warm winter, with a daily mean temperature of about 15 °C. The summer temperature being in the order of 30 °C is high and uncomfortable. The annual rainfall varies between 150 and 250 mm.

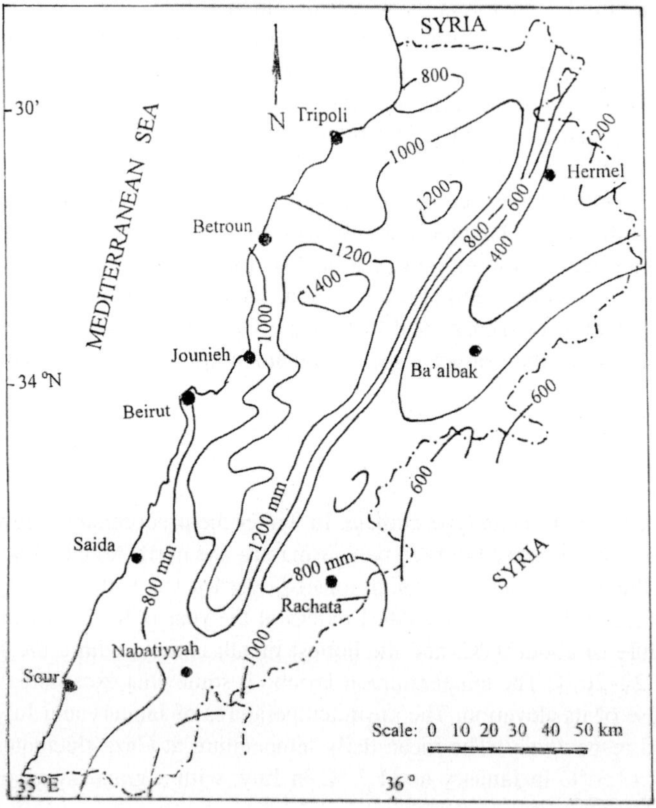

Figure 14. Isohyetal map of Lebanon (partly from the country report on Water Resources of Lebanon, 1986)

– The hilly area comprises both the western and eastern hills bordering the Jordan Valley. The dominant climate is dry and cool in summer, cold and rainy in winter. The winter temperature is in the range of 8 to 12 °C rising to say 25 °C in summer. The Salt station (station No.139), as an example, has a mean temperature of 7.9 °C in January and 25.5 °C in August. Close temperature figures can be found at Madaba (station No. 141).

The western hills receive annual rainfall in the range 600–800 mm, and the eastern hills 500–600 mm. The same station (139), which has an elevation of 796 m, receives an annual mean rainfall of 618 mm. At higher locations, e.g. Al Shoubak station (No. 142, more than 1300 m a.m.s.l.) the snow cover can remain several days (Takahashi and Arakawa, 1981).

– The desert zone is in fact an extension of Badiet Esh Shām (Syrian Desert). The main characteristics of this zone are the long duration of sunshine, hot and dry summer, and cold, rainless winter. The average annual rainfall at Qatrana (31°15′N, 36°02′E and 770 m high) is round 100 mm.

The monthly relative humidity in Jordan is lowest in summer, then it averages to 50% in the western hills and round 20% in the desert areas. Winter is the season of highest relative humidity of the air. It ranges between 60 and 77% in all parts of Jordan except for Aqaba on the Red Sea and the desert where it ranges between 47 and 55%. More information regarding the relative humidity in Jordan can be obtained from Table 6, Appendix I.
Rainfall occurs mainly in the winter season. It starts in September or early October and terminates around mid May. January and February are the months of heaviest rainfall. The rainfall particulars for As-Salt (station No. 139, 796 m a.m.s.l.) are: beginning of rainy season, October; end of rainy season, early may. Total rainfall = 618 mm; and sum of rainfall in January and February = 274 mm or 44.3% of the total rainfall. Figure 15 shows the isohyetal lines of annual rainfall over the territory of Jordan.

3.3.3.4 Palestine

The West Bank has a Mediterranean type climate. In winter the predominantly low pressure Mediterranean area centred between two air masses, the north Atlantic high of North Africa and the Euro-Asian winter high located over Russia, is the primary cause of winter weather of the area. The coldest month of the year is January, with mean daily temperature of about 9 °C, and the hottest month is July, with a mean daily temperature of 24–26 °C The temperature at Jericho is somewhat exceptional, most probably because of its elevation. The mean temperatures of January and July are 10.9 and 30.9 °C respectively. The mean daily temperature at Gaza fluctuates from a minimum of 13.3 °C in January to 31.7 °C in July, with a grand mean of 23.3 °C for the whole year.

The mean annual relative humidity varies from 53% at Jerusalem to 64% at Nablus, both in the West Bank and reaches 83% at Gaza in the Gaza Strip. The lowest relative humidity occurs between April and October and the highest in January/February. Because of its location on the eastern shore of the Mediterranean, there is hardly any significant change from month to month.

The hills in the west of Palestine affect the behaviour of the low-pressure area, resulting in westerlies, which force moist air upwards, causing precipitation on the hill ridges. The steep gradient of the Jordan Valley produces a 'lee' effect, which greatly reduces the rainfall in the Jordan Valley Rift area. In general, the annual rainfall in the West Bank increases from east to west. This can be seen from the isohyetal map in Figure 16 for the average rainfall in the period 1953–1993 (Husary et al., 1995). Occurrence of rainfall is limited, however, to winter and spring months, i.e. from October to May. This is clear from precipitation data for all stations as can be observed from Table 12, Appendix I. From the same Table of rainfall data it is clear that the annual rainfall at Nablus (station No. 143), Ramallah (station No. 144), Jerusalem (station No. 146) and Hebron (station No. 147) exceeds 500 mm. Because of their geographic location and low altitude, the annual rainfall at Jericho (station No. 145) is about 200 mm and at Gaza (station No. 148) 390 mm.

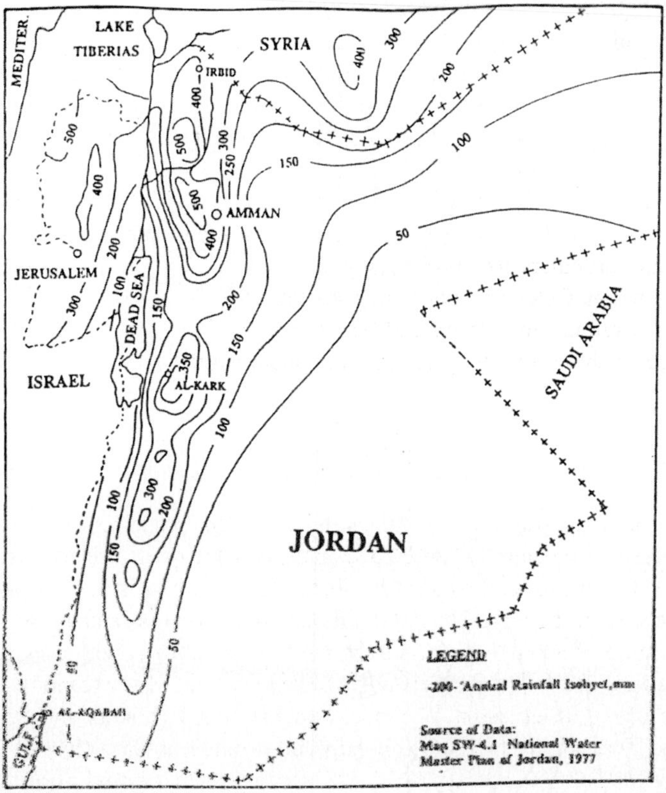

Figure 15. Isohyetal map of Jordan (partly from Rainfall in Jordan, 1982, and partly from the country report on Water Resources of Jordan, 1986,

3.3.3.5 Iraq

The country in general can be considered semi-arid, subtropical with hot dry summers (May-October) and cold winters (November to April). According to Köppens' classification for world climates, Iraq comprises three climatic types:

- The first class of climate is characterised by being warm, temperate and rainy with a dry summer in a relatively small area in the north of the country, e.g. Kirkuk (station No. 151).
- The second class is a dry, hot steppe covering the central and southern parts, e.g. Baghdad (station No. 152).
- The third class is simply dry, hot desert covering a vast surface in the west, e.g. Ar Rutba (station No.153).

Despite the differences between the three mentioned classes, the climate of Iraq is generally characterised by its extreme continentality. This can be evidenced

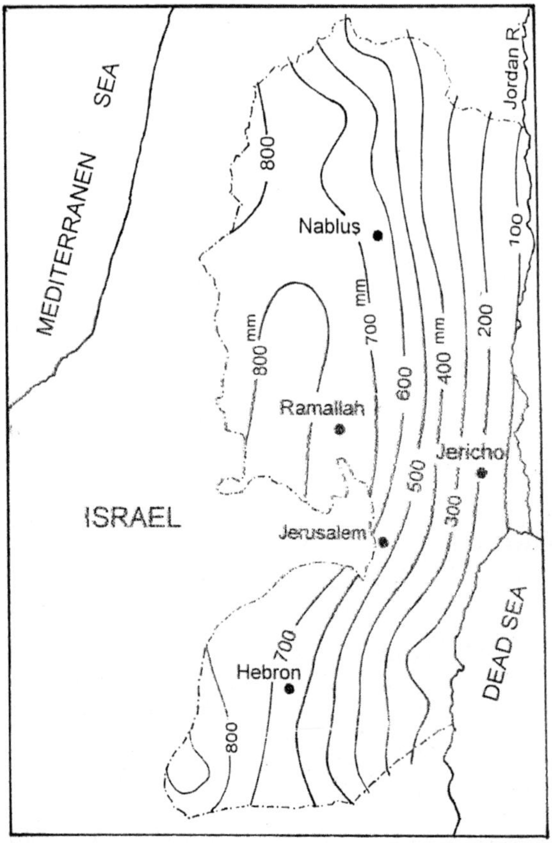

Figure 16. Isohyetal map of the West Bank (from Husary et al., 1995

from the large mean annual range of temperature; 27 °C (maximum) in the north, 24 °C in the central and western parts and 21 °C (minimum) in the Gulf area and in the extreme south. Shade temperature can reach above 50 °C in summer and falls to below −10 °C in winter. The exact values vary from location to location. In mountainous areas, however, summer heat is mitigated by the high elevation.

The air over Iraq is generally dry except in the extreme south. The considerable rise of temperature in spring and summer seasons causes the mean relative humidity to drop to 26% or less in these two seasons. The diagrams in Figure 17 show clearly the rise and fall of the mean air temperature and mean relative humidity throughout the year at Baghdad (station 152). In the western desert relative humidity figures of 0% have been reported several times.

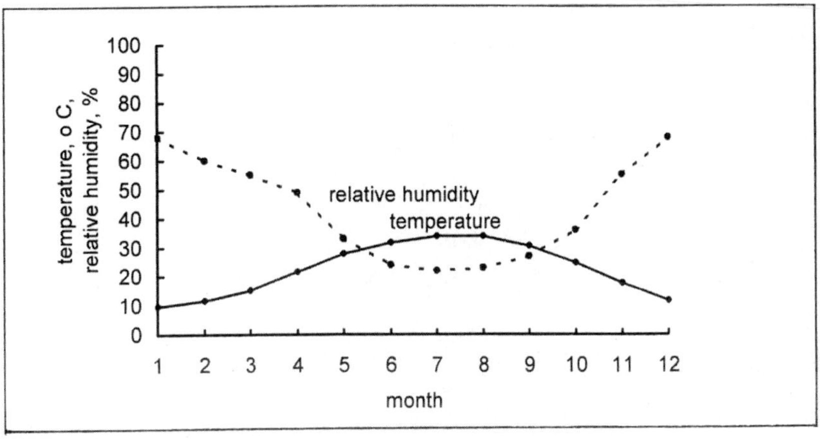

Figure 17. Mean daily air temperature and relative humidity of air at Baghdad, Iraq

Rain usually falls in the winter season, November-April. The annual rainfall is scarce in the plains of Iraq, usually around 150 mm. It increases up to 800 mm in the northeast, and reaches 1,000 mm in some localities. Snow falls heavily in winter on the mountains in the north. Snow melting contributes to the flood of Tigris and Euphrates in the spring season. The annual rainfall isohyets for Iraq are shown in Figure 18.

The weather of Iraq, like many more countries in the Middle East, is characterised by sand and dust storms. They occur in the central and southern parts, and can be expected any time of the year. However, the maximum frequency of occurrence is in the summer and the minimum is in the winter. Most of the area occupied by the central and southern parts is covered with sandstorms in the spring, whereas summer sandstorms are mainly confined to the southern extremity near the Gulf.

The prevailing winds in the wet season are mainly northwest, but they shift to the southeast ahead of the arrival of low-pressure systems. The northwestern winds are referred to as 'Shamali' (northern) winds contrary to the southwestern 'Suahili' (coastal) winds. The average annual wind speed is in the order of 12–15 km h^{-1} (Table 11, Appendix I). Annual evaporation from free water surfaces varies from to 2,000 mm to 3,000 mm depending on the location.

3.3.3.6 Kuwait

The climate of Kuwait is an extension to the climate of the southern part of Iraq, extremely continental. The climate, according to Köppen's classification is of the dry, hot desert-type.

The mean daily temperature drops from 36 °C in summer (June-September) to 14 °C in winter (December-February). The absolute maximum and minimum

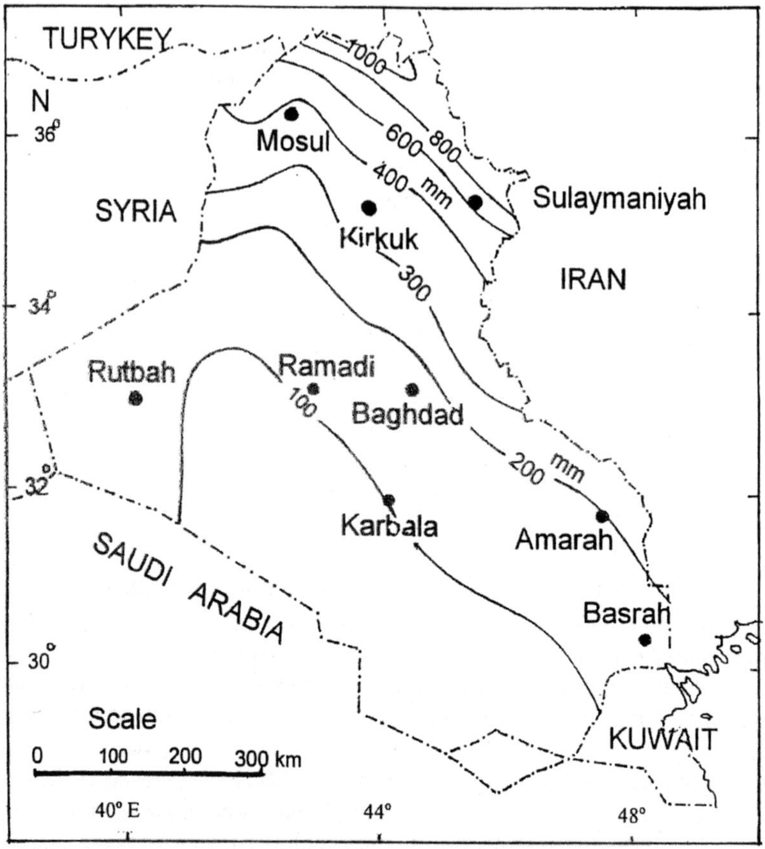

Figure 18. Isohyetal map of annual rainfall in Iraq (from Natural Resources/Series of Studies on Water, Report No. 9, UN Publications, 1982)

temperatures in these two seasons in their respective order are 50 °C and −6 °C. The diurnal range (difference between day and night) of temperature is 12 °C in the winter season increasing to 18 °C in summer.

The relative humidity is highest in December and January reaching 65% inland and 85% along the coast. These figures drop to about 20 to 30% respectively in the summer. The annual average precipitation is about 110 mm and exhibits considerable temporal and spatial variation. In fact, the Arabian Gulf is the source of moisture for the rain-bearing depressions. As such, no precipitation is expected in the summer (June September). In the transitional seasons, when upper troughs drive cold air from the northwest over warm air in the lower layers, the vertical temperature gradient in some upper air patterns is favourable for violent thunderstorms bringing hail and heavy rainfall (Takahashi & Arakawa, 1981).

The frequent winds from the northwest (Shamali) are cool in winter and early spring, and hot in summer. Southeasterly winds, usually hot and moist, spring up between July and October. Spring and early summer are dominated by hot and dry winds. Frequency analysis of surface wind data at Kuwait Air Port (station No. 163) has given the results shown in Table 4. These winds added to the high air temperature and low mean relative humidity are the main factors causing high rates of evaporation from the free water surfaces. In deep summer time (34–35 °C mean daily temperature and 40–45% relative humidity) the daily evaporation can be in the range of 20 to 25 mm.

3.3.3.7 Bahrain

The mean annual temperature in Al Bahrain is around 23 °C and the mean daily temperature in the warm period (June-September) is 32 °C. It falls down to 15.4–17.1 °C in the winter period (December-February). These figures bring the difference or range between summer and winter mean temperatures to 17–18 °C. It goes without saying that the absolute maximum temperature can climb up to more than 43 °C while the absolute minimum can be as low as 7 °C or less.

The mean annual relative humidity is high, especially when compared to tropical and subtropical areas. This is caused by the location and elevation of the islands with respect to the surrounding sea. The average relative humidity is 67% for the year, 60% and 74% for the summer and winter seasons respectively.

The annual total hours of sunshine are in the order of 3,200 h. The daily number of hours varies from 5 h or less in March to 11 h or more in August.

Nearly one half of the winds blowing on the Bahrain Islands are NNW-NW-WWN. The WWN winds are dominant in December and January. They blow with a high speed reaching up to 110 km h^{-1}. The next five months are under the influence of the NNW winds. These are replaced by the NW winds until end of November. The SSE winds blow only for short durations. The winds blowing on Bahrain bear different names depending on the month or period in which they blow, and the location from which they blow. These are: Northern winds (N), Southern winds (Kōs, S), As Sumūm, Al-Batīn (N, W), Al Agūz (N), Suhaili (S), B. As Sarayāt (N),

Table 4. Percentage frequency distribution of blowing wind at Kuwait Air Port (from Takawashi and Arakawa, 1981)

Month	Frequency distribution of blowing wind, %									Temp.,	Relative
	N	NE	E	SE	S	SW	W	NW	Calm	°C	hum., %
January	10.7	3.3	6.7	12.6	10.4	3.6	8.9	27.3	16.5	12.8	69
April	15.0	5.7	13.7	16.7	12.5	4.4	6.1	16.6	9.3	24.2	61
July	14.7	3.0	6.1	5.4	5.0	2.7	12.3	39.4	11.4	34.7	43
October	13.2	5.2	10.0	9.6	13.2	5.5	8.7	19.2	15.4	27.8	62

Explanation
N = north, S = south, E = east, W = west, NE = northeast, SE = south east, NW = northwest and SW = southwest.

Al Marzam (S), Gelibīn (S), Suhaiyl, Wasam (SE), Ahmīr (S, SE) and Al Marba'âniyāt (W, E).

The scanty amount of rain falling in Al Bahrain occurs in the season from December to April. This period coincides with the passage of atmospheric depressions coming from the Mediterranean area. As soon as any of these depressions become stable over the Gulf, air masses of different temperatures flow towards the centre of the depression. The collision between the cool northern winds and the southern warm air results in the formation of a front, which may lead to the fall of rain showers lasting between one and three days. The rainfall in Bahrain for the short period 1962–1974 is 59.3 mm y^{-1}. The long-term annual average is about 67.4 mm y^{-1}. The annual depth varies from about 200 mm in one year to less than 10 mm in another year. The four month period, June-September, is a period of absolute dryness.

3.3.3.8 Qatar

The State of Qatar lies within the branch zone of the northern desert belt. This geographic situation characterises the country by high temperatures throughout the whole year, early hot and dry summer winds, and high levels of air humidity in the major part of the year. As such, the country has a climate, somewhat modified by the surrounding Arabian Gulf.

Despite the small difference in temperature between the coastal areas and the inland, the climate of the former is usually less harsh. The same observation applies when the climate of the oases are compared to other inland areas. The daily mean summer temperature is above 35 °C and falls down to around 15 °C in winter.

The relative humidity varies considerably within the day, and so from day to day. The maximum relative humidity exceeds 90% during 150 days of the year in several parts of Qatar. Despite the drop of the mean humidity from say 75% in winter to 45–60% in summer, the summer humidity coupled with high air temperature renders the climate not agreeable. Figure 19 shows the variation of mean daily temperature and relative humidity at Utoriyah (station No.168). Despite the difference in geographic location between this station and Baghdad (station No. 152), Iraq, the similarity of the pattern of variation of both temperature and humidity at the two stations is quite remarkable.

The dominating winds are the northern and northewestern winds. However, the southerly and southwesterly winds blow generally in February. The speed of winds blowing by night is less than the speed of winds blowing by day. The summer winds have an average speed of 25–30 km h^{-1} in protected areas. The speed increases considerably in desert, unprotected areas.

The average annual rainfall is about 75 mm, although wide variations occur from place to place and from year to year. Most of the rainfall occurs between November and April, largely in the form of high-intensity storms, and daily totals can approach the annual average depth. Northern Qatar receives more rainfall than the south, although the difference is negligible, especially when compared to free water evaporation or potential evapotranspiration. In June, pan evaporation can

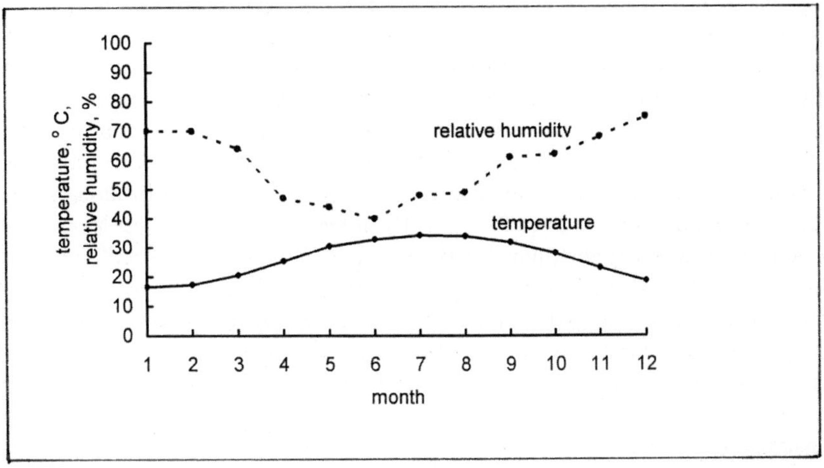

Figure 19. Mean monthly air temperature and relative humidity at Utoriyah (station No. 168), Qatar

reach 20 mm d^{-1}, but falls to between 2 and 4 mm d^{-1} in December and January. These figures, which are typical of an inland oasis site, may be increased by up to 50% in the open desert (Pike, 1979 and Country Report on Water Resources of Qatar, 1986).

3.3.3.9 Saudi Arabia

The country is situated in the tropical and subtropical region and its climate is linked to the East Mediterranean and surrounding countries. The air masses blowing from the north and west affect the climate of Saudi Arabia. On the contrary, the air blowing from the tropical zones in the south has a limited effect due to the sheltering effect of the mountain ranges extending from Oman up to the southern part of Yemen.

Saudi Arabia is characterised by a long, hot and arid summer for the greater part of the country. The mean daily temperature in summer is about 35 °C. This climate is due to the northerly winds moving from the eastern Mediterranean towards the Arabian Gulf. The winter season (December–February) is rather cold and short, with a mean daily temperature between 20 and 25 °C, but there are wide variations. The inland temperature ranges from below zero at night to a maximum of 50 °C during hot summer days. The summer temperatures in the northern part of Saudi Arabia are high, July being the hottest month. The coldest time in the southeast is January, whereas June is the hottest.

The relative humidity of air generally reaches its lowest level in June-July and its highest level in December–January. The variation of relative humidity from one month to another for locations along the seacoast is minor compared to inland locations.

Rainfall in Saudi Arabia is generally scanty, unpredictable and irregular. It varies in intensity and duration from year to year. Long periods may pass with no rainfall. When rainfall occurs, it is often local and sometimes very intense resulting in sporadic floods that cause severe damage and erosion of wadi beds and hill slopes. The mean annual rainfall generally ranges from 25 mm to over 200 mm. It varies from 30 to 50 mm in the north west of Saudi Arabia and increases to from 100 to 200 mm in the northeast. In the central areas, particularly the Riyadh area, the annual rainfall ranges from 85 to 110 mm. The total rainfall in the Hijaz Mountains and Asir region is fairly distributed between the months of the year with peaks in spring and autumn. Its mean annual depth is about 250 mm along the Red Sea coast south of Jeddah, but decreases to the north to reach less than 50 mm at Aqaba. The driest part of Saudi Arabia is the Rub Al Khali, which receives about 45 mm y^{-1} on its northern border and the rest is practically dry, rendering it lifeless. The areal distribution of rainfall in Saudi Arabia and the surrounding countries comprising the Arabian Peninsula can be seen from the isohyetal map in Figure 20.

Despite the generally high evaporation from free water surfaces, the temporal and spatial variations are quite considerable. Evaporation from the low areas in the west along the seacoast is relatively limited and remains at low rates while moving towards the mountainous regions. Contrarily, the evaporation rate increases eastwards while moving inland. The annual evaporation varies from about 2,000 mm at Ab'ha (on the mountain range east of the Tihama plain, about 550 km southeast of Jeddah) to 3,000 mm at the northern border of the Rub al Khali. The annual depth of evaporation in the region comprising Jeddah, Mecca and El Taif is confined to the range of 2,400–2,600 mm. Observations have shown that rates of evaporation along the coast of the Arabian Gulf are relatively small compared to other parts of Saudi Arabia.

3.3.3.10 United Arab Emirates

The eastern part of the country is mountainous, with elevation that reaches 1,000 m a.m.s.l. With the exception of the mountain ranges, the Emirates fall in the tropical, hot-desert climate zone. There is a marked difference between the climate of the coastal areas, inland, and the mountainous and highly elevated areas. The mean daily temperature, as given by the climatic normals, is 27.2, 25.8, 26.9 and 26.7 °C for the airports of Ras el Khaimah, Sharja, Dubai and Abu Dhabi, respectively, all along the Gulf coast. The deep summer mean daily temperature can be well above 37.5 °C coupled with high humidity, which frequently approaches the saturation level. The inland climate is characterised by the wide temperature range, which reaches 45 °C in July and August. Contrarily, the climate of the mountainous areas can be described as mild. The mean winter (December-February) temperature is about 20 °C.

The direction of the dominating wind can be southerly, southeasterly, westerly, northwesterly or northerly. The summer winds are usually laden with moisture. The rainfall is extremely variable with respect to time and space. The rainy season is from November to April. The United Arab Emirates is hit by thunderstorms due to

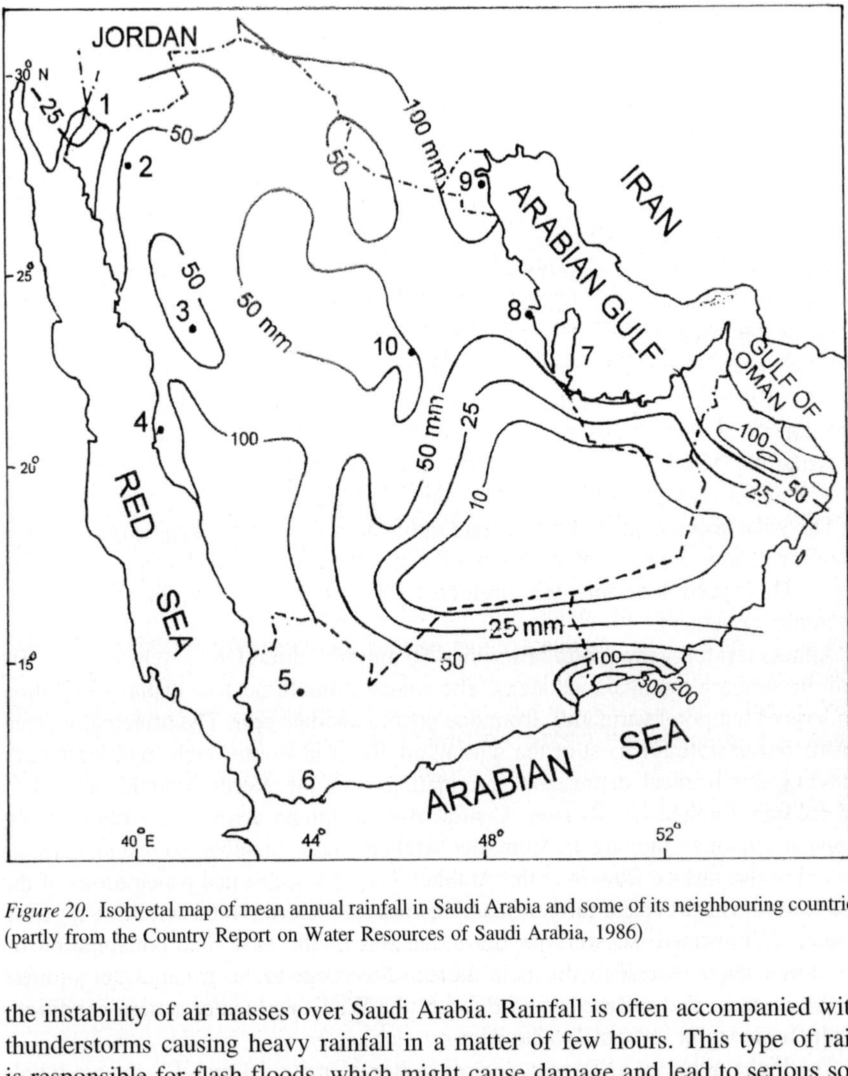

Figure 20. Isohyetal map of mean annual rainfall in Saudi Arabia and some of its neighbouring countries (partly from the Country Report on Water Resources of Saudi Arabia, 1986)

the instability of air masses over Saudi Arabia. Rainfall is often accompanied with thunderstorms causing heavy rainfall in a matter of few hours. This type of rain is responsible for flash floods, which might cause damage and lead to serious soil erosion. The average depth of rain varies from say 135 mm in the mountainous area to between 60 and 110 mm elsewhere in the country. The long-term mean rainfall at El Sharja (station No. 179) is 101 mm y^{-1}. This amount fell to just 11.5 mm in 1966–67, and rose to 206 mm in 1971–72.

Evaporation from open water bodies is generally high. However, it varies from one area to another depending on the distance from the coast. The Country Report on the Water Resources of the United Arab Emirates (1986) claims that the mean evaporation in the coastal strips is 2,980 mm y^{-1} increasing to 4,050 mm y^{-1} in Maliha, in the interior at a distance of 54 km from the coast. Potential evapotranspiration is high too and sometimes is taken as 75–85% of the free water evaporation.

3.3.3.11 *Oman*

The mean annual temperature is typically between 26° and 29 °C throughout the lowland area of Oman. The capital, Muscat (station No. 180), has a mean annual temperature of about 30 °C with maximum daily temperatures above 40 °C during three to four months each year. Compared to lowland areas, the temperature in the central desert plateaus tends to be 6–8 °C greater. In winter it sometimes snows on the Jabal Shams in the northern part of Jabal al Akhdar Range. There, the mountain height exceeds 3,000 m (Ministry of Water Resources, Oman, 1995).

The number of daily sunshine hours averages around 10 annually, with the exception of mountainous areas and the 'Khareef' (autumn)-affected south. The latter experiences much less sunshine hours between mid-June to mid-September.

Tropical cyclones, originating in the Indian Ocean area blow over the Arabian Sea quite regularly. Few of these cyclones cross the coast of the Arabian Peninsula, but not all of those cross the Oman coastline. The average number of cyclones crossing the Arabian Sea over the 20-y period, 1975–1994, was 2, 1, 7, 13, 1, 10, 5, 7, 8 and 3 for February, April, May through December.

The wind speed tends to be moderate throughout the country, typically averaging between 2 and 3 m s^{-1} at a height of 2 m above the ground level in lowland areas. This speed does not vary considerably during the year, except when there is cyclonic movement.

Annual rainfall varies from fewer than 50 mm in Central Oman to more than 300 mm in northern mountainous areas. The annual rainfall, next to spatial variability, undergoes temporal variability from one year to another year. The principal mechanisms responsible for most of the rainfall are the cold front troughs with eastward-moving extratropical depressions originating over the North Atlantic; advection of cold air southeastwards from Central Asia to Oman across the Arabian Gulf; tropical cyclones moving in from the Arabian Sea; on-shore monsoon currents linked to the surface flow over the Arabian Sea; and occasional penetrations of the shallow monsoon current further inland toward the United Arab Emirates triggering scattered convective showers on the mountains in the east. The precipitation on these mountains represents the main source of recharge to the groundwater aquifers in the country. The surface runoff flows to the Wadis and under certain conditions wadi flows can be extremely abundant.

The Piche gives an average evaporation rate of about 10 mm d^{-1} for the whole year at Muscat. This rate reduces to 7.0 mm d^{-1} at Salalah on the coast of the Arabian Sea, but increases to 16–17 mm d^{-1} in the interior of Oman.

3.3.3.12 *Yemen*

The topography and relief of the country added to its geographical situation have a strong influence on the factors affecting the climate of Yemen. As such, North Yemen has considerable variability in climate ranging from hot, humid coastal plain bordering the Red Sea to cool highlands with relatively low humidity and, towards the east, to hot, dry eastern desert area bordering Ar'Rub al Khali. Three basic climates and rainfall regions are distinguishable in the northern part of

Yemen: Western maritime, southern maritime and inland continental regions. South Yemen is characterised by a hot tropical climate throughout the greater part of the year. The coastal areas have more uniform temperatures and are more humid than the interior.

Locations along the coasts of the Red Sea, e.g. Al Hodeydah (station No. 189), and the Gulf of Aden, e.g. Khormaksar (station No. 198) experience high temperatures such as 33.0 and 32.2 °C respectively. Some of the highly elevated places such as San'a (station No. 186) and Ibb (station No. 194) enjoy a cooler temperature of 24.7 and 26.6 °C, respectively. January is the coldest month and June/July the hottest months of the year for all stations regardless of their locations whether on the coast or in the interior. The range between highest and lowest temperatures is less than 10 °C for all stations, except for Marib (station No. 185) where the range is 13.1 °C.

The relative humidity of air along the coasts of the Red Sea and the Gulf of Aden is characterised by its high value, between 70 and 80% for the whole year round. The second characteristic is that the difference between the monthly mean values of humidity is limited. The interior and mountainous areas are generally less humid. The annual figures for Marib (station No. 185), San'a (station No. 186), Yarim (station No.191) and Rihab (station No. 192) are 37, 44, 50 and 45% respectively. The maximum humidity occurs at these four stations in the period from January to July while the minimum occurs from June to November. So, it is difficult to specify a certain month for the occurrence of either maximum or minimum relative humidity.

South Yemen being located within the monsoon belt is basically affected by southeast monsoonal winds during the period October-April and northeast summer monsoonal winds during the period June-September. The period April-June is a transitional period between the two monsoons.

The rainfall in Yemen depends on two main mechanisms, the Red Sea Convergence Zone (RSCZ) and the monsoonal Inter Tropical Convergence Zone (ITCZ). The RSCZ, whose influence is most noticeable in the west of the country, is active from March to May and to some extent in the autumn, while the ITCZ reaches Yemen in July-September, moving north and then south again so that its influence lasts longer in the south. Both convergence zones produce precipitation in the form of convective storms of high intensity with short duration and limited areal extent, although the ITCZ storms have a larger extent than those produced by the RSCZ. The relative importance of the RSCZ and ITCZ in different parts of the country is reflected in the seasonal rainfall distribution.

The relationship between mean annual rainfall and topography is evident from the data of some of the stations listed in Table 12, Appendix I. Rainfall rises from less than 50 mm along the Red Sea coast to a maximum of 700–800 mm to the west of the main watershed west of San'a, and falls down steadily to below 50 mm along the gulf of Aden and also inland (Farquharson et al., 1996). The mean for the period 1985–91 is 1,510 mm (Ministry of Oil and Mineral Resources of

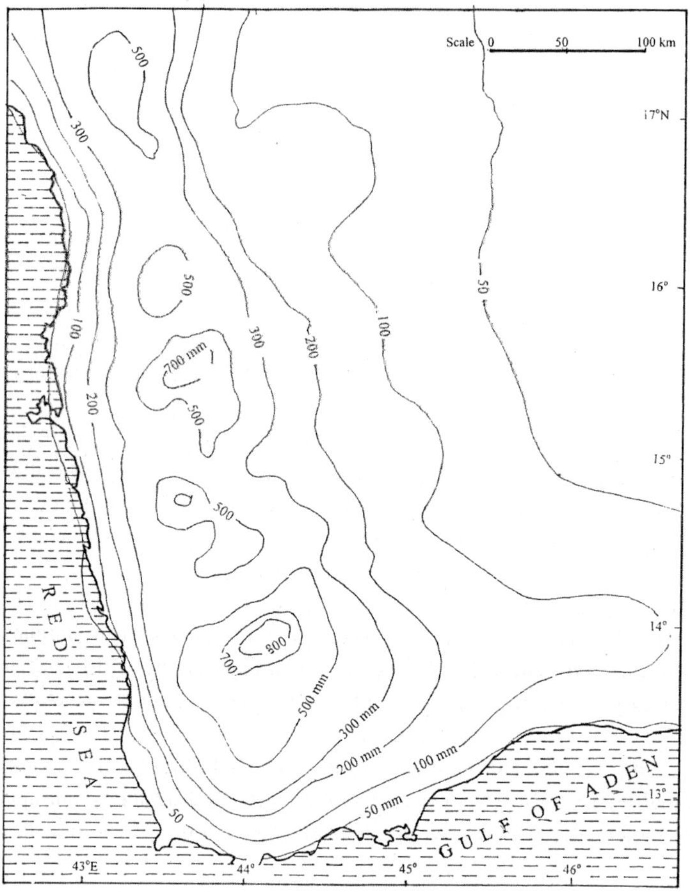

Figure 21. Isohyetal map of annual rainfall in Yemen

Yemen, 1987–91), and in Table 12, Appendix I, is 1,834 mm y^{-1}, both for Ibb (station No. 194). As many parts of Yemen are still without adequate rainfall data, one should regard the isohyetal map in Figure 21 as tentative.

3.4. CLIMATE CHANGE IN THE SOUTHERN AND EASTERN MEDITERRANEAN

Enormous volumes of research papers, scientific reports and books, all dealing with one or more aspect(s) of climate change, have been published in the last decades. In all those publications changes in temperature, precipitation, runoff, etc. are emphasised. For the purpose of the present chapter, temperature change in the southern Mediterranean will be reviewed here. Precipitation changes will be presented in Chapter 4 while other changes will be included in the corresponding chapters.

The increase of Carbon dioxide is claimed to cause the process known as greenhouse to take place. By doubling up the Carbon dioxide the greenhouse-gas-induced warming would add 2 °C in 70 years (Daniel, 1980). According to Jones et al. (1986) the near-surface air temperature averaged over the globe has increased by about 0.5 °C since the late 19th century. Parallel changes in the temperature of the lower troposphere have also taken place. This warming is not spatially homogeneous; trends have varied substantially from region to region. Jeftic et al. (1992(a)) wrote, "While the temperature trend in the Mediterranean over the past 40-years has been one of cooling, the trend over the past 20 years is quite different. Over the period 1967–86, the Mediterranean Basin has warmed in the west and cooled in the east. Clearly, with regard to future changes over the coming decades, one cannot simply assume that changes in the region (or any other region) will follow the global mean trend."

To verify those findings, the author considered annual temperature series at four stations; two in each of the western and central subregions. The time series plot of the mean annual temperature for the period of record, the 5-y moving average and the fitted trend line for the data at each of the four stations are shown in Figures 22(a) and 22(b). Te former represents the situation at the pair of coastal stations whereas Figure 22(b) represents the two inland stations. All four stations show that the mean temperature keeps fluctuating from year to year. The fluctuation assumes the form of a single or multiple cycle(s) mounted on a trend line. The cycles have neither equal amplitudes nor durations. The remarkable thing here is that the pattern of fluctuation of the data series of the two inland stations are almost identical. This is not the case at the two coastal stations: Oran (No. 23), Algeria, and Alexandria (No. 66), Egypt. The second intriguing feature is that the two inland stations, El Golea and El Minya, as well as the coastal station of Oran, show a rising trend while the trend line in case of Alexandria is a falling one.

The trends observed in the data series have been statistically tested to check whether the data are trend-free or not. In other words the null hypothesis that the trend is not significant can be rejected or not. To achieve this purpose the four data series have been tested using the Spearman's rank correlation test. The results obtained from this test are listed in Table 5. The tabulated results can fairly lead to the conclusion that the null hypothesis for the two Algerian series has to be rejected at 10% significance level (two-tailed normal test with $\alpha = 10\%$), and so for El Golea (Station 27) at $\alpha = 5\%$ where as for Oran (station 23) the result is very close to the critical value of the test statistic.

Wigley (1992) stated "While considering the future process of global warming, it is generally accepted that the equator-to-pole temperature difference will decrease, leading to changes in the atmospheric pressure field and atmospheric general circulation. It might be suggested that the subtropical high-pressure belt in the northern hemisphere will extend northwards as a consequence of a comparatively small amount of warming. The general northwards shift of the atmospheric circulation pattern will influence the path and frequency of mid-latitude cyclones over middle and eastern parts of the Mediterranean region. This general statement concerning the consequences of global warming may not hold everywhere on the micro scale where local factors prevail and can lead to inconsistencies. This conclusion is in agreement with an earlier statement given earlier by Lamb (1988) that "predictions of the magnitude of warming are still affected by a high degree of uncertainty and the implications of either positive or negative feedbacks are unknown, such that there is still no absolute confidence that climate will change, other than in the way it has changed in the recent past."

It may be of interest here to report on the study carried out by Al-Shalabei et al. (1996) on the implications of the expected climatic changes for the Syrian coast.

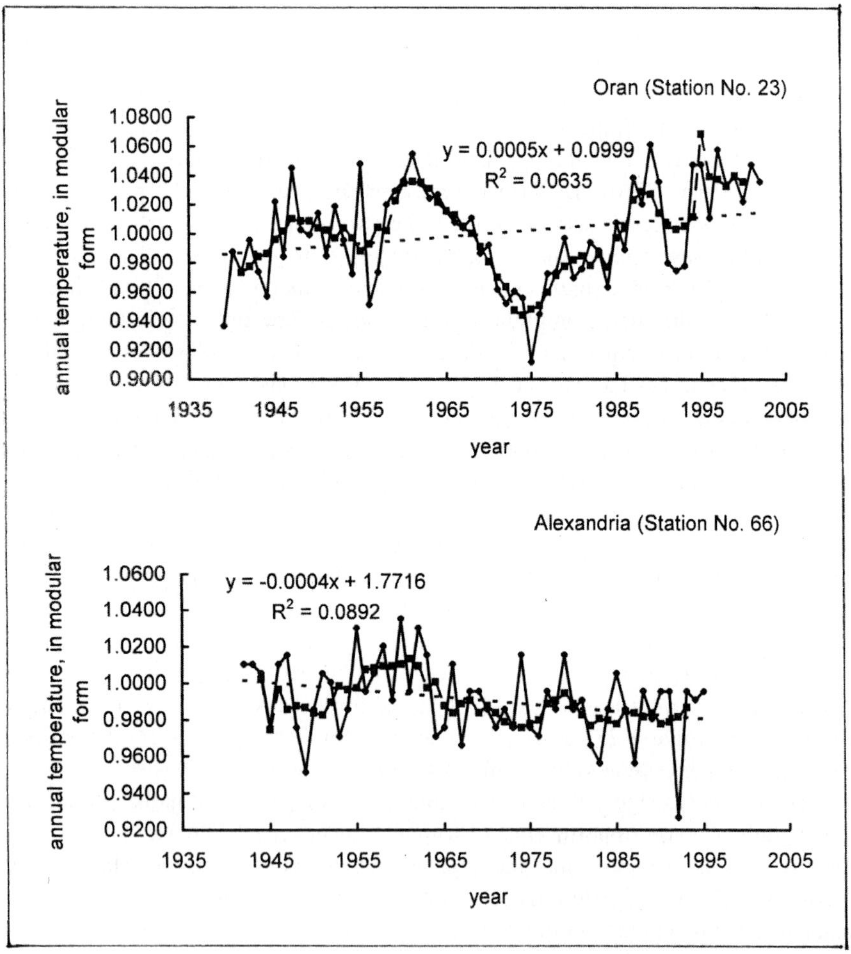

Figure 22(a). Mean annual temperature series at two locations along the Mediterranean Coast, North Africa

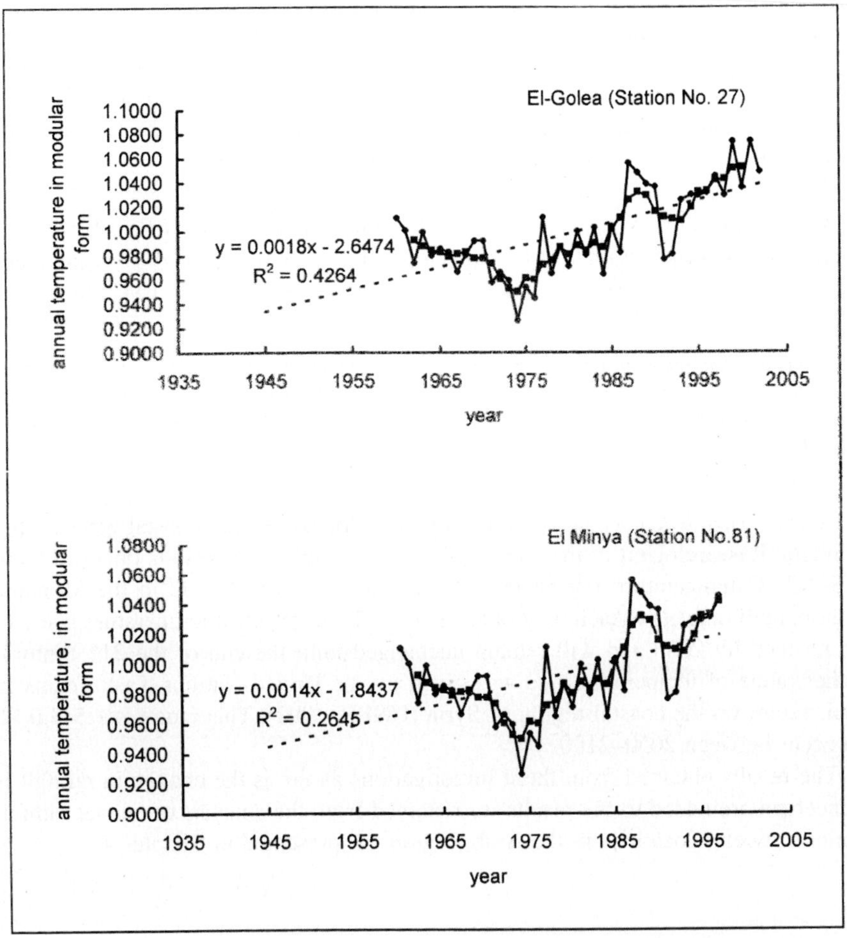

Figure 22(b). Mean annual temperature series at two interior locations in Algeria and Egypt

The coastal zone, which separates the Mediterranean shoreline from the arid zones of the interior, is transitional in both spatial and functional terms. The climate, as described earlier is Mediterranean of the humid or subtropical subtype, with cool, mild winters and hot dry summers. In general, temperature decreases from west to east and from south to north.

The assumptions used in that study have been based on the conclusions that, unless measures for greenhouse gas emission control are implemented by the entire world community, the global average temperature will rise by 1 °C by 2025, and by 3 °C before the end of the 21st century. The Global Circulation Models (GCMs), despite the uncertainty associated with the results they yield, can provide a means to investigate the consequences of CO_2 doubling through construction of

Table 5. Results of testing the null hypothesis that the mean annual temperature series are trend-free

Subregion/ country	Station Name	No.	Location	Record Period	N, yr	T_m, °C	s °C	z_{Sp}	H_o, given a 5%	10%
W.subregion										
Algeria	Oran	23	C°	1939–2002	64	17.69	0.596	1.933	NR	*R*
	El Golea	27	I*	1960–2002	43	21.50	0.764	3.941	*R*	*R*
C.subregion										
Egypt	Alexandria	66	C	1942–1995	54	20.10	0.430	1.271	NR	NR
	El Minya	81	I	1945–1995	51	21.22	0.409	1.54	NR	NR

Explanation

T_m = annual mean temperature for the given period of record, s = standard deviation, C_v = coefficient of variation, z_{Sp} = Spearman's z-statistic, H_o = null hypothesis of independence of the series, a = level of significance using 2-tail normal test, R = null hypothesis rejected and NR = null hypothesis not rejected. C° = coastal and ^{x} = interior

local scenarios. Such scenarios were produced for the Syrian coastal zone by the Climatic Research Unit of the University of East Anglia. The results obtained were 0.8–1.2 °C temperature rise in the coastal plain and 1.0–1.6 °C in the Montana region, both ranges for each °C global change. These ranges of temperature rise will be reached by 2030 and will remain unchanged until the end of the 21st century. Other range of temperature rise was given by the United Nations Task Teams in their report on the coastal region of Syria (UNEP, 1992). This range is 1.5–3.0 °C to occur between 2050–2100.

The results obtained from these investigations as far as the change in rainfall is concerned are added to the results we obtained from the analysis of annual rainfall series at twelve locations in the Arab Region, all presented in Chapter 4.

CHAPTER 4

ANALYSIS OF PRECIPITATION DATA

By virtue of its general characteristics, scarcity, high intensity and short duration, rainfall is an important water resource, at least for certain countries in the Arab Region. As such, denser and upgraded networks of rain gauges are strongly needed. Equally important are the care and interest in observing, collecting, analyzing and disseminating rainfall data.

The present chapter thus aims at using some of the available rainfall data to draw conclusions pertinent to land and water resources development, utilisation and management such as land conservation and rain-fed agriculture schemes. The latter have been in practice since early times in Yemen, Syria, Lebanon, Iraq, Palestine, Sudan, Libya, Tunisia, Algeria and Morocco.

4.1. ANNUAL RAINFALL

4.1.1 Homogeneity of rainfall series

Basic statistical descriptors of the annual rainfall records for about 120 rain gauging stations, some of which do not appear in Table 1, Appendix I, have been calculated and presented in Table 1 These are the arithmetic mean, X_m, standard deviation, s, coefficient of variation, C_v and skewness, C_s. The principal feature of these data is that three-quarters of the series have moderate to heavy skewness, i.e. having a coefficient of skew at or above 0.6, while nearly one-tenth is negatively skewed. The remaining 15% of the data series are lightly skewed. The magnitude and sign of skewness are important indicators to the frequency distribution function that may serve as a good fit to the observed data.

Twelve data series have been selected from the annual series included in Table 13, Appendix I, for further testing. The stations tested are: Nema (No. 6); Mauritania, Oran (No. 23); Algeria, Alexandria (No. 66) and Giza (No. 74); Egypt, Gedaref (No. 97) and Wau (No. 110); Sudan, Aleppo (No. 126); Syria, Amman A.P. (No. 140); Jordan, Beit Jala; Palestine, Al Manamah (No. 166); Bahrain and Aden/Khormaksar (No. 198); Yemen. The shortest record, 50 years, is that of Aleppo, Syria, while the longest, 90 years, is that of Djibouti Serpent, Djibouti.

Additionally, a group of five out of the twelve selected stations: No. 23, 66, 113, 166 and 198 are located along the seacoast and the other group comprises the

Table 1. Some statistical characteristics of annual rainfall series of certain stations in the Arab Region

Station No.	Name	Lat. °	′ N	Long. °	′ E	Z, m	n, year	X_m, mm	s, mm	C_v	C_s
MAURITANIA											
1	Atar	20	31	13	04 W	225	33	106		N.A.	
2	Tidjika	18	32	11	26	399	60	150	76.8	0.51	1.10
3	Nouakchott	18	07	15	36	21	61	107	60.8	0.57	0.62
4	Boutilimit	17	32	14	41	77	60	173	84.5	0.49	0.57
5	Kiffa	16	38	11	24	115	60	333	124.7	0.37	0.34
6	Nema	16	37	07	16	269	60	278	81.8	0.29	0.76
MOROCCO											
7	Tanger	35	43	005	54	21			N.A.		
8	Melilla	35	26	02	57	8			N.A.		
9	Oudja	34	47	001	56	470			N.A.		
10	Rabat	34	03	06	40	75	N.A.	503		N.A.	
11	Meknes	33	53	005	32	549	10	654	157	1.29	0.24
12	Casablanca	33	34	007	40	58	10	440	92.1	1.41	0.21
13	Ifrane	33	31	05	07	1640	N.A.	1112		N.A.	
14	Marrakech	31	37	008	02	466	10	259	84.6	0.41	0.37
15	Ouarzazate	30	56	006	54	1136	10	103	47.2	0.63	0.46
16	Aghadir	30	23	009	34	19	10	214	63.6	−0.18	0.30
ALGERIA											
17	Alger	36	46	03	03 E	60	57	668	182.5	0.27	0.32
18	Cap Cabron	36	45	05	06	230	10	758	140.8	0.19	−0.50
19	M. Blanche	36	43	003	15	22	10	746	136.1	0.18	1.62
20	Miliana	36	20	02	14	722	26	419	103.8	0.25	−0.64
21	S.B. Abbes	36	08	02	10 W	498	16	302	72.9	0.24	0.44
22	Biskra	34	48	05	44 E	56	31	123	51.4	0.42	0.36
23	Oran	35	38	00	37 W	99	61	369	108.3	0.29	0.34
24	Tebessa	35	26	08	08 E	816	31	325	95.7	0.29	0.83
	....						23	309	106.1	0.34	1.27
25	Laghouat	34	20	03	23 E	1180	44	337	106.3	0.32	0.73
26	C. Bechar	31	38	02	15 W	806	22	919	403.2	0.44	0.06
	....						16	627	294.6	0.47	−0.13
27	El Golea	30	34	02	52 E	393	43	35	29.9	0.85	2.18
28	In Salah	27	12	002	28	294			N.A.		
29	Ouallen	24	36	01	17	347	10	22.5	20.6	0.92	1.21
30	Tamanrasset	22	42	05	31	1405	10	73.4	43.5	0.59	0.17
TUNISIA											
31	Biezerte	37	14	009	49	3	10	684	116.1	0.17	−0.32
32	Kelibia	36	51	011	05	29	10	525	107.5	0.20	−0.64
33	Tunis	36	50	10	14	4	27	472	120.3	0.23	0.92
34	Jendouba	36	29	008	48	143	10	459	109.2	0.24	0.25
35	Kairaouan	35	40	001	06	60	10	355	105.2	0.30	0.82
							50	287		N.A.	
36	Monastir	35	40	010	45	2	10	387	177.6	0.46	1.33
37	Sfax	34	43	01	41	21	10	249	89.5	0.36	0.20
	....						50	196		N.A.	
38	Gafsa	34	25	008	49	313	10	198	71.5	0.36	0.64

39	Tozeur	33	55	008	10	45	10	96.4	16.7	0.17	1.09
40	Gabes	33	53	10	06	2	30	213	80.3	0.38	0.60
	....						50	170		N.A.	
41	Dj. Mellila	33	35	010	47	6	10	277	177.1	0.64	1.32
42	Remada	32	19	010	24	300	10	104	85.5	0.83	1.71
LIBYA											
43	Tripoli	32	57	13	12	25	39	309	107.7	0.35	0.20
	....						56	383		N.A.	
44	Shahat	32	49	21	51	625	16	577	153.3	0.27	−0.28
45	Garian	32	10	13	01	726	19	321	136	0.42	0.44
46	Derna	32	44	022	38	9	16	275	100.3	0.36	0.43
47	Aziziyah	32	33	013	02	145	10	252	101.5	0.40	0.86
	....						21	224		N.A.	
48	Misurata	32	19	015	03	32	10	285	56.7	0.20	0.26
49	Benina	32	06	20	16	132	36	255	90.0	0.35	0.09
50	Nalut	31	52	10	59	620	19	131	48.7	0.37	0.33
51	Nasser A.P.	32	06	023	55	51	20	83	34.7	0.42	0.81
52	Mizda	31	27	13	00	400	19	61.3	26.8	0.44	0.66
53	Sirte	31	12	16	35	22	39	188	83.0	0.44	0.97
54	Ajdabiyah	30	43	20	10	5	16	138	54.4	0.39	−0.05
55	Gadames	30	08	009	30	338	10	27	23.3	0.86	1.43
56	Hon	29	08	15	57	261	34	34	35.0	1.02	1.94
57	Gialo	29	02	21	34	52	30	9.1	8.5	0.94	1.32
58	Sebha	27	01	14	26	444	36	11	13.2	1.17	1.50
59	Kufrah	24	13	023	18	436	30	1		N.A.	
EGYPT											
60	S.Barrani	31	38	25	58	27	45	164	78.4	0.48	1.33
61	Sallum	31	33	25	11	7	47	100	211.0	2.12	1.93
62	Damietta	31	25	31	49	3	55	128	83.0	0.65	3.53
63	Rosetta	31	24	30	25	2	18	212	93.2	0.44	0.61
64	M. Matruh	31	22	27	14	7	42	136	59.4	0.44	0.98
65	Port Said	31	17	32	15	1	55	77	39.5	0.52	1.31
66	Alexandria	31	12	29	53	32	55	193	71.0	0.37	0.98
67	Al-Arish	31	07	33	46	10	49	105	44.0	0.42	0.52
68	Mansura	31	03	31	23	7	55	56	26.0	0.47	1.26
69	Damanhour	31	02	30	28	6	55	102	51.1	0.50	1.32
70	Tanta	30	47	31	00	14	55	55	27.7	0.51	1.20
71	Zagazig	30	35	31	30	11	51	35	26.3	0.75	3.70
72	Delta Barr.	30	11	31	08	20	27	22.6	14.6	0.65	1.19
73	Cairo	30	03	31	15	20	20'	23.8	11.73	0.49	1.67
74	Giza	30	02	31	13	21	55	21	17.5	0.83	1.56
75	Suez	29	56	32	33	10	30	25.6	19.2	0.75	0.85
76	Helwan	29	52	31	20	112	55	23	16.9	0.77	1.65
77	Faiyum	29	18	30	51	28	51	12	11.8	1.02	2.23
78	Siwa	29	12	25	19	17	51	9.8	11.5	1.17	2.36
79	Beni Suef	29	04	31	06	28	13	7.0	7.3	1.04	1.86
80	Tor	28	14	33	37	2	44	10	11.9	1.19	1.47
81	El-Minya	28	06	30	46	43	29	3	5.2	1.73	2.29
82	Hurghada	27	17	33	46	1	25	4.1	9.9	2.41	3.69
83	Assyut	27	11	31	13	71	39	3.0	4.6	1.53	1.78
84	Qena	26	10	32	43	75	28	5.0	11.3	2.26	3.90

(continued)

Table 1. (Continued)

Station No.	Name	Lat. °	′ N	Long. °	′ E	Z, m	n, year	X_m, mm	s, mm	C_v	C_s
85	Qusseir	26	08	34	18	6	55	8.8	13.7	1.56	1.87
86	Dakhla	25	29	29	00	122	32	1.0	2.2	2.20	4.93
87	El Kharga	25	26	30	24	70	51	1.7	4.1	2.43	4.06
88	Deadalus Is.	24	55	35	52	4	34	9.0	13.0	1.44	2.12
89	Aswan	24	02	32	53	108	55	1.1	2.2	2.02	2.54
The SUDAN											
90	W.Halfa	21	55	31	20	125	45	2.2	5.6	2.80	3.81
91	Port Sudan	19	37	37	13	6	45	87	54.5	0.63	0.69
92	Tokar	18	25	37	45	18	45	83	68.3	0.83	1.36
93	Atabra	17	42	33	58	348	45	64	40.8	0.64	0.55
94	Khartoum	15	37	32	33	385	45	158	70.7	0.45	1.15
95	Kassala	15	28	36	24	501	45	300	79.3	0.26	0.34
96	W. Medani	14	24	33	30	407	45	338	100.6	0.30	0.15
97	Gedaref	14	02	35	24	599	45	593	94.3	0.16	0.33
	…………..						89	578	119.1	0.21	0.31
98	Ed Dueim	13	59	33	20	379	45	276	101.4	0.37	0.50
99	Fasher A.P.	13	38	25	20	790	45	272	117.4	0.43	1.72
100	Sennar	13	33	34	37	419	45	460	105.6	0.23	−0.25
101	Geneina	13	29	22	27	779	45	515	134.3	0.26	0.36
102	Obeid A.P.	13	10	30	14	570	45	370	117.2	0.32	0.79
103	Singa	13	09	33	57	433	45	561	109.3	0.20	−0.32
104	En Nahud	12	42	28	26	540	45	411	117.7	0.29	1.29
105	Roseires	11	51	34	23	467	45	706	111.1	0.16	0.63
106	Renk	11	45	32	47	382	45	541	141.4	0.26	0.86
107	Malakal	09	32	31	39	389	45	770	137.6	0.18	0.25
108	Raga	08	28	25	41	460	45	1188	171.6	0.14	0.51
109	Akobo Post	07	48	33	03	403	40	934	199.3	0.21	−0.10
110	Wau	07	42	28	01	433	45	1127	171.4	0.15	0.09
111	Juba	04	51	31	37	462	45	988	166.5	0.17	0.34
112	Torit	04	25	32	33	625	45	1040	156.5	0.15	0.65
Djibouti											
113	Djibouti S.	11	30	43	09	7	90	137	97.8	0.72	1.48
Somalia											
114	C. Guardaffui	19 48	13 40	80	N.A.	N.A.		N.A.			
115	B. Cassim	11	17	49	11	6	12	18		N.A.	
116	Ergaivo	10	37	47	22	1730	18	434		N.A.	
117	Berbera	10	26	45	02	8	30	49		N.A.	
118	Hargeisa	9	31	44	06	1370	30	416		N.A.	
	……………						10	430	172.4	0.40	0.69
119	Galacaio	6	46	47	25	240	26	148		N.A.	
	……………						20	146	112.9	0.77	1.07
120	Obbia				N.A.		7	165		N.A.	
121	L. Ferrandi	3	45	42	35	193	39	309		N.A.	
122	Mogadiscio	2	02	45	21	17	46	399		N.A.	
	……………						12	368	142.4	0.39	0.47
123	Chismaio	00	22 S	042	26	10	10	467	247.6	0.53	0.67
CI	Comoros	11	42	043	14	12	18	2668	701.8	0.26	0.34

SYRIA											
124	Qamishliye	37	03	41	13 N	451	30	446	130.4	0.29	0.35
125	Hassakeh	36	31	40	39	N.A.	30	283	103.3	0.36	0.32
126	Aleppo	36	11	37	13	392	30	329	80.1	0.24	−0.30
127	Al Rakka	35	58	39	02	N.A.	30	215	77.8	0.36	0.33
128	Lattakia	35	33	35	45	8	30	807	182.0	0.23	−0.06
129	Deir ez Zor	35	45	40	48	204	22	155	74.1	0.48	0.40
	……………						48	157	68.5	0.44	0.41
UN	Safita*	34	49	36	09	369	30	1147	261.4	0.15	0.23
UN	El Kom	35	12	38	51	460	21	132	46.7	0.36	0.44
130	Hama	35	08	36	45	309	30	348	97.2	0.28	0.41
131	Palmyra	34	33	36	18	404	22	130	55.8	0.43	0.54
	……………						46	144	53.8	0.37	0.19
132	Abu Kamal	34	28	40	55	174	22	126	53.9	0.43	1.03
UN	En Nebk	34	02	36	44	N.A.	30	122	41.4	0.34	0.45
UN	Sad' Biar	33	47	37	41	820	22	114	89.6	0.79	2.21
UN	Jabal Ettenuf	33	29	38	40	722	22	100	57.2	0.57	1.39
133	Damascus	33	29	36	14	729	30	203	65.9	0.32	0.86
UN	Zuloff (Zilaf)	32	56	37	20	656	22	96	69.9	0.73	1.69
Lebanon											
134	Tripoli	34	35	36	00	8	30	765	189	1.54	0.25
135	Rayack	33	52	36	00	921	10	560	172	0.30	0.31
136	Kasara Obs.	33	50	35	53	918	50	617	144	0.08	0.23
137	Beyrouth	33	49	35	29	24	30	559	161	0.13	0.29
JORDAN											
UN	Kafr Som	32	33	35	51	585	39	515	154.1	0.30	0.10
UN	Kharja	32	40	35	53	455	39	450	133.5	0.30	0.35
UN	Ramtha S.	32	34	36	01	520	43	285	93.2	0.33	−0.08
138	Irbid Sch.	32	33	35	51	585	43	417	164.6	0.39	0.65
UN	Kafr Yuba	32	33	35	48	550	42	506	151.6	0.30	0.06
UN	Deir A.Sa'id	32	30	35	41	330	40	465	135.1	0.29	0.16
UN	K. Awwan	32	26	35	41	470	43	496	156.7	0.32	0.36
UN	Ajlun	32	20	35	45	760	42	641	198.1	0.31	0.01
UN	Kafrinja	32	18	35	42	640	43	620	176.9	0.29	0.12
UN	El Kitta	32	17	35	51	665	30	597	180.7	0.30	24
UN	El Jubeiha	32	02	35	58	980	42	472	146.1	0.31	0.25
139	Salt	32	02	35	44	796	43	625	218.6	0.35	0.20
140	Amman A.P.	31	58	35	36	790	43	291	131.4	0.45	1.78
141	Ma'daba	31	43	35	48	785	43	343	136.4	0.40	0.21
UN	Qatrana	31	15	36	02	770	42	100	41.3	0.41	0.71
UN	Buseira	30	45	35	36	1100	30	273	126.7	0.47	1.18
142	Shaubak	30	32	35	33	1300	43	319	139.7	0.44	0.84
PALESTINE (West Bank)											
UN	Qabatya	32	24	35	16	N.A.	28	569	208.3	0.37	2.31
UN	Meithalun	32	21	35	17	N.A.	38	602	150.0	0.25	0.67
UN	Burqa	32	18	35	12	N.A.	24	658	133.0	0.20	0.07
UN	Talluza	32	16	35	18	N.A.	29	641	174.2	0.27	0.61
UN	B.Dajan	32	12	35	22	N.A.	23	398	112.0	0.28	0.69
UN	Azoun	32	11	35	03	N.A.	34	543	144.0	0.27	1.01
UN	Beit Jala°			N.A.			85	591	172.4	0.29	0.27

(continued)

Table 1. (*Continued*)

Station No.	Name	Lat. °	′ N	Long. °	′ E	Z, m	n, year	X_m, mm	s, mm	C_v	C_s
143	Nablus	32	13	35	16	580	N.A.	652		N.A.	
144	Ramallah	31	54	35	12	883			N.A.		
145	Jericho	31	52	35	30	−276	3	198		N.A.	
146	Jerusalem	31	52	35	13	755	N.A.	627		N.A.	
147	Hebron	31	30	35	06	990	5	603	129.7	0.21	−0.21
148	Gaza	31	30	34	27	16	N.A.	313		N.A.	
IRAQ											
149	Mosul	36	19	43	09	223	20	377	106.8	0.28	0.61
150	As'Sulay-maniyah	35	33	45	27	853	10	717	125.9	0.95	0.18
151	Kirkuk	35	28	44	24	331	20	378	120.8	0.32	0.89
152	Baghdad	33	14	44	14	34	20	154	53.7	0.35	0.89
153	Ar'Rutba	33	02	40	17	615	20	117	59.6	0.50	0.51
154	Kut al Hai	32	10	46	03	15	20	152	62.9	0.41	0.24
155	Najaf	31	59	44	19	670	10	107	33.5	0.15	0.31
156	Ad'Dywaniyah	31	59	44	59	20	20	115	40.2	0.28	0.35
157	Amarah							179	76.5	0.07	0.43
158	Nassiriyah	31	05	46	14	3	20	114	44.2	0.39	−0.01
159	Basrah	30	34	47	47	2	20	129	43.7	0.34	−0.59
KUWAIT											
160	Failaka Is.	29	28	48	17	5	20	120	62.9	1.02	0.52
161	Shuwaikh	29	20	48	03	11	20	115	52.8	0.31	0.46
162	Omariyah	29	17	47	56	21	20	112	58.0	0.34	0.52
163	Kuwait A.P.	29	13	47	59	55	26	141	58.3	0.13	0.41
164	Al-Ahmadi	29	06	48	08	122	20	127	82.8	1.35	0.65
165	Ahmadi A.P.	29	04	48	09	16	12	100	68.4	1.26	0.69
BAHRAIN											
166	Manamah	26	12	50	30	5	64	75	50.2	0.67	0.93
QATAR											
UN	Al Ruwais	26	10	51	12	N.A.	21	81	56.9	0.70	1.13
167	R. Al Faras	25	49	51	20	14	21	100	67.3	0.67	0.83
NN	Um Saied	24	56	51	34	N.A.	21	60	42.3	0.70	0.91
UN	Um Bab	25	11	50	41	N.A.	21	63	52.7	0.83	2.03
168	Al Utoriyah	25	31	51	12	35			N.A.		
169	Abu Samra	24	44	50	50	3.1			N.A.		
SAUDI ARABIA											
170	Tabuk	28	24	36	35	2400	N.A.	76.1		N.A.	
171	Hail	27	31	41	44	988	20	77.3		N.A.	
172	Dhahran	26	16	50	10	21	22	71	56.6	0.80	1.30
173	Riyadh	24	39	46	42	591	20	108	71.6	0.66	0.70
174	Medinah	24	33	39	43	632	24	49	40.1	0.83	1.06
175	Yanbu	21	59	39	59	710	31	114	60.5	0.53	0.36
176	Jidda	21	28	39	10	6	31	69	48.4	0.70	1.50
177	Al-Taif	21	29	40	32	1457	20	193	86.5	1.12	0.45
178	Gizan	12	54	42	33	N.A.	N.A.	38.6		N.A.	
UNITED ARAB EMIRATES											
179	Sharjah	25	20	55	24	5	12	101.2		N.A.	

OMAN											
180	Muscat (El-Seeb)	23	35	58	17	16.6	6	91.8	53.4	0.58	1.34
181	Sur	22	35	59	31	13.8	6	86.7	63.4	0.73	−0.06
182	Masirah	20	40	58	54	16	15	54	80.5	1.49	2.35
183	Salalah	17	02	54	05	22	27	99	106.2	2.39	1.07
YEMEN											
184	Hajjah	15	41	43	36	1300	7	432	266.5	0.62	1.69
185	Marib	15	28	44	19	1100	N.A.	79.2		N.A.	
186	San'a A.P.	15	28	44	13	2190	N.A.	213		N.A.	
187	Karman Is.	15	20	42	37	6	21	86.4		N.A.	
188	Dahi	15	13	43	03	70	7	134	66.4	0.49	−0.07
189	Hudeydah A.P.	14	46	42	57	10	7	85.4	78.4	0.92	1.51
190	Riyan (Mukalla)	14	39	49	23	25	13	63.5		N.A.	
191	Yarim	14	18	44	23	2400	6	437	123.2	0.28	1.48
192	Rihab	14	13	44	11	1500	20	527	128.2	0.24	−0.07
193	Zabid Gerb.	14	09	43	26	240	7	353	133.6	0.38	0.25
UN	Taiz town	13	35	44	01	1350	24	642	149.1	0.23	0.52
194	Ibb	13	59	44	11	1800	6	1391	432.1	0.31	1.26
195	Al-Barh	13	27	43	42	600	7	244	92.5	0.38	−0.22
196	Al Mokha	13	21	43	16	5	N.A.	21.4		N.A.	
197	Al Khod	13	05	45	20	5	7	64.3	60.3	0.94	0.48
	....						34	65.2			
198	Aden-Khormakser	12	50	45	01	7	88	43	39.3	0.91	1.54
199	Barim Is.	12	39	43	24	27	9	50		N.A.	

Explanation
N.A. = Data are not available, inconsistent, unreliable or contain numerous gaps.
**UN = Unnumbered stations are those stations which are not included in the register of international stations listed in Table 1, Appendix I. Instead, they are included in the national register. ° Beit Jala is a small village in the West Bank, Palestine, about 12 km southwest of Jerusalem.*
X_m = arithmetic mean, s = standard deviation, C_v = coefficient of variation = s/X_m and Cs = Coefficient of skewness.

remaining seven stations, which are inland. Figures 1(a) and 1(b) show the 5-year moving averages versus years of record for each group of stations. They also show a straight trend line fitted to the moving averages of each data series using the method of least squares. All the series show rising or falling trends, though with different slopes.

The stations showing a rising trend are 5 in total, 3 coastal and 2 inland while 2 of the 7 series with falling trend are coastal and 5 are inland. To check whether these trends are apparent or true, i.e. their slopes are significantly different from zero, the respective data series have been tested using Spearman rank correlation technique (Gibbons, 1971) and the linear regression method (Himmelblau, 1969). The aim was to check whether the null hypothesis that the series are trend-free could be rejected or not.

The results obtained from the above-mentioned statistical tests are presented in Table 2. The listed values show that the Spearman's z_{Sp}-statistic for all stations, except the one at Wau (No. 110), is less than the critical values of the standard

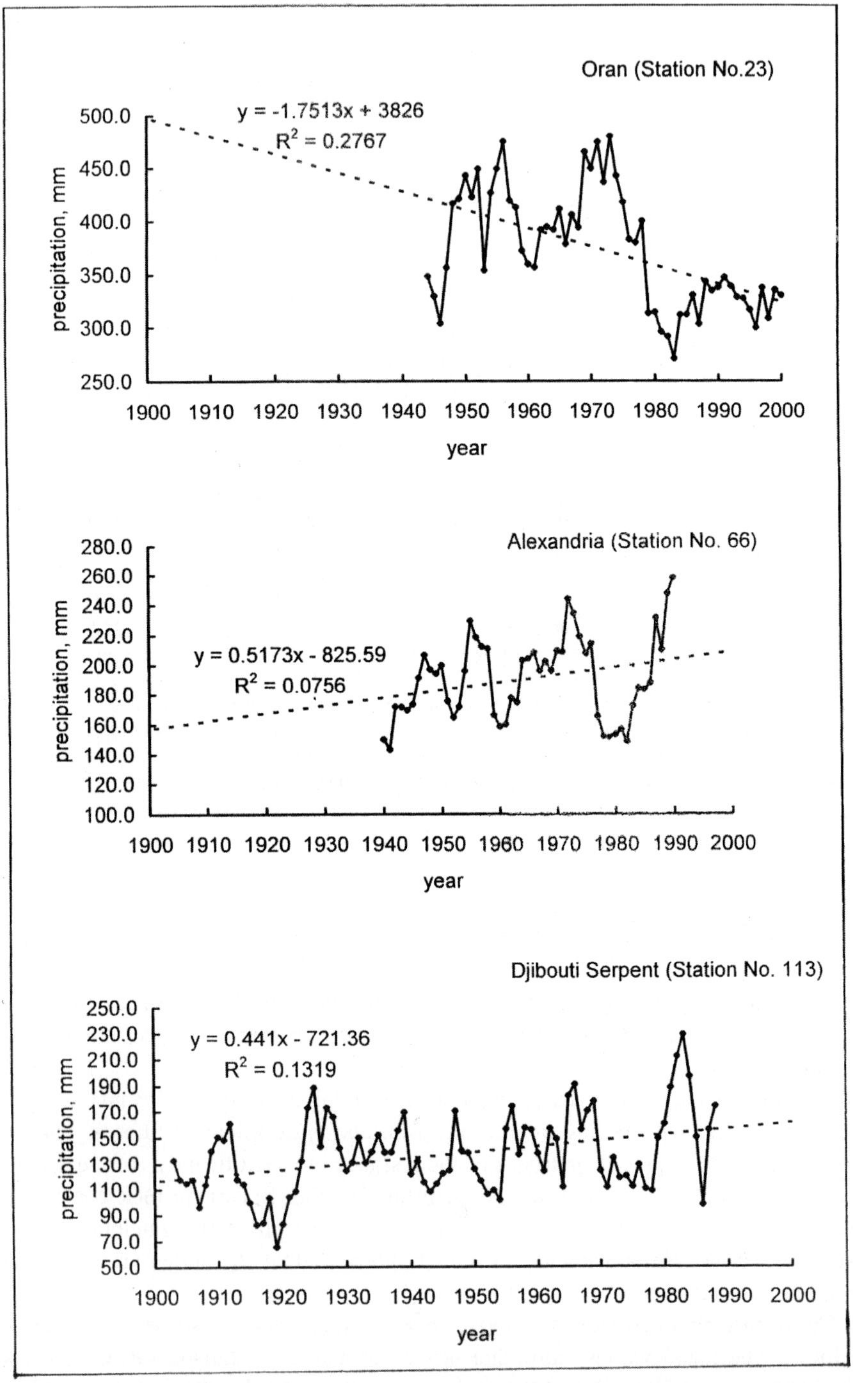

Figure 1(a). Fluctuation of annual rainfall series at coastal locations in the Arab Region

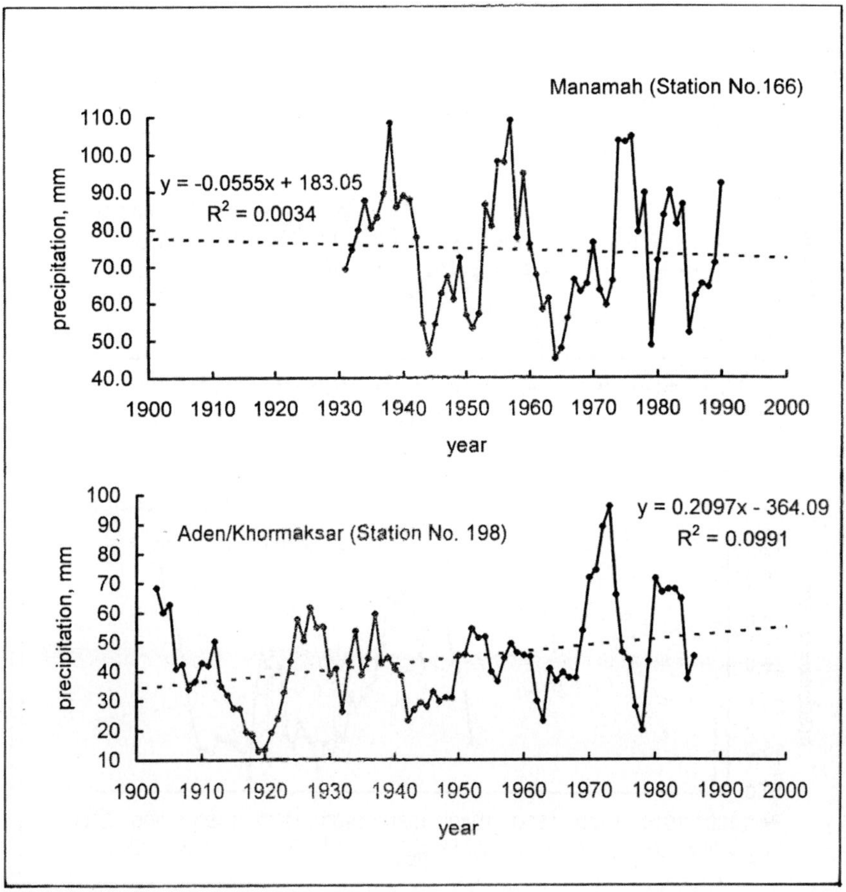

Figure 1(a). (Continued)

normal variate, z_{cr}, which can be obtained from the normal distribution at significance level, α, of 5% (two-tailed test). The same test shows, however, that the computed statistics for the precipitation series of Oran (No. 23) and Amman (No. 140) are very close to the critical limit, 1.645, when the significance α is at 10%. The regression test confirms the conclusion to reject the null hypothesis of having a trend-free series for each of the stations 23 and 110. As a further check, the homogeneity or stability of the means and variances of the sub-series formed by the pairs of segments of the full series before and after 1960 have been tested using the student-t and Fisher-F tests respectively. The results obtained from these two tests are presented in Table 3.

The variability of the annual precipitation is similar to that of the annual temperature presented in Chapter 3. The precipitation-time relationship assumes the form of periodic fluctuation(s) of different durations and amplitudes mounted on a rising

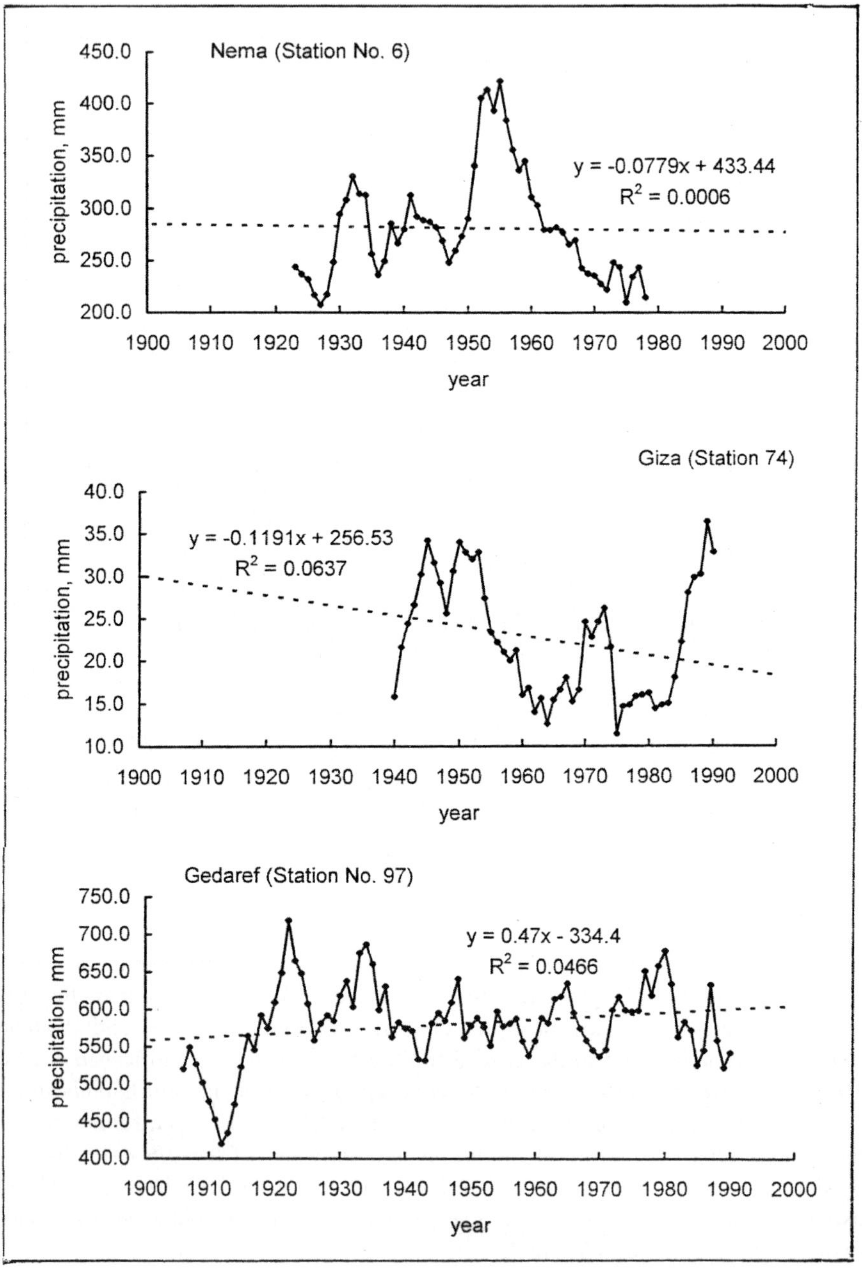

Figure 1(b). Fluctuation of annual rainfall series at interior locations in the Arab Region

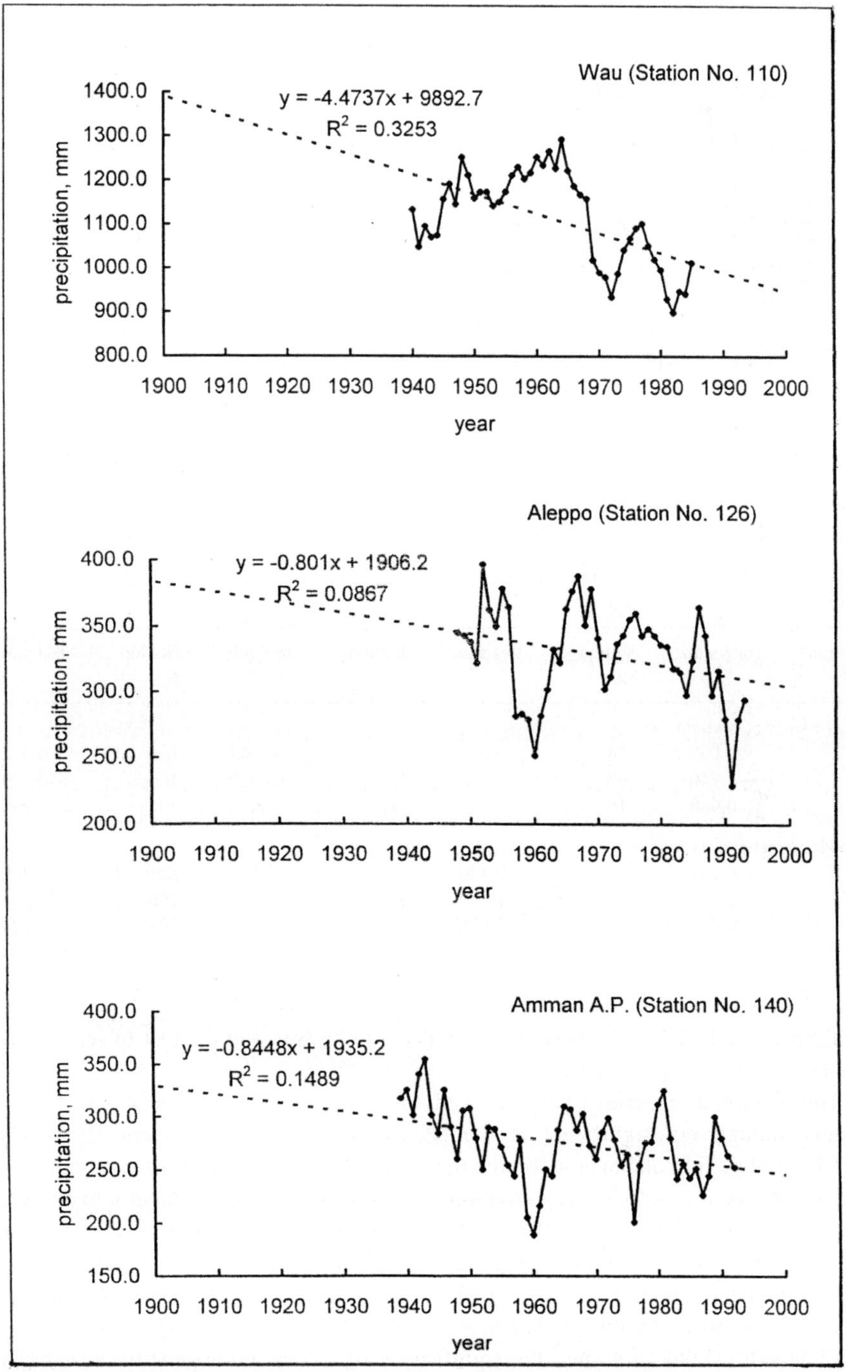

Figure 1(b). (Continued)

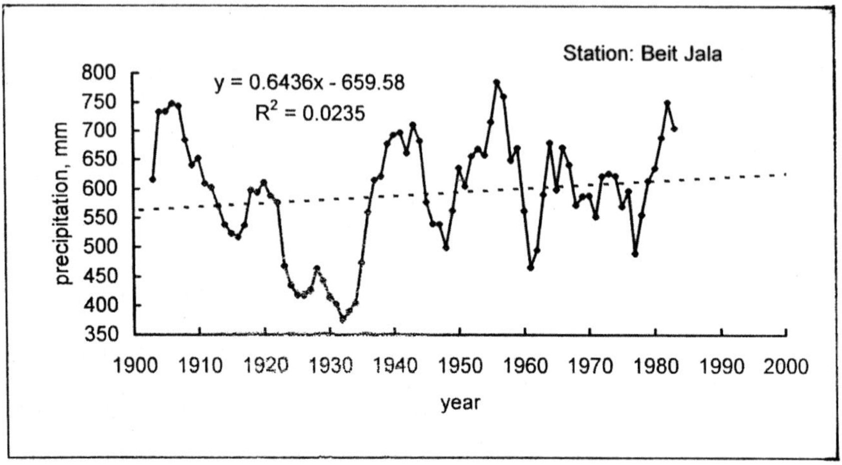

Figure 1(b). (Continued)

Table 2. Results of testing the significance of the trend line fitted to the moving averages of the annual rainfall series at 12 locations

Station No.	test stat.	Station No.	test stat.	Station No.	test stat.	Station No.	test stat.
Spearman's z_S-statistic							
6	−0.028	74	−0.216	113	0.560	Beit Jala	0.927
23	−1.540	97	1.390	126	−0.910	166	−0.335
66	0.968	110	−2.180	140	−1.615	198	0.948
Linear regression, t-statistic							
6	−0.141	74	−0.420	113	1.122	Beit Jala	1.240
23	−2.130	97	1.050	126	−0.871	166	0.111
66	0.833	110	−2.536	140	−0.989	198	0.294

or falling trend. A few of these trends appears to be true and the rest false, at least for the period of record used.

The precipitation series of Aleppo and Amman are among the ones with falling trends, though not significant at the chosen levels of significance. The study conducted by Al Shalabei et al. on the implications of expected climatic changes for the Syrian coast (1996) has reported that for each ° C global change in temperature the change by 2030 will be 0 to −2% and will remain so until 2100 for both the coastal plain and Montane region. The worst affected parts of the year are the summer and autumn seasons. The Global Climate Model results show that the respective seasonal rainfall will reduce by up to 22% and 18% respectively.

El Mahdi (1996) examined the significance of trends, homogeneity or stability of the mean and variance of annual rainfall series at 25 rain gauging stations. These

Table 3. Homogeneity (stability) of the means and variances of the subseries of annual rainfall before and after 1960

Station	Period	n, year	X_m, mm	s^2, mm^2	t_{comp}	t-test	F_{comp}	F-test
Nema	1931–60	30	313.03	8014				
No.6	1961–80	20	248.6	3742	2.790	*R.	2.035	**N.R.
Oran	1942–60	19	398.30	13155				
No. 23	1961–90	30	373.50	13079	0.739	N.R.	1.006	N.R.
Alexandria	1938–60	23	185.13	3981				
No. 66	1961–90	30	191.03	4670	0.322	N.R.	0.853	N.R.
Giza	1938–60	23	25.00	184				
No. 74	1961–90	30	19.67	167	1.459	N.R.	1.101	N.R.
Gadaref	1931–60	30	586.77	14062				
No. 97	1961–90	30	589.93	11295	0.109	N.R.	1.245	N.R.
Wau	1938–60	23	1168.60	23901				
No. 110	1961–87	27	1072.50	33854	1.938	R.	0.706	N.R.
Djibouti	1931–60	30	133.37	6356				
No. 113	1961–90	30	154.57	17369	0.571	N.R.	0.366	R.
Aleppo	1946–60	15	329.80	11965				
No. 126	1961–90	30	332.60	7770	0.093	N.R.	1.54	N.R.
Amman	1937–60	24	284.75	9429				
No. 140	1961–90	30	264.03	9866	0.776	N.R.	0.956	N.R.
Beit Jala	1931–60	30	597.20	32896				
	1961–85	25	612.32	32374	0.291	N.R.	1.016	N.R.
Manamah	1931–60	30	77.88	2048				
No. 166	1961–90	30	70.87	3261	0.527	N.R.	0.628	N.R.
Aden	1931–60	30	40.97	700				
No. 198	1961–88	28	50.89	2617	0.936	N.R.	0.277	R.

R = null hypothesis is rejected, N.R.** = null hypothesis is not rejected.*

stations are located within the area bounded by 4°52′ N and 19°35′ N latitudes, and 22°27′ E and 37°13′ E longitudes and elevation varying from 2 m to 805 m a.m.s.l.

The rainfall data series covering the period 1950–70 were compared to the rainfall data in the period from 1971 onward, assuming that the break in rainfall regime began around 1970/1971. Using the same statistical methods we mentioned earlier, i.e. the student t-test for the homogeneity of the means, the Fisher F-test for the homogeneity or the stability of the variances and the Spearman's rank correlation for testing the significance of the trend, the following results were obtained:

The hypothesis of homogeneity of the variances and means was not rejected in 21 and 9 out of 25 cases respectively. The null hypothesis that the data are trend-free was not rejected in 15 out of 25 cases, i.e. 40% of the series suffered from a significant trend.

It is an established fact that a state of severe meteorological drought, i.e. considerable reduction in precipitation, swept over vast areas in west and northwest Africa as from the second half of the nineteen-sixties. The drought moved gradually from west to east across the continent and became apparent in East Africa, especially Ethiopia and the Sudan, since the early 1970s to last 15 years or more. This state of affairs coupled with the results obtained from the statistical testing of the annual rainfall series have initiated to us to map the change in annual precipitation over the entire surface of the Sudan before and after the break in rainfall regime. The map in Figure 2 shows two sets of isohyetal lines, the solid lines represent the mean annual rainfall in the period 1950–67 and the dashed lines represent the mean for the period 1968–85. Most of the isohyetal lines show a general retreat of the annual rainfall from north to south, though not at the same rate at all locations. The example of the Sudan affirms to some extent the effect of climate change on precipitation in certain parts of the Arab Region in the second half of the 20th century.

4.1.2 Frequency distribution of annual rainfall

Frequency analysis of rainfall data is a useful tool for predicting rainfall parameters such as depth, intensity and duration given a certain probability of exceedance, P(x),

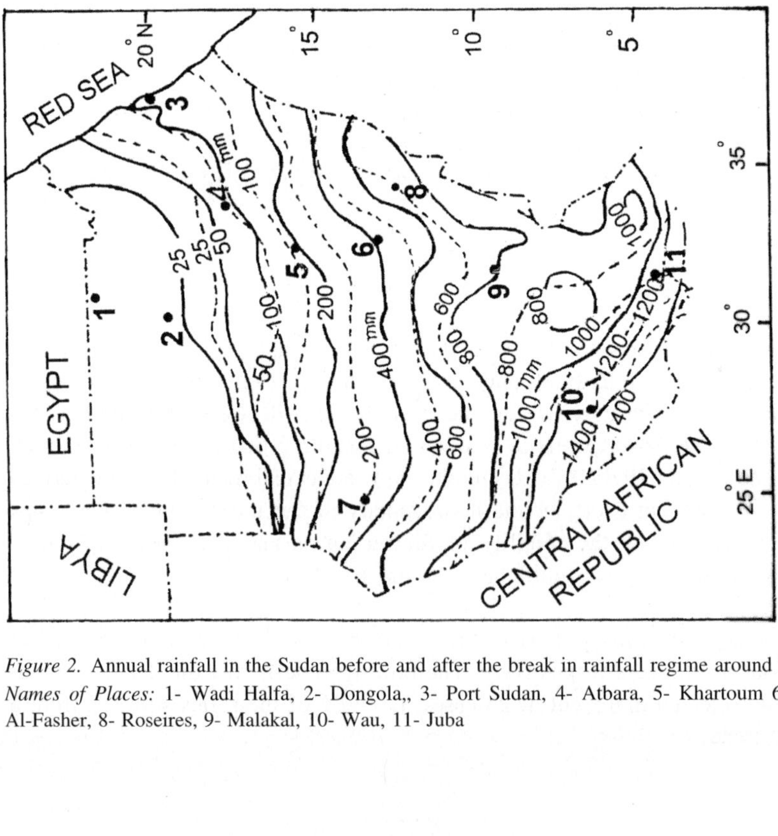

Figure 2. Annual rainfall in the Sudan before and after the break in rainfall regime around 1967/1968 *Names of Places:* 1- Wadi Halfa, 2- Dongola,, 3- Port Sudan, 4- Atbara, 5- Khartoum 6- Kosti, 7- Al-Fasher, 8- Roseires, 9- Malakal, 10- Wau, 11- Juba

non exceedance, $F(x) = 1 - P(x)$, or the corresponding return period, $T_r = 1/P(x)$. To do so, one has to search for the frequency distribution function that might serve as the best fit to the data in question. Several distribution functions are available for such a purpose. The two functions that have often proven to fit the rainfall data well are the log-normal and the Pearson Type III.

Upon excluding the rainfall data of Oran (No. 23) and Wau (No. 110), due to their failure in passing the trend-free test, we are left with 10 stations. From the literature we have borrowed the results of frequency analysis of rainfall data at seven additional stations in Syria (Al Saleh, 1983). As the variate values corresponding to P(x) of 99% and 95% were missing in the original results, we had to compute them using the same distribution function, Pearson Type III.

So, all in all, we have 17 series of trend-free annual rainfall. The Pearson III distribution showed to provide a better fit to the big majority of those series and the 2-parameter lognormal function to the remaining series. Examples of the two distributions are shown in Figures 3(a) and 3(b). The predicted annual rainfall is given in Table 4.

Before we leave this section, it might be of interest to add here the expressions of the Pearson Type III and the lognormal distribution functions.

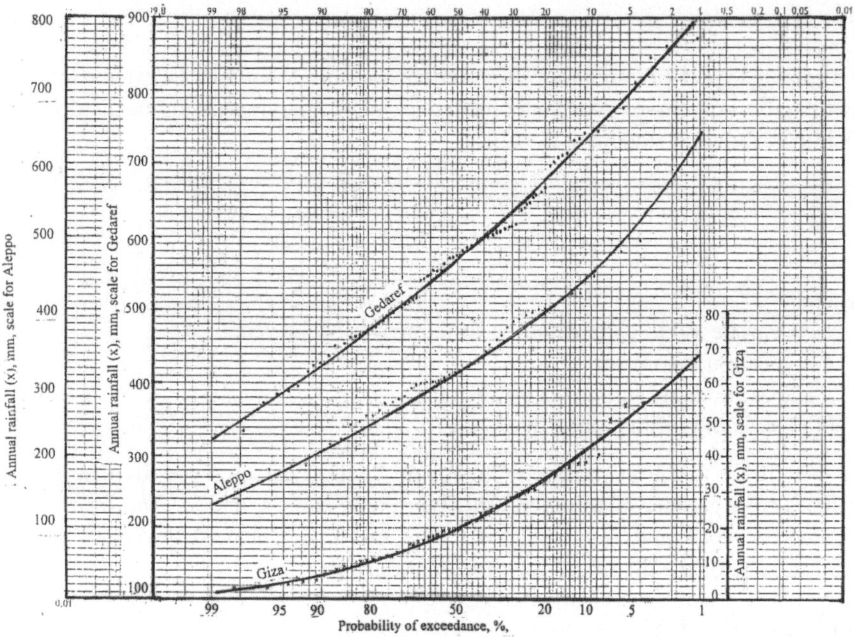

Figure 3(a). Fitting Pearson Type III function to stations No. 74, 97 and 126

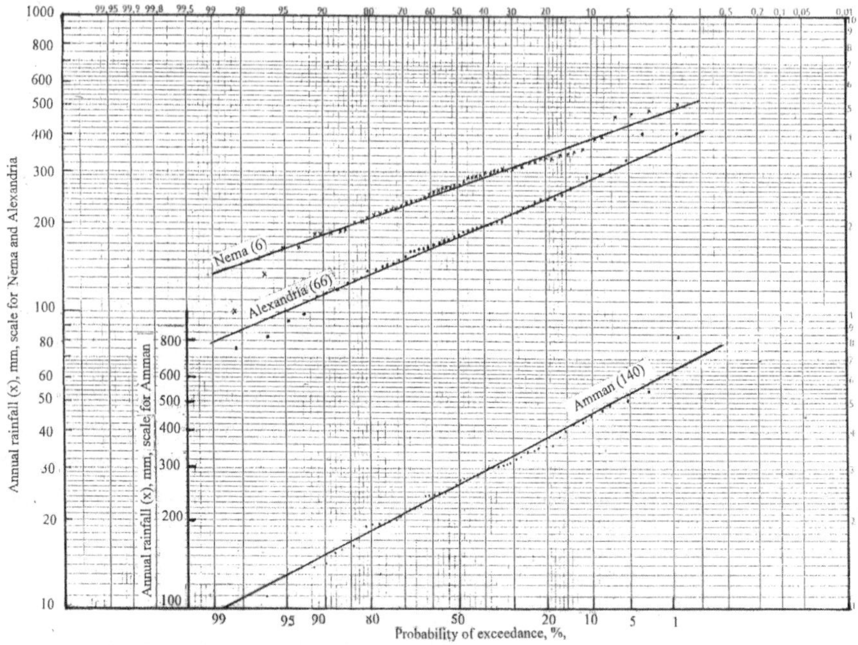

Figure 3(b). Fitting 2-parameter lognormal function to stations No. 6, 66 and 140

Pearson Type III distribution: If a random variable X follows a Pearson Type III distribution, sometimes goes under the name of three parameter gamma distribution, then the following statement is valid

$$(1) \qquad \Pr(X \leq x_T) = \int_A^{x_T} \frac{\lambda^\eta}{\Gamma(\eta)}(x-A)^{\eta-1}e^{-\lambda(x-A)}dx = 1 - \frac{1}{T}$$

where,

x_T = variate value corresponding to return period T, years,
A = location parameter,
η and λ are distribution parameters, and
$\Gamma(\eta)$ = Gamma function of η

In fact, if the location parameter A equals 0, the Pearson Type III distribution reduces to a gamma distribution with two parameters λ and η. The distribution parameters, η, λ and A are related to the mean, μ, standard deviation, σ, and skewness, γ, of the distribution by the relationships

$$(2) \qquad \eta = 4/\gamma^2$$

$$(3) \qquad \lambda = \sqrt{\eta}/\sigma$$

and

(4) $\mu = \eta/\lambda + A$

These relationships help to compute the frequency factor, k_T, needed for estimating the variate value, x_T, using the frequency factor method, which is given by the expression

(5) $x_T = \mu + \sigma k_T$

The frequency factor, k_T, is available in many pieces of literature (e.g. Water Resources Council Publication (WRC), 1967).

Lognormal distribution: The two-parameter lognormal distribution (zero lower bound) and, to somewhat less extent, the three-parameter distribution (arbitrary lower bound) are frequently used as probability models in frequency analysis of hydrologic data.

Table 4. Estimates of annual rainfall at different locations in the Arab Region corresponding to specified probabilities of exceedance

Station	Annual rainfall, mm, corresponding to given probabilities of exceedance, percent								
	99	95	90	80	50	20	10	5	1
Countries other than Syria									
Nema (6)*	132	164	182	210	268	342	388	438	535
Alexandria (66)	79	101	114	133	179	243	284	323	417
Giza (74)	2	4	7	10	20	33	42	50	68
Gedaref (97)	320	386	424	472	602	680	742	794	1000
Djibouti (113)	8	21	32	67	121	202	262	325	553
Amman (140)	94	130	151	182	263	380	462	542	825
Beit Jala	240	322	370	435	577	738	835	918	1090
Manamah (166)	1	14	19	32	67	115	150	182	255
Aden (198)	1	7	10	14	31	70	102	133	198
Syria									
Aleppo (126)	130	180	212	240	318	400	453	508	641
El Kom	38**	61	74	92	128	170	194	214	256
Deir ez Zor (129)	27	56	73	99	153	213	247	277	337
Palmyra (131)	33	61	77	99	143	189	214	236	277
Abu Kamal (132)	35	54	65	80	117	166	198	227	290
Sab 'Biar	23	31	37	46	85	164	226	290	446
J. Ettenuf	34	40	40	52	87	140	176	210	288
Zulof	8	20	28	40	78	142	187	232	336

**Figures between parentheses are the respective station numbers.*

***Rainfall figures for Syria written in italics are computed by the author*

If the logarithm of a random variable X is distributed like a normal, one can say that $Y = \ln X$ is normally distributed with parameters μ_y and σ_y^2, mean and variance of Y respectively. The distribution function $F(X)$ can be expressed by the relation.

$$F(X) = \Pr(X \leq x) = \frac{1}{\sigma_y\sqrt{2\pi}} \int_{-\infty}^{Y} e^{-\frac{1}{2}\left(\frac{Y-\mu_y}{\sigma_y}\right)^2} \tag{6}$$

Eqs. (6)–(10) express some of the important properties of the lognormal distribution.

$$\mu = e^{\mu_y + \frac{1}{2}\sigma_y^2} \tag{7}$$

$$\sigma^2 = \mu^2\left(e^{\sigma_y^2} - 1\right) \tag{8}$$

$$\gamma = \eta_v^3 + 3\eta_v \tag{9}$$

and

$$\kappa = \eta_v^8 + 6\eta_v^6 + 15\eta_v^4 + 16\eta_v^2 + 3 \tag{10}$$

where,

μ = mean of the original (untransformed) data, X,
σ = standard deviation of the raw data, X,
η_v = coefficient of variation = σ/μ,
γ = skewness of X-data, and
κ = kurtosis of X-data

Further elaboration of the relationships describing the lognormal function together with the substitution

$$z'/T = \frac{\ln x_T - \mu_y}{\sigma_y} = \frac{y_T - \mu_y}{\sigma_y} \tag{11}$$

leads to an expression of the frequency factor, k_T, as

$$k_T = \frac{1}{\eta_v}\left(\exp(-\frac{1}{2}\ln(1+\eta_v^2) + z'/T\sqrt{\ln(1+\eta_v^2)}) - 1\right) \tag{12}$$

Chow (1964) used Eq (12) for computing and tabulating the values of the frequency factor, k_T, for different values of return period, T, and coefficient of variation, η_v, or skewness, γ.

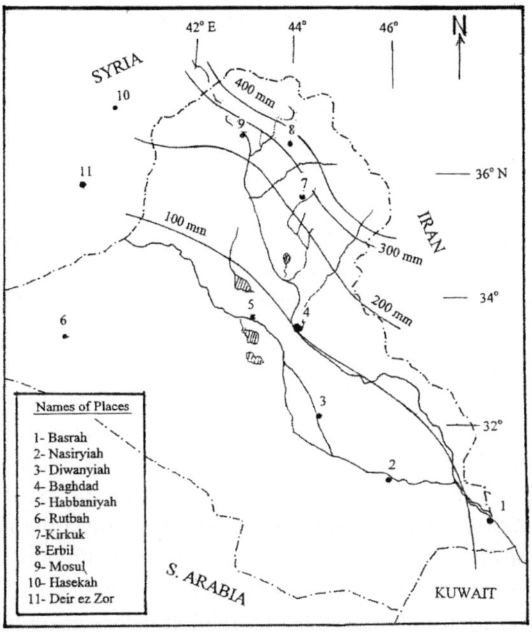

Figure 4. Map of Iraq showing isohyetal lines of probable mean total precipitation during four years out of five years (after Zaki et al., 1970)

Wallén (1967) investigated the water balance between annual rainfall and potential evapotranspiration figures in winter and summer in 12 subregions distributed between Jordan, Syria, Iraq, Iran and Azerbaijan. It is generally conceived that successful rain-fed agriculture is possible only in areas receiving adequate rainfall during four out of five years. With this aim in view, the Government of Iraq (1970) prepared a map showing the areas with probable mean total amount of precipitation in four years each pentad. Bearing in mind that the minimum precipitation needed to grow cereals is between 200 and 250 mm y^{-1}, rain-fed agriculture is thus confined to the northern and northeastern parts of Iraq (Figure 4)

4.2. SEASONAL AND MONTHLY RAINFALL

4.2.1 Wet season, wettest month and its contribution to total rainfall, and number of rainy days

The wet or rainy season has been arbitrarily taken as an uninterrupted period of successive months with monthly rainfall not less than 5 mm. A few, small parts of the Arab region exercise more than a single rainy season. The number of rainy days is the number of wet days with rainfall at or exceeding a certain preset value like 0.1, 0.25 or 0.5 mm. In our case the lowest limit has been specified as 0.1 mm. The reader should be well aware of the fact that the number of rainy days for the same location differs considerably depending on the period and number of years of record. It goes without saying that a period of record falling in a wet episode will have a lot more rainy days than if it had fallen in a dry episode. The number of rainy days given in Table 5 has been obtained from several sources (Ministry of War and Marine, Egypt , 1950, Meteorological Service of Algeria, 2002, Müller et al., 1996 and KNMI, 2002). It is not unusual to observe some discrepancies between one source and another regarding the mean number of wet days for the same location. As such, the number of wet days listed in Table 5 should be used with care.

The results given in Table 5 also include the wettest month of the year and its contribution in percent to the total annual rainfall. The stations located between 36° and 33° N latitudes, Morocco, Algeria and Tunisia in the west and Syria, Lebanon and northern and central Iraq in the east, show that the rainy season generally occupies the period from September to June. The wettest month is December/January, and its contribution to the annual rainfall ranges from say 14% to 18%. The duration of the rainy seasons decreases gradually in an easterly direction. It lasts from October to March/April in Libya and Egypt. Here too, the wettest month is December/January, but its contribution to the total rainfall increases up to 30%. The contribution of the wettest month increases further to more than 40% in central Sudan then falls down to an average value of 20% in southern Sudan where the annual rainfall is in the range of 1,000–1,200 mm. The wettest month in both central and southern Sudan is August.

The rainy season in Mauritania, extreme west of the Arab Region, lasts from May/June to October/November, with August or September as the wettest month. The mean percentage of rainfall in the wettest month to the annual rainfall is about 40%. The duration of the season in the Arabian Peninsula changes to Nov-Apr/May, with December or January as the wettest month, which is not usually the case in Yemen.

4.2.2 Variability of monthly rainfall, wet and dry spells

Due to intricate weather processes the rain that falls over a certain point is seldom uniformly distributed between the months of the year. In general it varies from month to month during the same one year causing one month to be wetter or drier than the month before or the month after. A succession of wet months is often referred to as a "wet spell" and a succession of dry months as a "dry spell". The term wet spell indicates to water excess, surplus, flood and humid period whereas the term dry spell indicates to shortage, drought and dry period. The rainfall in a certain month of the year is likely to vary from year to year. As such, monthly rainfall, given a certain period of record, shows a minimum as well as a maximum value. The ratio between these two extreme values varies from month to month, and for the same

Table 5. Number of rainy days, wettest month of the year and its contribution to the annual rainfall as recorded at some stations in the Arab Region

Stat. No.	Rainy season	No. R.D.	Wet. mon.	% of A.R.	Stat. No.	Rainy season	No. R.D.	Wet. mon.	% of A.R.
1	Jun-Nov	15	Sep	31	102	May-Oct	43	Aug	36
2	Jun-Oct	25	Aug	38	105	Apr-Oct	79	Aug	27
3	Jul-Nov	18	Aug	52[c]	107	Apr-Oct	77	Aug	22
4	Jun-Oct	24	Aug	37	108	Mar-Nov	88	Aug	22
5	Jun-Oct	N.A.	Aug	38	110	Mar-Nov	106	Aug	21
6	May-Oct	25	Aug	38					
7	Sep-Jun	77	Nov	16[c]	111	Feb-Dec	118	May	15
8	Sep-Jun	64	Dec	19c	113	Aug-Apr	29	Mar	19[c]
9	Sep-Jun	72	Dec	17	116	Jan-Nov	N.A.	Sep	26
10	Oct-Jun	69	Dec	17[c]	117	Nov-Jan,	2	Apr	23[c]
						Mar-May			
11	Sep-Jun	N.A.	Dec	15	118	Feb-Nov	61	Aug	20
12	Sep-May	77	Dec	20[c]	119	Oct-Nov	18	May	40
13	Jan-Dec	N.A.	Dec	15		Apr-May			
14	Aug-Jun	51	Nov	15	120	Oct-Jan	19	Oct	23[c]
15	Aug-May	35	Nov	17		Mar-May			
16	Oct-Apr	36	Dec	21[c]					
17	Sep-May	76	Dec	17[c]	121	Oct-Dec	N.A.	Apr	36
18	Sep-Jun	N.A.	Nov	17[c]		Mar-May			
19	Sep-May	96	Dec	17[c]	122	Mar-Dec	65	Jun	21[c]
20	Sep-Jun	77	Jan	18[c]	123	Mar-Nov	71	May	22[c]
					C.I.	Jan-Dec	170	Jan	13[c]
22	Sep-May	29	Nov	14	124	Oct-Jun	N.A.	Jan	17
23	Sep-May	44	Jan	18[c]	126	Oct-Jun	69	Jan	18
24	Sep-Jun	76	Mar	12	128	Sep-Jun	N.A.	Jan	23[c]
26	Sep-Feb	28	Jan	13	129	Oct-May	40	Jan	19
27	Nov-Jan	12	Dec	20					
					131	Oct-May	N.A.	Jan	16
31	Sep-May	101	Feb	17[c]	133	Oct-May	48	Jan	19
33	Sep-May	85	Jan	15[c]	134	Sep-May	74	Jan	26[c]
35	Sep-Jun	N.A.	Feb	13	137	Oct-May	66	Jan	21[c]
36	Sep-May	53	Dec	16[c]	138	Oct-Apr	52	Dec	26
37	Sep-May	41	Nov	16[c]	139	Oct-Apr	60	Dec	25
38	Sep-May	29	Oct	13	140	Oct-Apr	48	Feb	26
39	Sep-May	N.A.	Nov	14					
40	Oct-May	31	Nov	18[c]	141	Oct-May	38	Dec	28
					142	Oct-Apr	51	Jan	23
42	Sep-May	19	Mar	26	143	Oct-May	N.A.	Dec	24
43	Sep-Apr	47	Dec	23[c]	144	Oct-May	86	Dec	24
44	Sep-May	75	Jan	25	146	Oct-Apr	42	Jan	25
46	Sep-May	39	Nov	22[c]	148	Oct-Apr	34	Jan	27[c]
47	Sep-May	N.A.	Dec	27	149	Oct-May	54	Feb	18
48	Sep-Apr	46	Jan	26[c]					
					151	Oct-May	N.A.	Jan	18
49	Sep-May	57	Dec	26[c]	152	Oct-May	34	Jan	20

(continued)

Table 5. (Continued)

Stat. No.	Rainy season	No. R.D.	Wet. mon.	% of A.R.	Stat. No.	Rainy season	No. R.D.	Wet. mon.	% of A.R.
50	Sep-May	22	Mar	19	153	Oct-May	28	Dec	17
					154	Oct-May	N.A.	Jan	21
51	Oct-Mar	N.A.	Jan	24	156	Oct-May	N.A.	Jan	22
53	Sep-Mar	28	Jan	26[c]	157	Nov-May	N.A.	Dec	28
54	Oct-Mar	12	Jan	30	159	Nov-May	31	Jan	24[c]
55	scattered	11	Jan	18					
56	scattered	14	Oct	18	163	Nov-Apr	30	Dec	22[c]
60	Oct-Apr	22	Dec	26[c]	166	Dec-Apr	15	Jan	24[c]
61	Oct-Mar	30	Jan	19[c]	171	Nov-May	17	Nov	23
62	Oct-Apr	31	Jan	23[c]	173	Nov-May	17	Apr	29
63	Oct-Apr	34	Jan	30[c]	176	Nov-Jan	7	Dec	39c
64	Oct-Mar	35	Jan	24[c]	177	Jan-Dec	23	Nov	25
65	Oct-Mar	32	Dec	22[c]	179	Nov-Apr	9	Dec	30[c]
66	Oct-Mar	47	Dec	30[c]	180	Nov-apr	22	Jan	26[c]
67	Oct-May	20	Jan	22[c]					
68	Nov-Mar	17	Jan	20	183	Jun-Aug	26	Jul	32[c]
69	Oct-Apr	29	Jan	26	186	Jan-Oct	N.A.	Apr	22
70	Nov-Mar	18	Dec	21	187	Nov-Feb, Jul-Aug	14	Dec	25[c]
71	Dec-Mar	10	Dec	22	190	Feb-Apr, Oct-Nov	12	Apr	32
91	Oct-Jan	15	Nov	49[c]					
92	Oct-Jan	14	Dec	20[c]	192	Jan-Nov	47	Aug	19
93	Jul-Sep	11	Aug	48	193	Feb-Oct	27	Sep	28
94	Jun-Sep	19	Aug	46	195	Mar-Nov	32	Sep	25
95	May-Oct	35	Aug	36	198	Dec-May	13	Feb	14[c]
96	May-Oct	42	Aug	36	199	scattered	12	Sep	26[c]
97	Apr-Oct	59	Aug	31	195	Mar-Nov	32	Sep	25
99	May-Oct	35	Aug	45	198	Dec-May	13	Feb	14[c]
100	May-Oct	49	Aug	34	199	scattered	12	Sep	26[c]

Abbreviations

R.D. = Rainy days with at least 0.1 mm rainfall, Wet. Mon. = wettest month of the year and A.Rainfall. The superscript c refers to the respective station as being coastal.

one month it varies depending on the period of record. This situation is often dealt with statistically by computing the mean and variance for each month of the year.

In the next paragraphs we shall review two case studies from Libya and the West Bank, Palestine. Both cases describe some of the methods used for analyzing monthly rainfall data, and the results obtained therefrom.

Case study from Libya: Nine gauging stations located along or near the coast of the Mediterranean in Libya were selected for identifying some of the characteristics of wet and dry spells and their durations over the study area (Jadid and Şen, 1997a). Those nine stations comprise in addition to our stations No. 43, 48, 46,49, 53 and 54, the gauging stations at Zuwarah, El-Khums and Abugreen. Monthly mean and standard deviation of rainfall for 6 consecutive years, except for stations 43 and 48 where the record used was 54 and 37 years respectively, have been worked out for identifying the wet and dry spells. The procedure used is based on counting the number of spells at five levels.

These are; X_m, $X_m \pm s$, $X_m \pm 2s$, where X_m and s are the mean and standard deviation of the data respectively. The number of dry spells at all levels was found to exceed the number of wet spells. The probability of the wet and dry months over the study area is 0.742 for a dry month and 0.258 for a wet month. However, the probabilities of their respective durations are 0.428 and 0.572.

The same authors, i.e. Jadid and Şen (1997b), used the monthly-observed mean rainfall data for determining the parameters of the gamma distribution function. The number of months used for that purpose varied from 120 for stations No. 46 and 54 to 647 for station No. 43. The function has two parameters, the scale parameter, λ, and the shape parameter, η. The relationships linking these two parameters with the statistical descriptors of the data are as follows:

(13) $$f(x) = \frac{\lambda^{\eta}}{\Gamma(\eta)} x^{\eta-1} e^{-\lambda x} \text{ If } x \geq 0$$

$= 0$ elsewhere

(14) $$\mu_x = \eta/\lambda$$

and

(15) $$\sigma_x^2 = \eta/\lambda^2$$

where,

$f(x)$ = Probability density function,
$\Gamma(\eta)$ = Gamma function,
μ_x = Mean of x and
σ_x^2 = Variance of x

One should not forget that the gamma distribution function is a special case of Pearson Type III distribution function where the location parameter, A, is equal to zero. The shape and scale parameters were found to increase steadily in an easterly direction from 0.213 to 4.841 for λ and from 0.041 to 0.186 for η. As an example of the results obtained from this method, the probability that the monthly rainfall at Tripoli (Station No.43) exceeds 10, 25, 50 and 100 mm is 67, 36.8, 13.5 and 1.8% respectively.

Case study from the West Bank, Palestine: Husary et al. (1995) analyzed rainfall data in the northern West Bank, Palestine. Analysis of monthly rainfall was carried out using the data available at 15 gauging stations. One of those stations, (No. 143), is at Nablus. Results of statistical analysis of monthly rainfall for a period of 17 years, 1971–1993 with 6 missing years, are presented in Table 6. Despite the short length of record, 17 years, all months, except the dry ones, show a considerable

Table 6. Results of statistical analysis of monthly rainfall totals in the period 1971/93° at Nablus, West Bank, Palestine (Husary et al., 1995)

Item*	Mean monthly rainfall, mm, for Jan	Feb	Mar	Apr	May	Jun	Jul	Aug	Sep	Oct	Nov	Dec
X_m	155	135	90	34	5	0	0	0	2	17	60	158
s	83.0	107.0	33.1	53.9	9.0	0.1	0.0	0.4	5.7	20.3	50.1	120
C_v	0.54	0.79	0.37	1.60	1.73				3.05	1.22	0.83	0.76
C_s	1.79	1.96	0.63	3.42	1.92				3.75	2.61	0.94	1.42
C_k	7.03	7.95	3.61	15.40	6.00				16.9	11.2	3.99	5.92
X_{min}	55.4	20.4	41.3	0.0	0.0	0.0	0.0	0.0	0.0	0.0	2.9	9.5
X_{max}	389.3	444.7	155.5	224.6	28.5	0.0	0.0	0.0	22.4	82.6	169	472.2
P_{20}^{x}	85.1	45.7	62.6	zero	zero	0.1	zero	zero	zero	0.4	18.3	57.1
P_{80}^{x}	225	225	118	79	12.8	0.2	0.0	0.4	6.9	34	102	259

Explanation

° *Data from 1982 to 1987 are missing.*

* X_m = *mean*, s = *standarddeviation*, $C_v = s/X_m$, C_s = *skewness* = 3*rd moment*/(2*nd moment*)$^{1.5}$ *adjusted by* $n^2/(n-1)(n-2)$, C_k = *Kurtosis* = 4*th moment*/(2*nd moment*)^2 *adjusted by* $n^3/(n-1)(n-2)(n-3)$. X_{min} = *smallest value in record,*
X_{max} = *largest value in the record.*

P_{20} and P_{80}^{x} are monthly rainfall values corresponding to probabilities of non-exceedance *of 20% and 80% respectively, assuming the monthly rainfall to be normally distributed.*

difference between the largest and smallest values. The ratio X_{max}/X_{min} for April, May, September and October is infinite, as X_{min} for these months is zero. November and December show X_{max}/X_{min} in the range 50–60 whereas the same ratio for January-March is in the range 4–22. These values will undergo some change when the record becomes longer.

Estimates of monthly rainfall corresponding to 20% and 80% probabilities of non-excceedance, assuming the rainfall to be normally distributed, are also included in that table. The author upon revising the analysis found that the skewness and kurtosis of the data are too different from zero and 3.0 respectively. As such, the assumption of normal distribution cannot be justified. The relationships between the coefficients of variation, skew and kurtosis listed in Table 6 suggest that Pearson Type III will serve as a better fit to the observed rainfall. However, the discrepancy between the estimates using the normal and the Pearson Type III distributions is negligible in the central range between the chosen probability levels, i.e. 20% and 80%, beyond which the discrepancy grows larger. The normal distribution estimates for 1% and 99% non-exceedance probability levels, for example, are 437 and −121 mm respectively whereas the Pearson Type III distribution gives for these two probabilities in their order 550 and 0 mm.

Time series analysis and modeling consider chronological sequences of hydrological events. The importance of time series modelling lies in three major aspects: i- evaluation of irregularities in the record in question, ii- solving problems related to frequency analysis, and iii- data generation. Investigation of rainfall in the northern part of the West Bank, Palestine, included the application of the ARIMA models developed by Box & Jenkins (1976) to the monthly rainfall data. This class of models was basically adopted because of the non-stationarity of the series. The basic model used in fitting the time series is:

$$X_t = \frac{\theta(B)\theta_s(B)}{\phi(B)\phi_s(B)(1-B)^d(1-B^s)^D} Z_t \tag{16}$$

where,
X_t = rainfall at month t,
s = length of seasonality,
Z_t = Independent random term,
B = backward operator,
B^s = seasonal backward operator,

$$\theta(B) = 1 - \theta_1 B - \theta_2 B^2 - \ldots . - \theta_q B^q \tag{17}$$

$$\phi(B) = 1 - \phi_1 B - \phi_2 B^2 - \ldots .. - \phi_p B^p \tag{18}$$

$$\theta_s(B) = 1 - \theta_1^s B^s - \theta_2^s B^{2s} - \ldots . - \theta_Q^s B^{Qs} \tag{19}$$

$$\phi_s(B) = 1 - \phi_1 B^s - \phi_2 B^{2s} - \ldots . - \phi_P B^{Ps} \tag{20}$$

$\theta(B)$ = non-seasonal moving average operator,
$\theta_s(B)$ = seasonal moving average operator,
$\varphi(B)$ = non-seasonal autoregressive operator,
$\phi_s(B)$ = non-seasonal autoregressive operator,
p = order of non-seasonal autoregressive term,
q = order of non-seasonal moving average term,
d = order of non-seasonal differencing
P = order of seasonal autoregressive term,
Q = order of seasonal moving average term, and
D = order of seasonal differencing

The ARIMA model, often referred to as (p,d,q) (P,D,Q) model, when applied to the monthly rainfall data of Nablus has shown that (5,1,0)(3,1,0) fits adequately. The model in this particular case, owing to the absence of a moving average component, reduces to just an ARI (5,1)(3,1) model. Additionally, the order of the non-seasonal autoregressive component seems to outweigh the order of the seasonal component. It is necessary to mention here the need for a much longer series with no gaps in it in order to verify the extent of suitability of the model for generating data, and more importantly for estimating the model parameters accurately. The effect of length of record on some of the statistical properties of the data is presented in the next case study.

Case study from Algeria: The main objective of this case study is to present to the reader a realistic picture of the effect of the period of record and its length or duration on some of the statistical descriptors of the data. As an example we shall present here the variation of the mean, standard deviation, X_{max}, X_{min} and $X_{max} - X_{min}$ using the monthly rainfall data of Oran (Station No. 23), Algeria. The results given in Table 7 cover the months January-April and the periods 1993–2002, 1983–2002, 1973–2002, 1963–2002, 1953–2002 and 1943–2002.

The results given in Table 7 show that the arithmetic mean, X_m, is the descriptor with least variation or largest stability. The standard deviation, s, of the data, but not necessarily the coefficient of variation, C_v, follows the mean. Though not presented in Table 7, the skewness and kurtosis are likely to suffer the most. These two descriptors, next to the second statistical moment, depend on the third and fourth moments respectively. As such, upon selecting the distribution function that might serve as a good fit to the observations in the record, the two parameter distributions like the normal, lognormal or gamma are often more favourable than the three or four parameter distributions. Additionally, the reliability in parameter estimation decreases with the number of parameters. This is in contradiction with the flexibility of the distribution.

Table 7. Variation of some of the statistical properties of monthly rainfall in the period 1943–2002 with length of record at Oran (Station No. 23), Algeria

Month	Period of record	Length of record, y	Xm, mm	s, mm	Cv	Xmax, mm	Xmin, mm	Xmax-Xmin, mm
Jan.	1993–2002	10	36.54	98.98	2.71	82.1	0.9	81.2
	1983–2002	20	45.48	33.98	0.75	120.1	0	120.1
	1973–2002	30	42.66	32.25	0.76	120.1	0	120.1
	1963–2002	40	44.03	32.39	0.74	120.1	0	120.1
	1953–2002	50	49.68	36.08	0.73	158.8	0	158.8
	1943–2002	60	55.15	41.50	0.75	158.8	0	163.7
Feb.	1993–2002	10	49.41	39.81	0.81	107.6	0	107.6
	1983–2002	20	43.56	33.08	0.76	107.6	0	107.6
	1973–2002	30	45.9	32.56	0.71	107.6	0	107.6
	1963–2002	40	44.78	32.55	0.73	107.6	0	107.6
	1953–2002	50	47.69	38.03	0.80	144.0	0	144.0
	1943–2002	60	45.57	36.49	0.80	144.0	0	144.0
Mar	1993–2002	10	33.65	31.94	0.94	96.7	0	96.7
	1983–2002	20	41.67	35.2	0.84	119.2	0	119.2
	1973–2002	30	42.77	37.06	0.87	125.7	0	125.7
	1963–2002	40	45.12	35.21	0.78	125.7	0	125.7
	1953–2002	50	44.79	34.07	0.76	125.7	0	125.7
	1943–2002	60	42.04	34.96	0.83	125.7	0	125.7
Apr	1993–2002	10	28.84	16.76	0.58	57.2	0	57.2
	1983–2002	20	27.34	29.76	1.09	134.4	0	134.4
	1973–2002	30	37.10	38.02	1.02	159.6	0	159.6
	1963–2002	40	38.82	36.62	0.94	159.6	0	159.6
	1953–2002	50	36.96	34.13	0.92	159.6	0	159.6
	1943–2002	60	38.6	34.85	0.90	159.6	0	159.6

The largest and smallest values, i.e. X_{max}, X_{min} respectively, and thereupon the range $X_{max} - X_{min}$. are among the data descriptors that are likely to change considerably with increasing length of record. In many places in the Arab Region the minimum rainfall, X_{min}, can easily fall down to zero, even in months comprising the wet season. This fact is not exclusively confined to North Africa. It applies as well to many more parts of the region. One can find examples of this in Djibouti Serpent (Station No. 113), Djibouti, and Berbera (Station No. 117), Obbia (Station No. 120) and Chismaio (Station No. 123), Somalia. The maximum rainfall, on the contrary, will probably continue rising with number of years in the record. The chance that it remains unchanged is quite remote.

While describing the climate of humid, steppe and semi-arid regions of North Africa, Griffiths (1972) presented a summary of the monthly rainfall over 30 years in Jendouba (formerly Souk el-Arba, Station No. 34), Sfax (Station No. 37) and Gafsa (Station No. 38), Tunisia. As X_{min} happened to be often zero, the given rainfall data of the said stations, included X_m and X_{max} without any mention to X_{min}.

The ratio of the maximum rainfall to the long-term mean in case of stations No. 23, Algeria, and No. 34, Tunisia, both in relatively humid areas, varied between 2 and 4 for the months of the wet season, and. increased up to 6 in summer months. Values of the same ratio for the steppe, and semi-arid areas showed to be larger than those for the humid areas in the same region. In the steppe and semi-arid areas too, the ratio X_{max}/X_m for summer months was substantially greater than for winter months. The ratio in question increased up to 15–20 in July and August.

4.3. LONG AND SHORT DURATION RAINFALLS

4.3.1 Formulation of intensity-duration and intensity-duration-frequency relationships

The relationship between rainfall intensity, i, and duration, t, can be expressed as:

(21) $i = ct^{-n}$ for durations longer than 2-hours

and

(22) $i = \dfrac{a}{t+b}$ for a few minutes $< t <$ 2–3 hours

where a, b, c and n are constants to be determined locally, and i and t are in mm h^{-1} and h respectively.

The mean maximum intensity, i_m, of rainfall at Skikda, Algeria, has been described by (Belloum, 1993) for the period 1958–1987 using the law of Montana as:

(21') $i_m = at^{n-1}$

where a and n are parameters whose values depend on the frequency or return period, T, y. The values of a are 26, 28, and of n 0.31, 0.35 and 0.38 both for $T = 2$, 5 and 10 y respectively.

For countries without adequate rainfall observations, Bell (1969) has derived generalized rainfall-duration-frequency relationships.

The ratio of the rain depth, R_T^t, of t-min duration and T-y return period to rain depth, R_{10}^t, of t-min duration and $T = 10$-y return period is given by:

$$\text{(23)} \qquad \frac{R_T^t}{R_{10}^t} = 0.21 \ln T + 0.52 \qquad \text{for} \qquad 2 \le T \le 100 \text{ years}$$

Depth –duration ratios as a function of duration are given by:

$$\text{(24)} \qquad \frac{R_T^t}{R_T^{60}} = 0.54 t^{0.25} - 0.50 \qquad \text{for} \qquad 5 \le t \le 120 \text{ min}$$

where R_T^{60} is the T-y, 1-h fall.

Eqs. (23) and (24) can be combined to give the generalized expression for rainfall depth-duration-frequency:

$$\text{(25)} \qquad R_T^t = (0.35 \ln T + 0.76)(0.54 t^{0.25} - 0.50) R_{10}^{60}$$

The comparable generalized formula based on the 2-y, 1-h rainfall is:

$$\text{(25')} \qquad R_T^t = (0.35 \ln T + 0.76)(0.54 t^{0.25} - 0.50) R_2^{60}$$

Both Eqs. (25) and (25') are applicable for $2 \le T \le 100$ years and $5 \le t \le 120$ min. Eq. (25') is considered, however, less reliable, particularly for longer return periods (Shaw, 1983).

4.3.2 Daily and maximum 24-h rainfall

Before embarking on the analysis and discussion of 'daily rainfall' one should always distinguish between the fall in a calendar day and a period of 24 consecutive hours, which possibly occupies parts of two calendar days. The maximum 24-h rainfall is mostly referred to as the 'annual maximum daily rainfall (AMDR)'. Quite often it exceeds the fall in a calendar day and seldom equals it. In arid areas one should not be surprised if the AMDR is equal to or about the same as the total rainfall in the month in which it has occurred. Table 8 shows clearly that the observed annual maximum 24-h rainfall at several locations in the Arab countries comprises a considerable percentage of the maximum monthly rainfall in the same month of occurrence for a given sequence of years.

Case study from the northern part of the West Bank, Palestine- Frequency analysis of AMDR series at 13 stations has been investigated by Husary et al. (1995). The

Table 8. Maximum 24-hour rainfall, month of occurrence and corresponding maximum monthly rainfall at selected stations in the Arab Region

No.	Station Name	Maximum rainfall, mm 24-h	month	Month of occurrence	No.	Station Name	Maximum rainfall, mm 24-h	month	Month of occurrence
1	Atar	69	121	Sep.	102	El-Obeid	97	325	Jul.
3	Nouakchott	49	49	Oct.	105	Roseires	116	332	Aug.
6	Nema	125	312	Aug.	107	Malakal	176	345	Sep.
7	Tanger	122	374	Dec.	110	Wau	125	346	Aug.
8	Melilla	221	N.A.	Jun.	111	Juba	138	318	May
9	Oudja	70	109	Feb.	113	Djibouti	211	211	Mar.
10	Rabat	151	153	Apr.	117	Berbera	132	145	Mar.
12	Casablanca	72	76	May	118	Hargeisa	61	229	Mar.
13	Ifrane	163	524	Dec.				179	Aug.
14	Marrakech	63	139	Apr.	120	Obbia	70	N.A.	Mar.
16	Agadir	123	193	Dec.				N.A.	Dec.
17	Alger	147	284	Nov.	122	Mogadiscio	150	324	May
18	C. Carbon	66	N.A.	Jan.	123	Chismaio	>150	N.A.	May
23	Oran	81	175	Feb.				N.A.	Jun.
24	Tebessa	33	N.A.	Apr.	126	Aleppo	58	110	Feb.
30	Tamanrasset	48	95	May.				230	Nov.
31	Bizerte	135	N.A.	Oct.	129	Deir exZor	97	97	Jan.
33	Tunis	134	276	Dec.	133	Damascus	43	91	Dec.
34	Jendouba	85	143	Oct.	137	Beyrouth	140	N.A.	Oct.
35	Kairouan	112	N.A.	Oct.	140	Amman	79	142	Feb.
37	Sfax	95	144	Sep.	146	Jerusalem	99	418	Jan.
38	Gafsa	67	126	Oct.	148	Gaza	78	N.A.	Nov.
40	Gabes	105	320	Oct.				N.A.	Dec.
43	Tripoli	129	293	Nov.	152	Baghdad	56	149	Mar.
46	Derna	61	173	Nov.	153	Rutbah	46	46	Dec.
48	Misurata	78	222	Oct.	163	Kuwait	56	90	Nov.
49	Benina	43	200	Nov.	166	Manamah	71	147	Nov.
53	Sirte	67	67	Oct.	171	Hail	64	107	Jan.
54	Ajdabiyah	28	N.A.	Dec.	173	Er-Riyadh	61	61	Mar.
55	Ghadames	17	21	Mar.	176	Jidda	140	N.A.	Dec.
59	Kufrah	11	11	Aug.	179	Sharja	109	113	Nov.
61	Sallum	121	227	Nov.	180	Muscat	79	143	Jan.
64	M. Matruh	99	N.A.	Dec.	182	Masirah Isl.	53	N.A.	Dec.
66	Alexandria	63	154	Nov.	183	Salalah	43	N.A.	Oct.
73	Cairo	44	54	Dec.	187	Kamaran Isl.	76	N.A.	Aug.
91	Port Sudan	112	182	Nov.	190	Rayan	104	N.A.	Nov.
94	Khartoum	200	301	Aug.	198	Aden	25	N.A.	Mar.
99	El-Fasher	128	265	Jul.	199	Barim Isl.	43	N.A.	Dec.

historic rainfall data, which covered a period ranging from 7 to 39 years, has shown that the number of wet days range from 38 to 60 days. These numbers bring the ratio of wet days, WD, to dry days, DD, to 11.6–19.7%. The number of days with rainfall less than 0.5, 25 and 50 mm has been found to occupy 87, 98 and 99% of the total days year, respectively.

The Gumbel EV Type I distribution function and the exponential distribution function both have been worked out, and the results obtained therefrom compared to test their goodness of fit to the historical values. The comparison for 11 out of 13 stations has been in favour of the Gumbel EV Type I function. The AMDR value, x_T, estimated from this function can be written as follows:

$$x_T = \frac{1}{\alpha}(-\ln(-\ln(1-\frac{1}{T}))) + \beta \tag{26}$$

$$\alpha = 1.282/\sigma \tag{27}$$

$$\beta = \mu - \frac{0.5772}{\alpha} \tag{28}$$

where,

α and β are parameters of the distribution function as expressed by Eqs. (27) and (28), respectively,

μ and σ are the mean and standard deviation respectively. These can be estimated by the mean, X_m, and standard deviation, s, of the historic data.

T is return period in years.

Case study from the coastal zone of Algeria- While addressing the issue of agricultural hydrology in Algeria, Belloum (1993) suggested the fitting of an adjusted form of the lognormal frequency distribution (law of Galton) or Gumbel distribution to the observed maximum daily rainfall in the coastal zone. An example of fitting the latter distribution to the maximum daily rainfall in the period 1958–87 at Skikda is shown in Figure 5.

Case study from Badeyat esh Shām, Syria- Similar to the previous cases, Al Saleh (1983) applied Gumbel Type-I distribution for extreme events, fitted by the method of moments, to the maximum observed daily rainfall in the dry part of Syria, Badeyat (desert) of esh Sham. The estimates for the range of probabilities of exceedance from 1% to 99% are included in Table 9. The remarkable thing is that the observed rainfall and thereupon the estimated values given a certain probability vary in a narrow range.

Case study from Northern Oman- In the summer months (June-September), winds are mostly blowing from the east and rainfall may be formed along the line of the inter-tropical convergence zone when it penetrates the land areas from the Indian Ocean. Tropical cyclones form near the Indian coastline at any time between May and November as can be seen in Figure 1. These cyclones affect the coast of Southern Oman nearly once every three years. In 1977 rainfall of 424 mm was recorded at Masirah Island in a 24 h period. Type I Extremal (Gumbel) distribution has been fitted by the method of moments to 60 years (curve 1) of annual maximum daily falls and 67 years (curve 2) both al Muscat, and the results shown in Figure 6. The 90% confidence limits are also indicated on the same figure (Wheater & Bell, 1983).

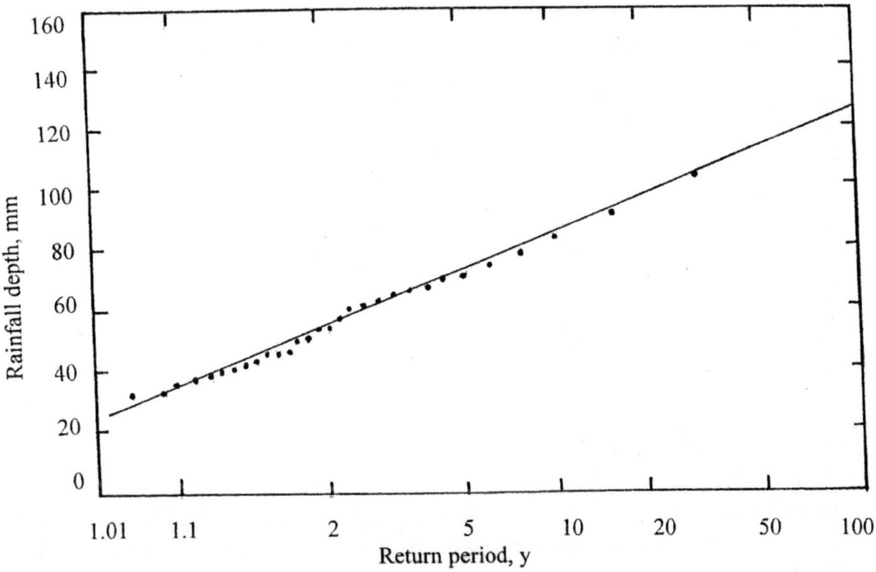

Figure 5. Fitting Gumbel distribution Type I to the observed annual maximum daily falls in the period 1958–1987 at Skikda, Algeria (Belloum, 1993)

Table 9. Estimates of maximum one-day rainfall in the Syrian Badyat (from El-Saleh, 1983)

Station name	X_m, mm	s, mm	C_s	Estimated maximum one-day rainfall, mm, corresponding to probabilities of exceedance,						
				1%	10%	20%	50%	80%	90%	99%
Deir ez Zor	23.55	12.31	1.22	62.2	39.6	32.4	21.5	13.4	10.0	3.3
Abu Kemal	25.54	11.34	1.15	61.1	40.3	33.7	23.7	16.2	13.0	6.9
Palmyra	23.20	11.32	0.84	58.7	38.0	31.4	21.3	13.9	10.7	4.6
J. Ettenf	21.06	10.80	1.23	55.0	35.2	28.8	19.3	12.2	9.2	3.3
Zulof	20.26	11.99	1.43	57.9	35.9	28.9	18.3	10.4	8.1	0.6
Rutbah	23.46	12.62	1.40	63.1	39.9	32.5	21.4	13.1	9.6	2.7

4.3.3 Short-duration rainfall (shorter than 24-h)

Some of the formulas expressing rainfall intensity-duration-frequency relationships have already been presented in subsection 4.3.1. The following paragraphs show the results obtained from case studies that have been conducted in a number of countries.

Case study from Algeria- The region of Skikda, Northern Algeria, is characterised by fractured, steep slopes that often cause the time of concentration to be less than 5 hours. In the absence of hyetographs, rainfalls of short durations cannot be determined by rigorous statistical methods. Instead, the method used here is

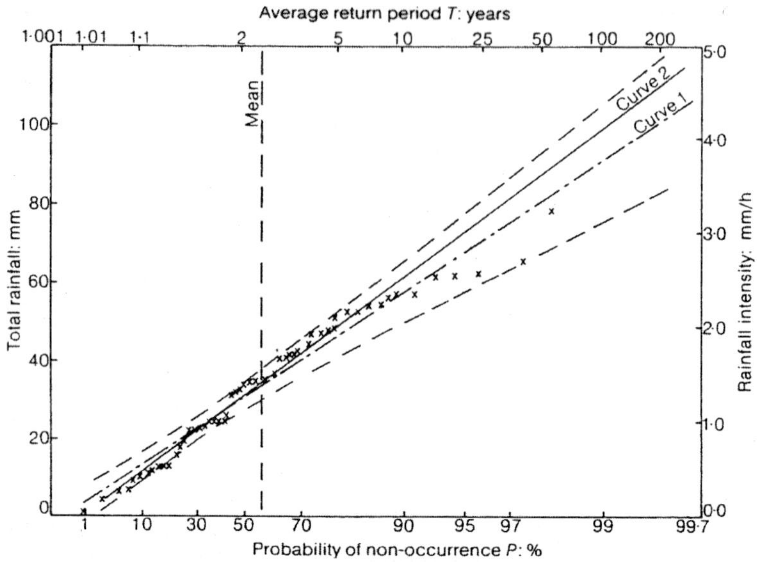

Figure 6. Frequency distribution of Muscat, Oman, 24-h historical rainfall (Type I Extremal (Gumbel) distribution) (Wheater & Bell, 1983)

Eq (22), known as Talbot's type formula. The findings obtained by Belloum (1993) for stations Azzaba and Bouati Mahmoud in the said region, are as follows:

Station	Average recurrence interval, y	a	b
Azzaba	2	431	10
	5	997	20
	10	1,266	25
Bouati Mahmoud	2	622	13
	5	735	16
	10	1,150	23

As such, the estimates of the mean maximum intensity of rainfall with duration of 60 min and average recurrence interval of 5-y are 12.5 mm and 9.8 mm h^{-1} for Azzaba and Bouati Mahmoud stations respectively.

Case study from Badyat esh Sham, Syria- Some of the statistical characteristics of the short duration rain storms have been reported by Al-Saleh (1983) for the Badeyat esh Sham. The results he obtained and those computed by the author are given in Table 10. The main problem with the tabulated results is that they are not associated with any return period. However, they show that the mean maximum intensity declines rapidly with the duration of the rainstorm. On the contrary, the coefficient of variation declines slowly with duration and reaches a stable value of about 0.5 as from 6-h duration until 24-h duration. It is remarkable that the coefficient of skewness remains almost constant, around 1.2 for all rainfall durations.

Case Study from Saudi Arabia- Point rainfall in the western part of Saudi Arabia has been investigated by Wan (1976). The depth-duration-frequency relationships

Table 10. Rainfall depth-duration and intensity-duration relationships for Badeyat esh Sham, Syria (based on results obtained by El-Saleh, 1983)

Duration, t, min	X_m mm	s mm	Cv	Cs	i, mm h-1	(Xm) t/(Xm)60
15	7.87	5.78	0.735	1.360	31.48	0.666
30	9.90	6.74	0.681	1.152	19.80	0.838
60	11.81	7.59	0.643	1.239	11.81	1.000
180	15.56	8.50	0.546	1.294	5.19	1.318
360	18.45	8.99	0.487	1.259	3.08	1.562
720	21.97	11.38	0.518	1.235	1.83	1.860
1,440	23.24	11.50	0.495	1.145	0.97	1.968

Explanation
X_m = mean rainfall depth, s = standard deviation, C_v = coefficient of variation, C_s = coefficient of skewness and I = mean maximum intensity.

for short duration storms were obtained and compared to those in a wide range of climatic regions of the world. Four rainfall zones, A, B, C and D, were considered. The mean annual rainfall for these zones in their respective order was 109, 176, 278 and 388 mm. As the period of record for any zone did not exceed 6 years, the lengths of the records were therefore extended using the station year concept. The series of annual maxima rainfall depths for selected durations –5, 15, 30, 60, 120 minutes and 24 hours were first abstracted from the rainfall charts. The ratios of the depths for the various durations to the 60-min rainfall of the same return period were next computed. The mean of the ratios and the standard deviation from the mean for each duration were obtained and have been summarized in Table 11(a).

Table 11(a). Depth-duration ratios for rainfall in the western part of Saudi Arabia (from Wan, 1976)

Rainfall zone	Rainfall, mm y^{-1}	5 min	15 min	30 min	120 min	1440 min
A	*90–140*					
X_m,		0.37	0.65	0.85	1.20	1.67
s_m		0.05	0.07	0.09	0.11	0.15
B	*140–200*					
X_m		0.31	0.57	0.78	1.19	1.70
s_m		0.03	0.05	0.07	0.11	0.81
C	*200–325*					
X_m		0.28	0.53	0.75	1.21	2.28
s_m		0.01	0.04	0.08	0.11	0.65
D	*325–450*					
X_m		0.29	0.57	0.81	1.21	2.08
s_m		0.04	0.04	0.06	0.11	0.03

Explanation
X_m = mean of ratios $(X)_t/(X)_{60}$ and s_m = standard deviation from the mean.

Table 11(b). Depth-frequency ratios of rainfall in Western Saudi Arabia (Wan, 1976)

Rainfall zone	Return period, years 2	5	25	50
A				
X_m	0.54	0.82	1.23	1.41
s_m	0.02	0.01	0.01	0.01
B				
X_m	0.61	0.85	1.25	1.34
s_m	0.02	0.01	0.01	0.02
C				
X_m	0.62	0.84	1.17	1.31
s_m	0.05	0.03	0.05	0.06
D				
X_m	0.63	0.85	1.19	1.33
s_m	0.04	0.01	0.02	0.02

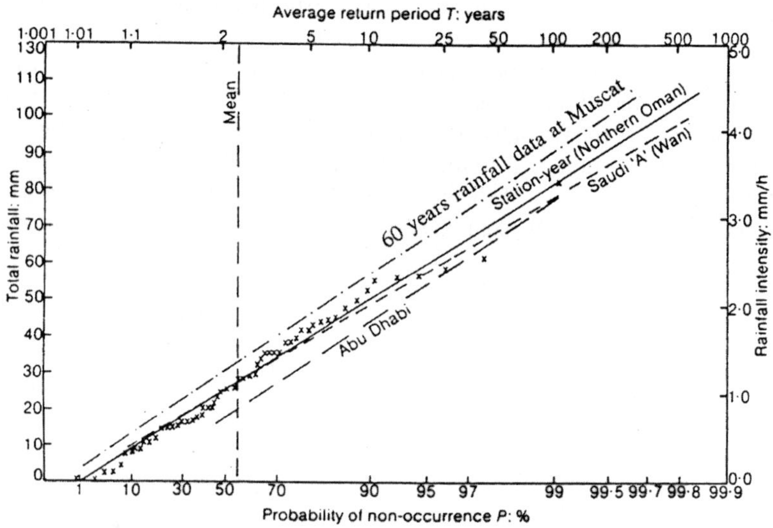

Figure 7. Gumbel Type I extremal distribution fitted to annual maximum 24-h rainfall in Northern Oman, Western Saudi Arabia (zone A), and Abu Dhabi (Wheater & Bell, 1983)

The depth-duration ratios listed in Tables 10 and 11(a) for durations of 15 min, 30 min and 1-d do not deviate substantially from one another. Additionally, it was found that the results in the latter Table are in fair agreement with those obtained from the USA, the former USSR and Australia.

The ratios of rainfall depths with specified return periods to the depths with a 10-y return period were computed for each duration of rainstorm from the derived

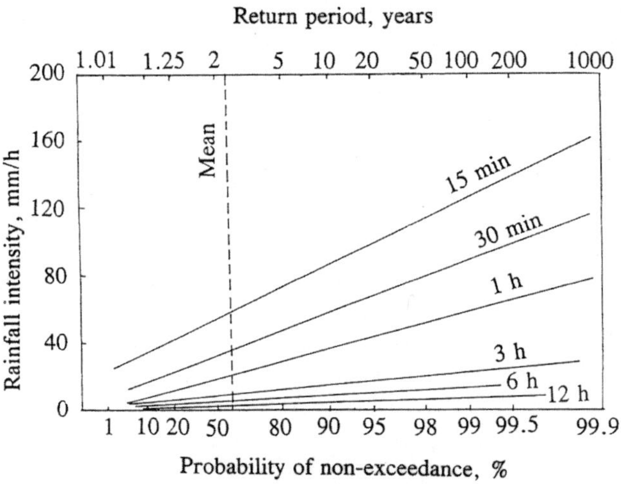

Figure 8. Gumbel Type I distribution fitted to annual series of maximum rainfall intensity with different durations, Northern Oman (Wheater & Bell, 1983)

frequency distribution. From the results included in Table 11(b) one can observe that although the ratios increased with higher return periods, they were practically independent of storm duration and of the rainfall zone, with the exception of zone A which produced slightly lower ratios than those from other zones.

Case study from Northern Oman- The annual maximum fall in Northern Oman was studied and reported by Wheater & Bell (1983). A brief summary of the results obtained has already been given in the previous subsection. Curve 1, Figure 6 was compared to the annual daily maximum rainfall in zone A, Western Saudi Arabia, obtained by Wan (1976) as well as the corresponding curve for Abu Dhabi. The comparison between the different curves can be seen from Figure 7.

Additionally, the Northern Oman study was extended to cover the short, up to 2-h, and medium, 3-12-h, duration rainstorms. The author has summed up the results obtained for these two ranges of duration in just one graph as can be seen in Figure 8.

CHAPTER 5

EVAPORATION AND EVAPOTRANSPIRATION

5.1. BACKGROUND AND DEFINITIONS

The Arab Region is traversed by a few major rivers and a lot more minor streams and wadis. These watercourses are mainly used to supply water for irrigated agriculture and to a less extent for producing hydropower.

The major rivers in the Arab Region are the Main Nile in Egypt; the Blue Nile, the White Nile, the Atbara and a certain reach of the Main Nile in the Sudan. Syria and Iraq are among the riparian countries sharing the Euphrates. Tigris is another international river shared by a number of countries. Compared to the Nile, Tigris and Euphrates, the Jordan River is a small river. Nevertheless, the Jordan is shared by a number of riparian countries. The wadis, which are seasonal or ephemeral streams, can be found in all countries the Arab Region comprises.

As the demand for water is not always met by the natural river flow or by the natural supply of the wadis, storage reservoirs are built at the appropriate sites on stream to regulate the natural flow. The Aswan High Dam (AHD), on the Main Nile in Egypt and Er-Roseires (Damazin) on the Blue Nile in the Sudan are just two examples to mention. In addition to storage reservoirs, there are natural lakes, depressions, wetlands and man-made pools in the Arab Region as described in Chapter 2. These water bodies when full of water provide large surfaces from which water can be freely evaporated. As evaporation depends on the prevailing climate, one should expect the free water evaporation to be quite substantial.

Cultivated fields, forests and woodlands loose water partly through evaporation and the rest through transpiration; the combined process is referred to as evapotranspiration. This can be actual, potential or reference evapotranspiration depending on the extent of availability of water and the plant raised or vegetal cover of the soil.

Evaporation and evapotranspiration constitute a major source of loss of water, at least locally, i.e. for the area from which they occur. In principle, the loss of water from a certain area can be a source of moisture gain to another area.

The definitions of the different terms relevant to evaporation and evapotranspiration as employed in this text are as follows.

5.1.1 Evaporation

It is a process through which the liquid state of water or solid state of snow, for example, is changed to vapour. Free water evaporation, E_o, takes place from open water systems such as lakes and reservoirs, running streams, barren soils with water table at or close to the land surface, and impervious surfaces such as building roofs and road surfaces. Penman (1948) defined open water evaporation as "The amount evaporated in unit time by a shallow layer of open water, for which reflection of radiation is determined by the surface only, and sufficiently extensive for edge effects to be negligible, under the atmospheric conditions measured above it." Stigter & Kisamo (1987) have argued this definition on the grounds that it does not apply to lake evaporation, since a lake is generally not a shallow layer of water. Evaporation can also take place from chotts, pools, tanks, pans and other containers of water. Evaporation measured from small size containers or evaporimeters often needs correction or adjustment to bring the measurement to its equivalent of lake evaporation under the same climatological conditions.

As established by the laws of physics, the transformation of water from liquid to vapour requires an exchange of approximately 600 cal for each gram of water evaporated. If the temperature of the surface is to be maintained, the large quantities of heat needed for evaporation must be supplied by radiation and conduction from the overlying air or at the expense of energy stored below the surface (Linsley et al., 1958).

5.1.2 Evapotranspiration

The World Meteorological Organisation (1966) has defined evapotranspiration, *ET*, as the difference between precipitation and/or irrigation minus surface runoff minus underground drainage minus change in water storage in the mass of soil concerned. This definition is strictly hydrological in nature. Evapotranspiration, as defined from the standpoint of physical meteorology is the combined process of evaporation from the soil and plant surfaces, and transpiration through the stomata of the plant surface.

The evapotranspiration, *ET*, from a specified plant or crop grown under well defined conditions including fertility and other cultural conditions is said to be actual, $(ET)_a$, whenever the supply of water at a designated time is below adequate. It becomes potential $(ET)_o$, or maximum, $(ET)_{mx}$, when the soil surface is completely covered by actively growing vegetation and supply of moisture is unrestricted.

The term reference crop evapotranspiration, $(ET)_r$, has been defined as the water requirements of a disease-free crop, growing in large fields under non-restricted soil conditions including soil moisture, fertility and achieving full production potential under the given growing environment (Doorenbos & Pruitt, 1977). According to Allen et al. (1994) the reference evapotranspiration approach to calculating crop evapotranspiration is widely accepted by engineers, agronomists and managers in field practice, design and research. For example, grass reference evapotranspiration is the maximum water loss from a clipped grass surface having a height of 0.12 m and bulk surface resistance of about 70 s m^{-1}, both at the same prevailing weather conditions. Burman & Pochop (1994) mentioned that $(ET)_r$ is similar to $(ET)_o$ except that it is applied to an identifiable crop, such as alfalfa or grass. As such, $(ET)_r$ is the evapotranspiration from a well-adapted local crop grown under the same conditions as for $(ET)_o$.

The term consumptive use of water by crops or any kind of vegetation is frequently used in irrigation practices and engineering. As defined by Blaney & Criddle (1966), it is the sum of the volumes of water used by the vegetative growth of a given area in transpiration and building of the plant tissue and that evaporated from adjacent soil, snow, and intercepted precipitation on the area in any specified time, divided by the given area. As such, the term consumptive use, *CU*, is synonymous to evapotranspiration, *ET*.

5.2. INSTRUMENTAL MEASUREMENT

5.2.1 Measurement of evaporation

Although instrumental measurements are neither simple nor straight forward as for rainfall, it is a compensating factor that evaporation quantities are less variable from one season to another, and therefore easily predicted than rainfall amounts (Shaw, 1983). Measurement of evaporation, mostly the evaporative capacity of the air rather than evaporation itself, began a long time ago.

The use of the Piche evaporimeter (Figure 1(a)) dates back to the beginning of the 20th century or earlier. As a consequence of its sensitivity to wind characteristics, it is strongly dependent on the elevation of the setup containing the Piche, degree of its exposure to the air and the nature of the surface surrounding it. These factors have rendered the use of Piche readings in practice quite limited. The former French colonies in the Arab Region are accustomed to use, next to the Piche, the sunken pan (Figure 1(b)). Additionally, the Wild evaporimeter, the Helwan and Abbassiya tanks were used for sometime in Egypt, and so the Sudan evaporation tank in the Sudan. Since the beginning of the second half of the 20th century most of the Arab countries began to use the USWB Class ‘A’ pan (Figure 1(c)). Currently it is the main device used for measuring evaporation, or evaporative demand in a certain environment better to say.

A detailed description of these devices as well as the relationship between the Class ‘A’ pan and other measuring devices is given in some of the hydrology literature, e.g. the Technical Report No. 105 on measurement and estimation of evaporation and evapotranspiration produced by the World Meteorological Organisation (WMO, 1966). A comprehensive description of evaporation pans can also be found in FAO Irrigation and Drainage Paper No 27 (Doorenbos, 1976).

The average monthly and annual evaporation measurements at selected stations in the Arab Region are listed in Table 14, Appendix I. Those measurements were obtained from the readings of the Piche evaporimeter, Colorado sunken pan and

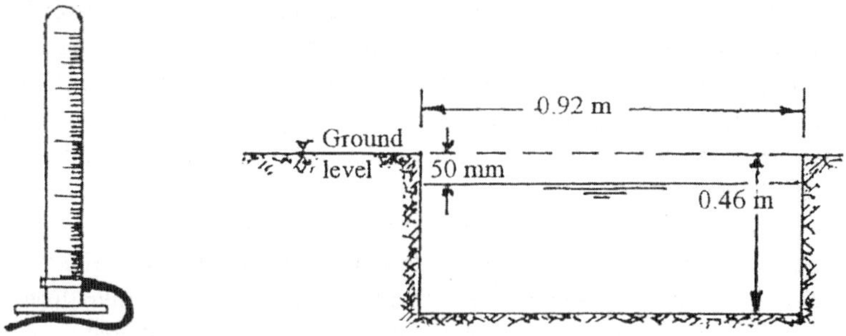

Figure 1(a). Piche evaporimeter. *Figure 1(b).* Sunken evaporation pan

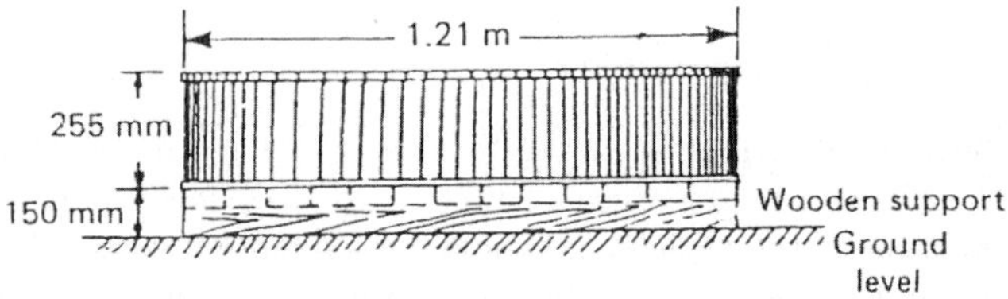

Figure 1(c). USWB Class 'A' evaporation pan

USWB Class 'A' evaporation pan. It goes without saying that the readings of the three devices are not in agreement with each other, even under the same climatic conditions. Additionally, the evaporation from any of them does not really represent the evaporation from an extended surface of free water. It is therefore a common practice to try to adjust or correct the readings of each evaporimeter. For example, it used to be a practice in countries like the Sudan, Egypt and Iraq to convert the Piche reading to free water evaporation by multiplying it by a coefficient of 0.50 for the year. The annual adjustment factor for the Class 'A' pan in Egypt was found to be around 0.60 (Mankarious, 1979) and in the Sudan 0.645 (WMO, 1966). The annual coefficient for a floating pan (1.0 m × 1.0 m × 1.0 m) installed in Lake Qaroun, Egypt, has been found to be close to 0.81. The monthly pan coefficient has been found to vary considerably from month to month.

5.2.2 Measurement of evapotranspiration

According to Israelsen (1956), the early measurements of crop water use were taken from irrigated field plots where the water table was at a sufficient depth below the soil surface. The depth of soil moisture removed from the root zone after each irrigation application throughout the growing season of the plant is determined from soil samples taken at specified intervals of time. Later, it has often become clear that percolation of water below the root zone was difficult to control. This difficulty has negatively affected the measurements taken from the experimental field plots. As a consequence, it was thought more practical to confine the body of vegetated soil from which evapotranspiration takes place in a container known as evapotranspirometer or lysimeter.

Lysimeters have been used for several years in numerous countries, and are still in use in some countries, to furnish data needed for issues related to evapotranspiration like irrigation water requirements and frequency of irrigation. According to Harrold (1966), interest in the use of lysimeters developed early in the 1900s. A lysimeter is basically a container in which soil is placed and water supplied through precipitation and/or irrigation. A plant is grown in the container and water table level as well as drainage, if any, are observed or measured.

Constant shallow-depth, water table lysimeters were used to obtain *ET* values by assuming that no change in water storage takes place during a certain period of time and equating *ET* to the total water added to the lysimeter in the same period. These limitations often caused a confusion about the measured *ET* values, whether they were actual or potential. If *ET* measurements have to be reliable, lysimeters should comply with certain requirements such as size (surface area and depth), crop cover, siting, material of construction and extent of isolation, water table and drainage conditions, etc.

Depending on their constructional design and operation, lysimeters can be classified as gravimetric and volumetric. The gravimetric lysimeters are either weighable or floating. The weighable are quite expensive, yet produce reliable *ET* values when the necessary precautions in design, operation and siting are taken care of (WMO, 1966). The lysimeter is provided with a system of balances (Figure 2) and any change in weight is recorded with time. The floating lysimeters compared to the gravimetric ones are relatively less reliable. In this type the system of balances with which the weighable lysimeters are equipped is replaced by a hydraulic system.

In volumetric lysimeters the supply of water and the outflow are measured. By maintaining the soil moisture content at the field capacity level at the beginning and end of designated periods of time, the difference between the supply and outflow represents the amount of evapotranspiration at the said period. The Mather and the Thornthwaite (Figure 3) are amongst the volumetric type of lysimeters.

Tanks and lysimeters are used in the Arab countries, though not to a large extent. Examples of case studies in which tanks and lysimeters were used are as follows:

Case Studies from Egypt:

i- Results obtained from tank experiments at the agrometeorological station, Giza, were reported by Omar (1960). During the 3-y period, 1957–59, potential evapotranspiration was measured from three evapotranspirometers of the type known as 'modified evapotranspirometer' developed by Mather. Each tank had an opening at the bottom where a tube was fixed in order to enable the measurement of

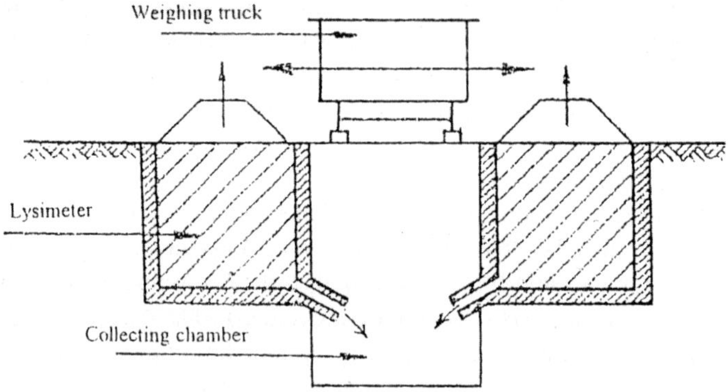

Figure 2. Simplified sketch of the weighable lysimeter

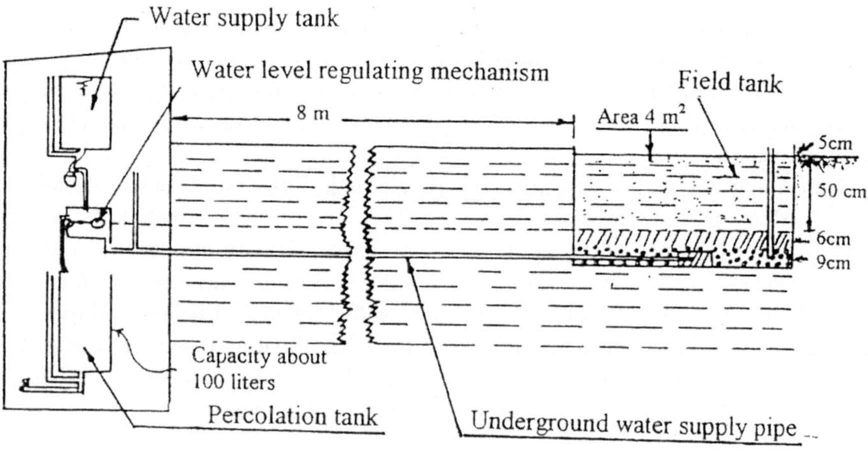

Figure 3. Diagrammatic sketch of Thornthwaite's lysimeter

percolation water. Two tanks were cylindrical, 60 cm in diameter each, and the third tank had a rectangular cross section 130 × 90 cm. The three tanks were planted with libya grass and installed in the station's grass field. The same vegetation level was maintained inside and outside the tanks by cutting the grass whenever necessary. The difference in area between the evaporating surfaces was claimed to be of no significant influence.

Other measurements were taken using Popoff lysimeters, which have smaller area (500 cm^2) compared to Mather's evapotranspirometers. The results obtained there from were, however, 10% larger than those obtained from the bigger tanks. The ET_o from cotton raised in the two sets of evapotranspirometers is shown graphically in Figure 4.

ii- Zein el-Abedine et al. (1967) compared evapotranspiration from field plots to that from weighable (hydraulic) lysimeters, all raising late corn 'early American', under two different irrigation treatments. The experiments were conducted at the experimental farm of the Faculty of Agriculture, Cairo University, Egypt. Results were obtained directly through daily weighing of the lysimeters and indirectly from measurements of soil moisture content using a neutron moisture meter. For each treatment one weighable lysimeter was used surrounded by five plots of the same surface area, 2 m × 2 m, and receiving the same levels of water and fertilizer. A cornfield of about one hectare in surface area surrounded the whole experimental setup. The average evapotranspiration from the two lysimeters and the field plots as calculated from the readings of the neutron moisture meter for both treatments are given in Table 1.

iii- Fathi (1995) compared the consumptive use and crop coefficient for some field crops grown in desert sandy soil as obtained from volumetric lysimeters to those obtained from the "CROPWAT" computer software (FAO, 1989). The two lysimeters were cylindrical in shape, each 2.25 m inner diameter and 1.0 m in depth, made of galvanized iron equipped with a drainage facility. The two lysimeters were surrounded by a buffer area of 2,500 m^2 cultivated with the same crop and receiving the same amount of fertilizer. Irrigation and drainage quantities were monitored and recorded over short time intervals (2–4 days). Climatological data covering a period of ten years were collected from a nearby agrometeorological station and fed into the CROPWAT computer program. The lysimeter-obtained

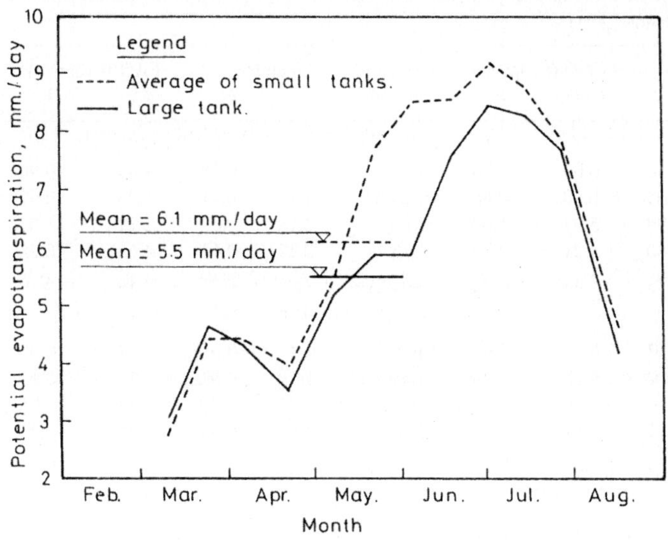

Figure 4. Comparison between potential evapotranspiration from small and large tanks raising cotton at Giza (Omar, 1960)

Table 1. Evapotranspiration from lysimeters and field plots raising maize under two different irrigation treatments, Cairo University Experimental Farm, Giza, Egypt (Zein el-Abedine et al., 1967)

Year 1966 from	to	Treatment 1 depth of root zone, cm	ET, mm Lys.*	Plot	Year 1966 from	to	Treatment 2 depth of root zone, cm	ET, mm Lys*.	Plot
24.07	1.08	40	34.4	31.8	24.07	7.08	40	52.6	48.8
2.08	9.08	60	40.8	39.2	8.08	20.08	80	50.6	64.2
10.08	20.08	80	57.6	59.2	21.08	7.09	100	75.2	73.8
21.08	4.09	100	76.6	62.4	8.09	25.09	130	119.6	67.2
5.09	17.09	120	80.2	67.0	26.09	11.10	140	119.2	88.2
18.09	1.10	140	72.8	77.0	12.10	5.11	140	141.4	112.0
2.10	11.10	140	82.8	72.8					
12.10	5.11	140	147.3	116.2					

Explanation
Lys. = Lysimeter*

crop coefficients were compared with the FAO suggested values (Doorenbos & Pruitt, 1977). Figures obtained from the lysimeters and CROPWAT are included in Table 2.

Semi-arid Mediterranean Region in Spain: The Province of Murica in the south-eastern part of Spain enjoys a semi-arid Mediterranean climate more or less similar to that in the northern parts of Morocco and Algeria. The test plot, located in

Table 2. Consumptive use of water and crop coefficients for some crops grown in the period 1987–90 at South Tahrir, Egypt (Fathi, 1995)

Stage of growth	Lysimeter cu	k_c	FAO (Cropwat) $(ET)_o$	k_c	Stage of growth	Lysimeter cu	k_c	FAO (Cropwat) $(ET)_o$	k_c
Sorghum					*Wheat*				
Init.[x]	3.04	0.50	6.09	0.70	Init.	1.07	0.58	1.84	0.90
Dev.	3.08	0.63	6.08	0.85	Dev.	1.23	0.71	1.73	0.97
Mid.	5.29	0.99	5.37	1.00	Mid.	2.18	1.03	2.12	1.05
Late	3.35	0.80	4.21	0.50	Late	3.12	0.67	4.66	0.25
Average	3.87	0.73	5.44	0.76	Average	1.9	0.75	2.59	0.79
Peanut					*Barley*				
Init.	3.53	0.57	6.14	0.65	Init.	1.1	0.6	1.84	0.90
Dev.	5.41	0.90	6.01	0.80	Dev.	1.19	0.69	1.73	0.97
Mid.	5.60	1.10	5.90	0.95	Mid.	2.31	1.09	2.12	1.05
Late	1.72	0.45	3.61	0.55	Late	3.03	0.66	4.66	0.25
Average	4.06	0.76	5.26	0.74	Average	1.91	0.76	2.59	0.79

Explanation
Values of cu and $(ET)_o$ *are expressed in* $mm\ d^{-1}$.
[x] *Init. = initial, Dev. = development and Mid. = middle*

Sangneora at about 45 km west of the Mediterranean Sea (Figure 5), forms part of a vast area irrigated by the Guadalentin River. The physiography and climate of the experimental area justify the applicability of the results obtained there from to the Mediterranean land strips of Algeria and Morocco. Experiments in Sangonera were run using a drainage lysimeter 4.0×7.0 m and 1.3 m deep, situated in a 4.2 ha square plot where an agroclimatic station is found.

Grass prairie was planted in the lysimeterunder investigation as well as the area surrounding it. The grass was maintained at a height of between 8 and 15 cm, in accordance with the recommendations of the FAO (Doorenbos & Pruitt, 1977). The monthly $(ET)_r$ from the lysimeter was determined for two growing seasons from September 1987 to August 1989 (Toribio et al., 1996) using the water balance method, the results of which are given in Table 3. From this Table one can easily notice that the ratio of the lysimeter $(ET)_r$ to Class'A' pan evaporation ranges from a little less than 0.6 to slightly above 1.0, with an overall mean of 0.83 and standard deviation of 0.091. The determination coefficient, r^2, between the FAO-Class 'A' Pan and lysimeter is 0.960 on a monthly basis and 0.935 on a ten-day basis.

Reference evapotranspiration was also computed using the methods of Penman-Monteith, Penman-FAO, radiation (Jensen)-FAO, Blaney-Criddle-FAO, Hargreaves

Figure 5. Location map of the experimental field in Murcia, south of Spain

Table 3. Water balance of the lysimeter used for measuring reference grass evapotransiration in the Sangonera Basin, Murcia, Spain, (Toribio, 1966)

Year and month	E_p mm	Rainfall mm	Irrigation mm	Drainage mm	Change in storage	$(ET)_r$ mm
1987						
Sep.	173.5	20.0	122.5	9.7	−10.4	143.2
Oct.	83.0	45.0	127.1	42.1	49.9	80.0
Nov.	53.0	62.2	0.0	64.8	−46.6	44.0
Dec.	24.5	20.7	0.0	16.3	−15.8	20.2
1988						
Jan.	37.2	35.5	0.0	17.6	−11.9	29.8
Feb.	53.0	60.0	15.7	28.6	2.4	44.7
Mar.	134.4	10.4	124.1	25.7	15.4	93.4
Apr.	148.5	28.3	159.0	76.7	−3.7	114.3
May	191.0	23.1	185.2	67.8	−16.8	157.3
Jun.	179.0	52.4	169.6	41.5	23.9	159.6
Jul.	277.0	0.0	284.6	31.4	17.2	236.0
Aug.	241.0	0.0	295.0	54.9	15.2	224.9
Sep.	166.0	10.0	163.5	25.1	7.6	140.8
Oct.	129.0	6.5	106.8	18.9	−18.2	112.6
Nov.	127.3	85.9	47.3	45.5	12.8	74.9
Dec.	42.0	3.5	2.3	9.3	−37.1	33.6
1989						
Jan.	26.5	36.7	33.6	25.2	16.9	28.2
Feb.	62.3	52.8	0.0	33.6	−32.6	51.8
Mar.	82.7	126.2	5.9	34.8	35.0	62.3
Apr.	136.3	23.0	88.6	21.6	−24.4	114.4
May	160.6	9.3	168.2	34.0	11.3	132.2
Jun.	206.4	11.4	168.6	29.0	−27.6	178.6
Jul.	288.0	8.6	261.8	5.0	20.0	245.4
Aug.	291.4	14.7	247.6	1.1	47.0	214.2

Explanation
E_p = Class 'A' pan evaporation and $(ET)_r$ = grass reference evapotranspiration.
All tabilated values are in mm mo^{-1}.

and Turc. A comparison between the estimates obtained from these methods and the lysimeter $(ET)_o$ is presented in sec. 5.4.3.

Case study from Lebanon: Abou-Khaled (1972) reported the results he obtained from two lysimeter setups raising grass; one at Tyr in the southwest along the Mediterranean coast and the second at Tal-Amara in Al-Beqaa Valley in northeast Lebanon. In addition to those experiments, evapotranspiration from citrus (orange and lemon) orchards were measured and the results compared to estimated values using the Penman and Blaney-Criddle methods. These methods will be explained in secs. 5.3.3. and 5.4.2 respectively. Table 4 gives the measured *ET*, evaporation measurements at T-yr using Colorado and Class 'A' pans. The pan coefficient, i.e. measured (*ET*/ *Ep*), for the period of measurement varied from 0.67 for Class 'A' pan to 0.79 for Colorado pan.

Table 4. Measured evapotranspiration *(ET)* for citrus orchards during the summer of 1969 and ratios to pan evaporation at T_{yr}, Lebanon (from Abou-Khaled, 1972)

Month	Decade	Measured *ET**		Measured *Ep**		*ET/Ep*	
		Orange	Lemon	Class A	Colorado	Class A	Colorado
	I	x	x	x	x	x	x
June	II	4.3	4.2	6.6	5.1	0.66	0.82
	III	4.2	4.0	6.7	5.8	0.61	0.71
	I	4.1	3.8	6.7	5.4	0.59	0.72
July	II	5.0	4.9	7.0	5.5	0.71	0.89
	III	4.3	4.8	6.5	5.3	0.70	0.85
	I	4.1	4.3	6.4	5.6	0.66	0.75
August	II	3.6	3.8	5.9	4.8	0.63	0.77
	III	3.6	3.6	0.6	0.5	0.64	0.71
	I	x	3.5	5.3	4.5	0.66	0.78
September	II	x	3.6	5.2	4.3	0.68	0.84
	III	3.3	3.8	4.3	3.8	0.83	0.92
Total, mm (112 days)		445.0		664	562	0.67	0.79

All measurements are in mm d^{-1}.

Case Study from Tunisia: Results obtained from an old investigation on water requirements in Tunisia have been summarized and reported by Combrémont (1972). That investigation comprised measurement as well as estimation of evapotranspiration. Some of the conclusions reached were as follows:

i Lysimeters gave correct values in the winter season. The advective heat in summer time affected the lysimeter readings, especially those was filled with heavy soil.

ii The Piche method greatly overestimated the requirements in autumn (October-mid December). The increase in summer time has been rather slight and became negligible in the January-April period.

iii Class'A' pan gives consistently higher values than those given by the Colorado sunken pan. The difference is in the order of 30% at Cherfech station near the Mediterranean coast (semi-arid climate) and falls to about 17% at the Chott (shallow, salt lake) in the desert at Tozeur (Station No. 39). Evaporation from the sunken pan in oasis environment is roughly 30% less than evaporation from the same pan in the Chott area.

iv All devices mentioned in i, ii and iii used the development of the crop's consumption of water rather than the actual value of this consumption. In addition to the measurements mentioned above, evapotranspiration was estimated using the methods of Thornthwaie, Blaney-Criddle, Turc and Penman. The monthly estimates as obtained from the last method together with the measurements obtained from the Piche, Class 'A' pan and Colorado sunken pan are shown in Figure 6.

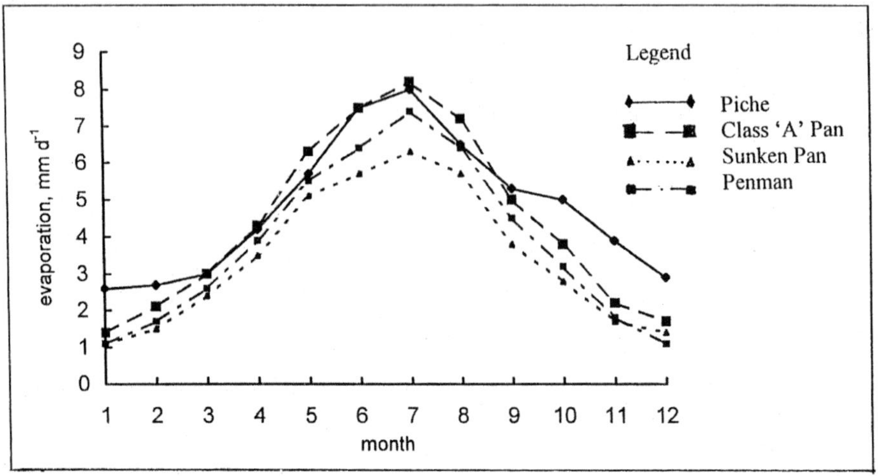

Figure 6. Evapotranspiration at Cherfech Sttaion, Tunisia (after Combrémont, 1972)

5.3. ESTIMATION OF EVAPORATION

5.3.1 Introduction

Estimation of lake evaporation depends largely on the availability of a certain amount of reliable hydrologic and climatic data as in the case of the water balance method. Most of the remaining estimation methods are of the climate-based. No wonder then if one talks about the temperature approach, the radiation approach, etc. The last seven decades or so have witnessed the development of a considerable amount of methods aiming at the estimation of free water evaporation, pan evaporation, potential, reference and crop evapotranspiration using methods based on climatic data.

It should be well remembered, however, that not all estimation methods require the same amount of climatic data. The results obtained from estimation methods requiring much data can be produced equally well, at least in certain cases, with less climatic data. According to Shih (1984), the lack of basic data and the difficulties in measurement required in the field methods have accounted for the great efforts made to develop evapotranspiration equations that can relate evapotranspiration to some readily available climatic data.

The common factors affecting evaporation and evapotranspiration are the latent heat required to change a liquid into its gaseous form, temperature of both air and evaporating surface, the vapour pressure of the air and wind speed and the nature of the evaporating surface that modifies the wind pattern. The latent heat of vaporization is provided by the solar (short-wave) and terrestrial (long-wave) radiation. The amount of water vapour in the air depends on the corresponding air temperature and its relative humidity. Additionally, evapotranspiration is affected by the kind of vegetal cover, its density and stage of growth. The effect of the quality of water, unless it is too saline, is usually neglected.

5.3.2 Calculation (estimation) of evaporation, E_o, using the water balance method

Estimation of evaporation from an open water body, like a lake, pool or a reservoir, depends largely on the availability and reliability of certain data. These comprise the direct precipitation on the surface of the body in question, tributary flow and surface runoff and/or surface water extraction or diversion from the body of water, groundwater flow to or from it and the volume of water in storage at the beginning and end of the balance period (Morton, 1979). Furthermore, seepage, ungauged local flow and bank storage are uncertain and often unmeasurable items (WMO, 1983). The balance can generally be written as:

$$(1) \qquad E_o + \delta = I + P + (G_i - G_o) - O - L \pm \Delta S$$

where,

δ = resultant error given by Eq. (2),
I = inflow of surface water to the water body,
P = direct precipitation on the surface of the water body,
G_i = ground water inflow,
G_o = groundwater outflow,
O = surface water outflow,
L = loss by leakage, and
ΔS = change in storage over the time interval, Δt, from the beginning to the end of the balance period.

$$(2) \qquad \delta = \sqrt{\delta_1^2 + \delta_2^2 + \ldots\ldots\ldots + \delta_m^2}$$

in which $\delta_1, \delta_2, \ldots, \delta_m$ are the estimated errors in the measurement or estimation of the items in the right-hand side of Eq. (2).

The average values of monthly and annual evaporation from the reservoir formed by the Aswan High Dam, Egypt, for two periods, 1968–89 and 1976–82 (period with relatively stable water level), as obtained from the water balance method are included in Table 5. Evaporation from the same reservoir using other estimation methods is presented and discussed in a forthcoming subsection.

5.3.3 Estimation of free water evaporation using climatic data

5.3.3.1 Mass transfer method

Denote the saturated vapour pressure of an air mass at temperature T_a by e_s and its average relative humidity by h. The actual vapour pressure, e_a, is equivalent to the

Table 5. The total losses from the reservoir formed by the Aswan High Dam, Egypt (after Sadek, 1992)

Total loss, mm d^{-1}, from Lake Nasser for												
Jan.	Feb.	Mar	Apr.	May	Jun.	Jul.	Aug.	Sep.	Oct.	Nov.	Dec.	Year
From 1968 up to 1989												
5.76	4.52	4.63	6.22	4.74	5.87	10.0	10.9	7.11	6.43	6.14	5.00	6.86
From 1976 up to 1982												
4.70	4.07	4.43	5.44	5.17	5.68	8.65	7.87	7.22	5.89	5.96	5.50	5.88

vapour pressure, e_d, at the dew point, T_d. These quantities are connected together by the bulk aerodynamic or Dalton's equation as follows:

$$(3) \qquad E = f(u)(e_s - e_a) = f(u)e_s(1-h)$$

where,
E = evaporation rate,
$f(u)$ = wind function, and
$(1\text{-}h)$ = saturation deficit

The wind function can take different forms. The one mentioned below is the first form obtained by Penman (1948) from his detailed studies in the United Kingdom. This, together with Eq. (3) resulted in:

$$(4) \qquad E_a = 0.35(e_s - e_d)(0.5 + 0.54u_2) = 0.35e_s(1-h)(0.5+0.54u_2)$$

in which the air measurements were taken at 2.0 m above the surface. The vapour pressure is expressed in mm mercury (Hg), u_2 in m s^{-1} and E_a in mm d^{-1}. The evaporation term E_a refers to the open water evaporation in the hypothetical case of air and water temperatures being equal.

5.3.3.2 Heat budget method

This method is based on the energy balance principle, i.e. the balance between incoming, outgoing and stored heat energy. This balance can be expressed as follows:

$$(5) \qquad (Q)_{E_o} = Q_{sh} - Q_{rsh} - Q_l - Q_s \pm Q_g \pm Q_v$$

where,
$(Q)_{Eo}$ = energy required for evaporation,
Q_{sh} = short-wave solar radiation,
Q_{rsh} = reflected part of short-wave radiation,
Q_l = long wave radiation from the water body,
Q_s = sensible heat transfer to the air overlying the water body,
Q_g = change in stored heat energy, and
Q_v = heat transfer between water, and bed and walls of the water body

Denoting the latent heat of vaporization of water by L and the air density by ρ, the rate of evaporation from open water can be given by:

$$(6) \qquad E_o = (Q)_{E_o}/\rho L$$

in which E_o is given in length (e.g.mm) divided by unit of time (e.g.day).

5.3.3.3 Combination method (Penman Formula)

Penman (1948) combined the energy budget with the Dalton bulk aerodynamic expression together. Using an approximate analytical solution he was able to develop the expression known as Penman formula. This formula can be expressed as:

$$(7) \qquad E_o' = \frac{\Delta H + \gamma E_a'}{\Delta + \gamma}$$

where,

E_o' = free water evaporation expressed in cal cm^{-2} d^{-1},

Δ = slope of the saturated vapour pressure-temperature curve at the given air temperature,

H = heat budget expressed in cal cm^{-2} d^{-1} as given by Eq. (8),

γ = psychrometer constant, depending on the rate of ventilation, and equals 0.49 if the temperature is in °C and the vapour pressure in mm Hg, and E_o'

is given by Eq. (4), except that the units used are cal cm^{-2} d^{-1}.

The net heat budget, H, left at the evaporating surface can be calculated from the expression:

$$(8) \qquad H = Q_{sh} - Q_{rsh} - Q_l = R_A(1-r)\left[a + \left(b\frac{n}{N}\right)\right] - R_l$$

in which Q and R are equal, R_A is the amount of radiation reaching the outer limit of the atmosphere, a and b are regression parameters, and r = reflection coefficient or albedo. R_A depends on the geographic location of the place and on the time of the year. The parameters a and b were originally given by Penman (1948) as 0.20 and 0.48, respectively. The notations n and N refer to the actual number and maximum possible number of sunshine hours per day respectively. The coefficient r is usually taken as 0.06 for water. The long-wave radiation, Q_l or R_l, can be computed from:

$$(9) \qquad Q_l = R_l = \sigma T_a^4(0.47 - 0.077\sqrt{e_d})(0.20 + 0.80\frac{n}{N})$$

where,

σ = constant of Lummer and Pringsheim = 117.4×10^{-9} cal cm^{-2} d^{-1}, and

T_a = absolute temperature = $273 + T$ °C

To transform E_o', cal cm^{-2} d^{-1}, into its equivalent E_o, mm d^{-1}, one can divide it by 60.

Numerical Example: Consider the mean climatic data of Baghdad station (No. 152) in July. As the latitude of the station is 33°14′ N, the radiation R_A in July is 988 cal cm^{-2} d^{1}. From the tables of climatic data in Appendix I, one finds $T = 33.9$ °C, $T_a = 306.9°$ A, $h = 0.22$, $n/N = 0.7875$ and $u_2 = 2.6$ m s^{-1}.

From Eq. (8) $H = 988 \times 0.94 \times (0.20 + (0.48 \times 0.7875)) - R_l = 530.4 - R_l$

Also, from Eq. (9) $R_l = 117.4 \times 10^{-9}(306.9)^4(0.47 - 0.077\sqrt{30.93})(0.20 + 0.80 \times 0.7875) = 36.1$ cal cm^{-2} d^{-1}. This figure brings H to be $530.4 - 36.1 = 494.3\ cal\ cm^{-2}d^{-1}$. From Eq. (4) $E'_a = 21(1 - 0.22) \times 40.41(0.50 + (0.54 \times 2.6)) = 1260.3\ cal\ cm^{-2}d^{-1}$

Substitution of these values together with 2.2125 and 0.49 for Δ and γ respectively in Eq. (7) yields E'_o to equal $633.2 cal\ cm^{-2}\ d^{-1}$. The equivalent evaporation rate, E_o, is $10.55 mm\ d^{-1} or \approx 327 mm\ mo^{-1}$.

It should be remembered, however, that by substituting a = 0.20 and b = 0.48 in Eq. (8), the parameters in Eq (9) substituted for by 0.47, 0.077, 0.20 and 0.80 and the wind function given by Eq. (4) can vary from experience to experience and from one area to another resulting in a difference in the end result. Radiation investigation in Egypt by Mehanna (1976) has shown that a and b can better be substituted as 0.22 and 0.55 respectively. The corresponding values for these two parameters in their respective order were found by El-Agib (1995) to be 0.350 and 0.368 for the Sudan.

Table 6 gives the mean monthly and annual evaporation rates as obtained from calculations similar to those used in the above example. As already mentioned earlier, it is possible to predict the Penman evaporation from a certain water body using fewer climatic factors than what is actually needed to calculate it using Eqs. (3) through (9). The climatic data available at Uqlat Suqur meteorological station (25°50′ N, 42°11′ E and 740 m a.m.s.l.), Saudi Arabia, were used to develop simple relationships for predicting Penman evaporation using fewer climatic factors (FAO, 1981). One of such relationships uses R_c, which is the radiation reaching the evaporating surface, and expressed as:

$$(10) \qquad E_o(predicted) = -2.403 + 0.0184 R_c$$

The climatic data of Baghdad (Station No.152) with a and b equal to 0.20 and 0.48 respectively give R_c as 571 cal cm^{-2} d^{-1}. This figure upon substitution in Eq. (10) yields (E_o) *predicted* as 8.1 mm d^{-1} or 77% of the value computed from Penman formula with more climatic data. It is not certain whether this difference is due to the approximate nature of Eq. (10) or because the values of a and b used in obtaining this equation are different from the ones used in the calculation of Penman's E_o.

At Khartoum Observatory, comparisons were made among evaporation from a 12-ft or 3.66 m diameter tank, 4-ft or 1.22 m deep, and a class "A" pan. The average ratio between the 12-ft pan and Class 'A' pan was 0.645, and between the 12-ft pan and Penman 1.20. If these ratios are correct, the ratio of Penman evaporation to Class 'A' pan evaporation must be around 0.54. It is unfortunate that these ratios were obtained from an experimentation period of just 2 years or 24 months and did not last much longer.

Table 6. Mean monthly and annual estimated Penman (1948) evaporation at selected stations in the Arab Region

Stat. No.	Mean monthly and annual Penman evaporation, mm d^{-1}, for J	F	M	A	M	J	J	A	S	O	N	D	Year
33	1.1	1.7	2.6	3.9	5.5	6.4	7.4	6.4	4.5	3.2	1.8	1.1	3.8
37	1.4	2.1	3.3	4.5	5.7	6.4	7.4	6.9	5.2	3.5	2.2	1.4	4.2
39	1.5	2.5	4.3	5.9	7.6	9.3	9.9	9.1	6.5	4.1	2.3	1.3	5.4
61	2.4	3.0	4.1	5.4	5.6	7.6	8.5	7.6	5.7	3.9	3.1	2.3	4.9
65	2.1	2.8	3.7	5.2	6.2	7.0	7.1	6.6	5.5	4.1	3.2	2.1	4.6
66	2.3	3.0	3.6	4.8	5.3	6.8	6.6	5.6	5.5	4.0	3.2	2.1	4.4
70	1.5	2.1	2.8	4.2	5.3	6.4	6.1	5.8	4.6	3.6	2.6	1.5	3.8
71	1.6	2.6	3.6	4.9	6.3	7.0	6.9	6.4	5.2	4.0	3.0	1.7	4.4
74	1.5	2.6	3.3	5.2	6.7	7.9	7.4	6.2	5.0	3.5	2.3	1.5	4.4
75	2.3	3.1	4.6	5.7	7.0	7.9	7.9	7.4	6.3	5.0	3.6	2.2	5.2
77	1.5	2.6	3.9	5.4	7.6	8.2	8.2	7.3	5.5	4.0	3.0	1.6	4.9
81	1.9	2.6	3.8	5.3	7.0	7.7	7.7	6.9	5.6	4.1	3.3	2.0	4.8
82	3.7	4.4	5.6	7.0	8.3	9.5	9.6	9.0	7.7	6.2	5.2	3.8	6.7
83	1.8	3.1	4.3	6.6	8.3	8.7	8.5	8.2	6.4	4.9	3.4	2.0	5.5
84	2.1	2.9	5.0	6.6	8.4	9.2	8.9	8.0	7.2	5.2	3.7	2.2	5.8
85	3.1	3.9	5.1	6.2	7.3	8.2	8.5	8.0	7.3	5.4	4.4	3.2	5.9
87	2.4	3.5	5.2	6.9	8.6	9.4	9.2	8.5	7.7	5.7	3.6	2.4	6.1
89	3.5	4.6	6.2	7.7	8.5	9.4	9.3	8.8	8.2	6.5	4.5	3.7	6.8
90	4.5	5.8	7.1	8.7	9.3	9.8	9.6	9.3	8.6	7.7	6.1	4.0	7.5
91	4.5	5.0	6.0	7.0	7.9	8.4	8.5	8.6	7.5	6.3	5.3	5.6	6.7
93	5.5	6.2	7.5	10.6	9.3	9.0	8.8	8.6	8.6	9.9	6.5	5.6	8.0
94	6.7	7.3	8.8	10.0	9.9	9.6	7.9	6.3	6.6	8.4	7.7	6.6	7.9
95	4.1	4.8	5.6	6.1	6.8	7.3	6.9	6.1	6.0	5.5	4.6	4.1	5.7
96	5.8	6.3	7.8	9.2	8.9	8.3	6.4	5.3	6.3	7.5	6.6	5.6	7.0
97	5.6	6.7	7.8	8.0	8.1	7.5	6.8	6.4	6.8	6.1	6.0	5.3	6.8
99	4.8	5.9	7.3	7.9	8.1	8.0	7.3	6.6	6.5	6.2	5.2	4.7	6.7
102	6.0	6.8	7.2	8.0	8.0	7.8	6.0	5.2	5.7	6.3	6.6	6.0	6.6
105	6.3	6.5	7.8	8.3	7.6	6.4	5.0	5.1	5.3	5.5	5.7	5.8	6.3
107	6.0	7.1	7.3	7.1	6.2	5.0	4.6	5.0	5.4	5.5	5.7	6.3	5.9
110	5.7	6.1	6.6	7.1	6.3	5.4	5.0	5.2	5.3	5.4	5.4	5.2	5.7
111	5.4	5.7	6.2	5.6	5.2	4.9	4.5	4.7	5.2	5.3	5.1	4.7	5.2
166	1.6	3.2	4.0	5.5	6.4	8.4	7.8	7.3	7.0	4.6	3.1	2.3	5.4
167	3.1	4.2	4.9	7.1	7.5	9.0	8.2	8.1	6.5	5.3	3.9	2.9	5.9
168	3.3	4.7	5.7	8.1	8.3	10.1	8.8	9.0	6.8	5.2	4.0	3.0	6.4
169	3.5	5.2	5.8	7.9	8.1	8.6	7.9	7.9	6.7	5.3	4.4	3.6	6.2
Najran^{o}	4.8	6.0	7.2	7.9	9.0	9.6	9.6	9.1	8.6	6.8	5.5	4.6	7.4
185	3.4	4.6	5.2	5.8	7.3	7.4	7.5	7.0	6.9	5.5	4.6	4.0	5.7
186	4.5	4.8	5.3	5.0	6.7	7.1	5.9	6.2	6.2	6.1	5.0	4.4	5.6
197	3.9	4.5	5.0	5.6	6.0	5.9	5.7	5.9	5.6	5.0	4.4	3.9	5.1

Najran^{o} = 17°37′N, 44°26′E and 1210 m height, Saudi Arabia.

5.3.3.4 *Estimation of free water evaporation using Morton's models*

Morton (1979, 1983 and 1986) developed a series of models under the title of 'Complementary Relationship for Lake Evaporation (CRLE)' for estimating evaporation from shallow and deep lakes.

In his models, Morton suggested the routing of solar and water-borne heat input through a hypothetical heat reservoir with time lag and storage parameters depending on the depth and salinity of lake water. The energy routing concept he used is similar to the concept of flood routing in lakes and reservoirs. Morton's model was run in four forms for estimating evaporation from the reservoir formed by the Aswan High Dam at both Aswan and Wadi Halfa, almost the two extremities of the reservoir. The first model was run assuming that the reservoir was a shallow lake. The second model included storage and advective heat terms employed by Omar & el-Bakry in an earlier study (1981). In the third model only the incoming radiation was routed and the advective heat was left unused. The last model, similar to the second model and different from the third one, included the routing of the net water-borne energy (advective heat). The first and third models yielded comparable results at the annual level; 2,081, 2,150, 2,080 and 2,244.mm in their respective order. Monthly evaporation rates as estimated from the model are included in Table 7.

5.3.3.5 Measurements and estimates of evaporation from large bodies of water using different methods

i *Case study of evaporation from the Qattara Depression, Northern Egypt*: The Qattara Depression is the second largest sub-sea-level depressions in the world. The total area b.m.s.l. is about 20, 000 km^2, with a total volume of about $1,200 \times 10^9$ m^3 and deepest point nearly 135 m b.m.s.l.. The depression is situated roughly halfway between the Nile Valley and the Libyan border.

The idea of filling the reservoir with sea water up to a certain level and allowing the stored water to fall down to the Mediterranean Sea for developing power is not a recent one. The hydrologic design and operation of the reservoir, which is supposed to develop within the depression, will depend largely on the evaporation. In view of the immensity of the volume of evaporation, many authorities became interested in carrying out the necessary calculations using a variety of estimation methods. Among those authorities that reported on those calculations were Flohn & Wittenburg (1980) and Sundborg & Nilsson (eds., 1985). The results reported by those two sources are included in Table 8. Despite the relatively wide differences in the monthly estimates between one method and another method, the agreement on the annual level is quite remarkable. The least estimate being 1,701 mm and the largest 1,768 mm, assuming water in every method to be fresh, bring the maximum difference to less than 4%. It is expected that the annual evaporation will drop somewhat below these figures depending on the salinity of water.

ii *Case study from Lake Qaroun, Middle Egypt*: The Fayoum Depression lies in the western desert of Egypt, about 100 km southwest of Cairo. The depression is separated from the Nile Valley by a narrow strip of desert land 25 km wide. At the northern edge of the depression lies lake (birket) Qaroun. The lake water surface is at about 44 m b.m.s.l., the corresponding surface area of 240 km^2.

Since the lake has no natural outlet, a delicate balance exists between inflow and evaporation. The lake receives each year about 30% of the irrigation water supply to Faiyum Province, i.e. a volume of say 700×10^6 m^3 as drainage water. The lake

Table 7. Mean monthly and annual evaporation from the reservoirs of the Old Dam and the Aswan High Dam using different estimation methods (Sadek, 1992)

Mean monthly and annual rate of evaporation, mm d^{-1}, for													
J	F	M	A	M	J	J	A	S	O	N	D	Year	Source
3.8	x	x	8.7	x	x	10.0	x	x	7.8	x	x	7.5	a_A
4.5	x	x	9.3	x	x	9.9	x	x	8.1	x	x	7.9	a_H
3.8	4.6	6.5	8.4	9.3	10.8	9.8	9.6	9.1	7.8	5.4	3.7	7.4	b_A
4.4	5.5	7.2	9.1	9.2	10.7	9.7	8.8	9.1	8.0	5.9	4.3	7.6	b_H
5.8	4.9	3.4	5.0	6.3	7.9	8.9	4.0	10.9	10.0	9.0	7.0	7.4	c
3.9	5.4	7.4	8.9	10.0	11.2	11.0	10.6	9.5	7.3	6.3	4.1	7.9	d_1
3.8	4.5	5.1	6.4	9.2	10.9	10.0	9.6	9.5	6.0	6.0	4.7	7.4	d_2
3.3	3.9	4.5	5.2	5.8	6.1	5.7	5.5	5.0	4.3	3.7	3.2	4.7	e_1
3.7	4.3	5.1	5.8	6.1	6.4	6.6	6.4	5.7	5.1	4.2	3.7	5.2	e_2
3.4	4.2	5.3	6.4	7.5	7.9	7.4	7.3	6.5	5.4	4.1	3.3	5.7	e_3
4.4	4.5	4.8	4.4	6.8	7.8	7.0	6.1	6.6	7.2	6.4	5.3	5.9	e_4
4.1	5.0	6.9	8.8	9.5	10.8	9.8	9.2	9.1	7.9	5.6	4.0	7.5	g
5.2	5.3	5.6	5.7	7.3	8.1	9.4	9.5	9.4	8.5	7.0	5.8	7.2	h_1
4.8	5.9	7.3	8.5	10.6	10.5	9.2	9.9	9.0	8.5	6.2	8.0	8.0	h_2
3.5	4.6	6.2	7.7	8.5	9.4	9.3	8.8	8.2	6.5	4.5	3.7	6.7	i_A
4.5	5.8	7.1	8.7	9.3	9.8	9.6	9.3	8.6	7.7	6.1	4.0	7.5	i_H
4.7	4.1	4.4	5.4	5.2	5.7	8.7	7.9	7.2	5.9	6.0	5.5	5.9	j_1
4.2	4.6	3.9	3.6	6.3	8.2	7.5	6.1	6.7	7.8	6.7	5.1	5.9	j_2
4.6	4.9	5.3	5.1	7.5	8.7	8.1	7.0	7.6	8.0	6.8	5.2	6.6	j_3
4.2	4.3	4.7	4.4	6.7	8.0	7.3	6.3	6.6	7.1	6.2	5.0	5.9	j_4

Explanation

a_A = *0.5 x Piche evaporation at Aswan in the period 1920–29 (Hurst & Philips, 1931)*

a_H = *0.5 x Piche evaporation at W.Halfa in the period 1905–29 (Hurst & Philips, 1931)*

b_A = *0.5 x Piche evaporation at Aswan in the period 1920–50 (Hurst et al., 1959)*

b_H = *0.5 x Piche evaporation at W. Halfa in the period 1905–50 (Hurst et al., 1959)*

C = *monthly evaporation distributed according to Dalton's law (Attia et al., 1979)*

D_1 = *bulk aerodynamic (Dalton) combined with heat budget (Omar & Bakry, 1970)*

D_2 = *estimate is based on lake measurements taken in 1970–71 (Omar & Bakry, 1981)*

E_1 = *results obtained from the original CRLE model (Morton, 1979)*

E_2 = *same as* e_1 *except that expression of net heat budget of Lake Mead, USA, has been used*

E_3 = *results obtained from Morton's modified CRLE model (Morton, 1983)*

E_4 = *same as* e_3, *except that heat input data mentioned in* d_2 *have been used*

H_1 = *aerodynamics of the air near the reservoir for the period 1981–84 (PTJC, 1982, 83 and 84)*

H_2 = *0.5x Piche evaporation from the floating station in the reservoir (PTJC, 1982, 83 and 84)*

i_A = *Penman evaporation estimated from the original (1948) formula for Aswan (Shahin, 1985)*

i_H = *Penman evaporation estimated from the original (1948) formula for Wadi Halfa (Shahin, 1985)*

J_1 = *from the water balance of the reservoir for the period 1976–82 (Sadek, 1992)*

j_2 = *heat budget method applied to the reservoir of the Aswan High Dam (Sadek, 1992)*

j_3 = *Penman evaporation using heat storage, advective heat and wind function as in* d_2

j_4 = *average of the estimates for Aswan and Wadi Halfa from the CRLE model and the same advective energy and heat storage as in* d_2 *(Sadek, 1992).*

Table 8. Calculated monthly and annual fresh water evaporation from the (projected Qattara Lake (from Flohn & Wittenburg, 1980 and Sundborg & Nilsson, eds. 1985)

Mean monthly and annual evaporation, mm, for													Method
J	F	M	A	M	J	J	A	S	O	N	D	Year	
59	62	103	137	177	228	201	190	190	164	119	71	1701	a
62	70	121	150	195	204	217	226	189	152	105	81	1772	b
49	75	129	175	220	237	236	221	178	128	76	44	1768	c
62	68	115	148	191	210	214	217	189	155	109	78	1756	d

Explanation
a = bulk aerodynamic method, b = energy budget method, c = Penman method and d = Mc Ilroy method.

has a capacity to evaporate 425×10^6 m^3 y^{-1} and the remaining 275×10^6 m^3 y^{-1} must therefore be spilled.

Table 9 includes evaporation measurements as well as estimates obtained from a number of methods. The observations from a pool 144 m^2 in surface area adjusted by a constant factor of 0.9 (method 'b') and from a floating pan adjusted by a variable factor, 0.75–0.90, depending on the month (method 'c') and Penman evaporation (method 'f') give nearly the same estimates. This is especially true on the annual level where the average is in the order of $1,700$ mm y^{-1} and the difference between the three values does not exceed 2%. The water balance (method 'a') compared to the average of the three methods 'b', 'c' and 'f' shows a difference of about 8% for the year. The same percentage difference can be obtained while comparing pen evaporation from the two methods, 'd' and 'e'. The difference between the annual evaporations should have been expected as one of the pans is painted and filled with saline water while the second is

Table 9. Measurements and estimates of monthly and annual evaporation from Lake Qaroun, Egypt (Ali et al., 2000, and Mankarious, 1979)

Mean monthly and annual evaporation, mm, for													Method
J	F	M	A	M	J	J	A	S	O	N	D	Year	
79	117	135	160	210	224	228	212	174	133	95	67	1834	a
60	77	121	152	199	226	229	224	187	147	101	62	1685	b
52	65	110	151	208	233	227	208	168	130	80	51	1683	c
96	125	193	247	317	350	349	312	273	226	150	103	2741	d
98	130	188	264	307	353	370	360	345	270	170	119	2974	e
55	82	121	162	199	227	220	202	172	134	90	58	1721	f

Explanation
a = water balnce of the lake 1990–95, b = measurements from a pond 144 m^2 *multiplied by 0.90, c = measurements from a floating pan* 1.0 m × 1.0 m × 1.0 m *multiplied by 0.75–0.90 depending on the month, d = Class 'A' pan filled with sweet water and evaporation estimated from Christiansen's method, e = evaporation from a standard Class 'A' pan painted in grey and f = Penman's evaporation. All estimated values from b to f are averages for the period 1969–75.*

obtained from calculation using Christiansen method (1966) in which water is assumed to be sweet.. According to Mankarious (1979), the lake evaporation can be obtained by adjusting the measurements from the painted pan by a constant factor of 0.59.

iii *Case study of the reservoir of Aswan High Dam (Egypt/Sudan)*: This dam is located on the Main Nile in Egypt. The dam construction began in 1960 and continued until 1968. The reservoir at its flood level is located between latitude 23°58′ N in Egypt and 20°27′ N at the Dal cataract in the Sudan. At its full level the surface area of the reservoir formed by the dam is in the order of 6, 500 km^2. The loss of water by evaporation is enormous, qualifying it for intensive amount of research and observation. Table 7 gives long-term mean monthly and annual evaporation.

The annual averages as obtained from the CRLE models, water balance, heat balance, Penman and bulk aerodynamic estimation methods are 5.7 mm d^{-1}, 5.9 mm d^{-1}, 5.9 mm d^{-1}, 6.6 mm d^{-1} and 7.1 mm d^{-1} respectively. According to Sadek et al. (1997), the limited data available for the purpose of calculation are generally inconsistent in time and inadequate as regards spatial extent. These constraints can justify not taking the results obtained from the aerodynamic approach into account. By doing so, the overall average rate drops to just 6.0 ± 0.3 mm d^{-1}. Even if the results of the aerodynamic method are not left out and taken into account, the overall average of the five methods becomes 6.25 ± 0.3. These figures are 20% and 17% less than the figure currently used in practice.

iv *Case study from the Sennar Reservoir, the Sudan*: The Sennar (Makwar) Dam has been built in 1925 on the Blue Nile, the Sudan. It is located at latitude 13°53′ N and longitude 33°35′ E and elevation 403 m a.m.s.l. The dam has been functioning since its construction, although its live storage capacity has reduced in the course of time from an initial volume of 0.93×10^9 m^3 to about 0.61×10^9 m^3 in 1986 (Shahin, 1993).

The surface area of the reservoir created by the dam is about 160 km^2 at the full retention level. The reports of the Permanent Technical Joint Committee for the Nile Water for the two water years 1982–83 and 1983–84, using the water balance method, give an average annual evaporation from the said reservoir as 317×10^6 m^3. This figure represents about 50% of the current live storage capacity. Additionally, it shows that the depth of evaporation is 2, 265 mm y^{-1} corresponding to an average of 6.2 mm d^{-1}, assuming the average surface area of the reservoir 140 km^2.

Table 10 gives monthly and annual estimates of evaporation from the Sennar reservoir. It appears from this table that the annual evaporation rate, 7.2 mm d^{-1}, adopted by the Ministry of Irrigation, Sudan, is 16% higher than that obtained from the water balance method. This difference is reduced to say 9% when the water balance figure is compared to that obtained from the Piche evaporimeter (method b) after adjusting it by the coefficients included in Table 10. It is remarkable that Abulai et al., (1990) have reached a similar conclusion though using a different approach. "..... Sennar evaporation data calculated with the Penman approach show an appreciably lower value for the lake evaporation than just correcting the

Table 10. Estimates of monthly and annual evaporation from the reservoir formed by the Sennar Dam on the Blue Nile, the Sudan

Mean monthly and annual rate of evaporation, mm d^{-1}, for													
J	F	M	A	M	J	J	A	S	O	N	D	year	Source
4.5	3.2	3.5	5.7	7.4	7.3	8.1	8.7	11.0	10.3	9.3	7.4	7.2	a
15.0	19.0	22.0	21.0	16.0	14.0	7.9	5.1	6.6	9.9	15.0	14.0	14.0	b
6.2	7.4	8.7	9.4	7.7	7.7	5.5	4.2	5.0	7.1	6.8	6.1	6.8	c

Explanation
a = evaporation rates adopted by the Ministry of Irrigation in the Sudan, b = Piche evaporation,
b = Piche evaporation, average for the period 1921–50 (Hurst et al., 1959) and
c = adjusted Piche evaporation using a monthly coefficient between 0.40 and 0.80.

Piche reading (placed within a Stevenson screen) by a coefficient equal to 0.5, already applied in the Sudan since long. This is true for all three stations used in the said investigation. The difference varies from 1.5 mm d^{-1} for the reservoir data input to 4 mm d^{-1} for the Sennar meteorological station data input, with its reduced air movement." As such, it is highly probable that the Sennar reservoir, despite its limited surface area, acts more or less similar to the reservoir of the Aswan High Dam so that the average evaporation from both reservoirs is in the range 6.0–6.3 mm d^{-1}. This means, conservatively, that the Ministries of Irrigation in both Egypt and the Sudan overestimate the reservoir evaporation by 15%. To the opinion of the author this matter needs a more profound investigation.

v *Case study of the Tharthar Reservoir (Central Iraq) and Lake Tiberias (Israel)*: The Tharthar reservoir is situated in the Tharthar region, Central Iraq. The climate of this region is continental subtropical. Very high temperatures, low atmospheric pressure, low atmospheric humidity and frequent periods of strong wind characterize the long summer of the region.

Wartena (1959), upon investigating the evaporation from the said lake concluded that due to the absence of cloud in Iraq the usual formulas are not suitable for calculating the total global radiation of the sun. As such, it was necessary to develop a new relationship between the degree of cloudiness and relative duration of sunshine. He estimated the evaporation once assuming the lake shallow (10 m deep) and another time assuming the lake deep (50 m). Additionally, he computed the evaporation using Penman (1948) formula. The annual evaporation from the three calculations in their respective order was 2,456 mm, 2,563 and 2,820 mm. The distribution of these amounts between the months of the year is shown in Figure 7(a).

Wartena (1959) attempted to calculate the temperature of the surface using elementary meteorological quantities. The method was checked on Lake Tiberias in the Jordan Rift, and a reasonable agreement found. Since the Tharthar Lake did not exist at the time that this calculation method was developed, it was not possible to compare the calculation results to actual measurements from the lake. The curves shown in Figure 7(b) represent the monthly evaporation from Lake Tiberias using the formula of Penman, the values obtained by Neumann (1953) and

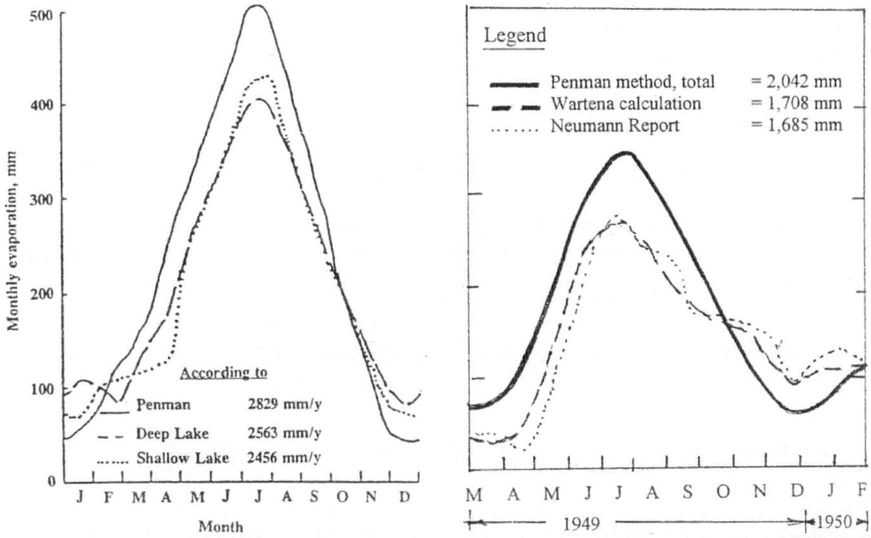

Figure 7(a). Lake evaporation in Central. (b). Evaporation of Lake Tiberias, Iraq (Wartena, 1959) Jordan Rift (from Wartena, 1959)

the calculations by Wartena according to the suggested course of temperature. As usual the formula of Penman gave the greatest annual evaporation, 2,042 mm. The annual evaporation as found by Neumann and Wartena was 1,685 mm and 1,708 mm, respectively. The average of these two figures correspond to 4.65 mm d^{-1}, which is about two-thirds the annual figure for Lake Tharthar in Iraq.

The above-reviewed case studies show clearly that the average annual rate of evaporation varies globally from of 4.65 mm d^{-1} for the Qattara Depression near the Mediterranean Sea as well as Lakes Qaroun and Tiberias, situated in rather wet environments, to 6.0–6.3 mm d^{-1} for the reservoirs of the Aswan High Dam and Sennar Dam and ultimately reaches 6.8 mm d^{-1} in Central Iraq.

5.4. ESTIMATION OF EVAPOTRANSPIRATION

5.4.1 Actual evapotranspiration $(ET)_a$

5.4.1.1 Empirical formula of Turc for computing catchment actual evapotranspiration

Assuming that annual precipitation, P, and mean annual temperature, T_m, are the dominant factors in evapotranspiration, Turc (1954) gave an expression for estimating the annual evapotranspiraion, *ET* The expression he developed is as follows:

$$ET = \frac{P}{(0.9 + (P/L)^2)^{1/2}} \tag{11}$$

where,

P and ET are in mm y^{-1}. L is a heat index given by:

$$(12) \qquad L = 300 + 25\ T_m + 0.05\ T_m^3$$

in which T_m is in °C.

Example: Consider an area with an annual precipitation of 200 mm y^{-1} and a mean temperature of 22 °C. From Eq. 12, the index L is equal to 1382.4. Substituting this value along with $P = 200$ mm in Eq. 11, the annual ET can be obtained as 208.4 mm. Obviously, unless there is a contribution from another source, e.g. high water table, the value of ET can not exceed 200 mm. Suppose that P for the same area was 600 mm instead of 200, the annual ET value should have become 575 mm.

5.4.1.2 *Theoretical approach based on depletion of soil moisture in the root zone*

In order to determine the actual evapotranspiration $(ET)_a$, the level of the available soil moisture in the root zone must be considered. According to the definitions given in subsection 5.1.2. $(ET)_a = (ET)_{mx}$ when the available soil moisture in the root zone to the crop is adequate. Doorenbos et al. (1986) have defined the available soil water as the fraction (p) to which the total available soil water can be depleted without causing $(ET)_a$ to fall below $(ET)_{mx}$. This fraction varies from crop to crop and probably with the stage of plant growth. The magnitude of $(ET)_a$ can be quantified for periods between successive irrigations or heavy rainstorms, and for monthly periods as well.

The total available soil water, *TASW,* is the difference between the soil moisture level at the field capacity and the moisture level at the wilting point, both expressed in mm/m. In the absence of actual measurements one can fairly assume that (*TASW*) equals 200 mm m^{-1}, 140 mm m^{-1} and 60 mm m^{-1} for heavy textured, medium textured and coarse textured soils respectively. The interval, I, between two successive irrigations for the condition $(ET)_a = (ET)_{mx}$ can be computed from the following equation:

$$(13) \qquad I = pD(TASW)/(ET)_a = pD(TASW)/(ET)_{mx}$$

in which D is the depth of the root zone, m. When $(ET)_a = (ET)_{mx}$ is given in mm d^{-1} the irrigation interval, I, will be in days.

The maximum evapotranspiration, $(ET)_{mx}$, will be maintained until the fraction (p) of the total available soil water has been consumed. Beyond this depletion level actual evapotranspiration, $(ET)_a$, becomes increasingly smaller than $(ET)_{mx}$ until the next irrigation application or rain spell takes place. In the phase of $(ET)_a < (ET)_{mx}$, the actual evapotranspiration will depend on the amount *(1-p) D (TASW)* as well as the maximum evapotranspiration. Values of $(ET)_a$ given p, *D (TASW)* and $(ET)_{mx}$ can be read directly or by interpolation from the values given in certain tables

Table 11. Variation of $(ET)_a$ over the irrigation interval given $(ET)_{mx}$, *p and D. (TASW)* (from Doorenbos et al., 1986)

D. (TASW), Mm	*p*, %	$(ET)_a$, mm d^{-1}, for the given irrigation intervals, days 4	6	8	10	12	14	16	18	20	24	30	40
$(ET)_{mx} = 4mm\ d^{-1}$													
25		4.0	3.5	2.9	2.4	2.1	1.8	1.6	1.4	1.3	1.0	0.8	0.6
50		4.0	4.0	4.0	3.8	3.5	3.2	2.9	2.6	2.4	2.1	1.6	1.2
100	60	4.0	4.0	4.0	4.0	4.0	4.0	4.0	3.9	3.8	3.5	3.0	2.4
150		4.0	4.0	4.0	4.0	4.0	4.0	4.0	4.0	4.0	4.0	3.8	3.3
200		4.0	4.0	4.0	4.0	4.0	4.0	4.0	4.0	4.0	4.0	4.0	3.8
$(ET)_{mx} = 6mm\ d^{-1}$													
25		5.2	4.0	3.1	2.5	2.1	1.8	1.6	1.4	1.3	1.0	0.8	0.6
50		6.0	5.9	5.2	4.6	4.0	3.5	3.1	2.8	2.5	2.1	1.7	1.3
100	60	6.0	6.0	6.0	6.0	5.9	5.6	5.2	4.9	4.6	4.0	3.3	2.5
150		6.0	6.0	6.0	6.0	6.0	6.0	6.0	5.9	5.7	5.2	4.6	3.6
200		6.0	6.0	6.0	6.0	6.0	6.0	6.0	6.0	6.0	5.9	5.4	4.6
$(ET)_{mx} = 8mmd^{-1}$													
25		5.8	4.1	3.1	2.5	2.1	1.8	1.6	1.4	1.3	1.0	0.8	0.6
50		8.0	7.0	5.8	4.8	4.1	3.6	3.1	2.8	2.5	2.1	1.7	1.3
100	60	8.0	8.0	8.0	7.6	7.0	6.4	5.8	5.3	4.9	4.1	3.3	2.5
150		8.0	8.0	8.0	8.0	8.0	7.7	7.4	7.0	6.6	5.8	4.8	3.7
200		8.0	8.0	8.0	8.0	8.0	8.0	8.0	7.8	7.6	7.0	6.1	4.8
$(ET)_{mx} = 10mm\ d^{-1}$													
25		6.0	4.2	3.1	2.5	2.1	1.8	1.6	1.4	1.3	1.0	0.8	0.6
50		9.5	7.6	6.0	4.9	4.2	3.6	3.1	2.8	2.5	2.1	1.7	1.3
100	60	10.0	10.0	9.5	8.5	7.6	6.8	6.0	5.4	4.9	4.2	3.3	2.5
150		10.0	10.0	10.0	9.9	9.5	8.9	8.2	7.6	7.0	6.0	4.9	3.7
200		10.0	10.0	10.0	10.0	10.0	9.8	9.5	9.0	8.5	7.6	6.4	4.9

prepared by Doorenbos et al. (1986). An example of such values can be found in Table 11.

Jensen (1968) suggested that the relationship between the evapotranspiration, *ET*, and the potential evapotranspiration, $(ET)_o$ can be represented as:

$$(14) \qquad K_{co} = ET/(ET)_o$$

However, when the soil water limits *ET*, the crop coefficient, K_{co}, has to be adjusted according to the available water *(1-p)D(TASW)*. One of the possible ways for adjustment is to use a reduction coefficient, K_a, along with K_{co} such that the reduced crop coefficient, $K'_c = K_{co}.\ K_a$. The reduction coefficient, K_a, takes the values of 0.852, 0.891, 0.924, 0.952, 0.977 and 1.00 when p = 50, 40, 30, 20, 10 and 0.0, respectively.

Wright (1982) developed a different procedure in which he replaced K_{co} by K''_c. He claimed that this procedure helps to establish more accurate estimation of the daily ET whenever the state of wetness of the soil surface can be specified, e.g. after rainfall or irrigation. The proposed coefficient can be expressed as:

$$(15) \qquad \boldsymbol{K}_c'' = \boldsymbol{K}_{cb}\boldsymbol{K}_a' + \boldsymbol{K}_s$$

in which K_{cb} is a basal crop coefficient, K_a' is a function of the available soil water, and K_s is an adjustment coefficient depending on the extent of wetness of the soil surface, $K_s > 0$ for a wet surface and 0 when dry.

The FAO Paper No. 56 (Allen et al., 1998) went another step further by changing Eq. (15) slightly to read

$$(16) \qquad \boldsymbol{K}_c'' = \boldsymbol{K}_{cb} + \boldsymbol{K}_e$$

in which K_{cb} can be defined as $ET/(ET)_r$ when the soil surface is dry but transpiration is occurring at a potential rate, i.e. water is not a limiting factor. The coefficient K_e serves to get values for surface evaporation. This is called the dual crop coefficient approach.

5.4.2 Consumptive use, and potential and reference evapotranspiration

The group of formulas presented in subsections 5.4.2.1, 5.4.2.2 and 5.4.2.3 is often referred to as the temperature group. Of the group usually referred to as the radiation group we shall confine our review and case studies to just one approach, i.e. the original Jensen & Haise (1963) formula and its subsequent development.

5.4.2.1 Blaney-Criddle formula

The Blaney-Criddle (B-C) consumptive use formulas (1950/52) can be expressed as:

$$(17) \qquad cu = k_c(0.4572T + 8.128)p/100$$

and

$$(18) \qquad CU = \Sigma cu = K_c \sum_{i=1}^{i=n} (0.4572T_i + 8.128)p_i/100$$

where,

$cu =$ monthly consumptive use, mm mo^{-1},
$CU =$ seasonal or annual consumptive use, mm,
$k_c =$ monthly consumptive use coefficient,
$K_c =$ seasonal consumptive use coefficient,
T, $T_i =$ monthly temperature of month i, °C, and
$p/100$, $p_i/100 =$ percent of day time hours of the month considered to daytime hours of the year,

Example; Consider the climatic data of Giza (Station No.74), Egypt. The mean temperature, T, is 26.9 °C and $p/100 = 9.93$. The consumptive use coefficient for cotton is 1.1. Substituting these data in Eq. (17) the consumptive use of water for the said month is 223 mm or 7.2 mm d^{-1}.

Blaney (1957) gave monthly values of the coefficient k_c for a variety of irrigated crops as obtained from the consumptive use studies in Arizona and California, USA. He distinguished between crops grown under coastal, interior and intermediate climates. Additionally, Blaney & Criddle (1966) gave monthly values of p for latitudes 26 to 48° north of the Equator, and the seasonal coefficients, K_c, for a number of crops.

In view of its simplicity and ease of application, the formula of Balney and Criddle has gained a wide success in many of the Arab countries since the 1960s. The values of k_c and K_c for a variety of crops growing in Egypt and listed in Table 12 are just an example.

Some adjustments to the B-C formula were introduced by Quackenbush & Phelan (1965) and shortly after that by the Soil Conservation service of the United States Department of Agriculture (USDA/SCS, 1967). The modified B-C (sometimes referred to as FAO/B-C) formula introduced by Doorenbos & Pruitt (1977) can be written as follows:

$$ET_r = a + b[p(0.46T + 8.13)] \tag{19}$$

where,

$$a = 0.0043H_m - \frac{n}{N} - 1.41 \tag{20}$$

and

$$\frac{n}{N} = 2(R_G/R_A) - 0.5 \tag{21}$$

where,

Et_r = reference evapotranspiration,
H_m = minimum relative humidity, in percent, quite often taken as the humidity at noon, H_n,
R_G = global short-wave radiation,
R_A = extraterrestrial radiation, and

B = parameter depending on n/N and wind speed, u_2 in m s^{-1}, measured at 2 m above ground surface. Values of b can be obtained from the tables prepared by Allen & Pruitt (1991).

i *Case study from the Arabian Peninsula*: Mean monthly and annual values of grass reference evapotranspiration at 9 stations in the Arabian Peninsula have been calculated by the author using the FAO-Blaney-Criddle formula and the results presented in Table 13 (Shahin, 1998). The stations used comprise 1 station

Table 12. Monthly and seasonal consumptive use coefficents for crops grown in Egypt (from El-Gibali et al, 1966, and Shahin & El-Shal, 1969)

Crop	Monthly use coefficients, k_c, for												K_c
	J	F	M	A	M	J	J	A	S	O	N	D	
Cotton			0.26	0.39	0.66	1.07	1.03	0.56	0.51				0.71
Wheat	0.49	0.59	0.66	0.49	0.31						0.46	0.47	0.51
E. corn*				0.36	0.71	1.06	0.87	0.60					0.79
L. corn**							0.51	0.91	1.13	0.80			0.88
Berseem+	0.31	0.42	0.72	0.82	0.64	0.25				0.31	0.37	0.62	0.56
Citrus orchards	0.35	0.51	0.55	0.57	0.60	0.64	0.70	0.64	0.59	0.56	0.42	0.32	0.54
Fenugreek and lupin	0.56	0.57	0.26								0.38	0.59	0.46
Chick pea and lentil	0.42	0.46	0.18								0.34	0.50	0.37
Sugar cane	0.29	0.72	0.87	0.84	0.90	1.00	1.35	1.37	1.32	1.00	0.79	0.41	0.91
Field beans	0.54	0.73	0.67	0.48									0.58
Potatoes S°		0.63	0.77	1.01	0.69								0.77
Potatoes F^x	0.86									0.19	0.49	0.98	0.59
Snap beans									0.37	0.80	1.14	0.78	0.77
Cow peas				0.30	0.35	0.80	0.99	0.60					0.64
Squash							0.44	1.02	1.01				0.87
Cucumber							0.30	0.63	0.87	0.79			0.69

Explanation
**E. = early*, ***L. = late*, + *Berseem = Egyptian clover*, S^o = *summer and* ^{x} *F = fall*.

Table 13. Estimates of mean monthly and annual grass reference evapotranspiration using FAO-Blaney-Criddle formula (Shahin, 1998)

Station Name	Mean monthly and annual grass reference evapotranspiration, mm, for														Period
	No.	Jan	Feb	Mar	Apr	May	Jun	Jul	Aug	Sep	Oct	Nov	Dec	Year	of record
El-Shuwaikh	161	80	106	144	178	259	297	306	311	250	200	125	81	2337	1965–89
Al-Manamah	166	83	93	122	179	252	296	291	263	225	180	136	86	2207	1989–93
R. Al-Faras	167	79	85	116	145	217	259	237	224	197	157	117	84	1917	1972–92
Abu Samra	169	82	90	120	154	217	230	226	207	189	153	109	95	1872	1976–92
Er Riyadh*	173	89	112	147	191	248	298	341	293	247	176	126	88	2356	1970–81
Muscat	180	112	131	167	200	258	252	258	228	218	169	147	121	2261	1981–92
San'a	186	135	115	149	157	209	238	230	227	211	166	152	126	2115	1982–86
Al-Hodeydah	189	113	116	142	154	171	173	182	180	170	152	124	110	1787	1983–90
Al-Khod	197	148	142	164	159	217	205	210	194	185	177	158	143	2102	1982–83

* *Results are quoted from Salih and Sendil, 1984, and Solaiman et al., 1987.*

(No. 161) in Kuwait, 1 station (No. 166) in Bahrain, 2 stations (No. 167 and 169) in Qatar, 1 station (No. 173) in Saudi Arabia, 1 station (No. 180) in Oman and 3 stations (No. 186, 189 and 197) in Yemen. The stations at Al Hodeydah, Yemen (2,382 mm), Riyadh, Saudi Arabia (2,356 mm) and Shuwaikh, Kuwait (2,337 mm) form the group of stations yielding the highest annual reference evapotranspiration. The stations in Oman (2,261 mm) and Bahrain (2,207 mm) form the second highest group, 4–5% smaller than the first high group. The group having the least $(ET)_r$ is the two stations in Qatar (1,917 and 1,872 mm). As such the minimum $(ET)_r$ constitutes about 80% of the maximum $(ET)_r$.

ii *Case study from Syria*: The Khabur Basin irrigated plains are located in the semi arid region of Syria as shown in the map, Figure 8. As explained in Chapter 3, the prevailing Mediterranean climate is characterized by a rainy winter and a dry hot summer. The mean annual rainfall in the area is about 275 mm. The annual total evaporation from Class 'A' pan established within an irrigated field in the said area averages 2,100 mm. The monthly pan evaporation varies from a maximum of 12.1 mm d^{-1} in June to 6.9 mm d^{-1} in October to a minimum of 1.4 mm d^{-1} in December-January (Wakil & Bonneli, 1996).

The monthly potential evapotranspiration from rainfed/irrigated wheat was estimated from the FAO-Blaney-Criddle formula (Doorenbos & Pruitt, 1977). The estimates thus obtained were: 45, 40, 10, 15, 70, 145, 130, 85 and 35 mm for October, November, ….., May and June respectively, with a seasonal total of 575 mm.

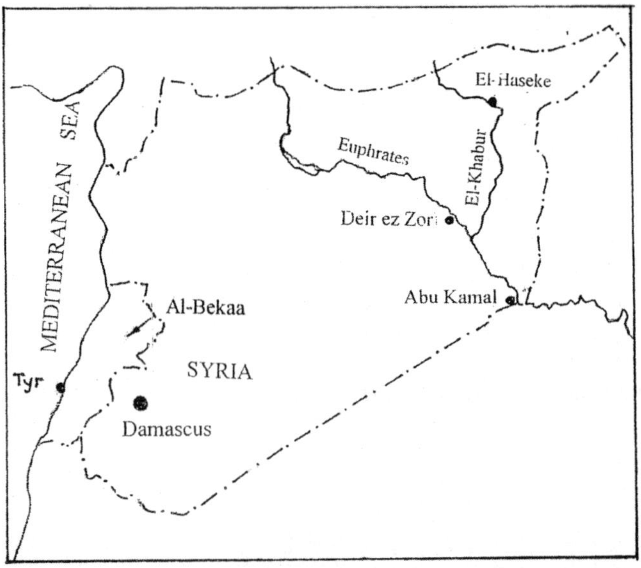

Figure 8. Location map of the Khabour River in Syria

5.4.2.2 *Hargreaves formula*

The original formula developed by Hargreaves (1956) for estimating evaporation was adjusted later by Hargreaves and Samani (1982) to yield estimates of potential evapotarnspiration using the expression:

$$(22) \qquad (ET)_o = 0.0135 R_A.(T + 17.8)\,[0.055 + 0.042(1 - h)]\,\sqrt{n/N}$$

Values of R_A in units of evaporation can be obtained for any given latitude and time from the standard tables available in current literature, e.g. FAO-Paper 24 (Doorenbos & Pruitt, 1977) or ASCE Report on Consumptive Use of Water for Irrigation and Water Requirements (eds. Jensen et al., 1990). The results obtained from 29 m^2 weighable lysimeters at Davis, California, raising Alta fescue grass, were used by Hargreaves and Samani (1985) to rewrite Eq. (22) as:

$$(23) \qquad (ET)_o = 0.0023 R_A.(T + 17.8)\sqrt{5.56 + 17.8(1 - h)}$$

Again, the mean monthly and annual reference evapotranspiration for the same stations in the Arabian Peninsula, appearing in Table 13, have been recalculated using the Hargreaves formulas. The results thus obtained are given in Table 14 (Shahin, 1998).

5.4.2.3 *Thornthwaite formula*

Thornthwaite (1948) used the results he obtained from certain parts of USA to develop a procedure for estimating ET_o, cm mo^{-1}, from short, dense vegetation with adequate water supply. The procedure employs the mean air temperature, number of daytime hours and days of the month as follows:

$$(24) \qquad ET_o = 1.6\left(\frac{N_m x N_{dh}}{30x12}\right)\left(\frac{10T_m}{J}\right)^a = \left(\frac{N_m x N_{dh}}{225}\right)\left(\frac{10T_m}{\Sigma_1^n (0.2T_m)^{1.514}}\right)^a$$

where,

N_m = number of days in the month considered,
N_{dh} = average number of daytime hours per day of the month considered,
T_m = mean air temperature, in °C, of the month considered,
J = heat index over the period considered,
n = number of months over which the summation is made, which can be a season or a year, and
a = a function of the heat index, J, given as:

$$(25) \qquad a = (675x10^{-9})J^3 - (771x10^{-7})J^2 + (178x10^{-4})J + 0.498$$

Referring to Serra, Wilson (1983) suggested that Eq. (25) could be approximated to:

$$(25') \qquad a = 0.016J + 0.5$$

Table 14. Estimates of mean monthly and annual grass reference evapotranspiration using Hargreaves formulas (Shahin, 1998)

Station	Mean monthly and annual grass reference evapotranspiration mm, for														Period of
Name	No.	Jan	Feb	Mar	Apr	May	Jun	Jul	Aug	Sep	Oct	Nov	Dec	Year	record
El-Shuwaikh	161	73	93	139	167	237	264	280	280	220	166	97	66	2082	1965–89
		69	*87*	*137*	*182*	*237*	*267*	*282*	*263*	*210*	*153*	*95*	*76*	*2058*	
Al-Manamah	166	75	82	112	148	193	221	234	222	184	151	108	82	1802	1989–93
		77	*84*	*123*	*163*	*207*	*222*	*233*	*212*	*170*	*138*	*101*	*79*	*1809*	
R. Al-Faras	167	71	83	106	155	200	219	212	203	174	143	99	74	1739	1972–92
	x	*79*	*91*	*134*	*177*	*231*	*234*	*247*	*232*	*205*	*162*	*109*	*82*	*1983*	
Abu Samra	169	67	73	101	128	167	176	174	183	158	135	93	71	1526	1976–92
	x	*76*	*81*	*123*	*164*	*194*	*192*	*206*	*196*	*177*	*147*	*99*	*83*	*1738*	
Er Riyadh	173	93	112	154	182	220	248	257	243	210	172	120	90	2101	1970–81
	x	*88*	*103*	*145*	*182*	*232*	*250*	*263*	*258*	*216*	*169*	*114*	*91*	*2111*	
Muscat	180	101	115	157	189	233	211	215	196	175	160	117	99	1968	1981–92
		95	*107*	*145*	*175*	*212*	*189*	*187*	*171*	*156*	*139*	*109*	*89*	*1774*	
San'a	186	121	114	138	142	148	166	149	160	166	143	132	109	1688	1982–86
	x	*129*	*130*	*154*	*162*	*190*	*188*	*190*	*170*	*165*	*154*	*131*	*122*	*1885*	
Al-Hodeydah	189	*113*	*116*	*142*	*154*	*171*	*173*	*182*	*180*	*170*	*152*	*124*	*110*	*1787*	1983–90
Al-Khod	197	121	122	143	141	177	163	156	149	142	155	133	114	1716	1982–83
		115	*116*	*140*	*146*	*163*	*163*	*175*	*164*	*149*	*136*	*119*	*113*	*1699*	

Figures in upper line refer to formula,1982; in lower line italic, 1982 and 1985 combined; and in lower line italic preceded by x, 1985 alone.

Table 15. Estimation of mean monthly and annual potential evapotranspiration at selected stations in the Arab region using Thornthwaite's formula

Stat. No.	Potential evapotranspiration, in mm mo^{-1}, for Jan	Feb	Mar	Apr	May	Jun	Jul	Aug	Sep	Oct	Nov	Dec	Year
1	30	41	120	159	191	200	207	200	181	167	101	31	1626
2	25	53	139	167	198	199	202	190	177	169	112	54	1685
3	58	60	92	119	153	158	157	160	157	148	101	56	1419
4	36	106	149	166	195	193	180	179	170	174	129	42	1719
5	61	107	163	184	203	200	199	187	161	174	138	35	1852
6	68	117	169	176	203	199	198	186	176	180	147	94	1913
7	30	31	42	53	79	102	123	123	97	72	42	32	826
8	32	32	45	51	82	104	138	137	104	73	46	32	876
9	17	24	36	54	76	115	164	152	106	68	32	21	885
10	29	32	44	56	79	100	122	123	97	77	46	35	840
11	22	26	39	49	81	108	147	142	103	69	38	23	847
12	30	31	43	53	79	102	122	119	99	75	45	34	852
13	6	12	24	43	59	93	131	118	78	48	26	8	646
14	19	23	43	60	95	125	185	179	129	84	37	23	1002
15	14	20	37	54	88	120	188	171	116	59	28	14	899
16	38	40	56	61	83	94	112	107	92	79	55	38	855
17	26	28	42	58	85	116	149	148	113	78	44	29	916
18	25	27	43	49	80	110	139	143	109	72	45	33	875
19	23	24	36	53	81	112	143	139	106	70	40	25	852
20	14	17	34	44	81	124	167	161	110	59	31	18	860
21	9	17	41	73	163	208	224	208	169	110	32	11	1265
22	14	19	39	63	115	192	214	200	151	66	31	15	1129
23	27	31	40	54	85	110	128	131	101	72	45	31	855
24	10	14	30	50	84	131	166	152	107	59	28	13	844
25	12	16	33	52	101	153	194	177	114	60	26	12	950
26	11	18	48	60	117	170	236	223	134	66	31	11	1125
27	10	16	38	67	121	191	235	213	142	71	28	11	1143
28	10	16	40	75	185	208	215	205	182	136	36	14	1320
29	14	24	50	88	146	230	253	230	174	97	41	16	1363
30	19	24	52	85	146	180	179	167	136	94	51	26	1159
31	24	25	35	47	78	115	114	145	108	70	38	27	826
33	19	22	36	54	88	129	168	161	118	76	41	22	934
34	20	22	32	44	86	128	165	159	111	67	33	21	888
35	15	22	35	57	91	142	189	174	125	76	38	20	985
37	22	26	39	52	88	121	148	148	115	76	39	25	899
38	10	16	35	66	113	165	201	185	134	77	33	16	1051
39	14	19	38	66	136	182	209	192	149	96	28	14	1143
40	21	26	41	59	91	124	162	161	120	79	43	24	951
41	24	27	41	58	92	123	162	161	119	80	42	28	957
43	24	28	33	66	93	125	151	152	132	93	55	29	981
44	21	22	34	53	88	117	127	123	96	70	44	27	822
45	14	20	35	55	96	142	161	157	109	71	39	19	918
46	28	28	43	63	87	114	140	140	114	90	56	35	938
47	17	22	38	64	106	166	184	174	149	87	46	21	1074
48	26	28	43	59	89	121	144	156	118	85	51	31	951
49	25	25	40	65	103	136	141	140	113	85	49	29	951
50	14	20	36	61	103	144	171	164	118	72	36	18	957

52	14	21	39	66	111	167	183	174	119	74	48	18	1034
53	24	28	47	71	99	129	150	157	131	103	60	32	1031
54	21	35	47	73	116	128	137	154	112	83	56	29	991
55	8	16	42	91	167	202	210	197	159	96	36	12	1236
56	16	22	41	75	119	173	172	163	120	78	39	20	1038
57	19	24	45	78	126	184	187	178	147	83	42	23	1136
58	14	22	44	84	165	191	189	181	155	94	40	17	1196
59	16	28	56	111	183	191	195	189	153	114	58	25	1319
60	27	30	41	58	84	108	127	128	106	81	51	32	873
61	25	38	44	65	94	123	145	138	110	92	58	33	955
64	27	29	43	59	83	113	132	132	114	90	56	37	915
65	26	26	42	62	96	138	153	159	119	94	57	35	1007
66	28	29	47	63	90	116	154	154	140	94	60	37	1106
67	24	25	41	61	94	110	134	139	115	91	50	30	914
68	23	24	40	65	108	135	173	165	126	92	57	30	1083
70	23	22	37	67	108	130	145	137	104	85	49	29	936
73	21	27	47	75	140	169	179	168	132	98	53	27	1146
74	19	22	41	68	111	139	163	153	106	82	49	26	979
75	27	26	46	72	138	168	186	177	143	112	56	29	1180
76	20	20	42	71	113	163	166	158	119	81	52	27	1032
77	20	21	42	79	133	168	171	138	131	85	51	20	1059
78	17	22	43	72	88	173	185	172	139	92	46	22	1069
81	18	17	43	75	134	169	181	171	126	88	49	28	1095
82	27	27	46	72	148	173	185	180	150	117	54	35	1214
83	16	20	43	85	153	175	184	174	137	97	45	22	1151
85	32	34	50	76	161	179	183	184	156	139	65	46	1305
86	15	20	43	103	170	186	192	183	150	113	44	20	1239
87	16	21	46	83	177	189	197	187	154	126	46	21	1236
89	18	23	49	139	187	198	204	195	169	148	81	35	1446
90	24	29	60	143	186	193	197	191	167	147	58	30	1425
91	74	61	88	141	176	191	204	198	176	157	132	101	1699
93	63	70	139	170	197	198	200	191	181	171	134	141	1855
94	70	82	149	170	179	179	172	166	174	173	153	90	1757
95	106	112	158	176	195	187	172	159	161	170	149	115	1860
96	88	115	153	173	191	191	172	157	156	164	138	97	1795
97	128	126	168	179	192	174	154	131	144	159	147	131	1833
99	51	63	113	155	177	176	168	153	150	147	84	57	1494
100	107	115	160	177	190	180	165	153	150	163	143	124	1827
102	58	68	123	161	180	173	161	152	153	155	114	72	1570
103	117	124	163	177	188	174	160	147	145	160	144	126	1825
105	134	130	165	179	151	161	146	137	136	151	144	135	1769
107	139	137	169	171	169	149	130	125	140	149	140	138	1756
110	140	137	166	166	165	147	135	129	137	150	145	141	1758
111	156	147	164	156	150	138	118	120	136	145	143	150	1723
113	99	108	144	157	182	190	212	193	177	160	135	113	1870
115	113	104	140	157	182	194	199	196	179	148	126	125	1863
116	52	56	67	77	89	86	88	86	78	63	52	48	842
117	86	90	132	153	181	193	199	195	179	156	132	124	1820
118	54	60	90	102	113	113	118	107	106	86	65	53	1061
119	120	128	153	155	160	152	153	151	152	147	137	130	1738
121	175	165	186	178	177	168	166	165	166	167	166	165	2044

(*continued*)

Table 15. (Continued)

Stat. No.	Potential evapotranspiration, in mm mo^{-1}, for Jan	Feb	Mar	Apr	May	Jun	Jul	Aug	Sep	Oct	Nov	Dec	Year
122	145	140	151	155	152	138	126	125	137	145	145	144	1703
123	143	132	152	148	144	125	119	120	121	143	140	146	1633
124	7	9	20	53	106	175	229	220	137	77	26	11	1070
125	6	10	21	51	105	176	231	233	141	80	27	12	1093
126	7	13	27	51	108	153	166	157	119	74	32	12	993
127	9	15	30	63	109	159	165	148	110	78	34	13	933
128	21	24	39	60	90	122	150	160	120	86	46	26	944
129	5	6	27	57	123	186	215	199	151	75	30	9	1083
130	7	11	31	61	114	183	211	198	145	83	35	12	1091
131	9	12	31	63	113	168	190	181	120	74	26	11	998
132	7	12	26	57	110	161	199	190	120	62	26	12	982
133	10	15	31	61	101	134	164	163	118	74	36	15	922
134	23	26	38	59	90	125	121	141	121	88	50	29	911
135	9	12	26	48	86	121	141	137	102	69	34	14	799
137	25	26	40	59	94	126	153	163	134	104	53	33	1010
138	12	20	38	46	95	121	144	140	112	79	42	19	868
139	13	20	41	37	101	119	145	142	113	83	43	19	876
140	14	19	38	33	109	124	143	139	113	82	44	21	879
141	13	19	39	40	105	129	150	140	110	82	46	20	893
142	10	17	38	49	105	120	137	129	99	78	37	20	839
143	16	15	31	60	99	134	140	129	100	80	53	31	888
144	15	19	42	67	106	140	145	137	102	82	47	28	930
145	10	15	33	75	159	192	199	186	153	112	65	31	1230
146	16	18	29	55	104	116	132	128	101	82	52	24	857
147	15	16	28	61	86	127	136	129	96	84	54	29	861
148	16	21	44	104	174	194	201	186	152	117	48	21	1278
149	5	10	24	54	118	196	219	202	141	66	30	8	1118
150	2	6	20	47	103	182	213	198	149	79	25	6	1030
151	6	10	22	67	157	208	222	209	167	102	30	10	1210
152	5	12	33	78	168	209	222	209	173	117	33	12	1271
153	7	12	29	64	124	168	196	186	141	81	34	11	1053
154	8	13	30	115	198	212	219	208	176	131	29	11	1350
155	6	13	35	96	188	211	219	208	178	116	27	9	1306
156	8	13	31	75	188	201	214	203	175	126	31	13	1268
157	7	14	34	85	190	211	219	207	177	116	30	11	1301
158	11	15	32	83	186	205	214	204	174	115	36	12	1287
159	11	15	38	94	183	203	212	201	174	120	38	15	1304
161	12	16	39	127	192	210	217	207	189	138	37	15	1394
162	11	15	39	125	193	210	216	207	179	140	39	15	1389
163	11	15	41	105	187	198	214	206	179	139	54	16	1365
165	10	15	40	138	196	211	217	207	183	149	42	18	1426
166	24	29	52	111	182	197	208	200	174	146	87	37	1447
167	21	22	43	126	183	200	209	200	175	146	53	27	1405
168	16	22	53	158	195	200	213	200	179	145	75	24	1480
169	18	25	58	144	182	189	206	186	171	130	55	27	1391
170	12	19	41	75	153	186	199	188	153	89	38	17	1170
171	11	14	39	68	142	186	195	185	152	108	49	15	1164
172	14	21	44	116	189	195	212	202	174	142	44	22	1180

173	17	22	60	111	185	204	210	197	169	116	55	21	1367
174	19	26	60	154	198	207	211	203	184	156	50	23	1491
175	34	43	58	139	176	184	191	183	163	149	69	44	1433
176	67	59	106	144	170	177	191	188	165	152	126	92	1637
177	30	32	55	73	134	169	171	160	143	77	42	31	1117
178	125	122	153	171	191	195	198	192	175	171	144	125	1962
179	31	37	56	100	171	185	205	200	170	144	86	47	1432
180	51	51	105	161	202	203	205	186	169	158	119	67	1677
181	34	37	123	168	204	204	205	188	170	155	109	40	1637
182	64	69	117	159	187	180	162	150	136	137	107	83	1551
183	76	80	118	144	161	173	148	129	134	122	109	91	1485
184	54	60	84	91	110	140	149	145	106	90	38	49	1116
185	28	47	128	136	161	185	197	191	161	115	44	32	1425
186	33	46	69	61	96	107	108	95	80	55	40	41	841
187	112	119	145	164	189	189	199	192	178	171	142	127	1927
188	116	112	149	162	191	194	199	194	179	167	139	132	1934
189	116	115	149	163	186	188	195	189	174	165	133	120	1893
190	81	88	120	146	171	180	177	166	158	136	108	95	1626
191	32	42	63	71	89	96	115	107	83	63	40	31	832
192	30	35	59	63	93	107	121	163	91	68	50	24	904
193	118	119	161	172	187	192	194	188	170	160	131	120	1912
194	51	57	71	83	88	79	97	87	74	56	44	40	827
195	120	112	148	155	167	173	184	183	164	159	138	124	1827
196	121	115	146	160	183	184	191	186	175	160	137	123	1881
197	120	119	135	155	173	182	189	181	170	147	129	122	1882
198	103	101	145	156	182	189	191	185	177	155	124	109	1817
199	114	108	143	158	183	185	192	188	176	164	140	131	1882

whereas $J = \sum_1^n j = \sum_{m=1}^{m=n} 0.09T_m^{1.5}$

For T_m greater than 26.5 °C, the corresponding value of Et_o can be obtained from a certain table after adjusting it by the factor $\frac{N_m x N_{dh}}{360}$.

In view of the availability of the temperature and daytime data at all stations in the Arab Region, Table 15 has been prepared to give mean monthly and annual Thornthwaite's potential evapotranspiration. The values included in this Table have been used while classifying the climate of the region, as shown in Chapter 3. In general, the values estimated from the formula of Thornthwaite underestimate the monthly as well as the annual potential evapotranspiration. It is very possible that the environment of that part of the United States in which the formula has been established does not resemble the environment of the Arab Region. Additionally, factors like wind speed, humidity, sunshine and radiation are totally ignored in the Thornthwaite's method. The number of daytime hours and the air temperature alone are not sufficient to represent the processes underlying evapo- transpiration.

5.4.2.4 The original Jensen-Haise (J-H) formula, and its modifications

Jensen & Haise (1963) began by collecting and re-assessing the existing measurements of potential evapotranspiration, $(ET)_o$, from several irrigated areas in the western states of USA. Those measurements were interpreted in the light of the corresponding temperature data and estimates of solar radiation. The relationship between these variables, expressed in CGS units, and the solar radiation in evaporation units, was expressed as:

$$(26) \qquad (ET)_o = R_{sh}(0.025T + 0.08)$$

in which R_{sh} is, as before, the global short wave radiation.

The original work of Jensen & Haise provides a procedure to calculate R_{sh} given the cloudiness and the extraterrestrial radiation, i.e. the solar and sky radiation flux on total clear days.

Upon reviewing the energy balance underlying the derivation of Eq. (26), Jensen (1966) developed other expressions for calculating the reference evapotranspiration, $(ET)_r$. These expressions are as follows:

$$(27) \qquad (ET)_r = [68 - (0.0036xZ) + (650/(e_{mx} - e_{mn}))]^{-1}\,(T - T_x)R_{sh}$$

and

$$(28) \qquad T_x = 27.5 - [0.25(e_{mx} - e_{mn}) - 0.001Z]$$

where,

Z = height in ft,
e_{mx} = saturation vapour pressure at the mean minimum temperature, mb,
e_{mn} = saturation vapour pressure at the mean minimum temperature, mb, and
T = mean air temperature, ° F.

The use of the above procedure has been recommended by Jensen et al. (1990) for time intervals not less than 5 days.

Example-Calculate $(ET)_r$ given $T_{mx} = 42.0$ °C, $T_{mn} = 31.2$ °C, $T = 36.7$ °C, $Z = 33$ ft and $R_{sh} = 12.37$ mm d^{-1}.

For the given temperatures $e_{mx} = 82.4$ mb, $e_{mn} = 45.4$ mb and $(e_{mx} - e_{mn}) = 37$ mb. From Eq. 28, $T_x = 27.5 - ((0.25 \times 37) - 0.033) = 18.28$. This brings $(T - T_x)$ to 79.78° F. From Eq. (27), $(ET)_r = [1/(68 - 0.12 + (650/37))] \times 79.78 \times 12.37 = 11.6$ mm d^{-1}

The method of Jensen & Haise (1963) and of Jensen (1966) have been used by the author for calculating the mean monthly and annual alfalfa reference evapotranspiration from a number of locations in the Arabian Peninsula (Shahin, 1998). The calculation results are presented in Table 16. These results show that for most of the investigated locations, the formula of 1966 yield annual values ranging from 77.4% to 110.4%, with a mean of 92.6% of the values obtained from the original expression of 1963. The monthly values show a wider range of variation from one expression to the other.

Table 16. Estimates of mean monthly and annual alfalfa reference evapotranspiration (Shahin, 1998) using the formulas of Jensen & Haise (1963) and Jensen (1966)

Station Name	Mean monthly and annual alfalfa refence evapotranspiration, mm for No.	Jan	Feb	Mar	Apr	May	Jun	Jul	Aug	Sep	Oct	Nov	Dec	Year	Years of record
El-Shuwaikh	161	73	97	158	207	309	356	382	382	293	209	112	68	2646	1965–89
		77	*101*	*159*	*200*	*292*	*333*	*355*	*4356*	*275*	*201*	*111*	*71*	*2531*	
Al-Manamah	166	77	87	126	179	249	292	312	297	242	191	132	92	2276	1989–93
		72	*81*	*115*	*159*	*217*	*253*	*272*	*256*	*210*	*168*	*117*	*84*	*2004*	
R. Al-Faras	167	75	90	122	191	260	291	284	272	229	182	118	82	2196	1972–92
		87	*104*	*136*	*205*	*272*	*300*	*292*	*279*	*238*	*192*	*129*	*93*	*2327*	
Abu Samra	169	71	79	117	158	215	230	232	243	207	170	113	78	1913	1976–92
		76	*83*	*126*	*158*	*213*	*226*	*227*	*238*	*204*	*170*	*115*	*82*	*1918*	
Al-Hofuf°	x	115	136	209	238	319	369	389	346	212	158	110	79	2680	1970–71
Er Riyadh		106	134	189	239	284	332	344	322	270	217	148	106	2691	1970–81
		109	*127*	*172*	*202*	*240*	*266*	*276*	*264*	*225*	*180*	*132*	*108*	*2301*	
Al-Kharj°		124	147	214	270	310	351	366	347	282	217	156	118	2902	1973–81
Dirab°		133	155	220	270	326	354	366	350	293	234	161	124	2986	1977–81
Muscat	180	119	140	196	245	309	280	286	260	230	192	146	118	2521	1981–92
		105	*122*	*170*	*211*	*264*	*239*	*244*	*222*	*197*	*166*	*127*	*104*	*2171*	
San'a	186	117	120	154	122	173	198	177	187	190	153	139	111	1841	1982–86
		126	*131*	*167*	*165*	*189*	*216*	*194*	*204*	*207*	*165*	*150*	*119*	*2033*	
Al-Hodeydah	189	133	155	188	210	228	202	200	218	209	219	179	153	2294	1983–90
		116	*136*	*164*	*182*	*196*	*173*	*172*	*187*	*159*	*159*	*156*	*134*	*1934*	
Al-Khod	197	149	152	179	172	229	214	205	194	186	195	166	141	2182	1982–83
		116	*118*	*139*	*133*	*176*	*164*	*157*	*148*	*143*	*150*	*134*	*110*	*1688*	

Figures in upright numerals the upper rows are obtained from Jensen & Haise (1963) and figures in italics in the lower rows are obtained from Jensen (1966). ° = results are taken from Salih et al. (1984) and from Solaiman et al. (1987) and x = measurements taken from lysimeters.

5.4.2.5 *Modified Penman formulas*

The original formula of Penman (1948) aimed at predicting evaporation from free water surfaces using a number of climatic data. That formula was later modified for the purpose of calculating grass reference evapotranspiration. The modified formula, which is often referred to as the FAO-Penman formula (1977), can be expressed as:

$$(29) \qquad (ET)_r = c\,[wR_n + (1-w)f(u_2)(e_s - e_d)]$$

where in Eq. (29) R_n is the net radiation, and c and w are adjustment factors that can be obtained according to the instructions given in the ASCE manual on Consumptive Use of Water and Irrigation Requirements (Jensen et al., 1990). The $(ET)_r$ calculated from Eq. (29) can be expressed in terms of free water surface E_o using the relationship:

$$(30) \qquad (ET)_r = \alpha E_o$$

in which α is a correction or adjustment factor that varies from month to month and has an average value between 0.65 and 0.85.

Eq. 25 was later updated by Allen et al. (1994) and since then has become known as FAO-Penman-Monteith formula. The formula can be written as:

$$(31) \qquad (ET)_r = 0.408\Delta(R_n - G) + [900\gamma u_2(e_s - e_d)/T_K]/[\Delta + \gamma(1 + 0.34u_2]$$

where,

$$(32) \qquad G = 0.07(T_{m=i-1} - T_{m=i+1}) or G = 0.14(T_{m=1} - T_{m=i+1})$$

and

R_n and G are in MJ m^{-2} d^{-1}, γ in kPa °C^{-1}, e_s and e_d in kPa, T_K is the temperature in degrees Kelvin and m is an arbitrary month, $m = i$ is the number of month in question, $m = i - 1$ is the number of preceding month and $m = i + 1$ is the number of the succeeding month. All parameters needed to use Eq. (27) for calculating $(ET)_r$ are available in the ASCE manual (Jensen et al., 1990) and Allen et al. (ICID, 1994).

Mehanna (1976) used the 1948 Penman formula with a locally found expression for the global radiation in Egypt. The results of calculation he ran over 23 stations covering the different parts of Egypt are given in Table 17.

Shahin (1998) applied Eqs. (29), (31) and (32) to the conditions at certain locations in the Arab Peninsula and the results obtained are presented in Table 18. These results show clearly that the FAO-Penman-Monteith expression yields for all locations reference evapotranspiration rates less than those given by the FAO-Penman. The ratios vary between about 88% and 95%, with an average 91.4%.

Table 17. Estimation monthly and annual potential evapotranspiration* over Egypt using the original Penman, 1948, formula with a local adjustment expression for the global radiation (Mehanna, 1976)

Station Name and serial No.	Monthly and annual evapotranspiration, mm, for J	F	M	A	M	J	J	A	S	O	N	D	Year
S. Barrani (60)	66	98	122	158	157	173	175	170	141	121	88	83	1552
Sallum (61)	93	103	132	154	175	194	202	192	150	150	102	96	1743
Damietta (62)	58	69	106	132	155	171	169	162	131	109	75	51	1388
M.Matruh (64)	80	93	122	141	161	178	180	176	146	124	86	85	1572
Port Said (65)	69	88	121	143	166	182	183	179	150	128	89	100	1598
Alexandria (66)	73	86	121	146	165	182	185	178	148	123	86	73	1566
El Arish (67)	65	79	110	133	158	167	174	175	142	112	61	69	1445
Damanhour (69)	64	76	108	143	173	186	178	166	135	112	74	65	1480
Tanta (70)	69	80	113	150	185	198	187	181	147	123	81	64	1578
Cairo (73)	93	110	151	185	218	230	210	208	170	152	103	95	1925
Giza (74)	70	94	131	170	204	217	204	194	159	137	88	75	1743
Suez (75)	98	91	151	203	217	236	257	253	198	180	134	95	2113
Faiyum (77)	80	101	142	180	219	235	229	218	178	154	103	76	1915
Siwa (78)	96	115	162	180	239	259	253	242	195	165	109	88	2103
Beni Suef (79)	79	97	131	180	215	231	221	212	171	151	99	78	1865
El Tor (80)	114	130	141	191	214	223	224	221	183	155	120	102	2018
El Minya (81)	86	105	166	198	238	250	237	219	181	162	106	77	2025
Hurghada (82)	131	141	175	208	247	265	269	260	219	186	138	120	2359
Asyut (83)	102	130	183	224	260	266	253	246	211	198	129	105	2307
Quseir (85)	123	134	165	187	218	229	227	228	191	170	135	114	2121
El Dakhla (86)	110	133	179	222	266	276	269	264	215	206	140	103	2383
El Kharga (87)	121	151	201	247	290	308	301	288	249	232	154	122	2664
Aswan (89)	152	173	221	265	304	313	307	302	262	253	182	144	2878

* *Reference grass is nodiflora*

5.4.2.6 USWB Class 'A' pan

One of the most systematic approaches regarding the prediction of pan evaporation using climatic data is the method established by Christiansen and his graduate students in the 1960s.

The expression they developed has made it possible to predict Class 'A' pan evaporation in case of absence of actual measurements as we have already demonstrated in Table 9 in connection with evaporation estimates from Lake Qaroun, Egypt. Additionally, the calculated pan evaporation, E_p, has been some time later adjusted so as to be used for computing the potential evapotranspiration. This can be expressed as:

$$(33) \qquad (ET)_o = CRC_T C_W C_H C_S C_M$$

where,

$(ET)_o$ = potential evapotranspiration, in mo^{-1},
C = a constant equal to 0.37,

Table 18. Estimates of reference evapotranspiration(Shahin, 1998) using FAO-Penman and FAO-Penman-Monteith methods

Station Name	No.	Jan	Feb	Mar	Apr	May	Jun	Jul	Aug	Sep	Oct	Nov	Dec	Year	Years of record
El-Shuwaikh	161	88	115	148	179	237	254	274	271	225	183	133	98	2205	1965–89
		79	*107*	*132*	*164*	*218*	*231*	*244*	*247*	*208*	*161*	*119*	*88*	*1998*	
Al-Manamah	166	89	94	148	205	283	295	332	307	247	180	144	86	2410	1989–93
		74	*81*	*120*	*172*	*247*	*263*	*305*	*278*	*231*	*177*	*122*	*76*	*2146*	
R. Al-Faras	167	80	92	131	178	231	243	234	224	188	161	101	76	1939	1972–92
		78	*91*	*125*	*150*	*210*	*223*	*227*	*205*	*178*	*154*	*108*	*88*	*1837*	
Abu-Samra	169	97	110	141	194	235	239	240	236	202	162	103	83	2042	1976–92
		96	*110*	*139*	*165*	*209*	*221*	*222*	*213*	*185*	*165*	*120*	*91*	*1936*	
Er Riyadh	173	95	123	167	194	231	263	278	243	210	158	102	81	2145	1970–81
Al-Kharj°		90	112	158	188	186	237	248	222	191	147	98	82	1959	1973–81
Najran		148*	170	223	237	279	288	298	282	258	211	165	143	2702	6 years ?
		0.65**	0.70	0.75	0.80	0.80	0.85	0.85	0.80	0.80	0.75	0.70	0.65	0.76	
		97***	119	166	189	223	245	253	226	206	158	114	93	2089	
Muscat	180	118	143	171	213	272	260	269	247	203	180	138	119	2333	1981–92
		115	*128*	*166*	*185*	*248*	*250*	*262*	*235*	*211*	*162*	*139*	*112*	*2213*	
San'a	186	124	139	161	141	200	212	188	190	180	181	150	125	1991	x
		140	137	165	151	209	212	183	192	186	190	149	136	2050	1982–86
		115	*118*	*137*	*141*	*171*	*198*	*189*	*181*	*153*	*156*	*125*	*105*	*1789*	
Al-Hodeydah	189	146	135	165	175	192	174	196	194	182	183	152	142	2036	1983–90
		109	*132*	*147*	*163*	*179*	*162*	*164*	*169*	*160*	*162*	*152*	*136*	*1835*	
Al-Khod	197	127	134	159	154	193	183	184	174	160	153	150	127	1898	x
		121	127	155	167	186	177	176	182	167	154	133	120	1865	1982–83
		119	*120*	*140*	*144*	*151*	*157*	*163*	*150*	*141*	*138*	*128*	*116*	*1667*	

Mean monthly and annual grass reference evapotranspiration, mm, for

Explanation:
*Figures in upright numerals in the upper rows are obtained from FAO-Penman (1977) and figures in italics in the lower rows are obtained from FAO-Penman-Monteith (1994), o = results are taken from Salih et al. (1984), *, ** and *** are Penman free water evaporation, adjustment factor and Potential evapotranspiration respectively(FAO, 1981). x = results taken from Farquharson et al. (1996) and years of record are not known.*

$R =$ extraterrestrial radiation, inches, received at the outer limit of the atmosphere,
$C_T = 0.048 + 0.014T$, T is the temperature in ° F,
$C_W = 0.790 + 0.0035W$, W is the wind velocity, mi d^{-1}, at the pan level,
$C_H = 1.140 - 0.0035H$, H is the relative humidity at noon,
$C_S = 0.550 + 0.00564(100 - 1.6SC - 0.84SC^2)$, SC is the sky cover, scale $1 - 10$

and

$C_M =$ monthly vegetative coefficient

With the wide spread of the number of Class 'A' pans in the Arab countries it appeared more convenient to use the direct measurement of the evaporation pan, E_p, instead of computing it from the method of Christiansen. The crop reference evapotranspiration, $(ET)_r$, can be estimated using a pan coefficient, k_p, that depends on the kind of crop or vegetal cover, pan used, wind run, humidity and stage of plant growth. The relationship between $(ET)_r$ and E_p can be expressed as:

$$(ET)_r = k_p E_p \tag{34}$$

As early as 1956–58, Boumans et al. (Dieleman ed., 1963) used pan coefficients of 0.42 and 0.56 for winter and summer seasons respectively in their consumptive use study at Dujailah, Iraq. The aim was to convert Class 'A' pan evaporation to the corresponding consumptive use of water by winter crops like barley and wheat, and green gram as summer crop. The figures in Table 4 clearly show that for the 112-day period of observation in Lebanon, the pan coefficient, k_p, varied from 0.59 to 0.83 with a mean of 0.67 for class A pan, and from 0.71 to 0.92 with a mean of 0.79 for the Colorado sunken pan.

The FAO Paper No. 24 (Doorenbos & Pruitt, 1977) and the ASCE Manual on evapotranspiration and water requirements (Jensen et al., 1990) are among the literature in use for obtaining values of the pan coefficient, k_p. These references show that k_p varies from a minimum of 0.40 to a maximum of 0.85 for Class 'A' pan and from 0.45 to 1.10 for the sunken pan. As will be seen from the next sub-section, the adjusted pan evaporation approach was one of the methods used for estimating the evapotranspiration in Egypt.

5.4.3 Comparison between the different approaches used for estimating reference evapotranspiration

The next subsections review in brief some case studies from the Arab countries in which a number of estimation methods have been used for estimating reference evapotranspiration.

5.4.3.1 Estimation of crop reference evapotranspiration in Egypt

In this study Egypt has been subdivided into three zones, i.e. Lower Egypt or the Nile Delta, Middle Egypt and Upper Egypt. The original Penman and the modified Penman were both applied in an old lands study as well as in a more recent

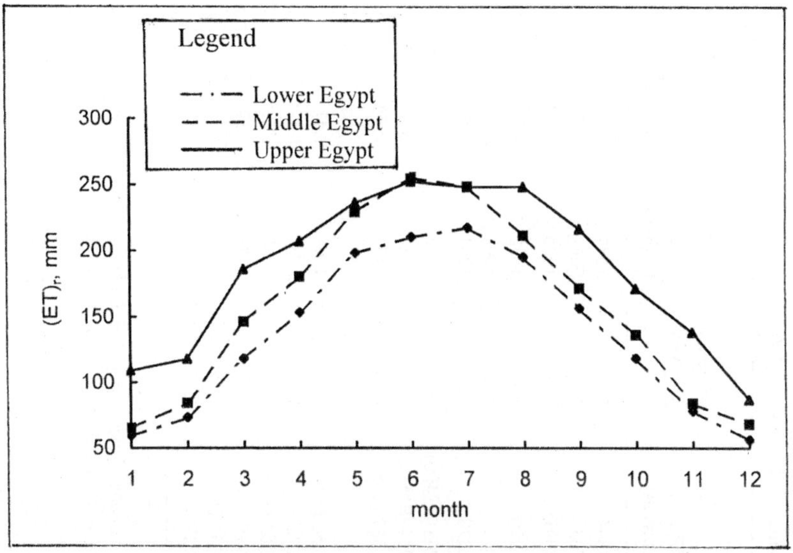

Figure 9. Monthly evapotranspiration in Egypt (data are taken from the Report of the Ministry of Irrigation and Water Resources, Egypt, et al., 1991)

study performed by the Water Researches Center. The Christiansen evaporation pan approach was applied too by the Water Researches Center while the modified Penman was used in a separate study referred to as Technical Report No. 17 (Ministry of Water Resources Egypt et al., 1991). A few adjustments to the results were felt necessary, as the climatic data of the station of Assyût (Station No. 83), Upper Egypt, seemed to be not sufficiently reliable and therefore had to be omitted. The calculation results before and after adjustment are given in Table 19.

The Christiansen evaporation pan method yields estimates that are closest to the average of all methods for Lower Egypt. The closest results for Middle and Upper Egypt are given by the modified Penman method. As such, one can fairly conclude that the modified Penman methods can generally be regarded as the most suitable method for estimating the reference evapotranspiration. Figure 9 shows graphical plots of the monthly reference crop evapotranspiration for the three zones using the modified Penman method as worked out by the Water Researches Center, Egypt.

5.4.3.2 Estimation of reference evapotranspiration for Saudi Arabia

Three locations, Riyadh, Kharj and Dirab, all in the same region situated in Central Saudi Arabia were chosen for the purpose of estimating alfalfa reference evapotranspiration. That investigation was supported by data from Al-Hofuf station in Al-Hassa area, about 350 km to the east of that region (Salih et al., 1984). The estimation methods used were those of Jensen & Haise, Hargreaves & Samani,

Table 19. Calculation results of reference evapotranspiration in Egypt (Ministry of Water Resources of Egypt et al., 1991)

Region	Period	Annual / seasonal reference crop evapotranspiration, mm							
		Original Penman		Modified Penman			Christiansen pan WRC	Average of all methods	Adjusted overall average
		Old lands study	WRC° estimates	Old lands study	WRC° estimates	TR 17⁺ estimates			
Lower Egypt	Annual	1311	1435	1630	1630	1654	1528	1541	1532
	Winter	401	482	560	536	563	492	506	508
	Summer	910	953	1070	1094	1091	1036	1035	1024
Middle Egypt	Annual	1500	1677	1948	1878	1950	2069	1837	1790
	Winter	461	561	645	627	672	672	606	593
	Summer	1039	1116	1303	1251	1278	1397	1231	1197
Upper Egypt	Annual	1713	2083	2346	2214	2481 (2400)[x]	2021	2143	2167
	Winter	594	799	868	844	924 (915)	727	793	806
	Summer	1119	1284	1478	1370	1557 (1485)	1294	1350	1362
Ratios									
Annual UE/LE*		1.31	1.45	1.44	1.36	1.50 (1.45)	1.27	1.39	1.41
Annual ME/LE*		1.14	1.17	1.19	1.15	1.17	1.3	1.19	1.17
Winter UE/LE		1.48	1.66	1.55	1.57	1.64 (1.63)	1.48	1.57	1.59
Winter ME/LE		1.15	1.16	1.15	1.17	1.19	1.37	1.20	1.17
Summer UE/LE		1.23	1.35	1.38	1.25	1.43 (1.36)	1.18	1.30	1.33
Summer ME/LE		1.14	1.17	1.22	1.14	1.17	1.27	1.19	1.17

Explanation

* UE = *Upper Egypt, ME = Middle Egypt and LE = Lower Egypt. WRC° = Water Researches Centre, Egypt, TR 17⁺ = Technical Report No. 17.*

[x] *Figures have been obtained after excluding the climatic data for Assiut. The last column excludes Christiansen pan and Assiut TR 17.*

modified Penman, original Blaney-Criddle and Class ‘A’ evaporation pan. All of these methods have already been reviewed in section 5.4.2.

The results of calculation averaged over the period 1970–81 for Er-Riyadh Station are shown graphically in Figure 10. From this Figure it can be noticed that the Class ‘A’ pan values, as one should expect, are the highest, then followed by the Jensen-Haise estimates, which are distinctly higher than those of the remaining methods. The methods of Hargreaves, modified Penman and the original Blaney-Criddle yield comparatively comparable results. These conclusions are in general agreement with the results listed in Tables 13, 14 and 16 about other locations in the Arabian Peninsula. The results from the different estimation methods used at three locations in Saudi Arabia were compared with measured $(ET)_r$ at Al-Hufouf station in Al-Hassa area, 350 km to the east of Riyadh. The comparison was based on the measured obtained from two studies. The first one was performed in the period November 1970-October 1971 and with alfalfa as a reference crop, and the second from December 1980-November 1981 using an experimental field plot, with alfalfa as a reference crop. The comparison has shown that Class ‘A’ pan and Jensen-Haise methods exchanged the first and second ranks. Hargreaves has consistently occupied the third rank, and modified Penman and Blaney Criddle the fourth and fifth ranks respectively (Salih et al., 1984).

More estimation methods were later used in connection with the above-described studies at Al Hufouf station. The total number of methods used for estimating alfalfa reference evapotranspiration reached 23 (Solaiman et al., 1987). Some evaluation criteria have been worked out so as to award each method its proper rank. The different ranking methods, however, did not show a satisfactory agreement. The differences were mainly caused by differences in the evaluation parameters used. Finally, the choice fell on an evaluation method that was established earlier by Salih and Sendil (1984). As such, the first rank went to Jensen-Haise, and the second and fifth to Class’A’ pan and adjusted Class ‘A’ pan respectively,

5.4.3.3 *Reference evapotranspiration in the Gezira region, the Sudan*

The Gezira region can be described climatically as hot, semi-arid with considerable number of sunshine hours, except for the autumn season (July-September). The Gezira vertisols are quite suitable for raising cotton, and long staple cotton is traditionally the main crop grown in the region.

Climatological data for the prediction of reference evapotranspiration were obtained from the Gezira Agrometeorological station (14°24′N, 33°29′E and 405 m altitude). The data were averaged over 10-day periods. A coefficient of 0.5 was used to adjust the readings of the Class ‘A’ evaporation pan available at the station. The methods used for calculating $(ET)_r$ were those of Hargreaves, FAO-Penman, FAO-Class ‘A’ pan, and Jensen-Haise. The first three methods estimate the evapotranspiration with grass as a reference crop while alfalfa is the reference crop in the case of Jensen-Haise method.

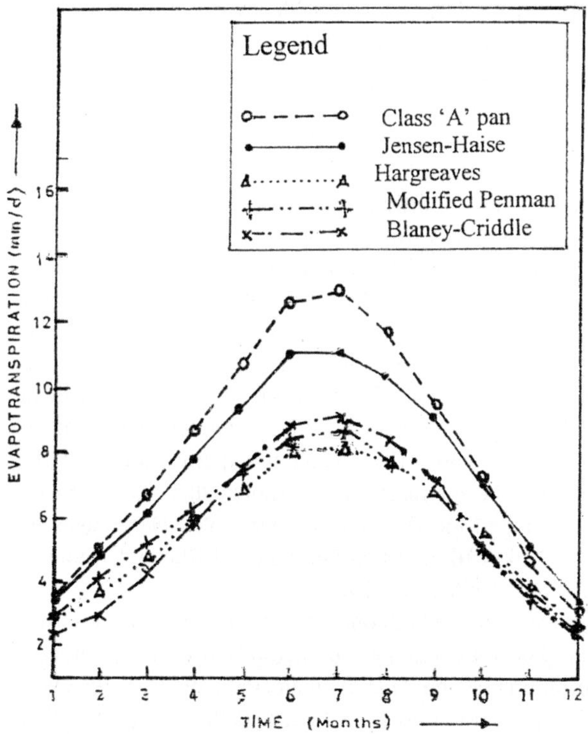

Figure 10. Reference evapotranspiration for Ryiadh, Saudi Arabia (Salih et al., 1984)

The 10-day measured and estimated evapotranspiration of warm-season grass (Cynodon Dactylon) using four different methods is shown in Figure 11(a). From this Figure it is clear that the Jensen-Haise method, originally developed for estimating $(ET)_r$ of alfalfa, consistently overestimated the measured data by at least 1 mm d^{-1} throughout. Despite using a pan coefficient of 0.5 in the FAO-Pan method, the estimated $(ET)_r$ remained still larger than the measured data. According to Hussein & El-Daw (1989), this result highlights the difficulties in using pan evaporation for estimating reference evapotranspiration because of the significant effect of local pan environment. As a consequence, the remaining part of the study in the Gezira region was limited to the FAO-Penman and Hargreaves methods.

Both expressions had to be modified locally to yield estimates in reasonable agreement with the measured reference evapotranspiration from warm-season grass. The modification dealt with the wind function and temperature in the case of FAO-Penman method while only a temperature correction had to be introduced to the method of Hargreaves. The modified estimates are shown graphically in Figure 11(b).

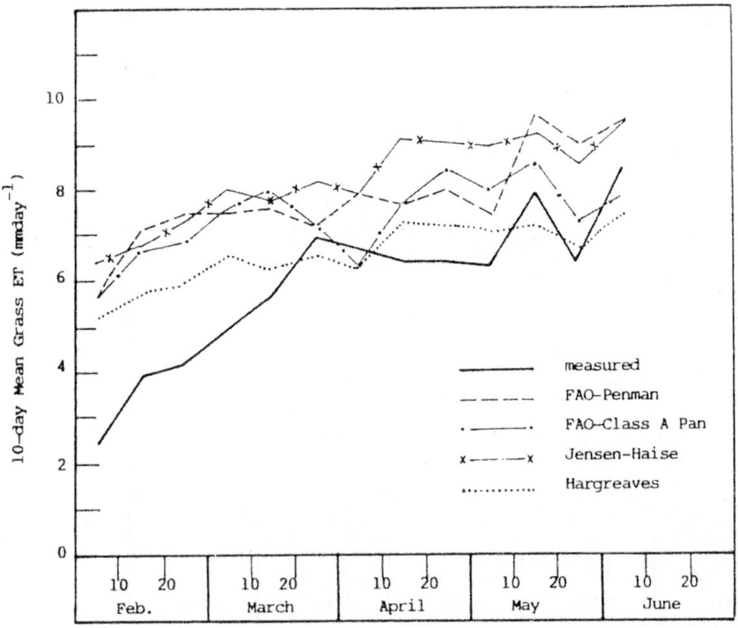

Figure 11(a). 10-day mean measured and estimated evapotranspiration of warm-season grass in Gezira region, the Sudan (Hussein & El-Daw, 1989)

A statistics-based comparison between the results of the modified methods and the measurements showed that the coefficient of determination (r^2) for the relationship between FAO-Penman and the measured $(ET)_o$ was 0.92 reduced to 0.82 for the Hargreaves method. The mean ratio of computed to measured $(ET)_o$, the standard error of estimate (*SEE*) and the coefficient of variation (C_v) for the two relationships in the same order were in favour of the FAO-Penman. On the contrary, the regression parameters (intercept, *a*, and slope, *b*, of the regression line) were in favour of the Hargreaves than the FAO-Penman method. According to Hussein & El-Daw (1989), when the same weight is accredited to each of these criteria, the FAO-Penman is ranked the first and the Hargreaves method ranks the second.

5.4.3.4 Reference evapotaranspiration in the semi-arid zone, south of Spain

This case study has already been presented in section 5.2.2. Despite its geographic location outside the Arab Region, the proximity of the area investigated to the southern shore of the Mediterranean Sea and the similarity of the climatic conditions might qualify the results obtained from that investigation to be transposed to the northern, coastal zones of Morocco and Algeria.

In addition to the measured (ET)o from the lysimeters, the reference evapotranspiration was calculated using each of the FAO-Penman-Monteith, Penman-FAO, Blaney-Criddle-FAO, Radiation-FAO, Class 'A'-FAO, Turc and Hargreaves

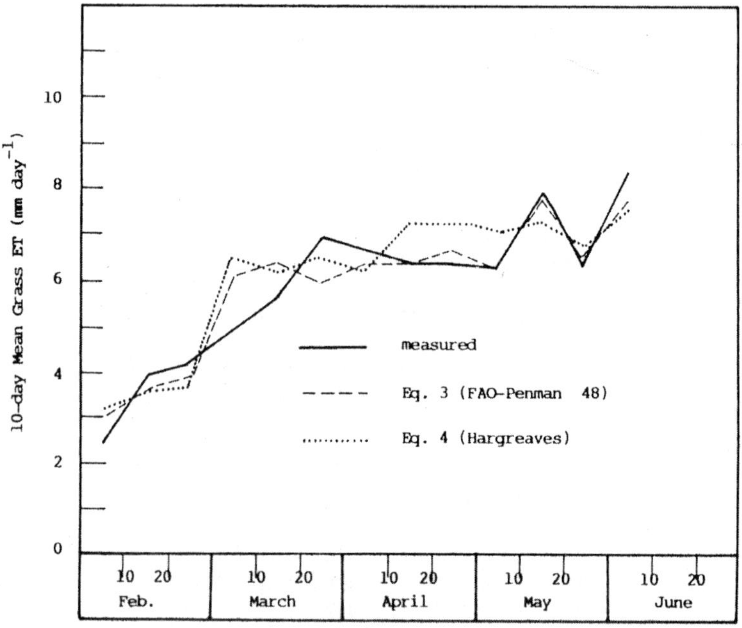

Figure 11(b). Temperature-corrected FAO-Penman and Hargreaves methods (Hussein & El-Daw, 1989)

methods. Linear regression relations were established between the average monthly and 10-day measurements and the corresponding estimates from each method. A correlation matrix was also built up between the results obtained from the different methods. The parameters of the regression lines and the elements comprising the correlation matrix are given in Table 20.

On a quantitative basis, the FAO-Penman methods provides the best adjustment equation with intercept, a, of −0.11 and regression coefficient, b, of 1.01. The estimates obtained from this method are in closer agreement with the measured values than the estimates from the FAO-Penman-Monteith or FAO-Class 'A' do. The predicted $(ET)_o$ using methods based on solar radiation and temperature are somewhat higher than the corresponding lysimeter measurements of $(ET)_o$. Despite the observed discrepancies, the estimates from these methods correlate strongly with lysimeter results and with other estimation methods regardless of having empirical or physical base. According to Toribio et al. (1996), this is possibly due to the fact that climate of the region studied, and so are the semi-arid coastal areas of Algeria, is characterized by light winds and medium to high relative humidity. As such, solar radiation and temperature are the fundamental climatic factors in control of the process of evapotranspiration.

This can be evidenced from the very strong correlation in the area investigated between the methods of Turc and Hargreaves.

Table 20. Parameters of regression lines between measured and estimated $(ET)_o$ for different estimation methods and the corresponding correlation matrix (from Toribio et al., 1996)

Time interval and estimation method	Intercept, Slope, Coefficient of determination (r^2) for								
	a	b	$(ET)_o$	1	2	3	4	5	6
Monthly									
Penman-Monteith	−0.16	1.08	0.962						
Penman-FAO	−0.11	1.01	0.965	0.933					
Blaney-Criddle-FAO	−0.20	0.88	0.969	0.967	0.969				
Radiation-FAO	−0.30	0.97	0.958	0.994	0.994	0.974			
Class 'A'-FAO	−0.13	1.07	0.979	0.933	0.932	0.940	0.925		
Turc	−1.03	1.41	0.961	0.990	0.995	0.979	0.995	0.926	
Hargreaves	−0.54	1.20	0.973	0.987	0.994	0.986	0.994	0.940	0.996
Ten-day									
Penman-Monteith	0.01	1.04	0.820						
Penman-FAO	−0.04	1.00	0.854	0.939					
Blaney-Criddle-FAO	−0.12	0.87	0.866	0.846	0.904				
Radiation-FAO	−0.25	0.96	0.848	0.924	0.976	0.938			
Class 'A'-FAO	−0.01	1.04	0.940	0.755	0.801	0.822	0.787		
Turc	−0.96	1.44	0.848	0.907	0.949	0.921	0.966	0.808	
Hargreaves	−0.50	1.19	0.876	0.918	0.975	0.954	0.986	0.836	0.975

Explanation
1 = Penman -Montieth, 2 = Penman-FAO, 3 = Blaney-Criddle-FAO, 4 = Radiation-FAO, 5 = Class 'A' FAO and 6 = Turc.

5.4.4 Estimating reference evapotranspiration under inaccurate data conditions

Since its development (Allen et al., 1998), the formula known as FAO-Penman-Monteith (FAO-P-M) has been widely used for estimating reference evapotranspiration and currently is regarded as a sort of standard. It is privileged by its physically based approach and being well documented on a wide rang of soft ware. It has been calibrated versus a variety of lysimeters and can be used globally without any need for additional parameter estimations.

The major drawback of the FAP-P-M method, however, is the relatively high data demand, as it requires accurate data regarding air temperature, relative humidity, radiation and wind speed. The number of stations where all these data are observed is limited in many parts of the globe, especially in the developing countries. Even in the case of availability of data, the question that arises every now and then is to what extent does the accuracy of the observed data affect the $(ET)_o$ estimates, especially for the FAO-P-M. The other question is whether more realistic $(ET)_o$ estimates can be obtained using a simplified approach such as the Hargreaves than with FAO-P-M given a certain level of inaccuracy in the meteorological observations (Droogers & Allen, 2002)

The expression used here is that of Hargreaves (1994), which is a function of the mean daily maximum and minimum temperatures. This expression, originally Eq. (22), can be rewritten in a slightly different form as given by Eq. (23):

$$(35) \qquad (ET)_o = 0.0023.0.408 R_A.(T + 17.8).TD^{0.5}$$

where,

R_A = extraterrestrial radiation in MJ m^{-2} d^{-1},
T = average daily temperature, °C, and
TD = difference between mean daily maximum and mean daily mean temperature, °C.

It should be noticed here that the constant 0.408 has been introduced in Eq. (35) to convert the radiation to evaporation equivalents in mm.

Inaccuracies in data were introduced into the FAO-P-M and MHG (modified Hargreaves) by assuming each climatic factor has a distribution like normal with the observed value as a mean and confidence limits equal to the mean and one standard deviation and the mean minus one standard deviation, i.e. the confidence interval equal to 95%. These assumptions yielded the following results:

Climatic factor	FAO-P-M	HG	MHG	2x standard deviation
Minimum temperature	x	x	x	1 °C(~ 5%)
Maximum Temperature	x	x	x	1 °C(~ 5%)
Humidity	x			25%
Wind speed	x			25%
Radiation	x			25%
Precipitation			x	10%

Monthly values of $(ET)_o$ were calculated using FAO-P-M and HG methods. The method of HG tends to underestimate FAO-P-M largely in the very dry regions and to overestimate FAP-P-M in very wet regions. After a number of trials Droogers & Allen (2002) arrived to the so-called Modified Hargreaves (MHG) equation, given by:

$$(36) \qquad (ET)_o = 0.0013.0.408 R_A (T = 17.0).(TD - 0.0123P)^{0.76}$$

in which P is the monthly precipitation in mm.

The above listed measurement errors were introduced into a certain database and worked out statistically. Deviation of predicted $(ET)_o$ values using the FAO-P-M and MHG methods including data errors were summarized and compared to FAO-P-M values based on error-free data. The coefficient of determination (r^2) was found to be 0.871 for the FAO-P-M and 0.915 for the MHG. The root mean square difference *(RMSD)* for the two calculation methods in their respective order was 0.93 and 0.72 mm d^{-1}, respectively. Based on these two evaluation criteria, i.e. r^2 and *RMSD,* the MHG method ranks first and the FAO-P-M ranks second.

5.4.5 Concluding Remarks

Review and discussion of a considerable number of reference evapotranspiration estimation methods and actual case studies, like the ones presented in this section, can fairly bring the user to the conclusion that until present there is no single climate-based method that is universally adequate for all climatic regions.

Jensen-Haise method has proven its adequacy in estimating reference (alfalfa) evapotranspiration in extremely arid areas such as Saudi Arabia. Differently, FAO-Penman has shown to be superior to other methods in warm areas under intensive irrigation such as Egypt and the Sudan. Probably the same method applies well to the small-size states of the Arabian Peninsula that are surrounded by water, like Qatar, Bahrain and Kuwait. It goes without saying that many of the formulas available in literature need calibration or adjustment so as to be applicable to local conditions. The modified Hargreaves formula followed by the FAO-P-M has proven, however, that none of them needs any adjustment. The fact that the modified Hargreaves requires less climatic data and the results are less sensitive to errors in these data compared to the FAO-P-M, renders its application for predicting reference evapotranspiration more feasible and the estimates more reliable.

Despite the relative scarcity of published material from the Arab Region in general, there are strong evidences that open water evaporation and evapotranspiration are measured and/or estimated in many parts of the region. The available information, some of which has been presented in this Chapter, can be useful for the purposes of water resources development and management.

CHAPTER 6

RUNOFF AND RIVER FLOW

6.1. RUNOFF

6.1.1 Runoff process

The ground rain, i.e. the rain that reaches the ground surface, evaporates, infiltrates or lies in depression storage. Rainwater retained in puddles, ditches, and other depressions in the soil surface is termed 'depression storage'. It does not always happen that some of the rainwater in every rain event is left on the land surface after evaporation and infiltration losses take place. However, if there is any water left, the land surface will become covered with a film of water known as 'surface detention'. Besides, almost immediately after the beginning of rainfall excess occurs, the smallest and shallowest depressions become filled and 'overland' flow begins to move downslope toward the nearest stream channel. This water, upon entering a channel is referred to as 'surface runoff'. The channels coalesce into rivers or wadis. Rivers eventually find their way down to the sea or ocean while wadis often disappear after some distance in the desert.

When the rainfall is intense or prolonged, or both, the network of stream channels in a certain catchment may not be able to carry the surplus surface runoff within the defined sections of these channels, and start to overflow and cause inundation as they become filled with water. After all, floods are not limited to humid areas only but can also happen in arid and semi-arid areas.

It should be noted that not all streamflows are a result of surface runoff alone. Streamflow may be the resultant of surface runoff and subsurface runoff (baseflow). "Since baseflow represents the discharge of aquifers, changes occur slowly and there is a lag between cause and effect that can easily extend to periods of days or weeks (Wilson, 1983)." In arid and hyper-arid zones there is hardly any groundwater-generated baseflow, since the water table is generally below the streambed. A distinction between arid and humid regions with regard to runoff and streamflow is illustrated schematically in Figure 1 (Margat, 1979).

It is customary to plot streamflow in units of discharge, e.g. $m^3\ s^{-1}$, as ordinate against time as abscissa. The resulting graph is known as discharge hydrograph or simply hydrograph. If, instead of the discharge, the water level in a stream is used the graph is then called the stage hydrograph. The water level and streamflow are linked by the so-called rating function.

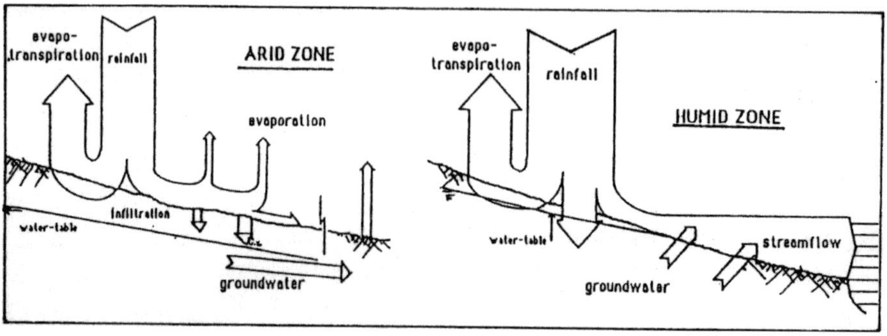

Figure 1. Schematic illustration of runoff formation and streamflow in arid and humid climates (Margat, 1979)

A perennial stream or river is a stream that never stops flowing throughout the year, thus has a continuous hydrograph. The discharge in the low-flow season is to a large extent sustained by baseflow. In this case the river functions as influent stream in the high-flow season and as effluent stream in the low-flow season. A good example of this is the River Euphrates in Iraq. The hydrograph of such a stream that is fed by both surface and groundwater is shown in Figure 2(a). The perennial rivers in the Arab Region are limited in number. Most of the perennial rivers in northwest Africa are national rivers, with relatively small basin areas and moderate

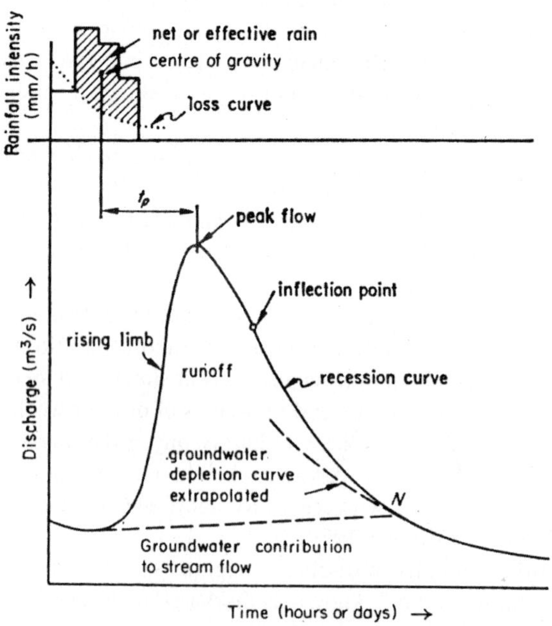

Figure 2(a). Natural hydrograph of a perennial stream (Wilson, 1983)

to fairly high average specific discharge. This is not the case with perennial rivers in the middle and eastern subregions of the Arab Region. The rivers in these two subregions are generally international shared by a number of riparian countries each. This state of affairs is a source of friction between one Arab state and another Arab state, and between some of the Arab States and their neighbouring countries. Water conflicts between riparian countries shall be discussed in a forthcoming chapter. Table 1, Appendix II, gives the catchment area, length of mainstream and the riparian countries sharing each river basin in the Arab Region.

Another type of streams, wadis (oueds is the French word used in the Western Sub-region) is the seasonal stream. In this type the streams do not run empty in the wet season. As such, the discharge hydrograph has two distinct seasons, in one season the discharge is larger than zero and in the second season it is just zero. The River Atbara, whose catchment is shared between Ethiopia and the Sudan, is a typical example of such a stream. Figure 2(b) represents the hydrograph of a seasonal stream.

A third type of streams is the ephemeral stream in which the river dries up completely in rainless periods. There are many examples of such streams in the Arab Region. In some countries as Abu Dhabi and Oman the word falaj (plural aflajes or aflaj) is used to describe a tunnel tapping undrground flow near a stream and bringing the flow to the surface for irrigation. In a few cases it is used to describe a surface channel fed from the main stream itself, upstream of a temporary embankment. Generally speaking, Falaj is the equivalent of qanat in Persian (Figure 2, Chapter 1) and foggara in al-Mghreb al-Arabi.. This case is illustrated graphically in Figure 2(c). In case of extremely intense precipitation the stream is described as a torrential stream (in Arabic makharr es' syoul).

Surface runoff may in certain situations cause soil erosion. Running water carries sediments, the natural products of erosion. The sediment yield of a basin varies from part to part and from time to time. A stream thus can be defined as a natural carrier of both water and sediments. In some cases there is some proportionality between the sediment load in the flowing water and the stream discharge.

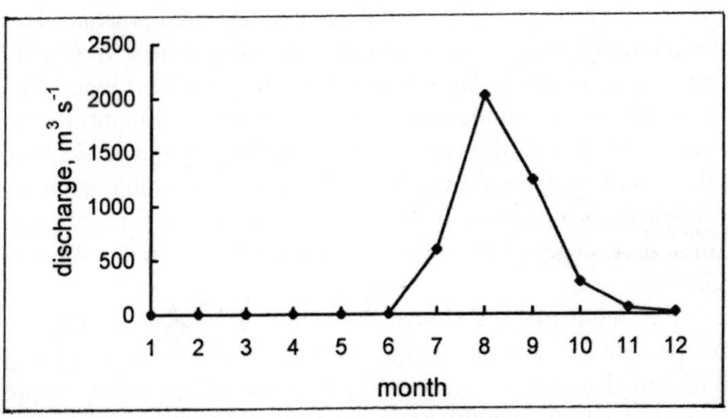

Figure 2(b). Hydrograph of a seasonal stream (Atbara K 3, the Sudan) for the year 1942

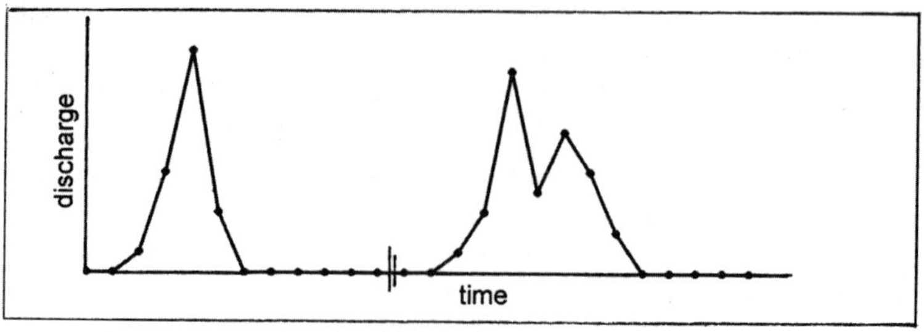

Figure 2(c). Hydrograph of an ephemeral stream

6.1.2 Rainfall-runoff relationships

The primary physical characteristics of a drainage basin are the surface area, geometric shape, topography, orientation, slope, soil type and vegetal cover, network of drainage channels and water storage capability. A large drainage basin is one where the storage effects in watercourses, lakes and aquifers dominate, thus rendering the response of the basin to precipitation slow. Additionally, the large catchment is not sensitive to short-interval variation of rainfall intensity, and often so to land use. A fairly detailed explanation of the role of each of these factors is given by Starosoloszky (ed., 1987).

Predicting runoff from existing rainfall measurements depends largely on the length of time interval considered. For short durations (hours, days or weeks) the intricate relationship between the rainfall and runoff renders prediction of runoff difficult and often the result unreliable. As the time period becomes longer the relationship between rainfall and runoff becomes better defined and prediction or estimation of runoff becomes less difficult and less unreliable. It is a generally accepted concept that after having the initial losses (mostly due to infiltration) covered, the relationship between the two variables tends to become linear.

Expressing runoff, as a percentage of the rainfall, is a widely used practice. This percentage is usually referred to as 'runoff coefficient'. It varies from a very small value, practically zero, to a relatively large value depending on the topography, average slope of the catchment, magnitude and intensity of storm rainfall, infiltration rate and the initial soil moisture condition as mentioned in the previous subsection. In humid climates the runoff coefficient may be, with some justification, assumed constant for the catchment in question. In arid zones the situation is different rendering the runoff coefficient widely variable from interval to interval, even within the same rainstorm.

The Soil Conservation Service of the United States Department of Agriculture has long since developed a set of curves giving the depth of runoff corresponding to a measured rainfall provided that the nature of land surface and its cover, etc are known. The user should check the conditions in the field to be able to select the appropriate curve and read the runoff given the rain depth. Generally, the reliability in predicting the annual runoff is higher than the seasonal, and the seasonal runoff than the runoff produced by a single storm event.

From the maps available in the Atlas of World Water Resources (Kurzon, 1979) one can conclude that the runoff coefficient for the larger part of the surface of Arab Region is in the range from 1% to 5%. Exceptions can be found in the northern parts of Algeria and Morocco where the coefficient reaches up to 20% along the Mediterranean Coast in Algeria and 50% in certain parts in the extreme northwest of Morocco. Exceptions can also be found in the mountainous regions of Yemen, northeast Iraq and northwest Syria. In these regions the runoff coefficient can grow up to 40% in certain locations. It is not surprising that the runoff coefficient increases with annual rainfall in the catchment of interest. The map in Figure 3, gives contour lines of equal annual runoff covering a large area of Africa and Asia including the Arab Region.

6.1.3 Modelling the catchment performance

Hydrological models deal in a quantitative sense with reconstructing the past rainfall-runoff behaviour and forecasting future runoff response to a given rain event. Three types of models can be mentioned, deterministic, stochastic and conceptual. The art of modelling comprises four basic stages. These are model identification, i.e. selection of the type of model to be used; estimation of model parameters; model verification, i.e. estimation of quality of model based on say independent experimental data; and application of model to situations of practical interest.

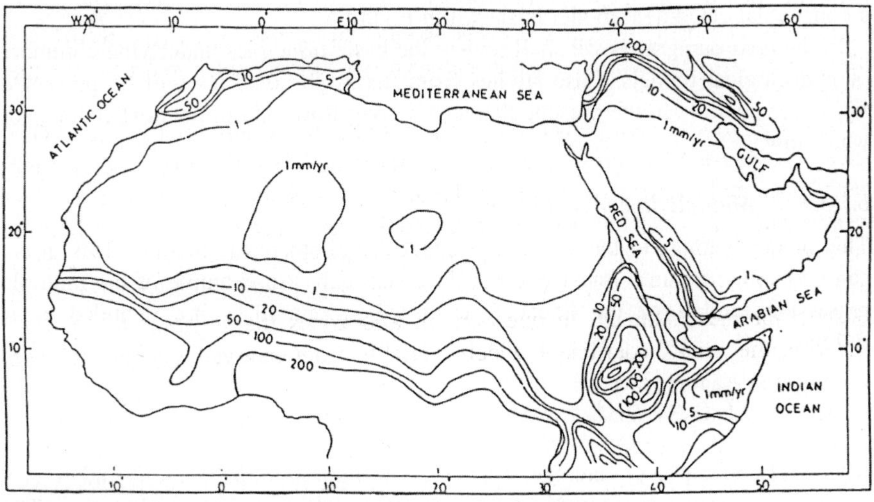

Figure 3. Contour lines of equal annual runoff in the Arab Region (from Kurzon, 1979)

Deterministic models are based on and account for the physical processes, storage and interactions. For a given set of initial and boundary conditions a deterministic model leads to a unique definable output, i.e. to any given input there is only one possible answer. One can also think of a deterministic model whether it gives a lumped or distributed description of the catchment considered, and whether the description of the hydrological process is physically based, conceptual or lumped. As such, deterministic models can be empirical (black-box), lumped models (grey-box) and distributed physically-based models (white-box).

Distributed models attempt to explain the actual process occurring within the catchment. They require a greater in-depth knowledge of all processes occurring within the catchment. They also attempt to account for a high degree of spatial variability of climate, soil, vegetal cover and topographical characteristics. On the contrary a lumped model ignores the spatial variation of variables and parameters within the catchment such as rainfall, evapotranspiration, soil properties and topography. Such models are designed on the basis of a simple arrangement of a relatively small number of components; each of them is a simplified representation of one process element in the system being modeled.

Stochastic models transform a given input to an output without considering the physical processes involved in the transformation. Traditionally, a stochastic model is derived from a time series analysis of a historical record. Then it can be used for generating long sequences of events having the same statistical properties as those of the historical record.

Conceptual models presuppose that the hydrological processes to be modelled are partly understood. For a complex hydrological system with considerable heterogeneity in input and output, both in time and space, several considerations have to be simplified or approximated to represent this heterogeneity. The model parameters are often averaged over temporal and spatial intervals. A schematic illustration of a conceptual catchment model is shown in Figure 4.

In the next paragraphs we shall review the basic principles underlying a number of hydrological models. Case studies from the Arab countries will be presented whenever possible in the present chapter for river flow, and in the next chapter for wadi flow.

6.1.3.1 Multiple linear regression

This model belongs to the group of statistical and stochastic models. This group can be subdivided into linear and non-linear models. An example of the multiple regression models of the linear type in which rainfall is not included is as follows:

$$Q_a = \alpha + \beta_1 Q_b + \beta_2 Q_c + \cdots + \beta_m Q_n + \varepsilon \quad (1)$$

where, Q_a, Q_b, Q_c, Q_n are the discharges at stations $a, b, c, \ldots, n$, and α is a constant, which can be viewed as representing baseflow and channel storage. The weights $\beta_1, \beta_2, \ldots, \beta_m$ are partial regression coefficients.

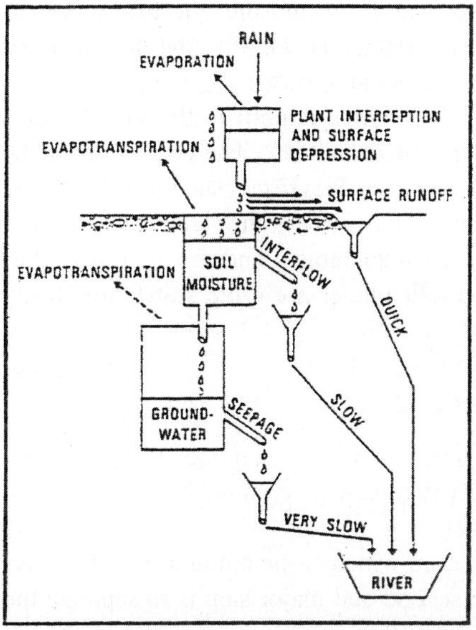

Figure 4. Schematic representation of a conceptual model (Breater, 1989)

Similarly an example of multiple regression models with rainfall included is:

$$(2)\qquad Q_{a,t} = \alpha + \beta_1 Q_{b,t} + \cdots + \beta_m Q_{n,t} + \lambda_1 P_t + \lambda_2 P_{t-1} + \cdots + \lambda_r P_{t-r} + \varepsilon$$

where, $\lambda_1, \lambda_2, \ldots$ and λ_r are additional model parameters to be associated with current and antecedent precipitation, $P_t, P_{t-1}, P_{t-2}, \ldots$ and P_{t-r}.

6.1.3.2 *Antecedent precipitation index (API)-model*

The antecedent precipitation index *(API)* expresses the state of dryness of the basin, which depends, one way or another, on all rains which have fallen prior to the day under consideration. Obviously, the effect diminishes with time that has passed since the last rainfall. The *API* can be expressed as:

$$(3)\qquad (API)_t = P_t + k(API)_{t-1}$$

in which k is a damping factor less than unity.

The discharge and rainfall are linked together by the functional relationship:

$$(4)\qquad Q = f(P, API, N, t_{dur})$$

where N is the month number, 1,2,3,….,12, or the week number 1,2,3,….,52, and t_{dur} is the duration of rain storm.

The procedure starts by plotting measured Qs versus the corresponding Ps. Lines of equal *API* are drawn through the plotted points (Q, P). Next, the residual discharges $(\Delta Q)_l$ are measured and plotted versus N and the line that serves as best fit is drawn through the plotted points $((\Delta Q)_l, N)$. Finally, the new residual discharges $(\Delta Q)_2$ are plotted versus the independent variable, t_{dur}.

In this way, to compute the direct runoff for a given rainfall P, the value of Q is read from the graphical plot for given values of P and AIP. The corrections to be applied are taken from the other graphical plots for N and t_{dur} shown in Figure 5.

The *API* model has been used on catchments from less than 200 km^2 to more than 2,000 km^2. It has been calibrated for each catchment and this requires a fair amount of recorded data. This renders the *API* model not quite suitable to many situations in the Arab Region.

6.1.3.3 Unit hydrograph

The unit hydrograph, one of the earliest models, is a black-box model. It incorporates the relationship between net rain and direct runoff. This relationship is one of the components of the general watershed model.

The method begins by assessing the net rain, which is the amount of rain equivalent to the volume of surface runoff. The second and major step is to separate the baseflow from the surface runoff and to check that its volume is equal to the net rain depth times the watershed area. The unit hydrograph, which is a semi-empirical technique, helps to correlate the surface runoff with rainfall that caused it. This is based on three postulates; equal base width of hydrographs resulting from different storms having the same duration, the discharges are proportional to the depths of net rainfall for all rains of the same duration, and the applicability of the principle of superposition.

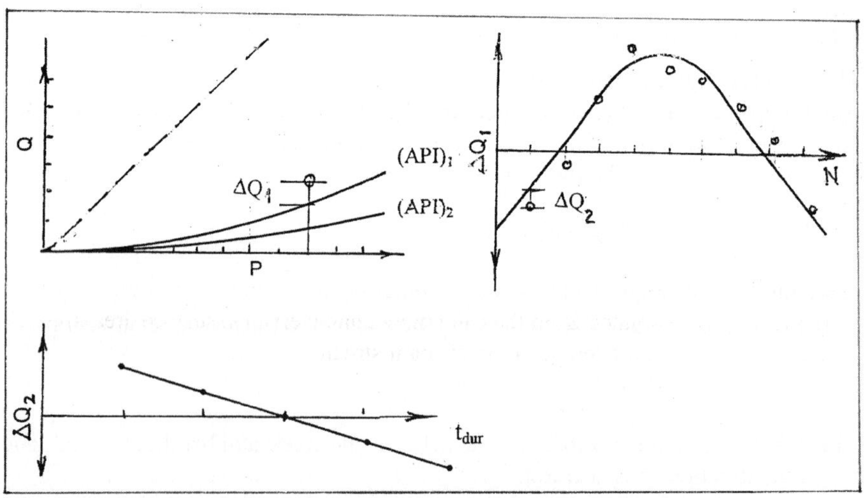

Figure 5. Procedure for the application of the Antecedent Precipitation Index method

The unit hydrograph gives the time distribution of the discharge from a watershed produced by a uniform net rain of given depth precipitated on the area. It shows how this rain is transformed through a linear process into discharge at the outlet. As such, the unit hydrograph can be regarded as a characteristic of any watershed. The ordinates of the unit hydrograph are often expressed by the symbol U:

$$U = U(t, t_{dur}) \tag{5}$$

where t_{dur} refers to the duration of the net rain producing the hydrograph, (sometimes referred to as the period of the unit hydrograph). It is possible to change this period to another period. Detailed explanation of the procedure and applications can be found a large number of hydrology textbooks.

A number of methods are available for the derivation of the IUH (e.g. Singh et al., 1981). Rodriguez-Iturbe and Valdes (1979) and later other researchers contributed to the development of the concept of geomorophpological instantaneous unit hydrograph (GIUH) as alternative methodology. In this concept, the shape and scale parameters of the catchment transfer function are related to the topology and other catchment's characteristics. More recently new developments have been introduced to the IUH allowing the peak and time-to-peak of the IUH to be expressed in terms of a group of catchment and channel characteristics and the intensity of rainfall excess. The hydrograph thus developed is referred to as the geomorphoclimatic instantaneous unit hydrograph (GCIUH).

6.1.3.4 The Nash model

In this storage model the effective rainfall is assumed to be transformed into a surface runoff through routing that rainfall down a series of equal reservoirs (Nash, 1957). The outflow, Q, is assumed to be a linear function of the reservoir storage, S, as:

$$S = KQ \tag{6}$$

in which K is a constant. The form of the instantaneous unit hydrograph (IUH) at time t after a cascade of n reservoirs can be written as:

$$U(0, t) = \frac{1}{K\Gamma(n)} \left[\frac{1}{K}\right]^{n-1} e^{-t/K} \tag{7}$$

in which $\Gamma(n)$ is the gamma function of n. For ungauged catchments the parameters K and n can be estimated from the catchment characteristics, surface area, overland slope, and length and average slope of main stream.

6.1.3.5 The runoff-routing model (RORB)

This model was developed by Laurenson (1964). It is based on the principle of rainfall excess through a non-linear catchment storage described by the equation:

$$S = K(Q)Q \tag{8}$$

where, S is the stored volume in the reservoir and the coefficient $K(Q)$ is a function of Q, the outflow rate from the reservoir. The relationship given by Eq (8) implies that the storage delay time is dependent on Q, which varies during the rainstorm. Thus the flow through the various storage processes of interception, surface detention and soil moisture detentjon is made essentially non-linear.

In its original form, the catchment area considered was divided into a number of contributing sub-areas separated by isochrones, starting from the outfall, and a relative time-area diagram constructed. Beginning at the head of the catchment, the effective rainfalls over the top sub-area were converted into an inflow hydrograph, $I(t)$, for the time increment Δt. The hydrograph, $I(t)$, was then routed using Eq. (8) along with:

$$K = aQ^{-b} \tag{9}$$

in which a and b were derived from the historical records. The excess rainfall inputs, $I(t)$, from the next sub-area were added to the resulting values of $Q(t)$, and the routing procedure was repeated. From the continuity and storage equations, $Q(t)$ can be calculated from the expression:

$$Q_2 = \frac{I_1\Delta t + I_2\Delta t + (2K_1 - \Delta t)Q_1}{2K_2 + \Delta t} \tag{10}$$

In the first instances K_2 is put equal to K_1 and an approximate value of Q_2 is used to obtain an improved K_2. By repeating, a final value of Q_2 is obtained by iteration. The Q_2 thus obtained serves as Q_1 for the next routing step. The operation of the runoff-routing model over a catchment beginning with the hyetograph of excess rainfall converted to an inflow hydrograph $I(t)$, follows the sequence:

$$I_l(t) \rightarrow [Q_l(t) + I_2(t)] \rightarrow [Q_2(t) + I_3(t)] \rightarrow [Q_3(t) + I_4(t)] \rightarrow Q(t)$$

where t is the increment of time of the excess hyetographs and hydrographs, and the routing processes are indicated by the arrows. $Q(t)$ is the final storm hydrograph at the outfall of the catchment.

The division of the catchment into sub-catchments of major tributaries instead of travel time areas has rendered the model application more flexible. The improved model again uses a non- linear catchment response, and is a distributed model since it can incorporate non-uniform rainfall and losses on the catchment (Mein et al., 1974, 1988).

6.1.3.6 *The Dawdy-O'Donnell model (1965)*

This is one of the earliest conceptual models, built up of four storages whose contents vary with time. These components are:

- surface storage, R- Augmented by rainfall, P; and depleted by infiltration, F, evapotranspiration, ET, and, when R exceeds a threshold, R^*, channel inflow, Q_1.

- Channel storage, S- Augmented by channel flow, Q_1; and depleted by surface runoff at the gauging station, Q_S.
- Soil moisture storage, M- Augmented by infiltration, F, and capillary rise, C; and depleted by transpiration E_M, and when M exceeds a threshold, M^*, deep percolation, D.
- Groundwater storage, G- Augmented by deep percolation, D, depleted by capillary rise, C, and baseflow at the gauging station, B; and if and while G exceeds G^*, M is absorbed into G, C and D no longer operate, but E_M and F now act on G.

A schematic representation of the above model is shown in Figure 6. The model has nine parameters appearing in the equations describing the way each component or interchange acts. Two parameters are associated with the storage- outflow expression given by Eq. (6) for Q_S and Q_B. Three parameters f_o, f_c and k come from Horton's equation used, where f_o is the maximum rate of infiltration, f_c the minimum infiltration rate and k is the exponent in infiltration equation. The three threshold levels R^*, M^*, G^* and the maximum rate of capillary rise, C_{max}, are the remaining four parameters. These nine parameters control the functioning of the model. The input data consist of P, ET and Q for each time interval of a known record together with the initial estimates of the nine parameters. Next, F is calculated according to Horton's equation using f_o, f_c and k.

A potential rate of infiltration, f_i, is calculated for the end of the current interval. Then Q_1 is determined to be the excess, if any, over R^* left in surface storage, after ET and F have been abstracted. The model then uses the P, ET data to calculate

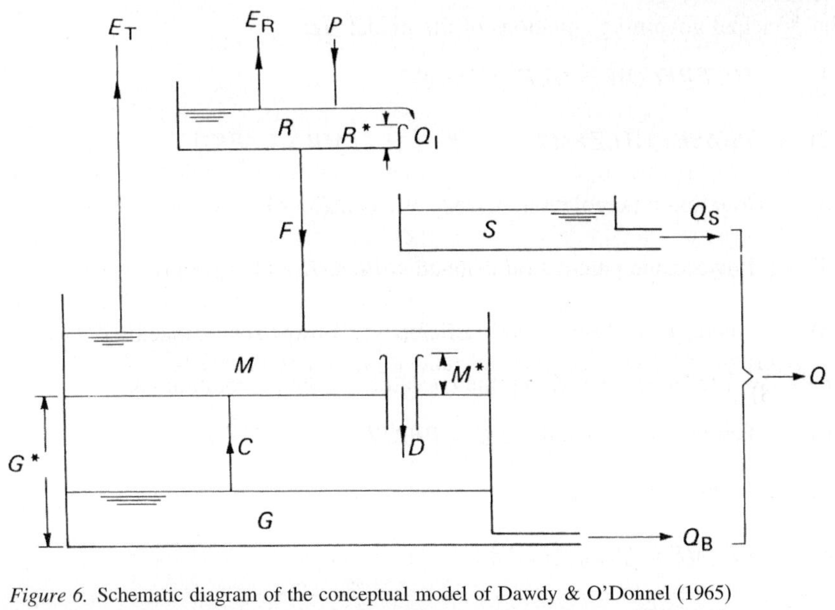

Figure 6. Schematic diagram of the conceptual model of Dawdy & O'Donnel (1965)

a runoff volume for each interval of the record. This often does not agree with the observed Q values. An optimization technique adjusts the initial parameter values such that the differences between the model and observed Q values are eliminated (Dawdy & O'Donnel, 1965).

6.1.3.7 Sacramento model

The model, developed by the Staff of the National Weather Service, River Forecast Center at Sacramento, California, USA, is a conceptual model. The primary elements involved in this simulation model are: (i)- determination of basin precipitation, (ii)- determination of basin evapotranspiration and (iii)- determination of the basin characteristics for the generation of streamflow. These characteristics include soil moisture storage, percolation and drainage. The whole system, as shown diagrammatically in Figure 7, is subdivided into two main zones, the upper zone *(UZ)*, which represents the surface water system, and the lower zone *(LZ)*, which represents the subsurface system. Each part has its storage capacities for tension and free water. Water entering any of these two zones is added to tension storage so long as the capacity is not exceeded, and any excess is added to free water storage.

A portion of any precipitation is diverted immediately to the channel system as direct runoff. This is the portion that falls on the channel system directly and on impervious areas adjacent thereto. The extent of this area is time variant in the model. Free water in the upper zone is depleted either as interflow or goes as percolation in the lower zone. If the rate of moisture supply to the upper zone is greater than the rate of depletion, the excess becomes surface runoff. Free water in the lower zone is divided between primary (slow drainage) storage and supplementary storage.

The principal governing equations of the model are:

(11) $INTERFLOW = UZK \times UZFWC$

(12) $PBASE = [(LZFMS \times LZSK) + (LZFMP \times LZPK)]$

(13) Maximum percolation capacity $= PBASE \times (1 + z)$

(14) Lower zone percolation demand $= PBASE \times (1 + z \times f)$

(15) where, $f = [(\text{Lower zone deficiency})/(\text{Lower zone capacity})]^{\text{REXP}}$

and $REXP$ = coefficient describing the curvature of the percolation curve.

(16) Percolation Demand $= PD = PBASE(1 + z \times f) \times u$

(17) $LZPK = (1 - k_p)$

(18) $LZFMP = Q_{p\max}/(LZPK)$

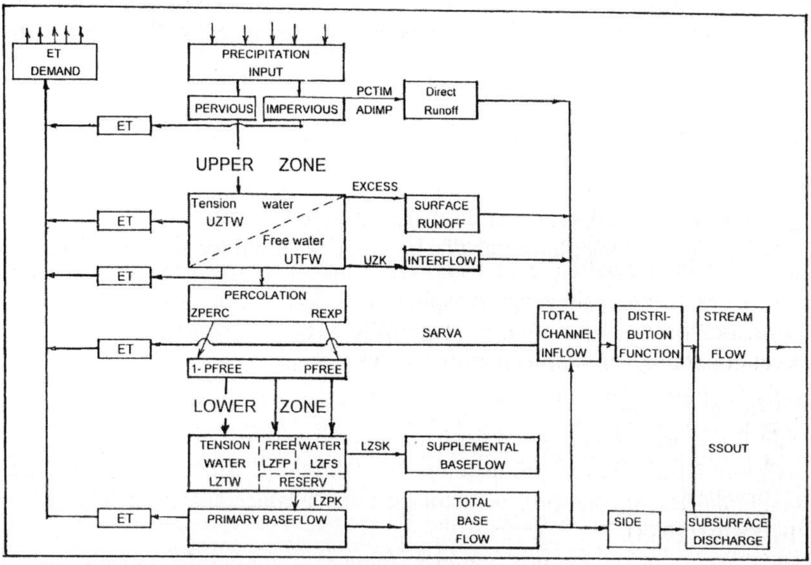

<u>Notations</u>

UZK = upper zone free water storage depletion coefficient,
UZFWC = upper zone free water in storage,
UZTWM = upper zone maximum capacity of tension water,
UZTWC = upper zone content of tension water
UZFWM = upper zone maximum capacity of free water,
UZFWC = upper zone content of free water,
LZFMS = lower zone free water maximum supplementary storage,
LZFPM = lower zone free water maximum primary storage,
LZFPC = lower zone primary content of free water in storage,
LZSK = lower zone supplementary storage depletion coefficient,
LZPK = lower zone primary storage depletion coefficient,
LZTWM = lower zone maximum capacity of tension water,
LZTWC = lower zone content of tension water,
PBASE = limiting rate of drainage from the lower zone storages,
z = a multiplier used to augment percolation from minimum, PBASE, to the maximum,
f = $[(\text{lower zone deficiency}) / (\text{lower zone capacity})]^{REXP}$
REPX = coefficient that determines the curvature of percolation curve,
u = (upper zone free water content) / (upper zone free water capacity),
PCTIM = fraction of impervious basin contiguous with stream channels,
ACTIM = fraction of impervious area that appears when tension water requirements are met,
ADIMP = maximum value for ACTIM,
POTI M = the potential impervious, is the sum of (PCTIM) plus (ADIMP),
SARVA = fraction of the basin covered with streams, lakes and wetlands,
ED = evapotranspiration demand,
Z = proportional increase in percolation from saturated to dry condition,
PFREE = percentage of percolated water that directly enters the lower free zone aquifers without a prior claim by lower zone tension water deficiencies,
SIDE = ratio of non-channel subsurface flow to channel baseflow,
SSOUT = discharge required by channel underflow,
PCTPN = dimensioning of evapotranspiration

Figure 7. Diagrammatic representation of the Sacramento model

where, k_p = recession coefficient of primary baseflow for the time unit used and Q_{pmax} = maximum primary baseflow that can be inferred from hydrograph analysis.

(19) $$LZSK = (Q_{sl}/Q_{\max})^{1/t} = (1 - k_s)$$

(20) $$LZFSM = (Q_{\max}/LZSK)$$

where,

$Q_{\max}$ = lower zone supplemental discharge at time t = zero,
Q_{sl} = lower zone supplemental discharge at an arbitrary time from time of peak,

and

k_s = rate of change of supplementary storage.

Some of the results obtained from the application of the Sacramento model to the Blue Nile Basin are presented in a forthcoming sub-section.

Many more models have been developed for different purposes, under different names and in different countries. Some of the notable conceptual models have been listed by Shaw (1983).

Another group of deterministic models, known as component models, seek to describe each part of the whole hydrological setup mathematically by means of appropriate and rigorous equations. Despite the scientific supremacy of this group of models, the complexities in the hydrological cycle make it doubtful whether all the model components can be linked and made operational.

6.2. RIVERS

6.2.1 Large river basins

Large river basins, as meant here, are those river basins with a total surface area exceeding $100,000\ \text{km}^2$ each. This class of river basins in the Arab Region comprises the Sénégal, Nile, Shebelle, Juba, Tigris and Euphrates. Catchment areas of these rivers and the gauging stations installed on them are given in Table 2, Appendix II.

6.2.1.1 The Sénégal and its tributaries

The Sénégal is an important river in West Africa. Four riparian states, Guinea, Mali, Sénégal and Mauritania share its basin. The river forms the Mauritania-Mali and the Mauritania-Sénégal international boundaries. Description of the Sénégal Basin and its hydrology has been reported by a number of authorities, e.g., Rodier (1964), Rochette (1974) and Shahin (2002).

What interests us here is a small part of the upper sub-basin and all of the lower sub-basin, which is drained by the river reach below the key measuring station at Bakel down to the river mouth on the Atlantic Ocean. The Karakoro, last tributary in the upper sub-basin, originates in the region of Kiffa, Mauritania, at altitude 130 m a.m.s.l. After flowing in a rather dry region for about 300 km it joins the

Sénégal River at Lami-Tounka. As the Karakoro originates and flows in a sahelian climate it adds but little to the flow of the Sénégal (Rodier, 1964). More important is that it increases the basin area at Bakel to $218,000\ km^2$.

The length of the river between Bakel and St. Louis on the Ocean is 784 km. The principal streams forming the drainage network in the lower sub-basin are the Nirode (Noiré), Ghofra and Savalel wadis and the Gorgol River, formed by the junction of the White (Blanc) Gorgol and the Black (Noir) Gorgol. The Ghofra originates in the Assab massive, dividing the two parts of the basin, at elevation 318 m a.m.s.l. It runs a distance of 190 km before joining the Sénégal at a point 40 km below Ouaunde. The junction of the Black Gorgol and White Gorgol forms the Main Gorgol. Again, as all tributaries in the lower sub-basin rise in a fairly dry environment, they are in fact wadi type streams, and their contribution to the flow of the Sénégal River is quite limited. Some of the hydrologic aspects of the Gorgols will be presented in the section dealing with wadi flow. Before finding its way to the Atlantic Ocean, the river reach in the lower basin is connected with Lakes Guiers and Rkiz, as shown in the map in Figure 8.

The channel of the Sénégal River is the international boundary between Sénégal and Mauritania rendering the river to be contiguous. The share of Mauritania in the basin area of the river is 50.2%, representing 23.7% of the surface area of Mauritania. The long-term (1904–84) mean annual discharge at Bakel given

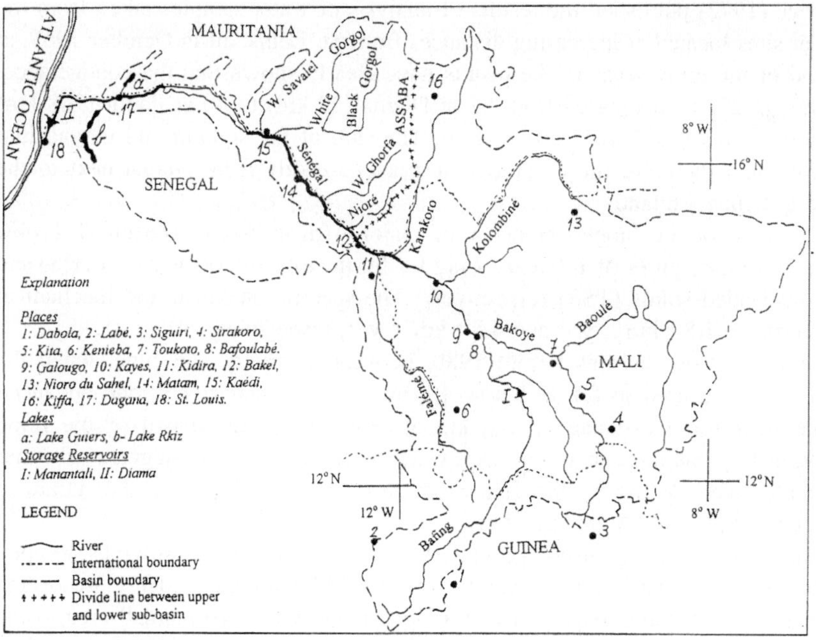

Figure 8. Sénégal River Basin

in Table 4, Appendix II, is 709 m^3 s^{-1}. As the basin area at St. Louis on the Atlantic Ocean is 270, 000 km^2, the mean value of the specific discharge becomes 2.63 l s^{-1} km^{-2}. Tables 3, 4, 4(a) and 4(b), Appendix II, give the mean monthly discharges, annual flow series, mean maximum and minimum monthly discharges, and extreme discharges, respectively. As the mean maximum (September) and mean minimum (May) river discharges are 3,138 and 5.2 m^3 s^{-1} respectively, one can conclude that the river is highly seasonal.

The maximum annual discharge records are available for the 63-y period, 1904–66, in the UNESCO Publication "Discharge of selected rivers in Africa" (1995). This period happens to represent the natural flow condition in the river before regulating it. The mean of the maximum discharge series is 4866.35 m^3 s^{-1} with coefficients of variation and skewness of 0.362 and 0.483 respectively. Pearson Type III- distribution function showed to be a good fit to the discharge data in that period. Discharge estimates corresponding to 95, 90, 80, 50, 20, 10, 5, 2 and 1% probability of exceedance are 2,240, 2,724, 3,357, 4,720, 6,290, 7,197, 7,991, 8,937 and 9, 598 m^3 s^{-1} respectively.

Tables 1(a) and 1(b), Appendix III, contain monthly and annual suspended sediment load and concentration figures for the years 1979–84 at Bakel. The tabulated figures show that the mean annual concentration of suspended sediment during the given period varied between 156 and 239 mg l^{-1}, with an average figure of 185 mg l^{-1}. The mean annual load is 1908 × 10^3 t and the corresponding specific load is 8.75 t km^{-2} y^{-1}.

Grove (1972) published the results of analysis of water samples taken from the river at sites located at increasing distances from St. Louis, all in October 1969, at the end of the rainy season. The results have clearly shown that the conductance, in mhos at 25° C, decreased from 52 at Diama (29 km) to 40 at Podor (268 km) and further to 33 at Tiguélré (610 km). The sum of the sodium and magnesium contents, and therefore the chlorinity decreases steadily from Diama next to the ocean to Sénégal inland.

In their report on biogeochemistry of major African rivers, Martins & Probst (1991) gave the figures of 140 and 57 mg l^{-1} for the total dissolved salts (TDS) and total suspended solids (TSS) respectively. The specific chemical and mechanical transports are 1.8 t km^{-2} y^{-1} and 4.4 t km^{-2} y^{-1}, respectively.

In the low-flow seasons before 1988, between February and June, when the mean discharge used to fall to below 100 m^3 s^{-1}, the seawater wedge advanced up the river for a distance of 300 km or more. With the arrival of the flood season in July, the large incoming freshwater helped to flush the stagnant saltwater down the river so that by the end of August, the Sénégal above the mouth at St. Louis became once more free of salt.

Considering the mean discharge as 750 m^3 s^{-1} and the concentration as 40 ppm, the dissolved load carried in an average year would be around 1 × 10^6 t. According to Grove (1972), the greater part of this load is derived from a small area in the headwater zone of about 75, 00 km^2, with over 1,000 mm rainfall, indicating a specific yield around 10–15 t km^{-2} y^{-1} from that area.

Since 1988, the Sénégal basin has been equipped with two major structures, the Diama Barrage to prevent or at least reduce the salt intrusion from the Atlantic Ocean and the Manantali Dam, which controls 50% of the of the river discharge. These two works are supposed to supply fresh water for large-scale land development and irrigation projects as well as power development.

6.2.1.2 The Nile and its tributaries

The Nile Basin (Figure 9) has a total surface area of about 2.88×10^6 km^2, of which 62.7 and 9.95% are in the Sudan and Egypt respectively (Krishna, 1986). Southern Sudan is a semi-arid to semi-humid country with high temperature the whole year round. The northern part of the Sudan, and so Upper Egypt, have hot, dry and long summers, and agreeable winters. The summer along the Mediterranean Coast is dry and pleasant and so is the winter, except for light showers.

After leaving Uganda, where the source of the river lies, the Upper Nile crosses the international Uganda-Sudan border at Nimule, and flows through southern Sudan under the name Bahr el-Jebel. The first key station is Mongalla, which is situated some 215 km below Nimule. The mean annual flow for the period 1905–83 is close to 33×10^9 m^3 with variation coefficient of 0.39 (Elwan, 1995). It is worthwhile to state here that discharge measurement at Mongalla has stopped since 1983 due the hostile situation between southern and northern Sudan.

North of Bor, 120 km below Mongalla, the river valley becomes wide and less defined. Extensive swamps spread out on each side of the river and continue down to Lake No. The river runs northwards between walls of papyrus and tall swampy grasses. The region is known as the Sudd (Sudd is an Arabic word meaning dam, barrier or blockage). Sutcliffe & Parks (1999) estimated the average monthly flooded areas below Bor for the period 1905–83 as 12,800 km^2. The net flow in the Bahr el-Jebel below the swamps is nearly one-half of the river flow at Mongalla. The Bahr el-Jebel is joined from the west by Bahr el-Ghazal, a river with a basin area of 520,000 km^2. This basin too is a source of huge losses that from a total rainfall of more than 400×10^9 m^3 the annual runoff that reaches Lake No is 0.6×10^9 m^3, i.e. 0.15%.

The Sobat joins the White Nile (the name of Bahr el-Jebel below the confluence of the Bahr el-Ghazal) at Hillet Doleib. The basin of the Sobat has a surface area of 225,000 km^2, part of which is in Ethiopia and the remaining part in the Sudan. The Baro (41,400 km^2) and the Pibor (10,000 km^2) are the principal tributaries of the Sobat. The yield of the Sobat Basin encounters heavy losses as a result of the excessive spillage from the river section added to the evaporation losses from the Machar Swamps in the basin of the Baro. The average annual flow of the Sobat over the period 1905– 83 measured at Hillet Doleib is 13.5×10^9 m^3 with coefficient of variation of 0.20. This amount when added to the annual volume of flow of the Bahr el-Jebel above its confluence with the Sobat brings the mean annual flow over the period (1905–95) at Malakal to 29.65×10^9 m^3, with coefficient of variation of 0.18. Malakal is the second key stream gauging station in the Sudan. The hydrograph of the White Nile at Malakal is less uniform than that of the Bahr el-Jebel at

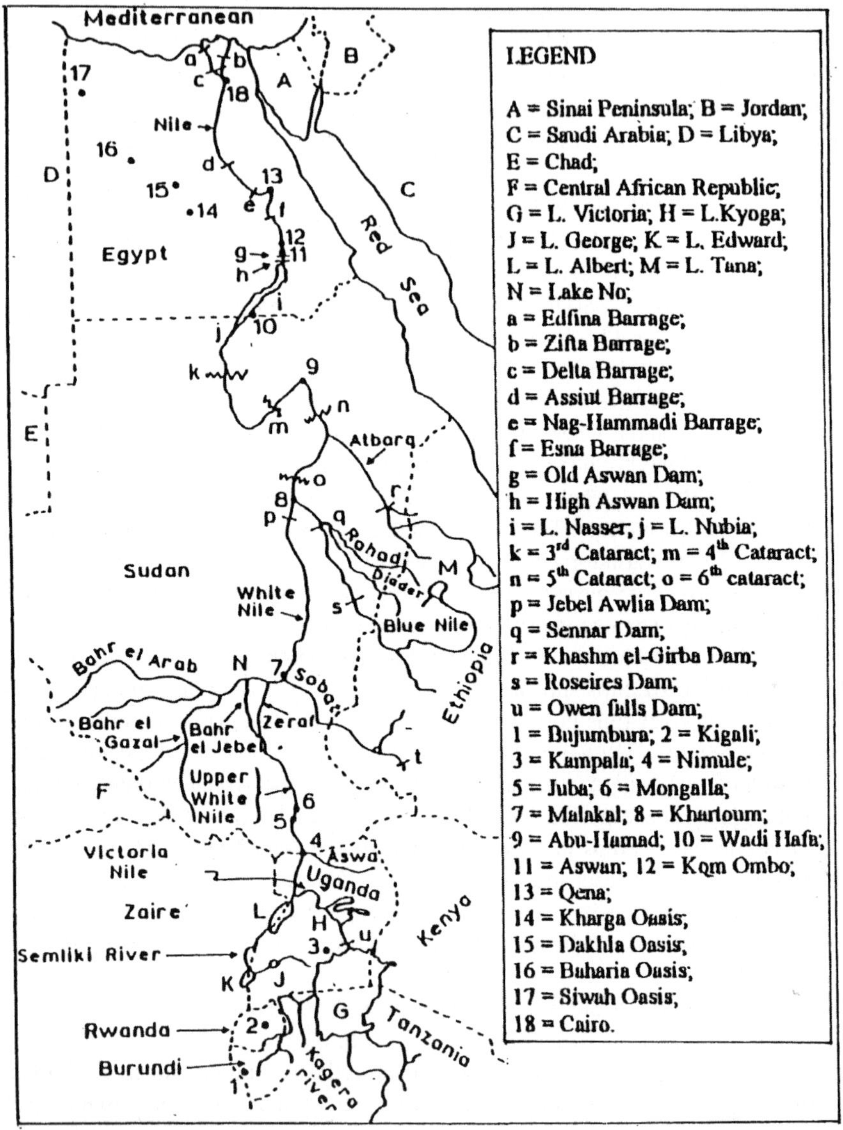

Figure 9. Map of the Nile Basin (based on Kashef, 1981)

Mongalla caused by the relative seasonality of the flow of the Sobat. The long-term (1905–83) mean flow at Hillet Doleib varies from a minimum of 232 $m^3\ s^{-1}$ in April to 1,992 $m^3\ s^{-1}$ in October. These two figures and the corresponding figures for Mongalla and Malakal bring the ratio minimum to maximum discharge to 0.686, 0.116 and 0.456 for Mongalla, Hillet Doleib and Malakal respectively. This ratio, sometimes its inverse, is referred to as seasonality index.

From Malakal the White Nile flows due north for a distance of about 800 km to just upstream Khartoum where it is joined by the Blue Nile. The White Nile between Malakal and Khartoum resembles a long, narrow lake without any tributary. As the average evaporation is much higher than the average precipitation, the White Nile is a source of loss. The annual loss in the river flow between Malakal and Khartoum can be fairly estimated as 1.5×10^9 m^3. Some 40 km above the confluence of the Blue Nile at Khartoum, a dam was built at Jebel-Aulia around 1937 to store water for the interest of Egypt. The rapid silting up of the reservoir formed by the Jebel-Aulia dam, and the construction of the Aswan High dam, 1960–1968, have stopped its function as a storage reservoir. The average annual flow in the White Nile for the period 1911–1995 at Mogren (below the Jebel-Aulia reservoir) is 26.03×10^9 m^3.

The Blue Nile and its tributaries all rise on the Ethiopian Plateau at an elevation of 2,000–3,000 m a.m.s.l. The total basin area, including Lake Tana and its basin is $324,530$ km^2. The Blue Nile, almost 120 km blow the point where it crosses the Ethiopian-Sudanese boundary reaches Er Roseires (Damazin). This too is a key stream gauging station. The natural river supply of the river at this station averaged over the period 1912–97 is 48.7×10^9 m^3. The Dinder ($16,000$ km^2) and the Rahad ($8,200$ km^2) are two tributaries joining the Blue Nile at Gwasi and Hawata, respectively. The flow season for these two tributaries lasts from June to December. Their natural supply averaged over the period 1908–1997 is 2.8 and 1.1×10^9 m^3 y^{-1}.

A short distance below the confluence of the Rahad, the Blue Nile traverses the reservoir formed by the Sennar Dam. This dam was built in 1925 mainly to guarantee the supply of adequate amounts of water to the Gezirah canal for cotton irrigation. Below Sennar, the Blue Nile travels a distance of 370 km before reaching Khartoum. Just above its junction with the White Nile at Khartoum, the average annual flow for the period 1912–1982 at Khartoum is $1,532$ m^3 s^{-1} or 48.35×10^9 m^3. The sum of the annual volumes brought by the two Niles together downstream their confluence is 74.38×10^9 m^3. The average flow in the Nile for the same period as measured at Tamaniat, the first key station on the Main Nile 41 km below the confluence of the Blue Nile, is 72.69×10^9 m^3. The difference between the two figures being 2.3% is quite acceptable.

The Main Nile flows 277 km below Tamaniat in a desert climate before it reaches Hassanab, a few kilometers above the junction of the Atbara River with the Main Nile. The conveyance loss between Tamaniat and Hassanab stations is about 0.7×10^9 m^3 y^{-1}, bringing the annual flow at Hassanab to 72×10^9 m^3. This estimate is close to the long-term average figure obtained from flow measurements, 72.34×10^9 m^3 y^{-1}.

The Atbara Basin has a total surface area of $100,000$ km^2. The runoff coefficient of the upper part of the basin, $68,000$ km^2, is 0.20 and for the lower part, $32,000$ km^2, 0.10. The Khashm el-Girba reservoir on the Atbara is operating since 1965–66 for the purpose of flow regulation. This is necessary as the natural supply is confined to the season June-December of each year as can be observed from the figures in Table 3, Appendix II. The mean annual flow for the period 1903–94 at the Atbara mouth (Kilo 3) is 11.05 10^9 m^3, bringing the total river flow downstream the Atbara junction to 83.4×10^9 m^3.

Dongola is the site of the first key stream gauging station on the Main Nile Below the confluence of the Atbara River. The distance between the two locations is 740 km. There is no gain whatsoever in this reach. On the contrary, the river loses some of its water due to strong evaporation caused by the predominantly hyper-arid climate in this region. The mean annual flow of the Main Nile in the period from 1911 up to 1995 at Dongola is 80.76×10^9 m^3. This volume is less than that just downstream the confluence of the Atbara by the conveyance losses (evaporation + seepage) plus any probable errors in measurement and/or calculation. As a matter of fact, the annual volume of 80.76×10^9 m^3 for the period 1911–95 at Dongola is on the lower side compared to other periods of time, e.g. 86.13×10^9 m^3 for the period 1911–60. One should not forget, however, that the mean for the period 1961–95 was 73.1×10^9 m^3, which is small indeed.

The Main Nile flows a distance of 450 km before reaching Wadi Halfa, and further 345 km before reaching Aswan, Egypt, the last key station on the Nile. The conveyance loss in this reach of about 800 km exceeds 2×10^9 m^3 y^{-1}. The annual natural supply at Aswan is traditionally taken as (84-85) $\times 10^9$ m^3, of which (10-11) $\times 10^9$ m^3 are lost by evaporation and seepage. The remaining amount is divided between Egypt (55.5×10^9 m^3) and the Sudan (18.5×10^9 m^3). Considering an average natural flow of say 85×10^9 m^3 y^{-1}(2, 695 m^3 s^{-1}) at Aswan and the corresponding basin area 2.9×10^6 km^2 gives an average specific discharge of the Nile as 0.93 l s^{-1} km^2. The mean monthly discharges at four key stations on the Nile and its tributaries in the Sudan are shown in Figure 10.

The annual flow below the Aswan High Dam, 55.5×10^9 m^3, is distributed between the different agricultural zones in Upper Egypt, Middle Egypt, and the Nile Delta according to their respective requirements for water. The water requirements for the Nile Delta area are supplied via the two river branches, the Rosetta and Damietta.

Old information about the quality of the Nile water has been collected from a number of sources, summarized and included in Appendix II, Part B, of the reference book "Hydrology and Water resources of Africa" by Shahin (2002). In 1995, the Ministry of Irrigation and Water Resources, Egypt, reported on the water quality assessment program of the Nile River, the main channel and the Damietta and Rosetta branches, which took place in the period 1976–91. Some of the results given in this report are included in Tables 2(a) and 2(c), Appendix III of the present book. The contents Table 2(b) of the same Appendix are quoted from the report on biogeochemistry of major African rivers: carbon and mineral transport by Martins & Probst (1991).

Dissolved minerals in water are composed of calcium, magnesium, sodium and potassium in combination with the anionic radicals, bicarbonate, sulfate and chloride. The Nile water is characterized by the dominance of calcium of the cations and bicarbonate of the anions. As any natural river, the Nile contains other

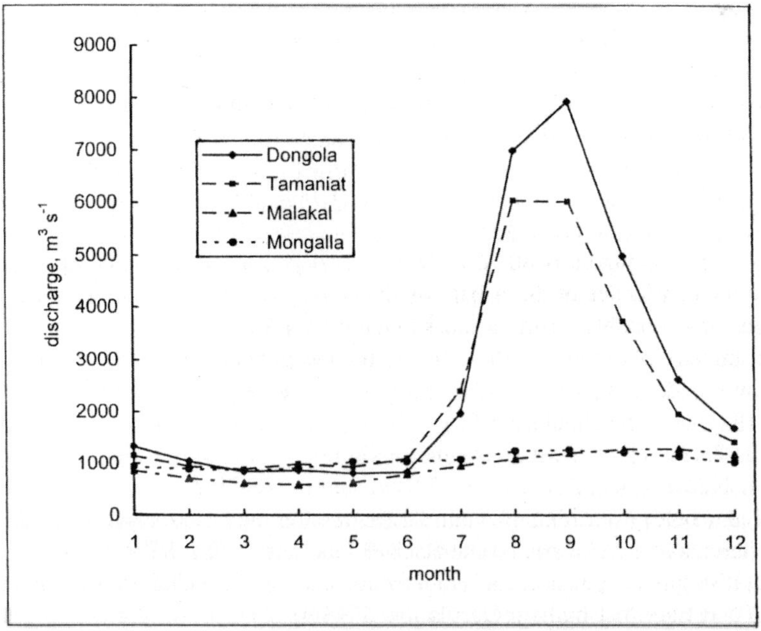

Figure 10. Long-term average discharge hydrographs of the Nile at four key stations

constituents of minor concentrations among which salts of phosphorous and nitrogen and silica are of importance. Salts of the major ions, except the bicarbonate radical, are of conservative nature. The concentration of the bicarbonates in natural water is controlled by the carbonate-bicarbonate equilibrium. In surface waters, however, the biological activities can disturb this balance resulting in changes in the bicarbonate concentration.

Comparison between the pairs of figures for TDS listed in Tables 2(b) and 2(c), Appendix III, show that former gives TDS as 318 mg l^{-1} while the figure for the same parameter in Table 2(c) is 177. The corresponding figures for TSS are 54 mg l^{-1} at Assyut and 15.8 mg l^{-1} as an average for water in the full length of the main channel, Table 2(c). These two pairs of figures show that the ratio TSS/TDS for the year 1983 was 0.18 at Assyut and 0.09 in July 1991 for the river channel. The corresponding river discharges for these situations in their respective order were 100×10^6 m^3 d^{-1} and 220×10^6 m^3 d^{-1}.

Furthermore, the generated data showed that the river system receives heavy loads of dissolved solids, nutrients, organochlorine pesticides and heavy metals from the investigated agricultural drainage canals and industrial outfalls. However, the loads are largely diluted by the river flow such that the impact on water quality is generally limited. Nevertheless, some effects appeared at some river sections especially in the downstream reach of the main channel and in the Damietta branch.

Modelling the Nile flow: Shahin (2002) reviewed some of the models applied to Nile flow. These models arranged in chronological order are:

- Balek (1977) modelled the maxima and minima series of the Nile levels at Rodha using autoregressive (AR) models. Hipel & McLoed (1978) reanalyzed almost the same series using autoregressive moving average (ARMA) models.
- Salas (1981) analyzed the series of annual outflows of the Equatorial Lakes, Victoria and Albert, and tried to model the abrupt rise in lake discharges around 1962 using an intervention model.
- Fahmy et al. (1982) modelled the river flow from Pakwach, at the outlet of Lake Albert, down to Wadi Halfa at the mouth of the reservoir formed by the Aswan High Dam using a Constrained Linear System (CLS) model.
- Shahin (1986) modelled the annual flows of the Bahr el-Jebel at Mongalla using AR models. Aguado (1987) reanalysed the same sequences used by Balek except that he used the ARIMA models.
- Reusing (1994) applied ARMA models for slightly shorter series than the ones used by Balek or Aguado.
- Elwan et al. (1996) used AR and Multiple Regression models to supplement the missing flow data at Mongalla on the Bahr el-Jebel after 1982–83.
- El-Tahir (1996) developed a linear relationship between the annual flow arriving at Aswan and El-Niňo Southern Oscillation (ENSO).

Ibrahim (1993) modelled the flow in the Blue Nile at Ed-Deim station on the Ethiopian-Sudanese border. The approximate catchment area at this station is $180,000\ \text{km}^2$ or 6.2% of the total area of the Nile Basin, yet it produces no less the 60% of the total Nile flow. The forecasts at Ed-Deim can be forecast about three days ahead from rainfall over the Nile Basin in Ethiopia.

The objectives of that study were (i)- estimating the basin rainfall using the so-called Cold Cloud Duration (CCD) and the Cold Cloud Coverage (CCC), produced from the METEOSAT TIR-images, (ii)- application of the SACRAMENTO model (complex conceptual model), (iii)- application of the RORB model (simple conceptual model) and (iv)-application of the ARMAX-type stochastic model.

The CCD and CCC data were available for the entire basin divided into four sub-basins as well as for the whole basin, for the flood seasons (July–September) from 1987 up to 1990. In this season the rainfall varies from 65% to 85% depending on the location of the sub-basin. The idea was to establish a regression relation between the rainfall, P, the fractional area average cold cloud cover, C (threshold temperature −40°C), and the cold cloud duration, D. Denoting the regression parameters by a and b, the relationship can be expressed as:

$$(21) \qquad P = aC + bD$$

The method of analysis that yielded the best estimates of monthly rainfall is that in which the mean rain depth per rainy day was regressed on the mean cold cloud duration, h d^{-1}. The estimated rainfall was 289, 275 and 179 mm and the observed 307, 314 and 215 mm for July, August and September, respectively.

The parameters of the Sacramento model were found through an iteration procedure. This required elaborate interplay and adjustment between the free water, primary and secondary, both in the lower zone as well as between the tension water in the upper and lower zones. After establishing the capacities and values of parameters such as *UZTW, UZFW, LZTW, LZFSW, LZFPW, UZK, LZSK, LZPK, ZPERC, REXP, PFREE*, etc. a pair of simulation approach runs were carried out for the period 1987–1990. These were referred to as single segment approach and multiple segment approach. The results of the two approaches were compared based on analysis of errors in monthly flows between predicted and observed values. The results obtained are summarized in Table 1.

From Table 1 it is clear that for the four years 1987–90 the model results were slightly to moderately different from the observed discharges for July and August. The results produced by the single and multi segment approaches are quite close to each other. However, the difference between the model-predicted and observed discharges for September in all four years was quite large and unacceptable, the model overestimated the river flow considerably. The difference in case of the multi segment approach was definitely larger than the difference obtained from the multi segment approach. Three trials were made; the first two dealt with the reduction of the September rainfall estimated from the CCD method for two of the

Table 1. Simulation results of The Blue Nile flow at Ed-Deim Station for the period 1987–1990 as obtained from Saceamento model (after Ibrahim, 1993)

Item and year	Single segment approach			Multi segment approach		
	Jul.	Aug.	Sep.	Jul.	Aug.	Sep.
A, 1987	0.89^x	2.05	1.65	0.94	1.93	2.03
B	0.90	1.87	1.32	0.9	1.87	1.32
C	−0.01	0.18	0.33	0.04	0.05	0.71
D	1	10	25	4	3	53
A, 1988	1.64	3.64	3.84	1.75	3.66	3.89
B	1.65	3.61	2.8	1.65	3.61	2.8
C	−0.01	0.03	1.04	0.11	0.06	1.09
D	1	1	37	7	2	39
A, 1989	1.37	2.48	2.48	1.32	2.34	2.70
B	1.19	2.24	2.01	1.19	2.24	2.01
C	0.18	0.24	0.47	0.13	0.1	0.69
D	16	11	22	11	4	34
A, 1990	0.97	2.03	2.27	0.97	2.14	2.30
B	0.88	2.35	1.86	0.88	2.35	1.86
C	0.09	−0.32	0.40	0.10	−0.21	0.44
D	10	−14	22	10	−9	24

Explanation
A = predicted flow, B = observed flow, C = difference between predicted and observed flows and D = percentage difference. A, B and C are in mm d^{-1}.

sub-catchments. The third trial went to raising the evapotranspiration in September by 25%. This reduced the overestimation to almost zero.

The detailed analysis of the results has led to the conclusion that the Sacramento model is quite sensitive to the rainfall over the system and to evapotranspiration from it. It is also sensitive to the unit hydrograph components, which indicate whether a proper time shift or travel time has been applied effectively.

The same source, i.e. Ibrahim (1993) while applying the RORB model to the Blue Nile basin used the same subdivision of the whole basin into four sub-basins as in the Sacramento model. In both models the criteria for subdivision was the homogeneity of the rainfall in each sub-basin. The application of the model was carried out using the rainfall data derived from the CCD and CCC values mentioned earlier. The procedure followed has already been described in an earlier sub-section. A comparison between the predicted and observed parameters of the runoff hydrographs for the period 1987–1990 is included in Table 2.

The model results for the four years look quite acceptable; they resemble the observed hydrographs in as much as the rising limb is concerned. The error in estimating time to peak, though small in percentage, yet the absolute difference between observed and predicted times varied from one to two days. The falling limb of the predicted hydrograph emphasized the fact that the derived CCD rainfall for September need to be adjusted as mentioned in connection with the Sacramento model. The error in estimating the time to centroid of the hydrographs varied between 7 and 22 hours, or between 0.6 and −1.8%, is quite acceptable.

Table 2. RORB model simulation results of the Blue Nile flow (Ibrahim, 1993)

Criteria	Hydrograph predicted	observed	Error absolute	%
Peak discharge, $m^3\ s^{-1}$, 1987	5296	5430	−134	−2.5
Time to peak, h	1344	1296	48	3.7
Volume, $10^9\ m^3$	22.3	22.7	−0.4	−1.8
Time to centroid, h	1175	1197	−22	−1.8
Peak discharge, $m^3\ s^{-1}$, 1988	9270	8690	580	6.7
Time to peak, h	1152	1128	24	2.1
Volume, 109 m^3	43.7	44.6	−0.9	−2.1
Time to centroid, h	1228	1221	7	0.6
Peak discharge, $m^3\ s^{-1}$, 1989	7312	7550	−238	−3.2
Time to peak, h	1334	1296	48	3.7
Volume, $10^9\ m^3$	29.2	30.2	−1.0	−2.1
Time to centroid, h	1226	1240	−14	−1.1
Peak discharge, $m^3\ s^{-1}$, 1990	5984	6150	580	6.7
Time to peak, h	1176	1152	24	2.1
Volume, $10^9\ m^3$	26.8	28.4	−1.6	−5.8
Time to centroid, h	1244	1259	−14	−1.2

The time-series approach to flow modeling of the Blue Nile was carried out using the micro-Captain 2-computer package. This package serves, among other purposes, the function of being a day-to-day tool for recursive analysis and forecasting of univariate time-series as well as modeling of bivariate time- series. There are two modes of operation of the main program: "A-R" for univariate analysis and forecasting, and "I-O" for input-output, transfer function. The approach used involves a two-stage procedure starting with AR identification and parameter estimation. The residuals, i.e. white noise (ε_k) from this stage are then used to construct an I-O data set with the estimated noise residuals forming the input series and the noise (ζ_k) forming the output series. The ARMA model is then obtained by performing I-O transfer function modelling exercises on this I-O data set using the so-called simple refined IV (SRIV) recursive estimation algorithm. In principle the analysis should provide the best ARMA model approximation to higher order AR models since the former may have fewer parameters and thereupon help to avoid over-parameterisation. For more details on time series analysis the reader can be referred to any of the specialized books, e.g. Young (1984).

The best identification model for the year 1987 was found to be (1,1,5). The physical interpretation is that the present day outflow consists of 0.93 of the previous day outflow and the rest is the contribution of rainfall with a lag from 2 to 5 days. Though the coefficients of determination, being in the range 0.71–0.80, seem to be reasonable for the whole period, all models failed to reproduce the peak flows occurring in the second half of August. It is likely that a much larger part could be explained by the I-O model would the periodic component in the series had been first identified and removed (Ibrahim, 1993).

6.2.1.3 The Tigris and its tributaries

Wadi Ar-Rafedein (valley of the two rivers) is an old Arabic description of the region that comprises a major part of Al-Arab Basin. The Tigris, Euphrates and Shat el-Arab Rivers and their respective tributaries are included in the said drainage basin (Figure 11). As such, the Wadi ar-Rafedein comprises the Tigris and Euphrates Rivers. The reaches of these two rivers, which traverse the Arab Region, are the main sources of water for Iraq and Syria.

The Tigris River springs from the southern slopes of the Torus Mountain Range in East Turkey. The total length of the Tigris River from source to mouth is about 1,718 km, of which 1,418 km (82.5% of total length) run inside Iraqi territory. The flow of the river depends on three sources. These are rainfall, snow- melt and water supplied by the tributaries flowing from Turkey and Iraq. The relatively high value of the runoff coefficient causes a considerable proportion of the rainfall on the upper part of the catchment to run off to the river channel. The rainfall in January and December causes the rise of the water level in the river and thereupon augments the discharge. The snowfall helps to maintain the same water level till more supply reaches the river as a consequence of spring rainfall coupled with snow melting in April and May. From the beginning of June the river discharge, consequently the water level falls rapidly until the minimum flow is reached in September. The second source is the supply flowing to the river channel as a result of melting of the snow accumulated on the mountain tops in the river watershed area. This source can yield in some years quite considerable volumes of water, which can be stored and used in the low-flow season in summer time. The third source is represented by the runoff to the Tigris carried by the tributaries springing from the upper parts of the river basins in Turkey. The flood of the Tigris River generally occurs in April.

The Tigris crosses the Turkish-Syrian-Iraqi border at Faysh Khabur, near Zakho, where it is joined by the River Khabur. The channel of the Khabur River is 160 km long and the surface area of its catchment at Zakho ($\varphi = 37°\ 08'$ N and $\lambda = 42°\ 42'$)

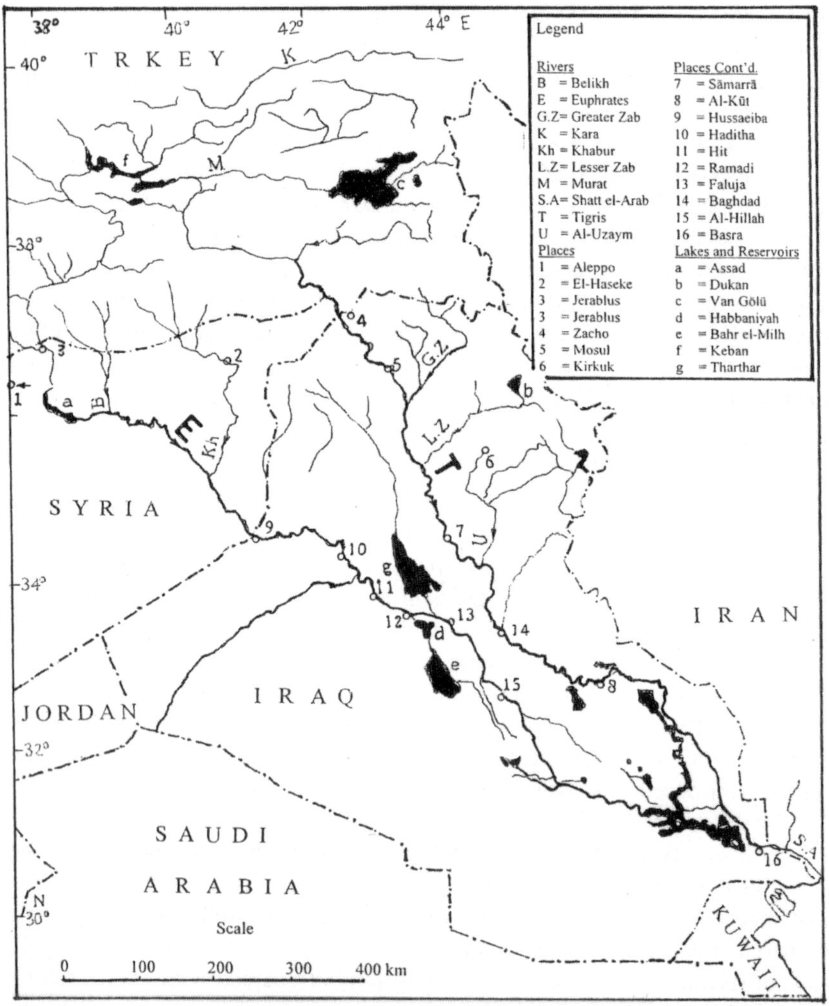

Figure 11. The Tigris and Euphrates Rivers, and their tributaries

is 3, 500 km^2. Assuming the average annual precipitation (rain and snow) as 1.0–1.2 m and an annual runoff coefficient of 0.5–0.6, the total river flow becomes about 66.5 m^3 s^{-1}. This figure agrees well with the arithmetic mean flow for the said river in the period 1958–75, that is 65.5 m^3 s^{-1}, with coefficient of variation of 105% (Kamal el-Din, 1981).

Mosul, the most important city in northern Iraq, is situated at a distance of 188 km below the junction point at Faysh Khabur. South of Mosul the Greater Zab, the larger tributary of Tigris within Iraq, joins the Tigris River. The Greater Zab is a perennial stream 392 km long with a catchment area of 40, 300 km^2 at Eski Kelek ($\varphi = 36°\,16'$ N and $\lambda = 43°\,39'$ E). It supplies the Tigris with a substantial volume of water each year in the order of 12×10^9 m^3. According to Kamal el-Din (1981), the mean annual flow for the period 1931–75 is 13.2×10^9 m^3, with coefficient of variation of 85%.

The Little or Lesser Zab, 400 km long river, joins the Tigris south of Ash Sharqat. The catchment area at Dukan ($\varphi = 35°\,53'$ N and $\lambda = 44°\,58'$ E) is 11, 700 km^2. Assuming the average runoff depth for the whole year to be the same as that for the Greater Zab, i.e. 0.55 m, the corresponding mean discharge becomes almost 203 m^3 s^{-1}. This figure too is close to the mean for the period 1931–75, which is 197.8 m^3 s^{-1} with coefficient of variation of 91%. From Al-Fatha, south of the junction of the Little Zab, the Tigris runs in a gorge through a mountainous region.

The Uzaym (Adhaim River) joins the Tigris south of the town of Balad. The catchment of the Adhaim River has a total surface area of 12, 965 km^2, of which 9, 840 km^2 can be measured at Narrows station ($\varphi = 34°\,30'$ N and $\lambda = 44°\,31'$ E). It is totally situated in Iraqi territory between latitudes 34°20′ and 35°54′ N and longitudes 44°00′ and 45°00′ E. The ground slope decreases from north to south, with an average of 1.5 m km^{-1}. The river has its sources in the mountainous region of the Sulaymanîya Province. The length of river channel from source to the junction with the Tigris River is 230 km.

The climate over the catchment area varies from semi-humid in the north to semi-arid in the south, where the annual rainfall ranges from 600 to 150 mm respectively. The rainy season is from November to May, with maximum rainfall occurring between January and March. The summer is hot and long (May–October).

For more information about the climate of the area the reader can be referred to the data of the meteorological station in Kirkuk (Station No. 151). The mean annual discharge of the Uzaym River for the period 1945–74 was 25.8 m^3 s^{-1}, reduced over the period 1945–84 to just 24.0 m^3 s^{-1} (Al-Jabbari and Mansour, 1986) and further to 18.6 m^3 s^{-1} for the period, 1975–84.

The last tributary of Tigris, the River Diyala has a total drainage area of 29, 700 km^2. It joins the Tigris at a point ($\varphi = 33°\,06'$ N and $\lambda = 44°\,58'$ E) situated 31 km south of Baghdad. The mean annual discharge over the period 1931–75 was 193.2 m^3 s^{-1}, with coefficient of variation of 95%.

The mean monthly and annual of the discharges of the Tigris River over the period 1931–75 at Mosul, Fatha and Baghdad and of the Uzaym at Narrows are included in Table 3, Appendix II. The annual series at the three stations for the period 1931–75

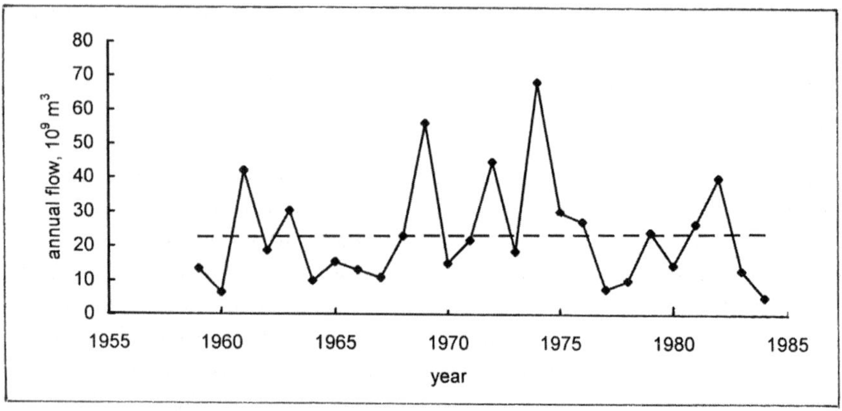

Figure 12. Mean annual discharge for the period 1959–84 at the Narrows gauging station on the Uzayem River (from Al-Jabbari and Mansour, 1986)

are included in Table 4, Appendix II, while the annual flows for the Uzayem River for the period 1959–84 at Narrows station are shown graphically in Figure 12.

Tables 3(a), and 3(b), Appendix III, contain monthly discharge, suspended sediment load, and dissolved solids data for the year 1983–84 in the Tigris River at two stations in the Baghdad area. It is important to mention here that the total transported load at Saddam Bridge station, being 6.5×10^6 t, is less than the total suspended load at North Baghdad station, 7.94×10^6 t, by an amount of about 1.44×10^6 t. The difference between the two sediment loads was deposited in the river reach between the two stations. Additionally, it was concluded that the suspended load transport was not always directly related to water discharge.

The maximum and minimum concentrations of the dissolved solids varied from a minimum of 245 mg l^{-1} and a maximum of 971 mg l^{-1}. At Saddam Bridge station, these two concentrations increased to 1,205 mg l^{-1} and 290 mg l^{-1} respectively. The total solute discharge at the latter station was 8.1×10^5 t greater than that transported through North Baghdad station.

Table 3(c), Appendix III, contains similar data for the same year in the Uzayem River, a tributary of the Tigris River. The transported sediment load during the 1983–84 year reached a total of 3.03×10^6 t. Almost 88% of this amount was transported during few days of heavy flood, especially on 26.03 and 24.04.1984. This observation confirms the role played by flood events in transporting suspended sediments. The concentration of suspended sediment for the given year varied from 5,549 mg l^{-1} during April to a just 21 mg l^{-1} during August, both in 1984, with an average annual of 605 mg l^{-1}.

6.2.1.4 *The Euphrates and its tributaries*

The Euphrates (Al-Furat/Firat, as written in Arabic/Turkish languages, respectively) originates in the high grounds northeast of Turkey, the elevation of which exceeds

3,000 m a.m.s.l. in some places. The river length is about 2,780 km, of which 1,200 km lie within Iraq. The total basin area is estimated as 444, 000 km^2, shared by Turkey, Syria and Iraq. The shares of Syria and Iraq are 64,100 and 197, 000 km^2 respectively.

The Euphrates has two principal tributaries (Figure 11), Murat Suyu (Eastern Euphrates) and the Kara (Western Euphrates). Other tributaries are the Biri River in Turkey, north of Keban Dam, Sajur, Belikh and Khabur in Syria. There are also torrential streams in Turkey, Syria and Iraq. The Euphrates crosses the Turkish-Syrian border at Jerablus, and the Syrian-Iraqi border at Hussaiba, which is situated at 161.8 m a.m.s.l.

The Euphrates is supplied in its upper reach with runoff generated by rainfall and snowmelt in the mountains of Armenia. The long-term average discharge of the Euphrates at the station of Jerablus is 26.8×10^9 m^3 y^{-1}. The mean monthly and annual discharges over the period 1976–79 at Qadahiyah ($\varphi = 36°32'$ N and $\lambda = 38°15'$ E), Syria are included in Table 3, Appendix II (UNESCO, 1993). The mean annual flow for those four years at the said station was 30.67×10^9 m^3. Using the annual flows for 1980–83, which are given in The Country Report of Syria (1986), the overall mean for the period 1976–83 becomes 31.12×10^9 m^3.

The Syrian part of the catchment of Euphrates receives an average areal precipitation of 278 mm y^{-1}. The runoff generated by this amount of rain, assuming an average runoff coefficient of 14%, is 2.5×10^9 m^3 y^{-1}. This annual volume when added to the contribution of the Khabur and Belikh., the tributaries of the Euphrates, brings the long-term yield of the Euphrates to 31.5×10^9 m^3 y^{-1} (Syria's Country Report, 1986).

Upon crossing the Syrian-Iraqi border, the Euphrates bends its course toward the east and southeast for about 246 km before it reaches Al Haditha station and flows further southeast for 100 km or so before reaching Al Hit station. The average slopes of these two reaches are 2.8×10^{-4} and 3.8×10^{-4} respectively. The river channel from Hussaiba to Hit is characterized by its large width and shallow depth. The river width at Hit is 3,400 m and the average depth 6.5 m at bankfull discharge. The Euphrates and the Tigris join together in southeastern Iraq to form the Shatt al-Arab River, which in turn discharges into the Persian Gulf. The river used to divide into many channels at Basra, forming an extensive marshland. The marshes (Ahwar) were largely drained by the Iraqi administration in 1990s before the invasion in 2003. Since the invasion the drainage policy has been reversed, but it remains to be seen whether the marshes will recover.

The Euphrates has no perennial tributaries from the end of its course in Syria up to its junction with the Tigris. The rate of flow of the Euphrates is widely variable not only from month to month but also from year to year. The discharge record of the Euphrates at Husseiba on the Syrian-Iraqi border dates back to 1954/55. The mean discharge over the period 1954–84 is 873 m^3 s^{-1} or 26.4×10^9 m^3 y^{-1}. The mean annual discharge reached a maximum of 1, 850 m^3 s^{-1} in 1968–69 and dropped to a minimum of 279 m^3 s^{-1} in 1973–74. At Hit gauging station (catchment area 264, 000 km^2), the mean annual discharge over the period 1932–84 is 894 m^3 s^{-1}, the maximum and minimum annual flows for the same period were 2,010 m^3 s^{-1}, 1969, and 286 m^3 s^{-1}, 1974, respectively.

The minimum and maximum annual discharges of the Euphrates during the period 1959–84 at Faluja gauging station were 284 m^3 s^{-1} and 1,530 m^3 s^{-1} respectively while the annual average discharge reached 920 m^3 s^{-1}. The mean monthly and annual discharges of the Euphrates at Hit for the period 1931–75 are included in Table 4, Appendix II.

In Syria, Al-Thawrah (revolution) or the Tabaqah Dam (completed in 1974) is creating a storage reservoir known as Lake Assad that is used for cotton irrigation. Syria has also dammed off the two tributaries of the Euphrates and is constructing another dam. These storage works plus the seven dams, which are in operation in Iraq, added to the impoundment of the Keban Dam, Turkey, have greatly altered the flow characteristics of the river Euphrates within Iraq.

Turkey's southeast Anatolia Project, better known as GAP, is a source of concern to both Syria and Iraq regarding their future shares in the water of the Euphrates. The implementation of the project that was launched in the 1990s includes the construction of 22 dams and 17 hydropower stations. When the Turks began filling the Ataturk Reservoir in 1990 they had to stop the flow of the Euphrates for a month without letting any drop of water flow downstream to Syria and Iraq. Professionals from Syria and Iraq are claiming that the said reservoir is likely to reduce flow from the Euphrates by 40% to Syria and by 90% to Iraq.

The mean annual discharge at Hussaiba for the period 1974–84 is 748 m^3 s^{-1} while that for Hit in the same period is 726 m^3 s^{-1}. Halfway the distance between Hussaiba and Hit a third gauging station was established in 1974 at Haditha. The flow record at this station indicates an average annual discharge of the order 676 m^3 s^{-1}. The maximum and minimum daily discharges at this station were 2,970 m^3 s^{-1} and 62.5 m^3 s^{-1} respectively.

Additionally, the mean monthly discharges for the period 1974–84 at the specified stations are shown in Figure 13. The maximum and minimum daily discharges at this station were 2,970 m^3 s^{-1} and 62.5 m^3 s^{-1} respectively.

A comparison between the discharge figures for Husseiba shows that mean annual discharge for the 20-y period, 1954–73, must have been 942 m^3 s^{-1} whereas the annual mean for the 10-y period, 1974–84, was 748 m^3 s^{-1}, i.e. a drop of 20% after 1973–74. It is possible that this difference is partly due to stochastic variation in the annual rainfall. Nevertheless, the major factor contributing to this difference is definitely the storage of river water in both Turkey and Syria.

Samples of water were taken from the river since the water year 1979–80 onwards. The total dissolved salts (TDS), and major cations and anions were analyzed and evaluated for each individual sample. These data were used to establish the relationships between water discharge and TDS for each of the four gauging stations in the upper reach of the river. The stations are located at Husseiba (drainage area = 200,000 km^2), Hit (264,000 km^2), Ramadi (271,300 km^2) and Faluja (272,300 km^2). The mean annual river flow in m^3 s^{-1}, rate of erosion in t km^{-2} y^{-1}

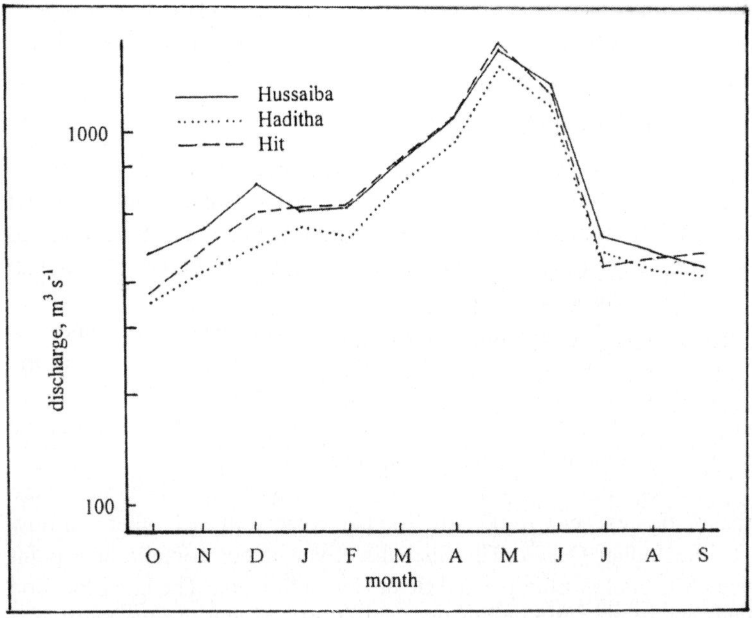

Figure 13. Hydrographs on mean monthly discharge of the Euphrates for the period 1974–84 Th Hussaiba, Haditha and Hit in Iraq. (from Al-Ansari et al., 1986)

and the annual total transported dissolved load in t km^{-2} y^{-1} are available in Table 3(d), Appendix III.

The sediment discharge in the reach of the Euphrates from Husseiba to Hit was investigated by Al-Ansari et al. (1986) for the period 1960–84. Similar investigation of the total transported dissolved load for the same period was carried out by Al-Jabbari & al-Ansari (1986). This investigation covered in addition to Husseiba and Hit stations the stations of Ramadi and Faluja. Some of the results obtained from those two investigations are included in Table 3(d), Appendix III.

The average sediment load at Husseiba for the period 1959–84 ranged from a minimum of 1.2×10^6 t y^{-1} to 53×10^6 t y^{-1}, with a mean of 13×10^6 t y^{-1}. These figures changed for the station at Haditha to become 1.9×10^6 and 21×10^6 respectively with an annual mean of 14×10^6 t. The corresponding figures in their respective order further downstream at Al-Hit station were 0.79×10^6, 480×10^6 and 68×10^6 t y^{-1}.

It goes without saying that all the above-mentioned figures should be used, whenever necessary, very conservatively and cautiously. Rates of erosion and thereupon sedimentation, including TDS and TSS are likely to have changed considerably as a consequence of storage and control works on the upper reaches of the river. The fact that Iraq since 1980 has been engaged in wars with Iran then Kuwait and the American invasion in 2003 has made it difficult to keep the gauging stations operating. The every now and then conflicts between Syria and Turkey added to the occupation of the Syrian Golan Heights by the Israelis have forced the Syrian authorities to conceal most of its hydrologic and hydrometric data of their water resources.

6.2.1.5 *The Shebelle and Juba rivers*

Each of these two rivers is a transboundary water resource whose water is shared by a number of riparian countries. The Shebelle is shared between Ethiopia and Somalia while the Juba is shared by Ethiopia, Kenya and Somalia. The total drainage area of the two river basins has been estimated as 805, 100 km^2. Figure 12 is a map showing the two rivers and their tributaries.

The eastern highland of Ethiopia, known as the Somali plateau, is somewhat lower in elevation compared to the western highland. Nevertheless, the tops of the Ahmar and Bale Mountains in the eastern highland can be up to 4,000 m a.m.s.l. The elevation of the eastern highland falls gradually in the direction towards the Indian Ocean in the south and the Gulf of Aden in the northeast.

The Shebelle and Juba Rivers originate in the eastern highlands of Ethiopia where the annual precipitation exceeds 1,000 mm. The two rivers join each other and flow as one river to the Indian Ocean. The Shebelle River enters Somalia at a point halfway between Mustahil in Ethiopia and Belet Uen in Somalia. The latter location is equipped with one of the key hydrometric stations of Somalia. Additionally, it is an important point as the river water quality in Somalia could be observed there. It is worth mentioning that all tributaries of the Shebelle are in Ethiopia and there is none in Somalia. Over 90% of the river flow at this point is supplied by catchments within Ethiopia.

The Juba is formed by three tributaries, the Gestro, the Genale and Dawa, all three meeting near the border between Ethiopia and Somalia. The rainfall at the source reaches 1, 500 mm y^{-1}, dramatically decreasing southwards to say 550 mm y^{-1} in the region where river enters the Somali territory. It enters Somalia at a place called Dolo some 50–60 km north west of the measuring station at Lugh Ganana and flows to the west, almost parallel to the Shebelle River at an average distance of about 400 km. It should be stated here that 90% of the flow in the Juba River comes from Ethiopia. Although Kenya is a riparian country, it does not contribute any water to the flow in the Juba and it has no significant interests in the river water.

The Shebelle River in the south, not far from Mogadiscio, bends its course and runs parallel to the coast of the Indian Ocean for more than 300 km before joining the River Juba near Gelib (sometimes written Jilib). The two rivers then flow jointly in one channel for about 120 km before discharging its water in the Indian Ocean at Kismayo city as can be seen from the map in Figure 14.

Mangrove trees and other wetland plantations, which cover the coastal strip where the rivers flow, form a source of considerable loss of water by evaporation. Additionally, the coastal strip causes the river water to be highly saline.

There is hardly any significant contribution to river flow from inside Somalia. Any supply that feeds the rivers there occurs only during heavy rain spells. Despite the fact that the Shebelle Basin is larger in size and the river course is longer than

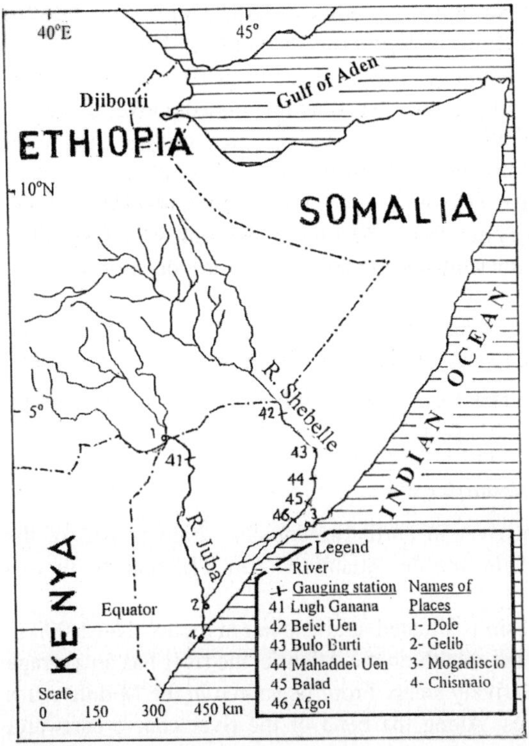

Figure 14. Map showing the Juba and Shebelle Rivers in Ethiopia, Somalia and Kenya

the Juba, the prevailing climatic and geological conditions cause the yield of the basin of Juba to be more than that of the Shebelle.

The catchment area of the Shebelle increases gradually from 211, 000 km^2 at Belet Uen to 231, 000 km^2 at Bulo Berti to 255,330 at Mahadell Uen to 272,700 at Balad and 278, 000 km^2 at Afgoi, all inside Somalia. The corresponding mean annual discharges for the period 1965–79 (with interruptions) were 68 m^3 s^{-1}, 68, 66, 52 and 60 m^3 s^{-1}, respectively. The mean annual discharge of the Juba River at Lugh Ganana (catchment area = 179, 520 km^2) for the same period was about 182 m^3 s^{-1} (UNESCO, 1993). In a more recent publication (UNESCO, 1995) the annual mean discharges of the River Shebelle over the period 1951–78 (with interruptions) at the same gauging stations in their respective order can be found as 66 m^3 s^{-1}, 61, 56, 43 and 43 m^3 s^{-1}. For the same period, i.e. 1951–78 the mean annual discharge of the River Juba at Lugh Ganana was 193 m^3 s^{-1}.

The estimated supply carried by the Shebelle and Juba Rivers from Ethiopia to Somali in a normal year is in the order of 7.5×10^9 m^3 y^{-1}. This figure corresponds to an average areal runoff of 10 mm y^{-1} or less (Figure 3). The distribution of this annual volume between the months of the year for the two rivers can be observed from Table 3, Appendix II. The distribution pattern shows the lowest flow to occur in January-April and the highest flow in September-November. The ratio of the mean maximum to the mean minimum discharge ranges from 10 to 15.

The flow in the lower reaches of the two rivers, as mentioned above, decreases considerably as a result of infiltration, evaporation, overbank spillage and water abstraction. In this region the annual rainfall ranges between less than 100 mm to more than 200 mm while free water evaporation is not less than 2,000 mm y^{-1} and potential evapotranspiration ranges between 1,500 and 2,000 mm y^{-1}. The soil in the coastal strip though suitable for wetland plantations is poorly developed.

6.2.2 Small river basins

These are basins with basin area less than 100,000 km^2 each and are drained by perennial rivers.

6.2.2.1 River Sebou and its tributaries

The Oued Sebou is an important river in northern Morocco, draining part of the Atlas Mountain and the Rharb Plain into the Atlantic Ocean. The total basin area is 37,000 km^2.

The upper part of the Sebou Basin is situated at an elevation higher than 1,000 m a.m.s.l. From the source down to the mouth on the Atlantic, the river has an average gradient of 2×10^{-3}, which is relatively steep. From its source in the Middle Atlas the stream flows northwest to Fés. Along the bend of the river course eastward, Oued Lebene joins the Sebou from the right side. At some 10 km above the confluence, the Oued Inaouène, after traversing the Idriss Reservoir, joins the Oued Lebene.

Several tributaries and feeders join the Sebou on both sides of its course while it is flowing northward to the ocean. Almost 3 km above the gauging station of Azib Soltane, the Oude Mikkes joins the Sebou. Below this confluence, the Oude Ouergha too joins the Sebou. The Ouergha is a major tributary of the Sebou and has its own drainage basin, which includes a number of smaller tributaries and the M'Jara Reservoir. Additionally, the Oued Sebou bends its course and assumes the shape of an arch as shown in the basin map, Figure 15. It receives on its right bank oueds Rhat and At-tnine while the left side receives the mightiest tributaries, Oude Rdom and Oude Beth. The mean monthly and annual discharges of these tributaries are given in Table 3, Appendix II.

The Sebou, at the end of a 500 km journey, reaches the Atlantic Ocean, where it discharges its water at Mehdiya. The Rharb coastal plain, which is formed by alluvial deposits of the Sebou, is often flooded by the heavy floods originating from the Sebou and its powerful tributary, Oued Ouergha.

The Colorado sunken pan at El-Kansera in the southeastern part of the Sebou Basin gives the average monthly evaporation as 47, 47, 77, 79, 104, 139, 183, 183, 144, 119, 82 and 79 mm for the months January thru' December. The sum of these figures gives an annual pan evaporation of 1,283 mm. As such, the equivalent free

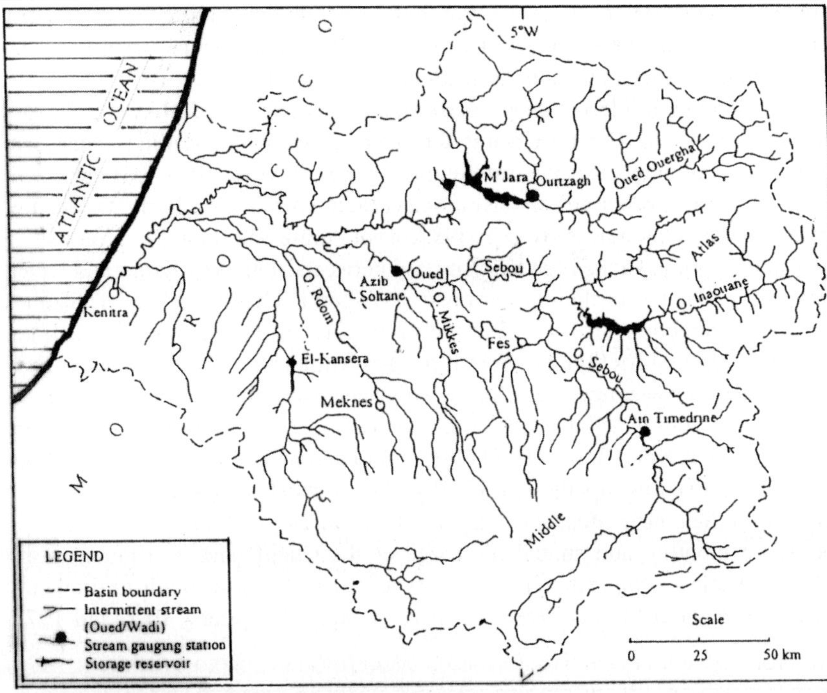

Figure 15. Drainage Basin of Oued Sebou, Morocco

water evaporation falls in the range of 1,020–1,090 mm (pan coefficient 0.80–0.85). The average specific discharge varies considerably from 3.64–3.83 $l\ s^{-1}\ km^{-2}$ for Ain Timedrine and Azib Soltane to 14.4–15.35 $l\ s^{-1}\ km^{-2}$ for Ouertzgha and M'jara. Additionally, the annual flow at Azib Soltane and probably at other stations as well suffered from a significant falling trend in the period 1934–1988. This will be illustrated in a forthcoming section using the data given in Table 4, Appendix II.

The suspended sediment yield of Oueds Ouergha and M'jarra are given in Table 4, Appendix III. The corresponding sediment yields of 3,500 and 2,910 $t\ km^{-2}\ y^{-1}$ are regarded amongst the highest in Africa and in the Arab Region. Heusch & Cayla (1986) gave the figures of 1,110 $t\ km^{-2}\ y^{-1}$ for the Inaouene catchment (3,384 km^2), 2,250 $t\ km^{-2}\ y^{-1}$ for the Lebene catchment (792 km^2), both as averages for the period 1933–62. These high rates have been attributed to the steep, unstable nature of the terrain there.

6.2.2.2 *Oued Chéliff*

The original name of the river as used by the Romans was Chenaleph. The Chéliff is the most important river in Algeria. It does not run dry in any season of the year while other streams run dry in all or part of the summer time. Due to geography, land physiography, relief, and areal distribution of precipitation, rivers and other streams in Algeria are characterised by their short length and small basin area.

The river originates in the Saharan Atlas at an elevation of about 1,940 m a.m.s.l. It has a total length of 725 km and a total catchment area of 43, 750 km^2 at the key gauging station of Sidi Belatar close to the river mouth on the Mediterranean Sea. The river has a number of tributaries supplying it with water (map in Figure 16). The two principal feeders are, the Mina (catchment area = 6, 635 km^2 at Oued el Abtal) and Rhiou (catchment are = 2, 398 km^2 at Ammi Moussa).

The Chéliff springs in the high land situated between Mounts of Ksour and Ouled Nail east of el-Bayadh in the northern part of Algeria. The river channel assumes a northeastern direction for about 250 km up to Kasr el-Boukhan. From there the river bends northwest for 90 km before it reaches Miliana. Below Miliana, the river flows parallel to the Mediterranean Sea at a distance of 50–60 km from the coast for about 130 km before its tributary, Rhiou, joins it. Some 40 km further downstream the other tributary, Mina, joins the Chéliff, and the combined stream flows another 40 km in its course towards the sea. Finally, the river debouches its water into the Mediterranean Sea, near Mostaganem, 100 km northeast of Oran.

The Mean monthly and annual discharge of the Chéliff and its tributaries are included in Table 3, Appendix II. The short record makes it difficult to draw any fair conclusion about the discharges of these streams. The years, 1976 and 1977,

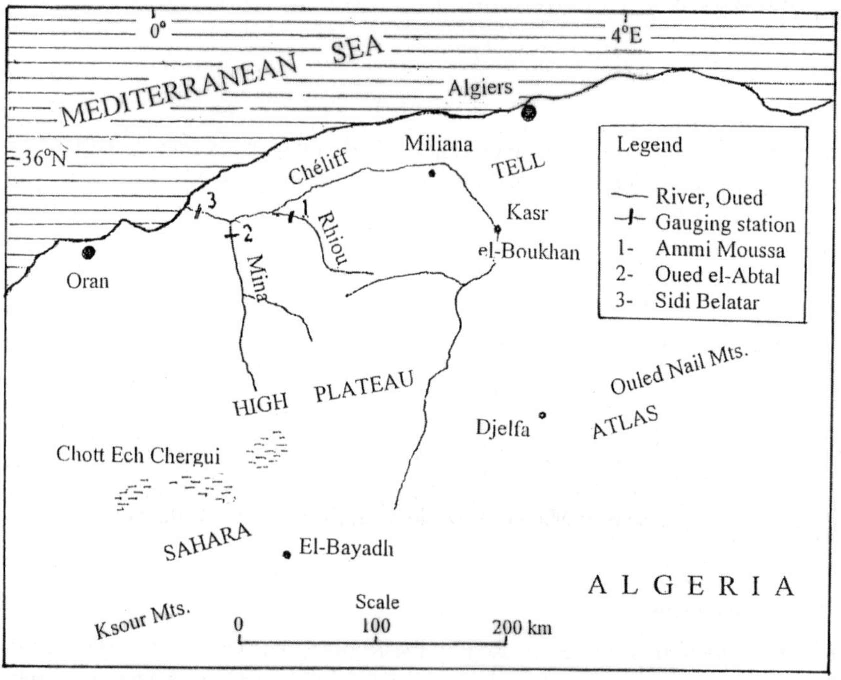

Figure 16. Map of the Chéliff River and its tributaries, Algeria

of record of the Chéliff near mouth have an average of 23.7 $m^3 s^{-1}$. This amount corresponds to $0.75 \times 10^9 m^3 y^{-1}$, is about 50% of the long-term average annual discharge of the Chéliff. The Country Report of Algeria (1986) estimates the annual flow of the river at mouth as $1.8 \times 10^9 m^3$, of which $1.6 \times 10^9 m^3$ represents the surface runoff and $0.2 \times 10^9 m^3$ baseflow. Assuming the areal average rainfall in a normal year to be around 350 mm and the runoff coefficient in the order of 10%, the surface runoff can be easily found as $1.53 \times 10^9 m^3 y^{-1}$, which is very close to the long-term average figure of $1.6 \times 10^9 m^3 y^{-1}$. The same report gives the long-term average yield, including baseflow, of the Tafna, Issar and Rhumel as 0.36, 0.53 and $0.63 \times 10^9 m^3 y^{-1}$ respectively. These figures are larger than those in Table 3, Appendix II, which were obtained from very short periods of record. Despite the short record, the observed extremes in the period 1976–78 were 1, 137 $m^3 s^{-1}$ and 0.31 $m^3 s^{-1}$ for the maximum and minimum discharges respectively.

Bouguerra (1986) presented the results obtained from experiments in micro basins of the catchment area of the Issar River. This catchment is adjacent to the catchment area of the Chéliff. Some of the results obtained are included in Table 5(a), Appendix III.

Table 5(b) in the same Appendix gives more recent results obtained from experiments on erosion in the basin of Oued Mina (Touaïbia et al., 1999). These results will be discussed in Chapter 8. The main conclusion that can be stated here is that heavy rainstorms can produce excessive amounts of erosion in short periods of time. The sediment concentration in water carried by some rivers or ephemeral streams exceeds 200 g l^{-1} and reaches 600 g l^{-1} or more. This state of affairs renders the life age of storage reservoirs quite limited and their operation rather complicated. One should not forget that at the beginning of the 1990s, Algeria had nearly 55 dams with a total live storage capacity of $4.4 \times 10^9 m^3$. The region known as Chéliff Zahrez comprises 12 dams in service with a total capacity of $1.88 \times 10^9 m^3$. Four dams, with a total initial capacity of $0.435 \times 10^9 m^3$, are projected. When all these storage works are completed the storage capacity within the basin of the Chéliff will rise to about $2.65 \times 10^9 m^3$. It is anticipated that the increase of water in storage will help to irrigate more land.

6.2.2.3 Oued Zerou

The Oude Zeroud is in fact a boundary case between a perennially flowing river and an oued (wadi). However, as the discharge during low flow season is relatively small yet hardly falls down to zero, we thought to include it among the group of small rivers rather than wadis. The high flow in the river usually, though not always, occurs in the season May-July while as during the rest of the year the river exercises low flow.

The basins of Oued Zeroud (8, 575 km^2), Oued Merguellil (675 km^2) and Nebhana together constitute a lager basin known as the Sabkha-Kelbia catchment (14, 400 km^2). Most of the sub-basin of Oued Zeroud penetrates into the Kairouan plain in the north central part of Tunisia. The Zeroud Basin at station Sidi Saad has a surface area of 8, 650 km^2. The Zeroud as can be observed from the map,

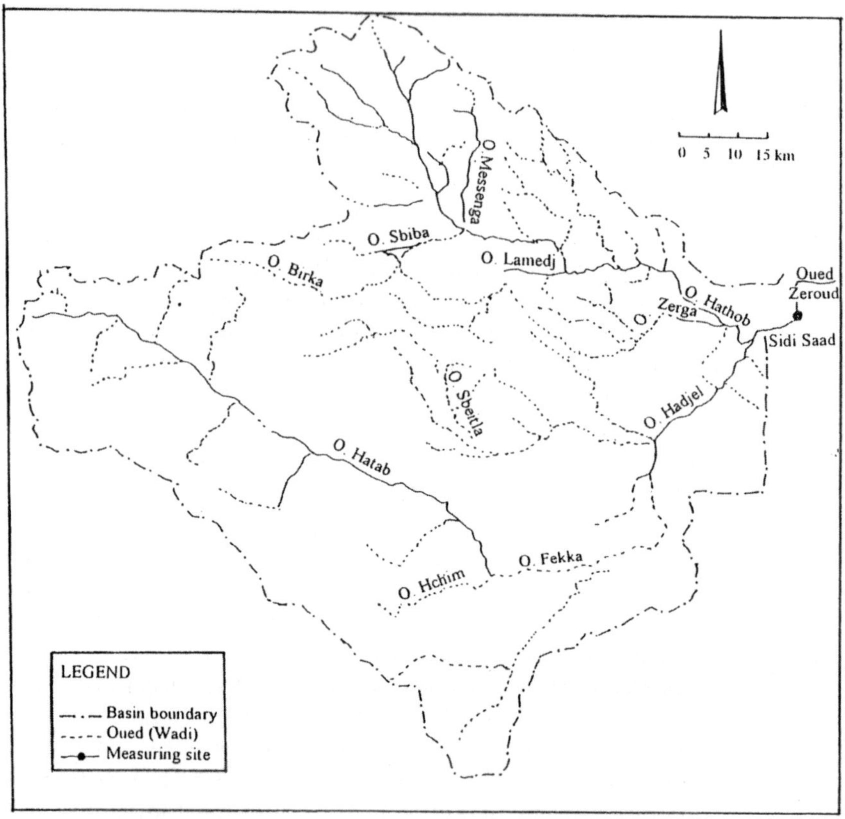

Figure 17. Drainage basin of Oued Zeroud, Tunisia

Figure 17, has two branches; the northern (Oued Hatoub) and the southern branch (Oued Hadjel). The yield of the Oued Zeroud counts for 55% of the surface water resources of the central part of Tunisia or 4% of the total Surface water resources of Tunisia. For a detailed description of the basin and its hydrology the reader is referred to Shahin (2002).

The report of Bouzaiane and Lafforgue (1986) contains a considerable amount of statistical analysis of direct flow, baseflow and total monthly and annual volumes of flow based on measurements taken between the years 1949 and 1974. It lists the mean monthly and annual baseflow two times, once including the extremely wet year 1969–70 and another time without having it into account. The mean total runoff (baseflow plus surface runoff) at Sidi Saad in the second case is about $95 \times 10^6\ m^3\ y^{-1}$ of which 18.5% comprise the baseflow component. If the exceptional year, 1969–70 is taken into consideration, the arithmetic mean jumps to $190 \times 10^6\ m^3\ y^{-1}$ with variation coefficient of 2.65. As the mean in this case is not a meaningful statistical descriptor of the flow, it is more appropriate to describe the river flow it by the total volume of water in a certain period of time. The flood studies over the 27-y period, 1948–78, have shown that the contribution of the largest flood counts, on the average, to 44% of the annual runoff volume of the oued. Figure 18 shows two graphs of monthly volumes of total flow corresponding to 10 and 50-y return periods. Additionally, it might be worth mentioning that the specific yield at Sidi Saad, without taking the year 1969–70 into account, is 0.35 l km^{-2} s^{-1}.

The discharge carried by the Zeroud during the low-flow season is practically free of any suspended sediments, which is definitely not the case with floodwater. The analysis of results obtained from samples collected during 27 consecutive years has shown that the most frequent range of sediment concentration is 20–50 g l^{-1}, which is the case with light to moderate floods. Severe floods have much higher sediment concentration, usually in the of 200–350 g l^{-1}. The concentration-time curve during heavy floods lags behind the discharge-time curve. The relationship linking the annual flood volume, V_f in 10^6 m^3, and the annual silt load brought by the flood, T_s in 10^3 t, is:

$$T_s = 37.41(V_f)^{1.1} \tag{22}$$

Eq. (22) estimates the sediment load carried by annual volume of flow of 80.2 × 10^6 m^3 as 4,651 t, with an average concentration of 58 g l^{-1}.

A series of devastating floods swept over the middle and south of Tunisia in Autumn of 1969. From the end of August till end of October 1969, the basin received an average areal rainfall of 678 mm. The corresponding runoff reached 2.636 × 10^9 m^3, of which the solid sediments occupied 0.240 × 10^9 m^3 and the remaining 2.40 × 10^9 m^3, water. According to Rodier (1989), a 50-y flood may transport a total sediment load far greater than the total transported load during the

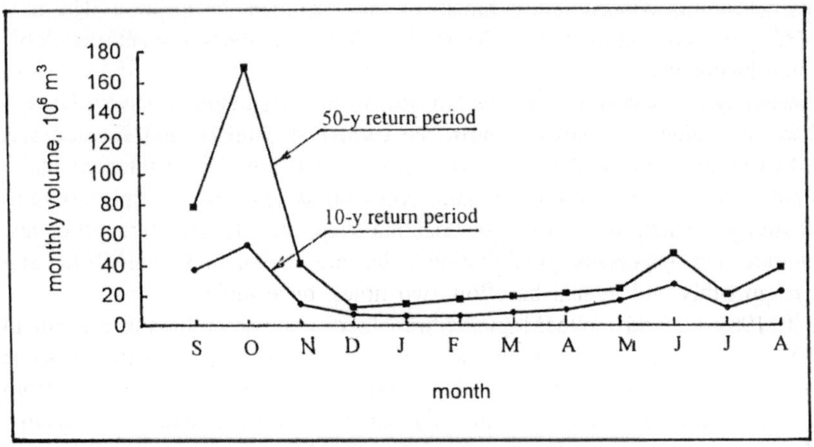

Figure 18. Estimated monthly flow volumes of Oued Zeroud, Tunisia, corresponding

preceding 50 years. Anyhow, it was reported that the total sediment load transported by Oued Zeroud during the 1969 floods was 450×10^6 t while the average sediment load transported by the river in an average year is about 6×10^6 m^3.

To evaluate the water quality, samples of river water were collected in the period from 1949/50–1979/80 for measuring the electrical conductivity. The average salinity as to 10 and 50 year return periods (Bouzaiane and Lafforgue (1986) obtained from the annual volume of floodwater and the corresponding annual salt load were found to vary from a minimum of say 1.0 g l^{-1} to a maximum of 2.54 g l^{-1}. The upper limit of this range was found to increase up to 3.8 g l^{-1} when the annual total flow volume had been considered instead of the annual flood volume alone. Other samples were used for determining the major ion constituents in the water, Ca^{++}, Mg^{++}, Na^{+}, SO_4, Cl^{-} and CO_3H^{-}. Notwithstanding the differences in chemical analyses between the flood flow and the low-flow, the percentages of cations and anions examined were as follows:

$$Cations: Ca^{++} = 30.8\%, \quad Na^{+} + K^{+} = 50\% \quad and\ Mg^{++} = 19.5\%$$

$$Anions: SO_4^{-} = 46.2\%, CI^{-} = 45.1\% \quad and\ CO_3H^{-} = 8.7\%$$

The above-listed figures show that water carried by the Oued Zeroud is highly charged with Sodium and Calcium sulfates. Besides, chemical analyses have shown that water is charged with Sodium and Calcium sulfates in larger amounts in floodwater than in low-flow water.

Application of Storm Water Management Model (SWMM) in Tunisia: The SWMM is a conceptual, deterministic and distributed model that contains three basic routines. These comprise the calculation of losses, i.e. depression storage and infiltration according to Horton's equation, surface runoff using Manning's formula, and conduit and channel flow using the kinematic wave equations. The model needs a number of parameters describing the physical properties of surface and channel systems pertaining to each subcatchment of the catchment in question. The data needed for parameter estimation can be obtained from field measurements, available maps and documents, etc.

Niemczynowicz (1989) applied the SWMM to the catchment of Rivers Guerab and Roriche, which is a part of Tunis, the capital of Tunisia, and situated less than 200 km northeast of the center of the Zeroud Basin. The catchment under investigation is 20 km^2 in surface area, receiving a long-term average rainfall of 460 mm y^{-1} (Station No.33). The location map, and the runoff coefficients, precipitation and antecedent precipitation index are shown in Figures 19(a) and 19(b), respectively.

The first runs of the SWMM model were performed after having the model parameters estimated according to the experience of the investigator in some European countries. The simulated runoff hydrographs, however, were far from being satisfactory. Thorough examination of the model results showed that the Manning roughness coefficient for impermeable surfaces needs significant correction. Additionally, the roughness of the ground surface, channels and conduits

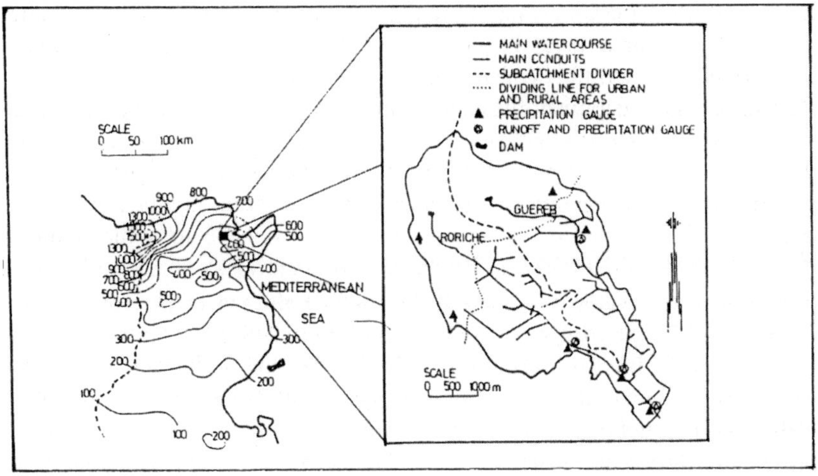

Figure 19(a). Location map of the Guereg-Roriche experimental catchment

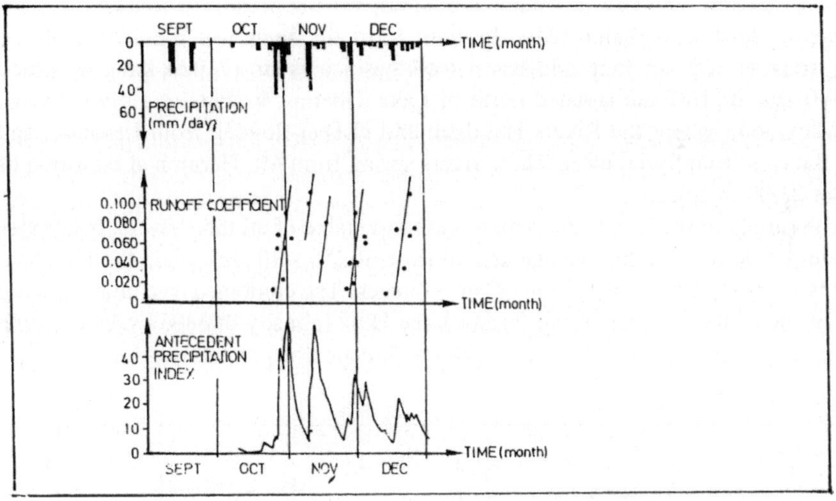

Figure 19(b). Precipitation, runoff coefficient and antecedent precipitation index

appeared not to remain constant, but rather variable depending on the season whether wet or dry. Another important reason appeared to be responsible to a large extent of the discrepancy between the model-predicted runoff and the observed runoff. Horton's representation of infiltration, successfully used in humid countries with time-invariant parameters, was unsuitable for the case of Tunis because it does not take into account the seasonal variations of infiltration. The spatial variation of rainfall in arid and semi-arid areas is so wide that it creates a source of difficulty.

This had to be overcome by increasing the number of rain gauges that each gauge had to serve a very small area, 1–2 km^2. This obviously rendered the process of data collection costly and time consuming.

Taking the above-mentioned difficulties into account, the model that was originally developed for humid areas had to be adjusted to the actual conditions in arid and semi-arid countries. The predicted runoff improved largely its agreement with the observed runoff became quite satisfactory as one see from Figure 19(c).

6.2.2.4 *The River Jordan and its tributaries*

Christians regard the water of the Jordan River as holy and of precious biblical value. They consider the river reach below Lake Tiberius (Sea of Galilee) to be the place where John the Baptist baptized Jesus Christ. The holy water of the Jordan, in this sense, is equivalent to the blessed water of Bir (well) Zamzam for the Moslems, who claim that Prophet Mohammad drank from it. Another example is the water of a certain well near Lake Tana, which is claimed by the Ethiopian priests to be the origin of the Blue Nile (Great Abbay), named in Genesis 14, of King James version of the Bible, by Gihon and regarded as one of four Holy rivers.

The Jordan River is an international river, shared by three riparian countries: Jordan, Palestine and Israel. A number of springs rising in the southern and western slopes of Jebel esh-Sheikh (Mt. Hermon) form the source of the Jordan River. The river is 320 km long and has a total basin area of 17,000 km^2, of which 2,740 km^2 or 16% are situated north of Lake Tiberias. It starts at a place known as Marjyaoûn where the Rivers Hasabani and el-Dan flowing from Lebanon, and the Baniyas from Syria, meet. These rivers spring from Mt. Hermon at elevation of about 2,000 m a.m.s.l.

The supply of the Dan is the richest and most stable of all three rivers. It supplies the Jordan River with an average annual flow of 258×10^6 m^3 while the Hasabani and Baniyas Rivers supply 157×10^6 m^3 y^{-1} each. The combined river after flowing out of the of the swamps of the former Lake Hula (already drained by Israel some years ago) continues its course in a southerly direction before entering Lake Tiberias

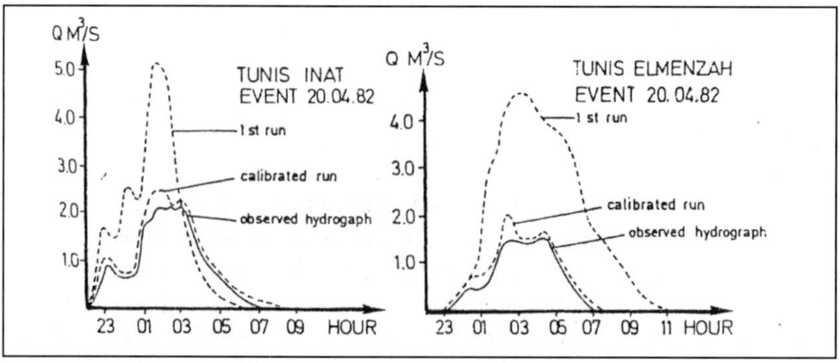

Figure 19(c). Observed and SWMM simulated hydrographs (Niemczynowicz, 1989)

continues its course in a southerly direction before entering Lake Tiberias. The mean annual flow in the Upper Jordan for the period 1965–72 was 646×10^6 m^3 measured at the Southern station (catchment area = $1,495$ km^2), Israel, and 502×10^6 m^3 over the period 1976–84 at the Obstacle Bridge station (catchment area = $1,376$ km^2), Israel. While drawing up the water balance sheet of the Jordan River, the figure of 605×10^6 m^3 has been opted for the long-term average discharge of the Upper Jordan. Of this amount an annual volume of 130×10^6 m^3 is used by Israel and the remaining 470×10^6 m^3 flows to Lake Tiberias. Additionally, the lake receives annually some 200×10^6 m^3 as direct precipitation and runoff from some local streams. The lake loses by evaporation an annual volume of water in the order of 300×10^6 m^3 (assuming the annual rate of evaporation equal to 1.8 m, see Chapter 5). The salinity of water of the Upper Jordan before traversing the Hula swamps is 15–20 ppm while that of Lake Tiberius itself is 300 ppm.

The river below the lake, known as the Lower Jordan River, runs in a twisted channel for a distance of 110 km through a low-lying valley, known as el-Ghawr (also written as Ghor) Valley, till it debouches its water in the Dead Sea at elevation of 392 m b.m.s.l. The Jordan Valley is usually the name given to the area encompassing the valley floor from Lake Tiberius to the Dead Sea.

The precipitation in the river basin decreases from about $1,500$ mm y^{-1} on the tops of Mt. Hermon to less than 100 mm y^{-1} at the Dead Sea, with an areal average of 432 mm y^{-1}. The river has a number of tributaries of which the Yarmuk, Syria and Zarqa, Jordan are by far the most important ones. The map in Figure 20 shows the basin of the Jordan River and its tributaries. The Yarmuk Basin has a total surface area of $6,800$ km^2, of which 1800 km^2 are within Jordanian territory and the remaining $5,000$ km^2 located in Syria. The long-term average discharge of the Yarmuk River used to be around 0.45×10^9 m^3 y^{-1}. The diversion of some of its water to the State of Israel, has contributed to the reduction of the mean annual flow for the period 1965–75 to 0.29×10^9 m^3 at Maqarin (catchment area = $5,950$ km^2), Jordan. The catchment area of the other tributary, i.e. River Zarqa, is $3,400$ km^2, all of which are in Jordan. The long-term average discharge used to be about 0.09×10^9 m3 y^{-1}. However, it fell to just 0.067×10^9 m^3 y^{-1} as average for the period 1965–75 at Jerash bridge station (catchment area = $3,100$ km^2) (UNESCO, 1993). The mean monthly discharges at this station, Maqarin stations and other two stations located in Israel are given in Table 3, Appendix II.

According to ECWA (1981), the water resources of the Jordan Valley consist of the streamflow of the Jordan River and its eleven tributaries (the east bank only) whose mean annual discharges, expressed in 10^6 m^3, are as follows: Yarmouk, 450; Arab, 36; Ziqlab, 13; Jurm, 12; Yabis, 4.5; Kufrinja, 13; Rajib, 4.0; Zarqa, 88; Shueib, 10; Kufrein, 9.5; and Hisban, 5.0. The sum of these flows is 645×10^6 m^3 y^{-1}, of which 538×10^6 m^3 y^{-1} or 83.4% is supplied by the Yarmuk and Zarqa alone. When this amount is added to the flow in the Upper Jordan the subtotal flow becomes 1.25×10^9 m^3 y^{-1}. The Country Report presented by the Palestinian Authority (1986) estimates the annual supply of the western wadis and springs rising in the Jordan Ghawr and flowing eastwards to the Lower Jordan River as

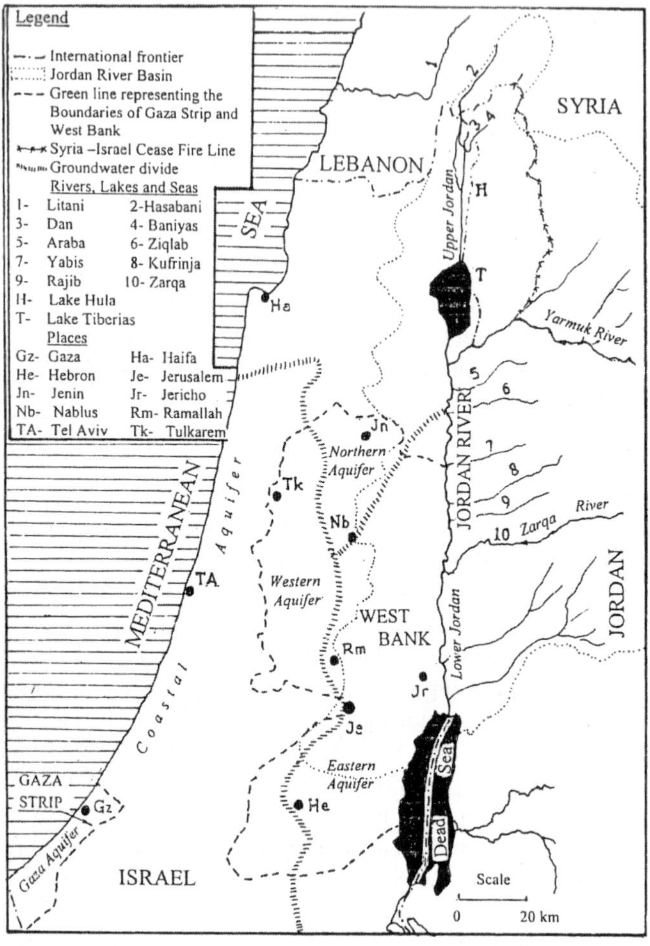

Figure 20. The Jordan River and its tributaries

505×10^6 m^3, of which $40–50 \times 10^6$ m^3 y^{-1} are surface water and the rest groundwater. Assuming the average annual precipitation over the West Bank to be around 0.5 m, the estimate of surface runoff seems to be too small to check the reliability of the flow estimate of the western wadis the author made a rough calculation of the runoff coefficient for two wadis in the northwestern part of the West Bank; namely, Alexander and Soraq (Khedera). The catchment areas of these two wadis at the corresponding flow measuring stations are 953 and 492 km^2 respectively. The corresponding mean annual flows for the period 1980–84, as given by UNESCO (1993), were 0.306 m^3 s^{-1}(9.66×10^6 m^3 y^{-1}) and 0.266 m^3 s^{-1}(8.39×10^6 m^3 y^{-1}) respectively. Assuming the annual rainfall as 0.5 m y^{-1}, the runoff coefficient must be in the range of 2–3%. As such, for a surface area of $2,540$ km^2, the total catchment area of the wadis flowing eastwards to join the Jordan River, the mean annual flow of these wadis should be close to 40×10^6 m^3.

Accepting 0.505×10^9 m^3 as an average for the total flow (direct and baseflow), one can fairly conclude that the average annual yield of the Jordan River Basin should be about 1.8×10^9 m^3.

6.3. ANNUAL FLOW SERIES

6.3.1 Independence of the series

Eight annual flow series are considered in this section. These series belong to the River Sénégal (Bakel), White Nile (Malakal), Blue Nile (Khartoum), Main Nile (Dongola), Euphrates (Hit), Tigris (Baghdad), Sebou (Azib Soltane) and Zurud (S.Saad). The annual flows of these rivers at the specified stations, amongst others, are listed in Table 4, Appendix II. Moreover, Figures 21(a) through 6.21(h) have been prepared with the aim of having a visual impression of the time variation of flow volume from year to year at the chosen stations. Additionally, on each plot of the time series, except the one in Figure 21(h), the trend line with its equation and coefficient of determination are given.

To check the null hypothesis of the independence of the series of time, the Spearman rank correlation test has been applied to each of the seven series, Figure 21(a) thru' 21(g). The series of annual flows at Sidi Saad on the Oued Zuroud represents an extraordinary case in which the flow in one year in a period of 27 years of record rose more than ten times the flow in any one of the other years. Any decision that will be made about the time-behaviour of the series with that particular year taken into account will be too different from that without taking it into consideration. Additionally, this series is too short, especially when compared to the other series.

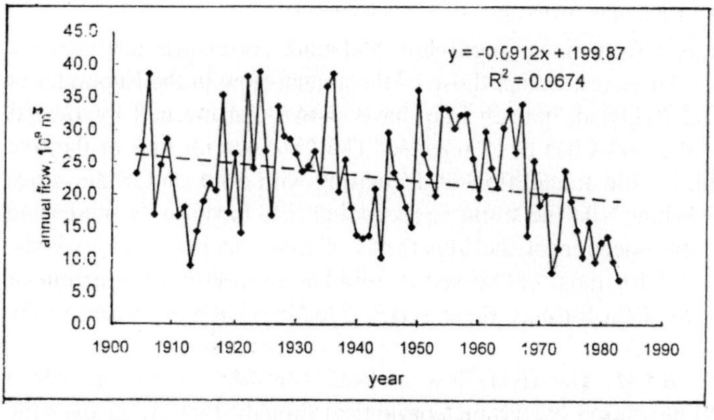

Figure 21(a). Time series of annual flow at Bakel, Sénégal

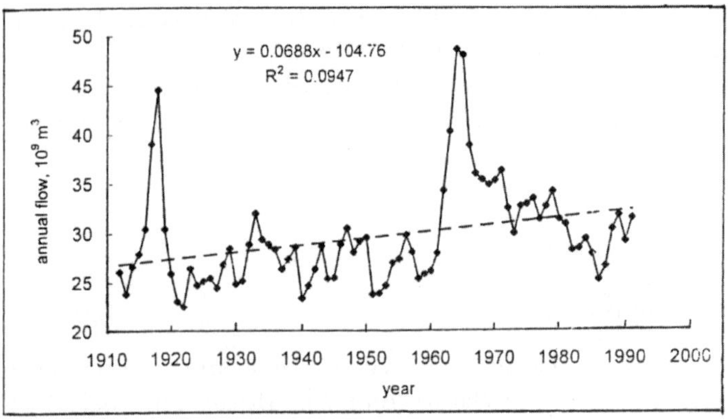

Figure 21(b). Time series of annual flow at Malakal, W. Nile

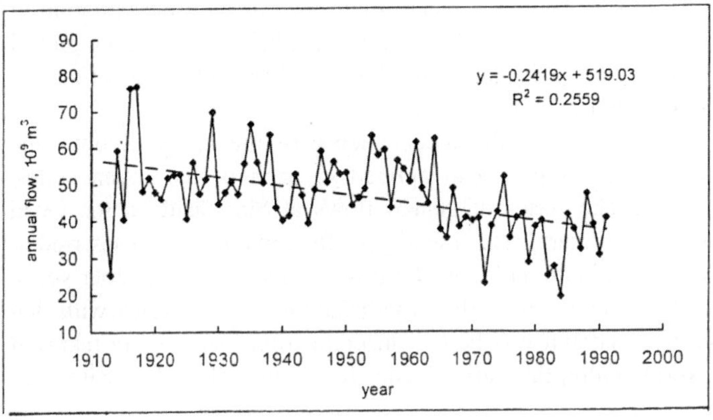

Figure 21(c). Time series of annual flow at Khartoum, B. Nile

The results obtained from linear regression and rank correlation analyses are included in Table 3. All series, except those of the annual flow in the Euphrates at Hit and the Tigris at Baghdad, both in Iraq, have shown that the null hypothesis of independence of the series has to be rejected. The trend line in four of the five series is descending, i.e. the annual flow is decreasing with time, and in the series of Malakal on the White Nile the trend is ascending, i.e. the flow is increasing with time. Despite the rejection of the hypothesis of time-independence over the used periods of record for most of the series, one has to question the extent of consistency among the data forming these series. This can be summarised in the following:

- *River Sénégal at Bakel-* The river flow at Bakel on the Sénégal showed a considerable decline caused by the meteorological drought that swept over the Sahel Region in the late 1960s and lasted more than 15 years in some areas. This

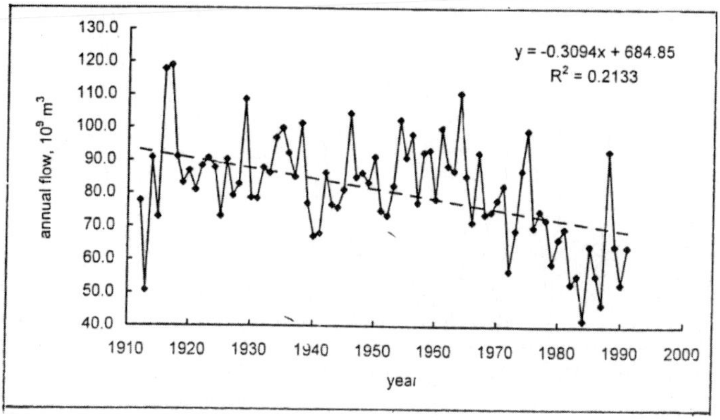

Figure 21(d). Time series of annual flow at Dongola, M. Nile

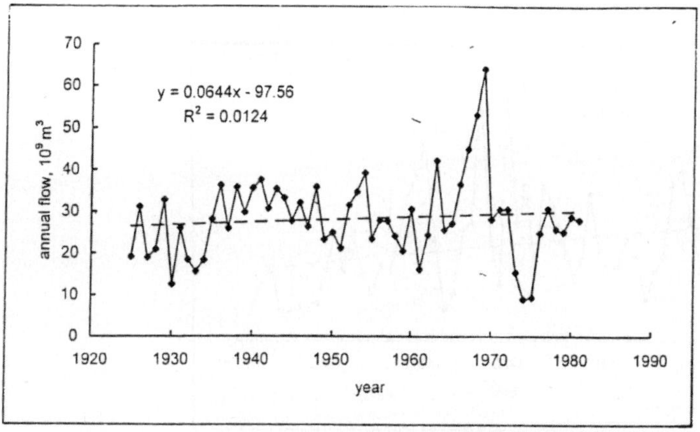

Figure 21(e). Time series of annual flow at Hit, Euphrates

is obviously a natural cause. Additionally, the flow in the Middle and Lower Sénégal is no longer natural since the 1980s onward, but rather under control the Manantali and Diama dams. As the tested series ends by 1983, the river flow in the period 1904–83 at Bakel is hardly affected by man-made activities and can be fairly regarded as natural.

- *The White Nile at Malakal-* The flow in the Upper White Nile below the Equatorial Lakes down to Malakal and below it for some distance, has undergone a significant increase caused by the abrupt, considerable jump in the lake levels that took place in 1961. The change in river discharge kept fluctuating up and down after 1962 till the end of record in 1992. Since then, the mean annual flow has never fallen back to its value for the pre-1961 period. The periodic change in the mean of the series can be observed from the following tabulation:

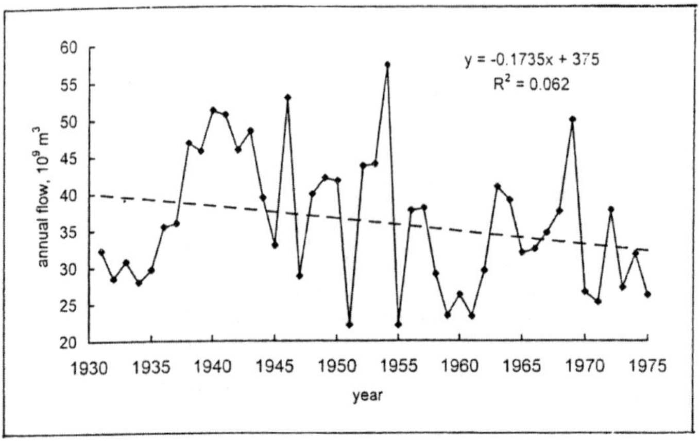

Figure 21(f). Time series of annual flow at Baghdad, Tigris

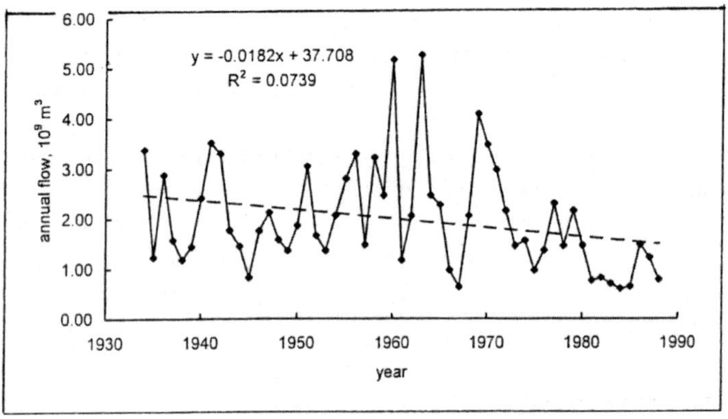

Figure 21(g). Time series of annual flow at A. Soltane, Oued Sebou

Mean discharge Pre-1961, 10^9 m^3 y^{-1}		*Mean discharge, post-1961, 10^9 m^3 y^{-1}*		*Mean discharge, full series, 10^9 m^3 y^{-1}*	
1905–73	*27.63*	*1962–73*	*37.52*	*1905–85*	*29.77*
1912–61	*27.48*	*1962–85*	*34.93*	*1905–95*	*29.64*
		1961–95	*32.85*	*1912–73*	*29.44*
				1912–85	*29.72*

The mean for the pre-1961 and so for the full periods of record are almost constant, though different by about 11%, emphasizes the role of the abrupt change in 1961 and the years after. The behaviour of the flow of the White Nile at Malakal is similar to that of the Sénégal at Bakel in that the time-dependence of the flow series in each case is caused by natural causes, i.e. significant rise and fall of the annual rainfall respectively.

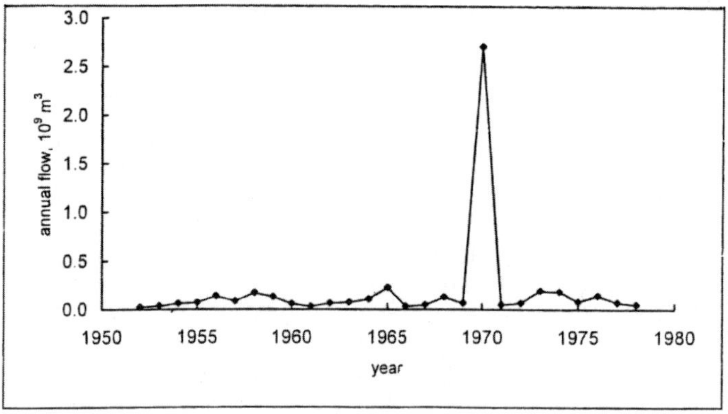

Figure 21(h). Time series of annual flow at S. Saad, O. Zurud

- *The Blue Nile at Khartoum-* The station at Khartoum is situated on the downstream reach of the Blue Nile. The flow series at Khartoum is affected by the reservoirs at Sennar, which has been operating since 1925 and Roseires since 1966. As the storage at the former is small compared to that at the Roseires, one can regard the series up to 1964, with little approximation, as if natural. This is not the case after 1965.

The drought that swept over west and northwest Africa extended to east Africa, and affected Ethiopia since the early 1970s for more than two decades. As such, it is fairly safe to conclude that since 1965 the flow in the Blue Nile at Khartoum is suffering from a significant decline due to reservoir storage (man-made activity) and drought (natural cause). The mean of the sub-series 1912–64 (53-y) was 51.85×10^9 m^3 while that for the sub-series 1965–92 (28-y) was 37.13×10^9 m^3, i.e. there is a negative jump in the mean of say 30% between the two sub-series. Because of the rapidly declining live storage in the reservoir, one can reasonably take the average effective storage capacity in the period 1965–92 as about 2.5×10^9 m^3. Accordingly, the reservoir storage can be held responsible for about 15% of the difference between the means of the pre-and post-1965 periods. The drought of the 1970s, 1980s and the early part of 1990s is responsible for the remaining 85%. This result led us to examine the record covering the 1912–91 period at ed-Deim, a stream gauging station upstream the location of the Reseires Dam, i.e. the natural flow of the Blue Nile. This caused the z-statistic in the Spearman rank correlation test to change from −4.52 to −2.20. Since the critical values of z for normal distribution (5% significance level) is ±1.96 removing the storage effect brought the computed- z close to the critical value, though the flow series remained time-dependant.

- *The Main Nile at Dongola-* Of all annual flow series dealt with in this section, the series for Dongola on the Main Nile is the most complicated one. The flow in the Main Nile is affected by the abrupt rise of the discharge of the White Nile, which began in 1961–62 and since then kept fluctuating up and down till present

Table 3. Some of the results obtained from statistical analysis of annual flow series of perennial rivers in the Arab Region

River	Station	Record, x	N, year	Linear regression		Spearman correlation			
				a	b	R^2	r_s	z	H_o
Senegal	Bakel	1904–83	80	199.9	−0.0912	0.067	−0.257	−2.28	R
				26.3	−0.0912				
W. Nile	Malakal	1912–91	80	−104.8	0.0947	0.095	0.466	3.96	R
				76.2	*0.0947*				
B. Nile	Khartoum	1912–91	80	519.0	−0.2419	0.256	−0.506	−4.52	R
				56.7	−0.2419				
M. Nile	Dongola	1912–91	80	684.9	−0.3094	0.213	−0.427	−3.80	R
				93.6	−0.3094				
Euphrates	Hit	1925–81	57	−97.6	0.0644	0.012	0.066	0.49	NR
				27.6	0.0644				
Tigris	Baghdad	1931–75	45	375	−0.1735	0.062	−0.267	−1.77	NR
				40.1	−0.1735				
Sebou	A. Soltane	1934–88	55	37.7	−0.0182	0.074	−0.335	−2.46	R
				2.5	−0.0182				

Explanation
x = Year in record, a = regression constant or intercept, b = regression coefficient or slope of regression line as in equation $y = a + bx$ used in the upper line, and $y = a + b\,n$ *used in lower line* where n is the year number starting with $n = 1$ for first year in record and N for the last year in record.
R^2 = coefficient of determination, r_s = Spearman rank correlation coefficient, z = standardised normal variable (for $n > 20$). H_o = null hypothesis of independence, R = reject null hypothesis and NR = null hypothesis is not rejected.
For the last column in table, the distribution of z is assumed to be two tailed with 5% level of significance.

as mentioned while reviewing the situation at Malakal. The effect of that rise is steadily attenuated along the river course from Malakal to Jebel Aulia Dam. Besides, the limited effect of storage in the reservoir formed by that dam has become negligible since the High Dam at Aswan, Egypt, was put into operation in 1968 onward.

The decline of annual rainfall in Ethiopia during the 1970s–1990s has definitely reduced the runoff from the basins of the Blue Nile and Atbara. The situation regarding the Blue Nile at Khartoum has been reviewed in the previous subsection. The onset of the drought was preceded by the storage at Sennar, 1925, Roseires, 1964, and Khashm el-Girba on the Atbara, 1968. The annual flow of the Atbara dropped from 12.3×10^9 m^3 for the period 1903–60 to 8.6×10^9 m^3 for the period 1961–94 (Sutcliffe & Parks, 1999).

The resultant influence of the above-mentioned factors shows, as should have been expected, the prevalence of the significant decline of the Blue Nile flow as illustrated by the series at Khartoum on the resultant flow of the Main Nile at Dongola. Again, according to Sutcliffe & Parks (1999), the mean annual flow at Dongola dropped from 86.1×10^9 m^3 for the period 1911–60 to 73.1×10^9 m^3 for the period 1961–1995.

- *The Euphrates at Hit*- The annual flow series at Hit station for the period 1925–81 has shown to be independent of time. From the results of analysis given in Table 3, one can see that the computed z-statistic is much less than the critical value of the standard normal variable using 5% or more than 5% as level of significance. It is likely that the effect of annual storage of certain amounts of river flow since the late 1970s in Turkey and Syria on the dependence of the series would have become evident if the series extended beyond 1981 for some years. Regrettably, the involvement of Iraq in successive wars and war-like troubles since 1980 till present has made it difficult to maintain data collection, analysis and dissemination in a regular fashion.

The monthly flow series at the same station, similar to monthly series of many more rivers in the world, contains a cyclic component repeating itself once every 12 months. This can be seen from the serial correlogram shown in Figure 22(a) (Kamal el-Din, 1981). The analysis further showed two remarkable features; firstly all harmonics after the second, i. e. the 3rd, 4th, 5th and 6th harmonics, combined have a small weight compared to the second harmonic alone. Secondly, the serial correlation of the residual series, which remained after removing all six harmonics, shows a highly significant autoregression between the successive members of the series. This can be realized from Figure 22(a) as the ordinates of the dashed line for all lags, k up to 24, fall outside the 95% confidence limits of independent series.

- *The Tigris River at Baghdad* – The results of statistical analysis of the available flow record for the period 1931–75 shows that the annual flow series of the Tigris River at Baghdad can be regarded as time-independent. The computed z-statistic as obtained from the Spearman rank correlation test was −1.77, which is slightly larger than the critical z, −1.96, for normal distribution with 5% significance level. As such, the null hypothesis of independence could not be rejected. This is partly due to the fact that the available period of record did not witness any effective storage of the river water.

Similar to the flow in the Euphrates River, the monthly flow series of the Tigris exercises a periodic component that recurs each year (Kamal el-Din, 1981). The combined weight of the 3rd, 4th, 5th and 6th harmonics together is small compared to the 2nd harmonic alone. Besides, the serial correlogram of the residual series, i.e. after removing the six harmonics, showed, similar to the Euphrates River, not to be totally independent of time. Instead, it assumes undulating form with serial correlation coefficient at many lags, k in months, falling outside the 95% confidence limits as can be seen from Figure 22(b).

- *The Oued Sebou at Azib Soltane*- The annual discharge data of the Oued Sebou at the station of Azib Soltane over the period 1934–88 were partly, 1934–75, provided by the technical report entitled 'flood report study' (NEDECO, 1975), and partly, 1960–88, by UNESCO Publications on "Discharges of Selected Rivers of Africa" (1996).

The annual flow series over the full length of record was statistically analysed and the results obtained are included in Table 3. These results show that the members composing the series are serially correlated and the hypothesis of time-dependence

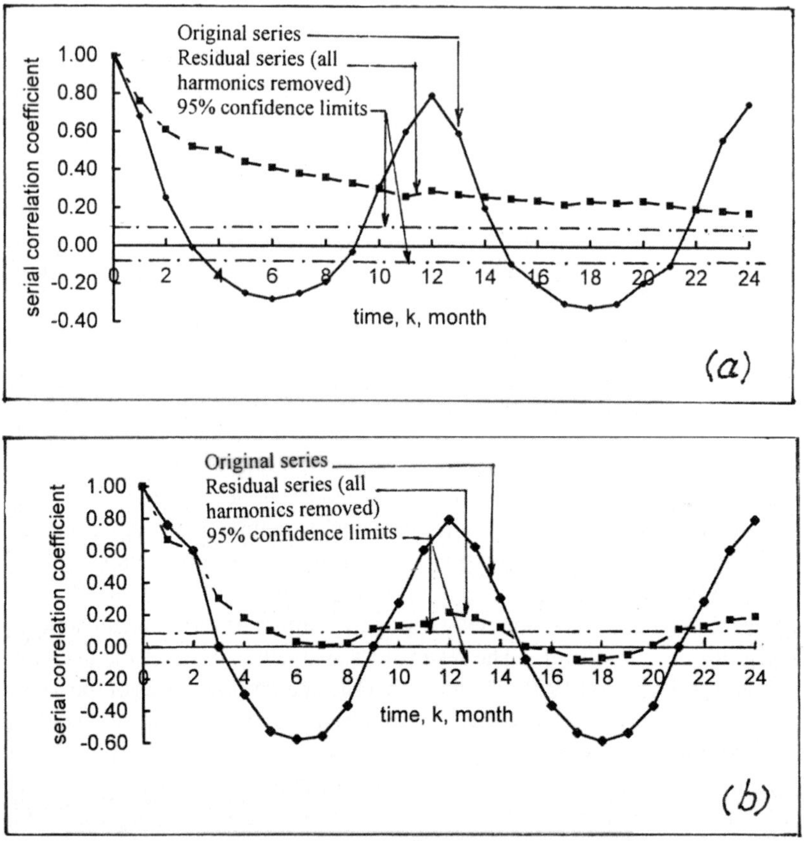

Figure 22. Serial correlograms of original series and of residual series after removing all harmonics, and the 95% confidence limits of monthly flows of (a) Euphrates River at Hit and (b) Tigris River at Baghdad (from Kamal el-Din, 1981)

has to be rejected. Further testing of the series was performed after splitting the full record into two sub-series, the first 1934–1970, and the second 1971–88. The difference between the means of the two sub-series has shown to be significantly different from zero and not a matter of chance.

Morocco, like many parts of west and northwest Africa, was stricken by a severe drought that occupied many years in the 1970s and 1980s. Besides, Al-Kansara storage dam, originally built in 1935, was heightened in 1969 thereby increasing the storage capacity of its reservoir to 272×10^6 m^3. In 1973 the Idris-I storage dam on the Oued Inaouène was completed and its reservoir put into operation. By the end of the 1980s, the sum of regulated flows in these reservoirs reached 1.2×10^9 m^3. The Country Report of Morocco (1986) mentioned that the projects planned for implementation in the 1990s and later will augment the regulated flow further to 5.2×10^9 m^3. These projects include the M'Jara and M'Dez storage dams, Ait

Ayoub diversion Dam, Matmata conduit and a number of small storage reservoirs. It is obvious that the storage works that have been completed and put into operation within the years of available record have contributed to the non-homogeneity of the discharge series as can be observed from Table 3.

6.3.2 Frequency analysis of annual flow

The information presented in the preceding sub-section about the series of annual river flows, their dependence/independence on time, and the probable underlying causes, have been worked out while performing frequency analysis. The only exception is that the flow of the Blue Nile at Khartoum has been abandoned and replaced by the annual flow series at Ed-Deim gauging station as the river supply there is natural and not influenced by any storage reservoirs. The results obtained from the statistical analysis are listed in Table 4. This Table gives for each of the considered flow series the mean, and the coefficients of variation and skewness. The Table also includes the frequency distribution function of best fit to the available flow data, and the estimates of annual flows for specified return periods using the frequency factor method. The distribution functions were the normal, 2-parameter lognormal, Pearson Type III and Gumbel Extreme-Value Type I. In view of the inconsistency and/or non-homogeneity of the data, one should use the estimated values cautiously.

6.3.3 Concluding remarks

The foregoing analyses of river discharge data are quite limited as most of the available data are not up to date. Consequently, the results obtained are of limited value too. This is basically due to lack of observations and/or data collection, processing or dissemination. This unfavourable situation can be attributed to the overall instability prevailing in many parts of the region. The troubles between north and south Sudan, the unsettled disputes over water between Turkey, Syria and Iraq, and so between Israel, Syria, Lebanon, Jordan and Palestine, and the local troubles between the different groups and factions in North Africa, have certainly contributed adversely to any serious efforts to develop the water resources in the region. In some Arab States the data are available but they are regarded as classified pieces of information and their dissemination outside pre-assigned circles is forbidden.

Climate change is identified by changes in temperature and precipitation. These changes have been reviewed earlier in those chapters dealing with climate, precipitation, evaporation and evapotranspiration. Unfortunately, there are no studies in the Arab Region for demonstrating the effect of climate change on river runoff.

Revelle and Waggoner (1983) carried out a study on some basins in Arizona, USA. The results obtained from the five wettest basins showed that a rise of 2° C and a drop of 10% in precipitation were equally effective; each resulted in a decline of 23% in basin runoff. The two changes combined produced a decline in the runoff of 41%. These results were strongly contradicted by those obtained from the then

Table 4. Basic statistics and probability functions of best fit to the data available in the listed periods of record, and estimates of annual flow values given certain return periods

River	Station	Period of record	Xm, 10^9 m^3 y^{-1}	C_v	C_s	D.F.	Annual flow, 10^9 m^3 for the given return periods, years									
							1.01	1.05	1.11	1.25	2	5	10	20	50	100
Senegal	Bakel	1904–83	22.65	0.36	0.16	N	3.7	9.2	12.2	15.8	22.7	29.5	33.1	36.1	39.4	41.6
Blue Nile	Ed-deim	1912–91	49.39	0.2	0.32	P-III	28.7	34.1	37.1	41.1	48.9	57.5	62.3	66.4	71.1	74.4
Euphrates	Hit	1925–81	28.26	0.34	0.95	LN-2	12.6	15.5	17.6	20.3	26.7	35.3	40.9	46.1	52.9	57.9
Tigris	Baghdad	1931–75	36.14	0.25	0.41	P-III	17.6	22.2	24.9	28.3	35.5	43.5	48.3	52.4	57.3	60.7
Sebou	A. soltane	1934–88	1.99	0.54	1.16	EV-I	0.08	0.48	0.72	1.05	1.82	2.9	3.55	4.2	5.1	5.71

Explanation
X_m = arithmetic mean, C_v = coefficient of variation, C_s = coefficient of skewness and D.F. = distribution function.
N = Normal, P-III = Pearson Type III, LN-2 = 2 parameter log-normal and EV-I = Extreme value type- I.

ongoing study by Macabe and Hay, which began in 1988 and reported in 1995, using the so-called 'Precipitation-Runoff Modelling System (PRMS)'. The United States Geological Survey (USGS) developed this model. The results obtained have shown that the effect of precipitation change is more effective than temperature change. A change of 4° C in the annual temperature produced the same effect as 5% change in annual precipitation, causing a change of 9–10% in the peak runoff. The extent of influence of precipitation change on runoff depended on the time of change. A change of 20% in precipitation produced a 25–26% change in the peak runoff. These results are just a sample of the large number of studies carried out in several locations for predicting the effect of climate change on runoff. Unfortunately none of them is situated in the Arab Region.

Last, but not least, building on-stream dams in the Arab Region for the purpose of land irrigation and power development began around the turn of the 19th–20th century by building the Old Aswan Dam, Egypt, between 1898 and 1900, followed by Sennar Dam, the Sudan, in 1925. Since then many more dams of different sizes have been built in the Sudan, Egypt, Turkey, Syria, Iraq, Sénégal/Mauritania and North Africa.

Most of the data pertaining to water balance of storage reservoirs on international rivers are not known with a high degree of accuracy. This state of affairs renders any calculations of the impact of impoundment on downstream flow of limited value.

CHAPTER 7

WADIS AND WADI FLOW

7.1. INTRODUCTION

The surface of the Arab Region, as already mentioned in some of the previous chapters, is intersected and dissected by a large number of intermittent and ephemeral streams. These streams (wadis) have drainage basins with surface areas ranging from a few square kilometers to tens of thousands of square kilometers. Likewise, the length of the wadi channel can range from a few tens of kilometers to hundreds of kilometers. A wadi comprises a single stream, but can also have a number of tributaries feeding the main stream. During or shortly after the offset of heavy rains, the peak flood carried by a wadi can reach some thousands of $m^3 s^{-1}$. Devastating floods frequently damage and destroy dwellings and other properties including farms and irrigation structures. They also inundate and erode agricultural areas, especially the high-elevated ones, which are subject to torrential rains every now and then. Products of erosion cause the reservoirs to fill with sediments, thus reduce their service lives.

Some of the wadis in Yemen, Saudi Arabia and Oman, as examples, have a long-standing history in supplying the population of their respective basins with water for domestic as well as irrigation purposes. As a matter of fact, spate (flood) irrigation has been practiced for millennia in the alluvial plains along the Yemeni mountain range and in many more Arab countries. It is a unique form of irrigation, predominantly found in arid and semi-arid regions where use is made of occasional heavy floods of very short duration carried by wadis. Agricultural yields are, in general, limited and may vary considerably from year to year depending on depth and frequency of spates. Spate water infiltrates into wadi beds and irrigated fields thus feeding the underlying aquifers.

At present the available information regarding the existing wadis in the Arab Region is quite limited. This information includes some of the characteristics of the drainage basins, like morphological, geological, physiographic, and most important of all hydrologic and hydrogeologic conditions including flow and water level measurements, and quality of water.

The growing interest in developing and better management of water resources so as to cope with the ever-increasing demand on water, caused mainly by increasing population, has recently led some of the Arab states to pay attention to surface flow brought by wadis as a source of freshwater. It might be of interest to report here that the Steering Committee of the Arab Center for the Studies of Arid Zones and Dry Lands (ACSAD) by its meeting in July 2004 has declared that the Center shall act as a coordinator of the Wadi Hydrology Network in the Arab Region.

The present chapter presents a rather elaborate review of the quantity and quality of surface water carried by major wadis in the Arabian Peninsula. Review of the hydrological characteristics of these wadis has been based on measured and/or estimated flow. Surface flow estimates have been obtained using either empirical rules or hydrological models as explained in Chapter 6. Wadi hydrology in the Arab countries located in the Eastern, Central and Western Sub regions are reviewed less extensively.

7.2. WADI BASINS AND WADI HYDROLOGY IN YEMEN

7.2.1 Drainage basins of Yemeni wadis

The main wadis of Yemen are shown in Figure 1 (Jacobi, 1991). Most of these wadis are ephemeral due to the short duration of rainstorms coupled with high rates of evapotranspiration. However, some of the wadis flow perennially in their upper reaches but disappear rapidly as soon as they traverse a porous medium such as the alluvium of the Tihama Plain. The factors affecting the ordinates of a wadi hydrograph are: basin precipitation (distribution in time and space), catchment factors such as the shape and extent of basin, climatological, geological and geomorphological factors, hydrological conditions and hydraulic characteristics of the wadi and human activities.

The territory of Yemen can be subdivided into four major drainage basins. These are: Red Sea basin, Gulf of Aden basin, Arabian Sea basin, and Rub el-Khali basin (Figure 1). It should be remembered, however, that the boundary between the Gulf of Aden basin and the Arabian Sea basin is somewhat arbitrary. The catchment areas, average areal rainfall and the mean annual runoff, whether measured or estimated, of the wadis belonging to these four subdivisions are listed in Table 5, Appendix II.

i *Red Sea Basin*: This basin is often referred to, though not accurately, as the Tihama Coastal Plain: It stretches some 400 km along the Red Sea from the Asir region in the north to Bab-el Mandab in the south. It ranges in width from 20 to 50 km, with its widest part stretching between the Mawr and Zabid Wadis. The plain is a flat to rolling terrain whose highest elevation is about 400 m a.m.s.l. The Tihama plain is intersected by almost east-west running wide, shallow wadis. These wadis can carry enormous floods from the western mountain slopes to the Red Sea. The drainage pattern is determined by the watershed dividing line, which runs over the peaks of the central mountain range. The major wadis draining the western slopes to the Red Sea are: Harad, Abs, Mawr, Surdud, Siham, Rima, Zabid, Hayes, Rasyam, Mawza and Ghubab. The well-defined

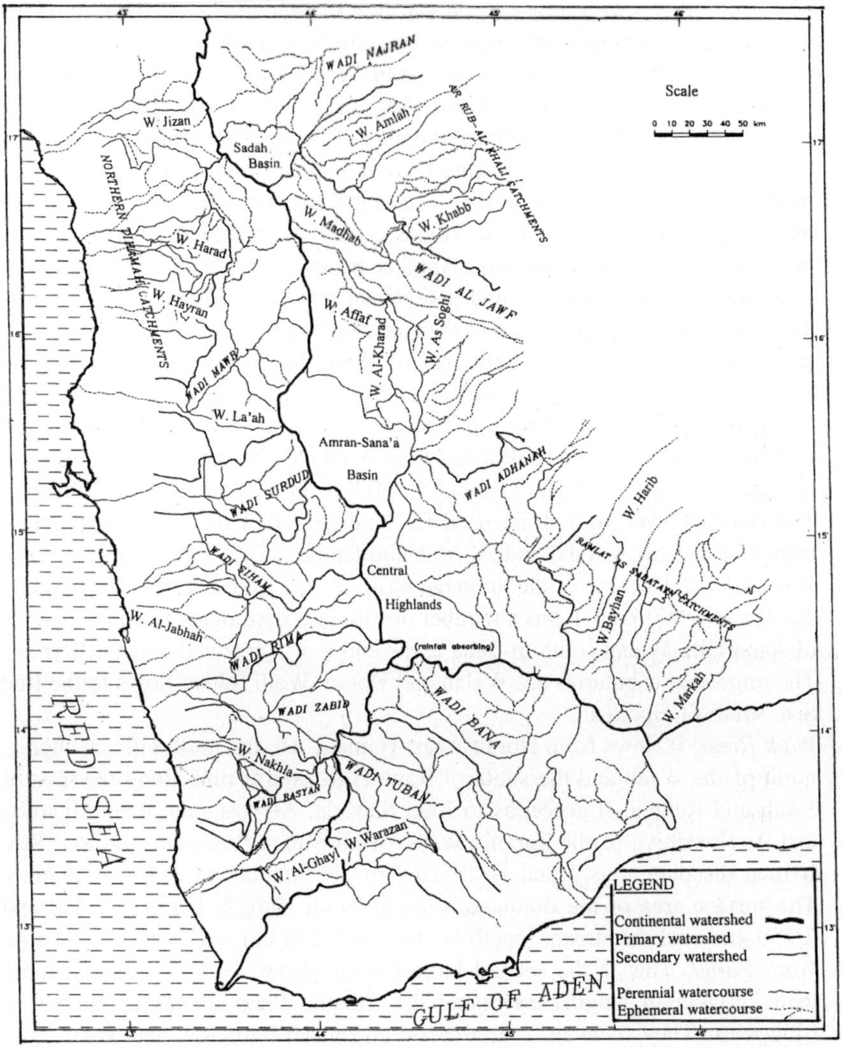

Figure 1. Map of the major wadis in Yemen (after Jacobi, 1991)

streambeds of these wadis degenerate rapidly westwards and become hard to trace as they approach the seacoast.

In view of its climate, the Tihama Plain is best described as a tropical, arid land. The runoff generated by the rainfall within the upper and middle catchment areas supplies the Tihama wadis. Two flood seasons occur, in spring and summer. They occasionally form a single prolonged season extending from April to September. The mean annual discharges of the individual wadis are commonly in the range

10–230 million m^3, with only exceptional flows reaching the sea. The following paragraphs describe briefly the majorwadis of the Tihama Plain (Al-Garoo, 1987).

a *Wadi Mawr*: With a basin area of about 7,500 km^2 and a length of about 300 km, the Wadi Mawr is by far the largest wadi traversing the western plains. The wadi springs from the heights of Al-Amshiya near Sa'da and is joined by some tributaries. Despite the continuous flow of its tributaries, the main stream runs only seasonally. The major tributary that joins the wadi in the south is called Wadi Zaa; its floods come from the Sawas Hill, Kokaban heights, and flows continuously. Another tributary, Wadi Mawr, runs north-south until it joins Wadi Zaa and then deviates to the west for about 25 km in Tihama Plain. The wadi flows to the Red Sea north of Al-Lhaia.

b *Wadi Surdud*: The upper tributaries of Wadi Surdud originate in Al-Nabi Shoaib and Alhasas region north of Haraz, and in Al-Mahwit in the south. The tributaries join in Khamis Bet Saad, and the main stream extends itself westwards to the Strait of Bab Anaka. Wadi Ahger is the most important tributary. The Surdud basin is about 2,450 km^2 in area and 230 km in length. The floods of W. Surdud are used for land irrigation of certain agricultural areas like Adahi, Azaidia and Al-Mahgam lands.

c *Wadi Siham*: The size of the drainage basin of this wadi is estimated between 3,200 and 4,050 km^2. It has a number of tributaries springing from the heights of Asod Hill, Walan south of Sana'a, the south of Haraz and north of Ryama. The important tributaries are Walan and Doran. Wadi Siham flows to the Red Sea, south of Hudayda.

d *Wadi Rima*: It flows from Doran Anes, Hamam Ali, the north hills of Otoma, north of the Wsab and the south of Rayma. Its course runs along the hills of Swab and Rayma. It descends to Bani Sawada, Al-Mashrafa, then Algaroba and Al-Husainiya of the Zaranik lands. Wadi Siham flows into the Red Sea. It runs discontinuously and is affected by the amount of seasonal rainfall. The surface area of the drainage basin of Wadi Rima is between 2,550 and 2,750 km^2 and the stream length is more than 250 km.

e *Wadi Zabid*: This is the second largest wadi (4,500 km^2 in area and more than 250 km in length) traversing the Tihama Plain. It has a number of tributaries. They meet in a strait between the Hills of Ras and Wsab. The main stream runs continuously to the northwest then to the west heading to the Red Sea.

f *Wadis Rasyan and Mawza*: Wadi Rasyan is situated to the south of Wadi Zabid. It has a basin area of 1,750–1,900 km^2. It reaches the Red Sea at Maha Port. Wadi Mawza (1,300–1,600 km^2) is situated to the south of Wadi Rasyan. These two wadis, i.e. Rasyan and Mawza, originate from the heights of Saber near Taiz.

ii *The Gulf of Aden Basin*: The major wadis, each with catchment area larger than 1,000 km^2 are, Wadi Tuban, W. Bana, W, Hassan, W. Ahwar, W. Maifa'h, W. Hajar and W. Huwayrah. There seems to be a great deal of similarity with the Red Sea Basin. However, the surface water outflow into the Gulf of Aden occurs more frequently. This may be attributed to the fact that the coastal plains along the Gulf of Aden are less developed and often steeper than those along the Red Sea.

Rainfall in the mountainous catchments is heavier than near the coast. Apart from this, the catchments in the western part of the basin receive more rainfall than those in the central and eastern parts do.

iii The Arabian Sea Basin: This drainage basin is rather complex. Apart from the catchments of the wadis of the Al-Ghayda Depression (Haghawat, Tinhalin, Al-Jiza, Fauri and Idunut and a few minor wadis), it includes the 'Wadi Hadramout System'. The large distance between the upper reach of the wadi and the sea coupled with the small amounts of rainfall, it is unlikely that surface water travels over the entire distance. Instead, most of the surface water easily infiltrates the soil and joins the groundwater. As such, the wadi from upstream to downstream comprises alternating zones between runoff producing and runoff absorbing. Entrapped as groundwater, it takes long travel time for water to cover the distance from the eastern slopes of Yemen to the Arabian Sea. The Arabian Sea Basin is drier than the Red Sea Basin and the Gulf of Aden Basin. Outcropping of bare rocks is predominant in the runoff producing zones.

Wadi Hadramout (Al-Ah'qāf. as it appears in the Holy Qu'rān) has one of the largest drainage basins of Yemen. It is situated between latitudes 15° 03′ to 16°00′ N and longitudes 48°00′ and 49°15′ E. The altitude of the basin ranges between 495 and 645 m a.m.s.l. The basin has two high plateaus at elevations ranging from 700 to 850 m high. The southern plateau, from which eight tributary wadis spring, slopes towards the Arabian Sea. The northern plateau slopes towards the Rub el-Khali. Ten tributary wadis spring from the northern plateau, all of them debouch their discharges into the main wadi.

The main wadi is 115 km long, running from west to east with an average slope of 8×10^{-4}. The wadi covers an area of 1,200 km, and located about 200 km from the Arabian Sea. The average annual rainfall is 76 mm of which no less than 57 mm y^{-1} infiltrates the soil and helps to replenish the groundwater. The daily evaporation varies from a minimum of 5.4 mm in December to a maximum of 11.7 mm, with an average rate of 9.4 mm d^{-1} or 3.4 m y^{-1}. The total basin area of Wadi Hadramout is estimated as 32,000 km^2; the southern plateau occupies 15,200 km^2 and 5,800 km^2 are located in the northern plateau. The tributary wadis of Amd/Do'an, Al-Ayn, Ben-Ali, Tarbah, Idim discharge their waters on the right side of the main wadi. The Heynan, Sirr, Juaymah and Thibi, and many small tributaries, discharge their runoff on the left bank of the main wadi. Figure 2 shows the location of the tributaries of the Hadramout wadi. The catchment area, the average rainfall and the annual basin yield of each tributary are, as mentioned earlier, included in Table 5, Appendix II. The following paragraphs give a short description of some of these wadis (Al-Garoo, 1987):

a The Abyan Delta: The delta is situated in the Abyan Governrate. The two major wadis that emerge from the mountainous hinterland and discharge regularly into the Abyan Delta are Wadi Bana and Wadi Hassan. Wadis Suhaybiah and Maharia approach the Delta from the west and join Wadi Bana. The catchment

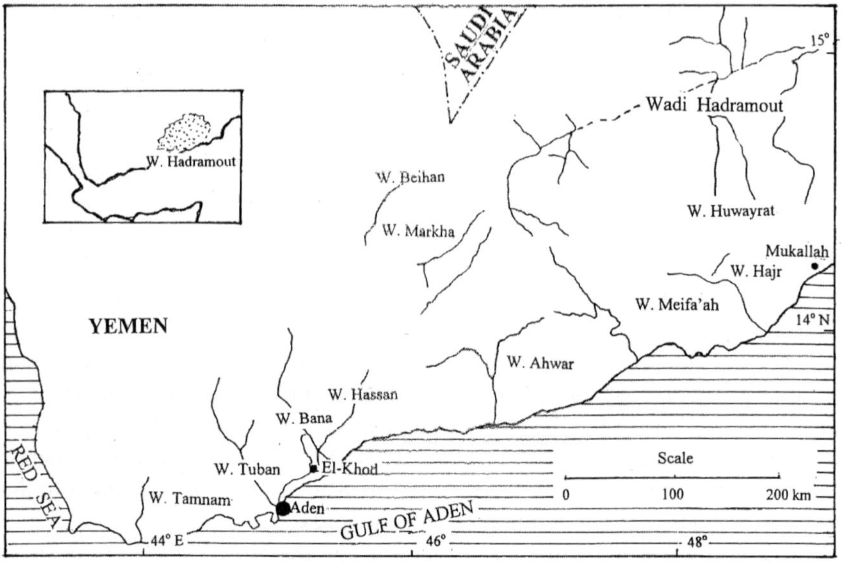

Figure 2. Map showing the tributaries of Wadi Hadramout

areas of Wadi Bana and of Wadi Hassan are 7,200 km^2 and 3,300 km^2 respectively. These two catchments in their respective order receive annual rainfall of 360 mm and 200 mm. The total annual runoff has been estimated at between 171 and 206×10^6 m^3.

b *The Wadi Tuban Delta*: This is located in Lahej Governrate. It has a catchment area of 5,600 km^2, receiving at its head part an annual rainfall of about 1,500 mm. The annual flow of Wadi Tuban and its tributaries averages about 125×10^6 m^3 most of which is diverted in the upper part of the Delta for irrigation purposes. However, about 70% of the flow goes as groundwater recharge before it reaches the fields.

c *The Wadi Ahwar Delta*: The catchment area of this wadi, which is situated in the coastal zone of the Abyan governrate, is about 6,325 km^2. The average annual runoff recorded from 1955 to 1982 was about 69×10^6 m^3, while the maximum annual runoff recorded in 1982 reached 290×10^6 m^3 with a peak discharge of 5,340 m^3 s^{-1}.

d *The Wadi Beihan*: The basin of this wadi varies from 800 to 1,600 m a.m.s.l. The wadi has two tributaries-Wadis Nahr and Khirr, which join to form the Beihan. The catchment area of Wadi Beihan at Naqub is 3,300 km^2 and that of Wadi Nahr and Wadi Khirr 2,500 and 530 km^2 respectively. Surface water is in the form of floods when the rainfall exceeds 17 mm, providing an average annual inflow of 48.4×10^6 m^3. About 44×10^6 m^3 infiltrates the soil and is stored in the aquifer and the rest of it is mostly lost in the desert.

e The Wadi Hajar Delta: This area is situated in the Hadramout Governrate and has a catchment area of 9,160 km^2. According to Girgirah et al. (1987) Wadi Hajr is the only wadi perennial flow, discharging into the sea 72 km west of Mukalla. The average annual runoff has been estimated at between 230 and 470×10^6 m^3.

iv The Rub el-Khali(empty quarter) Basin: The northward sloping northern zones of the Yemen Mountain Massive and Eastern Plateau Region are dissected by numerous, almost parallel wadis. The most important wadis descending from the Yemen Mountain Massive arranged from west to east are: Wadi Najran, W. Atfyan, W. Khubb, W. Amwah and W. Ghummur. Those draining the Eastern Plateau are: Wadi Hadi, W. Aywat as'Sayer, W. Makhya, W. Khadra, W. Hardah,W. Qinab, W. Aywat, W. Harthuth, W. Rumah (or Arma), W. Dahyat Ba'ut, W. Arabah, W. Rakhut, W. Mitan and W. Shihan. These wadis together with a large number of minor wadis drain the excess water of the exceptional heavy rainstorms into the sands of the Rub al-Khali Basin, and thus recharge the groundwater. So far, it is not known whether this groundwater evaporates or finds its way to drain into the Arabian Gulf.

The Rub al-Khali Basin, having a hyper-arid desert nature, is the driest of all four drainage basins. Nevertheless, some green oases are locally present, e.g. in Wadi Najran and Wadi Khubb.

7.2.2 Wadi hydrology and surface water quality in Yemen

7.2.2.1 Wadi runoff

Before embarking on the hydrology of some wadis in Yemen, it is worthwhile mentioning that until a few years ago the wadis that have been gauged for at least four uninterrupted years without any gaps did not exceed twenty. These are, Mawr, Surdud, Siham, Rima, Zabid, Rasyan, Tuban, Rabwa, Bana, Adhana, Ahwar, Juwaymah, Sarr, Amd/Du'an, Idim, Thibi, Al-Ayn, Ben-Ali and Masila (Ministry of Oil and Natural Resources of Yemen, Yemen & TNO Institute of Applied Geosciences, the Netherlands, 1995). These data, despite their being limited, have thrown some light on the hydrologic characteristics of wadis.

- Wadi basins have shown that a threshold depth of 5 to 8 mm of water is needed before they start producing runoff. The runoff coefficient has shown to increase with increasing rainfall until a constant value is reached. Steeply sloping basalts receiving heavy rainfalls can have runoff coefficient up to 70%. Oppositely, flat, highly pervious soils hardly produce any surface runoff.
- Abruptly rising limbs till flood peaks are reached represent a general characteristic of the discharge hydrograph of wadis in Yemen. A falling limb at a rate that varies from rainstorm to rainstorm and from wadi to wadi follows the flood peak. In between the irregular floods during and after the rainstorms, wadis run either empty of water or carry only minor base flows. Flow measurements at Faj el-Hussein station in Wadi Surdud basin, Tihama Plain, have shown that the discharge increased from about 1.0 m^3 s^{-1} to 87m^3 s^{-1} within 15 minutes.

A short record of floods observed at Qarn Attah in a small wadi west of the town of Rada is presented in Figure 3. The wadi is situated in the headwaters of Wadi Adhana system, and has a catchment area in the order of 100 km^2. It is obvious that short rainstorms generated those floods. Again, the time to the peak is in the order of 15 minutes. The wadi at Qarn Attah is intermittent, hence permanent baseflow such as Faj el-Hussein is lacking.

- The coefficient of runoff has been determined for the above- mentioned wadis by dividing the measured runoff expressed as depth by the areal average rainfall over the surface of the catchment area, both for the same year. This operation has been done for the available years of record and the mean annual runoff coefficient and its variability obtained. From the results listed in Table 1 one can observe that the mean annual runoff coefficient ranges from 0.011 for Wadi Rasyan to 0.11 for Wadi Rima. The coefficient of variation of the annual runoff coefficient varies from 0.22 for Wadi Surdud to 1.37 for Wadi Ahwar. Both statistical descriptors, i.e. the mean and coefficient of variation, have a wide range of variation.
- The measured monthly and annual flows of five wadis are listed in Tables 5(a) thru' 5(e), Appendix II. These five wadis have been chosen because of their relatively long record, 17–27 y. This does not eliminate the fact that both Wadi Mawr and Wadi Bana have some interruptions. Not withstanding the length of record, the mean monthly and annual flows for all gauged wadis are included in Table 5(f), Appendix II.
- The estimates listed in Table 5, Appendix II, show that the surface water carried by the wadis in Yemen is in the order of 2×10^9 m^3 y^{-1}. This figure is subdivided

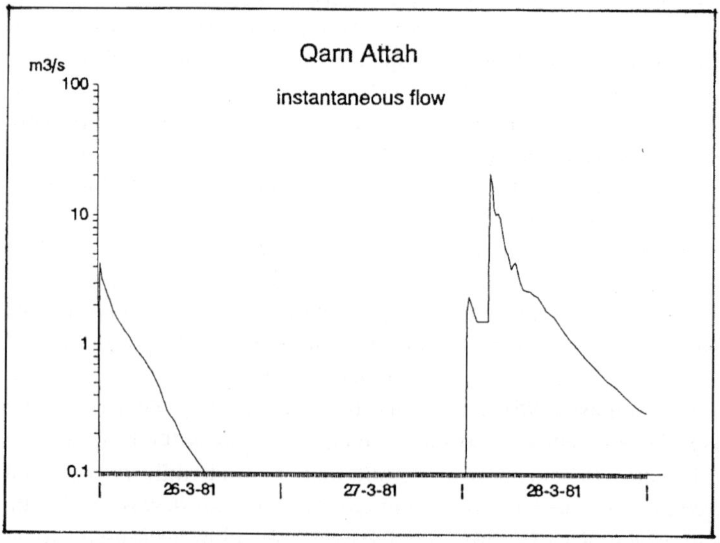

Figure 3. Hydrograph of instantaneous flow at Qarn Attah, Wadi adhana, Yemen (from WRAY Report, Yemen-The Netherlands, 1995)

Table 1. Mean annual rainfall, runoff and runoff coefficient for a number of wadis in Yemen (from WRAY Report, Ministry of Oil and Mineral Resources of Yamen and TNO Institute of Applied Science, The Netherlands, 1995, and Farquharson et al., 1996)

Wadi	No. of years of record	Annual runoff, mm	Annual rainfall mm	Runoff coefficient
Mawr	13	20.5	475.0	0.043
Surdud	5	29.2	440.0	0.066
Rima	8	44.0	400.0	0.110
Zabid	23	27.0	550.0	0.049
Rasyan	7	6.0	550.0	0.011
Tuban	8	21.6	465.0	0.046
Rabwa	17	12.5	320.0	0.039
Bana	16	27.4	370.0	0.074
Ahwar	18	11.0	190.0	0.058
Adhana	8	10.5	180.0	0.059
Amd/Du'an	4	3.1	80.0	0.039
Al-Ayn	4	6.4	75.0	0.086
Sarr	4	1.2	45.0	0.026
Ben-Ali	4	5.8	65.0	0.089
Juaymah	4	1.0	35.0	0.028
Idim	4	7.5	70.0	0.108
Thibi	4	2.6	40.0	0.066
Masia (Qassam)	4	2.3	68.0	0.033

into 36, 27, 28 and 9% between the western slopes, Gulf of Aden catchments, Arabian Sea catchments and the Rub el-Khali catchments respectively.

- The available discharge measurements have been used for estimating the wadi discharges and annual flows corresponding to given return periods or probabilities of exceedance. The estimates obtained are listed in Table 2. It is necessary to emphasise here that not only the data used but also the frequency distribution fitted to these data have a strong influence on the flow estimates. This can be seen from the two estimates found for Wadi Zabid. The check dams built for irrigation purpose in the basin of Wadi Zabid have been designed on the basis of an estimated flow of 1,400 m^3 s^{-1}, and 4,000 m^3 s^{-1} for those in Wadi Mawr.
- Records of instantaneous peak flow rates are rare in Yemen. Table 3 presents the maximum floods observed as given by different sources (cited in WRAY Report, Ministry of Oil and Mineral Resources, Yemen and TNO Institute of Applied Science, The Netherlands, 1995). The observations have been plotted versus the respective sizes of catchment areas as shown in Figure 4. The figure also includes the so-called Creager curves, which represent the envelopes of world floods. They give estimates of probable maximum floods (PMF) as a function of the catchment size A, km^2, using the following expression:

$$Q_{PMF} = 1.3C(0.385A)^{0.935A^{-0.048}} \quad (1)$$

Table 2. Estimates of wadi flows for given return periods or probabilities of exceedance as given in Country reports of (formerly) Yemen Arab Republic and the People's Democratic Republic of Yemen (1986)

Zone and Wadi	T_r, yr pexc, %	Discharge corresponding to return period 100	50	10	5	2	1.25	1.11
		1	2	10	20	50	80	90
Tihama Plain								
Zabid		2500	2200	1350	980	580	350	270
				350	*210*	*150*	*100*	*75*
		256.2	*238.5*	*191.1*	*166*	*121.6*	*82.1*	*63.2*
Mawr		1550	1205	*353*				
				388	*330*	*237*	*170*	*142*
Rama'a		1800	1625	1200	1000	725	500	450
				135	*115*	*60*	*60*	*50*
Arabian Sea								
Amd/Du'an		1540		750		130		
Al-Ayn		750		400		150		
Sarr		710		330		85		
Ben-Aly		600		320		110		
Thebi		550		280		100		
Idim		1600		850		260		

Explanation
Figures in upper lines (upright numerals) are in $m^3\ s^{-1}$ *and in lower lines (italics) are in* $10^6\ m^3$. *Underlined values in the third line opposite Wadi Zabid have been estimated by the author Using the data in Table 6(b) and Pearson Type III distribution.*

Table 3. Observed maximum discharge as observed for some wadis in Yemen and their dates of occurrence (from WRAY Report, Ministry of Oil and Mineral Resources of Yemen, and TNO Institute of Applied Geoscience, the Netherlands, 1995)

Wadi	Catchment area, km^2	Q_{max}, $m^3\ s^{-1}$ (observed)	Date of occurrence
Amd/Du'an	6553	985	March, 1981
Al-Ayn	1500	500	October, 1977
Sarr	2540	2160	March, 1989
Ben-Ali	720	200	April, 1977
Juaymah	760	559	March, 1989
Thibi	5485	1314	March, 1981
Masila	718	350	July, 1978
Hadramout	22500	975	October, 1977
(Shibam)	12800	974	April, 1987

Baban & Gaish (1987), using the data obtained from recorder stations at Mahala on Wadi Nahr (tributary on Wadi Beihan), found strong correlations, R, between the volume, V, of flow generated by a rainstorm and the peak of the discharge hydrograph, Q_p, or the peak height of water, h_p, and the time base or duration of the flood, T. The relationships, which are claimed to be valid

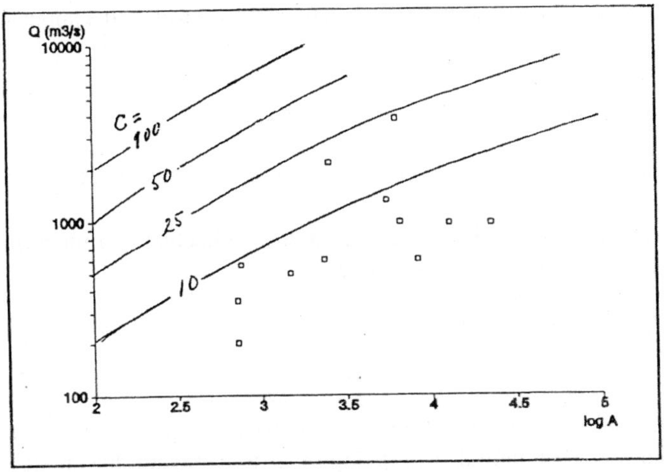

Figure 4. Observed maximum floods versus catchment area, km^2, and the theoretical curves as obtained from Creager's equation for different values of the coefficient C.

for single peak hydrographs, in which by far the majority of floods fall, can be expressed as:

$$V = 8.712 \times 10^{-4} Q_p^{0.653} T^{1.306} \text{ with } R = 0.985 \tag{2a}$$

and

$$V = 0.0105 h_p^{1.2936} T^{1.3297} \text{ with } R = 0.966 \tag{2b}$$

where, V is in 10^6 m^3, Q_p in m^3 s^{-1}, h_p in m above zero datum of the gauge and T in h.

Numerical example: A rainstorm of 50 mm depth falling on a catchment area of 200 km^2 with and average runoff coefficient of 0.10 generates a runoff volume of 1×10^6 m^3. Assuming the base width of the runoff hydrograph to be 6 h, the peak discharge, according to Eq. (2a), is equal to about 1,350 m^3 s^{-1}.

7.2.2.2 *Runoff estimation*

In view of the unclear dependence of the runoff coefficient on rainfall alone, the WRAY Report (Ministry of Oil and Mineral Resources of Yemen and TNO Institute of Applied Geosciences, the Netherlands, 1995) assumed that the relationship between the mean annual rainfall, P mm, and mean annual runoff, R mm, is linear and can be expressed as:

$$R = 0.055P \tag{3}$$

A few years earlier, the author in his unpublished lecture notes developed a slightly different relationship:

$$(4) \qquad R = 0.08(P - 100)$$

It goes without saying that both Eqs (3) and (4) are empirical and easy to apply for predicting approximate runoff values.

The same report gives also the results obtained from the application of a multiple regression model based on the following algorithm:

$$(5) \qquad R_t = a + b_0[P_t - P_o] + b_1[P_{t-1} - P_o] + b_2[P_{t-2} - P_o] \ldots + b_i[P_{t-i} - P_o]$$

in which a is a regression constant, $b_0, b_1, \ldots,$ and b_i are regression coefficients, P_t, P_{t-1} and P_{t-i} are rainfall at month t, one month back and i months back, and P_o is a threshold value of rainfall (mm). Figure 5 is a plot of the model-calculated monthly runoff versus measured monthly runoff. The correlation between the observed and calculated values was found as 0.8. Further, it has been suggested that the straight line drawn between the points indicates that other parameters should be taken into account.

In 1996 Farquharson et al. published a paper describing the combined use of two models developed to synthesise both storm rainfall and runoff in certain parts of Yemen. The model can be summarised as a distributed one that takes account of the wide variations of rainfall input (50–800 mm y^{-1}) and land type over each basin. The northern part of the country was classified into runoff-producing, P, types and runoff-absorbing, A, types as far as surface runoff was concerned. An example of the latter can be found in the Tihama Plain. A runoff-absorbing zone is generally

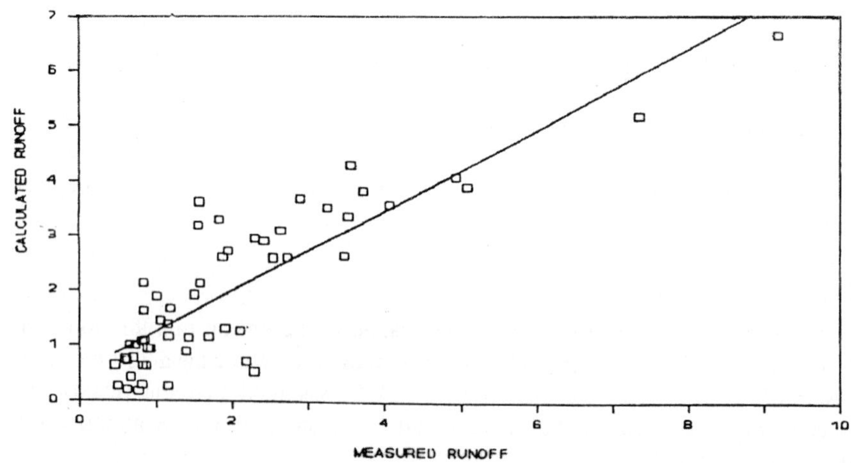

Figure 5. Calculated versus measured runoff for Wadi Surdud (WRAY Report, Ministry of Oil and Mineral Resources of Yemen and TNO Institute of Applied Geosciences, The Netherlands, 1995)

a flat area that may contribute baseflow runoff, but generate negligible surface runoff. Oppositely, runoff-producing zones are generally in the steeper rocky parts of catchments with thin soils.

The essential part of the model comprises the conversion of daily rainfall data within each separate rainfall zone to runoff, with the response of each runoff zone being modeled separately in each case. The model flow chart is shown diagrammatically in Figure 6. The model uses the Soil Conservation Service (1964) curve number (CN) method to convert the rainfall within each zone to runoff. The CN is adjusted dynamically during the year according to antecedent precipitation indices for each runoff zone. The computed runoff from each zone is summed and a proportion of this runoff is routed through a storage representing interflow and terrace storage 'storage'. The remainder, together with the excess from the interflow store, is routed through a baseflow store, which divides runoff into surface runoff and baseflow (Farquharson et al., 1996).

The model was calibrated for the six wadis flowing towards the Tihama, using about five years of rainfall and runoff data in each case. The calibration was based on comparisons of observed and model predicted monthly flood flows, baseflow and total flows. The use of the model was extended to cover two wadis near Aden, which have been observed for a reasonably long period, and many more where runoff measurements cover shorter periods. The estimated runoffs for the wadis used in the model are listed in Table 5, Appendix II. Figure 7 illustrates a graphical plot of the mean annual runoff as obtained from the model versus the mean annual rainfall. It also shows the curve obtained from Turc-Pike equation (Pike, 1964) assuming the open water evaporation, E_o, to remain constant, $2,250$ mm y^{-1}, for all wadis. The equation is written as:

$$E_a = PE_o / \left(P^2 + E_o^2\right)^{1/2} \tag{6}$$

where, E_a = annual actual evaporation, mm and P = annual precipitation, mm.

From Figure 7 it is evident that the calculated runoff, R (mm), taken as $P - E_a$, falls far below most of the plotted points. As such, there is no way to accept the results obtained from the Turc-Pike equation as a comparable substitute to the model results. The runoff, R, which in case of arid and semi-arid climates has small values, is numerically the difference between two larger values. As such, any small relative error in estimating E_a and/or observing P, can result in large absolute errors in estimating R. As the value of E_a calculated from Eq. (6) depends on the value of the denominator, any small change in the power to which the amount between brackets is raised, can lead to considerable change in the estimate of R. With this idea in mind, a different (improvised) curve has been obtained by the author through a gradual change of the power of $(P^2 + E_o^2)$ from 0.500 for P = 800 mm to 0.503 for P = 200 mm. Though the improvement in the fit is clear, the author does not have any theoretical justification for changing the power with decreasing rainfall. The empirical relationship expressed by Eq. (3) too provides a better fit to the model-calculated runoff than the Turc-Pike curve.

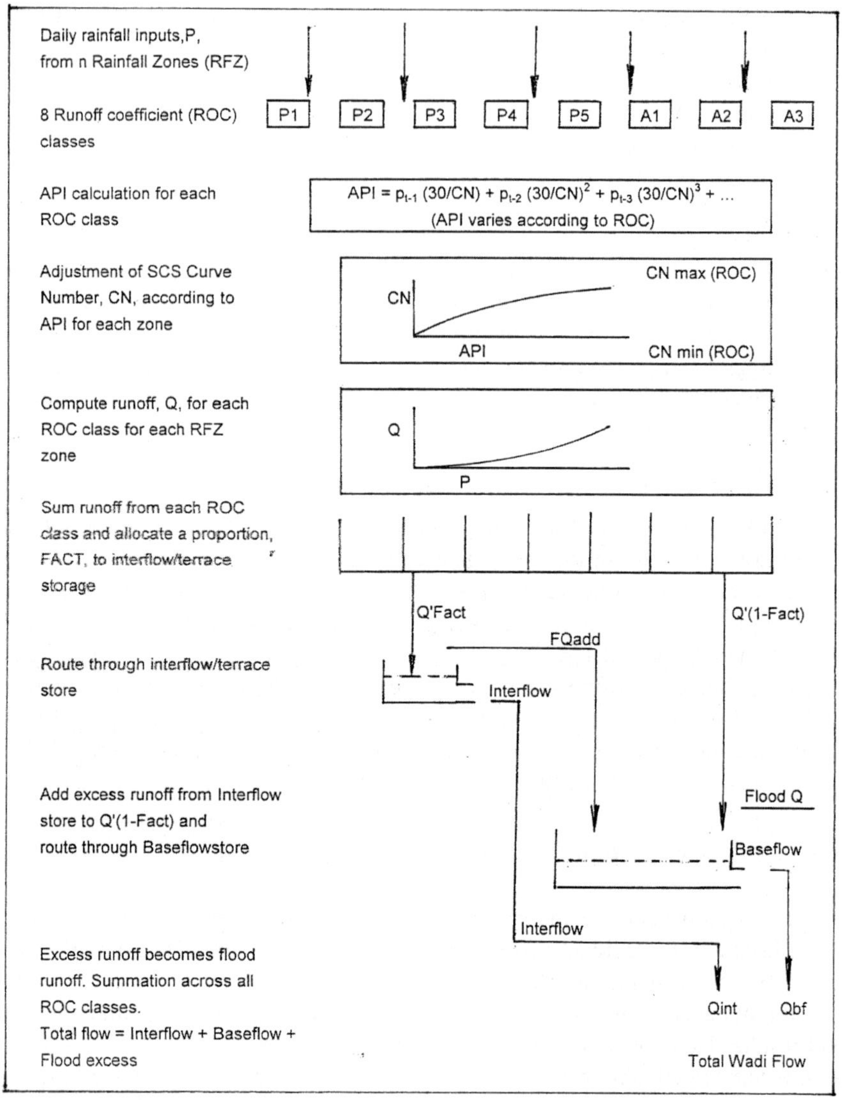

Figure 6. Schematic description of distributed daily rainfall- runoff model based on the SCS method (from Farquharson et al., 1996)

7.2.2.3 'Water Quality

Electric conductivity: Surface water in Yemeni wadis is generally fresh. Old measurements report the total dissolved solids (TDS) were 290 mg l^{-1} in Wadi Rima and 500 mg l^{-1} in Wadi Zabid. Since 1.0 mmho cm^{-1} is equivalent to about 2/3 EC_{25} (electrical conductivity at 25 °C), the reported values correspond to 440

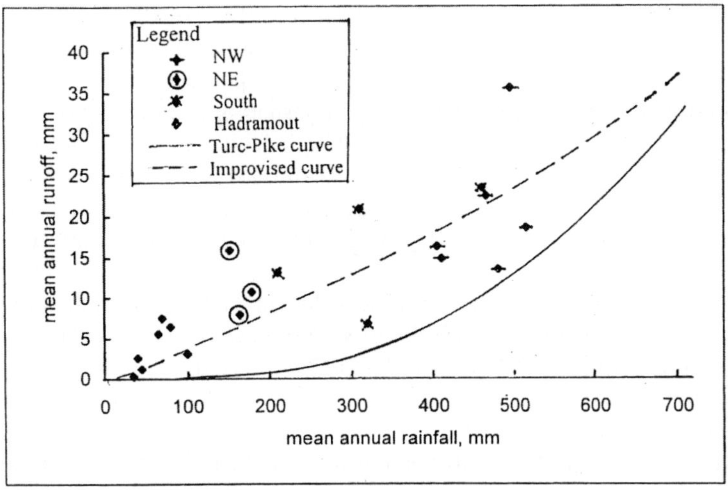

Figure 7. Mean annual basin rainfall and runoff (mm) showing Turc-Pike runoff estimate (from Farquharson et al., 1996)

and 740 mmhos cm^{-1} respectively. The difficulty with these two figures is that there is no specification as to whether they represent the TDS content in flood flow, baseflow or average flow.

Mu'Allem (1987) referred to an old report in which it was mentioned that irrigation water from the wadis in the Abyan Delta has a salt content of 6-10 meq l^{-1}. The principal cations are sodium (50%), calcium and magnesium. The principal anion, bicarbonate, invariably exceeds the total of calcium and magnesium ions.

The WRAY Report (Ministry of Oil and Mineral Resources in Yemen, and TNO Institute for Applied Sciences, The Netherlands, 1995) mentions that the EC_{25} of water of Wadi Surdud is relatively low, varying between 500 and 600 mmhos cm^{-1} during low-flows, and drops to 300 mmhos cm^{-1} during floods. The corresponding figures for Wadi Bana are 1,500-1, 800 mmhos cm^{-1} in the low-flow season and 400-500 mmhos cm^{-1} for floods. Table 6, Appendix III, gives some old measurements of EC_{25} in the southern part of Yemen.

Sediment transport: Sediment transport by wadis in Yemen, especially during flood periods, is substantial. The colour of water, as an indicator of its sediment charge, changes from yellowish-brown in flood to crystal clear in low-flow periods. Bedload is substantial too, especially in wadis with a steep gradient. High concentrations of bed load sediments, approaching 100,000 ppm, were recorded in flood flows of Wadi Zabid. Storage reservoirs in such wadis can be quickly filled with sediments.

Girgirah et al. (1987) reported that nearly all the transport of sediment occurs in wadis during heavy floods. It can carry along course sediment and gravel. The finest sediment comes to rest at the end of the flood and may clog the wadi bed. The heavy flood, however, destroys this mud cake and renews the wadi bed infiltration ability. The suspended solid load concentration in Wadi Tuban was found to vary between 11.4–55.7 gm l^{-1}, and supposed to reach an upper limit of 100 gm l^{-1}. An average value of 25 gm l^{-1} was taken for estimating the sediment transport of Wadi Tuban, bringing its annual load to 3.5–4 × 10^6 t. The corresponding annual volume of flow is 150 × 10^6 m^3. Furthermore, the bed load transport is assessed as 10–20% of the suspended solid load. The same source, i.e. Girgirah et al. (1987), added that the sediment load concentration in spate flows of Wadi Bana was investigated and found to be in the range of 3.2-13.6 gm l^{-1}, and the concentration of suspended sediment load in spate flow in Wadi Hajr was found to fall within the range 6-8 gm l^{-1}.

The above paragraphs show that the available sediment data are far from being adequate. In attempt to overcome the scarcity of kind of data in Yemen, data from mountainous, semi-arid to arid catchments in Jordan have been used to derive some rules of thumb for estimating sediment volumes carried by Yemeni wadis. The general results obtained were as follows:

- The total average volume of transported sediments is in the order of 1–2% of the volume of floodwater.
- Suspended load counts for two-thirds of the total sediment load.
- Annual volume of sediments has a much wider range of variation than the range of variation of flood volume. The sediment load in a wet year can easily reach ten times as much as in dry years.

A case study of Wadi Adhana and Ma'rib describing the sedimentation processes in the Ma'rib Reservoir is included in Chapter 8 'Erosion and Sedimentation'.

7.3. WADI BASINS AND WADI HYDROLOGY IN OMAN

7.3.1 Drainage basins of Omani wadis

The surface of Oman is crowded with a large number of wadis each having a drainage area of a few hundred of square kilometers. Different approaches for dividing the surface of the country into geographic units have been adopted while considering the drainage basins of those wadis. In this text we shall adopt the approach used in the Country Report of Oman (1986). This can be seen from the approximate map of Oman in Figure 8. The regions into which the surface is subdivided are as follows:

i- Musandam Peninsula: This region is located in the northwestern extremity of the country. There is a limited number of wadis, all springing from the mountain tops passing through the center line of the Peninsula. Examples of these wadis are, Khous, Maqlili and Khabb el-Shay. As the amount of surface water carried by these wadis is quite small, 1.5–2.0% of the total surface water carried by the major wadis in Oman, it is often not taken into account.

ii- Sahel Al-Batinh: This region comprises not only the plains intersected by the wadis running down the mountain slopes but also the mountain heights and slopes from and along which the wadis originate and flow before reaching the low-lying areas. The surface area of the region exceeds 12, 000 km^2 subdivided between 25

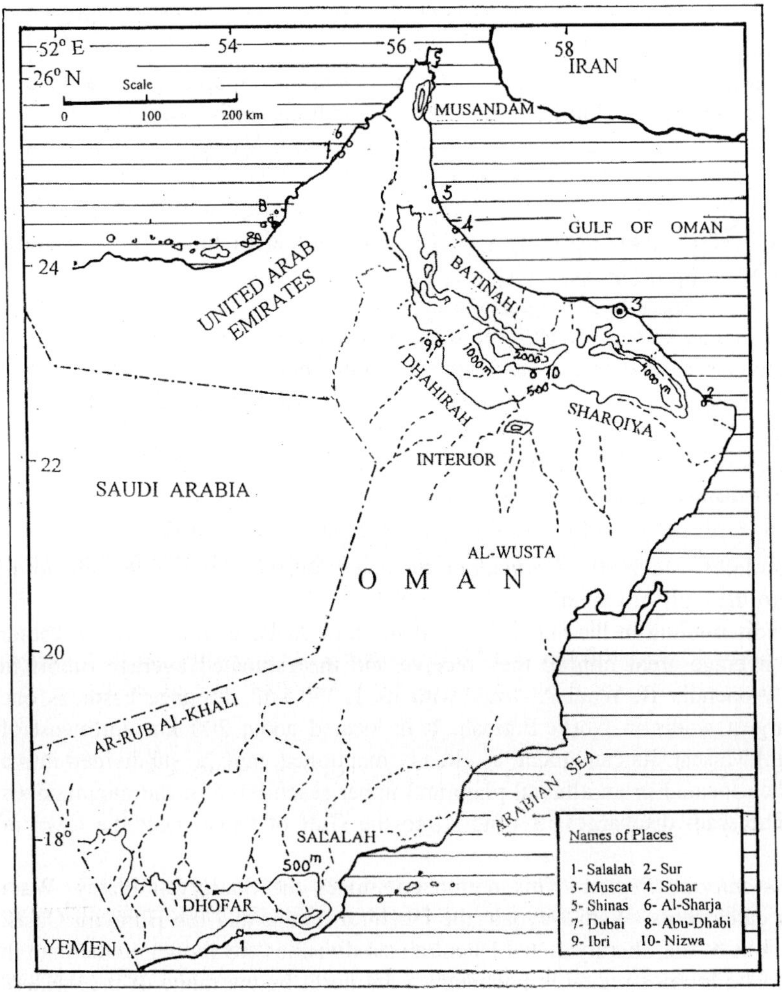

Figure 8. Map showing the division of Oman into wadi regions

wadi basins as can be seen from the map in Figure 9. The northern part of Al-Batinah comprises the basins of Wadi Hulu, Qar, Hatta, Faydh, Rajimi/Zabin, Fizh, Ban-Omar, Souq, Al-Jizzi, Yanbu/Salahi, Al- Hilti, Ahen, Saken/Sheda, Sarami, Shafan/Kanut and Al Hawasinah. The central and eastern parts of Al-Batinah comprise the drainage basins of Lansab, Al-Rasyl, Samail, Taww, Al-Ma'awel, Bani- Kharus, Al-Far', Bani-Ghafer, Al-Hajer and Mabrah.

The southern parts of Al-Batinah from which wadis spring are mountainous in nature with elevations ranging from 200 and 1,500 m a.m.s.l., and receive relatively heavy rainfall 200-250 mm y^{-1}. These upper reaches of wadis run full for a short distance, below which the slope of the terrain declines rapidly and so does the

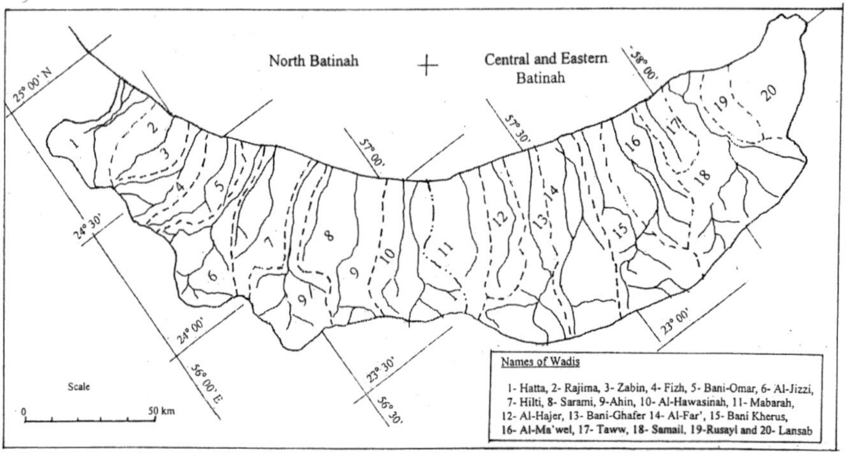

Figure 9. Map showing the drainage basins of wadis in Al-Batinah Region

annual rainfall to reach 50–75 mm along the coast. Wadis discharge the remaining runoff into the Gulf of Oman.

The wadis running in the north, east and south of Al-Batinah are listed together with the average areal rainfall they receive and the estimated average runoff in Table 8, Appendix II. Wadi Al-Jizzi, with its $1,250\ km^2$ drainage basin, is one of the largest wadis in North Batinah. It is located about 200 km northwest of the Capital Muscat. Its catchment, as already mentioned, may be subdivided into a lower reach located in an alluvial plain and upper reach lying in the mountainous region. The wadi discharges its runoff into the Gulf of Oman near the town of Sohar.

iii- The Interior Region: This region comprises the Sharqiyah Plains, Wadi Al-Batha catchments, Wadi Dayqah, the Dhahirah Plains and the Buraymi Oasis. The locations of the Sharqiyah and Dhahirah subdivisions are shown on the map in Figure 8. The Interior region is bounded on the north by the range of the western Al-Hajer Mountains and from the west the international boundary between Oman and the United Arab Emirates and Saudi Arabia. It is joined in the south by the Central Plateau region (Al-Wusta) and from the east and south by the Gulf of Oman and the Arabian Sea respectively.

The Sharqiyah subregion lies on the flanks of Al Hajar ash Sharqi Mountain range, and extends over 42,2000 km^2 of rock ridges, wadis, alluvial plains and the sand sea of Ramlat Al-Wahaybah. It includes three major drainage basins, the Batha (8,300 km^2), the Adnan/Samad ($3,170\ km^2$) and the Dayqah (2,000 km^2). The remaining wadi basins are much smaller in surface area.

The $8,300\ km^2$ Wadi Al Batha basin is unusual among the interior catchments in having a continuous channel from the mountains to the coast. Five sub-basins form the headwaters of the Batha Drainage basin. These are: Ibra, Adh Dahir, bu Taymah, Aghda and Bani-Khalid. The grand total catchment area of Al-Batha is claimed to be close to 29,900 km^2 (Singh and Cherchali, 1995). Each of these sub-basins has its own characteristics that affect the surface runoff, recharge to the groundwater, evaporation and other losses.

Wadi Dayqah is a perennial wadi less than 100 km east of the Capital Muscat. It is well known with surplus flood flows most of which are wasted to the sea. Dam construction for storing floodwater is already under consideration for quite some years. The long, high gorge of the lower reaches of the wadi breaks through the Eastern Hajar Mountains to drain some 1,500 km^2 of the Interior region.

The Dhahira plains form a part of the Interior region of Oman. It comprises the slopes extending from the extreme north of Oman Mountains to the western frontier of Oman with the United Arab Emirates near Al-Buraymi Oasis. It also includes the southwestern slopes of Al-Aswad and Al-Abyadh Mountains. The major wadis of Dhahira are Yanqul (1,460 km^2), Lusayl (1,726 km^2), Tumaymah (1,590 km^2) and Dank (1,250 km^2).

The Buraymi Oasis, which forms another part of the Interior region, lies in the northwestern corner of Oman near to and extending into the United Arab Emirates. Table 6, Appendix II, gives the catchment area of the Buraymi/border wadis as 2,750 km^2. Aday/Mayh, Mah'dha, Al Ain, Chiq and Sumayni are some of these wadis.

iv- Central Region (Al-Wusta): The Al-Wusta Region of Central Oman covers about 80,000 km^2 or one quarter of the surface of the country. It is a hot dry desert with an annual average rainfall of between 10 mm to 50 mm. This region may be sub-divided into five principal geomorphic and hydrologic provinces-Central Plateau, Ar-Rub al-Khali, Al-Huqf, Eastern Wadi Province and Bajada. In view of the scanty runoff produced by the meager annual rainfall, this region has not been included separately in Table 6, Appendix II. The thin aquifer underlying the bed of Wadi Rawnab is probably the only significant freshwater body so far known in the Wusta Region.

v- The Southern Region: This region is bounded on the east and south by the Arabian Sea, on the west by the frontier with Yemen and by Al-Wusta Plateau on the north and Ar-Rub Al-Khali Desert on the northwest. The Southern Region includes the Salalah coastal plain. The Dhofar Mountain slopes from east to west and further northward up to Najd Plateau where they terminate in the desert. A number of relatively small wadis, e.g. Dahnat, Andour and Hallouf, spring from the northern basins in the Interior Region and discharge their water in Najd.

7.3.2 Wadi hydrology and surface water quality in Oman

Wadis in Oman do not behave differently from wadis in Yemen or any other arid or semi-arid land. The upper reaches flow almost continuously and become ephemeral upon traversing relatively flat country where the ground has been described as rainfall-absorbing. The peak runoff is often reached during or shortly after the off set of rain. The ascending limb of the flow hydrograph is rising rapidly from zero or a small value to reach large values in short time. Having the peak flow reached, the descending limb of the hydrograph falls rapidly or slowly depending on several factors. The flood flow usually has a high concentration of sediment in suspension as well as bed load. The next paragraphs review in brief a number of case studies from Oman

7.3.2.1 Al-Batinah Wadis

Surface water fills the wadis during and after heavy rains or successive rainstorms. It flows in the upper wadi reaches on the highly elevated belt and in the hilly areas. In the lower wadi reaches and in the plain, however, surface water can be observed just for a few hours or days. Preliminary investigation has shown that between 15% and 40% of the runoff from the high ground is lost in the sea and the rest of the torrential water is absorbed by the fluvial soil while flowing and joins the groundwater body all along the plain. The effect of the light rain showers on the gravely parts of wadi channels is limited. The hydrographs of Wadis Bani-Ghafer at Al-Hawqyn and Al-Far'a at Al-Mazahiet stations are shown in Figures 10(a) and 10(b) respectively.

From the old data included in the Country Report of Oman (1986) one finds that a certain rainstorm on wadi Bani- Ghafer brought a total volume of rain of 3,037,880 m^3 of rain and 595, 980 m^3 as surface runoff. This pair of figures tells of a rainfall coefficient of 19.6%. Likewise, another rainstorm on Wadi Al-Far'a yielded 5,753,400 m^3 rainfall and 817, 200 m^3 surface runoff. This pair of figures shows that the runoff coefficient for that storm was 14.2%. These two values for the runoff coefficient were used by the authorities responsible of the water resources in Oman for estimating the annual surface runoff of the said wadis.

As wadi runoff depends principally on the characteristics of the rain falling on the catchment area, it has been found imperative to analyse rainfall events regarding their spatial and temporal distributions. Hofman & Rambow (1995) defined a rainfall event as a period of rainfall exceeding 0.2 mm preceded and followed by one hour of rainfall less than or equal to 0.2 mm. They carried out probability analysis using the rainfall events in Wadi Al-Jizi, the northern part of Al-Batinah Plain. The mean conditional probability of having rain at a certain gauge given rainfall at another station was found to be 0.21 in summer (April–September) and 0.35 in winter (October–March). Generally the probabilities decrease with increasing distance. Figure 11 was obtained for the stations of Daqiq (840 m a.m.s.l.) and Sohar (15 m a.m.s.l.) and gauge separation up to 70 km. Whether the joint occurrence of hourly rainfall is dependent or not dependent shall be discussed in the next section while reviewing wadi hydrology in Saudi Arabia.

The Wadi Adai, which is situated in northern Batinah, has a catchment area to Wattayah near the seacoast of about 370 km^2. The catchment consists of an area of gravel hills surrounded by more steeply sloping, bare rock areas. The combined area of the gravely and rock covered surfaces is 321 km^2. The flow from the combined area is concentrated through a gorge 10 km upstream of Wattayah. A series of tributaries flow from a further bare rock area over the length of the

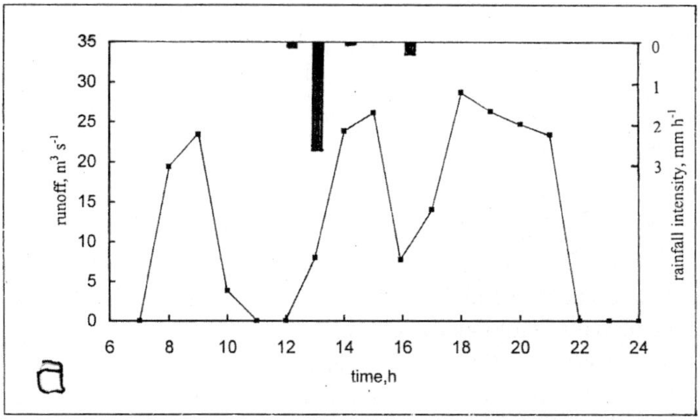

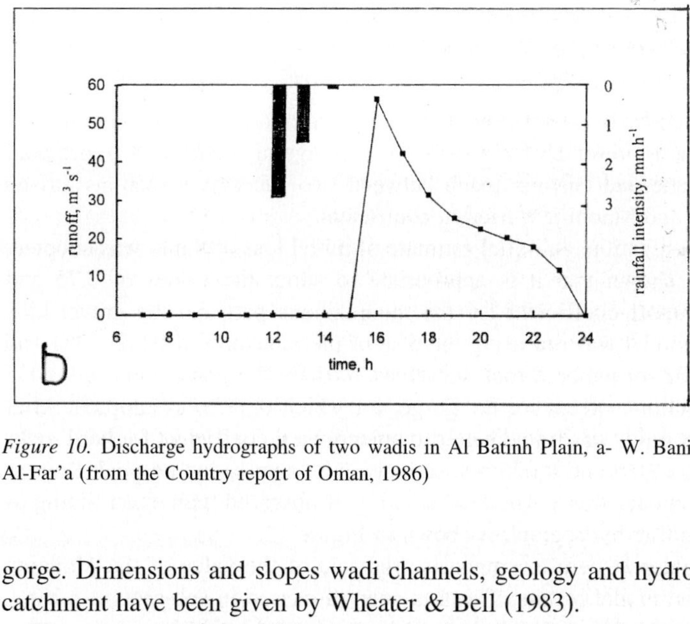

Figure 10. Discharge hydrographs of two wadis in Al Batinh Plain, a- W. Bani-Ghafer and b- Wadi Al-Far'a (from the Country report of Oman, 1986)

gorge. Dimensions and slopes wadi channels, geology and hydrology of the wadi catchment have been given by Wheater & Bell (1983).

The steepness of the Adai catchment when coupled with intense rainfall results in severe floods that cause damage to properties and infrastructures in the area and part of the surroundings. The wadi flow data and rainfall-runoff relationship were investigated and modelled by Wheater & Bell, and the results reported. (1983). The peak discharge of the wadi was estimated applying the slope-area method to four sections. The value of Manning's n of 0.032 was adopted. The travel time between two successive sections ranged from 4.5 to 3.6 m s^{-1}. To simulate the flood hydrograph with the limited data available, a physically-based rainfall-runoff model had to be chosen such that the model parameters can largely be estimated from the catchment measurable characteristics. As flood runoff from arid areas is

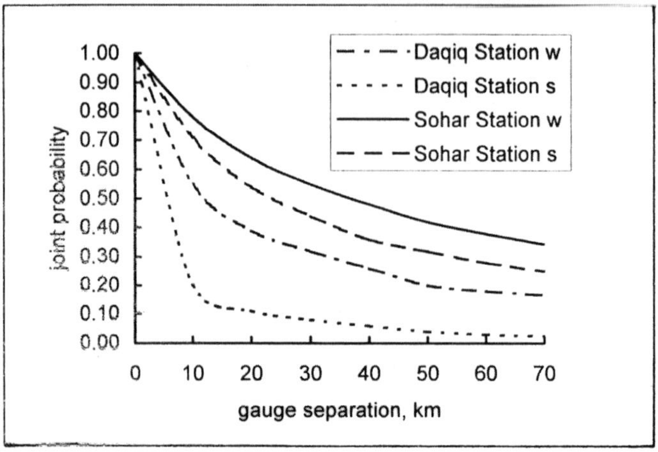

Figure 11. Probability versus gauge separation for Wadi Daqiq and Wadi Sohar, Al-Batinah, for both winter (w) and summer (s) seasons (after Hofman & Rambow, 1995)

essentially generated by overland flow, the choice went to kinematic wave model. The catchment was represented by a series of rectangular planes and trapezoidal channels. Each plane and channel reach between two successive sections has its dimensions, slope and Manning's friction coefficient.

For the said investigation, an initial estimate of initial loss of 8 mm was adopted. Calculations have shown that it is appropriate to adopt the values of 0.75 and 0.50 for the net runoff coefficient for the mountainous part and the gravel hills respectively. The model was run using the data of the rainstorm of 03.05.1981 and Manning's n of 0.02 for the hard rock sub-areas, 0.03 for the gravel hills and 0.035 for the channel sections except for the gorge, for which 0.032 was adopted. After the first simulation run it was found better to amend the n-coefficient for hard rocks and gravel hills to 0.016 and 0.025 respectively. The corresponding peak flow as obtained from the model was 1, 1630 $m^3 s^{-1}$, 1% of observed with exact timing as occurred. The resulting hydrograph is shown in Figure 12.

It might be of interest here to emphasise the role of the value of the Manning coefficient n on the model results. This has been discussed in sub-section 6.2.2.3, Chapter 6, while reviewing the hydrology of Wadi Zeroud in Tunisia.

7.3.2.2 Al-Sharqiyah Wadis

The upper wadi reaches in the Sharqiyah region are of barren hard rock, impermeable and very steep without vegetal cover. The elevation exceeds 600 m a.m.s.l. The annual rainfall ranges between 250 mm and 300 mm and the runoff coefficient is in the order of 60%. The elevation of the lower reaches is below 600 m a.m.s.l. They traverse alluvial plains, and have relatively mild slopes, less than 1%. The annual rainfall is in the order of 110 mm. As such, the lower wadi reaches help to recharge the groundwater aquifers in the region.

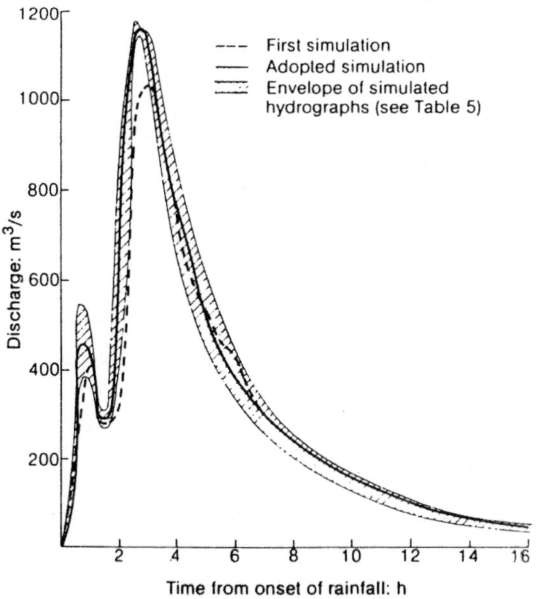

Figure 12. Synthetic Wadi Adai flood hydrographs for the rainstorm of 3 May 1981

Hydrology of Wadi Dayqah has been investigated for quite some time in view of its potential for storage dams, and for estimating water use by crops. This was supported by meteorological investigation of rainfall and evapotranspiration. Open water evaporation, for example, was computed using Mazara (main gauging station) climate data with FAO modified approach of Penman, which is already reviewed in Chapter 5.

The resulting average annual open water evaporation was about 2.8 m or 7.7 mm d^{-1}. Flows in W. Dayqah are perennial at Mazara, though highly variable from one year to another as far as total runoff is concerned. At Mazara, which is the most downstream wadi gauge measurimg perennial flows before they are lost in the karst, the recorded annual flows from 1978–1993 ranged from 3 to 192×10^6 m^3. The average annual flow is 45×10^6 m^3 for the 15-y record period, but 37×10^6 m^3 (equivalent to 22 mm of runoff from catchment average annual rainfall of about 150 mm) if the extremely wet year 1982 is excluded. In this case the average annual runoff coefficient is in the order of 15%. The flow duration curve for the said period at the station of Mazara is shown in Figure 13 (from Kaul, 1995)

Wadi Dayqah's floods can be enormous and pose a danger to any water resources development structures not properly flood-proofed. In 1982, the flood peak reached $6,000$ m^3 s^{-1} at Mazara, and in 1927 has been estimated as above $9,000$ m^3 s^{-1}, both high by world standards. The probable maximum precipitation (PMP) with duration of 24 h set at 470 mm point rainfall was found to be critical. For the 1,700 catchment area at the dam site, the probable maximum flood (PMF) was computed as $14,300$ m^3 s^{-1}, which is large by world standards.

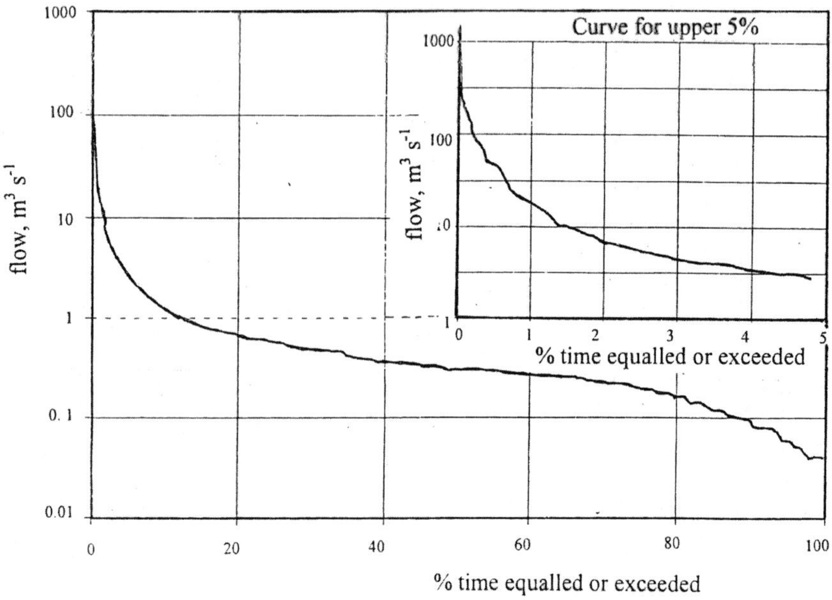

Figure 13. Wadi Dayqah flows at Mazara 1978–93, flow-duration curve: daily flows (Kaul, 1995)

7.3.2.3 Wadis of the interior region

Some of the hydrological characteristics of eight wadis, namely; Halfayn and Al-Muaydin (Dakhliyah, i.e. Interior), Al-Ghalaji (Sharqiyah), Al-Aridh, Yanqul, Al-Kabir, Al-Hajr and Al-Asil (Al-Dhahirah/Ibri Region), were studied by the Ministry of Water Resources, Oman, in the period 1992–94. The studies were associated with the plans to construct groundwater recharge dams, and their possible impacts on the water resources. The major items included in those studies and the results obtained therefrom were:

Catchment characteristics such as surface area, channel length and slope, surface coverage by rock outcrop and alluvium were described by Mohsin et al. (1995). The smallest catchment area is 128 km^2 for Wadi Aridh at the A-I site and the largest 1,416 km^2 for Wadi Asil at AsSulayf Gap. The channel length ranges between 24 and 67 km and the slope between 0.01 and 0.06. The largest percent of total rock outcrop is 95 (5% alluvium) for Wadi Al-Muaydin and the smallest 52 (48% alluvium) for Wadi Ghalaji. The catchments and sites of the eight wadis are shown on the map in Figure 14.

The rainfall in the mentioned eight basins was found to be highly localized, with an average annual depth ranging from 70 mm to 200 mm. Potential evapotranspiration was estimated from 1,500 to 2,300 mm, with the lower values in the higher elevations. Rainfall for each one of the five sub-catchments in the Ibri area was estimated by weighing nearby daily rainfall records by the inverse square of the distance between the station and the centroid of the corresponding sub-catchment.

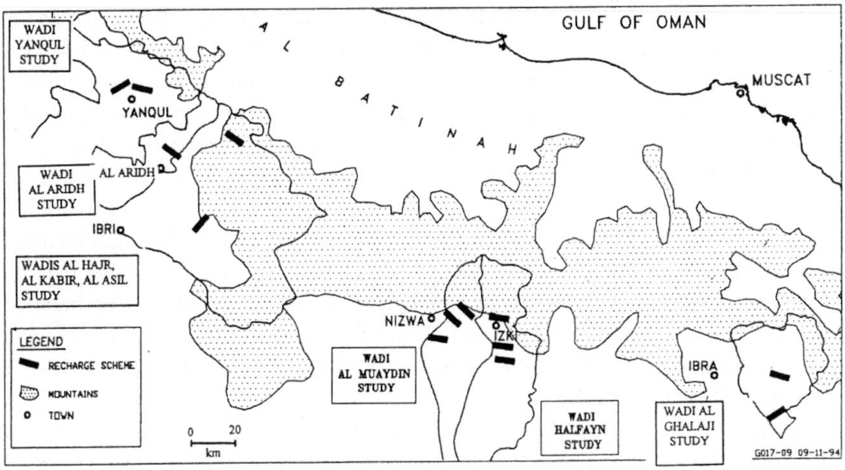

Figure 14. Map showing the location of eight Interior catchments Oman and the recharge schemes, Oman (from Mohsin et al., 1995)

Runoff was based on the area of sub-catchment, and then calibrated to the Yanqul flow gauge and the recharge model.

Initial losses and losses of remaining rainfall were required for input to the groundwater model. The initial losses were found to fall in the range 7–15 mm while the latter ranged from 70 to 80% for the mountainous areas to 50% in alluvium.

A distributed runoff model for Al-Ghalaji Wadi catchment was developed using the Soil Conservation Service method. Runoff curves were selected by experience from similar arid area studies and calibrated to the only storm event with rainfall and runoff data in the study area. Calibration of runoff also involved analysis of groundwater response to recharge. The limited surface water data was a constraint to that study.

Composite rainfall/runoff/recharge relationships have been established from the model results. Runoff from the eight wadi catchments ranged from less than 1% to as much as 14% in the normal rainfall years. Recharge ranged from 2 to 20% of average annual rainfall. The higher rates of recharge were in the basins with less steep slopes and with higher areal percentage of Tertiary limestones.

Two approaches to the water balance were used in the Halfayn and Al Muaydin studies. In the first approach the natural annual volume of recharge was estimated and the assumed volume checked against the available aquifer characteristics for the period 1982–86. The second approach was a certain operation model, which sized the reservoir sizes and quantified their impacts. The inflow to this model was a 10-y synthetically generated series derived from the main flow gauges.

An integrated water resources model was developed for Wadi Ghulaji to facilitate the coupling of a distributed rainfall model, a distributed water balance model and a groundwater model (Wheater et al., 1995). The model comprised a rainfall generator supplying input to the Jebel (mountain) Plane, Alluvial Plane and Wadi Bed. The water balance model distributed the input (rainfall) to the first and second planes between evaporation, surface runoff and input to the groundwater model. The Jebel Plane produced surface runoff only without recharge to the aquifer while the Alluvial Plane produced both surface runoff and aquifer recharge. The curve number CN for the steep Jebel region varied between 90 and 95 and for the alluvial plain between 77 and 85. The input (rainfall) to the Wadi Bed was distributed between evaporation and input to the transmission losses model. The wadi channel carried the runoff produced by the Jebel and Alluvial Planes and recharged the aquifer (Figure 15). The groundwater recharge simulated by the water balance model served as input to the groundwater model. Groundwater flow and water levels were modelled using the 'Modflow' software package, which will be reviewed in Chapter 9 on Groundwater Resources.

Calibration of the water balance model using the data of the rainstorm of 5–6 April 1993 at three stations showed a reasonable agreement between the simulated and observed surface runoff volumes. The maximum difference between the model-simulated runoff and the observed runoff, 10%, occurred at Al-Qabil station. Besides, from the results obtained from the water balance of Wadi Ghulaji, the runoff coefficient ranged from 3.2% for a dry year to 4.6% for a wet year, with an average of 4.0%.

7.3.2.4 Surface water in the southern region

From the available data it appeared that a seasonal (Khareef) rainfall of 200 mm represents the limit above which recharge to groundwater occurs. A considerable amount of rainwater goes to the soil moisture reservoir and the consumption of water by the vegetal cover of the ground. As such, there is hardly any surface

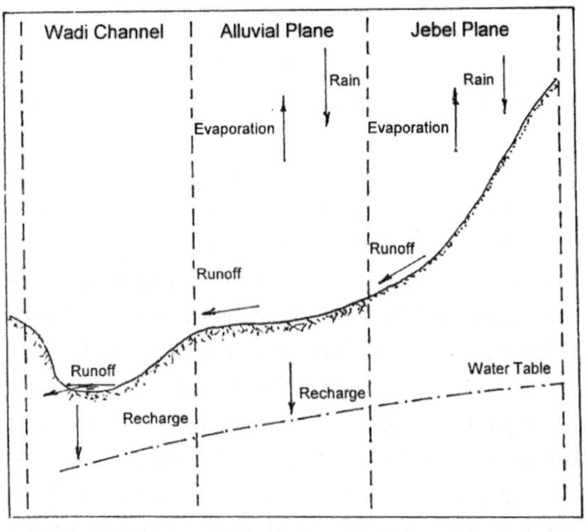

Figure 15. Schematic representation of the water balance model (Wheater et al., 1995)

runoff. However, the rainfall in May, June, October and November, which takes the form of intensive rainstorms, produces runoff that flows swiftly to the wadis and eventually to the sea. This process, as mentioned earlier, occurs during a short period of time.

Local experiences show that rainfall during the rainy seasons is about two-thirds of the total annual rainfall, and the remaining one-third represents the amount of torrential rainstorms, which occur sporadically. One should bear in mind that torrential rain (sayl) might not happen for a sequence of years and then unexpectedly comes a year with excessive torrents.

The water balance of the Salalah Coastal Plain is based on an average annual (7-y) rainfall of 310 mm and a catchment area of 910 km^2. These figures bring the average annual volume of rainfall to 282×10^6 m^3. It has been estimated that about 10% of this amount, i.e. 28×10^6 m^3, moves annually through the groundwater reservoir and supplies the springs by an amount of 9×10^6 m^3 y^{-1}. The remaining 19×10^6 m^3 keep flowing through the groundwater reservoir. To this amount there is an additional amount of 7×10^6 m^3 from the rainfall produced by torrential storms and other circulating air masses bringing the total to 26×10^6 m^3 in an average year. The considerable difference between the 282×10^6 m^3, annual rainfall, and the 26×10^6 m^3, groundwater flow, can be attributed to the losses by evaporation from the pools and watercourses, and evapotranspiration from the vegetal cover of the soil. The percentage of losses to the total rainfall increases from 80–85% in the mountainous parts and reaches up 98% of the annual rainfall in the Salalah Plain.

Before we end this sub-section it is worthwhile mentioning that the Country Report of Oman (1986) estimated the total wadi surface runoff as 917.8×10^6 m^3 y^{-1} distributed as follows (all figures are in 10^6 m^3 y^{-1}):

Musandam Peninsula	*23.4*
East and Central Batinah	*180.2*
North Batinah	*168.1*
Al-Dhahirah (northern basins, interior region)	*121.7*
Interior	*143.2*
Al-Sharqiyah	*95.9*
Qurayyat and Sur	*90.1*
Dhofar and northern wadis	*27.3*
Salalah and sourthern wadis	*67.9*
Total	*917.8*

The ECWA Report (1981) estimated the total annual volume of surface runoff and groundwater as $1,409 \times 10^6$ m^3 distributed between Northeastern Oman (485×10^6 m^3), Northwestern Oman (500×10^6 m^3), the Interior (389×10^6 m^3) and Dhofar (35×10^6 m^3). Comparison between these two estimates shows that the surface water alone comprises about two-thirds the total volume, and the groundwater recharge comprises one-third only. The sum of the subtotals adopted in

Table 6, Appendix II, is $1,150 \times 10^6$ m^3 y^{-1}, almost 25% larger than the 1986 estimate appearing in the ECWA Report. The difference may be attributed to the greater number of wadis consequently the larger area of the drainage basins on one hand and the marked increase in the research tools and amount of studies carried out in recent years.

It is important to remember that intensive rainstorms coupled with the poor vegetation cover can result in extreme runoff peak flows. Discharge observations and estimates show that Omani wadis are capable of providing flood peaks of the order of 20 m^3 s^{-1} km^{-2}, and possibly more during rare and extreme events. Muscat Flood Alleviation Dam has a design flood peak of over 100 m^3 s^{-1} km^{-2} (Al Moqbali et al., 2000).

7.3.2.5 Water quality

In general, the surface water carried by the Omani Wadis, caused by intensive rainstorms, can be described as of high quality. This is especially true for wadi channels traversing the rocky parts. When the lower reaches traverse the coastal plains there is a chance that the quality of water becomes inferior.

Huge quantities of sediments are moved by water during major floods. Sediment transport in general, and particularly bed loads, are too difficult to measure when floods are extreme as access to key stations is difficult and velocities are so high that they would damage measuring equipment. Generalised sediment-flow curves have been developed from bed material surveys of Omani wadis. Additionally, valuable information has been collected from routine annual measurement of accumulated sediment within existing recharge dams, similar to bathymetric surveys of lakes and other storage reservoirs. A combination of this information, together with bathymetric survey of Wadi Dayqah, resulted in an estimated average sediment flow in the range 0.1-0.3 mm y^{-1} of equivalent erosion runoff from the catchment (Kaul, 1995).

Assuming an average rate of sedimentation of 0.2 mm y^{-1} of equivalent erosion runoff from the total area of drainage basins of the wadis listed in Table 6, Appendix II, the estimated annual volume must be in the order of 15×10^6 m^3. According to Kaul (1995), the total sediment inflow to Wadi Dayqah caused by the severe flood of 1982 was in the order of 9×10^6 m^3, most of which would have been trapped. This leads to the conclusion that large storage reservoirs are needed not only for storing floodwater but also to cover potential sediment inflow.

7.4. WADI BASINS AND WADI HYDROLOGY IN SAUDI ARABIA

7.4.1 Drainage basins of Saudi wadis

The territory of the Kingdom of Saudi Arabia has been subdivided for quite some time on the basis of the main geomorphologic and hydrogeologic features into eight provinces as indicated in Figure 16. As this subdivision generally applies to surface water runoff it will be adopted here.

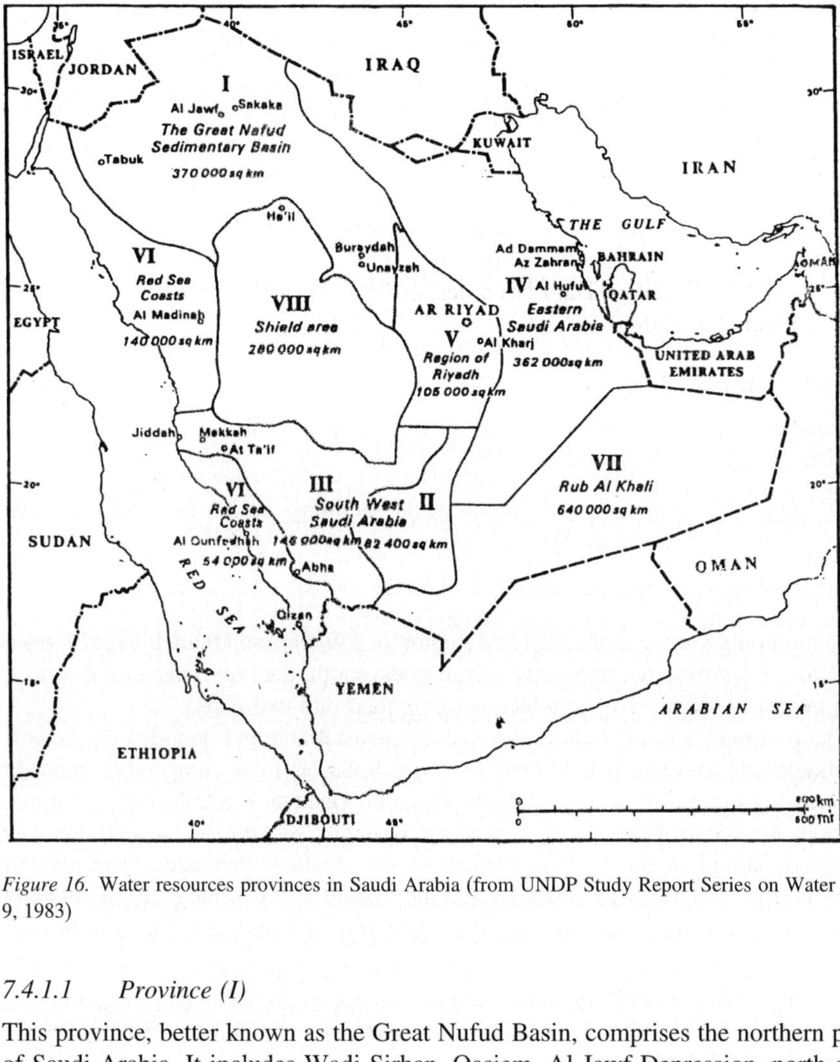

Figure 16. Water resources provinces in Saudi Arabia (from UNDP Study Report Series on Water No. 9, 1983)

7.4.1.1 Province (I)

This province, better known as the Great Nufud Basin, comprises the northern part of Saudi Arabia. It includes Wadi Sirhan, Qasiem, Al-Jawf Depression, north and northeast Hail, Skaka, Tabuk, Taima', Al-Ula, Al-Qaryat and Ar'ar areas up to the international boundary with Jordan. The surface area of this province is about $370,000\ km^2$.

7.4.1.2 Province (II)

It has a surface area of $84,400\ km^2$ in south central Saudi Arabia. This province, often referred to as Najd Basin, lies in the south of Saudi Arabia. It is bounded on the west by the Hijaz Plateau and the Najd Pediplain, on the north by the south Tuwayq Mountains and Ar-Rub al-Khali on the east. It covers the Wajid Plateau, Aflaj and Wadi Dawasir until Wadi Tathlieth.

7.4.1.3 Province (III)

This area is situated in eastern Asir and has a surface area of 144,000 km^2 extending from north of Taif to Najran in the south. The most important wadis in this province are: Turabah-Khurnah (15,475 km^2), Ranyah (11,950 km^2), Bishah (35,780 km^2), Tathlieth (29,782 km^2), Wadi Idemah (6,475 km^2), Wadi Tablah and Habawnah (7,275 km^2).

7.4.1.4 Province (IV)

It comprises the northeastern and eastern areas of Saudi Arabia, covering a surface area of 362,000 km^2. This province is the most important region for the economy of Saudi Arabia as it contains the country's oil fields and refineries.

7.4.1.5 Province (V)

It is commonly referred to as Er-Riyadh region, comprising the northern part of Aflaj, Al-Kharji up to Sdair in the north. The total surface covered by this province is 105,000 km^2. This province includes the drainage basins of Wadis Fatima, Al-Nu'man, Khulays, Wijh, Turabah, etc.

7.4.1.6 Province (VI)

It is commonly known as the Red Sea Region of Tihama and Hijaz. It extends from the Gulf of Aqaba in the north up to Jizan in the south, thus covering a total surface area of 194,000 km^2. This province is subdivided into two parts.

The first part is located along the Red Sea coastal belt and bounded on the east by the basaltic plateau. It is known as Harrat Rahat and has comparable drainage patterns from east to west, e.g., Rabigh, Thimrah, Qudayd, Khulays, Usfan, Fatima, Numan, Abyad, Ad-Dam, Sadiyah, Markub, Kaderah, Al-Lith. The second drainage system is found in the Central Plateau of the Arabian Peninsula. The general trend is from southwest to northeast and the system drains principally in Al-Jarad Depression, with the wadis Al-aqia, Wajj, Qayum, Liyyah, Um Salam and Busal.

7.4.1.7 Province (VII)

This is Ar-Rub al-Khali area, with a surface area of 640,000 km^2. It is the south-eastern Empty Quarter.

7.4.1.8 Province (VIII)

This province also called the Arabian Shield has a surface area of 260,000 km^2. It encompasses the central Najd area.

7.4.2 Wadi hydrology and surface water quality of Saudi wadis

The rainfall in Saudi Arabia, as already described in Chapters 3 and 4 is scanty and unpredictable. As such, the resulting runoff is irregular and storage of surface water is very limited. As a consequence to limited precipitation there are no perennial streams. However, surface runoff from the high parts of wadi drainage basins, especially those draining the Red Sea Mountains, can continue for several weeks and months.

Torrents are generally rare in the eastern province and increase in frequency and magnitude in Asir and the southwest of the country.

7.4.2.1 The Great Nefud Basin (Province (I))

This area is characterised by high intensity storms, which result in runoffs over a large surface of the province. However, data about the amount, duration and frequency of these floods are still inadequate.

7.4.2.2 Asir-Wadi-Dwasir (Province (II))

The most important runoff producing areas are located in the median rainfall belt (200–300 mm y^{-1}), away from the mountain region. Najran discharge observations showed that Wadi Bishah, Wadi Ranyah and Wadi Turabah are the most important catchments in this area in terms of total discharge. In the sedimentary and desert areas to the east of the Arabian Shield area, runoff appears to be localized and of infrequent occurrence.

7.4.2.3 Asir-Wadi Najran (Province (III))

In the Asir Highlands where comparatively high rainfall occurs, and the ground relatively impervious and sloping steeply, considerable volume of surface runoff occurs. Some perennial flows occur in the higher parts of the region but rarely reach the Red Sea. Most of the wadis in this area drain their waters in the interior. Examples of these are Wadis Tabalah (1,270 km^2) and Habwanah (7,275 km^2).

The total annual surface runoff from Provinces II and III was estimated in the Report on "Food Security in the Arab States (1980)" as 273×10^6 m^3.

7.4.2.4 Northeastern and Eastern Region (Province (IV))

There is hardly any surface runoff to mention in this region.

7.4.2.5 Er-Riyadh Region (Province (V))

There are no perennial streams in Riyadh Region. Analysis of rainfall over this region shows that the average annual rainfall ranges from 85 to 110 mm. Heavy showers accompanied with steep slopes of the Tuwayq Mountains, Marrat and Aruma formations can produce considerable runoff.

The province is characterised by an average annual rainfall of 100 mm. However, considerable runoff occurs in the Tuwayq Basin creating favourable conditions for feeding Wadi Hanifah, one of the most important wadis in Saudi Arabia. Several other wadis are also directly fed by local rainfall, e.g. the Marrat, Aruma, Nisah, Jubaylah and Al-Kharj Plain. According to the ECWA Report (1981), SOGREAH (French Society of Agronomic and Hydrological Researches) estimated the average annual runoff from the region as 300×10^6 m^3 distributed between Tuwayq

Mountains (200×10^6 m^3), Marrat formation (15×10^6 m^3), Aruma formation (70×10^6 m^3) and other formations (15×10^6 m^3).

7.4.2.6 *The Red Sea (Province (VI))*

The Biyash, Yiba and Hali are the main wadis in this area. The estimate of the combined runoff is around 100×10^6 m^3 y^{-1}. Al Lith, Itwad, Khulah, Qanunah, Ahsihah, Shaqah and Al-Shamiyah are somewhat smaller wadis and their combined runoff is estimated as 50–100×10^6 m^3 y^{-1}. The flood runoff of the wadis in this province has been estimated as 39.8 m^3 s^{-1}. Of this amount 27.0 m^3 s^{-1} occurs in the area south of Jeddah, 4.6 m^3 s^{-1} in the surroundings of Jeddah and 8.2 m^3 s^{-1} in the area north of Jeddah.

Again, according to the ECWA Report (1981), SOGREAH estimated the total runoff from the Red Sea Province as $1,610 \times 10^6$ m^3 y^{-1}. Peak floods have been estimated at 10,000 m^3 s^{-1} for Wadi Biyash and Hail, and for the other wadis between 4,000 and 5,000 m^3 s^{-1}, though generally do not last more than a few hours or a few days, and disappear before reaching the sea.

7.4.2.7 *Rub al-Khali (Province (VII))*

There is hardly any surface flow to mention as all rainwater practically disappears in the desert.

7.4.2.8 *Arabian Shield (Province (VIII))*

There is a limited information about surface water in this province.

Table 7, Appendix II, includes the names of major wadis in Saudi Arabia, surface area of their respective drainage basins, average annual rainfall and measured or estimated annual runoff. From these data the annual values of runoff coefficient have been computed for some wadis in the southwest and presented in Table 4. The average value of 11.5% for the annual runoff coefficient is certainly high compared to those of most wadis in Oman already presented in sec. 7.3.

Table 4. Annual runoff coefficientfor some wadis in southeast Saudi Arabia

Name of wadi	Surface area, km^2	Annual rainfall		Annual runoff		Runoff coefficient,%
		depth, mm	volume, 106 m3	depth, mm	volume, 10^6 m^3	
Sabya	843	275	232	35.6	30.1	13
Khulab	869	500	434	72.5	60.8	14
Dhamad	1084	346	375	34.1	37.5	10
jizan	1100	550	605	72.7	78.7	13
Liyyah	1293	266	344	29.4	37.8	11
Itwad	1553	163	253	33.5	54.1	21.4
Baysh	5164	350	1807	45.5	235.0	13
Najran	4800	250	1200	18.8	90.0	7.5
Aveage						11.5

The above sub-sections show clearly that surface water resources in Saudi Arabia are limited. There is no perennial streamflow in the proper sense. Floods occur but they are local and flow for just a few kilometers and then disappear into the valley fills and fractured rocks. Investigations have shown that runoff is generally subject to wide variation in both quantity and quality. Estimate of surface floods as given by different authorities are in the order of $2.1–2.2 \times 10^9$ m^3 y^{-1}. As an example to mention, specification of the said estimate, as given by Ukayli & Hussein (1988), is included in Table 5.

7.4.2.9 Hydrologic modeling

The inadequacy of the water resources data has urged the authorities in Saudi Arabia to request the assistance of highly specialised professionals to carry out a variety of research works, especially in wadi basins in the Red Sea Province. The main target eventually is to reach reliable estimates of surface runoff including flood peaks, time of occurrence and quantity. The researches conducted include analytical and model studies of rainstorms and rainfall characteristics.

- One of the early studies was that performed by Wan (1976) on the characteristics of point rainfall in Saudi Arabia. Some of the findings obtained from that study have been highlighted in sub-section 4.3.3 of Chapter 4 "Analysis of Precipitation Data".

Table 5. Estimated runoff from certain wadis in saudi Arabia (Source: Ukayli & Hussain, 1988)

Region[x]	Surface area, km^2	Main Wadis	Estimated runoff, 10^6 m^3 y^{-1}
Red Sea Coastal Belt	241,600	Jizan, Dhmad, Baysh, Hali, Yiba, Qanunah and Al-Lith	1,265
Asir-Wadi ad-Dwasir	180,000	Turabah, Ranya, Bishah, Dawasir, and Tathlieth	330
Asir-Wadi Najran	38,400	Najran	135
Wadi Birka-Nisah-Sahaba	162,300	Hatwa, Nisah, Hanifa and Sahaba	100
North Tuwayq	152,800	Sudair, Meshgar/Namil	95
South Tuwayq	48,300	Jadwal	55
Taif-Fadat al-Misbah	43,200	Wajj, Liyyah, Aqiq	25
W. ar-Rimah al-Batin	174,000		(–)
Ar-Rub al-Khali	756, 100*	Not known	(–)
An-Nafud	161,000		20
Al-Jawf	192,300	As-Sirhan	(–)
Total	2,150,000		2,025

Explanation

[x]*Some of the regions mentioned in Table 5 are somewhat different from those listed in Table 7, Appendix II. * This figure represents the difference between the surface area of Saudi Arabia and the sum of all other areas combined.*

– The spatial-temporal characteristics of rainfall in southwest Saudi Arabia were analysed, modelled and their long-term performance predicted by Wheater et al. (1991(a) and 1991(b)). Five study basins were chosen: Al-Lith, Yiba and Liyyah, draining towards the Red Sea, and Tabalah and Hawabnah, draining from the mountains to the interior towards Ar-Rub el-Khali. Figure 17 is a map showing the locations of the five basins.

Rainfall analysis has shown that rainfall occurs regionally in all months, but the highest frequency of occurrence can be classified in three main periods: spring (March-April), autumn (September) and winter (November-December) with individual wadis displaying some or all of these wet seasons.

Overall, the rainfall shows the characteristics associated with convective rainfall, namely short duration, high intensities and a high degree of spatial variability.

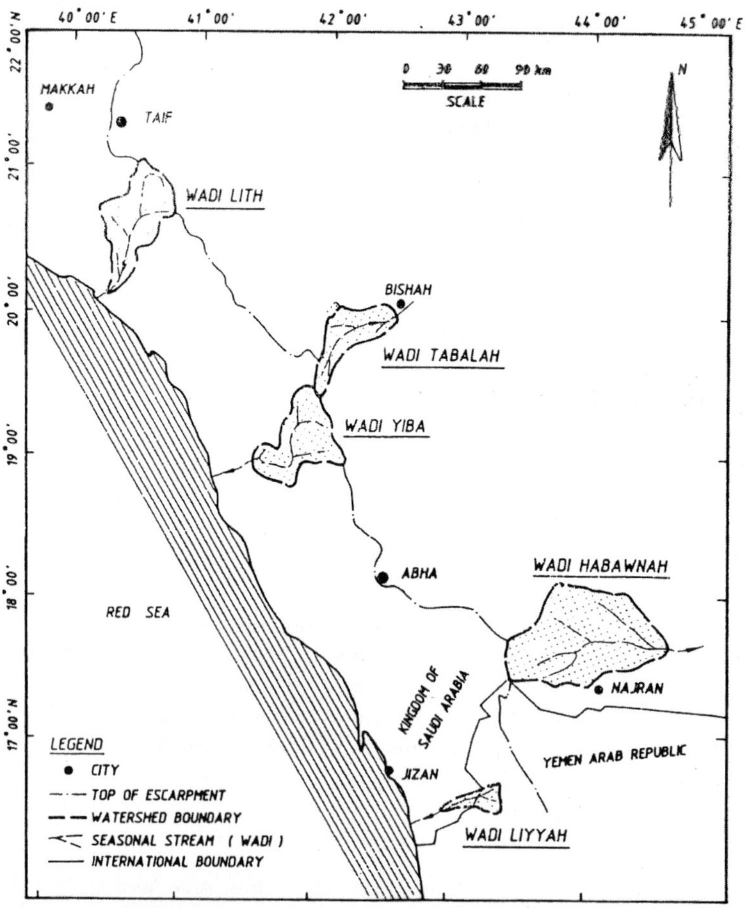

Figure 17. Location map of study basins in southwest Saudi Arabia (Wheater et al., 1991(a))

These characteristics led to the adoption of a multivariate approach, strongly based on observed rainfall characteristics.

The conditional probability of rainfall for a given hour at station B given rainfall at station A as a function of gauge separation at Wadi Yiba was calculated. The overall trend is decreasing probability with increasing gauge separation. The mean value of the relative probability for a stretch of 90 km as maximum gauge separation was 0.12, with only 5% of values greater than 0.3 and a significant number of gauges giving close to zero probability. As a consequence, overall correlation analysis was of limited value, but the population of simultaneous occurrences was analysed for cross correlation. The mean correlation coefficient was 0.01, and the scatter of results substantial. These figures are not in full agreement with the results obtained by Hofman & Rambow (1995) for Wadi Al-Jizi Basin, Oman, sub-section 7.3.2.1. The study of Wheater et al. (1991(a)) led to the conclusion that, at least as a first approximation, the data could be considered to be spatially independent. In the Oman study the mean values of the probability and correlation coefficient for winter rainfall in Wadi Al-Jizi are high. As no seasonal figures were given by Wheater et al. (1991(a)) comparison between the results obtained from southwest Saudi Arabia and northern Oman is limited.

Further analysis of rainfall showed that 37% of catchment raindays for Wadi Yiba arise as a result of rainfall at a single site. It became evident that once a convective rainfall has been triggered, the point rainfall properties such as mean hourly rainfall depth, mean duration of rainfall, mean number of events per day and mean event inter-arrival time on any given days are independent of location.

Based on 10 simulations of 2-y periods, it was demonstrated that the Markov model for catchment rainday occurrence successfully reproduces the observed seasonality of monthly occurrence, distribution of rainday inter-arrival times and other features such as the frequency of consecutive catchment raindays (Figures 18(a) and 18(b)). With this end similar analysis was carried out for the remaining four wadis.

In the second part of the study (1991(b)), an analysis was made of the regional variability of annual and seasonal characteristics of the available long-term rainfall records to provide the context for investigation of long-term model performance.

Well-defined relationships were observed for the individual study basins, linking the occurrence frequency of raindays with gauge elevation. A distinct difference of regime occurs between the Red Sea coastal catchments of Lith, Yiba and Liyyah, and the two inland catchments of Tabalah and Hawabnah. The coastal wadis appear to show a consistent geographical variation from northwest (Lith) to southeast (Liyyah). Likewise, Tabalah and Hawabnah, although similar in the regional context, also show a distinct difference between the two wadis.

The model results were compared with long-term rainfall data which could not be found except at gauges outside the boundaries of the study basins. The overall conclusion is that the comparison of model results with long-term data must be made with caution; the results can sensibly be interpreted in the context of the pattern of regional elevation relationships.

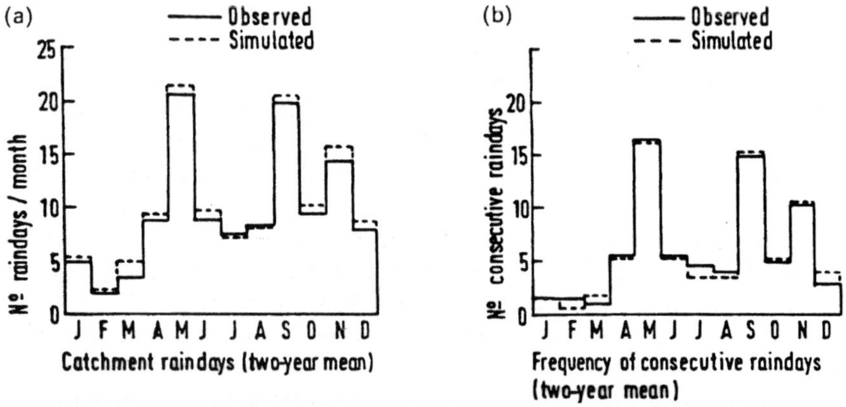

Figure 18. Two observed and simulated point rainfall characteristics, Wadi Yiba (Wheater et al., 1991(a))

A consistent period of record (1970-85) was adopted to define long-term mean annual rainfall. The 1984-85 data for all five basins are significantly lower than the long-term data. In the light of the discrepancy between the long-term and short-term annual rainfalls, it was considered necessary to adjust the model to accommodate the long-term results. This objective necessitated further examination of the link between differences in short-and long-term rainday occurrence, frequency and differences in annual rainfall.

The 15-y (1970-85) and the 2-y (1984-85) data for those gauges in regional proximity of the five basins under study gave depth-weighted ratios of 1.43 for Yiba, 1.45 for Al-Lith, 1.53 for Liyyah, 1.65 for Tabalah and 1.45 for Hawabnah. These values indicated general regional consistency of annual variability. These were applied to modify the model structure of the basins by adjusting the respective Markov chain parameters. The resulting increase in synthetic average annual rainfall ranged from 1.50 to 1.65.

An important property of the model is its ability to reproduce extreme value statistics and hence a comparison has been made of hourly and daily extreme values based on a sample of six long-term gauges in the vicinity of Wadis Lith and Hawabnah to represent the coastal and inland basins, respectively. This was done using the long-term (12-20 y) and 50-y simulations of hourly rainfall. An EV I distribution was fitted to the annual maximum series using the method of moments. The average values of the ratio of 100-y to 5-y for durations of 1-h and 1-y ranged from 1.83 to 2.06 for the long-term gauges and from 1.68 to 1.82 for simulated data.

– Wheater and Brown (1989) investigated the limitations of design hydrographs in southwest Saudi Arabia. They used for that purpose Wadi Ghat (sometimes written as Khat). The catchment of this wadi, 597 km^2, is a major subcatchment of Wadi Yiba. The Ghat has two principal tributaries (Wadi Ghat and Wadi A'qas), which approximately bisect the catchment. As the flows were ephemeral, baseflow separation for unit hydrograph analysis was not required. Eleven storms were selected for detailed analysis from the 1984 and 1985 data. The rainfall-runoff event characteristics showed the peak flow to range from 36.5 $m^3\ s^{-1}$ to 205.7 $m^3\ s^{-1}$ and the runoff depth from 0.28 mm to 2.41 mm. The runoff coefficient for those 11 events ranged from a minimum of 5.9% to 79.8%. The median value of the coefficient was 11.95% and the arithmetic mean 21.2%.

One millimeter, 10-minute unit hydrographs were derived for each of the events individually. The unit hydrographs thus obtained has proven that the unit hydrograph was generally an excellent model. The coefficient of efficiency of the model ranged from 0.954 to 0.998. However, the individual unit hydrographs showed high variability. For example, peak discharge varied from 57 to 358 $m^3\ s^{-1}$ and rise time from 6 to 55 min. The unit hydrographs of all events except for one event are superimposed in Figure 19.

An overall unit hydrograph was obtained from the combined set of eleven events (Figure 19) and tested in simulation for the individual storms (assuming observed rainfall loss). The results were described as poor. The high degree of variability in the unit hydrograph shape and runoff coefficient is entirely consistent with the very high spatial variability of rainfall observed in the region. Runoff generation is thus extremely localized and, hence, the time of travel of the flood to a gauging station, and the opportunity for transmission loss, is dependent on the spatial location of rainfall.

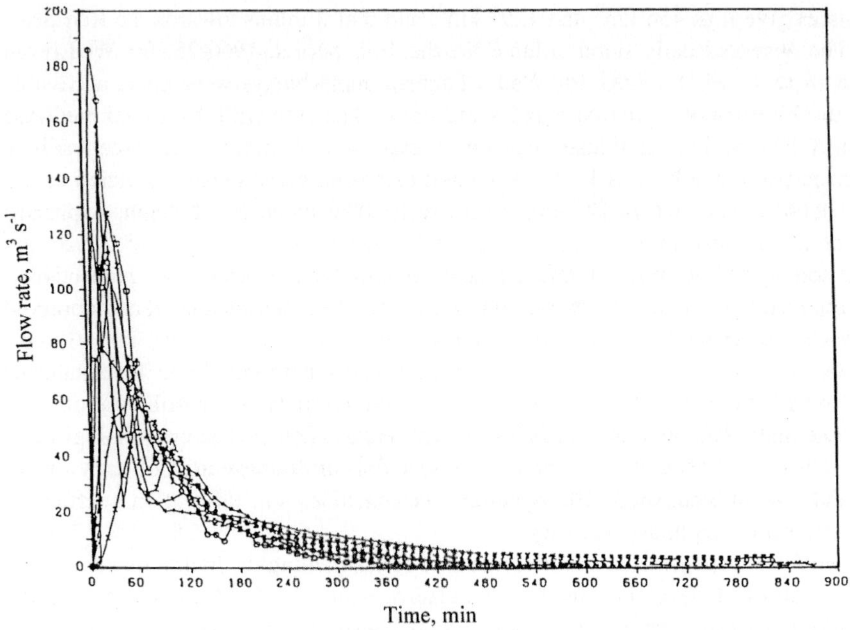

Figure 19. Unit hydrographs – all events, excluding the event of 1 may 1985 (Wheater and Brown, 1989)

– In the same year, i.e. 1989, Allam developed a physically-based flood volume distribution model. In that model the mechanics of the infiltration process, soil moisture, ponding time were taken into account. Approximately 80% or more of the watersheds in Saudi Arabia, particularly in western and southwestern regions, are mountainous. As Philip's expression does not apply for the infiltration process occurring in such a situation, a runoff coefficient for mountains was assumed in that study. Finally, The probability distribution of the flood volume was obtained. The input data to the model were:
  - Climatic data: Mean rainfall intensity, mean rainfall duration, mean annual evaporation rate, and mean value of inter-time between storms.
  - Soil data: Saturated hydraulic conductivity, porosity and pore size distribution index.
  - Vegetation data: Density of vegetation canopy, and ratio of potential rates of transpiration and soil surface evaporation.
  - Watershed Parameters: Total area of watershed, size of alluvial area in watershed, depth to water table, and runoff coefficient of the mountainous part of the basin.

The sensitivity of the flood volume distribution to each of the above listed variables was tested and reported (Allam et al. 1989). Three gauged wadis in southwest Saudi Arabia, namely W. Liyyah, W. Turabah and W. Khulays were used as case study. It is quite strange that the surface area of the watershed of Liyyah is given here as 174 km^2 and that it flows towards the east and drains into the desert whereas other sources give it as 456 km^2 and 1293 km^2, and that it drains towards the Red Sea!!

The observed daily flood volume for the 10-y period, 1966-75, for W. Liyyah and 14-y period, 1967-90, for Wadis Turabah and Khulays were made accessible by the Ministry of Agriculture and Water for comparison with the model predicted runoff. The mean annual historical runoff, expressed in mm, for the three wadis in their respective order was 17.23, 16.88 and 12.55 compared to the calculated values of 18.66, 17.11 and 14.10 mm, respectively. The mean flood depth, expressed in mm, for those three basins in the same order was 0.23, 0.36 and 0.1.26 and the corresponding model predicted depth was 0.24, 0.37 and 1.41, respectively. These results show that the maximum discrepancy between calculated and observed runoffs was about 12% for Wadi Khulays.

– The procedure known as 'Hydraulically-based Geomorphologic Instantaneous Unit Hydrograph (HGIUH)' is a tool for predicting runoff with limitations and potentials. The input data of HGIUH may be classified into geomorphologic data including number of watershed streams, length of drainage area of each stream and size of watershed. The hydraulic data includes soil infiltration coefficient, streamflow depth and velocity.

Allam (1990) suggested a simple procedure for estimating the hydraulic parameters of the HGIUH developed by Diaz-Granados et al. (1986). A simple infiltration expression approximated by an infiltration coefficient multiplied by the stream discharge was introduced in the original model. This procedure was applied to three gauged watersheds in southwest Saudi Arabia, namely Wadi Khat (597 km^2)

mentioned earlier in connection with the work of Wheater and Brown (1989), W. Al-Jawf (305 km^2) and W. Midhnab (20 km^2). It should be remembered that the first two wadis are tributaries of Wadi Yiba.

Wadi khat and Wadi Jawf receive approximately the same amount of precipitation, which is in the order of 400 mm y^{-1}. Rainfall data that were available then covered the 3-y period, 1984-86. Wadi Midhnab lies in the northern region of Saudi Arabia and receives annually rainfall in the order of 150 mm. The rainfall-runoff data, which were then available, were for the period 1982-84. Additionally, the runoff coefficient for Wadis Khat and Jawf was computed and found to be 10% and 3% respectively. The three study basins have a mountainous nature, with mountains covering over 80% of the basins of Wadi Khat and Wadi Jawf. The streams represent the only alluvial portion. In Wadi Midhnab, the alluvium covers 60% of the wadi drainage basins.

Out of the available storms, six storms for W. Khat, four for W. Jawf and 5 for W. Midhnab were selected for the HGIUH applications. The selection criteria are: rainstorm should be covering the whole surface of the watershed, rainfall-runoff events must show an obvious relationship, and runoff depth must be at least 0.1 mm. Simulation runs were carried out for the selected storms. Model predicted peak discharges and times to peak were compared to the observed ones, and simulation errors computed. The average absolute errors in estimating the peak discharge were 4.3%, 35.2% and 35.7% for Wadis Khat, Al-Jawf and Midhnab respectively. The average absolute error in simulating the time to be for the three wadis in their respective order was 44.9%, 61.8% and 27%.

All simulation errors, except that of the peak discharge of the hydrograph of Wadi Khat, are relatively large. However, the simulated runoff volumes seem to be the same as the observed ones. This result suggests that the proposed approach for computing the infiltration coefficient is warranted. According to Allam (1990), the large error in the peak discharge is mainly due to the incorporated linear infiltration expression as the effective rainfall is assumed to start from the beginning of the storm with duration equal to the storm duration. The incorporated linear infiltration expression is probably also the main reason behind the differences in the shape between the observed and simulated hydrographs as well as the simulation errors in the peak discharge. Last but not least the condition of the spatial rainfall distribution in the study areas.

Sorman & Abdulrazzak (1993) investigated the transmission loss (cumulative infiltration) from an ephemeral stream in Saudi Arabia. They suggested simplified procedures for the estimation of infiltration-recharge amounts, requiring only limited information on flood flow and channel characteristics. The Tabalah Basin, located on the eastern side of the escarpment of the mountain range that runs parallel to the Red Sea (Figure 17) was chosen for that investigation. The drainage area of the basin is $1,270$ km^2 and the elevation ranges from 1,200 to 2,400 m a.m.s.l. The basin was, by the time of investigation, equipped with a network of recording rainfall stations, climate station, runoff measuring stations and observation wells.

The basic equations describing the transmission losses are:

$$(7) \qquad TL = V_{UR} - V_{DR} + C_{TR}V_{TP} = V_{UR} - V_{DR} + [V_{TP}/(V_{TS} + V_{TP})V_{TP}]$$

and

$$(8) \qquad V_{TS} = V_{US}\frac{A_T}{A_U}$$

where,
- TL = cumulative volume of transmission loss,
- V_{UR} = cumulative upstream inflow volume,
- V_{DR} = cumulative downstream outflow volume,
- C_{TR} = coefficient of tributary runoff,
- V_{TP} = volume of tributary precipitation,
- V_{TS} = volume of tributary storage abstraction capacity,
- V_{US} = volume of upstream storage capacity,
- A_U = upstream basin area, and
- A_T = tributary basin area.

Sorman & Abdulrazzak (1993) tabulated the values of V_{UP}, V_{UR}, V_{TP}, V_{TR}, V_{DR} and TL they obtained from investigating 27 events in the period 1985-87. They classified the runoff information into four groups based on runoff volumes recorded at the upstream and downstream sections. This grouping helped them to interpret the variation in soil moisture replenishment, recharge and transmission loss. The author used the precipitation-runoff data to compute the runoff coefficient event-wise and for the 3-y period as a whole. The events with zero precipitation and zero runoff were not taken into account. Figures 20(a) and 20(b) show the runoff coefficient versus the precipitation volume for the upstream reach and the tributary respectively. Despite the inconsistency in the relationship between the precipitation and runoff coefficient, Figures 20 show a general tendency of decreasing runoff coefficient with increasing precipitation. The long-term magnitude of the coefficient for the drainage basin under study is 8.8% for the upstream reach and 10.1% for the tributary.

Further analysis of the available information has shown that the hydrograph characteristics strongly control the transmission loss and groundwater recharge. The hydrograph determines the width, depth and time of inundation of the wadi channel. The relationship between the hydrograph, channel and alluvial characteristics, and transmission loss-recharge process was examined through regression analysis methods using data from the 27 observed events. This led to the development of two regression relations for *TL*; one with the product of maximum stage height and flow duration, and the other with the product of the maximum flow width and the flow duration.

Groundwater recharge is dependent on the magnitude of water lost from the wadi bed and soil profile characteristics. Three regression relations were developed for estimating the recharge using the antecedent dry index (*ANTEC*) and/or depth to

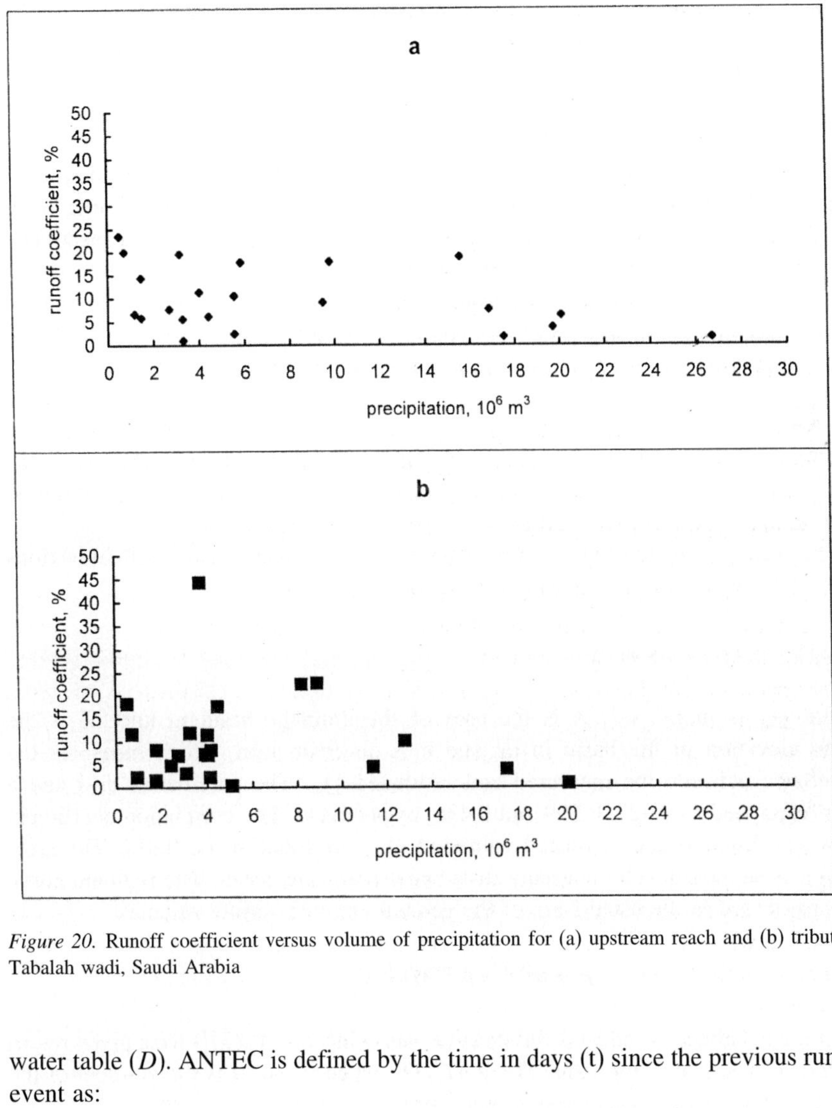

Figure 20. Runoff coefficient versus volume of precipitation for (a) upstream reach and (b) tributary, Tabalah wadi, Saudi Arabia

water table (D). ANTEC is defined by the time in days (t) since the previous runoff event as:

$$(9) \qquad ANTEC = 1.0 - 0.9^{t}$$

Regression relations for estimating transmission loss and groundwater recharge as obtained from this study should be used cautiously and when limited information on stage height and runoff duration, and antecedent conditions and depth to the water table is available.

– The author collected from the current literature a number of sets of measured peak discharges, Q_p, the flood volumes, V, and their respective durations, T.

The sets he used were taken from Yemen, Saudi Arabia, Oman and Iran. An empirical, easy to use relationship was developed and can be expressed as:

$$(10) \qquad V = CQ_p^a T^b$$

The values of the parameters C, a, and b that provide the best fit to the data used were found to be 13.6×10^{-4}, 0.865 and 1.0 respectively. One should bear in mind that the quantities V, Q_p and T are expressed in 10^6 m^3, m3 s^{-1} and h respectively. The coefficient of correlation is in the order of 0.98.

– The flood peak discharge corresponding to a given return period can be estimated using the appropriate frequency distribution function(s). Nouh (1988) on prediction of flood frequency of wadis in the southwest of Saudi Arabia suggested the use of regional frequency curves. These curves represent the frequency distribution of $(Q_p)T_r/Q_m$, where T_r is the return period and Q_m the arithmetic mean of the annual maximum instantaneous flows. This approach associates a return period T_r with $(Q_p)T_r/Q_m$ and the relation between them is assumed to be valid for all drainage basins in one region.

The mean flood flow, Q_m, can be estimated for drainage basins with elevations at or above 8 m a.m.s.l. using the expression

$$(11) \qquad Q_m = cA^d Z^e + \varepsilon$$

where Q_m is in m^3 s^{-1}, A is the area of the drainage basin in km^2, Z is the mean elevation of the basin in m, and ε is an error term, which measures the difference between the measured and estimated Q_m. The parameters c, d and e were estimated as 0.322, 0.599 and 0.436, respectively. The correlation coefficient between observed and estimated values of Q_m was found to be 0.812. The error term ε is supposed to be normally distributed with zero mean. The regional curve was expressed in the usual form of the general extreme-variate value as:

$$(12) \qquad (Q_p)T_r/Q_m = \mu + \alpha[(1 - e^{-ky})/k]$$

where, y is Gumbel's reduced variate ($y = -\ln - \ln[(T_r - 1)/T_r]$) for a given return period, T_r years, μ is the value of $(Q_p)T_r/Q_m$ when y = 0, α is the gradient of the curve at $y = 0$ and k is a curvature parameter.

– Sorman et al. (1995) examined the issue of flood peak estimation using satellite information on Wadi Itwad in southwestern Saudi Arabia as a case study. Information on lithology, drainage pattern, frequency and spatial distribution of the fractures were obtained from processed Landsat Thematic mapper data in coloured form.

Wadi Itwad ($1,350$ km^2), as mentioned in the previous sub-sections, is one of the major wadis in southwestern Saudi Arabia, which drain its water towards the Red Sea. Wadi Hali and Wadi Baysh surround it from the northwest and the southeast direction respectively. The study covered only the 14 km reach of the wadi channel and the small catchments, which drain between km 19 and km 33 measured from the town of al-Darb.

The peak discharge was estimated using the parameters obtained from the information collected by the Landsat-TM data and incorporated in the model known as TR-55. The topographic parameters include the slope and channel length and the geomorphologic parameters include the fracture length and density and node intersection. The input parameters of model TR-55 are the storm type, curve number and time of concentration. The rational formula, with runoff coefficient varying between 0.1 and 0.9 depending on the area cover and land use, was applied as well to the small wadis. Sorman et al. (1995) concluded that further studies of the kind described above are required to assess objective criteria to select the methodology and model application so that a better estimate for spatially variable data can be from remote sensing data in order to couple them with the models after several model calibration studies. The real objective is "trying to simulate complex rainfall-runoff processes with a simple hydrologic modeling approach".

7.4.2.10 Surface water quality

As most of the surface water is floodwater brought by torrential rains, there is no reason to doubt the goodness of its quality. Analysis of water samples for detecting its quality is generally confined to groundwater. This will be described in Chapter 9 "Groundwater Resources".

7.5. WADI BASINS AND WADI HYDROLOGY IN THE UNITED ARAB EMIRATES

7.5.1 Drainage basins in the United Arab Emirates (UAE)

Groundwater and desalinated water are the primary and secondary sources of water in the UAE respectively. As such, surface water discharged by wadis is not equally important. With an areal average rainfall of about 102 mm y^{-1}, the precipitation in UAE ranks after both Yemen and Kuwait, 135-140 mm y^{-1}. It should be noted that annual rainfall is widely variable. In a period of 15 years it varied from less about 12 mm in 1966–67 to 293 mm in 1981–82. In that last year the annual rainfall reached more than 450 mm in some locations.

Like many more arid lands, the annual rainfall is accompanied by thunderstorms causing the fall of heavy rain in a short period of time. This can lead to severe floods and intensive soil erosion.

Referring to the map in Figure 20 (Chapter 2), the United Arab Emirates has been subdivided into a number of hydrogeological units. This subdivision can also apply to surface water resources of the country as follows:

7.5.1.1 The Central Mountains (Province (1))

This province extends southwards from the Dibba Fault Line and forms the watershed that separates the drainage systems to the Gulf of Oman and to the

Arabian Gulf. The highest elevation of the Central Mountains barely exceeds 1,100 m. This province receives the highest precipitation that falls over the country. Several wadis have been formed as a result of fractures and faults that took place in this area long ago. Examples of these are Wadi Ham, Wadi Dibba, Wadi Al-Sadar, Wadi Al-Siji. The wadis intersecting the western slopes flow northwestwards and those intersecting the eastern slopes flow eastwards.

7.5.1.2 The Limestone Mountains (Province (2))

This province extends northwards from the Dibba Fault Line into the Musandam Peninsula of Oman up to Hurmoz Strait where evidences of strong marine erosion can be observed. The mountain tops reach 2,000 m a.m.s.l. Several wadis draining into the interior such as Wadi Bieh and Wadi Hgalielah intersect these tops.

7.5.1.3 The Batinah Coastal Plain (Province (3))

It lies between the Central Mountains (Province 1) and the Gulf of Oman. This province is an extension of the Batinah Plain of Oman. This province too is dissected by Wadi Dibba and Wadi Ham.

7.5.1.4 The Central Gravel Plain (Province (4))

It lies between the Central Mountains and the western hilly area where Jabal Faiyah, Jabal Aqabat and Jabal Buhats exist. The plain extends from opposite Dibba Gap to Wadi Maiha. It is about 171 km long and has a maximum width of 29 km. The plain is covered with alluvial deposits of increasing fineness towards the west. It also includes smaller plains like al-Zayd, Gharif, Madham, Goa, etc.

7.5.1.5 Buraymi Oasis (Province (5))

It comprises the oasis proper and the Jawa Plain to the west. The oasis lies in a region where the foothill folds west of the Central Mountains have been subjected to differential erosion giving chances for possible surface and groundwater occurrences.

7.5.1.6 Ras al-Khaimah-Jiri-Plain (Province (6))

This province is located between the Limestone Mountains and the Arabian Gulf. It extends northwards from opposite the Dibba Gap to where the mountains reach the seacoast in high cliffs.

7.5.1.7 Central Desert Foreland (Province (7))

This is the hinterland of the Dubai-Umm Al-Qaiwain coast with sand dunes through which some major wadis flow towards the sea draining the surface runoff from the Central Gravel Plain. In this province, the coastline comprises sabkhas, lagoons and tidal flat areas. The principal wadis in this province are Wadi Lamha, which drains the runoff flowing from the Al-Zayd Plain and discharges it near Al-Mudakak between Ras Al-Khaimah and Umal-Qaiwain in case of heavy rainstorms. Another wadi, Wadi al-Maliha, carries the runoff flowing from al-Madham area for a certain distance before it gets dry or seeps into the ground.

7.5.1.8 Dubai-Buraymi-Abu-Dhabi Plain (Province (8))

It is mainly a sand dune area of increasing size and diminishing vegetation.

7.5.1.9 The Southern Desert (Province (9))

It comprises sand dunes inland and major sebkhas along the coast westwards from Abu-Dhabi mainland.

7.5.2 Surface water hydrology in U.A.E.

It is claimed that there are 27 major wadis in the U.A.E. The central area drains its water to the west; the southern area and the east coasts of the country drain to the east; the western slopes drain to the north and the rest drain their waters to the south. The principal drainage areas are shown on the map in Figure 21. The Eastern Drainage (East Coast) covers the coastal area immediately north and south

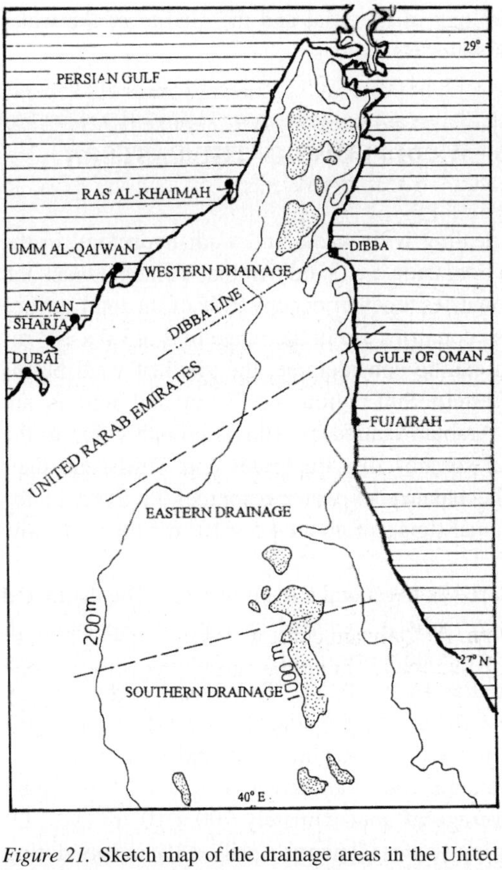

Figure 21. Sketch map of the drainage areas in the United Arab Emirates

of Fujairah, Wadi Ham, Wadi farfar and Wadi Farah. The southern area, which drains southward, is made up of two wadi catchmnts, the Wadi Ghor and the Wadi Hatta and their tributaries.

The average annual volume of water carried by wadis in the northern Emirates was estimated as 110×10^6 m^3. A later estimate of the volume of water carried by all wadis is about 150×10^6 m^3 y^{-1}. The author collected from different sources (e.g. UAE Country Report on Water Resources, 1986, ECWA Report, 1981, Report on Traditional Water Irrigation Systems in the U.A.E., 1982, etc.) approximate figures of the runoff carried by wadis in the Emirates. These figures are included in Table 8, Appendix II.

Most of the surface water in UAE is discharged into the sea, seeps through the ground and joins the aquifers and collected in aflaj (plural of falaj). The falaj is an underground conduit reaching the water table and collecting groundwater in its higher part and conveying it to the lower part. This is made possible by making the slope of the falaj smaller than the slope of the ground surface. The United Arab Emirates used to have some 150 falaj in the past. Currently this number has fallen to 30% of the old figure, i.e. 45 used for irrigating small farms and agricultural schemes. We shall discuss the hydrology of the aflaj and the quality of the water they carry in the Chapter on Groundwater Resources.

7.6. A SHORT ACCOUNT OF WADI FLOWS IN THE EASTERN SUB-REGION

The preceding sections have been dealing with wadis and wadi hydrology in the largest four countries of the Arabian Peninsula, i.e. Saudi Arabia, Yemen, Oman and United Arab Emirates. These four countries together occupy 99% of the total surface of the Peninsula. Wadi flows in these countries are in the range of $4.5\text{-}5.0 \times 10^9$ m^3

Despite their annual flow, which can be considerable, the yield of wadi basins in the two largest states of the Eastern Sub-region, i.e. Syria and Iraq is not better known than the wadis in the Arabian Peninsula. This is probably due to the availability of water from perennial streams like Euphrates and Tigris and their tributaries. Rough calculation of the balance of water resources in Syria in the period 1979–83 gives an average annual figure of about 4.5×10^9 m^3 for wadi flow (Country Report of Syria, 1986).

The Yarmouk in Jordan is considered as perennial stream or river. The estimated surface water provided by the major wadis is included in Table 9, Appendix II. From this Table it can be seen that the total storm (surface) runoff from the wadi basins in an average year is estimated as 140×10^6 m^3. The figures corresponding to dry and wet years are 50 and 280×10^6 m^3. The average total yield (storm runoff and baseflow) is close to 350×10^6 m^3 (ECWA Report, 1981 and Biswas, 1994).

The principal watercourse in Lebanon is the Litani River. The flow of the Litani varies from year to year with an average of approximately 700×10^6 m^3 y^{-1}. Of the annual value 60–65% occurs during winter (January-April); 15% occurs during May and June; 12% from July through October; and 10% during November and

December. From these figures it is clear that the Litani and so other streams in Lebanon cannot be regarded as wadis, instead they are perennial streams.

Dense networks of wadis dissect the ground surface of Iraq. Some of the wadis west of the Euphrates traverse the Western Desert in Iraq and some others extend to Syria, Jordan or Saudi Arabia. Several of these wadis disappear in the desert like the Hamir and Ar'ar. Others, like Wadi Amij, discharge their water into the mud flats or the salt lake (Bahr el-Milh) between Habbaniyah and Karbala like the Ubbayid, Al-Ghudaf, Al-Furukh and Al-Arjawi wadis do. Wadi Ghazilah drains its water into the Tharthar Depression, and Wadis Hawran and Khubbaz are amongst the wadis draining into the Euphrates directly.

The Mesopotamian Plain bounded by the Euphrates on the west and Tigris on the east is crowded by many wadis. Some of these wadis flow from east to west to feed the Euphrates while other wadis flow from west to east to feed the Tigris. The major wadi in this plain, Wadi ath-Tharthar runs from north to south and serves as major feeder to Tharthar Deprssion (Mileh Tharthar).

The third group of wadis spring from the high grounds along the Iraqi-Iranian border in the east of Iraq. The N.Wadi is an example of wadis which join the Tigris directly, while the Awhad discharges into Diyala, a tributary of the Tigris. Other wadis end into the mud flats (Ahwar) and swamps filling the area from Al-Kut down south to Basra, particularly west of the Tigris in its northern part, and the triangle between the two rivers north of Basra up to the line connecting An Nasiriyah on the Euphrates and Qal'at Salih on the Tigris.

It is unfortunate that there is hardly any data available on the hydrology of the drainage basins of the wadis of any of these groups. Assuming that the total surface area covered by wadi catchments is in the order of 200, 000 km^2, receiving an average annual precipitation depth of 230 mm (average rainfall for Iraq) and an average surface runoff coefficient of 5%, the average surface flow of the wadis becomes $2.3 \times 10^9\ m^3\ y^{-1}$. This estimate, according to the author's personal experience working in Iraq, is a modest one.

The above-mentioned figures show that 7-8 $\times 10^9\ m^3\ y^{-1}$ will be a fair estimate for the surface runoff of wadis in the eastern part of the Arab Region.

7.7. A SHORT ACCOUNT OF WADI FLOWS IN THE WESTERN SUBREGION

7.7.1 General remark

The names of the streams flowing in the Western Sub-region, as can be seen from geographic maps or given in French literature, are preceded by the French word Oued, which is always translated to wadi in English text. The word wadi, which is originally an Arabic word, refers to an intermittent or ephemeral stream. Oppositely, a river is a stream that never dries up or runs empty at any time of the year.

Some of the Oudes that carry the runoff flowing from the wet or relatively wet are as in North Africa and never dry up the whole year round cannot be considered as wadis. This is the case of streams in northern Tunisia like Mejerdah and Zeroud, and Sebou, Oum er Rbia or Moulouya in Morocco, and Chéliff in Algeria. The figures listed in Table 3, Appendix II, show, as an example, that the mean monthly discharges of (Oued) Mejerdah at Sloughia, station No. 23, never fell to zero during the stated period of record. The same can be said about (Oued) Sebou at Azib Soltane in Morocco and many more examples.

The drainage basin of (Oued) Mejerdah is 23,700 km^2 in area. It receives its water from the richest tributaries in Tunisia and carries a volume of water of about 1×10^9 m^3 in a normal year. Remarkable enough is that its flood carrying capacity reached 3,500 m^3 s^{-1} during the flood of March 1973. In that extreme flood, the Mejerdah discharged 943×10^6 m^3 in six days only. Likewise, the (Oued) Zeroud carried in the same year a volume of water $2,500 \times 10^6$ m^3 in just two months or about 160 times the average quantity; the average annual volume being 95×10^6 m^3. By reason of being perennially flowing streams, hydrology of Oudes Like Sebou, Morocco, Cheliff, Algeria and Zeroud, Tunisia, have been reviewed and discussed in Chapter 6 dealing with Rivers.

Wadis in the true sense of the word, i.e. intermittent and ephemeral streams, can be found in several parts of North Africa like central and southern Tunisia, east and south of Morocco and Algeria. They can be found also in different locations in Mauritania.

7.7.2 Summary of wadi flows by country

7.7.2.1 Tunisia

To get a rough idea about the runoff supplied by the wadis in Tunisia, it is claimed that the annual yield of the northern, central and southern parts are about 2.1×10^9, 0.30×10^9 and 0.20×10^9 m^3 respectively. The two last figures are supposed to be brought exclusively by wadis. Assuming arbitrarily that 30% of the annual yield of the northern part is brought by wadis, the share of perennially flowing streams in the surface water resources in Tunisia can be estimated as 1.5×10^9 m^3. The remaining 1.1×10^9 m^3 y^{-1} is the approximate share of wadis in the surface water resources.

7.7.2.2 Morocco

Most of the Moroccan streams draining their flows into the Mediterranean Sea, especially those having their drainage basins in the west, yield considerable surface runoff. Likewise, the streams springing from the Haut Atlas (High Atlas) and Moyen Atlas (Middle Atlas), often referred to as North Atlantic basins, are relatively rich in their supply. Examples of these streams are the Sebou, Oum Rbia, Grou, Bou Regreg and El-Abid. The Atlas Mountains act as a huge reservoir sustaining the flow of these rivers in the low-flow season.

The average annual flow of the eastern drainage basins, with the exception of Moulouya, such as Kies, Eslay and other wadis has been estimated as 0.164×10^9 m^3. The average surface runoff of the South Atlas eastern basins, such as those of Wadis Ziz and Gheris, Ghair and Dr'aa, has been estimated as 1.465×10^9 m^3 y^{-1}, and from the South Atlas western basins like Souss, Massa and other wadis as 0.570×10^9 m^3 y^{-1}. Runoff from the Desert basins like Tarfayah, As-Saquia al-Hamra and Wad el Dahab are practically negligible (Country Report of Morocco, 1986). All these wadis and many more run empty during certain periods of the year. A rough figure for the sum of the surface flows of the said wadis is 2.2×10^9 m^3 y^{-1}. The same reference estimates the average annual surface flow volume as 23.2×10^9 m^3. This means that the flow brought by the wadis alone represents approximately 10% of the total surface water resources.

7.7.2.3 Algeria

Most of the surface runoff in Algeria is generated in the northern part of the country, which occupies over 300,000 km^2. There are 17 principal drainage basins belonging to three distinct groups. The first and most important group covers a total area of 134,000 km^2 and receives annual precipitation ranging from 400 mm to 1,500 mm. This group comprises the Mediterranean basins which draw their supplies from the Tell Atlas. The total annual runoff flowing to the Mediterranean Sea is in the order of 12×10^9 m^3. All streams of this group are perennial.

The second group of drainage basins covers a total area of 100, 000 km^2 receiving an average annual precipitation in the order of 300 mm. The basins of this group are situated on the high plateaus of Constantine. The streams carry to the Mediterranean Sea an annual flow of about 0.75×10^9 m^3.

The third group also covers a surface area of about 100, 000 km^2 of Saharan Basins. It receives annually between 100 mm and 300 mm rainfall. The generated 0.7×10^9 m^3 y^{-1} is discharged into the Mediterranean Sea.

The three groups of drainage basins produce annually a total surface runoff of about 13.5×10^9 m^3 (Rapport National de la République Algérienne Démocratique et Populaire, 1983). The streams belonging to the second and third groups are all seasonal streams, i.e. wadis. Hydrological data relevant to these two groups are listed in Table 10, Appendix II. The supply brought by the wadis in Algeria is 1.48×10^9 m^3 y^{-1} or approximately 11% of the total surface water resources.

7.7.2.4 Mauritania

The important wadis in Mauritania like the Gorgol and Ghorfa discharge their flows into the Sénégal River thus adding to its flood, though not considerably. The junction of the White Gorgol (346 km long and 336 m fall) and the Black Gorgol (194 km long and 95 m fall) forms the main Gorgol. The Wadi Ghorfa springs in the Assaba massive at elevation of approximately 320 m a.m.s.l. It runs for a distance of about 190 km before joining the Sénégal as can be seen from the map in Figure 8.

Figures 22(a) shows the hydrograph of the Black Gorgol at Foum Gleita resulting from a rainstorm which occurred in July 1979 while Figure 22(b) illustrates the hydrograph of Wadi Ghorfa during the wet season of 1979. More flow data of the Gorgol and Ghorfa Wadis from the late 1970s are listed in Table 11, Appendix II. This Table shows that the annual flow sum of the Gorgols and Ghorfa is about 560×10^6 m^3, and the average runoff coefficient is 10%. The Country report of

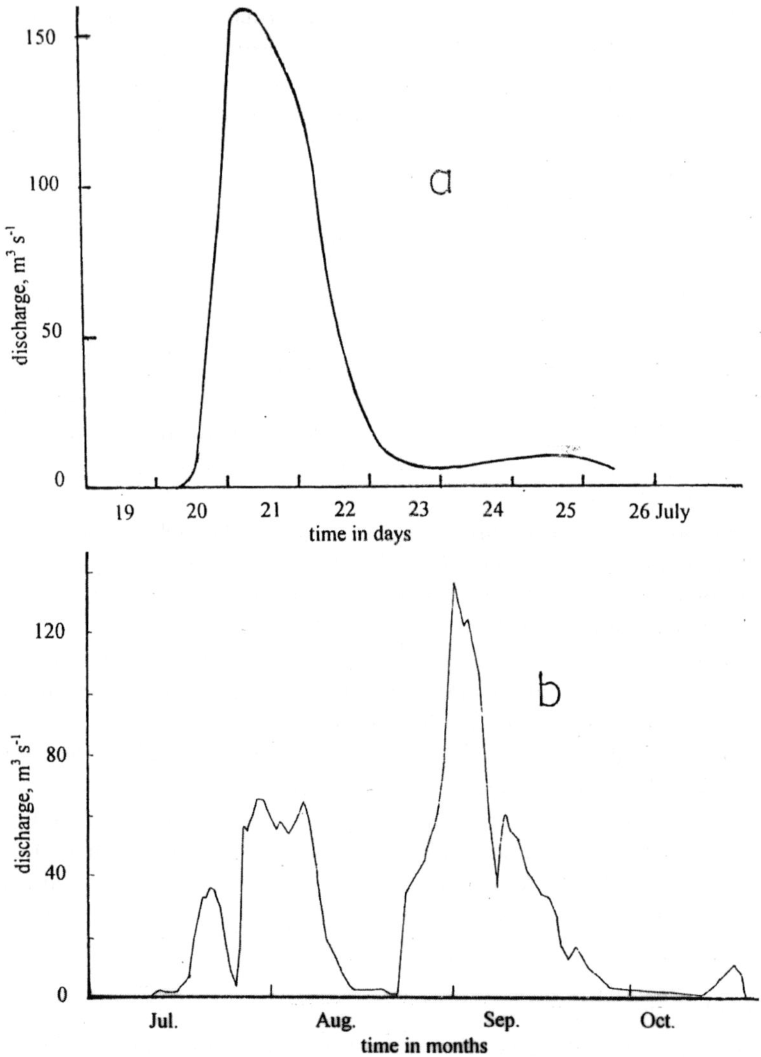

Figure 22. Hydrographs of (a) the Black Gorgol, and (b) Wadi Ghorfa, Mauritania

Mauritania (1986) estimates the annual contribution of the Mauritanian wadis to the flood of the Sénégal as 1.0-1.5 $\times 10^9$ m^3. One should not forget that other than the Gorgols and Ghorfa there are the Noiré and Savalel. The total annual supply of these wadis exceeds 1×10^9 m^3 y^{-1}.

7.7.2.5 Libya

In an unpublished report (1978) to the Food and Agriculture Organization of the United Nations on the Water Resources of the Socialist People's Libyan Arab

Jamahiriya, Pallas mentioned that the total mean annual runoff water calculated or measured at the entrance of wadis in the plains is roughly estimated as 0.2×10^9 m^3. A certain portion of this amount reaches the Mediterranean Sea, and a much larger portion is either evaporated or infiltrates the wadi beds and recharges the underlying aquifers. In an internal report (1983) prepared by the Department of Land and Water, a preliminary estimate of the annual surface runoff of 0.22×10^9 m^3 was given. This amount is distributed between Jabal Nafusa (140×10^6 m^3) in the northwest and Jabal Al Akhdar (80×10^6 m^3) in the northeast of the country.

A more conservative estimate of wadi flow in Libya is given in the Country Report of Libya (1986). In this report the country has been divided into seven major hydrologic provinces. These are: The Jefara Plain in the extreme northwest; Jabal Nafusa south of Jefara Plain; Jabal Al Akhdar in the northeast (Wadi al-Kof and Wadi Qattarah are the most important wadis); Hammadah al Hamra; Murzuq, Kufrah and Sarir in the southeast; and Jabal Al Haruj al Aswad in the middel of the country. The first three provinces receive annual rainfall much more than the other basins. Assuming a total catchment area of 30, 000-35, 000 km^2 receiving an average annual rain depth of 120 mm and a runoff coefficient between 5 and 6%, the generated volume of surface water is in the order of 170×10^6 m^3 y^{-1}, which is somewhat smaller than the previous estimates of 200 and 220×10^6 m^3 y^{-1}.

From the above figures one can fairly estimate the total annual surface water yield in the Western Sub-region as 6.3×10^9 m^3.

7.8. A SHORT ACCOUNT OF WADI FLOWS IN THE CENTRAL SUB-REGION

7.8.1 General remark

With the exception of Djibouti, the Central Sub-region of the Arab World compared to other sub-regions is the least in dependence on wadi flows. Egypt and the Sudan and Somalia, so far, cover the greatest bulk of their water demands from the Nile and its tributaries. The rest is covered from groundwater resources and recycled agricultural drainage water, as is the case in Egypt. As already mentioned in Chapter 6, the Juba and Shebelle Rivers, so far, cover the water needs for irrigation, drinking and domestic purposes.

However, for the sake of completion, we shall try in this section to present a brief description of wadi flow in the Arab Central Sub-region within the limits of the available information.

7.8.2 Summary of wadi flows by country

7.8.2.1 Egypt

According to the Report on Water Resources of Egypt (1986), the surface of the country, with regard to intermittent and ephemeral streams or wadis can be subdivided into five distinct hydrological provinces. These are roughly indicated on the map in Figure 23 and can be listed as follows:

- The Mediterranean Basin *(Province No. 1)* comprises more than 200 wadis draining into the sea. In the rainy season (November-February) there is always the chance that these wadis receive considerable volumes of surface runoff in short periods of time. Wadi Al Arish in Sinai, and Wadis Al-Garawlah, Ar Raml and Al-Kharroubah in the north coastal plain near Mersa Matruh are among the largest wadis in this province. Wadis intersecting the coastal strip in northwest of Egypt range between 10 and 15 km in length and between 1 km^2 and 240 km^2 in catchment area. The surface runoff these wadis carry is estimated as 11×10^6 m^3 y^{-1}, a small portion of which is used for irrigating an area of about 1,000 ha.
- The Red Sea Basin *(Province No. 2)* comprises the sea itself, the gulfs of Suez and Aqaba and the Isthmus. There are over 100 wadis draining the basin runoff.

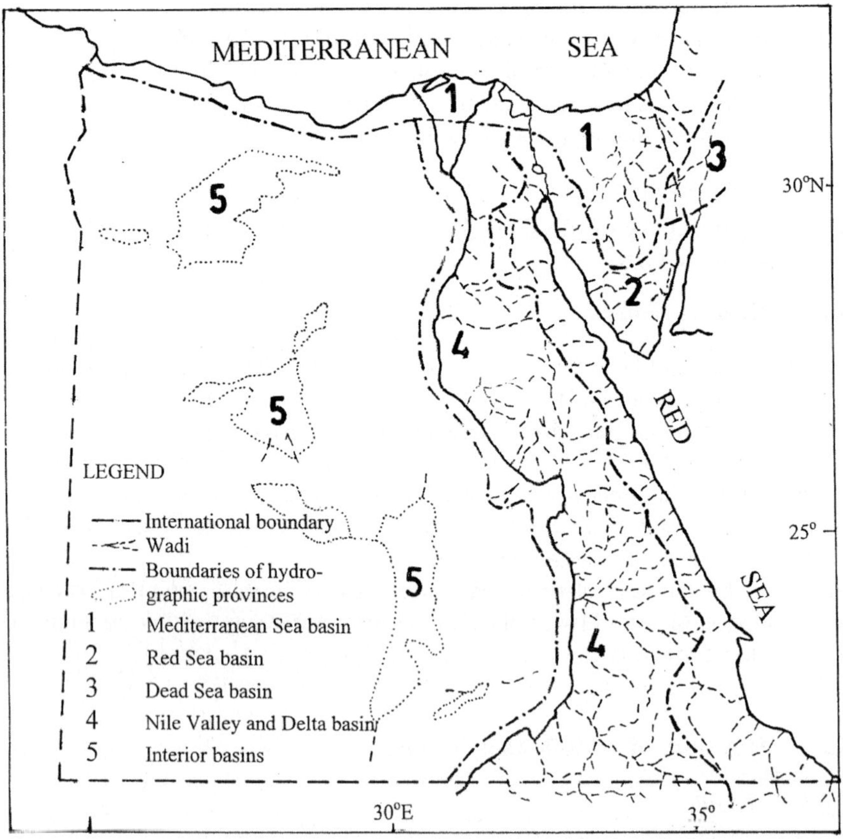

Figure 23. Map showing the boundaries of surface water provinces in Egypt

The major wadis in this province are Wadi Hodhein near Ra's Banas, Wadi Lahmi and Wadi Karim near Qusseir and Wadi Mallaha north of Safagah. They also include Wadis Asla, Al-Qa'a, Mizan, Ba'ba, Farandal, Sidr and Abu Souair, all discharging their flows into Suez Gulf, while Wadis Kabir, Watar and An Nasb drain their water into the Gulf of Aqaba.

- The Dead Sea Basin and Wadi Araba *(Province No. 3)* which receives the runoff brought by Wadi Al Jirafi from the rainfall in East Sinai.
- The Nile Valley and Delta Basin *(Province No. 4)* is the largest hydrographic province in Egypt as far as the geographical extent is concerned. The basin include many important wadis such as Al Alaqi, Karkar, Kalabsha, Qena and Assyut,
- Interior Basin *(Province No. 5)* includes Wadi-Natrun, Fayum, Qattara Depression, Siwa, Farafra, Bahaiyah, Dakhla and Kharga Oases. This province comprises several short wadis.

The total surface runoff of the above-mentioned provinces is estimated as $1.0\text{-}1.5 \times 10^9$ m^3 in a normal year. It increases to more than 2×10^9 m^3 and 5×10^9 m^3 in a wet year corresponding to 20% and 10% probabilities of exceedance, respectively.

The surface runoff of the five provinces is estimated as $1.0\text{-}1.5 \times 10^9$ m^3 in a normal year. This amount increases to more than 2×10^9 m^3 in a wet year, 1 in 5 years or with 20% probability of exceedance, and to 5×10^9 m^3 in a wet year, 1 in 10 years or with 10% probability of exceedance.

7.8.2.2 *The Sudan*

The total supply carried by the wadis in the Sudan has been estimated as between 3×10^9 and 4×10^9 m^3 y^{-1} (Country Report of the Sudan, 1986). Some of these wadis, like the Gash and Baraka originate in Ethiopia, i.e. outside the boundaries of the Sudan itself. However, they flow through the northeastern part of the Sudan and deposit the sediments they carry thus forming fertile deltas. The surface areas of the deltas are 2,950 km^2 for Al- Gash and 2,550 km^2 for Baraka.

The flow season of the Gash and Baraka extends from July to November and lasts between 80 and 110 days. The average annual flow of the Gash has been estimated, based on a few measurements, ranges from 0.2×10^9 m^3 to 0.8×10^9 m^3, with an average of 0.55×10^9 m^3 y^{-1}. It is exceedingly difficult to measure the flow of Wadi Baraka due to the rapid and frequent shifting of its channel within its wide flood plain. Instead of the standard methods of calculation, the flow of Baraka was estimated from the measurements of the area of surface annually inundated in its delta. In this way the estimate of the average annual flow has been found as $0.3\text{-}0.7 \times 10^9$ m^3, with an average of 0.5×10^9 m^3 y^{-1}.

Other than the wadis originating outside the borders of the Sudan, there are wadis originating within the borders of the country like Kathabati, which is fed by runoff generated from precipitation over the Amnong Mounts and the runoff carried by smaller wadis running down slope the Mara Hills, Red Sea Hills and Nuba Mountains. The flow season of these wadis lasts between 40 and 80 days, with total annual yield not exceeding 2×10^9 m^3. The annual rainfall in the western provinces of the Sudan is estimated as 600–1,000 mm. The generated surface runoff yield carried by the wadis along the middle slopes of Mara Mountains is about 120×10^6 m^3 increasing to 535×10^6 m^3 along the lower slopes. The importance of wadis in the Sudan lies in recharging the sedimentary groundwater basins which are regularly used for drinking water supply.

The Ministry of Water Resources in the Sudan has already installed a number of flow measuring stations on wadis in Kordofan, Central and Eastern, and Darfur Provinces. So far, there are no definite results available as to the total runoff carried by wadis in the Sudan. As such, one has to suffice with the early estimate of 3×10^9-4×10^9 m^3 y^{-1}.

7.8.2.3 Somalia

As already mentioned in the chapters on climate and analysis of rainfall data, the overall average rainfall in Somalia ranges from less than 200 mm y^{-1} in the north to 400 mm y^{-1} the south and 700 mm y^{-1} in the south west. On average, the country receives annually some 250 mm. It falls in two seasons, from mid April to June and from October to December.

Along the Gulf of Aden, a mountainous zone with rugged relief is subject to torrential flows, causing considerable erosion (FAO, 1995). The land slopes down towards the south, and the watercourses which flow southwards peter out in the sands of the desert. The total water resources produced internally, i.e. within the borders of the country are estimated as 6×10^9 m^3. Assuming that the intermittent streams or wadis carry half this amount, the annual wadi flow becomes 3×10^9 m^3. As already stated in the previous chapter the annual incoming water resources, Juba and Shabelle Rivers, supply Somalia by 7.5×10^9 m^3.

The above summary shows that wadi flow in the Arab Central Sub-region can be estimated to fall in the range 8-9 $\times 10^9$ m^3. This amount when added to wadi flows in the other sub-regions brings the total wadi flow in the Arab region to 25-30 $\times 10^9$ m^3 y^{-1}.

CHAPTER 8

EROSION AND SEDIMENTATION IN DRAINAGE BASINS AND IN STORAGE RESERVOIRS

8.1. GENERAL BACKGROUND AND DEFINITIONS

'Erosion' phenomena acting on the earth's surface by running water, commonly referred to as wet erosion, is the outcome of complicated natural processes together with consequences of human activities. As such, erosion can be subdivided into natural erosion and accelerated erosion. Climate, hydrology, morphology, geology, soil and vegetation prevailing in a certain region are the main factors causing natural erosion. Land management and use for agricultural, urban and other purposes are considered responsible for accelerated erosion.

Erosion might cause damage to land and water resources. It causes agricultural soil to lose its fertility and become unable to sustain crop production. Running water brings sediments to watercourses, and reduces the life age of impoundments and water storage reservoirs. Additionally, it affects adversely the function and operation of hydraulic structures and increases their maintenance costs.

Erosion can be defined as the detachment of soil particles from their position in the soil mass and their movement to any channel where they may be transported by water for the rest of their journey. Erosion can be subdivided further into 'soil erosion' and 'gully erosion'.

Soil erosion involves the removal of weathered loose soil materials from ground surface with the attendant transport and deposition of sediments elsewhere. Three erosion types of soil erosion exist; namely, 'sheet', 'rill', and 'channel and bank' erosion. Sheet erosion can be defined as the removal of a relatively uniform layer of soil from the ground surface. Rill erosion involves the removal of soil from small channels and rills by running water during floods. Channel and bank erosion occurs along flood, stream and river channels. Channels and banks collapse due to the instability of soil during fast flows and peak discharges. Figure 1 shows schematically the different types of erosion.

Gully erosion occurs where a small rivulet forms during heavy rains (e.g. torrential rain, in Arabic sayl, sail or seil). Turbulence in the flowing water creates local forces capable of dislodging soil particles from the bed and banks of the

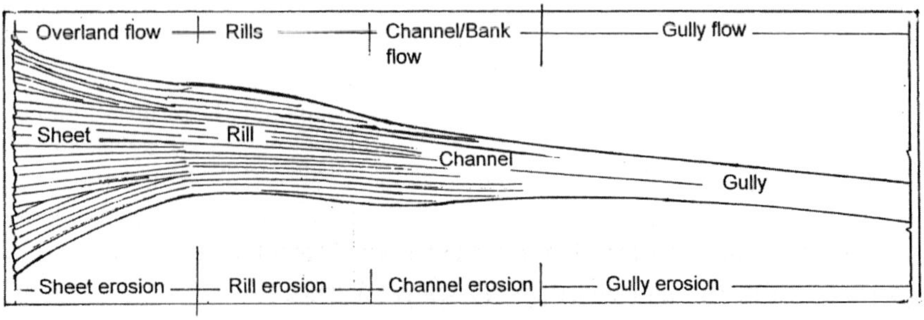

Figure 1. Schematic illustration of the different types of erosion

channel. As the gully deepens, the profile being steepest at head, it grows headward (Linsley et al., 1958). Gully erosion tends to be the last stage of erosion before the eroded surface attains some sort of stability. It may be tectonically-generated as ancient gullies, ravines, canyons or wadis. There may also be recent gullies that developed during the gradual or abrupt off-loading of sediments. Gully erosion with attendant slope stability or engineering foundation problems has bedevilled many Arab countries.

In a document prepared by Food and Agriculture Organisation (FAO, 1986) an evaluation of the severity of erosion and land degradation in Africa is given. Ten Arab States in the Western and Central Sub-regions are in Africa. Some of the results obtained from that source as well as other sources appear in various Tables included in the present chapter as well as in Appendix III.

'Sediments' are the natural product of erosion. The 'sediment yield' of a certain drainage basin can be correlated with the amount of stream runoff, and runoff basically depends on precipitation and temperature. Flood water triggers off sheet, rill and channel/bank erosion, thereby carrying away soil materials and nutrients depositing them elsewhere and increasing surface elevations with time. Sediment is transported by flowing water in different ways. Depending on a number of factors such as the physical characteristics and grain size composition of sediment, as well as the conditions of flow, the sediment grains may be moved by saltation (bouncing), rolling or sliding on or near the stream bed, or may be swept away from it and kept in suspension. The different phases of sediment transportation generally occur simultaneously in natural streams, and there is no sharp line of demarcation between them.

'Sediment discharge' is divided into three categories; 'wash load', 'suspended load' and 'bed load'. Wash load comprises the finest particles of silt and dust ($< 50\,\mu m$) that are brought to the stream and remain suspended until they reach the sea or reservoir. Wash load, therefore, has little or no influence upon the river behaviour, although it may quantitatively be by far the largest category of sediment transported by the stream flow. The second mode of sediment transport consists mainly of the fine particles that are continuously supported by the flowing water. The sediments in this category do not sink down because the upward force of the turbulence of the flow compensates their weight. Over a short stretch of river, the suspended load can be

assumed to remain in suspension, so that it does not have much influence upon the bed either. Considered over the full length of river, however, the suspended particles can settle and other particles can be picked up from the bed. This state of affairs can result in some sort of relationship between bed and suspended load. Bed load transport refers to the movement of the larger grains of material carried by the river through sliding and rolling over each other but hardly rising from the channel bottom. Sometimes a fourth mode of transport is distinguished. 'Saltation load', is a sediment movement in between the bed load and the suspended load. Intermediate-size grains are small enough to get loose from the bottom by sudden high drag impacts, but are not small enough to remain in suspension. Often bed load and saltation load are considered together as one type of transport.

While discussing the various aspects of sedimentation and sediment transport it is customary to refer to each aspect a number of times. 'Bed material' can be defined as the sediment mixture of which the channel bed is composed. The 'total sediment load' refers to bed load plus suspended load or bed load material plus wash load. Additionally, 'sediment concentration' can be defined as the mass of solid material in a given volume of water. It is generally expressed in grams per liter (gm l^{-1}) or in parts per million (ppm). Sediment load terminology using two classification systems is presented in Table 1 (Cooper and Peterson, 1970).

According to Walling & Kleo (1979) the available data from a number of countries do not indicate that sediment yields of river basins in arid and semi-arid regions are significantly higher than those of river basins in humid temperate and tropical regions. Hadely (1986) added that the percentage of material eroded from the upland areas within drainage basins in arid and semi-arid regions is higher than that in more humid environments. He attributed this conclusion to the meager precipitation and erratic distribution of runoff -producing storms.

8.2. MEASUREMENT/ESTIMATION OF EROSION

Soil loss and sediment yield can be studied at different scales, each of these having relevance to certain aspects of the erosion problem. Plot studies are perhaps the most

Table 1. Sediment load terminology (Taken from Jones et al., 1981)

		CLASSIFICATION SYSTEM	
		Based on Transport Mechanism	Based on Grain Size
TOTAL	Wash Load		Wash Load
	Suspended Bed Load Material	Suspended Load	
SEDIMENT			Bed Material
LOAD	Bed Load	Bed Load	Load

widely used method to yield information on soil loss. This method has led to, among others, the development of the Universal Soil Loss Equation by Wischmeier & Smith (1978).

8.2.1 Formula of Wieschmeier & Smith (1978)

This formula estimates the rate of soil loss, A (t ha^{-1}), in a certain period of time based on 6 factors as given by Eq. (1).

$$(1) \qquad A = 1.3R.K.L.S.C.P$$

where,

$R =$ the index of rainfall agressivity,
$K =$ the index of soil erodibility,
$L =$ length factor;
$S =$ slope factor,
$C =$ rainfall index, and
$P =$ index of water and soil conservation.

Touaïbia et al. (1999) upon investigating the (wet) erosion on different spatial scales in Wadi Mina, Algeria, calculated the value of the agressivity index, R, using rainfall data from 399 charts covering two consecutive years (1989–90 and 1990–91) and 902 rain events with intensity varying between 0.2 and 22 mm h^{-1}. The calculated value of R showed to be relatively small at 16.53 m t m ha^{-1} h^{-1}. The remaining 5 indices, i.e. C, K, L, S and P, were obtained using certain nomographs. The calculated soil loss, A, by erosion for the 16 experimental plots used in the investigation was found to vary from 0.12 and 1.09 t m^{-2}.

That investigation has led to a simple regression model for estimating the index R using the depth of rain, Lp. The model can be expressed as follows:

$$(2) \qquad R = m(Lp)^n$$

where,

$m = 0.0361$, $n = 1.3211$ and $r^2 = 0.87$ for the year 1989–90,
$m = 0.0214$, $n = 1.3248$ and $r^2 = 0.96$ for the year 1990–91, and
r^2 is the coefficient of determination (square of coefficient of correlation).

8.2.2 Experimental plots

Touaïbia et al. (1999) reported the results obtained from a series of experiments carried out in 10 plots each 1 m^2 in surface area. The plots are situated within three small experimental basins in Wadi Mina, Algeria. The slope of the terrain ranges from 12.5% to 50%. The land is mainly dominated by brown calcareous soil covered with temporary or permanent grass, else left barren. The experiments lasted from less than 30 minutes to more than 130 minutes. Rainfall was supplied via rain simulator (type ORSTOM) with an average intensity of 30 mm h^{-1} and duration from less than 30 minutes to more than 130 minutes. The actual duration, intensity and rain depth, water and sediment discharges were measured. Those experiments showed that the specific erosion under the given conditions varied enormously between a minimum of 9 g h^{-1} m^{-2} to a maximum of 2,805 g h^{-1} m^{-2}. In 7 out of 23 experimental trials the runoff coefficient exceeded 60%, and in 6 out of those 7 trials the specific erosion was between 1,279 and 2,805 g h^{-1} m^{-2}.

Another set of experiments was conducted on four plots each 87 m^2. The slope, nature of soil and the rain simulator used were different from those in the first set. The slope of the plots varied from mild to steep. The runoff volume varied widely from trial to trial resulting in a wide range of the runoff coefficient, from 0.34% to 63.74%. The corresponding specific erosion values were 0.1 and 184 gm m^{-2} h^{-1} respectively. Additionally, a simple regression model was developed for predicting the sediment discharge, Q_s, given the water discharge, Q. The regression model can be expressed as follows:

$$(3) \qquad Qs = aQ^b$$

where,

$a = 21.85$, $b = 1.227$ and $r^2 = 0.81$ for bare soil,
$a = 8.74$, $b = 1.194$ and $r^2 = 0.90$ for vegetated soil, and
$a = 15.40$, $b = 1.293$ and $r^2 = 0.84$ for mean values of Qs and Q.

8.2.3 Erosion maps and sediment discharge

In view of the strong erosion in Tunisia and its grave consequences on many aspects of life, Sida (1978) was requested by the Food and Agriculture Organisation to carry out a field study to determine the erosion rates there.

The investigator mapped erosion on 4 catchments delineating three erosion classes. Assuming a soil loss of 10,500 m^3 km^{-2} y^{-1} for severe erosion, 5,000 m^3 km^{-2} y^{-1} for medium erosion and 1,100 m^3 km^{-2} y^{-1} for light erosion, the overall soil loss was calculated and compared with observed siltation rates in the reservoir of the catchments. The calculated annual erosion expressed in m^3 km^{-2} was 1,920 for Nebhana, 5,050 for Masri, 3,520 for Chiba and 2,050 for Bezrik. The corresponding observed erosion for the four catchments in their respective order was 1,880, 4,830, 3,660 and 2,070 m^3 km^{-2}. For more details the reader is referred to Table 8, Appendix III.

In the marly catchments, under average annual rainfall of 500 mm, erosion is generally high and correlates well with monitored siltation rates. Thorough examination of the results obtained shows that the major contribution comes from the small, yet severely erodible areas. The erosion map of Tunisia has been later updated and could be used to evaluate the sediment discharge of ungauged catchments.

Table 2. Erosion forms and annual erosion rates in different parts of Morocco (from Heusch et al, 1986)

Region	Erosion form	Sediment discharge t km^{-2} y^{-1}	% of total surface area	soil loss
Anti-Atlas, Sahara, arid Moulouya and Souse plains	Local alteration	25	66	8
Tensift, Oum er Rbia, Bouregreg and Sebou plains	Limited areal erosion	150	17	10
Eastern Rif, Central High-lands and High Atlas	Limited gully erosion	400	11	20
Folded Middle Atlas and Atlantic coastal Rif	Frequent gullyerosion	700	2	7
Mediterranean coastal Rif and Lower Rif	Generalised gullyerosion	1700	2	16
Western Rif	Landslides	3500	2	39

Heusch combined the information on an old erosion map of Morocco with the available sediment discharge measurements and used similar methods to those used in Tunisia. Some of the results obtained are presented in Table 2.

It might be of interest to mention in this respect that monitoring of stream bank and gully erosion has recently been carried out in certain catchments of the United States by airborne laser. The investigation concluded, "Laser altimeter measurements of the physical properties of the surface of the landscape provide unique and rapid measurements of gully and channel morphology." Such measurements of macro scale and micro scale topography can be used to measure channel and gully degradation and aggradation and to study their relationship to the landscape. The method further offers the potential to measure landscape properties over large areas quickly and easily. Such large area measurements can provide valuable information for understanding and managing of the erosion cycle" (Ritchie et al., 1993).

8.2.4 Experimental basins

Bouguerra (1986) reported the results obtained from experimental work on three pairs of basins within the drainage basin of the Issar River, Algeria. The erosion caused by rainfall and running water was determined from land survey and from the experimental basins. The surface area, slope, type of soil and soil cover differed from basin to basin. The observations taken were precipitation, runoff and sediment yield. The distribution of total erosion between the sheet, channel and gully types of erosion were given. Some of the results obtained from that study have been summarised and presented in Table 5(a), Appendix III.

Touïbia et al (1999) expanded the investigation mentioned under 8.2.2 to comprise erosion measurements in experimental basins. The surface area of the six basins put under investigation ranged from 21.7 ha to 76.2 ha and the corresponding mean channel slope from 9.9% to 19.0%. The measured variables were the rainfall depth and volume, the runoff depth and volume and the sediment load, all on annual basis. Some of the results obtained are summarized in Table 5(b), Appendix III. The same measurements were taken over the period of the storm event of 11 May 1990. The latter set of measurements provided values for the mean and maximum sediment concentrations.

The observed erosion over the year 1989–90 varied from 12.38 to 48.72 t ha^{-1} y^{-1}. These figures appear to be rather alarming, especially when one realizes that the rainfall in the year of observation was less than that in an average year. The rain event of 11 May 1990 alone produced erosion more than 50% of the erosion produced in a whole year. That rain event produced a 30-min rain intensity of 44 mm h^{-1} over basins No. 5 and 6. However, the considerable difference in runoff coefficient between basins 5 and 6, 69% and 94%, respectively, caused the sediment load to increase from 557 t for basin No. 5 to 997 t for basin No.6.

As before, the investigators developed a simple regression model between the sediment discharge Q_s in t h^{-1} and the discharge, Q l s^{-1}. The model has the same form as the one expressed by Eq. 3, except that, for the available results, a ranged from 0.0668 to 1.3261 and b from 0.9024 to 1.9094. The coefficient of determination r^2 ranged from a minimum of 0.81 to a maximum of 0.95.

8.2.5 Measurements taken over large areas

8.2.5.1 The Western Sub-region countries

Heusch & Cayla (1986) presented a paper dealing with sediment discharge measurements in the countries of the Western Sub-region (Maghreb). In that paper the results obtained by a number of investigators in the period from 1960 up to 1982 were analysed and discussed. Some of those results appear in Table 4, Appendix III, and can be summerised as follows:

Tunisia- It may be of interest to hint here to the effect of rainfall and land use on the rate of erosion and sediment concentration. The annual erosion rate increased from 287 t km^{-2} for areas with 1% cultivation and 0–199 mm rainfall to 573 t km^{-2} for 12% cultivation and rainfall 300–399 mm and further to 1,765 t km^{-2} for 24% cultivation and rainfall larger than 600 mm. As the change in annual erosion is caused by increased percentage of cultivated area and rainfall, it is difficult to quantify the share of each one of these two variables. Likewise, the contribution of each variable in the reduction of the sediment concentration from 29 gm l^{-1} to 19 and further to 6 gm l^{-1} can be estimated.

According to Colombani et al. (1984), the catchment of Wadi Zeroud (8, 950 km^2 at Sidi Saad), Tunisia, received in the period from 30.08.1969 up to the end of October of the same year an average rainfall of 678 mm. The total flow produced by the rain in those two months was 2.64×10^9 m^3, of which 10% represented the sediment load and the remaining 90% represented the flow of water. One should remember that the median annual flow is 77.5×10^6 m^3 and the annual flow with a return period of 10 years is in the order of 200×10^6 m^3.

The observed peak flood discharge was 17,050 m^3 s^{-1} and the corresponding specific sediment yield 1.9 m^3 s^{-1} km^{-2}. The observed erosion, which occurred in just 2 months, was about 40,000 t km^{-2} or 26.8 mm. As a matter of fact this extreme depth of erosion, which happened in 2-month time normally occurs in 10–20 y.

The severe flood of 1973 in the basin of the Mejerda River in Tunisia, and the subsequent erosion is another example. The drainage basin of Mejerda River, Tunisia has a surface area of 23,000 km^2. The basin was hit by continuous rainfall, which after 3–4 days amounted to a depth of 100–300 mm depending on the location within the basin, causing complete saturation of the soil. This was followed by a severe flood with peak discharges estimated at return period ranging from 50 to 150 y depending on the location. The recurrence interval of the flood volume was even greater; it ranged from 100 to 200 y. The consequences of that flood were grave, high loss in human life, destruction of roads and bridges and inundation of a surface area of 473 km^2 or 2% of the river basin area.

The volume of sediments deposited on the inundated surface was estimated as 42.5×10^6 m^3 or 75×10^6 t and the corresponding specific yield was 1, 600 t ha^{-1} or 160 kg m^{-2}. The minimum amount of sediments discharged into the sea was estimated as 25×10^6 t, bringing the total sediment yield to no less than 100×10^6 t. The average concentration during the 6-day flood duration was in the order of 106 gm l^{-1}. One can, therefore, expect that the maximum concentration must have been somewhere between 150 and 200 gm l^{-1}. These figures show that the described flood must have caused an erosion of 4,250 t km^{-2} or 2.5 mm over the entire surface of the drainage basin (Colombani et al., 1984).

Morocco- Erosion estimates relevant to Morocco as suggested by Heusch (cited in Heusch & Cayla, 1986) are summarised in Table 2.

8.2.5.2 *The Eastern Sub-region*

Syria- Soil erosion is a general problem in the coastal region of Syria due to the combination of torrential rainfall, deforestation, forest fires, and inadequate reforestation. Approximately 75% of the region is likely to be affected by high or severe rainfall-caused (wet) erosion. Figure 2 illustrates the extent of different classes of potential soil loss. According to Shalabi et al. (1996) four zones can be recognized as far as their vulnerability to erosion is concerned. The area with least susceptibility covers the large plain south of Lattakia to 100 m altitude with slopes of less than 3% and surface area of 90,000 ha. The second zone, of slightly higher susceptibility, covers most of the areas north of Lattakia up to 50 m altitude (about 50,000 ha). The land slope here too does not exceed 3%. Annual sediment yield in these two zones is generally below 30 t ha^{-1}. The third and fourth zones are areas of high susceptibility to erosion and together cover a large surface of 330,000 ha, which comprises more than two-thirds of the coastal region of Syria. They include the northwest corner of the country and the slopes of Jebel as Sahel from 100 to 1,500 m altitude with gradients of 45% or more. Sediment yield ranges between 30 and 60 t ha^{-1} y^{-1}. The areas that are most vulnerable to erosion cover

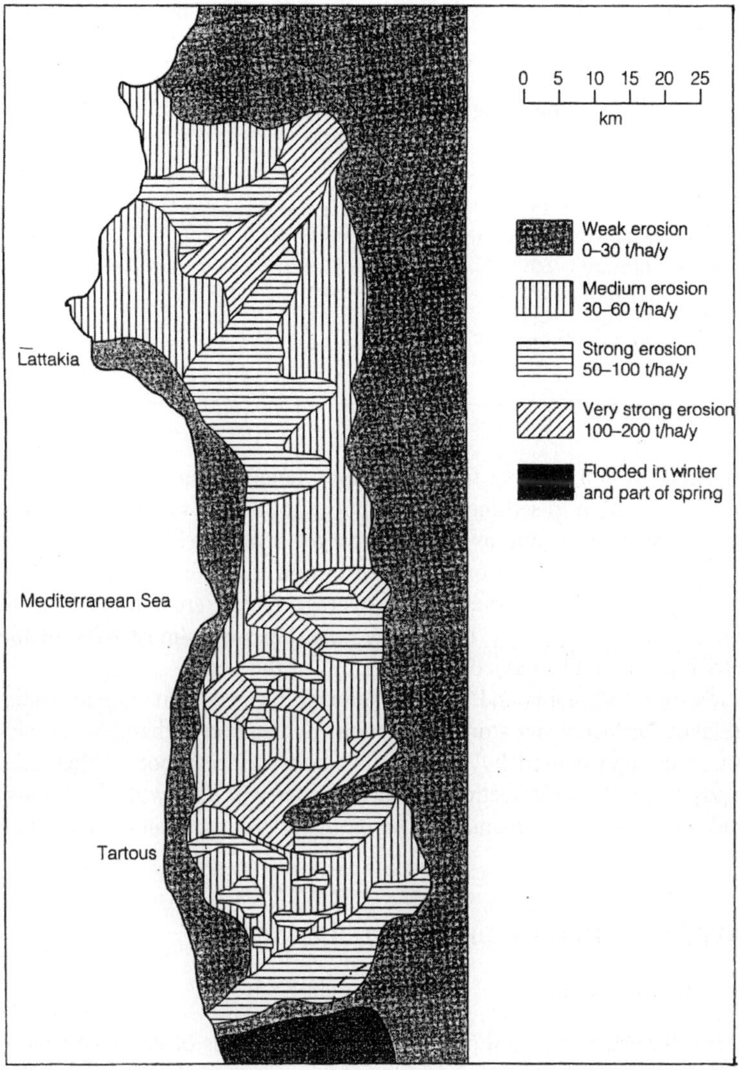

Figure 2. Soil erosion map of the coastal region of Syria (from Shalabi et al., 1996)

approximately 32,000 ha. They are located mostly in the hilly part of the coastal plain, northeast of Latakia and between Banias and Tartous where the altitude ranges from 100 to 600 m and the slopes vary from 15 to 35%. The annual rate of sediment loss in these areas can amount 50-100 t ha^{-1}.

Iraq- Seven sediment events have been monitored in 1983–84 at north Baghdad station on the Tigris. The water discharge, suspended sediment load and concentration, and computed rate of erosion for those events are listed in Table 3. The

Table 3. Erosion and suspended sediment data at North Baghdad Station on River Tigris (Al-Ansari et al., 1986)

Event No.	Duration from	to	Mean Discharge, m3 s-1	Suspended load weight, t	% annual total	Concentration, gm 1-1	Rate of erosion, t km-2 d-1
1 (20 days)	18-Nov	07-Dec	936	2115395	26.6	996	0.79
2 (10 days)	11-Dec	20-Dec	425	385740	4.8	488	0.29
3 (14 days)	12-Feb	25-Feb	402	417608	5.2	402	0.22
4 (33 days)	09-Mar	10-Apr	639	938698	11.8	491	0.23
5 (20 days)	14-Apr	03-May	1035	2732229	34.4	1371	1.02
6 (7 days)	09-May	15-May	911	194587	2.5	347	0.21
7 (10 days)	10-Jun	19-Jun	816	112539	1.4	159	0.08
Total = 114 days				6896796	86.7		51.49

mean concentration, as can be seen from the Table, ranges from 0.158 gm l^{-1} to 1.371 gm l^{-1}. The variation in sediment load, concentration and erosion rate is not always in direct proportion to the average discharge. It goes without saying that neither the rate of erosion nor the total erosion is uniformly distributed in time. From Table 3 one can easily observe that the average rate of erosion varies from event to event. For example, in the hydrologic year 1983–84 almost 87% of the total erosion took place in 114 days or about 31.2% of the year.

Sedimentation on flood plains and internal drainage basins, in rivers and wadis, and in natural lakes, surface water storage reservoirs and aquifer recharge reservoirs is the product of erosion caused by precipitation and surface runoff. Other than what has already been, the next sections of this chapter will deal with estimating, measuring and monitoring sedimentation in drainage basins, watercourses and reservoirs.

8.3. SEDIMENT YIELD OF DRAINAGE BASINS

8.3.1 Predictive equations

There are several for estimating and predicting sediment yields of drainage basins. These are basically statistical equations developed from easy-to-measure watershed parameters. More or less similar to erosion, these parameters include rainfall depth and intensity, soil type and geologic formation, ground cover and land use, runoff, and slope and geometry of the drainage basin. Examples of these equations are those developed by Brune (1951), Modified Wischmeier and Smith, Fournier (1960), Jansen & Painter (1974), Dunne (1979), etc.

Denote the annual runoff per unit area by Q, the mean elevation of the drainage basin by E, the annual precipitation by P, the average annual temperature by T, and the index representing the soil vegetal cover by C. The annual sediment yield per unit area of the watershed S can be expressed according to Jansen & Painter (1974) as the product $Q^{\alpha}.E^{\beta}.P^{\lambda}.T^{\delta}.C^{\eta}$, where α, β, λ, δ and η are coefficients.

8.3.2 Gross erosion and sediment delivery ratio computations

This technique of computation of sediment yield is a two-step procedure. At first, the gross erosion in a drainage area is computed from the universal soil loss equation proposed by Wischmeier and Smith (1965); then, a sediment delivery ratio is determined. This ratio is a fraction of the gross erosion, which is expected to be delivered at a certain cross section in the drainage network of the basin under consideration. Sediment yield is the product of gross erosion times the delivery ratio. The remaining part of sediments not transported from the drainage basin by the streams is deposited in channels, on flood plains, terraces and alluvial deposits of all kinds.

8.3.3 Generalised erosion and sediment yield maps

A convenient, though approximate, means of obtaining sediment yield information is by collecting and analysing all available sediment data and extending them over a national or regional scale. Equal erosion potential areas are delineated without a need to assign yield figures to them. A map of sediment yield may then be prepared and may be calibrated by all available information on sediment yield. Once such a map is available, the area under investigation can be projected on it and sediment yield predicted.

According to Walling (1986) the provisional maps of contemporary rates of soil degradation produced by FAO for North Africa and the Middle East provide some basis for comparing current estimates of erosion rates with measurements of sediment yield. Despite the limitations and general nature of such maps, it is certain that there is a lack of direct correspondence between rates of soil erosion and downstream sediment yield. This can be attributed to sediment redistribution within the drainage basin, thus rendering the assessment of sediment delivery ratio difficult and sometimes unreliable. The application of a predictive method, especially the modified universal soil loss equation with appropriate coefficients, can overcome some of the problems of determining a delivery ratio.

8.4. SEDIMENT DISCHARGE OF FLOWING STREAMS

8.4.1 Sediment measurement

8.4.1.1 Samplers

Due to differences in mechanism of bed-load and suspended-load, there are different samplers for each type of transport. Samples of sediment in flowing streams are taken at the discharge measuring cross sections. In lakes and reservoirs, locations of sampling verticals are scattered over an area. The samplers are suspended in the water on a rod or on a cable.

Most of the bed-load samplers are of the box or basket type, and consist of a pervious (mesh material with an opening in the upstream end) container. Water sand sediment enter the sampler; the mesh should pass the suspended material, but retain the sediment moving along the bed. A bed-load trap for research purposes on small streams consists of grated opening in the streambed into which the bed material falls. The trapped material is later excavated or sluiced out and measured. Turbulence-producing weirs have also been designed which throw the bed load into suspension locally so that it can be sampled with a suspended-sediment sampler. Comparison of the samples thus obtained to those from a section upstream of the weir indicates the quantity of bed-load. The method is suitable only when the bed material is relatively fine.

Suspended-load samplers are concentration samplers where a volume of water is sampled at a certain level or as an average over the depth of water. The mousetrap and the depth-integrating samplers are among the types often used. With a suspended-load sampler suspended-load is measured at a point or by integration over the depth of water. Mounting a number of samplers above each other on a vertical bridge pier, samples at different depths below water surface can be taken simultaneously. Samples are usually taken during the flood season at several pre-determined water levels.

The guide to hydrological practices, Volume 1 gives a more detailed description of the measurement techniques and procedures including computation of a continuous record of suspended-sediment discharge as well as of bed-load discharge (WMO, 1981). Additionally, Lawrence (1987) based on his long-standing practical hydraulic research of wadis in Yemen established a list of minimum requirement for checking sediment- transporting capacity of wadis. The listed items are as follows:

- i Concentration and size range of bed material sediments as a function of wadi discharge. It is necessary to measure the discharge and sediment concentration at the same time.
- ii Bed material size grading curves, where flows and sediment loads are measured, and at the site(s) of offtake structures. Samples should be collected from at least five trial pits at each location. The samples should be collected in such a way as to assemble a "representative" bed material size-grading curve.
- iii Bed material specific gravity.
- iv Wadi bed cross sections and stage-discharge curves at the measurement and offtake locations.
- v Water surface slope corresponding to known discharges at the measurement locations.
- vi Channel bed slope at the measurement and offtake locations.
- vii Water temperature.

8.4.1.2 Tracer techniques

Other than sediment samplers, sediment transport can be measured using tracers. Artificial grains are marked and added to the flow in small quantities and their displacements are determined. The measured displacements will help in computing the sediment transport. Several types of tracers are currently used:

i *Fluorescent (luminofores):* The marked grains can be detected upon sampling under ultra violet light.
ii *Activation analysis:* Particles are marked and radiated after sampling. This procedure is difficult, and mostly applied to silt only.
iii *Radioactive:* Natural sand is coated with a radioactive material and its movement can be detected in situ. It should not be used in excessive quantities to avoid health and environmental problems.

The constant injection method and the point-injection method are two among several methods used for interpretation. In the first method a constant amount of tracer material (rate τ) is distributed over the profile and injected during a long time interval. Samples are collected at a downstream cross section and the concentration as a function of time is determined. After some time the concentration reaches a stable value, C_o. The rate of transport s can be obtained as:

$$s = \tau / C_o \tag{4}$$

In the point-injection method a certain amount of tracer material is injected at a certain time. By observing the flow at a number of cross sections in the downstream, the concentration can be determined as a function of distance. From the displacement of the concentration-distance curves the average transport velocity can be computed. Multiplication of this velocity by the effective depth of transportation gives the rate of transport. Often, the effective depth is taken as half of the height of the bed forms.

8.4.1.3 Nuclear concentration gauges

These instruments sense the decrease of radiation intensity originating from a low-activity isotope such as Cadmium 109 or Cesium 137 due to absorption and/or dispersion by suspended sediment particles. The lowest detectable limit is 0.5–1.0 gm l^{-1} making this type of gauges inapplicable under moderate climate with low erosional activity. On the other hand, the upper limit of detectable concentration is around 100 gm l^{-1}. The isotope is placed into a sealed container, thus cannot contaminate the environment. Its lead housing makes it possible to send only a thin beam of radiation toward the detector and the radiation can be totally shielded when the instrument is not in use. As such the storage, transport and handling of the device are quite safe.

Laboratory calibration of nuclear gauges is made either in salt solutions having similar radiation-absorbing characteristics to the water-sediment mixtures, or in sediment suspensions maintaining a known and constant concentration at each step (Rákóczi, 1986).

8.4.2 Sediment transport functions

In the absence of reliable measurements or where sediment concentrations measured at low flows have to be extrapolated, then sediment functions have to be used.

The total sediment load of a stream can be determined by adding the bed load and suspended load as is the case in the Einstein Procedure (1950). Additionally, there are many more direct empirical relations like the procedure described by Englund & Hansen (1967), Ackers & White (1973). A comparison between eight of the most widely known transport formulas was conducted using 840 flume data and 260 field experiments in natural water resources. If the percentage of all data with a ratio T of calculated to observed transport in the range $1/2 < T < 2$ is taken, the following result is obtained:

Ackers & White	68%
Englund & Hansen	63%
Einstein	46%

It is not surprising that the formula of Ackers & White gives relatively good results in view of the large number of tuning parameters, all functions of grain size. It is surprising, however, that the much simpler formula of Englund & Hansen gives nearly equally good results.

According to FAO Paper 37 (Jones et al, 1981) the amount of bed load transported by an alluvial channel in arid zones is of the order of 5 to 25% of the suspended load. It cannot be measured satisfactorily and may therefore be accounted for only by using empirical formulas based on field and laboratory research data.

8.4.3 Sediment discharge and concentration in rivers and wadis of the Arab Region

In general, available data on sediment loads are less extensive and less reliable than data on flows. Ideally one requires detailed measurements at two or three-day intervals during low flows and more regularly during high flows, with not only sediment concentration but also particle size distribution being recorded. Such information is expensive to collect and rarely available.

The next sub-sections review the current situation with regard to sediment load and concentration in a number of rivers and wadis in the Arab Countries.

8.4.3.1 Sediment concentration of the Blue Nile water, the Sudan

It is generally conceived that sediment concentration in a river increases with discharge. The underlying reason is that the runoff increases with precipitation, and increase in precipitation leads to increase in erosion. However, this appears not to be always the case.

Sediment of the Blue Nile was noticed to increase during the drought years, which hit Ethiopia and the Sudan in the 1980s. This can be seen from the graphs in Figure 3. The river discharge in August 1982 was much less than that of the same month in 1975. Nevertheless, the sediment concentration was clearly higher for 1982. This had been attributed to the deterioration of the vegetal cover and subsequently loosening of the top soil in the catchment area of the river in Ethiopia due to succession of drought years. Gulleying and stream channel erosion in the uplands of the Blue Nile was claimed to be another source.

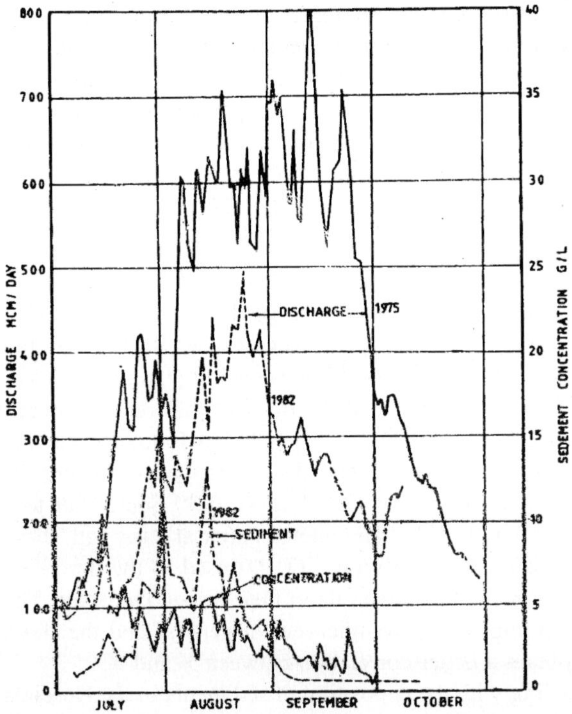

Figure 3. Sediment concentration of the Blue Nile, the Sudan (Hamad & Mohamed, 1986)

The increased amount of silt was carried to the canalization system of the Gezira Scheme. Although the system is designed as quasi non-silting non-scouring under normal conditions of operation, those abnormal conditions left their marked influence on the hydraulic performance of the system. Two aspects developed as a result; an annual effect caused by the succession of drought years, and a seasonal effect within the same irrigation season. It was not difficult to observe the annual effect through the change of the canal cross section over the years due to sediment deposition along the canal sides. The seasonal effect was reflected on the coefficient of roughness. It showed some variation between the flood and dry periods especially for high water levels (Hamad & Mohamed, 1986).

The important remark that has to be mentioned here is that the peak sediment concentration occurs about 4 weeks earlier than the peak discharge.

The sediment load in the Blue Nile and sediment deposition in the Rosiores Reservoir were analysed and discussed by El-Sheikh & Kaikai (1991). The investigators considered the flow in the period 1966/67–1987/88 at Ed'Deim station on the Blue Nile in Ethiopia, 120 km upstream of the Roseires Reservoir. The available daily sediment concentration in the high flow season, July-October, for the years 1954, 1955, 1970, 1973, 1975, 1987 and 1988 at the same station was also considered. Linear and parabolic regression relationships between $Y = logQ_s$ and $X = logQ$ were considered, Q is the water discharge and Q_s sediment discharge.

(5) $$Y = C_o + C_1 X$$

and

(6) $$Y = C_o + C_1 x + C_2 X^2$$

The annual discharge hydrograph and sediment discharge data were divided into four intervals; namely, 1–31 July, 1–20 August, 21 August – 30 September, and 1 October-30 June. The least correlation ($R^2 = 0.36$) between log Q_s and log Q was found when the annual data were correlated with hardly any difference between linear and polynomial regression relations. The two regression functions yielded almost the same correlation for the intervals. The determination coefficient R^2 ranged from a minimum of 0.63 for the interval 21 August to 30 September to a maximum of 0.90 for the interval 1–31 July. The author is not at ease with these results because the number of pairs of observations (Q_s, Q) used in the described analysis was 29, 33, 67 and 25 for the four intervals in their respective order. The principal merit here is that dividing the hydrologic year in general and the flood season in particular helps to obtain a better correlation between Y and X.

The part of investigation dealing with the deposition of sediments in the Roseires Reservoir will be reviewed in a forthcoming section dealing with siltation in reservoirs.

8.4.3.2 *Sediment concentration of the Main Nile at Gaafra, Egypt*

The suspended load of the Atbara River, Blue Nile and the Main Nile at Wadi Halfa/Kajinarti in the Sudan for the period from August 1967 to July 1969 was discussed by Shahin (1986). Due to its downstream location, the occurrence of peak concentration of the Main Nile at Halfa lags about 25 days the maximum concentration of the Atbara and Blue Nile Rivers.

The flow in the Main Nile is the sum of the flows brought by the White Nile, Blue Nile and tributaries, and the Atbara less the abstractions, conveyance and storage losses. This item has already been discussed in details in Chapter 6 and earlier by Shahin (2002). Until 1964 the storage of water in reservoirs on the Nile was limited and so its effect on the sediment concentration of water. The period 1964–68 witnessed the construction of the first phase of the Roseires reservoir on the Blue Nile, the Khashm el-Girba on the Atbara and the Aswan High Dam. The most upstream sediment measuring station in Egypt is Gaafra, located 35 km downstream of Aswan. To avoid the complexities introduced by these storage works on the concentration of the quasi-natural flow of the river, the season August-November of the period 1955–63 is considered for the purpose of analysis of suspended sediment concentration in the Main Nile. The concentration before August and after November is too small.

The principal objective of the study is to check whether a relationship between the sediment concentration (x) and the river discharge (y) does exist, and if there is any does it hold for the rising limb of the flood hydrograph as well as the falling limb? The graphical plot of y versus x for the rising limb of the flood, i.e. August-September is shown in Figure 4. The scatter of the points as can be seen from the figure is extremely wide, and at the upper end of the curve, where Q exceeds 650×10^6 m^3 d^{-1}, the concentration declines with further increase in the discharge. Above the discharge figure 650×10^6 m^3 d^{-1} the concentration decreases sensibly till the flood peak discharge, which is about 870×10^6 m^3 d^{-1}, is reached. The corresponding concentration as obtained from the equation of best fit to the plotted points is 3,340 ppm, i.e.75% of the maximum concentration. As such it has become obvious that there is no sense in trying to develop a single relationship describing the association between these two variables over the whole range of the plotted points, and was thought necessary to fragment the graph in Figure 4 into two segments, 4(a) and 4(b).

Figure 4(a) shows the concentration (y) as a function of the discharge (x) for the rising limb of the flood hydrograph until it reaches 650×10^6 m^3 d^{-1}. At this discharge the average concentration is 4,450 ppm. Figure 4(b) describes the $y - x$ relationship for x from 650×10^6 m^3 d^{-1} up to the flood peak at 870×10^6 m^3 d^{-1}. The equation of best fit covering the entire range of the flood rising discharges, 200–870 $\times 10^6$ m^3 d^{-1}, can be expressed as:

$$(7) \qquad y = -0.0243x^2 + 31.95x - 6061.2, \quad R^2 = 0.5172$$

where R^2 is the coefficient of determination.

The relationship for the first segment represented by Figure 4(a), covering the range of x from 200 to 650×10^6 m^3 d^{-1} and the second represented by Figure 4(b) for the range of x from 650 to 870×10^6 m^3 d^{-1} can be described by the following equations:

$$(8) \qquad y = 0.0936x^{1.6798}, R^2 = 0.8536$$

and

$$(9) \qquad y = -0.0283x^2 + 35.429x - 6278.9, \quad R^2 = 0.2164$$

respectively.

Eq. (8) yields $y = 4{,}970$ ppm for $x = 650 \times 10^6$ m^3 d^{-1}, a concentration that is 11.6% higher than that given by Eq, (7), and in the mean time much closer to the average of long-term observed values.

Eq. (9) is similar to (7), except that the values of the coefficients of x^2 and x, and the regression constant are somewhat different. Additionally, as one should expect, the coefficient of determination, R^2, is quite different. Furthermore, Eq. (7) underestimates the value of y for most of the given range of x compared to Eq. (9).

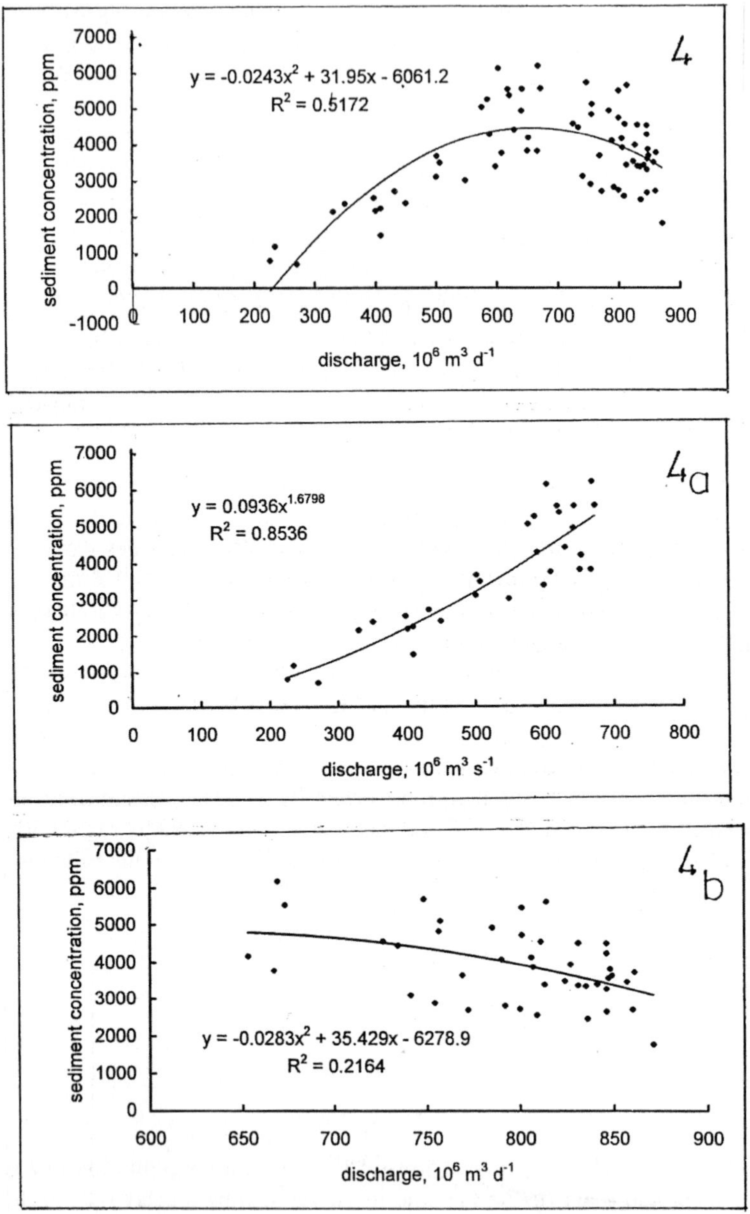

Figure 4. Suspended sediment concentration (y)-discharge (x) relationship at Gaafra on the Main Nile for the rising limb of the flood hydrograph ($200 \leq x \leq 870$ mill.m^3 d^{-1}); (a) y vs. x for $200 \leq x \leq 650$ mill m^3 d^{-1}, and (b) y vs. x for $650 \leq x \leq 870$ mill m^3 d^{-1}

The last part, which represents the concentration (y) versus discharge (x) for the falling limb of the flood hydrograph, i.e. below 870×10^6 m^3 d^{-1}, is illustrated by graph in Figure 5. The power function described by Eq. (9) provides the best fit for the plotted (x, y) points.

$$(10) \qquad y = 0.032x^{1.6505}, \ R^2 = 0.7436$$

Comparison between Eq. (8) and Eq. (10) shows that the power of $-x$ in both equations is almost identical, and the only difference is limited to the value of the regression constant. This yields a value of y for a given x, in the range from 200 to 650×10^6 m^3 d^{-1}, on the falling limb of the flood hydrograph to about 0.287 the value of y for the same x on the rising limb of the hydrograph. This is very likely due to the reduction in flow velocity during the falling limb of the flood compared to the higher velocity during the period of rising flood.

8.4.3.3 *Sediment concentration of the Tigris River in Iraq*

While discussing suspended load and solute discharge in the Tigris River within Baghdad, Al-Ansari et al. (1986) presented two graphs showing the daily sediment load versus water discharge (m^3 s^{-1}). These two graphs, (A) for the river at North Baghdad Station and (B) at Saddam Bridge, are shown in Figure 6.

In the said discussion the consequence of having more than one source of sediment supply on the relationship between the sediment load and water discharge in the main stream was reported as: “In both stations the data showed that the suspended load transport was not always directly related to water discharge. An example can be recognized where the mean daily discharge during February reached 640 m^3 s^{-1} and

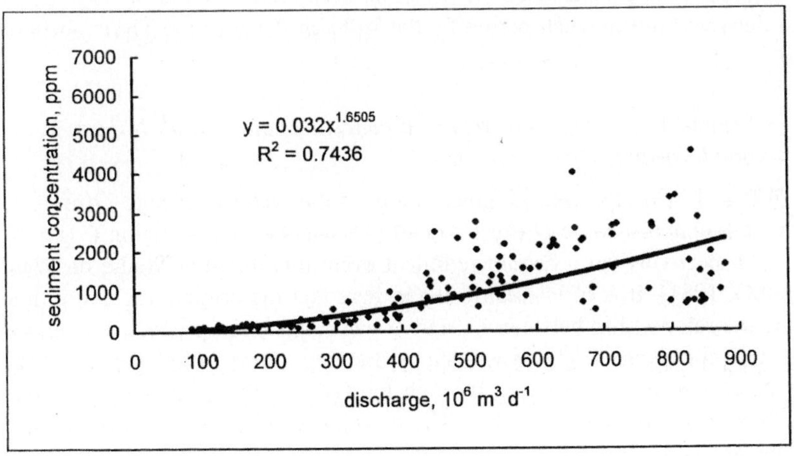

Figure 5. Concentration-water discharge relationship for the falling limb of the flood hydrograph of the Main Nile at Gaafra

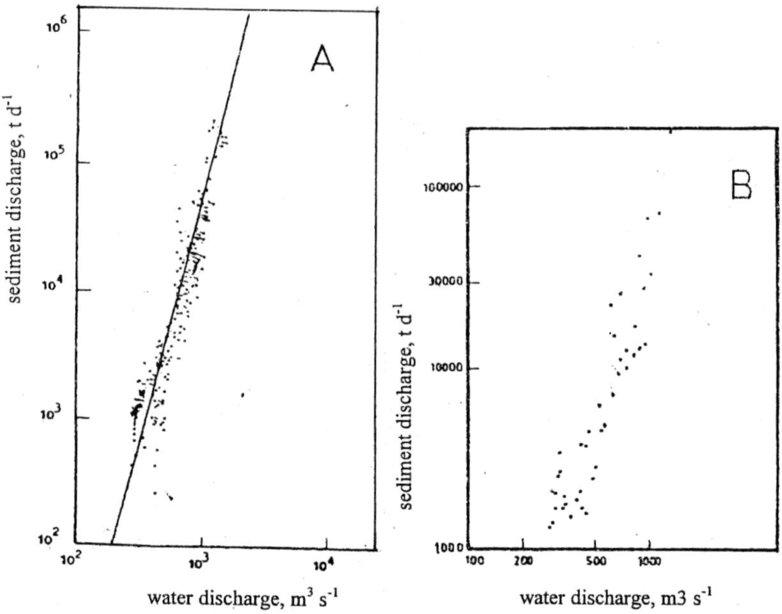

Figure 6. Sediment concentration versus discharge of the Tigris River in Iraq; (A) at North Baghdad Station, and (B) at Saddam Bridge Station (Al-Ansari et al., 1986)

the sediment discharge reached 466,984 t during that month while during March, when the mean daily discharge decreased to 598 m^3 s^{-1}, the sediment discharge increased to 743,190 t. A similar trend was noticed during April and May. During October and November, however, the increase in water discharge was accompanied with an increase in sediment load." This can be partly attributed to the significant sediment delivered during such period by the Adhaim River to the Tigris north of Baghdad.

8.4.3.4 Sediment discharge and sediment concentration in Wadi Najran, Saudi Arabia

Figures 7(a) and 7(b) represent graphical plots of the water discharge Q(m^3 s^{-1}) versus the sediment discharge Q_s(kg s^{-1}) and Q versus the concentration C (kg m^3 or gm l^{-1}) respectively for a certain sediment event monitored at Mudiq on Wadi Najran (FAO, 1981). It is of interest to report here that the original regression line expressing the relationship between Q_s and Q was adjusted to produce a weighted estimate of Q_s for given Q. This was done by drawing a straight line parallel to the original regression line and passing through the $(Q_s)_m - (Q)_m$ intercept, m refers to a weighted estimate of the variable. The original (unadjusted) and the adjusted expressions are as follows:

(11) $$Q_s = 1.14Q^{1.3334}$$

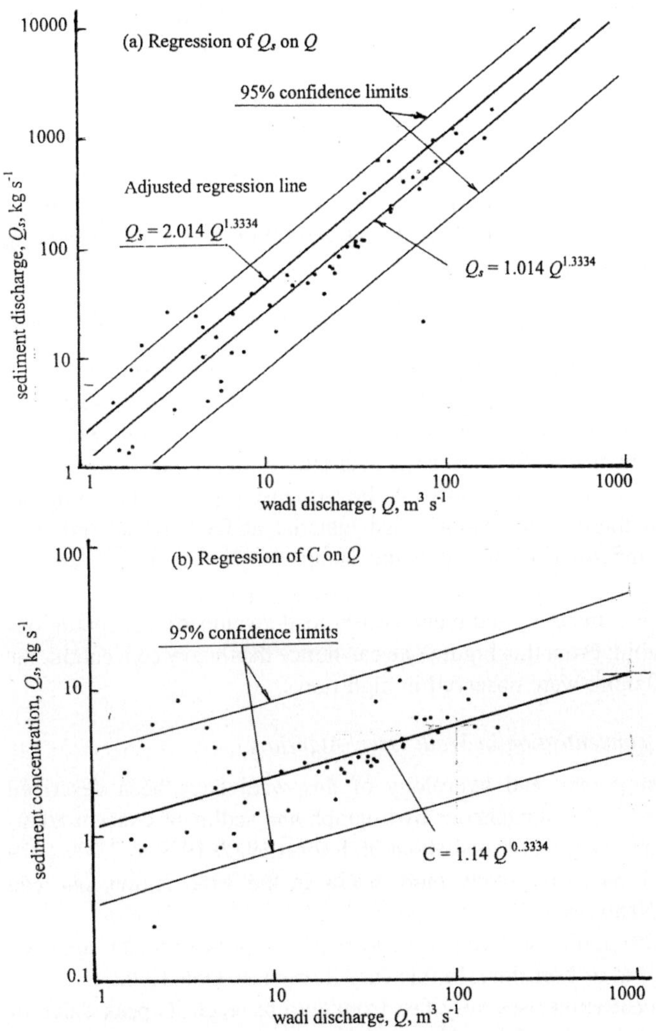

Figure 7. Sediment discharge and sediment concentration versus water discharge for Wadi Najran at Mudiq, Saudi Arabia (from FAO Paper No. 37, 1981)

and

$$Q_s = 2.014Q^{1.3334} \qquad (12)$$

respectively. The last two expressions, Eqs (11) and (12) indicate that the adjusted sediment discharge is 76.7% more than that estimated from the original or unadjusted regression for all water discharges. The scatter of points and the confidence interval are proportionately the same for C as for Q_s, with respect to Q.

Since $Q_s = CQ$, the slope of the regression line of C on Q differs from that for Q_s on Q in that the power to which Q is raised becomes less by 1, i.e.

(13) $C = 1.14Q^{0.3334}$

The sediment concentration in Wadi Najran at Mudiq, S. Arabia when compared to that of the Tigris at North Baghdad, Iraq, shows that the concentration in the former is nearly ten times that of the Tigris for $Q = 1000$ m^3 s^{-1}, and increases up to more than 5,000 times for Q = 100 m^3 s^{-1}.

8.4.3.5 Bed material concentration in Wadi Zabid, Yemen

Lawrence (1987) reported the difficulty encountered in getting the bottle samplers to the near-bed zone at high flows. It seems that the same difficulty had been encountered while sampling sediment-laden water from Wadi Najran when the discharge exceeded 200 m^3 s^{-1}. Besides, large quantities of trash, transported by the leading edge of a flood wave, are trapped on the suspension gear of the sampling equipment. To avoid these complications, bed material and wash load sediment concentrations were measured using automatic pump samplers in Wadi Zabid in 1982–83.

Figure 8 shows bed material sediment (wash load excluded) concentrations measured in Wadi Zabid. From this Figure one can notice that heavy concentrations, approaching 100,000 ppm, were observed in high flows.

8.4.3.6 Sediment concentration in Wadi Isser, Algeria

The catchment physiography and hydrology of this wadi have been described earlier. Figure 9 shows the instantaneous hydrograph and sediment concentration-time graph of a certain event that lasted about 90 h form 19.09.1979 to 23.09.1979 (Pitt & Thompson, 1984). Like many more wadis in the Arab region, one can observe three salient features:

- The discharge hydrograph rises from zero to reach its peak ordinate, approximately 600 m^3 s^{-1}, in a short time. In this case it was less than 12 h.
- The sediment concentration rises very fast from zero to reach its peak value of about 340 gm l^{-1} in less than 2 h.
- The peak concentration occurs much earlier than the peak discharge. The sediment concentration gets two values for the same discharge; one for the rising limb of the discharge hydrograph and one for the falling limb. This state of affairs, which is known as loop form, makes it impossible to get a single relationship between C and Q for all values of Q.

8.5. RESERVOIR SILTATION

The rate of sediment deposition in a reservoir depends on the rate of sediment supply by the inflowing stream(s), and hence on the sediment yield of the catchment area(s).

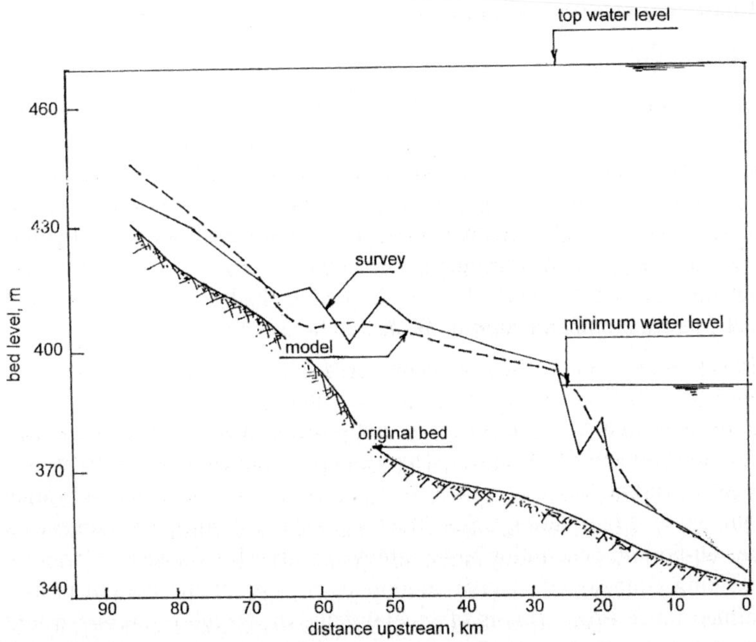

Figure 8. Bed material concentration-discharge relationship for Wadi Zabid, Yeman (from Lawrence, 1987)

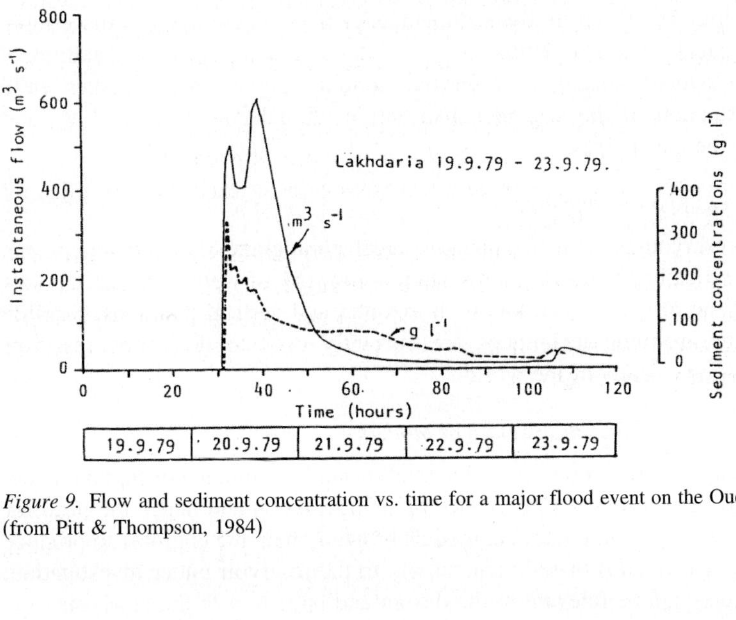

Figure 9. Flow and sediment concentration vs. time for a major flood event on the Oued Isser, Algeria (from Pitt & Thompson, 1984)

8.5.1 Monitoring reservoir siltation

8.5.1.1 Simple sounders

A graduated pole is used to measure the water depth. More typically is to lower a certain weight attached to a vertical line until it rests on the reservoir bed and hence measure the depth below the water surface down to the bed. By repeating the measurement at known time intervals it becomes possible to assess the sediment accretion in the reservoir. This method can be used in shallow water bodies. However, since it is a point measurement, local high or low bed areas can yield misleading results. As such the method can be limited to shallow reservoirs bearing in mind that the result obtained is rough and has to be used cautiously.

8.5.1.2 Echo-sounding

The need for a more accurate result than the one provided by the simple sounders and/or monitoring sediment deposition in deep reservoirs, is better served by means of repeated echo-sounding traverses on well-defined range lines. This method yields details of depth and bed morphology. Selection of suitable transmission frequencies permits some sub-bottom penetration to be achieved with echo-sounders. Correctly calibrated recoding echo-sounders give continuous records of the bottom profile along the defined range lines. This is of special value in storage reservoirs where the maximum thickness of deposits does not exceed 5 m. In this situation it may be possible to identify the pre-reservoir bed, thus making it possible to construct isopach maps of the deposits. However, for reliable sub-bottom profiling it is essential to use high-powered signals. Instruments, like the ones developed for marine exploration, permit deep penetration measured in tens of meters, depending upon the bed sediments. As observed in cores from reservoir floor deposits, irregularities may be present in the sediment sequences. These are also discernible using geophysical methods. Utilisation of side-scan sonar provides additional dimension to remote sensing of reservoir siltation. The acoustic images yield detailed information of the size and distribution of underwater mass flows and landslides on the basin sides.

8.5.1.3 Areal photogrammetry

If a storage reservoir is drained annually, areal photogrammetry techniques can be applied practically provided that a sufficient network of well-marked fix points exists around the reservoir with known horizontal and vertical positions. Satellite remote sensing mapping techniques are currently used to determine reservoir volumes in different parts of the world.

8.5.1.4 Conceptual and mathematical models

As a preliminary step in a reservoir sedimentation study, a simple conceptual model can be used to determine the significance of sedimentation and hence the need for more detailed studies. Computational models basically help the engineer to produce upper and lower estimates of sediment supply to the reservoir under investigation. Other advantages can be relevant to the design and operation of the reservoir.

Mathematical models are designed to simulate water flow and sediment behaviour in a river system containing one or more reservoirs. Hydrological data and reservoir surveys are unavoidable to provide the necessary calibration and validation data for such models, which can then be used to forecast the future life age of the modelled reservoirs.

As an example, sediment routing in the model developed by Okabe et al. (1993) was divided into three sub-regions as shown in Figure 10. The non-uniform flow routing equation and so the mass conservation relationship needed for calculating bed level change were derived. The latter was adjusted to the sediment condition in each sub-region. In this simulation model, the calculation of the variation in the grain size composition due to bed material load was carried out using another model. The numerical solution of the equations was obtained with the aid of a finite difference method. The model was applied to a certain reservoir used for waterpower generation in Japan. Simulated variation of stream-wise profile of the reservoir bed was drawn and compared with the observations. Furthermore, the simulated vertical distribution of the mean grain size in the top sediment layer was presented for the representative cross sections. The comparison led to the conclusion that the simulation results were in satisfactory agreement with the observations, although the alteration of the strata in the sediment layer was not as clear as the observations (Okabe et al., 1993).

One of the well-known models is called HEC 6BP. The model was applied to the Tarbela Dam on the Indus River, Pakistan. The dam was designed to impound 14×10^9 m^3 of the mean annual inflow of 79×10^9 m^3. A survey carried out in 1979 indicated that 6% of the reservoir storage volume had been lost due to sediment deposition since it was put into operation in 1974. This period was simulated using the said model and some of the results obtained are illustrated in Figure 11. This figure shows that the agreement between the model and surveyed bed profiles is satisfactory.

8.5.2 Trap efficiency of reservoirs

This is the ratio of the volume of sediments trapped in a reservoir to the volume of sediments entering it, both over the same period of time. This ratio is necessary to

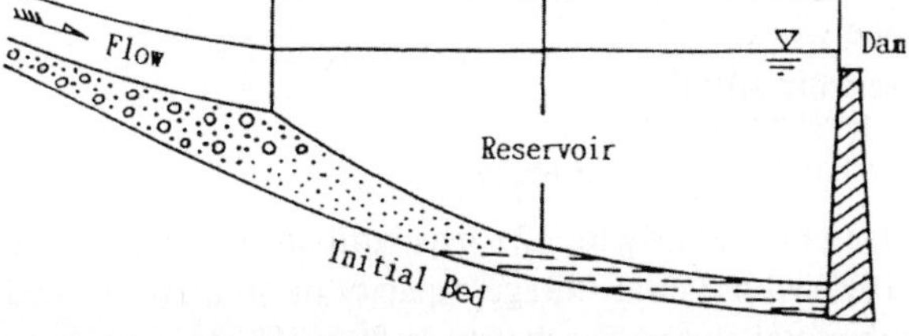

Figure 10. Schematic diagram of the three sub-regions for sediment routing (Okabe et al., 1993)

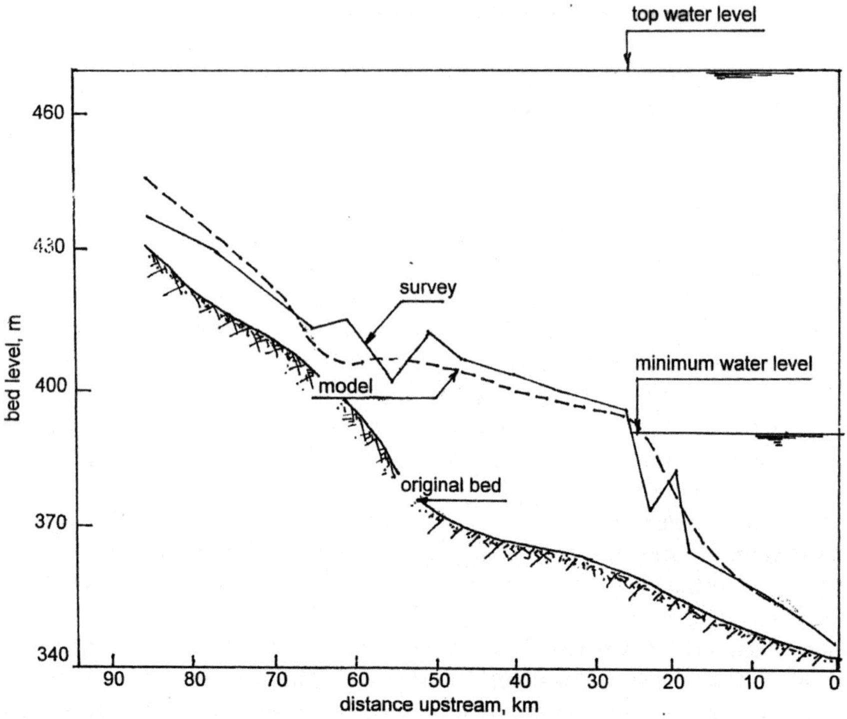

Figure 11. Proving of HEC 6BP with Tarbela Reservoir data (Thompson, 1986)

estimate the length of time needed before a certain volume of a planned or existing reservoir will silt up. The number of years at which 50% of the storage capacity is filled up with sediments determines the effective life age of a reservoir.

There are several formulas, e.g. the formula of Brune (1951) that can be used for estimating the trap efficiency. Temporal variability of inflow; detention time of water; and grain size of sediment load are the most paramount factors rendering the application of this formula for arid zones favouarable, especially when the upper envelope curve (coarse sediments) is used. This envelope curve gives a trap efficiency of 100% for reservoirs with storage capacities at or larger than 0.02 times the size of watershed.

8.5.3 Case studies of reservoir siltation

8.5.3.1 King Talal and Kafrein Reservoirs, Jordan

The King Talal Dam, on the Zarqa River (Figure 12), was built on two stages. The first stage was completed in 1977 with a total storage capacity of 56×10^6 m^3 and the second in 1987, bringing the total reservoir capacity to 80×10^6 m^3.

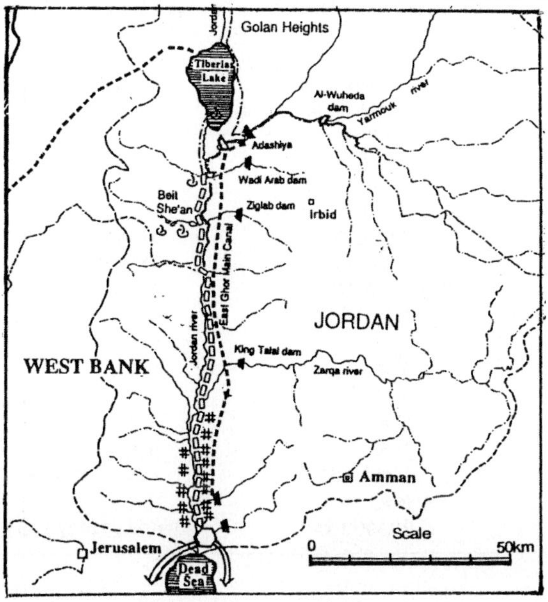

Figure 12. Location map of King Talal Dam, Jordan

Since the first stage, the sediment deposition in the reservoir has been monitored whenever possible. Table 4. lists the annual water inflow to the reservoir together with the volume of accumulated sediments for the period 1979/90–1993/94 (Malkawi & Abdulla, 1997). The irregularity of the annual sediment deposition in the reservoir and its disproportionality to the annual volume of inflowing water has been largely affected by the necessary excavation for heightening the dam body and the development of the lower Zarqa River Basin for flood protection and forest preservation.

The Kafrein Dam was built in 1968 on the Wadi Kafrein as a component of the Jordan Valley Scheme. The initial capacity of the reservoir was 4.75×10^6 m^3 declined to 2.3×10^6 m^3 by 1990, i.e. at an average annual rate of sedimentation of 107×10^3 m^3 in those 22 years. Since 1990 the annual rate of reservoir sedimentation increased significantly to reach an average of 265×10^3 m^3 for the period 1990–94, leaving only 1.39×10^6 m^3 for live storage by 1994. In view of its importance for water supply, the dam has been heightened to provide a storage capacity of 8.5×10^6 m^3.

8.5.3.2 *Reservoirs in Oman*

Until now, the available data of sediment yield in Oman are too little and estimates of the necessary storage capacity of reservoirs are therefore not as accurate as should be. However, it is anticipated that de-siltation of the reservoirs cannot be avoided after numerous flood events. The settling of sediments in the reservoir

Table 4. Annual series of inflowing water and deposited sediments in King Talal Reservoir, Jordan (from Malkawi & Abdulla, 1997)

Water	Annual flow, 10^6 m^3	Accumulated sediment, m^3	Annual sediment, m^3
1979–80	161.246	3,150,771	
1980–81	78.743	3,841,810	691,039
1981–82	63.507	4,629,582	787,772
1982–83	99.619	5,073,709	344,127
1983–84	57.604	5,407,624	333,975
1984–85	65.579	5,758,713	351,089
1985–86	43.148	N.A.	N.A.
1986–87	65.683	N.A.	N.A.
1987–88	116.075	N.A.	N.A.
1988–89	69.068	7,431,120	1,672,407 (4 years)
1989–90	60.862	N.A.	N.A.
1990–91	60.220	N.A.	N.A.
1991–92	234.340	10,745,852	3,314,732 (3 years)
1992–93	118.677	N.A.	N.A.
1993–94	79.458	10,990,100	0,244,248 (2 years)

itself is desirable as it protects the recharge area from being silted up, which would reduce the infiltration rate.

Reservoirs that have been under operation for more than five years have been desilted mechanically. For these reservoirs sediment yield figures have been derived and listed in Table 5. Since the periods of record are too short and the sediment measurements are rather approximate, the data in the said table should be dealt with cautiously (Al Muqbali & Schmid, 1995).

8.5.3.3 Marib Reservoir, Yemen

Marib Dam is built on Wadi Adhana in Yemen. The wadi, morphologically speaking, belongs to the eastern escarpment of the country. Most of the catchment

Table 5. Measured sediment volumes in reservoirs in Oman and derived sediment yields (from Al Muqbali & Schmid, 1995)

Recharge dam	Time period	Storage volume, m^3	Sediment volume, m^3	Catchment area, km^2	Sediment yield, m^3 km^{-2} y^{-1}
W.Jizzi	1990–92	5,400,000	160,000	812	66
W. Hitti Salahi 1989–92	1989–92	550,000	173,000	362	119
W. Quryat	1985–92	140,000	50,000	375	19
W. Tanuf	1990–92	600,000	40,000	168	79
W. Ghul	1990–92	400,000	30,000	179	56
W. Fulayj	1991–92	700,000	25000	685	18
W. Al-Kabir	1991–92	500,000	35000	971	18

area consists of Precambrian basement rocks; the eastern part around Ma'rib consists of a thick succession of carbonate rocks, the north western of sandstone and the southwestern part of volcanics. The annual precipitation is in the order of 100 mm and the Penman free water evaporation around 2,110 mm.

The Marib survey area is receiving surface runoff from the following catchment areas (map in Figure 13, WRAY Report No. 8.2, Ministry of Oil and Mineral Resources of Yemen & The Dutch Research Institute of Applied Geoscience, TNO, The Netherlands, 1988)):

– *Wadi Adhana (Marib Reservoir)*	
Runoff-absorbing zone	3,300 km^2
Runoff-producing zone	8,200 km^2
– *Wadi Adhana at gauging site I)*	7,700km^2
– *Wadi Masil*	823km^2
– *Wadi As Saila*	168km^2
– *Wadi al Mil*	82km^2
– *Remaining areas*	67km^2
– *Marib survey area*	600km^2
Total catchment	13,240km^2
– *Net runoff-producing catchment*	9,940 km^2

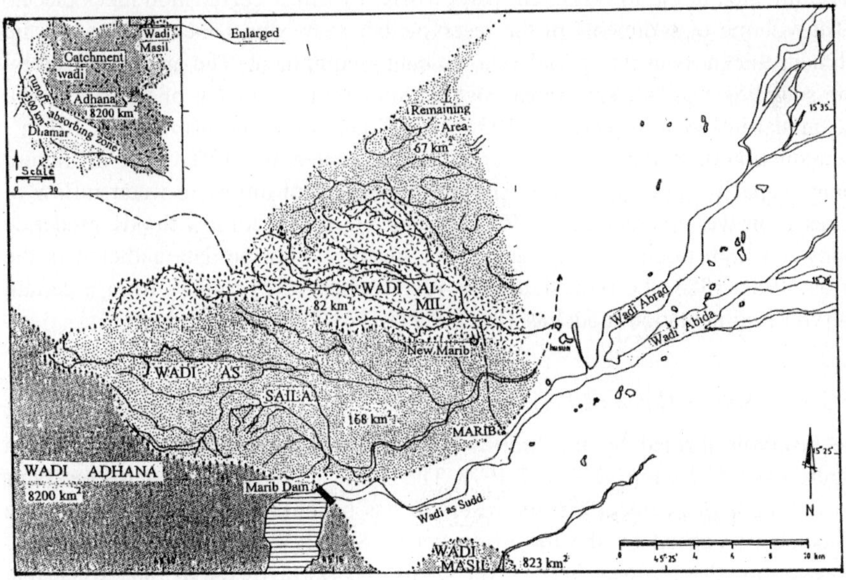

Figure 13. Location map of catchment areas producing surface runoff to Marib Reservoir, Yemen (WRAY Report No. 8.2, Ministry of Oil and Mineral Resources of Yemen & The Dutch Institute of Applied Geoscience, TNO, The Netherlands, 1988)

The gradient of the wadi and its tributary channel network varies from steep (5.4-6.5 m km^{-1}) to intermediate (2.8-3.8 m km^{-1}) and mild (0.5 m km^{-1}). The steep gradient results in low net sedimentation due to its high flow velocity. Contrarily, the mild slope results in higher net sedimentation due to its low flow velocity.

Four observation locations were chosen; one at the edge of the reservoir where the valley width is 500 m, the second at a point where the width narrows down to 150 m, the third at the location of the stream gauging station and the fourth is at a place called Sherwab, where a downstream widening of the valley takes place. Wadi floods were examined and three classes were distinguished; high (depth of water > 3 m), moderate (2–3 m) and low (depth < 2 m). The flood of April 1986 can be described as low with depth of water of 1.5–2.0 m, maximum discharge of 225–300 m^3 s^{-1} and a total runoff of 81×10^6 m^3. A year later, the floods of April 1987 ranged from moderate to high, with water depth of 3.4–5.0 m, maximum discharge 540-750 m^3 s^{-1} and a total runoff of $120–177 \times 10^6$ m^3. During high floods erosion power is quite strong causing deeper wadi channels, deposition processes are in the form of debris flows and flood sheets. Sedimentation takes place across the full width of the valley resulting in chaotic structure and a conglomeratic body. Most of the sand is flushed into the Marib Reservoir. Moderate floods result in a sheet flood deposition process. Sediments, which are mainly sand grains, are deposited within the full extent of the valley especially where the gradient is relatively low. The flow in a low flood has a normal fluvial pattern. Fine sandy-silt is accumulated in the low-gradient parts where a vertical aggradation takes place.

The volume of sediments in the reservoir began by direct measurement of the sediment thickness using special echo-sounding equipment. The area surveyed by echo-sounding was 8.2 km^2 (area covered after initial flooding of the reservoir). The annual inflow to reservoir is 103×10^6 m^3 and sediment inflow 1.5×10^6 m^3. The reduction of storage in a 3-y period was estimated as 1.12% of the maximum storage capacity (400×10^6 m^3). The ratio of sediment inflow to water inflow to the reservoir was assumed as 1.45%, 2.0% and 5.0 to 14% for low floods, moderate floods and high floods. Using these ratios together with the data gathered in the period, 1986–1988, the reduction over 40-y service was obtained from a certain reservoir sedimentation model as 15×10^6 m^3.

8.5.3.4 *Sidi-M-Hammed Reservoir, Algeria*

The reservoir formed by the Sidi-M-Hammed Benaouda (SMB) Dam, Algeria (Figure 14), was put into service in 1978. The initial storage capacity of the reservoir was 335×10^6 m^3. Since then the reservoir has become threatened by deposition of sediments carried by the inflowing water. Shortly before the exploitation of the reservoir, the National Service for Dams and Barrages (ANB) suggested that the average annual deposition of sediments could be assumed as 1×10^6 m^3. As the density of sediments is 1.6 t m^{-3} and the surface area of the drainage basin 4,900 km^2, the specific erosion was then estimated as 3.26 t ha^{-2} y^{-1}.

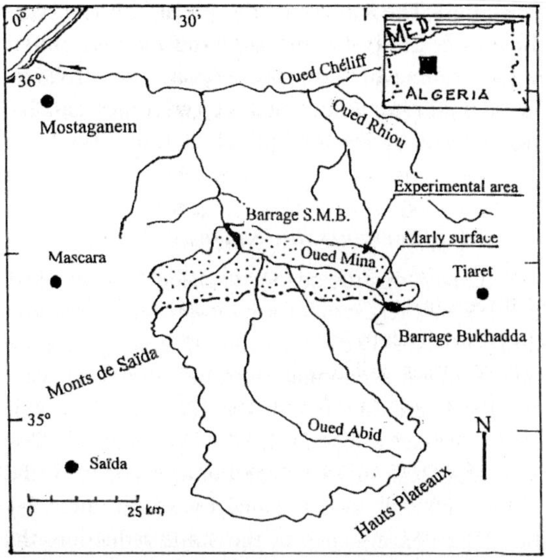

Figure 14. Location map of Sidi M. H. Benaouda Dam in Oued Mina, Algeria

When the calculated erosion rate obtained from the micro-basins is taken into account, and the surface area covered with marl in the basin (erosion surface) taken as 1,000 km^2, it should contribute to the quasi-total sediment deposition in the reservoir an annual rate of 16 t ha^{-1}. This does not exclude that the sedimentation figure given by the National Agency for Barrages had been underestimating the actual sedimentation in the reservoir (Touaïbia et al., 1999).

8.5.3.5 Reservoirs in Morocco

Table 7, Appendix III lists some data regarding water and sediment inflow to 15 reservoirs scattered in different parts in Morocco (sources: Country Report of Morocco, 1986, and personal communication with University of Liège, Belgium, 2000). The sum of the catchment areas supplying water and sediment to the reservoirs is $383,215$ km^2. The annual runoff is 9.25×10^9 m^3 corresponding to an average depth of 24.1 mm. The initial capacity of the reservoirs varied from 13×10^6 m^3 to $2,724 \times 10^6$ m^3. From the listed data it is evident that, with the exception of Abdel-Karim, Lalla Takerkoust and Mohammed V Reservoirs, the annual rate of reservoir siltation as percent of the reservoir capacity is less than 1%. In general, this percentage decreases with increasing reservoir capacity. The weighted average for all 15 reservoirs is almost 0.5%, i.e. the average life age (50% of the reservoir initial capacity is silted up) can be fairly estimated as 100 years.

8.5.3.6 Reservoirs in Tunisia

Table 8, Appendix III, lists some hydrological data of six storage reservoirs located in several parts of Tunisia. Four reservoirs have small catchment areas thereupon small initial storage capacities. The largest of the six is the reservoir on Wadi Mellegue with an initial capacity 268×10^6 m^3, bringing the total initial capacity to 383.5×10^6 m^3 Since the volume of sediment deposited in all reservoirs is about 8.5×10^6 m^3 y^{-1}, the annual rate of sedimentation becomes 2.025% of the initial capacity. In fact, this percent varies for individual reservoirs from 0.83 to 2.81.

8.5.3.7 Roseires and Khashm el-Girba Reservoirs, the Sudan

The Khashm el-Girba and Roseires dams, constructed on the Blue Nile in 1964 and 1966 respectively, created multi-purpose reservoirs for hydropower generation, irrigation and flood control. Within the same decade of commissioning of each dam, serious problems arising from sediment accumulation in the reservoirs were encountered. The analysis of two bathymetric surveys of the Roseires Reservoir in 1976 and 1981 showed that there has been a loss in the total capacity (3.02×10^9 m^3) of 18% in 10-y of operation and 21% in 15-y. The initial capacity of the Khashm el-Girba reservoir (1.3×10^9 m^3) was also reduced by 25.4% after 7-y and by 43.2% in 12-y of operation. As a consequence of this large reduction, the situation regarding the supply of water for irrigation reached critical dimensions and the surface of cropped area progressively reduced. Hydropower generation too decreased drastically, affecting industrial and domestic water supplies in the Sudan (El-Sheikh et al., 1991). The change of the effective reservoir capacity against water level in the period 1964–90 for the Khashm el-Girba, and in the period 1966–81 for the Roseires are shown in Figures 15 and 16 respectively (Shahin, 1993).

El-Sheikh et al. (1991) applied Eq. 5 and 6 to the historical discharge series together with appropriate trap efficiencies to generate sediment volumes deposited

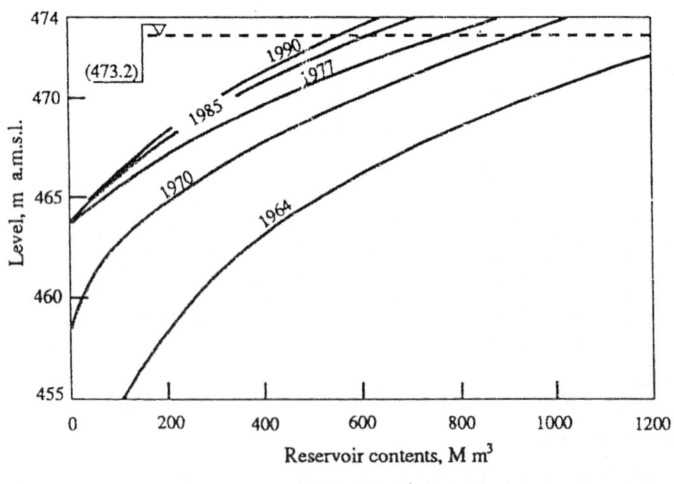

Figure 15. Reduction of live storage of the Khashm el-Girba Reservoir with time (Shahin, 1993)

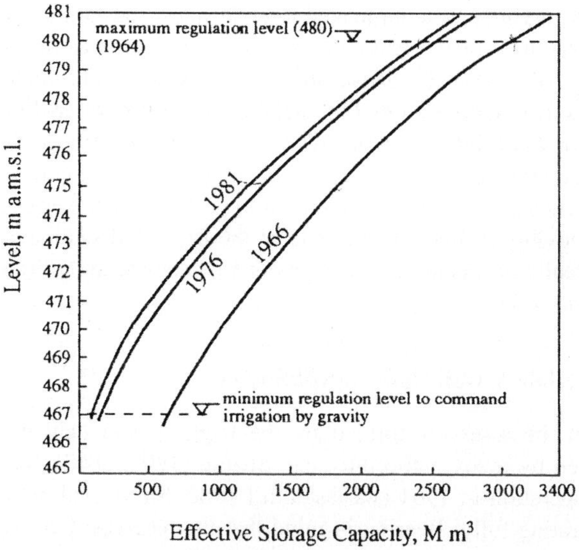

Figure 16. Reduction of live storage of the Khashm el-Girba Reservoir with time (Shahin, 1993)

in the reservoir. Three values of trap efficiency, corresponding to three sediment grain types, were computed for the various phases of the flow. Calculation was based on a formula expressing the trap efficiency as a function of the reservoir capacity-streamflow ratio, fall velocity of sediment particles, mean depth of reservoir, and duration of spilling. The accumulated deposition as obtained from the two equations was compared to the results of the bathymetric surveys of 1976 and 1981. The results given by the polynomial expression were generally higher than those obtained from the linear expression. Computations using the linear regression showed that the average annual siltation in the 10-y period, 1966/67–1975/76, was 43×10^6 m^3, with the subsequent five years decreasing to 41×10^6 m^3 and further to 30×10^6 m^3 in the period 1981/82–1987/88. The bathymetric surveys showed a rate of siltation of 54×10^6 m^3 in the first ten years followed by a sharp drop to reach an average of 20×10^6 m^3 in the subsequent 5-y period, 1967/77–1981/82. Both calculated and measured rates are far higher than the design rate of 15×10^6 m^3. It is worth mentioning that the calculated siltation rate corresponding to the excessive flood of 1988/89 reached 82.5×10^6 m^3. This happened to be contrary to what was thought of that the annual siltation has reached a stable rate of 20×10^6 m^3 y^{-1}.

The percentage of annual deposition was obtained from the computed annual sediment loads and deposited volumes. Denoting this percentage by $Q_{s\%}$, the following relationship was developed:

(14) $$Q_{s\%} = 5.25 \times 10^3 Q^{-0.48}, \mathrm{R}^2 = 0.83$$

The case of the Roseires Reservoir is a complicated one. Besides, the rainfall in the Ethiopian part of the catchment of the Blue Nile is generating the streamflow that reaches the reservoir in the Sudan. Likewise, the sediments deposited in the reservoir, leading to continuous reduction in its life age, are the product of erosion caused by the same rainfall on the Ethiopian highlands and transported by the same streamflow. Control of this erosion is, at least for the time being, beyond the control of the Sudanese authorities. As such there is no option left to prolong the life age of the reservoir other than removing as much as possible of the deposited sediments through mechanical or hydraulic means until an agreement between the authorities in the Sudan and Ethiopia is reached.

8.5.4 Reservoir of the High Aswan Dam, Egypt/Sudan

The sediment deposition in the reservoir formed by the High Aswan dam has been presented and discussed by many authorities, e.g. Shahin (1993, 2002). The reservoir began its partial operation in 1964 and lasted till 1968. Since 1968 until present the reservoir is operating fully. Bearing in mind that the reservoir being a huge storage facility, its trap efficiency is practically 100%. This means that all inflowing sediments are deposited in the reservoir.

The annual series of the volume of sediment deposited annually in the reservoir for the period 1964–89 is listed in Table 9, Appendix III. The annual deposition in that period varied from a minimum of 18×10^6 m^3 to a maximum of about 184×10^6 m^3, with an average value of 80×10^6 m^3 y^{-1}. It goes without saying that the annual volume of sediment inflow, subsequently settling in the reservoir, is depending on the volume of the flood in the same year. Assuming the volume allocated to storage of sediments (dead storage) to be 30×10^9 m^3, one can fairly expect this volume to be totally filled with sediments in a matter of 375 years after

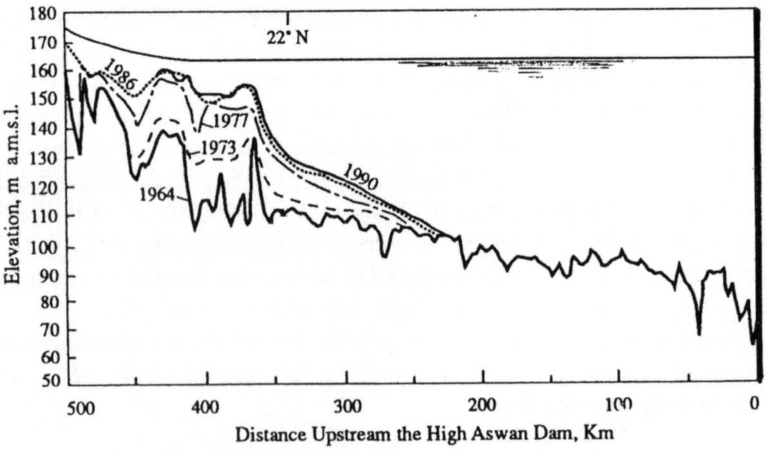

Figure 17. Accumulation of sediments in the reservoir of the High Aswan Dam

the reservoir has been operative. The old figure declared by the High Dam Authority, Egypt, was 500 years based on annual volume of sediments of 60×10^6 m^3. Later, the Ministry of Irrigation and Water Resources of Egypt, using a mathematical model describing the process of reservoir sedimentation reached a different figure, 408 years (Shalash & Makary, 1986), which is about 8.5% more than 375 years.

Bathymetric survey of the reservoir sides and bottom is carried out once every few years. The bottom profiles as observed in the years 1964, 1973, 2977, 1986 and 1990 are shown in Figure 17. The figure shows that after 25 years since the reservoir has been put into operation, the first 200 km upstream the location of the dam are free of any deposited sediments. The sediments are accumulating in the most upstream part of the reservoir and are slowly advancing with time towards the dam. The reservoir in certain reaches has become full with sediments, which have reached more than 40 m in thickness. Additionally, the reservoir in the most upstream 120–130 km or so has changed in the course of years, compared to its remaining length, to a shallow lake.

CHAPTER 9

GROUNDWATER RESOURCES

9.1. INTRODUCTION

Aquifers (water-bearing and transmitting formations) serve within water resources systems as 1) water resources, 2) water storage facilities, and 3) means of water quality upgrading. Groundwater as a source of water is extremely important in arid and semi- arid zones. It can also be used as a supplementary source in humid areas where the distribution pattern of precipitation does not match the pattern of demand on water.

Groundwater, unless contaminated, is generally of better quality than surface water. In many areas in the Arab Region ground water is obtained from basement rocks, fissures, and joints of the igneous and metamorphic rocks, and from layers of weathered basement rocks. The quality of this water is often unsuitable. Water is also extracted from alluvial deposits and aquifers in coastal areas. Excessive withdrawal of groundwater from these coastal aquifers can lead to salt water intrusion and to deterioration of the pumped aquifers.

In view of the scarcity of surface water resources in the Arab Region, groundwater is the primary source of water in some countries and the secondary source of the other Arab countries. It is regarded as the primary source in the United Arab Emirates, Bahrain, Qatar, Saudi Arabia, Libya and Djibouti. Groundwater in the remaining countries like Morocco, Algeria, Tunisia, Egypt, Jordan, Kuwait, Oman and Yemen is still regarded as a secondary source of water. Large-scale irrigation with groundwater from deep well fields and boreholes is becoming widespread in countries like Libya and the Sudan. Recent estimates show that groundwater comprises 5–8% of the total water use in Egypt, a country that is traditionally depending on river water. A much larger percent of demand for water in Tunisia, Jordan and Syria is covered from the groundwater resources in these countries.

Groundwater resources are generally subdivided into three zones. These areactive zones in the upper earth's crust with a depth not exceeding 200 m below land surface and above the erosion base; less active zones above the sea level with a depth of up to 400 m, interacting with various, extensive depressions and deep valleys; and zones below the sea or ocean level with a depth of 2,000 m. Over 90% of groundwater abstraction is from the top zone and the remaining 10% from the middle zone. There is hardly any abstraction from the zone below the ocean surface.

Geology, hydrogeology and groundwater quality are the three major issues one has to encounter while dealing with groundwater as a source of water, and should reckon with. In attempt to reach the goal of this book, Chapter 2 has been prepared with the aim of providing a reasonable, though approximate, coverage of the geological aspects of the region. A more detailed coverage can be found in specialised reference books of geology of the Arab Region. Raw data of the quality of groundwater are presented in the Tables included in Appendix III.

While presenting and discussing the storage in groundwater basins, yield of aquifers, natural flow and recharge, groundwater extraction and artificial recharge and similar issues, some of the available data of hydrogeological characteristics of aquifers are discussed in this chapter. Raw data such as measurements and estimates of the quantitative aspects are included in Appendix II.

9.2. BRIEF DESCRIPTION OF GROUNDWATER BASINS IN THE ARAB REGION

A groundwater basin or reservoir is a body that usually consists of a number of groundwater formations. A basin by definition is body that has a single or multiple natural recharge area(s) and a single or multiple discharge area(s). Hydrogeological properties of the individual aquifers largely influence the basin characteristics. The quantities and patterns of groundwater extraction, and natural and/or artificial discharge dictate the state of equilibrium of the basin. Moving from the extreme west to the extreme east of the Arab Region, one finds the following groundwater basins:

Region	*Groundwater basin*
Atlas Region	*Tel Atlas, Upper Plateau, Reef, Atlas coastal plain, Upper and Middle Atlas, and Desert Atlas.*
Great Desert Region	*Tarfaya-Dakhla, Nouakchot, Senegal, Taoudeni, Great Erg, Niger, Marzuq, Hammada Al-Hamra and Jifara.*
Nile Basin Region	*Western Desert, Nile Delta and coastal plain, Red Sea and Sinai, Um Rawaba and Western Sudan, and African Horn*
Eastern Region	*Coastal plain, Middle zone, and Tigris and Euphrates, and*
Arabian Peninsula	*Western, Eastern, Omani and Yemeni Mountains.*

The lithology of a certain water-bearing formation is connected to its geologic age. The geological succession introduced by Lloyd (1990) is as follows:

Geologic Age	*Lithology*
Quaternary	*Lacustrine deposits, Sebkhas, sand dunes and alluvium*
Miocene	*Sands, interbedded clays and sandstones* (2×10^6 *y*)
Oligocene	*Sands, sandstones, clays, limestone and basalts* (54×10^6 *y*)

Eocene	*Limestone, dolomites, marls and sandstones*
Upper Cretaceous	*Shale with some sandstone and limestone* (136×10^6 *y*)
Lower Cretaceous	*Sandstones, interbedded shale and clays (Nubian)*
Undifferentiated Paleozoic	*Shale and sandstone*
Cambro-Ordovician	*Sandstone and quartzite* ($> 300 \times 10^6$ *y*)

The crystalline rock (Pre-Cambrian) is exposed in large surfaces in the Arab region. Examples can be found in Mauritania, south and west of Algeria, southern part of Libya, and southwest and east of the Sudan and east of Egypt. It is also exposed along the coast of the Red Sea. Thick layers of sand and sandstone suitable for groundwater storage were deposited on the basement rock some 300 million years B.P. The thickness of the sandstone deposited in that period is estimated at about 720, 2,900, 3,400 and 2,500 m in Jordan, the Arabian Peninsula, Siwa Depression in Egypt and Marzouq Basin in Libya respectively. Later, additional layers of sand and Nubian sandstone covered a huge surface extending from northern Sudan to the Western Desert in Egypt and Libya. By the end of the Mesozoic Period the thickness of sandstone reached a total of 1,220, 5,100, 4,250 and 2,900 m for the mentioned locations in their respective order. The thickness of the sandy formation in the Sudan reached 3,000 m. Besides, in a later part of the Mesozoic Period thick layers of poorly permeable limestone were deposited in the Arabian Peninsula.

In the Tertiary and Quaternary Periods alternating series of lacustrine calcareous material and sand were deposited in many areas thus forming Al-Hamada in Algeria and Morocco, and together with limestone in Egypt, Syria, Iraq and the southern part of the Arabian Peninsula. These limestone formations can store groundwater under special conditions. The Atlas Region belongs to the broader Mediterranean Region. It is characterised by formations dominated by clays calcareous rocks and dolomites of the Lower Jurassic time. These formations can be found in the Upper Atlas in Tunisia, and in the high Plateaus of Morocco and Algeria. As a matter of fact, North Africa contains huge reservoirs. The locations of these reservoirs are shown on the map in Figure 1 (Margat & Saad, 1984)).

Subsections 9.2.1–9.2.6 describe briefly the groundwater basins in the Western and Central Subregions of the Arab Region while subsections 9.2.7 deal with some of the basins in the Eastern Subregion. The available hydrogeological characteristics of the basins in these sub-regions are summarized in Tables 12 and 13, Appendix II.

9.2.1 Grand Occidental Erg

This reservoir, which is often referred to as 'Continental Intercalaire', is located south of the Atlas in Algeria. Its surface area is estimated as 330,000 km^2, of which an area of 180,000 km^2 forms an artesian basin. The average thickness of the reservoir is between 250 and 600 m. The average value of the transmissivity is 43.2 m^2 d^{-1} and effective porosity 0.2. The annual natural recharge to the reservoir is estimated as 150×10^6 m^3 (Gischler, 1979).

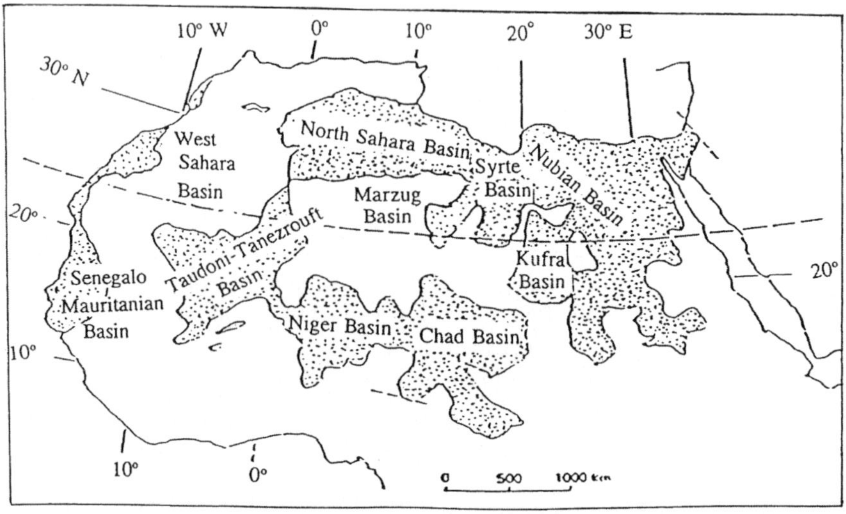

Figure 1. Location map of groundwater basins in North Africa (Margat & Saad, 1984)

9.2.2 Grand Oriental Erg

This basin, which is often referred to 'Complex Terminal', is located east of the Occidental Erg. The eastern edge of the Complex Terminal runs along the Algerian-Tunisian frontier. Its surface area is estimated as 375,000 km^2, 90% of which forms an artesian basin. The depth of the reservoir ranges from 100 to 400 m and its annual recharge as 375×10^6 m^3.

9.2.3 Murzuq Basin

This basin covers certain parts of Algeria, Libya, Chad and Niger. It covers a total surface area larger than 350,000 km^2. According to Pallas (1978), the saturated thickness of the reservoir can be up to 1,400 m with transmissivity varying from from 240 to 1200 m^2 d^{-1} and storativity from 0.002 to 0.02. The total annual recharge to this basin has been estimated as more than 140×10^6 m^3. Later investigation has shown that the basin consists of two systems; a lower Devonian aquifer and an upper Jurassic cretaceous aquifer. The two systems are separated by a carboniferous aquitard. The lower aquifer alone has transmissivity from 500 to 2, 500 m^2 d^{-1} and storitivity values from 0.005 to 0.00005.

The part of Murzuq Basin situated in Libya is known as Fazzan (sub) Basin. It covers a surface of about 175,000 km^2 in the southwest of Libya. This basin is recharged from the neighbouring heights at an estimated rate of 60×106 m^3 y^{-1}. The piezometric head in the artesian part of the basin is between 400 and 600 m a.m.s.l.

9.2.4 Tadouni-Tanzrouft Basin

This basin lies south of the Occidental Erg in Algeria. It covers a surface area of 240,000 km^2 and receives an annual recharge estimated as 20×10^6 m^3. The ground water level is at 150–200 m a.m.s.l.

9.2.5 Nubian sandstone Basin

With a total surface area of 1,800,00 km^2 covering northeastern Chad, south and east of Libya, the Western and Eastern Deserts in Egypt and the northwest of the Sudan, the Nubian sandstone basin in Africa is one of the most extensive artesian basins of the world. Of the total surface covered by the basin, 245,000 km^2 situated in Libya are centered in the Murzuq basin already described in 9.2.3. the thickness of aquifer in the Libyan part easily reaches 500 m and the total depth to water table below the ground surface is widely variable. The transmissivity coefficient at the northern edge of the aquifer ranges from 250 to 1,200 m^2 d^{-1}.

Generally, the entire surface covered by the basin is located in a hyper-arid region. The only pockets of life are encountered in a series of depressions (100–300 m deep) in the desert plateau, which slopes gradually towards the north. The most important oases are the Kharga, Dakhla, Farafra, Bahariya and Siwa in Egypt, and the Kufrah in Libya. The annual abstraction from the wells installed in the Egyptian Oases, according to Hefny (1991), has reached 570×10^6 m^3 in the late 1980s. The discharge figure increased from 14×10^6 m^3 for the first year of operation, 1965, to 208×10^6 m^3 in 1965, i.e. 10 years since the abstraction of groundwater from the Dakhla and Kharga Oases began. The discharge was maintained stable at about 205-210 $\times 10^6$ m^3 y^{-1} for the decade 1966–75. As a consequence of the drastic drop in piezometric head, some 40 m in 20 years of operation, it has been decided not to depend any longer on artesian flow. Since then groundwater is pumped out from deep wells so as to cope with the demand on water for expanding the agricultural area under irrigation.

The Nubian sandstone in the Sudan has a storage capacity estimated as 5.5 × '10^9 $m^3$with annual recharge in the range between 4×10^6 and 140×10^6 m^3. The aquifer thickness in northern Sudan ranges from 140–375 m and the transmissivity from 1, 040 m^2 d^{-1} to 2, 250 m d^{-1} (Hesse et al., 1987).

In view of the limited natural recharge, any abstraction in excess of the natural recharge will lead to mining the basin, a situation that should be avoided unless it is economically and/or socially justified.

9.2.6 The Kufra and Sarir Basins

These two groundwater basins combined cover a surface of about 675,000 km^2 situated in southeast Libya. The upper aquifer of the Kufra Basin, which is a Nubian aquifer of the Triassic to Lower Cretaceous, is currently under excessive exploitation. The transmissivity ranges from 300 to 3, 500 m^2 d^{-1} and the storativity from 0.015 and 0.0001. The Sarir Basin lies to the north and west of the Kufra Basin and has two main aquifers. The upper aquifer has a depth ranging from just a few meters to 210 m of graded sand to calcareous limestone with thin interbeds of clay. The lower aquifer is too thick, with a thickness that reaches 4,000 at some places. According to Salem (1991), the transmissivity of the upper aquifer has been found to range from less than 500 $m^2\ d^{-1}$ to more than $6,000\ m^2\ d^{-1}$, and the storativity from 0.05 to 0.0005.

9.2.7 Hadramaut Basin

This basin covers almost 40% of the total surface area of southern Yemen. The aquifer material consists of sand and gravel mixed with clay and sandy clay. In view of the poor quality of water in some of its parts, the productivity of the Hadramout Basin is rather limited. The basin recharge is claimed to be about $257 \times 10^6\ m^3\ y^{-1}$ (Gischler, 1979 and ECWA, 1981).

9.2.8 Az-Zarqua Basin

This is a rather small basin with a surface area of just $13,000\ km^2$ situated in Jordan. The aquifer is made up of basalt rocks and carbonatic formations. The annual basin recharge is estimated as $20 \times 10^6\ m^3$.

9.2.9 Amman Az-Zarqua Basin

This basin is relatively small compared to the basins described above. Nevertheless, it is one of the most developed sources of water in Jordan. Its water bearing strata are essentially made up of pebbles. Estimates of annual recharge are $20 \times 10^6\ m^3$ for the upper aquifer (free) and $5 \times 10^6\ m^3$ for the lower one (confined). Since the annual abstraction (more than $40 \times 10^6\ m^3$) is far in excess of the recharge, the water table and the piezometric levels are both dropping down with time.

9.2.10 Groundwater formations in the Arabian Peninsula

In the Arabian Peninsula there are several formations that by definition cannot be described as basins, yet they constitute important sources of water.

9.2.10.1 Formations of sandy facies

i- 'Al-*Saq' Formation*: This is a sandstone formation belonging to the Paleozoic Era. It covers a surface area of $160,000\ km^2$ and extends 1,200 km along the Arabian Shield from Jordan in the north to Saudi Arabia in the south. The average well yield in this formation ranges from 15 to $22\ l\ s^{-1}$, with a maximum yield of $100\ l\ s^{-1}$.

Pumping tests have shown that the transmissivity of the Saq formation varies from $35\ m^2\ d^{-1}$ to more than $23,000\ m^2\ d^{-1}$. The storativity decreases from 0.0013 for the confined aquifer to 0.000025. Since the average salinity is about 1,000 ppm, it has been considered as the best groundwater resource in Saudi Arabia. The groundwater storage above the 300 m depth has been estimated as 475×10^9 m^3.

ii- *'Al-Wajid' Formation*: It exists in the middle and south of Saudi Arabia, has a strong resemblance to Al Saq Formation and the quality of its water is also good. The formation outcropping has a maximum width of 100 km and thickness ranging from from 200 to 420 m. The well yield in this formation varies from 15 to 40 l s^{-1}, with a maximum of 80 l s^{-1}. The aquifer transmissivity has been estimated as from 50 to 1,750 m^2 d^{-1}. The annual recharge to the aquifer can be fairly assumed as 115×10^6 m^3, and the volume of storage above the 300 m depth as 430×10^9 m^3.

iii- *'Al-Tabuk' Formation*: This formation was deposited in a geologic time after that of the Saq formation. Nevertheless, the two formations seem to have similar characteristics. The extent of the outcropping of the Tabuk is 77,000 km^2 and the aquifer thickness varies from 20m to 1050 m. The maximum well yield is 127 l s^{-1}, and the transmissivity ranges from say 13 m^2 d^{-1} to 130 m^2 d^{-1}.

iv- *'Al-Manjour' Formation*: This is a formation that developed in the Mesozoic Era, and outcrops over a surface of 6,500 km^2 in Saudi Arabia. The thickness of the formation ranges from 350m to 400 m. The formation is situated at a depth of 1,200–1,400 m below ground surface in the Ryadh area, thus maintained under excessive piezometric pressure. The aquifer transmissivity ranges from 40 m^2 d^{-1} to about 1,400 m^2 d^{-1} and the storativity is about 0.00013. The aquifer recharge and storage above the 300 m depth are 80×10^6 m3 y^{-1} and 487×10^6 m^3 respectively.

v- *'Al-Bayadh' and 'Al-Wassie' Formation*: These two formations belong to the Mesozoic Era, and together form one source of groundwater extending with unequal proportions into Saudi Arabia, southern Yemen, Kuwait, Iran, Qatar and Jordan.

The thickness of the two formations combined together ranges from 230 m to 1,230 m, and their combined annual recharge is 420×10^6 m^3. The range of transmissivity is from 112 m^2 d^{-1} to 8,400 m^2 d^{-1}, and the storativity from 0.0002 to 0.12. Information about the hydrogeologic constants of each formation individually is listed in Table 13, Appendix II. The salinity of Al-Bayadh formation ranges from 550 ppm to 990 ppm increasing to 6,000 ppm or more depending on the location. The water of Al-Wassie formation is, generally speaking, less saline than that of Al-Bayadh. It varies between 1,000 ppm and 3,000 ppm.

9.2.10.2 Formations of calcareous facies

These formations appear extensively in the deposits of the Mesozoic and Tertiary Eras. Their thickness in the Arabian Peninsula is about 1,900 m. The flow of groundwater in these formations is under pressure and the potential as a water resource is high. The geographic distribution of the water-bearing formations in the Eastern subregion including the Arabian Peninsula are shown country-wise in the next sections.

i- *'Al-Dammam' Formation*: This formation consists of carbonatic rocks. The outcropping, which extends over a surface of 20,000 km^2, can be found in Kuwait, Qatar, Bahrain and Oman. The formation comprises five layers of which 'Al-Ulat' and 'Al-Khobar' are the principal water-bearing layers. The former has an overall average thickness of 83 m and can be found at depths ranging from 140m to 230 m. The thickness of Al-Khobar is about 57 m and can be found at a depth ranging from 200 m to 300 m. Despite the limited thickness of the water-bearing layers, the high permeability of their material helps the well yields to be fairly high; 4–10 $l\ s^{-1}$ for Al-Ulat and 7–22 $l\ s^{-1}$ for Al-Khobar. The value of the coefficient of transmissivity of the aquifer is widely variable from one location to another. Average values for the coefficients of storativity and transmissivity for both layers can be found in Table 13, Appendix II. The salinity of the groundwater increases from 2, 500 mg l^{-1} in the southwest to 100, 000 mg l^{-1} in the north.

ii- *'Umm er-Radhuma' Formation*: This formation can be found extensively in large areas in the Arabian Peninsula. Its outcropping, which extends over a stretch of 1,200 km, can be found in Dhofar, Oman, Hadramaut and south east of Yemen. It disappears in the United Arab Emirates and Qatar to reappear in Bahrain, Saudi Arabia, Jordan and Syria. The average thickness of this formation reaches 300 m in certain areas of Saudi Arabia. The upper aquifer yields the most potable water of all aquifers in the Umm er-Radhuma formation. The Rus and Dammam formations lie above the Umm er-Radhuma aquifer. The depth to water table ranges from 60 to 100 m, and the maximum thickness of the formation is 700 m.

The coefficients of transmissivity and storativity of the Umm er-Radhuma formation are included in Table 13, Appendix II. The wide range of values of the hydrological properties of the aquifer has resulted in a wide range of values, 10–70 $l\ s^{-1}$, for the well yield. As the salinity of water is at or greater than 1, 000 mg l^{-1}, the quality of groundwater can be described as fairly good to saline. A typical figure for the water salinity in Bahrain is 12, 000 mg l^{-1} and occurs in the form of a large lens. The recharge is estimated as $2.4 \times 10^9\ m^3\ y^{-1}$, of which $400 \times 10^6\ m^3\ y^{-1}$ are supplied by rain showers and torrents, and $2.0 \times 10^9\ m^3\ y^{-1}$ by leakage from other layers.

The information in the text and in Appendix II, are drawn from Gischler (1979), ECWA (1981), Khory et al. (1986), Hesse et al. (1987), Lloyd et al. (1990) and Sayed & Al-Ruwaih (1995).

9.3. BRIEF ACCOUNT OF GROUNDWATER SALINITY IN THE ARAB REGION

The salinity can be classified into three distinct categories. These are: freshwater with salinity less than 1, 000 mg l^{-1}, slightly saline to brackish with salinity ranging between 1,000 and 3, 000 mg l^{-1}, and saline with salinity exceeding 3, 000 mg l^{-1}.

In Mauritania, brackish to saline groundwater generally prevails. Freshwater can be found in the northern parts of Morocco and Algeria. The salinity of groundwater in these two countries increases towards the south. There is hardly any fresh groundwater in Tunisia. The situation in Libya is different than that in Tunisia. Fresh groundwater in Libya can be found in the Nubian sandstone formation, e.g. the Kufra, Sarir and Marzuq Basins. It can also be found in the northeast of the country.

Egypt, like Libya, gets fresh groundwater from the Nubian sandstone. The aquifers underlying the Nile Valley and Delta yield fresh groundwater originating from seepage from the river and irrigation canals. The salinity of water abstracted from the Nubian sandstone in the Sudan is generally less than $1,000\ \mathrm{mg\ l^{-1}}$. The Sinai Peninsula and the Eastern Desert in Egypt hardly contain any freshwater.

Similar to the Atlas Mountains in Morocco and Algeria, the relatively high precipitation in the mountainous areas in northwest Jordan, northern Iraq and Lebanon helps to maintain a good quality for the groundwater sources in these countries. In Iraq, the quality of groundwater deteriorates in a southerly direction due to the presence of evaporates (gypsum). In Syria, the salinity is generally higher than $1,000\ \mathrm{mg\ l^{-1}}$ in the interior and reaches $5,000\ \mathrm{mg\ l^{-1}}$ in certain areas, e.g. the Jezirah area, due to the presence of evaporites.

The Dammam formation in Kuwait contains groundwater ranging from brackish in the southwest ($2,000$–$5,000\ \mathrm{mg\ l^{-1}}$) to brine in the southeast ($5,000$–$10,000\ \mathrm{mg\ l^{-1}}$ or more). The quality of groundwater in the Tihama Plain, Yemen, can be described as high in the high land and declines in a westerly direction towards the Red Sea. Groundwater in the northern part of Qatar occurs as freshwater floating lenses within the limestone-dolomite succession of the Dammam-Rus formation overlying the brackish and saline water of the older rock units. The hydrogeological conditions in southern Qatar are more complex than in the north. There is hardly any freshwater body as brackish water covers the entire area (ECWA Report, 1981). Likewise the aquifer underlying the surface of the United Arab Emirates is of very poor quality (10,000 ppm) decreasing with depth to about 4,000 ppm. Better quality water (1,000–1,500 ppm) has been encountered in the shallow wells tapping the in the Liwa Oasis, Al-asab and Habshan (Barber & Carr, 1976).

The Riyadh-Wassie-Aruma aquifer, stretching diagonally across Saudi Arabia, has excellent to good quality groundwater with salinity in the range of 300–$900\ \mathrm{mg\ l^{-1}}$. The quality deteriorates eastwards reaching $6,000\ \mathrm{mg\ l^{-1}}$ and more than $20,000\ \mathrm{mg\ l^{-1}}$ (brine) near the border with the state of Kuwait. Groundwater quality of the Umm-Radhuma aquifer in the eastern part of Saudi Arabia increases from west to east in the direction of the Arabian Gulf, with the existence of a tongue of good-quality water in Al-Dammam area. The salinity of the Dammam aquifer there ranges from 1,500 to $2,000\ \mathrm{mg\ l^{-1}}$, and deteriorates rapidly towards the south and the east where it reaches far more than $5,000\ \mathrm{mg\ l^{-1}}$. In eastern Saudi Arabia, the overall regional potential of the quaternary alluvial aquifer is limited and of poor quality too.

9.4. GROUNDWATER RESOURCES BY COUNTRY

9.4.1 Mauritania

The water-bearing formations in Mauritania are generally made up of Basement Complex and Sedimentary deposits. These are distributed among 15 reservoirs shown on the map in Figure 2, which contains a brief description of each aquifer.

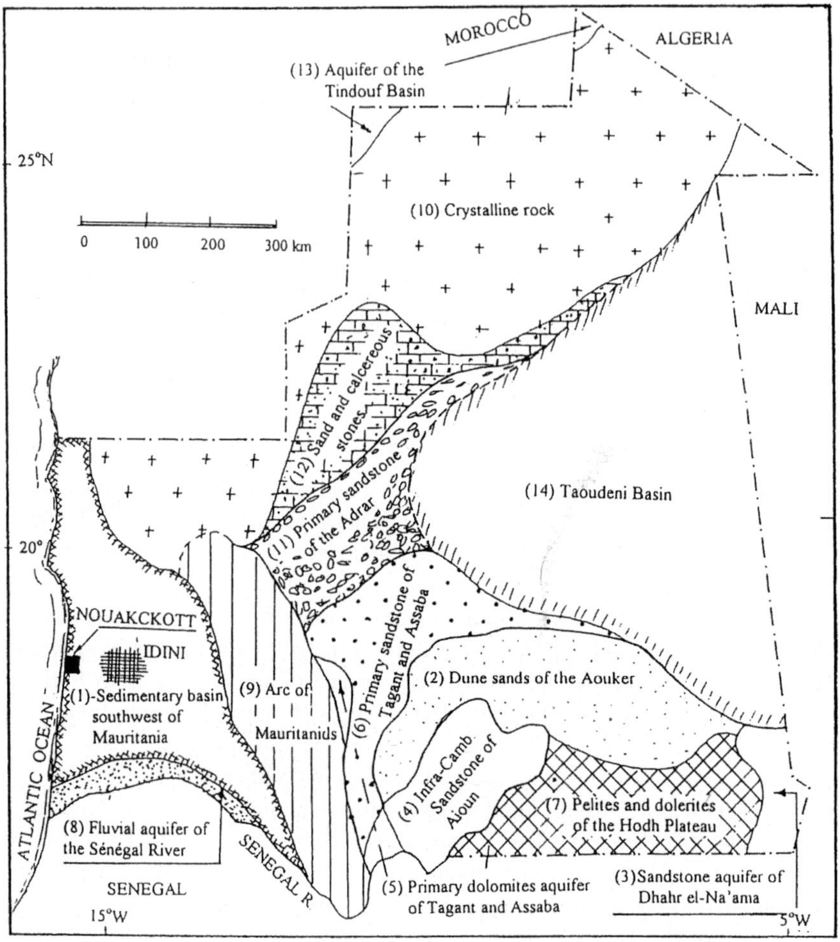

Figure 2. Map showing the water-bearing formations in Mauritania (from the Country Report of Mauritania, 1986) and the study area of Idini

The average annual rainfall volume in Mauritania is estimated as 100×10^9 m^3. Of this volume an amount varying between 0.5 and 1.5×10^9 m^3 recharge the groundwater aquifers. Little is known about the hydrogeological parameters of the different aquifers. The available information including well discharges in some parts of Mauritania are listed in Table 14, Appendix II.

The Idini area shown in Figure 2 is underlain by a sedimentary aquifer having a porosity of 5%, and a storage coefficient of 1×10^{-7} to 1×10^{-9} for the clay aquitard separating the phreatic and semi-confined aquifers of the reservoir. The average figure for the transmissivity of the Idini aquifer is 260 m^2 d^{-1}. Moving to the east, one finds a limited part of the aquifer with a relatively high transmissivity of up to 1,760 m^2 d^{-1} and storativity of less than 2%.

9.4.2 Morocco

In Morocco, most economic, agricultural and industrial activities are based on use of groundwater resources. These resources are encountered in six groundwater provinces.

i- *El-Rif Province*: This province is located in the northwest of Morocco adjacent to the Mediterranean Sea coast. It comprises mainly impermeable or poorly permeable erosion material and clay formations. It also comprises masses of calcareous stones, wadi fills, plains and small, isolated basins containing groundwater. So far, there is no reliable or accurate information about the calcareous masses, which are assumed to contain huge groundwater reserves in the range of 350 to 400×10^6 m^3. The importance of these reserves lies in that they sustain the surface water resources in the Rif region in the low-flow period. They also help in recharging the groundwater in the wadi fills.

The extent of the sedimentary coastal plains is variable. The Ghies and Nechor plains, which are the most important ones in this province, are made up of various sedimentary layers of different depths. The maximum depth in the middle exceeds 400 m, consisting essentially of erosion products. The aquifer in these two plains of Ghis and Nicor Wadis is recharged from the surface water resources in these wadis, thus accumulating a considerable groundwater reserve. The quality of water along the seacoast is saline.

The groundwater resources in the Rif Province are generally drained through the flow to the wadis or to the sea.

ii- *The Atlantic Province*: This province contains almost two-thirds of the groundwater resources in Morocco. These resources are recharged from the runoff along the mountain slopes, generated by precipitation on the Atlas Ranges, bordered on the north by the Rif Province, on the west by the Atlantic Ocean, and the Middle and High Atlas on the south and east respectively.

The province includes a number of rich basins of different sizes like Sayes Plateau, Al-Gharb, Tadelah, El-Houz, Al-Buhaira and Burashid Plain. The locations of these basins are shown on the map in Figure 3. Some of the aquifers in this province are deep while others are shallow. Additionally, some aquifers are of the unconfined type and the rest either artesian, confined or semi-confined. A certain portion of the surface runoff generated by rainstorms infiltrates the ground surface and recharges the aquifers. Recharge is also effected by land irrigation and underground seepage from the high lands to the low-lying plains.

iii- *The Atlas Province*: This province is made up of highly-permeable calcareous formations (table and rugged Middle Atlas and the calcareous High Atlas). The water bearing formation and the large number of spring serve as major contributors helping to sustain the perennial flow, characteristic of the wadis in the Atlas Mountainous region.

The old massif of the High Atlas plays a major role in retaining the water produced by melting snow, which in a ways counter balances the low permeability of the formations composing the massif. The western part of the Haut Atlas is characterised by its limited groundwater resources.

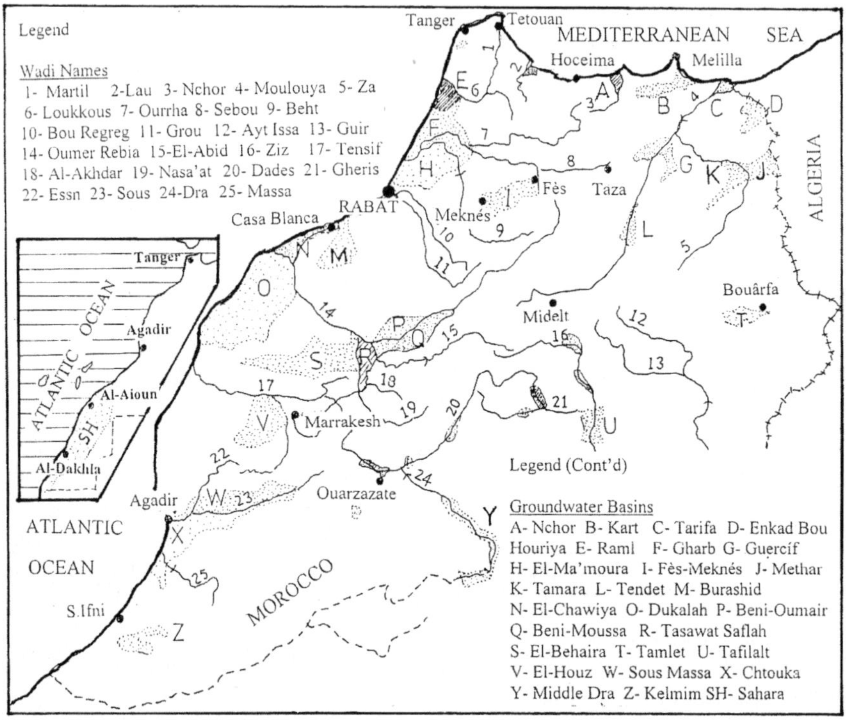

Figure 3. Map of Morroco showing the locations of groundwater basins (from Morocco's Country Report, 1986)

iv- *East Morocco Province*: This geographical unit is generally characterised by its aridity. The only exception can be found in the coastal strip whose climate is strongly affected by the Mediterranean Sea. The formations in this province are made up of sediments resulting from the water erosion of the mountains surrounding it (High and Middle Atlas, Eastern Rif, Beni-Yeznasin and Horst chain). The formations drain their groundwater almost completely to Wadi Moulouya. The drainage basin of the middle reach of Moulouya comprises a number of water-bearing formations discharging their waters through a large number of springs. The deep formations in certain locations of the said reach are artesian.

High plateaus can be found in the northeast region, they include a deep artesian aquife in the calcareous formation (e.g. Ain Methar) which is exploited via deep bore holes yielding a total amount of 1.2 $m^3 s^{-1}$. Between the plateaus in the east and the middle reach of Moulouya in the west lies the Rukam basin where aquifers are deep and have a limited yield.

The Horst chain is a mountainous belt 20–30 km in width and 100 km long. It extends from the Guercíf Plain to the Algerian frontier in the east. Only the resources in the eastern part are exploited and their yield supplies the neigbouring townships with potable water.

v- *South Atlas Province*: The climate of this province is hyper-arid, and geologically has unfavourable formation for either regular groundwater flow or exploitation. However, there are some important water-bearing formations, e.g. Sous, Tiznit, Oued Massa and Ifni (Figure 3). According to the Country Report on Water Resources of Morocco (1986), the groundwater reserve in the Sous Plain can be estimated as 10×10^9 m^3, yet the volume of water annually rejuvenated does not exceed 300×10^6 m^3. The groundwater yield of the free aquifer in the Shtouka Plain is limited. It is expected to increase as a consequence of the seepage of irrigation water supplied by Yusuf Ben Tesfin Reservoir on the Wadi Massa.

The South-Atlas Province comprises a number of wadi basins such as the Dra, Ouarzazate, Tafilalt and Guir. The Dra has a number of shallow, free groundwater aquifers, which are recharged from the flood events that are basically controlled by the operation of Al-Mansour Eddahbi Reservoir. The Ouarzazate drains the southern slopes of the Haut Atlas. The flood flows of Wadi Dades and Wadi Makoun provide a considerable proportion of the recharge of the groundwater in this group.

The Tafilalt Basin encounters a number of free aquifers, most important of which can be found at the junction of wadis Ghris and Ziz. So far, the groundwater resources of the Guir Basin are neither complete nor definitely known. However, it is suggested that the basin comprises some free, shallow and discontinuous aquifers as well as deep aquifers.

vi- *The Sahara (Desert) Province*: It consists of El-Saguia el-Hamra and Wadi el-Dahab Basins. The geology of the water-bearing formations in the eastern and southeastern parts of the province is essentially fractured crystalline rock belonging to the Paleozoic Era. The aquifers are of limited extent; poor yield and fairly moderate quality. The geology of the remaining parts of the province belongs to a sedimentary basin of the Mesozoic and Tertiary Eras. The most important water-bearing formations were formed in the Lower Cretaceous period and are situated at different levels. The aquifers are artesian yielding thermal water that contains sulphuric gases. Salinity of this formation increases from south to north: Al-Dakhla 2,000 mg l^{-1} and El-Aaiún 8,000 mg l^{-1}.

The formations deposited in the Pliocene and Eocene Periods are of little importance as they yield water of relatively high salinity (4,000–7,000 mg l^{-1}). The aquifers belonging to the Quaternary Period are phreatic with limited yield. The aquifer known as Fumm el-Wad (mouth of the wadi) yields relatively freshwater, which is exploited to supply El-Aaiún with potable water (30 l s^{-1}).

The total annual yield of the above-described hydrogeological provinces reaches 10×10^9 m^3 distributed as follows:
- 2.5×10^9 m^3 y^{-1} are drained naturally via the wadis,
- 2.5×10^9 m^3 y^{-1} are lost by evaporation and runoff to the sea, and
- 5.0×10^9 m^3 y^{-1} represent the volume of potable water

Other than these volumes, there are brine-water resources yielding an estimated annual volume of 2×10^9 m^3.

The most important aquifers in Morocco have been fairly well investigated to a degree that enables one to draw a fairly reasonable assessment of their yield. This can be specified as follows:

– *El-Gharb Plain*	$935 \times 10^6\ m^3\ y^{-1}$
– *Water-bearing formations of Tedelet*	$540 \times 10^6\ m^3\ y^{-1}$
– *Fès-Meknés Basin and Fès-Taza reach*	$450 \times 10^6\ m^3\ y^{-1}$
– *Sous-Shtouka aquifers*	$365 \times 10^6\ m^3\ y^{-1}$
– *Dukala aquifers*	$250 \times 10^6\ m^3\ y^{-1}$
– *El-Houz aquifer*	$170 \times 10^6\ m^3\ y^{-1}$
– *Burashid and El-Shawiya aquifers*	$150 \times 10^6\ m^3\ y^{-1}$
– *Lower Moulouya and El-Kart Plains*	$150 \times 10^6\ m^3\ y^{-1}$
Total =	$3.01 \times 10^9\ m^3\ y^{-1}$

9.4.3 Algeria

In Algeria one can identify six categories-types of groundwater aquifers. These are as follows (National report of Algeria to UNESCO/Regional Meeting of Arab States on Water Resources, 1983):

Type I : This is a succession of alluvial and colluvial deposits, generally perched, weathered by erosion processes that took place in the Pliocene and Quaternary Periods. It is argued whether this type was developed by faulted formations of limited extent, and is poorly drained by the near drainage channels. As such, the yield of wells drilled in this type of aquifers is low. The storage capacity of the aquifers in this type is almost nil.

The aquifers within the fissured crystalline rocks of the Hoggar can be included in this type. The potential of these aquifers has not been estimated but is supposed to be relatively weak.

Type II : The aquifers of this type are phreatic and of large extent. They can be found in alluvial formations of the thick plateaus or in the endoeric depression fills of the high plains. The total potential of the aquifers included in this type are quite high. The drainage of the aquifers is realised either through the drainage network (coastal aquifers) or by subterranean flow towards the chotts (closed depressions in the high plains of Central Algeria) where water evaporates.

The extent and thickness of aquifers are generally important, especially in the regions located in the high plains. In the meantime there are exceptions where the yields of wells are too little and do not allow to rely on substantial volumes of water. Aquifers are exploited by means of a large number of wells scattered all over their extent.

Type III : It comprises those narrow alluvial aquifers situated along the important streams and are in direct hydraulic connection with them. This type is well represented by the Kabylia Valleys: Wadis Isser, Sebaou and Soumam.

The aquifers transmissivity, consequently the well yields are high. However, the storage capacity is small. The aquifers can be intensively exploited provided that the relation between the groundwater and the surface water of the wadis is known. The use of storage capacity of the alluvial mass should be maximised whenever groundwater abstraction is planned for seasonal agriculture.

Type IV : Extended calcareous or dolomite massifs form this type of formation. Aquifers are generally the core where infiltration is high (in the order of 20% of the precipitation). The exploitation of these aquifers, however, is somewhat difficult in view of the heterogeneity of the fractures that renders the specific yield of wells widely variable. Accordingly, it is not possible to exploit the aquifers directly like those in the privileged zones (tectonic or fractured). Additionally, the free parts of the aquifers are generally located in areas not easily accessible.

The confined part, bordering the rock outcropping, is generally of much interest to exploit. Nevertheless, practical considerations might limit the aquifer exploitation. The tectonic movements may wash away important aquifer material, and thus the depth of wells becomes quickly prohibitive. The aquifers of this type are generally well drained by means of high-flowing streams, which do not permit the regulation of storage capacity through pumping except rarely.

Two important aquifers are considered as exception from the above-described general set up. The first is the karst aquifer in the region of Zibans, west of Biskara, which is exploited uniquely by means of artesian wells in its confined part. The second is the dolomitic aquifer of the Chergui Chott. The resources of this aquifer can be fully exploited in a certain zone where the aquifer is overlain by an extremely permeable calcareous lens, which acts as a drain to the entire dolomitic aquifer.

Type V: This type comprises the aquifers located in the alluvial fills of large depressions. They are by far the most important aquifers in northern Algeria. The aquifer in Algeria Plain can be regarded as an excellent example of this type of aquifers. The thickness of sand and gravel fills exceeds 100 m. In general, these aquifers are unconfined in the largest part of their areal extent. They are essentially located in the Tel zone where precipitation is usually heavy and hydraulically connected with important wadis.

The specific yield of wells is variable due to the heterogeneity of the fill material. Accordingly, the conditions of some zones only allow intensive exploitation of their catchment areas. The two fields of groundwater abstraction at the base of Mazafran supply about 2 $m^3\ s^{-1}$ to the city of Algiers through 35 wells. The balancing capacity of the aquifers is quite high; this does not only help to regulate the sharp differences in flow within the year but also tends to attenuate the interannual variability in supply.

The over-exploitation of coastal alluvial aquifers that resulted in the intrusion of seawater and the intensive use of fertilizers have contributed to the deterioration of groundwater quality. The rapid mineralisation in chlorides between 1989 and 1992 has resulted in concentrations the jumped from 460 mg l^{-1} to 3, 650 mg l^{-1} in the Mazafran aquifer. Beside detectable nitrates and increase in sulfate concentrations due to the excessive use of fertilizers, a noticeable inland migration of certain chloride constituents was also observed (Messahel et al., 1997). The change in the chemistry of groundwater in the neighbouring Oued Nador aquifer over the period 1982–1993 can be observed from the figures listed in Table 10, Appendix III.

Type VI : It comprises aquifers of extremely large extent as well as storage capacity. This type is uniquely represented as Sahara by the aquifers of the Terminal Complex and the Intercalary.

Management problems of the above-described resources, basically the balance between discharge and recharge and the economic aspects of water abstraction, limit the abstraction of groundwater from the mentioned aquifers. Proper knowledge of the aquifers geohydrology and the economic conditions of their exploitation are of prime importance for determining the water balance of these resources.

The Country Report of Algeria on Water Resources and their Utilizations (1986) estimated the groundwater potentials in northern Algeria as 1.7×10^9 m^3 y^{-1} and as 2.0×10^9 m^3 y^{-1} in the south. The annually exploitable volumes are 1.3×10^9 m^3 in the north (80% of the potential volume) and 0.7×10^9 m^3 in the south (35% of the potential volume). The annual volume abstracted by means of shallow wells and drawn from wadi beds, mainly for small irrigation schemes, is estimated as 0.9×10^9 m^3. If these estimates can be regarded as reliable, the total amount becomes 2.9×10^9 m^3 y^{-1}. More detailed information about water-bearing formations and groundwater aquifers in Algeria can be found in Table 15, Appendix II.

9.4.4 Tunisia

Groundwater reservoirs in the north and middle of the country, depending on their hydrogeological characteristics, can be subdivided into three major types: sedimentary, calcareous and Miocene erosion material reservoirs.

9.4.4.1 Sedimentary Basins

There are several sedimentary basins in the north and middle parts of Tunisia in the form of sand-pebble plains, 20–400 m in thickness, belonging to the Quaternary Era. The sedimentary reservoirs often comprise multiple water-bearing formations located at different depths below the ground surface, of which Ghar el-Dem'a Plain and the fertile castle in the north, and the Kairouan Plain in the middle can be mentioned.

The reservoirs are recharged directly from the rainwater, but mainly from the floods of the neighbouring wadis that inundate the plains containing such reservoirs. The reservoirs drain their flow either to the depressions (sabkhas/sebkhas) where groundwater is evaporated or to the perennially flowing streams. The quality of groundwater is good in the north and middle where the salinity does not exceed 1,500 mg l^{-1}. The quality becomes worse in the coastal strip where the salinity increases to 3,000 to 5,000 mg l^{-1}. Groundwater exploitation is accomplished via too close, shallow wells with diameters between 3 and 5 m and median flow ranging between 4 and 8 l s^{-1}. There are much deeper wells between 100 and 300 m depth with more intensive yield ranging from 20 to 50 l s^{-1}.

9.4.4.2 Calcareous reservoirs

These reservoirs are usually of limited areal extent and distributed between the north and middle of the country according to their geology and tectonic activities. The water-bearing formations belong mostly to the Eocene Period.

Depending on the development of fractures, the formations can turn into karstic networks associated with the presence of springs of different degrees of importance. In view of the rather heavy rainfall in the north, the recharge of the reservoirs is generally adequate (20–50% of the rainfall) causing the rate of flow to be high, in the order of 100 l s^{-1} or more.

Most of the calcareous reservoirs have good quality water (500 mg l^{-1}) thus making its abstraction for domestic purposes necessary. Additionally, a certain amount of this water is bottled as mineral water for drinking. Abstraction of water is done through tunnels collecting spring water, also by means of drilled wells.

9.4.4.3 Reservoirs formed of Miocene erosion material

The material forming these reservoirs can be found in the middle of Tunisia in the form of marine formations separated by clay layers or in the form of thick river erosion deposits. The aquifer storage capacity in either case is quite high. When these formations are directly overlying the cretaceous limestone layers, seepage water recharges these calcareous reservoirs thus bringing their pattern of water level change into uniformity. The case of Sidi Murzuq-Sbieba basin is an example of such a situation. It is common to find these Miocene formations at the base of the Paleo-Quaternary sediments in the form of bent, discontinuous layers.

The recharge of these reservoirs occurs through direct precipitation on their top surface. Besides, some recharge is accomplished through water flowing in the adjacent streams and/or seepage from close bye reservoirs. Groundwater flow in these reservoirs is discharged at the low-lying reaches of the wadis or to existing springs. The supply to the wadis during the low-flow period was estimated as 200 l s^{-1} for Wadi Besbieba in the Sidi Murzuq-Besbieba Basin and 400 l s^{-1} for Wadi el Darb in the Kasserine Heights. The quality of water of the erosion material reservoirs is generally good, with salinity not exceeding 1.5 l s^{-1}. Groundwater is intercepted by the existing wadis after having their floodwater diverted by means of certain type of diversion dams like the case of Sbieba dam across the Wadi Darb. However, water in more substantial quantities is abstracted using deep wells (200–1,500 m).

The groundwater reservoirs in southern Tunisia are portions of the two huge Saharan reservoirs known as Continental Intercalary and Terminal complex. It is possible to add to these two reservoirs the Jifara formation, which runs along the seacoast. The Jifara is replenished by seepage water from the Middle intercalary while the other two huge reservoirs used to be recharged with rainwater during the rainy episodes of the Quaternary.

9.4.4.4 *Terminal complex*

The rocks composing this formation were deposited during the Cinonian and Miocene Periods. Only small parts of the extended surface (350, 000 km^2) of this formation exist in Tunisia; Nefzawa (calcareous reservoir from the Cinonian Period) and Chott el-Djerid (sandy reservoir from the Miocene Period). The mean thickness of the formation at Nefzawa ranges from 100 to 300 m and from 200 to 600 m at Djerid. The groundwater flows in the direction south north. As such, the formation terminates at the Chotts, and the groundwater that rises to the ground surface by artesian pressure is lost by evaporation.

Drainage of this formation takes place through the springs nearby the Chotts; the discharges of the springs at Nafta and Touzeur were measured in 1983 and found as 134 $l\ s^{-1}$ and 165 $l\ s^{-1}$ respectively. However, recent intensive exploitation of these springs by artesian flow and/or pumping has resulted in a marked drop in their productivity. Additionally, there is a growing tendency to depend more on pumped wells for groundwater abstraction.

Groundwater salinity varies from location to location, for example 1, 500 mg l^{-1} at Nefzawa (in French Nefzaoua), 2, 500 mg l^{-1} at Djerid (El-Jerid), 4, 000 mg l^{-1} in Kibli Island and 8, 000 mg l^{-1} at Hamet el-Djerid. The areas with high salinity are strongly affected by intrusion of saline water from Chott Djerid. The given locations are shown on the map in Figure 4 (Country Report of Tunisia, 1986).

Feasibility studies have shown that the depth to water level in the Complex Terminal should not be lowered to a depth exceeding 60 m below the ground surface to safeguard the economy of groundwater abstraction. Model studies have yielded the following quantities that can be abstracted from the different formations (Country Report of Tunisia, 1986):

Location	*Abstraction, $m^3\ s^{-1}$ from*			
	*C.T.** *Year 1970*	*C.I**.*	*C.T*.* *Year 2010*	*C.I**.*
Djerid	*2.0*	–	*4.8*	*1.0*
Nefzawa	*2.3*	–	*6.5*	*0.9*
Chott Al-Fedjaj	–	*0.2*	–	*1.1*
Jiffara	*4.2*	–	*4.7*	–
Extreme south	–	*1.0*	–	*1.0*
Subtotal	*8.5*	*1.2*	*16.0*	*3.0*
Total	*9.7*		*19.0*	

* Complex Terminal and ** Continental Intercalary

9.4.4.5 *Continental Intercalary*

This reservoir comprises continental sandy deposits brought by erosion in the period between the end of the Jurassic and beginning of inundation by seawater in the Cenomanian (Upper Cretaceous) Epoch. The areal extent of this reservoir, as mentioned earlier, is about 600, 000 km^2. Those parts of the reservoir within the Tunisian territory can be found in the areas of Al-Zaher, Nefzawa, Djerid,

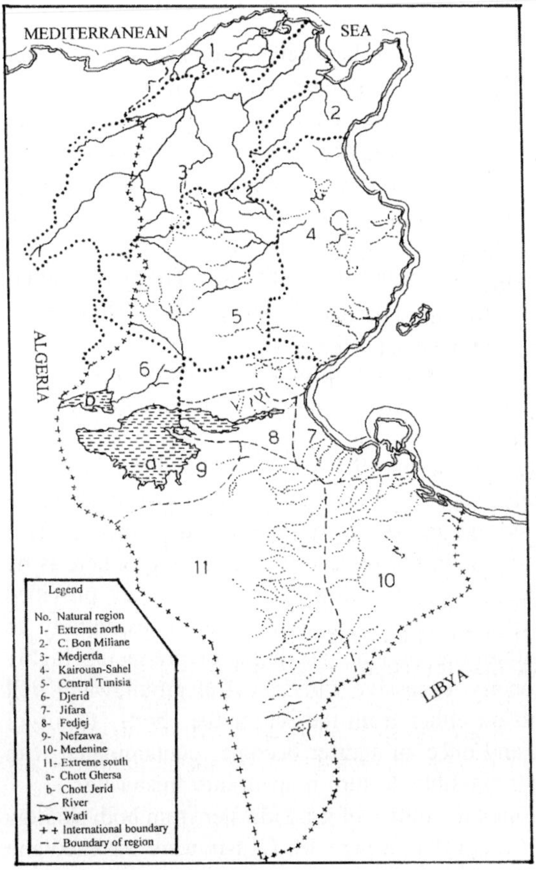

Figure 4. Natural regions of Tunisia (from Zebidi, 1980)

Chott el-Fedjaj and the extreme southern corner of Tunisia. The aquifer can be found at a depth from 1,000 to 2,000 m. As a considerable part of the Continental Intercalary runs under the desert of the Eastern and Western Grand Ergs in Algeria, the part of the formation in Tunisia has the privilege to act as an outlet.

The groundwater flow assumes a northeasterly direction and terminates at Chott el Fedjaj. Part ($3.5\ m^3\ s^{-1}$) of the reservoir discharge seeps to the Jiffara reservoir and a smaller amount ($135\ l\ s^{-1}$) at Hamet Gabès appears on the ground surface in the form of springs. The salinity of water ranges from 2,500 to $3,000\ mg\ l^{-1}$ in the areas of Chott el-Fedjaj, Nefzawa and Djerid and increases to $5,000\ mg\ l^{-1}$ in the southern extremity of Tunisia at el-Borma. Groundwater exploitation from this reservoir is witnessing intensive development as a result of the increase in number of deep production wells.

9.4.4.6 Al-Jifara Formation

This formation extends along the coastal plain between Gabès and Ben Gardane in southern Tunisia. It assumes the form of two reservoirs; a sandy reservoir of the Miocene Period and a calcareous reservoir of the Cinonian Period. Leakage of groundwater connects hydraulically the two reservoirs. The average depth to aquifer ranges from 100 to 300 m. The aquifer is recharged mainly with leakage water from the Continental Intercalary through the existing faults and fractures. It is also partly recharged with rainwater.

The groundwater flows eastward and terminates at the sea. Discharge points are represented by the natural springs, e.g. springs of Gabès. The abstraction of groundwater has led the discharge of these springs to decline from 800 $l\ s^{-1}$ to less than 150 $l\ s^{-1}$ from the beginning to the end of the 20th Century. The salinity of water ranges from 3,000 to 4,000 mg l^{-1} at Gabès to 6,000–8,000 mg l^{-1} at Djerba and Zarzis.

The annual recharge to the three reservoirs 9.4.4.4, 9.4.4.5 and 9.4.4.6 is too small compared to their potential resources. Nevertheless, the large areal extent of these reservoirs brings their storage to several billions of cubic meters. This state of affairs is very tempting to exploit the reservoirs. The difficulty here is by doing so most of the artesian springs will stop flowing unless they are provided with pumps capable of withdrawing the required amounts. This is essentially an economic problem unless the withdrawn amounts are used for producing highly marketable commodities. Additionally, excessive withdrawal of groundwater will eventually lead to saltwater intrusion either from the sea or the chotts. This will lead to environmental problems, and once an aquifer becomes contaminated with saline water, it is difficult, if not impossible, to turn it again into freshwater.

In 1985, the available and exploited quantities of groundwater from both shallow and deep groundwater aquifers in the different parts of Tunisia were assessed and the results obtained then were as follows:

Part of the country	*Shallow groundwater, $10^9\ m^3\ y^{-1}$*	*Deep groundwater, $10^9\ m^3\ y^{-1}$*	*Sum, $10^9\ m^3\ y^{-1}$*
North	*0.287*	*0.151*	*0.438*
Middle	*0.162*	*0.224*	*0.386*
South	*0.037*	*0.656*	*0.693*
Total	*0.486*	*1.031*	*1.517*

9.4.5 Libya

Over 90% of the surface of Libya is covered by deserts. In view of the limited surface water resources, groundwater resources supply more than 98% of the total water consumption. The five main basins are Jifara (200), Jebel L'Akhdar (200), Hamada (230), Murzuk (1,200) and Kufra-Sarir (1,600), all figures in parentheses are in $10^6\ m^3\ y^{-1}$. The last two basins, i.e. Murzuk, in the southwest of Libya, and Kufra-Sarir, in the southeast, are the principal basins as they together contain 81.6% of the total volume of groundwater.

– *The Murzuk Basin*: The basin covers a surface area of more than $35,000\ km^2$. It contains two aquifer systems. The lower system comprises aquifers belonging to the Cambrian, Ordovician and Divonian (Paleozoic Era). The Cambro-Ordovician and Divonian aquifers vary in thickness from 200 to 1,400 m and 40 to 60 m, respectively. The known figures for the aquifers transmissivity range between 500 and $2,500\ m^2\ d^{-1}$.

The storage coefficient ranges from 1×10^{-5} to 5×10^{-3}. There is hardly any current recharge to the aquifer, and the current depletion of the aquifer system is at a rate of 0.1–1.0 mm d^{-1}. The total solid salts are generally below 1,000 mg l-1 (Salem, 1991).

The upper system is made up of Permo-Teriassic and Upper Jurassic-Lower Cretaceous aquifers (Nubian Sandstone of the Mesozoic Era). The two aquifers are separated by a Carboniferous aquitard, allowing each system to behave as a group of independent aquifers. The Nubian Sandstone aquifer covers an extent of $190,000\ km^2$ situated in the middle of Murzuk Basin (Figure 5(a)). It consists of a thick series of sandstones intercalated with clay. The maximum thickness varies from 1,000 m in the middle to 200 m at the edges. Aquifer transmissivity at the northern edge varies from 250 to $1,250\ m^2\ d^{-1}$. The total dissolved solids in the shallow layers fall in the range of $1,000$– $4,000$ mg l^{-1} and 160–480 mg l^{-1} in the deep layers.

– *The Kufra-Sarir Basin*: This basin covers about $675,000\ km^2$ in southeast Libya. The water-bearing formations of the Kufra Basin belong to the Cambrian-Lower Cretaceous ages and consist of Continental sandstones with intercalated clays and silts. The saturated thickness of the aquifer system is 3,000 m in the middle as shown in Figure 5(b). The top middle part of the basin is occupied by Nubian Sandstone of the Triassic-Lower Cretaceous ages. Due to a topographic low in the regionally unconfined area the groundwater emerges at the surface as an oasis. The aquifer transmissivity varies from 300 to $3,500\ m^2\ d^{-1}$ and the storativity from

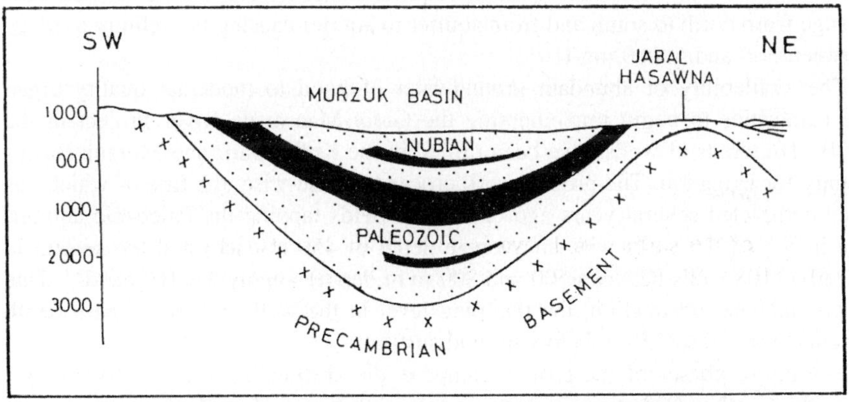

Figure 5(a). Southwest-northeast cross section through the Murzuq Basin, Libya

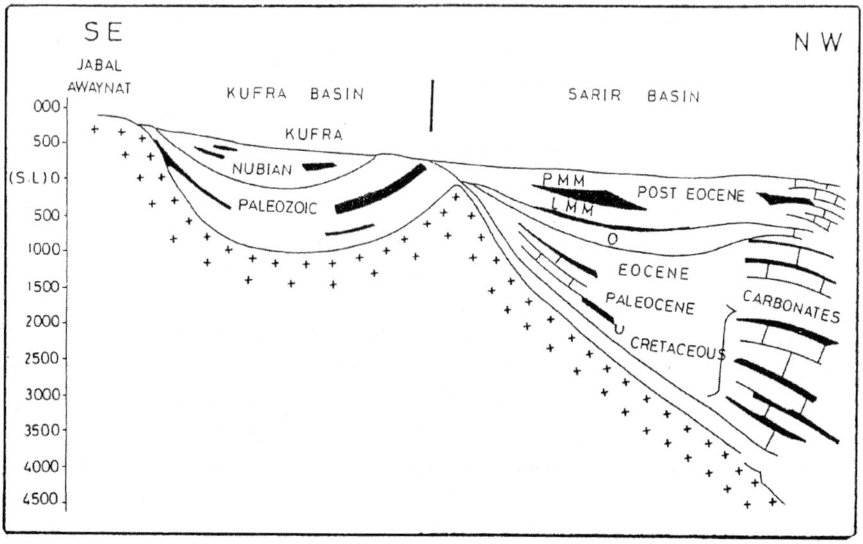

Figure 5(b). Southeast-northwest cross section through the Kufra Basin, Libya (Salem, 1991)

1.5×10^{-2} to 1.1×10^{-4}. The total dissolved salts in the water from deep wells fall in the range of 180–300 mg l^{-1}.

The Sarir Basin is situated north and west of the Kufra Basin. The waterbearing formations of this basin belong to the post Eocene with two main aquifers: the Post-Middle Miocene (PMM) and the Lower and Middle Miocene (LMM) and Oligocene. The thickness of the PMM can reach 200 m, consisting of coarse sand and calcareous sandstones with interbedded clays. The LMM is thicker and can reach 800 m. The LMM consists of interbedded clays, fluvial sands and sandstones. The transmissivity varies from 500 to 6,000 m^2 d^{-1}, and the storativity from 5×10^{-2} to 5×10^{-4} depending on the aquifer material. The total dissolved salts change from north to south and from aquifer to aquifer causing the salinity to range between 500 and 5,000 mg l^{-1}.

The availability of abundant groundwater of good to moderate quality urged the authorities to begin implementing the Great Man-made River Project in the 1970s. Hundreds of wells have been drilled in the Kufra, Sarir and Murzuk Basins mainly for irrigation. The project work proceeds phase-wise, the first of which has been completed several years ago. Two well fields tapping the Paleozoic aquifer, one in Sarir (126 wells with an average depth of 450 m each) and the second in Tazerbo (108 wells between 500 and 800 m in depth), supply 2×10^6 m^3 d^{-1}. Due to groundwater abstraction, the pumping level in the wells will be reach a depth ranging from 70 to 150 m below ground surface.

The future phases of the project comprise the abstraction of 2.5×10^9 m^3 y^{-1} from the Cambro-Ordovician aquifer in the northeast of Murzuk Basin using over 500 wells. The pumped groundwater will be transported to the Jifara Plain in the northwest along the Mediterranean Sea coast. Model results have shown that the expected static water level will be between 80 and 175 m below ground level. Figure 6 is a location map of the well fields feeding the Great Man-made river in Libya.

In a study area situated 16 km west of Tripoli many wells have been dug and groundwater excessively withdrawn. Originally, the quality of groundwater was good, with an average total dissolved solids of 500 mg l^{-1} and chloride concentration of less than 150 mg l^{-1}. The observation well at Ben Gashir, in the heaviest groundwater extraction zone of the Jifara Plain showed a drop of 75 m in the water level from 1958 to 1989 (Pallas, 1980 and Salem, 1991). As a result of aquifer over-drafting the quality of water is showing signs of gradual deterioration. The worst affected wells are the deep ones and those excessively pumped. Seawater intrusion into the Quaternary aquifer west of Tripoli was also investigated by Belaid (1995). The conclusion to be drawn from that investigation is that salinisation by seawater is in fact affecting many areas of the Jifara Plain as a result of intensive exploitation of the aquifer added to the meager recharge it receives. Major anion interpretations showed that chloride dominates in most of the water points studied. As far as cations are concerned, sodium showed depletion against enrichment in calcium content. This result can be attributed to the reverse ion exchange reaction that is taking place within the clayey and silt lenses in the aquifer as the saline waterfront advances landwards. Magnesium and potassium are also involved in the process of reverse ion exchange.

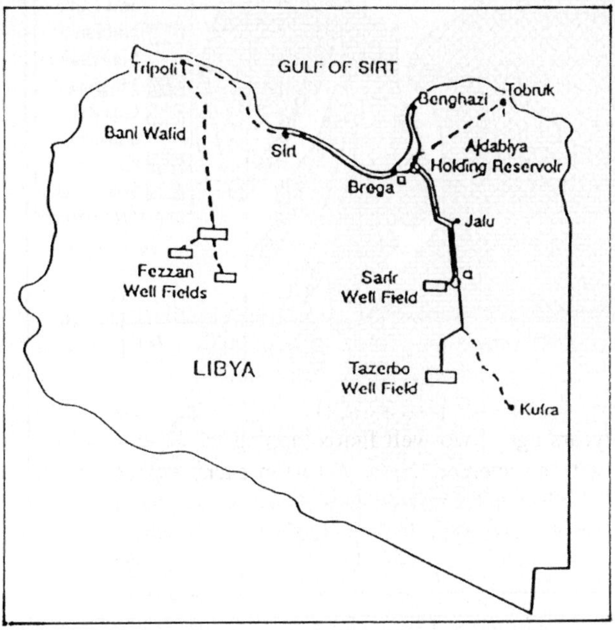

Figure 6. Location map of the well fields feeding the Great Man-made River, Libya (from Salem, 1991)

9.4.6 Egypt

9.4.6.1 Groundwater systems in Egypt

The main aquifer systems in Egypt are approximately indicated on the map in Figure 7. They can be grouped into six groups as follows:

– *The Nile Aquifer System*: This system is basically formed of the alluvial deposits of the Quaternary and Late Tertiary (Oligocene and Eocene), where sand and gravel beds are intercalated with clay lenses. The water-bearing formation is confined between an impermeable bed at the bottom and a relatively thin semi-impermeable clay layer at the top. The Nile Delta occupies a great tectonic depression, and is bounded on both sides by gravelly plains rising to more than 100 m a.m.s.l. The plains on the eastern and western sides of the Delta merge into the elevated tablelands (higher than 200 m a.m.s.l.), which act as watershed areas. The eastern tablelands are dissected by a number of wadis, which act during rainy seasons as drainage arteries. The sedimentary section in the Delta has an expected thickness of more than 10,000 m, at least in some locations (Shata & El-Fayoumy, 1969).

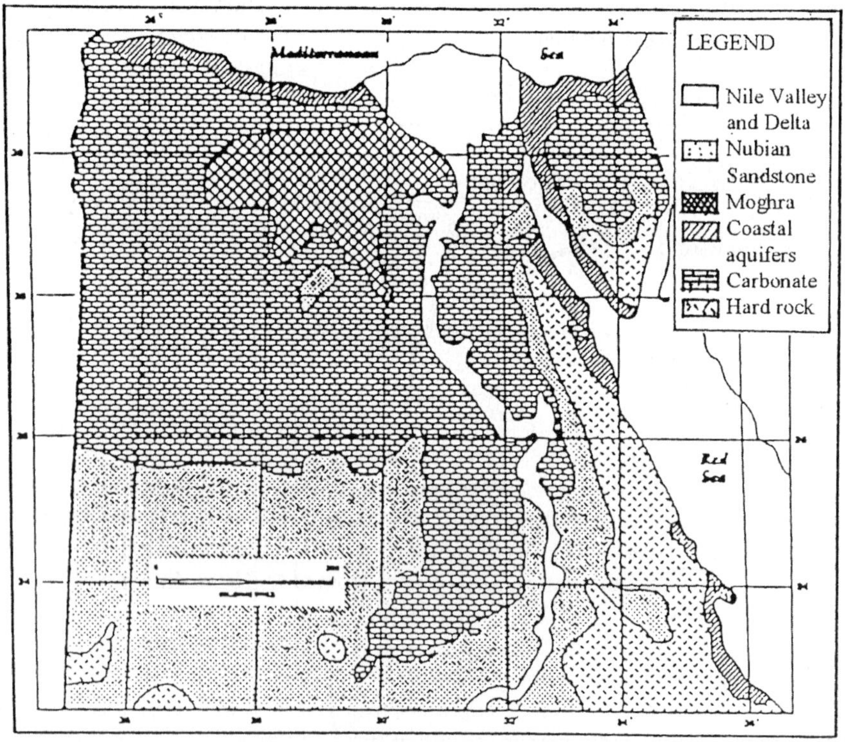

Figure 7. Aquifer systems of Egypt

The aquifers underlying the Nile Valley and Delta can be regarded as of similar properties. Both are composed of Quaternary and Late Tertiary sands and gravels with intercalated clay lenses and underlain by Pliocene clays. The thickness of the aquifer decreases from 300 m or more in Upper Egypt to around 50 m near Cairo. It increases in a northerly direction and reaches 1,000 m along the Mediterranean Sea coast. The hydraulic properties of these aquifers as obtained from the pumping tests are included in Table 16, Appendix II. From these properties one can fairly conclude that the aquifers enjoy reasonably high transmissivity and the water quality is generally good.

Kashef (1983) applied the leaky aquifer theories to the results of the available pumping tests and obtained the following values:

- *In central and southern Delta:*
  Vertical permeability of clay $= 0.385\ md^{-1}$, *permeability of aquifer* $= 10.4\ md^{-1}$, *Transmissivity of aquifer* $= 4,160\ m^2d^{-1}$, *and storativity* $= 6.1 \times 10^{-4}$.
- *In northern Delta:*
  Vertical permeability of clay $= 0.655\ md^{-1}$, *permeability of aquifer* $= 6.6\ md^{-1}$, *Transmissivity of aquifer* $= 3,828\ m^2d^{-1}$, *and storativity* $= 3.78 \times 10^{-4}$.

The study of the water budget by the same reference showed that an annual volume of $6.4 \times 10^9\ m^3\ y^{-1}$ represents the infiltrated recharge to the aquifer. This is due to seepage from irrigation canals as well as application of irrigation water in excess of the plant water requirements. Since the withdrawal of groundwater then, around 1982, had been about $2.4 \times 10^9\ m^3\ y^{-1}$, the said study was concluded by assuming that the difference, i.e. $4 \times 10^9\ m^3\ y^{-1}$ should represent the potential water surplus for future development, unless unaccounted losses or estimation errors would appear in the future.

The Nubian Sandstone: It belongs to the Paleozoic-Mesozoic eras (Cretacous, Jurassic and Triassic ages, $135–225 \times 10^6$ y BP), and occupies a thin long slice of the Eastern Desert and a large part of the Western Desert in Egypt. Erosion is claimed to have removed 200–300 m of the limestone deposited in the Cretaceous and the Eocene on the plateau bordering the Kharga from the north and east to expose the underlying Nubian formations. They contain the important aquifer which supplies water to the oases. Water levels in the wells scattered in the oases of the Western desert suggest that they belong to one huge body of groundwater. Maps indicating the equipotential lines of groundwater under the desert were prepared using the results obtained from several investigations.

Results of pumping tests have shown that the coefficients of permeability, transmissivity and storage are in the ranges of 2–20 m d^{-1}, $1,000–1,800\ m^2\ d^{-1}$ and $2–4 \times 10^{-4}$, respectively. The depth below ground surface to the top face of the aquifer in and around the desert oases varies from 0 to 20 m. In some locations water is even flowing freely. The total abstraction from the oases of the Western Desert in Egypt around 1975 was estimated as $0.38 \times 10^9\ m^3\ y^{-1}$. The inventories carried out in 1984 and 1990 were $0.45 \times 10^9\ m^3$ and $0.54 \times 10^9\ m^3$ respectively. The total abstraction from the Nubian Sandstone in Egypt is expected to reach $3.5 \times 10^9\ m^3$ by the year 2025. This corresponds to a total withdrawal of about $100 \times 10^9\ m^3$ from the volume in storage.

– *The Moghra Aquifer System*: The system belongs to the Lower Miocene (post Nubian). It occupies a vast area between the western edge of the Nile delta and Qattara Depression (Table 17, Appendix II). The aquifer is composed of sand and sandy shale with a thickness varying from 500 to 900 m. The aquifer is phreatic with water table level between 10 m above mean sea level (amsl) at the side bordering the Nile Delta and 60 m below mean sea level (bmsl) at the Qattara Depression in the west. The aquifer water is thought to be partly non-renewable. The salinity increases from 1,000 parts per million (ppm) near the Delta in the east to 5,000 ppm near the depression in the west. The aquifer is recharged, among others, by the lateral seepage from the aquifer underlying the Nile Delta at an annual rate of $50–100 \times 10^6$ m^3.

– *The coastal Aquifer System*: This belongs to the Late Tertiary and Quaternary, and extends along the northern and western seacoasts of Egypt.

– *The Carbonic Aquifer System*: It belongs to the Upper Cretaceous Period, and consists of fissured and karstified Carbonic rocks. The aquifer covers the northern half of the Western Desert and northern Sinai, thus occupying more than 50% of the surface area of Egypt.

– *The Fissured and Weathered Hard Rock Aquifer System*: This aquifer system belongs to the Precambrian Epoch. It covers most of the surface of the Eastern Desert and southern Sinai.

Weathered hard rock acts occasionally as a water-bearing medium through cracks and fractures. The water abstracted from this formation is used on a small local scale for domestic purposes as well as supply for tourism.

According to the Research Institute for Groundwater in Egypt (Hefny et al., 1991), the annual total withdrawal from all aquifer systems in Egypt was estimated then as 3.32×10^9 m^3; 0.97×10^9 m^3 from the Nile Valley, 1.67×10^9 m^3 from the Nile Delta and 0.68×10^9 m^3 from the desert and coastal aquifers. The reference estimated the possible increase of the total extraction to reach 8.3×10^9 m^3 y^{-1}. Again, the same authority, i.e. RIGW (Attia et al., 1995) gave the then pumped groundwater as $5–14 \times 10^9$ m^3 y^{-1} and its potential volume as $11.35 \times 10^9 m^3$ y^{-1}.

9.4.6.2 Hydrochemistry of groundwater

There are abundant determinations of groundwater quality in Egypt. As most of groundwater withdrawal occurs in the Nile Delta area for land irrigation, its salinity content (total dissolved salts) has been examined and classified as described below. The distribution of the salinity classes over the surface of the Nile delta area can be seen from the map in Figure 8 (El-Ghandour et al., 1983).

Class	*Salinity, ppm*	*Electrical conductance, μmhos cm⁻¹*
Low saline water	> 160	*0–25*
Medium saline water	*160–480*	*25–750*
High saline water	*480–1,440*	*750–2,250*
Very high saline water	*1,440–3,200*	*2,250–5,000*
Excessive saline water	< 3, 200	< 5, 000

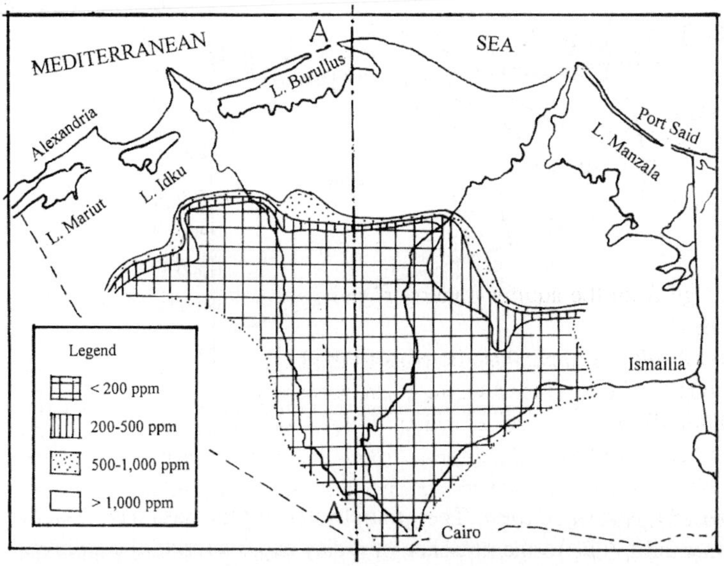

Figure 8. Lines of equal chloride content in the groundwater, the Nile Delta area (from El-Gandur et al, 1983)

Sherif et al. (1990) studied the salinity of groundwater along a longitudinal section in the Nile delta. They modelled saltwater intrusion under the dispersion zone approach using a two-dimensional finite element model. The seawater intrusion was assumed not to migrate inland more than 150 km from the Mediterranean Sea. Additionally, the depth of the domain under consideration varied from 680 m at the sea boundary to 240 m at the landside and the thickness of the upper semi-confining layer was assumed to vary from 15 m to 52 m depending on the location. The entire domain was subdivided into five sub-domains; each sub-domain in turn subdivided into a number of elements representing small areas. The entire domain was finally represented by a non-uniform grid of large numbers of nodes and elements.

The lines of equal concentration along the section line A-A extending a distance of 150 km inland from the seacoast are shown in Figure 9. The contour line of $35,000 \text{ mg l}^{-1}$ concentration (seawater concentration), as can be seen from this Figure, hits the bottom boundary at a distance of 62 km from the seacoast whereas the contour line of $1,000 \text{ mg l}^{-1}$ hits the bottom at about 110 km from the seacoast. The distance between these bounding lines represents the transition zone between seawater and freshwater.

Attia (1996) discussed the problem of vulnerability of groundwater pollution in the Nile Delta. Four classes of vulnerability were presented as shown in the map in Figure 10.

- *The reclaimed desert fringes*: This class has a moderate to high vulnerable groundwater due to the presence of sandy formations having high infiltration and low adsorption capacities despite the relatively deep groundwater.

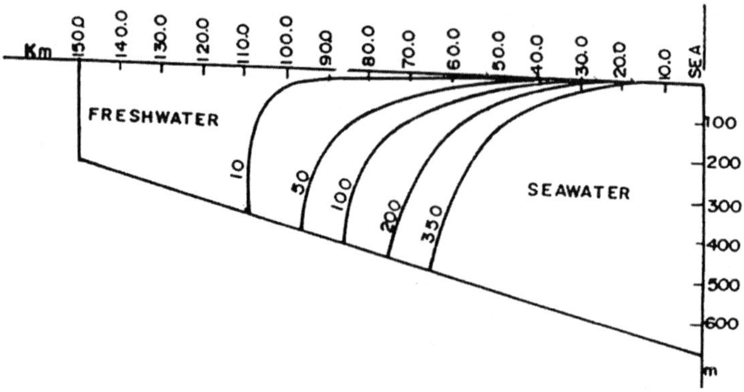

Figure 9. Lines of equal concentration of groundwater along the section line shown in Figure 8 (from Sherif et al., 1990)

- *The traditionally cultivated area*: The groundwater in this class has moderate to low vulnerability due to the presence of a clay cap overlying the aquifer.
- *The transition between the old land and the newly reclaimed land*: It has a highly vulnerable groundwater due to the presence of sandy soils and relatively shallow groundwater.
- *The northern part of the delta area*: This strip of land has a with very low vulnerability due to the presence of a thick clay cap overlying the aquifer coupled with the vertical upward flow of groundwater.

Recently, Margane et al. (2004) produced groundwater vulnerability maps of the Western part of the Nile Delta using the Classical DRASTIC and Modified

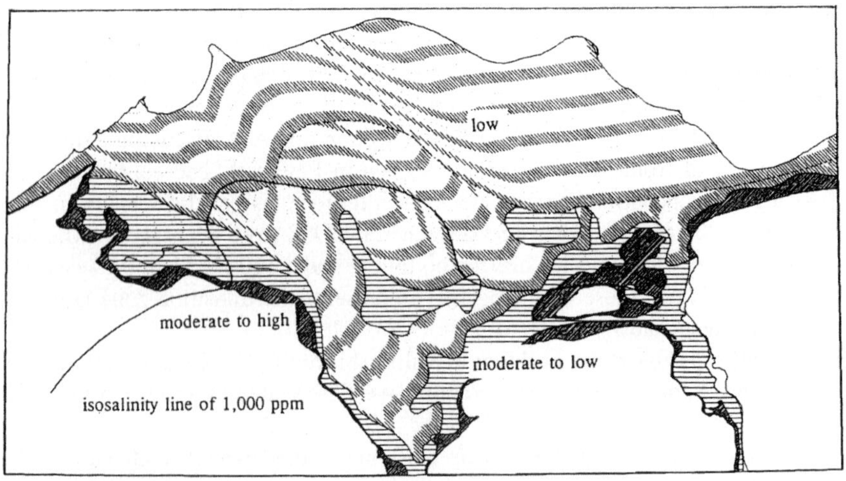

Figure 10. Groundwater vulnerability to pollution in the Nile Delta (from Attia, 1996)

DRASTIC Approaches. For the Classical Approach the reader can be referred to Bennett et al. (1985).

9.4.7 The Sudan

Water-bearing formations in the Sudan are diverse. These formations can be approximately classified as follows:

– *The Basement rock of the Precambrian age*: It covers the areas occupied by the Nubian mountains along the Red Sea, a total of about 1.08×10^6 km^2 or 43% of the total surface area of the Sudan. There is hardly any water to be abstracted from these rocks, except where there are fractures and faults.

– *The volcanic masses*: The surface area covered by thee masses is 228×10^3 km^2, i.e. 9.1% of the surface of the country. Groundwater in the volcanic formations is generally saline with a total concentration of dissolved salts varying from 1,500 to 3,000 ppm. This level of salinity is high for drinking and domestic purposes.

– *The Umm Rawaba Series of the Pliocene–Pleistocene ages*: The deposits of this series can be found south of latitude 14° N. The series covers 515×10^3 km^2 or 20.5% of the country's surface area. The water is of relatively good quality, as the total dissolved salts range between 400 and 600 ppm.

– *The Nubian Sandstone*: This water-bearing formation covers about 740×10^3 km^2 or about 28.1% of the surface area of the country north of 10° N latitude. The total volume of groundwater in storage in the Sudan has been estimated as 41.8×10^9 m^3, receiving an annual recharge of 1.38×10^9 m^3 (Country Report of the Sudan, 1986). The annual abstraction until a few years ago was estimated at less than 0.4×10^9 m^3.

There is, however, a growing tendency to augment the abstraction to meet the increasing demand on water for future domestic and small-scale irrigation purposes. A few more details about the properties of some of the water-bearing formations and the local groundwater reservoirs are included in Table 18, Appendix II.

Examples of groundwater abstraction from alluvial and Nubian sandstone deposits in the Sudan are as presented in the next subsections.

1. The town of Nyala in the western region of Darfur depends for its water supply on the groundwater in the alluvial sediments of the Wadi Nyala and those of Wadi Boloudanga, Bulbul and Gendi and others as shown on the map in Figure 11 (Khadam & Salih, 1986). The aquifer there consists mainly of medium to coarse sand of an average thickness not exceeding 10 m. Water is supplied by means from hand-dug wells and sometimes from boreholes. The alluvial aquifers at Wadi Nyala and the neighbouring wadis are of the unconfined type and are underlain by the impervious Basement Complex. The aquifer is recharged in the wet season from infiltration of the wadi surface runoff. A model study has shown that an annual abstraction of 4×10^6 m^3, reduced to 3×10^6 m^3 in a dry year, will help to cover the local needs for domestic and local industrial purposes (Linden, 1986).

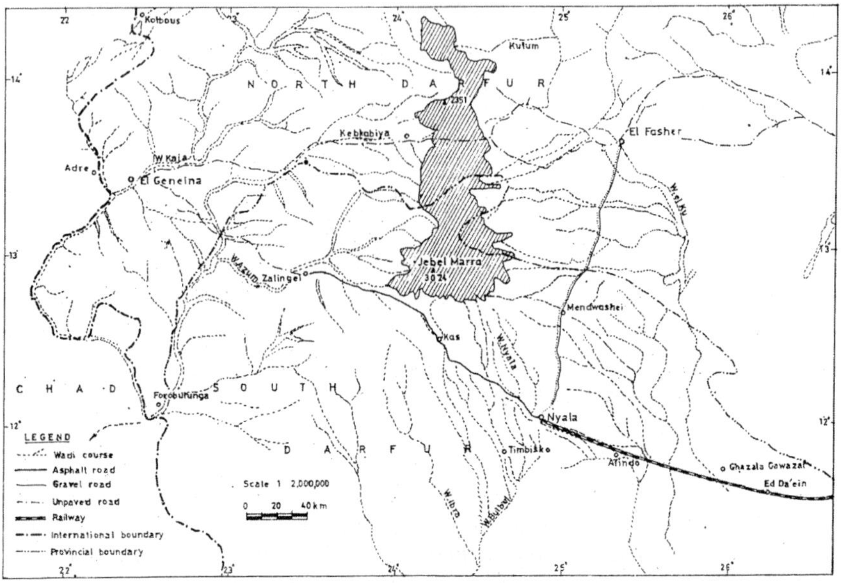

Figure 11. Map of Central Darfur, The Sudan, showing the location of the study area at Jebel Mara (Khadam & Salih, 1986)

2. Another example is the abstraction of groundwater from the Kassala-Gash in the eastern part of the Sudan. Water supply by wells is widely practiced for domestic purposes, supplementary irrigation and for raising livestock in the Kassala Province. The aquifer system there is underlain by impervious Precambrian Basement rock and laterally supported against the impervious Tertiary-Pleistocene clays of the Sudan Plain.

   The water is usually of excellent quality. The most productive part of the aquifer is that where deposits are generally coarse and the recharge is most intensive. Some of the hydrogeological parameters of the aquifers of the Eastern part are listed in Table 18, Appendix II.
3. As mentioned earlier, the northern desert region of the Sudan is underlain by a permanent groundwater reservoir in the sandstone of the Nubian series. It usually exists at a fairly large depth below the land surface. The available data shows that this aquifer system comprises a number of basins with an estimated storage capacity of 16×10^9 m^3 and annual recharge ranging between 43.5×10^9 and 257×10^9 m^3. Figure 12 shows the groundwater heads in certain parts of the Nubian Sandstone in the Sudan (Country Report of the Sudan/Hedayat, 1986), after some adjustment to fit the groundwater heads in the neighbouring countries. This map has been complemented by the findings of Salaam (1966) for the Gezira area between the White Nile and the Blue Nile, and by the results recently obtained by Hussein (2004) for the area between the Blue Nile and the Dinder River.

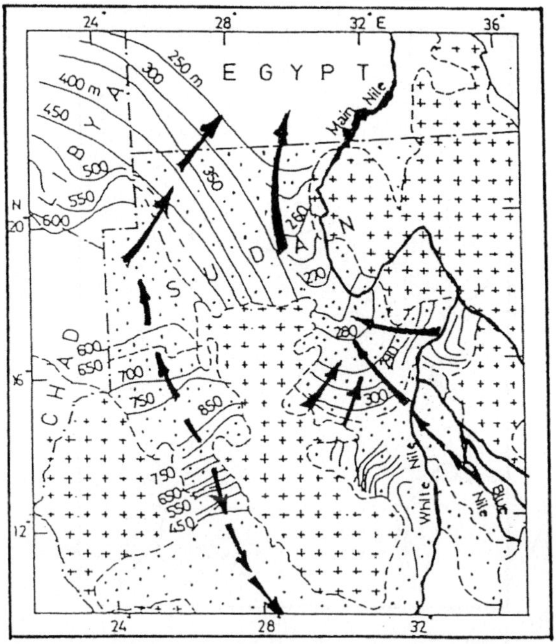

Figure 12. Groundwater heads in the Nubian Sandstone in the Sudan (Hedayat, 1986)

4. Hussein (2004) evaluated the hydrogeology and hydrochemistry of groundwater in the Blue Nile Basin. The aquifer system in the study area consists of the Precambrian Basement Complex in the east, Nubian Sandstone in the west and Ashtan formation in the middle. The Ashtan formation can be found locally in the Blue Nile Basin and consists mainly of fluviatile clays, sand and gravel. Groundwater occurs at an average depth of 25 m from the ground surface with a hydraulic gradient of 0.001–0.0001, and north-northwest as a general direction of flow in the vicinity of the river. Some of the seepage flows, however, laterally from the Dinder to the west and from the Blue Nile to the east where the two flows meet in the groundwater trough between the two rivers.

Water well inventory was conducted using 162 wells located mainly in the Nubian and Ashtan aquifers The TDS in the Nubian aquifer shows variations between 300 and 600 mg l^{-1} except in the central area of the Nubian aquifer where the TDS exceeds the upper limit of this range. Values above 800 mg l^{-1} were only encountered in wells tapping the Ashtan formation in the area to the east of the River Dinder. A simple relation was found to exist between the potentiometric levels and groundwater salinity. The lateral flow causes the relatively high salinity in the central part of the study area where the groundwater trough occurs. Table 11, Appendix III gives a summary of the major ionic constituents in the Blue Nile Basin.

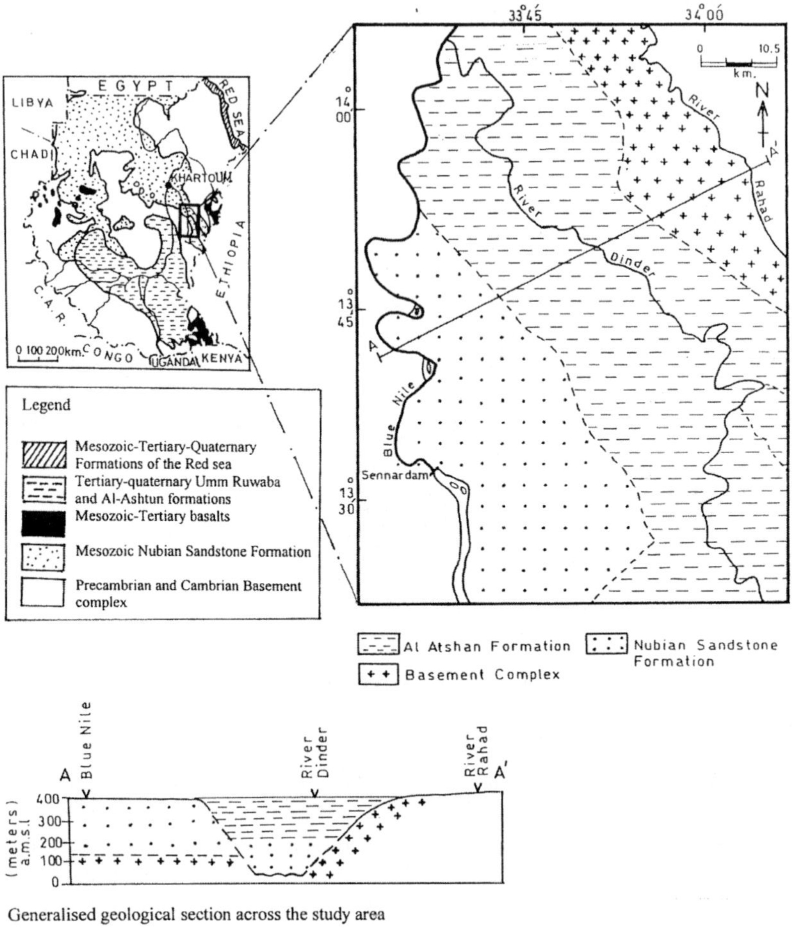

Figure 13. Map and geological section of the study area in the Dinder Basin (from Hussein, 2004)

9.4.8 Djibouti

Müller (cited in Verhagen et al., 1991) suggested that groundwater in Djibouti is recharged almost exclusively during periods of flow in the oueds, which may occur several times during the year. As aquifer recharge happens through infiltration in wadi basins, the water pumped will be mainly the volume that is locally recharged after reducting the evaporation losses.

The most important consumer of groundwater in Djibouti is the capital of the country. Water is supplied from the basalt aquifer, which is situated along the coast. Withdrawal during 1982 amounted as 10×10^6 m^3, and was expected then to rise to 15×10^6 m^3 during 1992. This withdrawal has led to a significant fall of the water table level and increase in the water salinity in the north, mainly as a result of seawater intrusion.

For the rest of Djibouti, no reliable withdrawal figures are available and one has to rely on estimates. Other important consumers are principal villages such as Al-Sabieh and Dikhil using 180,000 m^3 y^{-1}, and Tadjourah and Obock using 220,000 m^3 y^{-1}. The locations of the mentioned places can be seen on the map in Figure 14.

Generally speaking, most of the groundwater in Djibouti is highly mineralised. Electrical conductivity for tested samples ranges from 350 to over 6,000 $\mu S\ cm^{-1}$. The available hydrochemical data show that the lowest mineralisation level, usually below 1,000 $mg\ l^{-1}$, can be found in the deposits traversed by the oueds. The further the wells are located from the oueds, the worse the quality of water is. Groundwater from wells was sampled in the 1980s. Nitrate appeared in all samples, often up to 145 $mg\ l^{-1}$. For more information about hydrochemistry and results of isotopic examination of groundwater the reader can be referred to Verhagen et al., 1991.

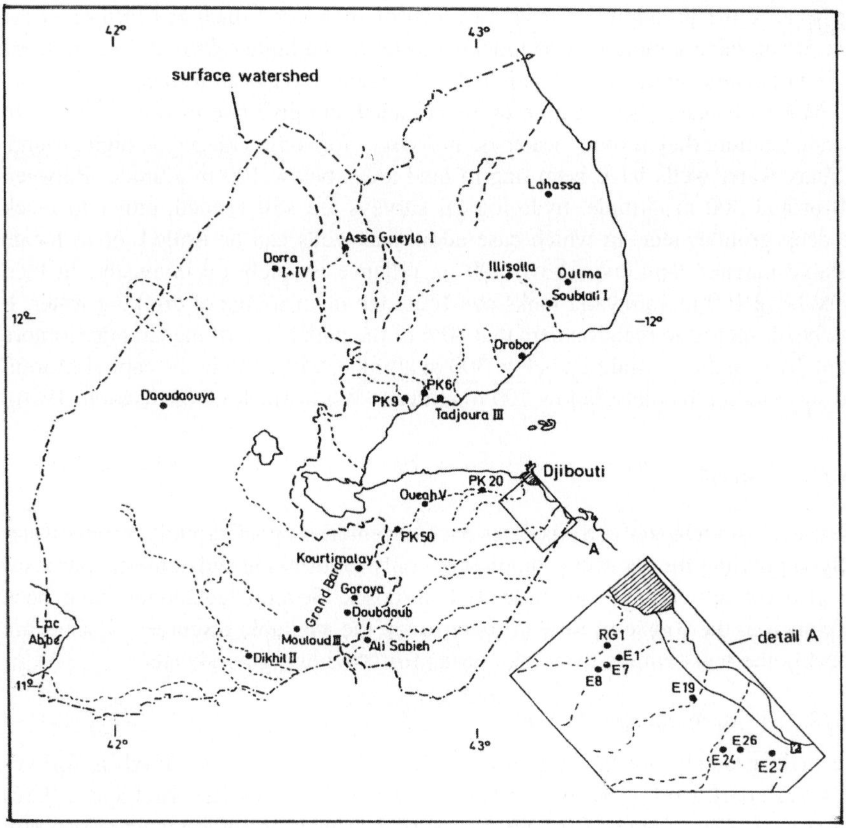

Figure 14. Locaton map of wells sampled for chemical and isotope analyses during 1980 and 1986 (from Verhagen et al., 1991)

9.4.9 Somalia

There is hardly any information on groundwater resources in the country. However, old technical reports mention briefly that pumped water had been used for irrigating small areas during the Italian occupation.

The Report of the Food and Agriculture Organisation (1995) entitled "Irrigation in Africa in Figures" mentions that the groundwater potential is limited because of the limited potential for recharge. It adds "In the northern region, some subsurface flows in wadis are tapped for small (1–25 ha) irrigated farms." The total withdrawal was estimated as 0.81×10^9 m^3 in 1987. Agricultural withdrawal, about 0.79×10^9 m^3 y^{-1}, is mainly for the full or partial control of irrigation schemes.

9.4.10 Comoros

The Comoro Islands are made up of recent volcanic rocks. Because these rocks are porous, groundwater can be tapped. This is particularly true in Grande Comore, but because the groundwater rests on a bed of seawater, which has infiltrated the rock, it can have a more or less brackish taste. In the higher grounds, where there are impermeable strata of clay, perched water tables have been formed. In Anjouan and Mohéli Islands, these can be more extended and give rise to watercourses. In Grande Comore they provide water-point sources for some villages on high ground.

Many water wells have been dug in land areas below 100 m altitude. Between 100 m and 300 m altitude, hydrological surveys are still needed, either to reach the deep groundwater, in which case additional wells can be drilled, or to locate possible perched aquifers. Above 300 m, reliance is solely on rainwater. In fact, even above 100 m, rainwater tanks constitute the main source of drinking water. It may be of interest to mention here that 80% of the population of the Grande Comore Island lives in the coastal belt below 300 m altitude, with 50% in the capital Moroni and surrounding hamlets, below 100 m altitude (World Health Organisation, 1996).

9.4.11 Syria

The use of groundwater in Syria dates back to centuries ago. Currently, groundwater in Syria provides the necessary supplies not only for drinking and domestic purposes but also for agricultural and industrial purposes. Several techniques have been developed in the course of time in order to tap the available resources. Springs are probably the most widely available means for providing groundwater.

9.4.11.1 Main aquifers in Syria

The main groundwater bearing formations in Syria can be classified as follows (ECWA Report, 1981, UNDP Report No. 9, 1982 and Country Report of Syria, 1986)

- *Jurassic Limestone and Dolomite formation*: This is one of the most resourceful formations. It is a karstic carbonate rock aquifer occurring mainly in the coastal highlands and the anti-Lebanon Range. The main springs gushing from this formation are the Banias on the southern slope of the Hermon Mountain; Barada Spring, the main feeder of River Barada; a group of springs scattered along the western bank of Al-Ghab Plain, springing from the eastern slopes of the Alaouite Range, most notable springs of this group are Al-Barid and Al Na'our; Al-Senn Spring, which is the head source of the Al-Senn River. The average flow of these springs combined is in the order of 15.1 $m^3\ s^{-1}$. The average discharge per spring is listed in Table 19, Appendix II. The salinity of the encountered groundwater from the mentioned springs is in the range of 360 to 380 ppm of total dissolved salts (T.D.S.).

– *Lower Cretaceous Sandstone*: It consists of sandstone layers with intervals of limestone and shales. Low-yield springs issue from this aquifer, particularly in Khafroun area of the Alaouite Range and Saghaya area northwest of Damascus. The average yield does not exceed 10 $l\ s^{-1}$. Groundwater quality is generally good and the T.D.S. ranges from 200 to 400 ppm.

– *Middle Cretaceous Limestone and Dolomite*: This formation too is one of the most important water-bearing formations in Syria. The main springs gushing from it are Al-Fija Spring, which supplies potable water to the capital Damascus; Fasraya Spring, north of Damascus; Al-Tannûr Spring, south east of Homs, Al-Sukhna Spring, south the city of Homs; Tel-Oiûn Spring, between Hama and Al-Ghab Plain; a group of springs originating east of Al-Ghab Plain; Banias Spring, north of Tartûs; and the group of Himma Springs, located in south west Syria. The total discharge of these springs has an average of 26.7 $m^3\ s^{-1}$. Groundwater salinity ranges from 420 to 600 ppm in the spring waters.

– *Upper Cretaceous Paleocene Chert and Chalky Marl*: This formation consists of alternating beds of chert, marl and chalky marl. Generally the aquifer is fair to low yielding and water quality is fair too (700–4,000 ppm) and has a high temperature that can be up to 35 °C.

– *Paleogenic Limestone and Miocene Marine formation*: This formation is generally karstic, from which many springs originate. The main springs are Ra's el-Ain, which form the head-water sources of Al-Khabour River; Al-Arûs Spring, source of River Bleikh; Uri Spring in Al-Rouj Plain west of Idlib; Healan Spring, north east of Aleppo (Halab); Qarina Spring, in Yabrud; and Maneen Spring in the outskirts of Damascus. The average total discharge of these springs combined amounts to 52.3 $m^3\ s^{-1}$. The salinity of groundwater falls in the range of 360–665 ppm of T.D.S. for both the springs and well waters.

– *Miocene gypsum, Anhydrites and Sandstone*: These formations originally formed in closed marine environment, include karstic sub-formations where groundwater is stored and flows similar to the limestone formations. They can be found spread in the basin of the Euphrates where the main springs rising from them are Al-Hool, Al-Khatûniyyah and Tel Tabban. The average total flow of these springs amounts to 1.65 $m^3\ s^{-1}$. The quality of this water is hardly suitable for drinking and/or irrigation in view of its high T.D.S. content and sulphuric taste. As such the use of this water is localised to those areas where no better water can be found at a reasonable depth below ground surface.

– *Volcanic formations*: These water-bearing formations are frequently found in Syria, especially in the upper Jordan Valley Basin in the southwest. They are permeable, saturated with water due to their being extensively cracked. Most important of the springs that issue from these formations are the Mizeirib; Zayzûn; Al-Sakhinah; Al-Ash'ary; Al-Azzûly; Al-Bandak; Al-Agamy; Al Thûrayy; Ummud-Dananier; Ain Zakar; As-Sayyada, Al-Buraygbt and others in the Golan area; and Juleibina in Huran district. These springs are characterised by their relatively stable discharge and good quality water. The average total discharge of these springs is about 12.33 $m^3 s^{-1}$. The groundwater quality of this aquifer is generally less than 500 ppm of T.D.S.

The distribution of the different aquifers over the Syrian territory can be observed from the map in Figure 15. From the above estimates of spring flows it is evident that the average totals to about 108 $m^3 s^{-1}$ or 3.41×10^9 $m^3 y^{-1}$.

9.4.11.2 Groundwater conditions

From the preceding subsection one can fairly conclude that karstified carbonate rock units form the most productive aquifers in Syria as well as in Lebanon. Besides,

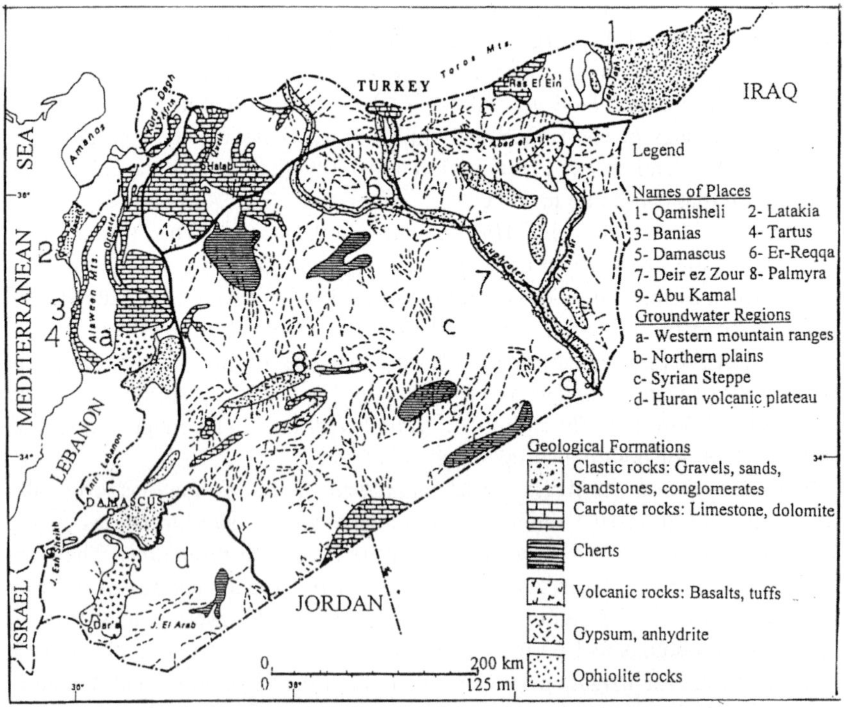

Figure 15. Groundwater regions in the Syrian Arab Republic (UNDP Report No.9, 1982)

these aquifers are mostly productive in zones where high tectonic activities, such as faults, folds, fractures and other geological structural patterns have occurred.

The internal low-land dislocated plains formed by the Rift Valley System; namely, Al-Ghab Graben constitute a huge interceptor drain to the aquifers that occur in the eastern and western flanking uplifted blocks. The total average spring flow along these dislocated plains is estimated as 27 $m^3\ s^{-1}$ or $0.85 \times 10^9\ m^3\ y^{-1}$.

In the southern coastal plain, good potential groundwater occurs in the Middle Cretaceous Limestone. However, most of the flow of this plain goes to the sea similar to the situation in Lebanon. In the northern coastal plain, the groundwater potential is low as a result of the predominating low permeable formations in the area. Groundwater of fair quality and quantity may be encountered within the alluvial deposits along the flood plains of streams and rivers originating from the coastal mountains and crossing the coastal strip of land.

Different authorities and individuals undertook detailed hydrochemical investigation covering the whole territory of Syria. The result of that investigation is represented by the map in Figure 16 (UNDP Report No. 9, 1982). Additionally,

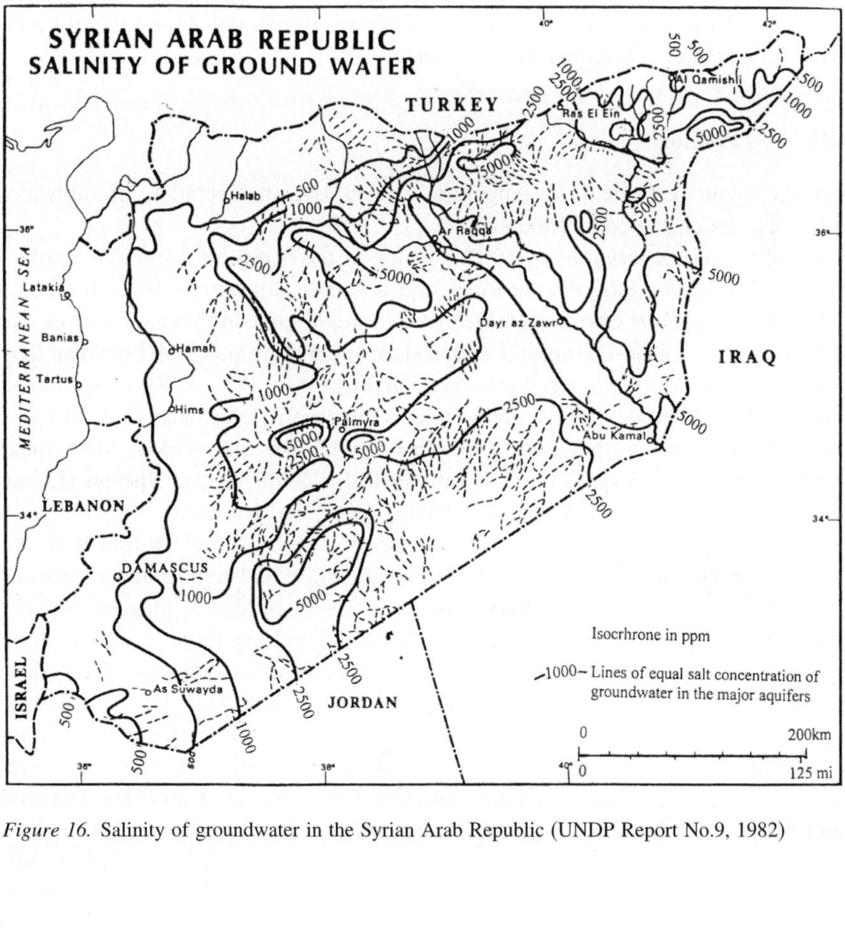

Figure 16. Salinity of groundwater in the Syrian Arab Republic (UNDP Report No.9, 1982)

the results obtained indicate that the groundwater quality is a function of rainfall distribution, geological structure, rock types and groundwater movement. In general groundwater of 1,000 ppm of T.D.S. and less may be encountered throughout western Syria. As one moves eastwards, water quality deteriorates and the salinity can exceed 5,000 ppm of T.D.S.; sulphate water in the northeast and middle Euphrates Basin; and chloride water in the southeastern part of the steppe Basin (Daw, Tadmur and Teniff areas).

Groundwater withdrawal in Damascus area and the Ghouta Plain has been met with increasing difficulties in the last years due to the rapidly increasing demand on water and the decrease in recharge caused by several years of drought. Currently most of the domestic wastewater of Damascus is treated before discharging it into the northern and central parts of the Ghouta and using it for irrigation. Nevertheless, pollution poses an ever-increasing risk. In many locations the Nitrate content in the groundwater already exceeds the limiting value for drinking water, 40 mg l^{-1}. A groundwater vulnerability map of the Ghouta Plain has been recently compiled by Hobler & Rajab (2002). Based on the results obtained from the vulnerability mapping, recommendations for reducing the risk of groundwater contamination have been suggested and listed. As an example the use of abandoned limestone quarries for waste disposal is posing a very serious threat and should therefore be stopped immediately (Margane et al., 2004).

9.4.12 Lebanon

From the hydrogeological investigation carried out in Lebanon, groundwater aquifers can be classified as follows:

– *Cenomanian-Turonian aquifer*: This is one of the two most important aquifers in Lebanon. It consists of limestones and dolomitic limestones with thin veins of marl. The aquifer covers a surface area of 4,290 km^2 and recharged from the high precipitation on the coastal and eastern mountain ranges of Lebanon. The average thickness of the aquifer is estimated as 800–1,000 m. Karstification in this aquifer is well developed giving rise to high well yields exceeding 50 $l\ s^{-1}$. Some of the springs that issue from this aquifer are Mar Sarkis; Mar Abda; Qadisha; Sannine; Afqa, Bakish; Yarmouneh; Bardouni; Faraya Springs (Laban and Asal); Anjar; Ras el Ain; Yunin Fakha; Inata; and Hasabani.

– *Jurassic aquifer:* It occurs in the cores of the coastal uplifts and the Hermon Mountain. The aquifer consists of dolomite and limestone successions interbedded with marl and volcanic rocks. It is well bedded to massive fossiliferous at top and locally interrupted towards the top by Lava flows. Laterally these volcanic rocks grade into a series of marls, limestones and shales of very low permeability. The extension of its outcrops is about 1,290 km^2. Hardly anything is known of the hydrogeological characteristics of this aquifer. However, judging by the large springs issuing from it, the aquifer seems to enjoy a high groundwater potential. The Seir; Dimniya; Jaita; Antelias; Safa; Barouk; Kub Elias; Deckoni and Ain el Delbeh are some springs to mention.

Other than the above two aquifers, there are other aquifers of local importance and limited occurrence and potentials. These are:

– *Eocene aquifer*: It occurs in the south of the country as well as in Al Beqa' area. A number of springs originate from this aquifer. The Hjair Spring, which produces about 8 $m^3 s^{-1}$ in high-flow season (winter) and dries up in summer, is one of the major springs that originates from it.

– *Miocene aquifer:* This aquifer exists in Tripoli Depression and in Al Beqa' Valley. Several productive wells and springs are fed by the aquifer in the Tripoli Depression while that other part of the aquifer is less productive.

– *Jezzine aquifer*: This is another aquifer existing in central Lebanon and consists of limestone, sand and argillaceous limestone.

– *Quaternary aquifer*: It consists mainly of seashore sand, alluvium and talus deposits. It occurs in a continuous manner along the flood plains of the major streams and rivers originating from both the coastal and eastern highlands. It also occurs in the Beqa' area and in the Tripoli Depression. The estimated total surface area covered by this aquifer is a little above 1,000 km^2. The aquifer permeability is widely variable and its thickness can be up to 600 m. Well yields fall in the range of less than 10 to 30 $l\ s^{-1}$.

The potential volume of potable groundwater has been estimated as $0.60 \times 10^9\ m^3\ y^{-1}$. The amount actually used in the 1980s was limited to say $0.16 \times 10^9\ m^3\ y^{-1}$ and increased to more than $0.25 \times 10^9\ m^3\ y^{-1}$ by the end of the 1990s. The recent studies have shown that the recharge to the aquifers can increase the available volume by $0.16 \times 10^9\ m^3\ y^{-1}$ distributed as follows: 0.015 for the Akkār Plain; 0.010 for Beirut District; 0.050 for Al Barûk Range/Niha; 0.020 for the coastal area; and 0.065 for the southern Beqa'a, all figures are in $10^9\ m^3\ y^{-1}$ (UNDP Report No. 9, 1982).

Groundwater in Lebanon is generally of good quality. It is fresh at the foothills on both slopes of the coastal and eastern highlands. In Beqa'a area, water quality is generally of less than 1,000 ppm T.D.S. The water quality in the coastal plains is inferior due to saltwater encroachment from the sea as a consequence of over-pumping the aquifer. Table 12, Appendix III gives the chemical analysis of groundwater in some areas of Lebanon.

9.4.13 Palestine

The geological build up in the area dictates the availability of groundwater. The mountain heights are the source of recharge of many a groundwater basin in the West Bank and Israel. A certain amount of the precipitation on these highlands penetrates the rocky surface and flow under ground to feed the water-bearing formations.

Quite often the groundwater resources are subdivided into the Mountain aquifer system and the coastal aquifer system. These two systems lie in Palestine and Israel. The former extends from the foot of the Carmel Mountain near Haifa in the north to Bir Al-Sabe' (Bir Sheva) in the south, and form the Jordan River and the Dead sea in the east to the tip of the coastal aquifer in the west. The Mountain Aquifer

System consists mainly of limestone and Dolomite of Cretaceous to Quaternary age. The current annual withdrawal from this system is estimated as 0.118×10^9 m^3 y^{-1}. The Coastal Aquifer extends from the slopes of the Carmel Mountain in the north to the northern Sinai in the south, and from the foothills of mountains in the east to the Mediterranean Sea in the west. The aquifer consists mainly of sand, sandstone and pebbles of the Pliocene to Pleistocene age. The safe yield is claimed to be around 0.3×10^9 m^3 y^{-1} (Naji, 2004).

In the West Bank, the principal basins where groundwater is flowing towards the west are the Ouja-Temsah and Al-Khalil (Hebron) Basins. The basins in which groundwater is flowing towards the north are the Nablus-Jenin-Gilboa Basin feeding the Bisan and Al-Far'b areas and the Gilboa-Ta'na Basin feeding the hot springs and aquifers in the northern and northwestern plains. The basins where groundwater is flowing to the east are Bardala, Al-Maleh, Fasayel, Ram Allah, Al-Quds, etc. The locations of the groundwater basins in the West Bank are shown on the map in Figure 17.

The annual amount of groundwater sustainable in the aquifers of the West Bank is estimated as 0.71×10^9 m^3. This amount is the sum of the volume in storage plus the return flow. The average annual return flow is claimed to be 0.60×10^9 m^3, of which 0.45×10^9 m^3 feed the basins with groundwater flowing to the west and the rest feeds those basins with groundwater moving towards the east. In contrast, the annual return flow to the aquifers of the Gaza Strip is estimated as 75×10^6 m^3 originating from precipitation, ground and surface water resources, and excess sanitary and agricultural drainage water (Country Report of Palestine, 1986).

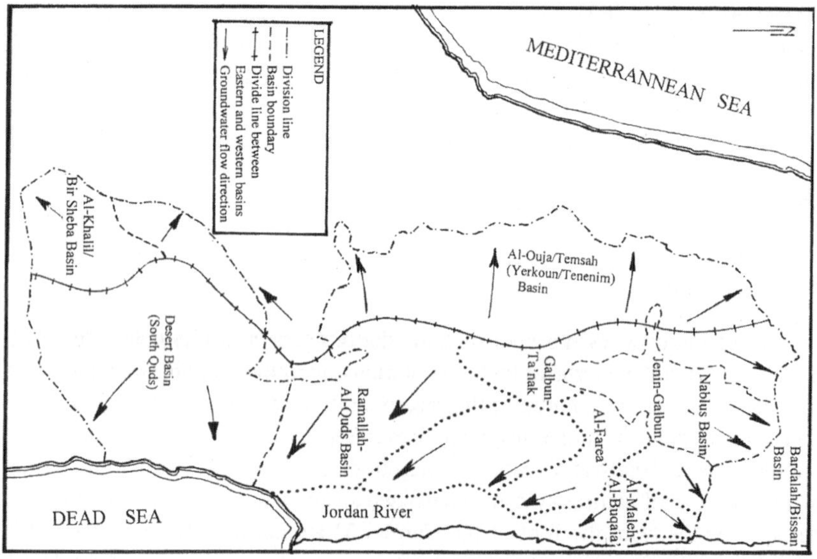

Figure 17. Groundwater basins in the West Bank (Country Report of Palestine, 1986)

Table 20, Appendix II, gives the yields of production wells (32 wells) used for drinking purpose in the West Bank in the 1980s. Youssef (2004) mentioned that the total annual available groundwater quantities in the West Bank from the existing groundwater sources are around 123.6×10^6 m^3. Of this amount 62.8×10^6 m^3 come from pumped wells and 60.8×10^6 m^3 from springs.

Groundwater in the Gaza Strip occurs in a system of shallow sub-aquifers made up mainly of Quaternary sand, calcareous limestone and pebbles with interbeds of impervious and semi-pervious clay. The top of the system consists of recent sand dunes in the western part of the strip and finer deposits (sands and loess) in the eastern part and beyond. The aquifer depth varies from 10 m in the east to 120 m in the west. Pumping test results have shown that the permeability varies from a minimum of 1.0 m d^{-1} for clay and chalk to over 2,000 m d^{-1} for coarse sand, causing the transmissitivity to range from a minimum of 30 m^2 d^{-1} to a maximum of 5,000 m^2 d^{-1}. Accordingly, well yields vary from 4 l s^{-1} to 65 l s^{-1} depending on the aquifer material.

Groundwater heads fall from say 15 m a.m.s.l. along the Strip's eastern border to mean sea level in the west along the seashore line. At locations where groundwater is extensively abstracted the groundwater level drops to more than 2 m b.m.s.l. According to Youssef (2004), the total quantity of fresh groundwater in Gaza fell from 1.2×10^9 m^3 in 1972 to around 0.7×10^9 m^3 in 1998, i.e. a drop of at least 40% in a period of 26 years caused by the increasing abstraction with time. Oppositely, the annual volume of abstraction increased from 65×10^6 m^3 in 1967 to 100×10^6 m^3 in 1995 (Mogheir, 1997). The same source estimates the annual recharge to the sandy areas in Gaza as 26×10^6 m^3 and for the clayey areas as 20×10^6 m^3, bringing the total annual recharge to 46×10^6 m^3. These estimates have been based on two approaches: the mass balance of the chlorides and the seasonal fluctuation in water level.

9.4.14 Jordan

9.4.14.1 Hydrogeology

Various hydrogeological studies have been carried out and more are under way. From the results of those investigations, the major aquifer systems in Jordan can be classified as follows

– Quaternary-Tertiary aquifers:

i- The shallow aquifer system consists of sedimentary and alluvial deposits of the Tertiary and Quaternary ages. Examples can found in the valley fills of the Jordan Valley and Wadi Araba, and the limestone, chalk, marl, sand and gravel layers in the Jafer, Azarq and Sirhan areas. Groundwater in this system originates from wadi floods and from the subsurface runoff flowing from the highland areas. The quality of groundwater in this system is strongly variable and depends essentially on the recharge characteristics, salt content within the aquifer and rate of evaporation. Likewise, aquifer parameters are widely variable and depend on the percentage of fine material in the soil.

ii- Basalt aquifer system: This system extends from the Syrian-Jordanian border southwards to Al-Azraq and Wadi Dhuleil areas, and occupies 1,100 km^2 in northeastern Jordan. The quality of water of this system is known to be good, and the aquifer material in some localities is high enough to allow the withdrawal of large quantities of groundwater. The aquifer discharge augments the baseflow in the basins of the Upper Yarmouk River, the Wadi Dhuleil-Wadi Zarqa and Al-Azraq River.

The aquifer parameters are extremely variable depending on the thickness of the scoriaceous basalt within the aquifer. The transmissivity ranges from almost zero to 11,300 $m^2\ d^{-1}$, and the specific capacity of wells from 0.07 to 3352 $m^3\ h^{-1}\ m^{-1}$. The quality of water is described as good with T.D.S. ranging from 500 to 1,000 ppm.

– *Carbonate Rock aquifers:*

This system of aquifers is known locally as Amman-Wadi Sir aquifer system. The aquifers are made up of limestone, chert limestone and sandy limestone of the Upper Cretaceous. As the system has the largest extent in both the east and west banks of the Jordan River, it forms the most important aquifer in Jordan. The groundwater contained in the aquifers originates from the wet zones of the highland areas. The degree of fracturing of the carbonate rocks (karstification) is widely affecting the values of the hydrogeologica parameters. The aquifer transmissivity is estimated to range between 1 and 46,000 $m^2\ d^{-1}$, the specific yield of drilled wells 0.2–2,100 $m^3\ h^{-1}\ m^{-1}$ and the storativity 1–10% (ECWA Report, 1981).

The aquifer recharge is supplied directly from the rainfall and indirectly by seepage from the neighbouring aquifers. Depth to groundwater is often less than 150 m below surface and the quality of water is generally good, less than 1,000 ppm of T.D.S.

– *Sandy-facies aquifers:*

The aquifers belong to two systems: the Disi group and the Kurnub group. Together they form a huge groundwater reservoir of considerable extent in east Jordan. There is hardly any information about it in West Jordan.

i- The Disi group aquifer: It consists of sandstone and quartizite, 350 m thick on the average, belonging to the Paleozoic age. Its main outcrops occur in the southwestern part of Jordan in a strip 20 km wide on the eastern edge of the basement outcrop between Ras el-Naqb in the north and the Umm Sahm Mountains in the south. The most important freshwater producing part of the aquifer lies in southern Jordan where the depth to water ranges from 60 to 80 m. The average transmissivity is 720 $m^2\ d^{-1}$ and the mean storage coefficient from 0.01 to 0.03.

ii- The Kurnub-Zarqa group aquifer: This aquifer extends almost over the whole surface of Jordan. The aquifer material consists of sand, sandy limestone with clay and shale of the Cretaceous-Jurassic age. The aquifer crops out in the Lower Zarqa River and along the eastern flanking escarpment of the Jordan Valley-Dead Sea-Wadi Araba Graben. The Kurnub sandstone formation in south Jordan overlies the Disi Group aquifer system described in i. The yield of production wells is limited and the quality is poor, more than 2,000 ppm T.D.S. The two-aquifer systems are separated by layers of sandstone, siltstone and shales.

The locations of the basins and wadis mentioned in the above description are shown on the map in Figure 18. Besides, Table 21(a), Appendix II, includes some of the statistics relevant to groundwater situation in Jordan (cited in ECWA Report, 1981 from the National water master Plan of Jordan, 1977). It has been added that an estimated subsurface amount of 33.5×10^6 m^3 y^{-1} flows towards the Red Sea, Dead Sea and Saudi Arabia leaving an annual amount of 169.6×10^6 m^3 as internal underflow within the basins in Jordan. As such, the recharge to East Jordan is in the order of 585×10^6 m^3 y^{-1}. Al-Weshah (1992) gave a different set of data, a summary of which is available in Table 21(b), Appendix II. Assuming that the data listed in both Tables are reliable, it can be easily observed that the baseflow, after excluding the Yarmouk, has dropped from 353.2×10^6 m^3 y^{-1} to 267×10^6 m^3 y^{-1}, i.e. 25%. On the other hand, the amount of groundwater annually exploited rose from 232×10^6 m^3 y^{-1} to 304×10^6 m^3 y^{-1}, i.e. an increase of about 31%. Furthermore, the total amount of water exploited and used by the different sectors in 1993 was 983×10^6 m^3. Renewable and non-renewable groundwater resources contributed about 534×10^6 m^3 (Shantanawi & Jayousi, 1995). According to the same source, the annual volume (10^6 m^3) used for irrigation alone increased from 138 in 1980 to 183 in 1986 to 318 in 1989 to 321 in 1991.

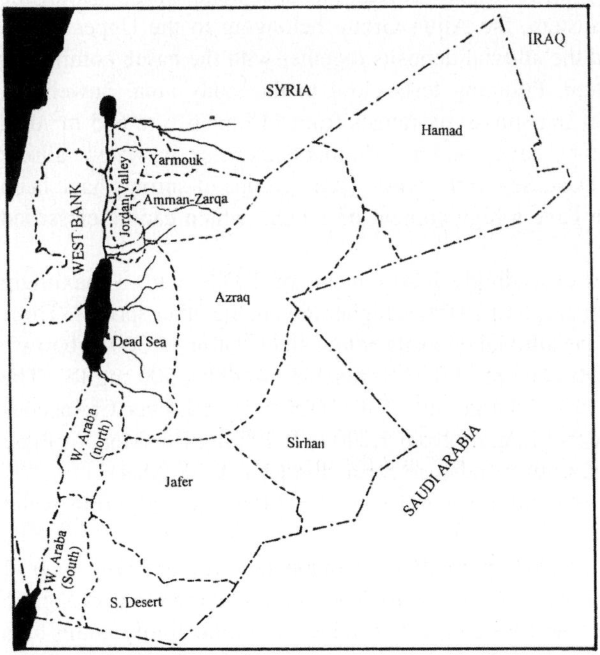

Figure 18. Groundwater Basins of Jordan (UNDP Report, Study Series on Water No. 9, 1982)

Adjacent to the northern Jordan Rift Valley (JRV) are the Golan Heights and the Aljoun Mount. The relief and active tectonic movement have led this zone to host tremendously prolific artesian aquifers. Natural discharge from the zones along the fault contributes to baseflow in the Yarmouk River. At Mukhabeh, the Maqla and Balsam freshwater thermal springs (31 and 43 °C, respectively) indicate a discharge of deeply circulating groundwaters. The neighbourhood of this area is well reputed with its thermal springs, which have been used since biblical times for curative purposes. A certain well drilled to a depth of 350 m for exploring the groundwater potential at Mukhabe, yielded a discharge of 6,000 $m^3\ h^{-1}$. Twelve wells were drilled in the 1980s in the neighbourhood of the thermal springs and the temperature of groundwater was found to range from 29 to 56 °C.

The combined discharge of the JRV wells (4) is about $11 \times 10^6\ m^3\ y^{-1}$ and the total yield of the Mukhabe well field (8 wells) in the Yarmouk Valley is $30 \times 10^6\ m^3\ y^{-1}$. The quality of water is fresh with up to 500 mg l^{-1} T.D.S. (Bajjali et al., 1997).

9.4.14.2 Hydrochemistry

The range of water salinity in each of the groundwater areas is already listed in Table 21(b), Appendix II. Lately 15 springs in the Suweimeh area have been hydrochemically investigated. This area is located in the northeastern part of the Dead Sea area. Hydrogeologically, this area comprises the three regional important aquifer systems in Jordan; the Zarqa-Ma'in and the Kurnub Sandstone Group within the deep Sandstone Aquifer System, the Aljun Group belonging to the Upper Cretaceous Aquifer System, and the alluvial deposits together with the basalt composing the Shallow Aquifer System. Pumping tests close to the study area, Suweimeh, have shown that the aquifer transmissivity ranges from 115 $m^2\ d^{-1}$ to 685 $m^2\ d^{-1}$. Hydrochemical investigations have shown a dramatic increase in water salinity from the east towards the Dead Sea in the west. Three groups of springs have been distinguished, all of which have a high content of T.D.S., which can be classified as medium saline to brine.

Two springs showed an exceedingly high content of T.D.S. with a maximum value of 40,670 mg l^{-1}, which is 5 to 10 times higher than in the other springs. Three springs discharging out of the alluvial deposits below an elevation of 300 m b.m.s.l. They also show a very high content of T.D.S. reaching 4,000–4,500 mg l^{-1}. The remaining 10 springs, which discharge out of the Triassic and Lower Cretaceous aquifers, have a T.D.S. content ranging from 1,200 to 3,400 mg l^{-1}. Most of these springs are located in the Kurnub Sandstone formation (Ali et al., 2004).

The investigation of stable isotopes confirmed the differences in spring water composition.

Recently, Margane et al. (2004) presented a groundwater vulnerability map of the Irbid area in Jordan. The water vulnerability has been found high to very high in much of the outcrop of the limestone aquifers. The level of vulnerability in this area correlates with the level of bacteriological contamination and with nitrate content. It is feared that the current water supply becomes polluted by sewage water from the city of Irbid and surrounding communities. The establishment of a groundwater protection zone and recommendations for restricted land use are considered to be top priorities for implementation in the near future.

9.4.15 Iraq

Groundwater in Iraq, generally speaking, is neither well developed nor totally identified. Since the beginning of the 1980s until the time of drafting these lines Iraq has become involved in a number of wars with neighbouring states, occupied by foreign powers, suffering from political and administrative unrest, and serious financial difficulty. These circumstances did not encourage any comprehensive investigation of the country's groundwater resources for the sake of completing and/or updating the rather old information and data. As a result, we shall base our presentation here to the available data, which date back to the end of the 1970s – beginning of the 1980s.

Groundwater in Iraq is locally used for irrigation and domestic purposes in areas where surface water resources are remote, inaccessible or not available. Five major groundwater areas can be found in Iraq: the mountains, the foothills, the Delta Plain, Al-Jazira and the desert areas. The locations of these areas are indicated on the map in Figure 19 (from Jibrael, 1973). The first two areas are known to store and to provide good groundwater potential in quantity and quality while as the remaining areas are claimed to be moderate to fair groundwater-producing areas. A descriptive assessment of the hydrogeological characteristics of these areas is summarised in Table 22 Appendix II. Additionally, estimates of the available groundwater resources were given almost three decades ago by the then Iraqi Ministry of Irrigation as $1.0 \times 10^9 - 1.2 \times 10^9$ $m^3\ y^{-1}$. The amount annually utilised at that time was only 40%.

Good quality groundwater has been found at the foothills of the mountains in the northeast of the country and in the area on the right bank of the Euphrates. The aquifer in the northeast of the country is estimated to have a sustained discharge between 10 and 40 $m^3\ s^{-1}$, lies at a depth of 5 to 50 m. Its salinity increases towards the southeast of the area until it reaches between 0.5 and 1.0 mg l^{-1}. The aquifers on the right bank of the Euphrates River are entrapped between gypsum and dolomites at a depth increasing towards the west. Water can be found at 300 m depth and the estimated total discharge of the aquifer is about 13 $m^3\ s^{-1}$. In the western part of the said area the salinity of water is 0.3 mg l^{-1} compared to 0.5–1.0 mg l^{-1} in the eastern section. In other parts of the country good quality water is limited due to high levels of salinity.

The Ministry of Municipalities drilled 103 wells in the Samara-Tikrit area to provide groundwater for several purposes. The area is located along the eastern bank of the Tigris River nearly 160 km north of Baghdad. The Tigris River borders the area on the west and south, the Hamrien Hills on the north and east, and the Adhaim River on the east and south (Figure 19). Haddad et al. (1970) listed the static water level and water level during pumping for all drilled wells. The same

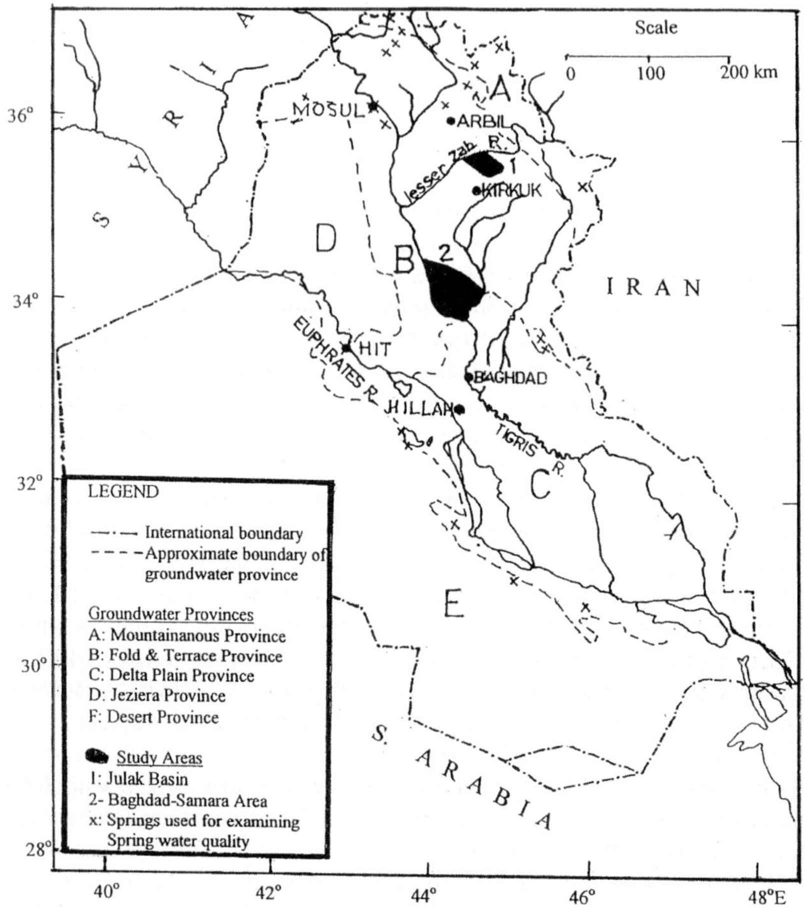

Figure 19. Groundwater Provinces and study areas in Iraq

source of information also includes the results of chemical analyses. The depth to water table, prior to pumping, ranged from 5 to 50 m below land surface. The groundwater quality showed a considerable variation from well to well. Some wells showed an excellent quality of water with total soluble salts in the range from about to 200 to1,200 ppm. Other wells showed inferior water quality with total soluble salts up to about 7,400 ppm. Water chemistry of the 10 wells with least salinity and the 10 wells with highest salinity is given in Table 13, Appendix III. Of the 103 wells analysed only 3 wells showed an electric conductivity less than 1.0 and 11wells up to 2.0 mmhos cm^{-1}. Those wells are the ones located in areas adjacent to the southern part of the Adhaim River and to Tigris River south of Al-Dour. The remaining samples showed electrical conductivity between 2 and 9 mmhos cm^{-1}. One should not forget that irrigation with water having total soluble salts exceeding

2,000 ppm leads sooner or later to soil salinisation with the obvious consequence of deterioration in land's productivity unless certain drastic measures are taken.

Hydrogeological data were also compiled from the area between Baghdad and Hillah (Figure 19) as a component of the investigation of the irrigation scheme of the Greater Mussayeb. The examination of the available logs of water wells showed that in the Baghdad area layers of sand and pebble overlain by a sequence of clayey silt and sand occur at an average depth of 13 m below ground surface. Water wells in the Hillah area are in general shallow, only 43% of the wells exceeded 13.5 m. Of these wells only one-third hit beds of pebble. The data collected from that investigation suggest that the sand-pebble beds between Baghdad and Hillah, considered to belong to the Older Alluvium, may occur at a depth of 10–15 m below surface. The water table in general lies between 3.0 and 4.5 m below surface. The measured well discharges and thereupon their specific discharges were described as not highly reliable. Measurements, however, indicated that the discharge ranged from less than 0.01 $l\ s^{-1}$ to 6.8 $l\ s^{-1}$. The salinity figures ranged from less than 100 ppm to over 5,000 ppm, total soluble salts. It was stated that the reliability of these data too should not be overestimated (Hassan & van der Sluijs, 1970).

Hydrogeological investigations were also carried out in the Jolak Basin with the aim of evaluating its groundwater resources. The basin is located about 20 km north of Kirkuk city, bounded on the northeast by the Khasah Su River and on the north and northwest by the Lesser Zab (Figure 19). The basin lies in the foothills province of Iraq, within the semi arid zone and receives annually some 380 mm of precipitation. The Jolak Basin can be considered as a typical synclinal valley at the low folded hydrogeological province of northern Iraq. The rock formations, which are exposed and cover the basin and the surrounding areas, generally range in age from Miocene through Recent. The water-bearing formations are thick sediments of the Bakhtiari, old and recent alluvium, and terraces resting on impermeable base of clay beds of the Upper Fars formation. The depth to groundwater varies within the areas of the different hydrogeological conditions. It ranged from 10 to 50 m in the plain area, less than 30 m in the foothills area and between 0 to more than 20 m in the area occupied by the valley of the lesser Zab. The thickness of water-bearing horizons in the plain exceeds 300 m.

However, in the foothills area each of the three water-bearing horizons has a thickness of 30 m on the average. Generally the groundwater in the basin flows towards the Lesser Zab River. Nevertheless, the basin has a discharge zone, discharge-recharge zone and a recharge zone. Direct precipitation and surface drainage are the main sources of aquifer recharge.

The groundwater and springs water have a total soluble salts ranging from 200 to 1,000 ppm. Bicarbonates are the dominant anion with very low sodium chloride concentration throughout most of the basin area. In general, the quality of water for drinking, domestic and irrigation purposes is rated as good (Haddad et al., 1971).

Water springs are numerous; probably reaching thousands. Despite their scatter all over Iraq, they are mainly confined to mountainous and foothill regions. According to one of the classification systems, springs are either freshwater springs or mineral water springs. The water of the first class is primarily used for drinking and domestic purposes, and only if water is abundant, it can be used for irrigation as well. As an example, the spring of Sarchinar supplies Sulaymaniyah with the necessary water needed for domestic use and the remainder goes to irrigating farms, gardens and woods. Mineral water springs are fewer in number though spread over a vast area. Some of these springs are well known to possess medicinal properties for curing certain ailments and skin diseases. The water is characterised by a relatively high temperature (35–50 °C) and H_2S content. Hammam Al-Alil (bath of the ailing person), 26 km south of the city of Mosul on the western bank of the River Tigris, is the most reputed spa in Iraq (Baghdadi, 1973). The discharge of the four springs forming the spa was measured in the period July-December 1972 and found to range from 16 $m^3\ h^{-1}$ to 45 $m^3\ h^{-1}$. The gas in these springs contains light hydrocarbons; mainly CH_4, C_2H_6, (36–63%), H_2S (24–46%)and CO_2 (13–17%). The average water temperature of the 4 springs is 42.5 °C. The water temperature is coldest in January and hottest in July-August (Al-Sawaf, 1973).

9.4.16 Kuwait

9.4.16.1 Hydrogeology

Two main geological formations are known to contain suitable groundwater in Kuwait. They are: Kuwait and Hasa Groups. The former comprises the Dibdibba, Fars and Ghar Formations. The second comprises the Dammam, Rus and Um er Radhuma Formations. The two aquifers are connected in a way that provides a continuous hydraulic system from the western edge of the outcrop areas of the Dammam Formation in Saudi Arabia to the Arabian Gulf and Shatt el-Arab. Table 23, Appendix II, summarises the lithology, water-holding characteristics and exploitation area(s) of each water-bearing formation.

The natural water resource in Kuwait is the brackish groundwater, the salinity of which ranges from 3,000 to 10,000 mg l^{-1}. The isosalinity line of 1,000 mg l^{-1} divides the country almost diagonally from northwest to southeast into two halves. The northern half, bounded by the frontier with Iraq and the coast of the Arabian Gulf, is more saline than the southern half located adjacent to the frontier with Saudi Arabia. In this half there are potential areas with estimated salinities from 3,500 to 6,500 mg l^{-1} total dissolved solids and with life expectancy of more than 25 years (ECWA Reoprt, 1981). The recharge the aquifers receive from infiltration into the outcropping zone is the origin of their water. Potable water in Kuwait consists of a blend of brackish groundwater with desalinated water in a ratio 1 to 9. The Dammam limestone aquifer, which underlies the surface of Kuwait, is the most productive source of brackish water.

The State of Kuwait extracted almost 1.6×10^9 m^3 of groundwater in the period 1957–89, at an average annual rate of 70×10^6 m^3. Actually the rate increased from 5×10^6 m^3 in 1957 to almost 120×10^6 m^3 in 1987 abstracted from well fields with more than 800 wells. The historical development of groundwater abstraction from 1957 to 1992 is shown in Figure 20 (Sayed & Al-Ruwaih, 1995). The decline in the water production around 1990 was due to the first Gulf war. Shortly after that, the production capacity increased to reach 160×10^6 m^3 in 1995, and expected to increase further to reach double this amount in the near future. Water is withdrawn mainly from well fields A, B, C, D, and E, located in the Shagaya area southwest of Kuwait. The location of these well fields is shown on the map in Figure 21.

The hydrogeology and hydrochemistry of groundwater extracted from the Umm Gudair area (450 km^2) in the southeast corner of Kuwait was examined and the results reported by Al-Ruwaih (1995). The two aquifer-groups from which the brackish groundwater is extracted were considered as a single reservoir penetrated by 41 production wells. The study has shown that the aquifer is semi-confined to confined. The average transmissivity, which increases towards the north and northeast, is 312 m^2 d^{-1}. The water quality of the two aquifers is relatively poor as the T.D.S. ranges between 3,130 and 4,750 mg l^{-1}. The cation facies are calcium-sodium and the anion facies chloride-sulphate. The type of groundwater is generally as being Na Cl. However, two genetic types have been identified: the Mg Cl_2 and Ca Cl_2 are of marine origin and old marine formations. The inflow entering the aquifer along the border with Saudi Arabia is about 1×10^4 m^3 d^{-1} while in 1995 it exceeded 6.1×10^4 m^3 d^{-1}. The prevalence of the imbalance between the inflow to and the extraction from the aquifer is expected to lead to further decline of the water level and deterioration of water quality.

Well field B, one of the oldest in Kuwait, is withdrawing its water from the Dammam limestone aquifer. It has been recently put under investigation with the aim of pursuing its hydrogeology and chemical evolution of its water. The investigation has shown that the Dammam aquifer has a higher piezometric head than the overlying aquifer of the Kuwait Group. However, as pumping water from the Dammam aquifer reduces its head, the flow of water between the two aquifers is reversed and leakage takes place from the Kuwait Group aquifer to the Dammam

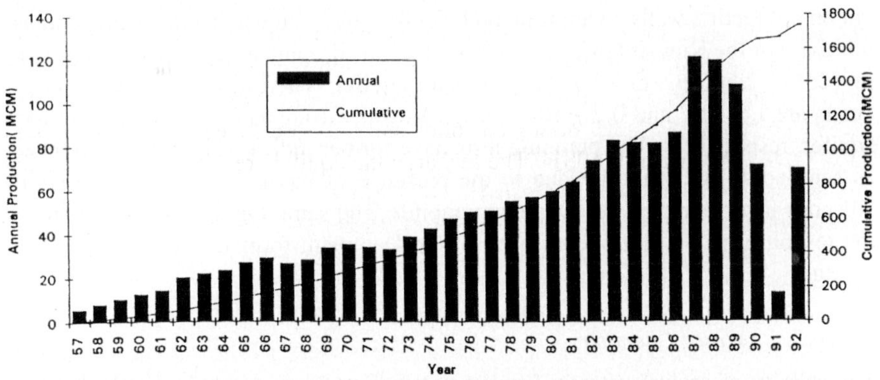

Figure 20. Annual groundwater production in Kuwait (Sayed & Al-Ruwaih, 1998)

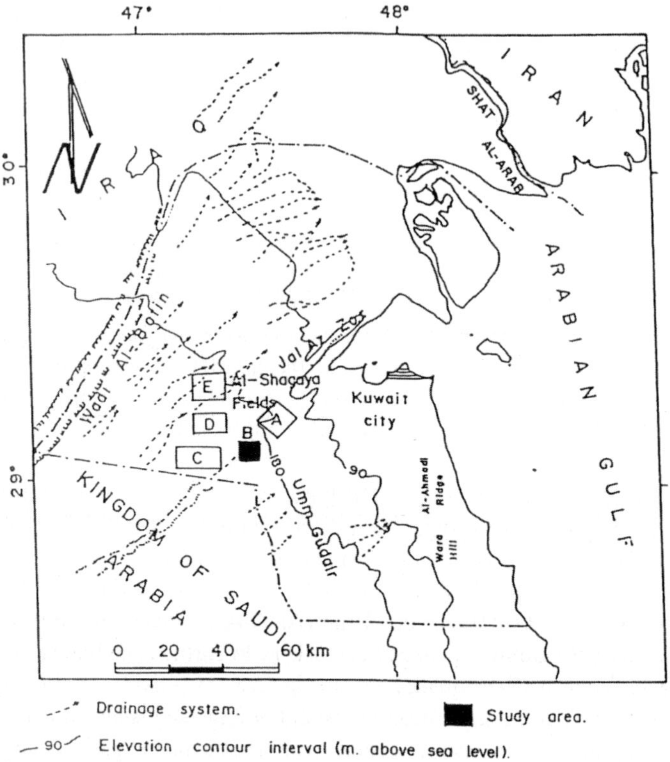

Figure 21. Location map of the study area, well field-B (Al-Ruwaih & Shehata, 1998)

aquifer. The variation of the piezometric heads in the two formations duting the period 1980–1990 can be seen from the graphs in Figure 22.

According to Al-Ruwaih & Shehata (1998), the results of pumping tests carried out on production wells penetrating both aquifers have shown that the average transmissivity of the Kuwait Group aquifer is 312.8 $m^2\ d^{-1}$ and a range of transmissivity of 32.9 to 2,606 $m^2\ d^{-1}$ for the Dammam aquifer, The estimated storage coefficients are 1×10^{-3} and 0.2×10^{-3} for the Kuwait Group aquifer and the Dammam aquifer, respectively. The pumping tests have further indicated that the aquifers act as semi-confined, primarily due to the presence of basal clay and discontinuous siliceous layers capping the Dammam aquifer. The same Figure shows also shows the response of the piezometric head in the Dammam formation to the abstracted amount.

9.4.16.2 Water Chemistry

The salinity of groundwater in Kuwait varies from basin to basin. Fresh or sweet groundwater with total dissolved salts (TDS) in the range 500–1,500 ppm can be

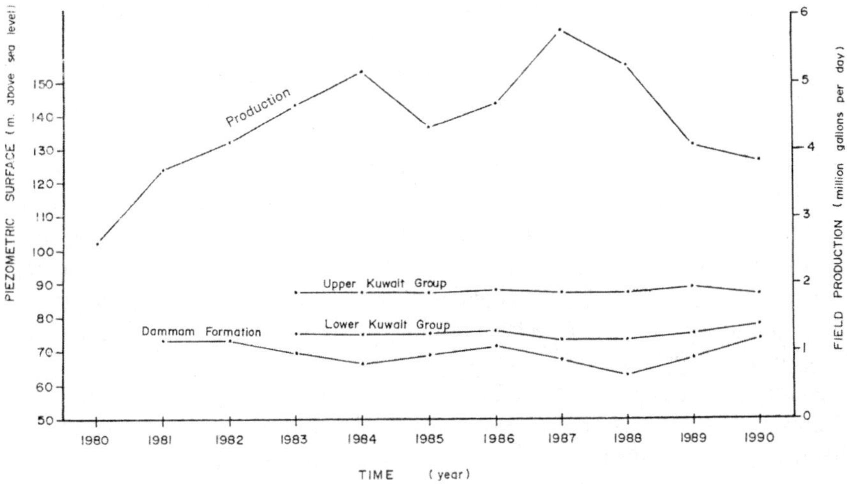

Figure 22. Variation of piezometric head with groundwater abstaction from the Dammam aquifer during the period 1980–1990 (Al-Ruwaih & Shehata, 1998)

found in the well fields of Ar-Rawdatain and Umm Al-Aish in northern Kuwait. The daily production of these fields has been reduced in the course of time so as to avoid their being exhausted. The production rate in 1984 was 25% of that in the year 1975. Slightly brackish groundwater is abstracted from the principal well fields, which are located in Salibiya, Shagaya and Umm Gdair. The TDS of this water falls in the range 1,000–10,000 ppm.

Water chemistry of the Kuwait Group and of the Dammam limestone as well as the water pumped from the two aquifers has been studied to assess the geochemical processes involved, and to identify the type and genesis of water. Samples of groundwater were collected in the period 1978–89 and analysed for the basic cations; Ca^{2+}, Mg^{2+}, Na^{+} and K^{+}, and anions, SO_4, Cl^{-} and HCO_3. The salinity of the Dammam aquifer in 1978 ranged from 2,725 to 3,575 mg l^{-1}. The range of variation became 2,685–4,897 mg l^{-1} in 1983. Details of chemical analysis of water from selected wells in Al-Shagaya Well Field B (Al-Ruwaih & Shehata, 1998) are given in Table 14, Appendix III.

9.4.17 Bahrain

Prior to 1925, freshwater was derived from 15 land and 6 offshore springs. Since 1924 the flow of springs has been surveyed at intervals and found to decline in the course of time from 57×10^6 m^3 y^{-1} in 1924 to 42×10^6 m^3 y^{-1} in 1953 and further to 28×10^6 m^3 y^{-1} in 1966 and finally to 8.1×106 m^3 y^{-1} in 1979–80 (Country Report of Bahrain, 1986). The year 1925 witnessed the drilling of the first artesian well in Bahrain, and since then the dependency on wells abstraction in meeting the country's water demand has been steadily growing.

The Dammam aquifer in Bahrain provides more than 75% of the country's local consumption. It forms a small part of the eastern Arabian Aquifer that extends from the Central Saudi Arabia, where its main recharging area is located, to the Arabian Gulf. It is this large aquifer that supplies the aquifer in Bahrain by lateral flow.

The Dammam aquifer consists of three aquifer zones; A, B and C. Zone A (Ulāt) is formed of limestone with an average transmissivity of 350 $m^3\ d^{-1}$ and its quality has significantly deteriorated in the course of time. Zone B (Al-Khobar), also made up of limestone, represents the principal aquifer for supplying water due to its high transmissive and storage properties. Zone C (Umm er-Radhuma) occurs as a large lens located beneath Zone B, and contains water with poor to very poor quality. Its salinity increases with depth from 8,000 mg l^{-1} to more than 40,000 mg l^{-1}, and is used mainly for feeding desalination plants.

Development activities in Bahrain added to the rise of water consumption per capita from 100 $l\,d^{-1}$ in 1952 to 457 $l\,d^{-1}$ in 1990 and the increase in number of consumers from about 90,000 in 1941 to 520,000 in 1991 have led to a steady increase in the abstraction rate from the Dammam aquifer from 63×10^6 m^3 in 1952 to 180×10^6 m^3 in the early 1990s while the safe yield is estimated at just 100×10^6 $m^3\ y^{-1}$. The deterioration of water quality due to the over-exploitation of the groundwater resources has urged the authorities in Bahrain to turn to alternative resources. These are water desalination and treated wastewater. Nevertheless, groundwater remains the principal source of freshwater. As an example, in 1984 it supplied an annual amount of 153×10^6 m^3 of the 170×10^6 m^3, total demand on water, i.e. 90%.

The fact that the discharge from the aquifer is being in excess of the recharge has resulted in the lowering of the piezometric head by 0.5 to 4.0 m depending on the location. It also caused the increase in water salinity due to seawater intrusion and upward leakage of the brackish/saline water from Zone C to the overlying aquifers (Zubari et al, 1994). The increase in salinity level (mg l^{-1}) of water from 1965 to 1990 in the Dammam aquifer, Bahrain, can be observed from the two maps in Figure 23 (Zubari et al., 1993).

9.4.18 Qatar

The two water-bearing formations comprising the principal aquifer system in Qatar are the Rus and the Dammam. They outcrop over the entire peninsula and overlie the older rock units that contain highly saline groundwater. The areal extent of the aquifer in northern Qatar is influenced by seawater intrusion along the coastal areas, particularly along the eastern coast. The depth to water table varies from close to zero along the coast to 80 m below ground level at some locations in southern Qatar. Free water table conditions generally prevail in Qatar, except in the southwest where confining beds occur and possess artesian conditions. The deepest groundwater levels in northern Qatar are up to 40 m below ground surface and over 30 m in the southern part. The elevation of groundwater is generally high in central Qatar and decreases radially.

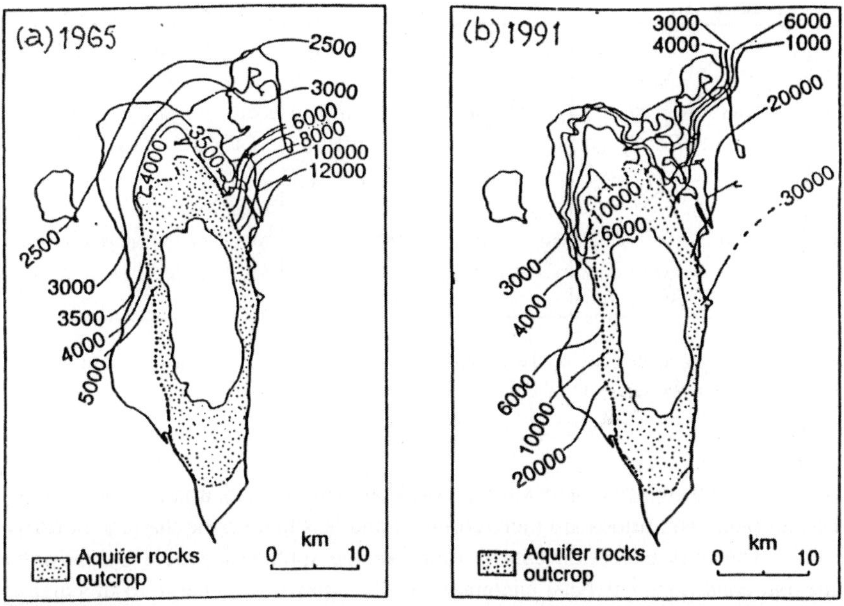

Figure 23. Salinity levels of groundwater, mg l^{-1}, in the Dammam aquifer, Bahrain, for the years 1965 and 1990 (Zubari et al., 1993)

Hydrogeological investigations of the aquifers including analysis of pumping test data from selected wells at different locations have yielded the following ranges of values (ECWA Report, 1981):

Saturated thickness = *1.7–51.1 m*
Discharge = *0.6–19.7 l* s^{-1}
Transmissivity = *2–4,500* $m^2\ d^{-1}$
Storage coefficient = *0.001–0.007*

The wide range of variation of aquifer parameters has been attributed to the varying extent of karstification of the limestone-dolomite aquifer sequence. Upon comparing the transmissivity values obtained for the Qatar aquifer system to those of the corresponding aquifer members in Saudi Arabia and Bahrain, one can observe a general decline in transmissivity from north to south along the adjacent Saudi coast.

The northern groundwater province, which comprises the northern half of Qatar, constitutes the most important source of fresh and potable groundwater in the country. Here, groundwater occurs in the form of freshwater lenses within the limestone-dolomite succession of the Dammam-Rus formation. These lenses are floating over a body of saline water contained in the older rock units as mentioned before. The estimated freshwater volume in storage was about $2.5 \times 10^9\ m^3$ and the surface area with T.D.S. from 400 to 2,000 ppm was 2, 180 km^2, both in the late 1970s. These lenses overlie the brackish and saline water of the Umm er-Radhuma formation.

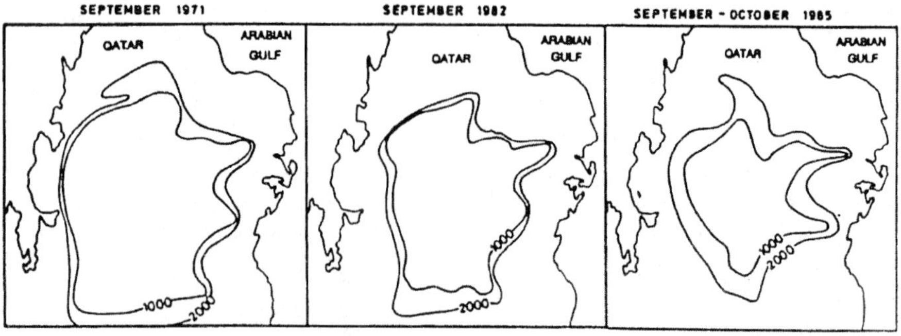

Figure 24. Reduction in the size of the main freshwater lens in Qatar with annual abstraction of groundwater; contour lines show the T.D.S. content, mg l^{-1} (Lloyd, 1992)

The southern province covers a little more than half the land area of Qatar, and its hydrogeological conditions are more complex and less favourable than the northern province. The only exception can be found in the southwest where groundwater of artesian conditions has been encountered. The source of recharge to this area is claimed to be from Saudi Arabia.

In Qatar, quantitative determinations of recharge were undertaken over a period of some years. Observations of hydrological parameters showed that 15% of annual rainfall, after elimination of insignificant rainfall from the record, could be adopted as an appropriate recharge rainfall ratio. Using this ratio, the annual recharge to the northern Qatar lens-type aquifer has been estimated as 27×10^6 m^3 (Pike, 1983).

Groundwater quality is controlled by two main factors. These are seawater intrusion along the coastal margin and the upward leakage from the underlying connate saline water body in central Qatar. The composition of the host rock affects also the groundwater quality.

Similar to what has happened in other countries in the Arab Region, the annual abstraction of fresh groundwater rose from zero in 1924 to about 20×10^6 m^3 in 1966, 63×10^6 m^3 in 1979 and further to more than 130×10^6 m^3 in 2000. Computations indicated an average rate of depletion of the aquifer storage of 20×10^6 m^3 y^{-1}. With such a rate of loss of aquifer storage, it has been estimated that the volume of water in storage will be depleted by the year 2025. The present rate of seawater intrusion is estimated as 1 km y^{-1}. Lloyd (1992) developed three maps showing the extent of the fresh groundwater lens in Qatar in the period 1971–85. These maps; 1971, 1982 and 1985, are reproduced and presented in Figure 24. By comparing the map of 1971 to that of 1985 it can be readily observed that the areal extent of the lens has been reduced by no less than 50% in 15 years. Several options have been recently considered to deal with the growing salinity problem so as to be able to fulfill the future demands on fresh water. The options include water transfer, increasing the production of desalinated water, water reuse and better demand management.

9.4.19 Saudi Arabia

Based on the main hydrogeologic and geomorphologic freatures, the country has been divided into eight units or provinces. Description of these provinces and their geographic locations are presented in subsection 7.4.2 and Figure 16, respectively. Additionally, a general lithological description of the water-bearing formations and the water-holding characteristics of the aquifers they include are given in Table 24, Appendix II. Average values of some of the hydrogeological parameters and salinity of these aquifers are as follows:

– Saq aquifer:	*Location*	*Provinces I, III and V*
	Depth of wells:	*1,000–2000 m*
	Yield:	55–100 $l\,s^{-1}$
	Transmissivity:	150–250 $m^2\,d^{-1}$
– Tabuk aquifer:	*Location:*	*Province I*
	Transmissivity:	*(lower aquifer)* 250 $m^2\,d^{-1}$
		(middle aquifer) 3 $m^2\,d^{-1}$
– Minjur aquifer:	*Location:*	*Provinces II, V*
	Depth of wells:	*1,200–1,500 m*
	Depth to water table:	*60–100 m*
	Transmissivity:	180–1,500 $m^2\,d^{-1}$
	Volume in storage:	$460 \times 10^9\,m^3$
	Salinity:	*400–5,000 ppm*
– Riyadh-Wasia-Aruma aquifer:	*Location:*	*Provinces II, III, IV, V*
	Transmissivity:	450–1,500 $m^2\,d^{-1}$
		Salinity:300–900 ppm reaching much higher values eastwards near the Kuwait border.
– Umm er-Radhuma Aquifer:	*Location:*	*Province IV*
	Thickness:	*50–300 m*
	Transmissivity	390–45,000 $m^2 d^{-1}$
	Salinity:	*1,000–6,000 ppm from west to east*
– Dammam aquifer:	*Location:*	*Province IV*
	Transmissivity:	730 $m^2\,d^{-1}$ *(Al-Khobar)*
		30–540 $m^2\,d^{-1}$ *(Al-Ulat)*
	Salinity:	*1,500–2,000 ppm*
– Quaternary aquifers	*Location:*	*Provinces II*
	Storage:	$15 \times 10^9\,m^3$
	Salinity:	*790–24,000 ppm*
	Location:	*Province III*
	Discharge:	$315 \times 10^6\,m^3\,y^{-1}$
	Salinity:	*600–1,500 ppm (rarely exceeds 6,000)*

Groundwater recharge and mining: Hydrogeological and hydrochemical investigations have proven that the groundwater resources in the Arabian Shelf are of meteoric origin resulting from direct recharge from rainfall and/or surface runoff along the major wadis (Country Report, 1986). The recharge occurs through direct infiltration into the shallow alluvial aquifers to the deeper carbonate and sandstone aquifers or directly in the areas where the aquifers outcrop. This recharge mechanism was proven by the general trend of the groundwater movement and salinity distribution. Both of them increase in northeasterly and easterly directions away from the major water divide, which coincides almost with the crest of the western Arabian

Shield. The major springs which exist in the Qatif and Has'a Oases inland, and the submarine springs along the coast of the Arabian Gulf are other indicators of the groundwater movement in the Arabian Shelf. Along the Red Sea Coast, Province VI, groundwater moves away from the Arabian Shield area into the Quaternary alluvium in the Tihama and Hijaz coastal plains.

The total runoff in Saudi Arabia has been estimated as 2.025×10^9 m^3 and the mean annual recharge as 0.94×10^9 m^3 (quoted from Al-Mohorjy, 1988). The latter figure is more conservative than the 1,145 $\times 10^9$ m^3 given by Farooq et al. (1985) for the renewable groundwater resources. The same authors estimate the non-renewable groundwater resources in as 3.45×10^9 m^3. The mining schemes from confined-aquifer areas in northwestern Saudi Arabia, have led to the development of major cones of depression. In these areas, excessive drawdowns occur because of the combined influences of small storage coefficient, small transmissivity, and large abstractions. From Figure 25 it can be seen that decline in the potentiometric surface of several meters occur over a large area in Saudi Arabia (Lloyd, 1994).

9.4.20 United Arab Emirates (UAE)

9.4.20.1 Hydrogeology

Hydrogeological studies have reached two main conclusions. The first is that groundwater is the main, and will remain, the major natural water resource in the UAE. The second conclusion is that resources of good water quality are limited.

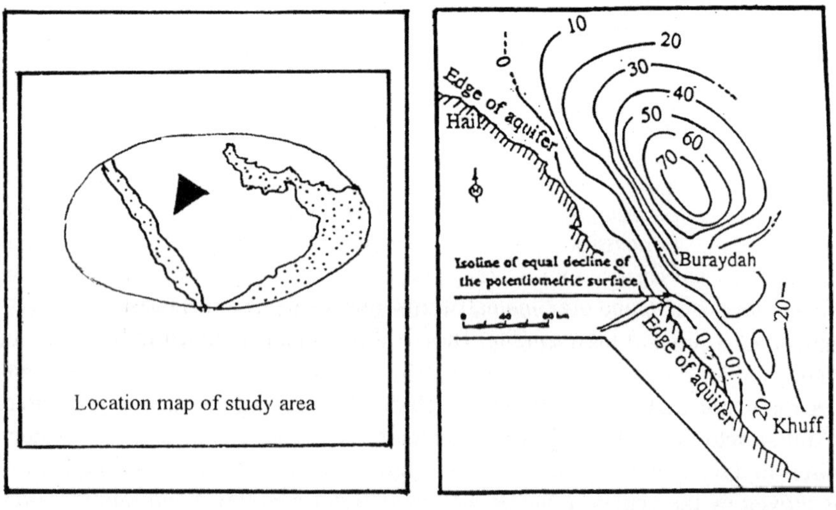

Figure 25. Decline of potentiometric surface in the Cambro-Ordovician sandstone aquifer as a result of groundwater mining, northwestern Saudi Arabia (BRGM, 1986)

While discussing basin drainage, subsection 7.5.1, the country was divided into nine water provinces; the Central Mountains, Limestone Mountains, Batinah Coastal Plain, Central Gravel Plain, Buraymi Oasis, Ras el-Khaimah-Jiri Plain, Central Desert Foreland,, Dubai-Buraymi-Abu Dhabi Plain and Southern Desert. Three main aquifer systems constitute the groundwater resources of the country. These systems are:

– *Alluvial aquifer system*: It is composed of extensive terrestrial deposits of boulders, cobbles, pebbles, and sand of the Quaternary Age. The productive part of the system is restricted to the coarse alluvium for reasons of both aquifer permeability and water quality. The saturated thickness of the productive zone exceeds 10 m in the Central Gravel Plain (Province IV). Aquifer sections increase below Province VI and northwestwards towards the coast. Along the coast, the saturated zone penetrates upwards in the overlying sand dune sediments.

The coarse alluvium is hydraulically connected with the underlying mudstone evaporitic sediments which contain groundwater of poor quality. The areal distribution of water quality within the aquifer system is a function of recharge mechanism and the transmissivity. As such, water of good quality can be found in the upper reaches of the major wadis. The quality deteriorates westwards towards the coast. The storage capacity of the aquifer system is claimed to be large. Surface runoff from the western slopes of the mountains is the main contributor to the aquifer recharge. The water budget of the stretch between Al-Ain and Ras Al-Khaima (170 km long) indicated to overdraft conditions. This has been confirmed by the steady decline of potentiometric heads, overdraft conditions in Al-Ain Oasis as well as overall water quality deterioration. This situation calls for the undertaking of a more careful abstraction. The ECWA Report (1981), referring to Halcrow, 1969, gives the total amount of groundwater in storage in these alluvial gravels capacity as $5{,}280 \times 10^9$ m^3. The same sources, referring to Carr & Barber, 1976, gives the recharge to the aquifer system through a 170 km recharge front extending from Al-Ain Oasis to Ras el-Khaimah at 124×10^6 m^3 y^{-1}.

– *The Batinah Coastal Plain aquifer*: This aquifer comprises the alluvial deposits, which form an almost littoral strip, extending from the Musandam Peninsula in the far north to the frontier between Oman and UAE along the Batinah. The aquifer is composed of alluvial fan material consisting of coarse sand, gravels and boulders, ending up with fine-grained sebkha deposits in some localities along the coast of the Gulf of Oman. The surface runoff from the wadis adjoining the western mountains and draining eastwards towards the Batinha coast is the major source of recharge to the aquifer. The yield of wells is claimed to be high. The depth to water table is usually shallow and a few meters above the sea level. The total renewable resources are reported to be adequate for development. Water quality is generally fair but deteriorates eastwards.

– *The Deep Carbonate aquifer system*: This aquifer system is composed of thick carbonate rock sequences underlying the region to the west of Oman Mountains and extending to the area of Al-Dhafrah in southern Abu-Dhabi. The aquifer is under artesian conditions with a hydraulic head at or near the ground surface in

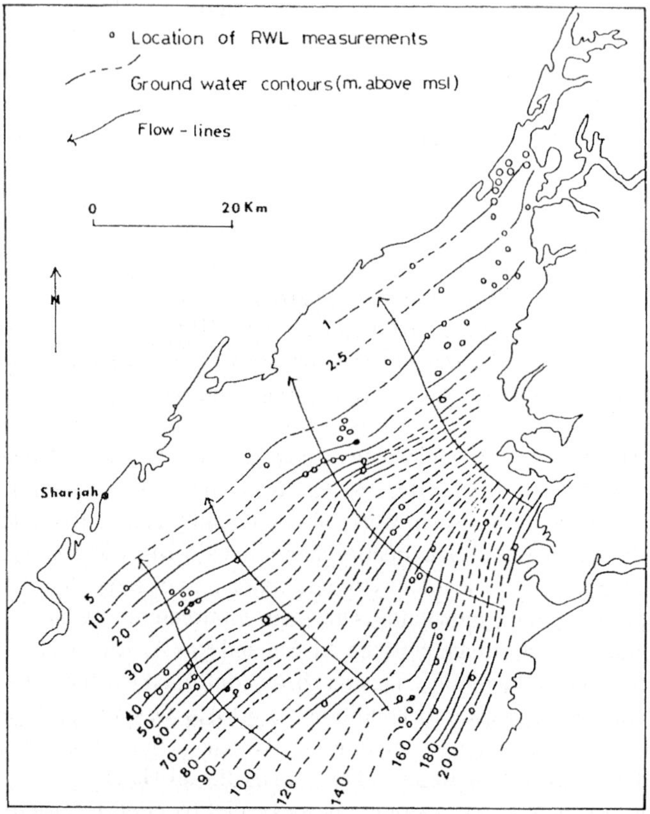

Figure 26. Contour map of groundwater potentials in the United Arab Emirates (Naqash & Marrei, 1983)

Al-Ain or desert foreland areas. The condition changes to unconfined in the area along the Omani frontier south of Al-Ain. Groundwater flows from the recharge zone over the Oman Mountains westwards, and the aquifer drains naturally in the sabkhas region where the Liwa Oasis is located. It is believed that the aquifer system offers a major water resource in the area between the Liwa Oasis and the coast.

The quality of groundwater encountered in this aquifer is generally poor (about 10,000 ppm) and decreases with depth to reach 4,000 ppm. Better quality of water (1,000–1,500 ppm) can be encountered in shallow wells tapping the aquifer in the Liwa Oasis, Al-Asab and Habshan areas.

The depth to water table was measured in numerous wells and was found to range from less than 2 m at Ras Al-Khaimah to a maximum of 55 m in the wadi gravels at Burairat Mountain front. The aquifer system is generally unconfined, having a free water surface. Figure 26 is a map showing the average steady state

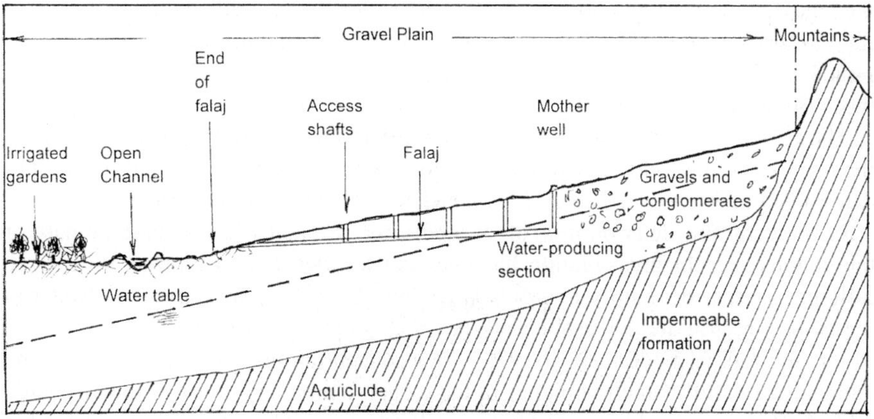

Figure 27. Typical section of one of the falages used for collecting groundwater from the gravel plains in the Gulf countries

configuration of this water table (from Naqash & Marrei, 1983). The groundwater contour lines in the southern part of the area, as can be seen from the map, run roughly parallel to the mountain front and bend to northeast-southwest alignment as one moves towards the coast. From the flow net shown on the map, the hydraulic gradient is found to vary from 1 in 175 to 1 in 1,200.

It should be noted that the exploitation of groundwater aquifers in some parts of the UAE is beyond the safe yield as dropping of water levels in the western part of the country is quite noticeable. As an example to mention, the water level in the main well fields in the Sharja Emirate dropped in the period 1975–80 at a rate of 1.5 m, and in the early 1980s the annual rate increased to more than 2.0 m. This was caused by increasing the abstraction from Bedai well field from about 140 m^3 h^{-1} from 3 bore holes in 1967 to 1,380 m^3 h^{-1} from 25 boreholes in 1978. In 1980, the extraction of 1,800 m^3 h^{-1} was reached from 35 boreholes in the same well field. As a consequence, some of wells became dry (Naqash, 1983). In Dubai, the total production of well fields in 1980 was close to 6,450 m^3 h^{-1}. These figures show that UAE has since long reached and exceeded its safe yield. New methods and techniques for control and management of water resources should be developed.

The UAE has a long-standing history in using falages (Figure 27) as a means of conveying water from one location to another. Flank spates are the major contributor to to the yield in the main falages. The numerous wadi bed seepages are also included which are generally subterranean flows forced to the surface of a wadi bed. In some literature, falages are associated with surface water resources. In others, and here too they are associated with groundwater resources.

Falages in UAE vary in length from 1 to 6 km, depending on the ground slope and the depth to water table. The deepest and longest falages can be found in the Buraimi area. Falages belong to two major types: Dawoodi, which are high potential falages emerging from groundwater and are continuous sources of water supply, and the Ghayl, which are low potential flages dependent on rainfall, and their flows last for shorter durations. Table 25, Appendix II, gives the annual discharges over the period 1977–82 for 46 major falages in different drainage areas. The total including those that dried off is 26×10^6 m^3 y^{-1}.

The above-mentioned figure drops down to about 21×10^6 m^3 upon excluding the falajes that dried off. The quality of water sampled from 40 selected falajes was determined and listed in Table 15, Appendix III (UAE Country Report, 1986).

In view of the deterioration of groundwater resources in the eastern part of the country, recent investigation has been carried out using Chlorine-36 to trace different sources of salinity in groundwater systems. ^{36}Cl data from groundwater in the eastern region are consistent with previous studies indicating that the recharge to the aquifers is modern (post-1950). ^{36}Cl concentrations could be related to the amount of rainfall, which is higher in the recharge area and lower in the discharge area. Eastern Gravel Plain aquifer has chloride concentrations in excess of what could be ascribed to evaporation or seawater mixing. Instead, it can be attributed to sources such as fertilizers and pesticides in this heavily agricultural area. The Chlorine-Barium data have shown that animal waste appears to be one of the sources. In addition, the chemical imbalance between chloride and other major cations as well as higher silica concentrations indicate the release of ions such as sodium and calcium from the host rocks in several groundwater systems (Murad & Krishnamurthy, 2004).

9.4.21 Oman

Oman, like many countries in the Arabian Peninsula, depends heavily on groundwater resources in meeting its water demands. Most of the groundwater used in Oman is withdrawn from the sedimentary deposits, old and recent, which exist in wadi basins.

The thickness of the fresh water bearing formations increases in the coastal areas, and reduces gradually as it moves towards the upper reaches of the wadis far from the sea. The thickness can be as much as 100 m at a distance of 5 km from the sea coast and reduces to almost zero at a distance of 12 km from it. Groundwater can be found at a depth of 1 m below ground surface in the manually operated wells close to the sea. The depth to groundwater increases to say 30 m in land.

The quality of groundwater in the aquifers underlying the flowing wadis is superior to the water that is available in the consolidate sediments along the wadi borders. The latter formations receive less recharge compared to the formations underlying the wadi channels. In general, the quality of groundwater improves with distance from the sea. This has to be expected in view of the possible seawater intrusion.

The yield of any water-bearing formation depends, among others, on the geology and physical setting of the formation as well as the recharge supplied directly through precipitation and indirectly on the seepage from the neighbouring formations. The overall annual recharge to the water-bearing formations in Oman has been estimated as 564×10^6 m^3 y^{-1}, corresponding to 61.4% and 38.1% of the surface runoff and the total runoff (surface and subsurface) respectively. Accordingly, it seems reasonable to present the groundwater situation in Oman region wise.

– *Musandam Peninsula:* The total recharge to the wadi basins of the Musandam Peninsula was estimated as 18.1×10^6 m^3 y^{-1} (Oman Country report, 1986). This amount represents some 75% of the annual surface water resources. Accordingly the yield of the water-bearing formations in this Peninsula is quite limited. The quality of water too is inferior.

– *Sahel (coastal strip) Al-Batinah:* This strip of land can be subdivided into two belts parallel to the sea. The first belt extends from the mountain bases downwards in the direction to the sea. It comprises highlands and hills, and is characterisd by the existence of an extensive network of falages mainly used for irrigation while dug wells are of secondary importance. The belt functions as a collector of seasonal rainfall in the defined, deep wadis. Some of the collected water penetrates the ground surface and joins the underlying body of water. As a matter of fact, the carbonate rocks outcropping in the middle of the belt helps to influence the movement of groundwater in it. These outcroppings force the groundwater in the underlying deposits, which form the terraces and sand dunes towards the low-lying areas between the hills and highlands. The Omanis took advantage of this situation, i.e. availability of abundant, yet shallow groundwater and built a chain of falages to intercept the groundwater and use it in land irrigation. The second belt is a coastal strip of land composed of wadi deposits, coastal sand dunes and sebkhas. This belt is characterised by depending on the system on dug wells. The thickness of the water-bearing formations in this belt is greater than the first belt and the quality of its water is superior.

It has become apparent from the available results of preliminary investigations that between 15 and 40% of the runoff produced by rainfall in the mountainous area is lost by flowing to the sea. The remaining amount infiltrates through the soil surface and joins the existing groundwater body. Recharge of groundwater in the second belt occurs at a relatively large scale shortly after the occurrence of surface runoff in wadi channels. The highest elevations of groundwater occur close to the wadi channels and decline far from them. The response of groundwater in storage to flood events almost diminishes in the middle of Sahl al-Bathinah where the depth to groundwater is quite large. It is worth mentioning that a short period of few years with reduced rainfall has a slight effect on the flow of groundwater. If such a drought period lasts longer and the abstraction of groundwater is not cautiously planned and managed, the drought effect is likely to become rapid and pronounced, especially in the coastal belt (Figure 28). A detailed description of the groundwater recharge processes in Eastern Batinah is presented in a later section.

Examination of groundwater samples using isotope tracers has shown that the abstracted is rejuvenated, young water dating back to 25 years only. The annual recharge of the aquifers in the eastern and central parts of Al-Batinah was estimates as 105.5×10^6 m^3 and in the northern part as 119×10^6 m^3.

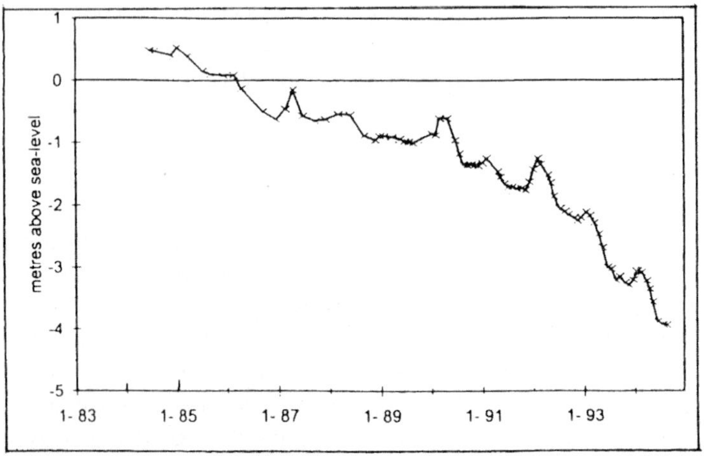

Figure 28. Hydrograph of high storage aquifer with over abstraction in the Batinah coastal belt (from Poulter, 1995)

i- *Ash-Sharquiya Province*: This region as has been described earlier, is bounded on the north by the Western Hajer range of the north Oman Mountains, from the west the international boundary with the United Arab Emirates and Saudi Arabia, from south the Central Plateau and from the east the Gulf of Oman and the Arabian Sea.

i- The eastern (Sharquiya) subregion is mainly (over 70%) covered by mountains and deserts. It also comprises a number of the major wadi basins. Coastal land strips cover the remaining part of the subregion. Groundwater is basically abstracted using the falages' system, which produces more than 3 $m^3\ s^{-1}$ enough to irrigate over 2,000 ha.

The falages' system can be classified into two classes; one consisting of an upper and lower falages and the second comprising upper, intermediate and lower falages. Most of the falages in Wadi Batha Basin are included in the first class. This is because the depth to groundwater in the distance between the upper and lower falages is relatively large, not less than 50 m. Figure 29 is a typical section showing a three-falages' system in the sedimentary deposits of the Al-Batha Basin and the Mudhaibi area. From the available information one can classify the groundwater reservoirs in the eastern (Sharquiya Province) as follows:

Al-Batha Plain: The groundwater reservoir covers a total area of 1, 300 km^2 with depth to groundwater ranging between 5 and 50 m. The quality of water at the outlet of wadis is good. The electrical conductivity of water ranges between 500 and 1, 000 mmhos cm^{-1}. These small figures are due to the large supply of freshwater flowing down the mountains. The high salinity figures (electrical conductivity at or larger than 3, 000 mmhos cm^{-1}) in some areas are caused by increased evaporation from the groundwater table near the ground surface, poor drainage condition of agricultural land, reuse of water after having it used for irrigating the upper reaches of wadis and seepage of saline water from neighbouring areas.

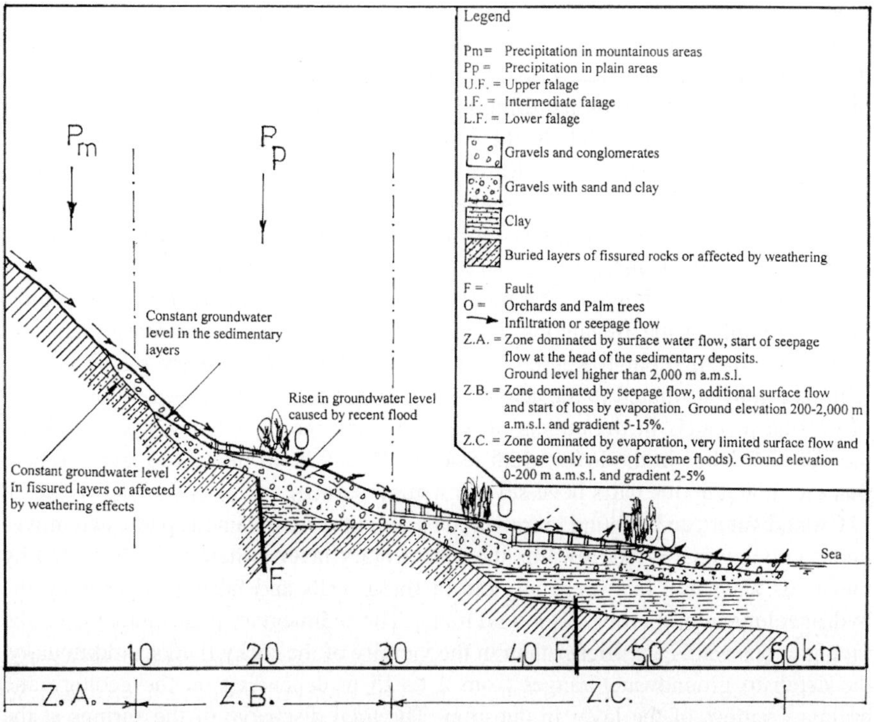

Figure 29. Cross section of a three-falages' system running in the sedimentary deposits of the plain of wadi Batha and al-Mudhaibi area (Oman Country Report, 1986)

Sahel Quryat: The reservoir area does not exceed 40 km^2 with an average thickness from 2 to 10 m and average depth to the groundwater table of 5 m. The quality of water in those areas far away from the seacoast, at the outlet of major wadis upon entering the plains, is good with electrical conductivity of about 1,000 mmhos cm^{-1}. Oppositely, the quality of water in those areas close to the seacoast is less due to seawater intrusion into the groundwater reservoir where abstraction exceeds the safe limit.

As-Sur Area: The groundwater reservoir is extremely small, 10 km^2 in surface area with an average thickness of 5 m and depth to the groundwater table ranging from 2 to 10 m. The problem of seawater intrusion in the groundwater body is acute as a result of the decline of groundwater level caused by excessive abstraction.

Andām/Saad Wadi Basin: There is hardly any information about this area. However, the depth to groundwater table is believed to be less than 10 m. The land strips parallel to wadi channels are thought to have groundwater of suitable quality (electrical conductivity less than 1,500 mmhos cm^{-1}). In areas located far away from the wadi channels and in those areas receiving limited quantities of surface flow and where water reuse is applied, the salinity increases due to evaporation.

Wadi Addey and wadi Mieh: The quality of groundwater in the hattat Plain is generally ranging between 1,500 and 2, 000 mmhos cm^{-1}. In some locations it deteriorated and reached 4, 000 mmhos cm^{-1}. As this wadi supplies a large proportion of the demands for drinking water of the capital of Oman it has been putunder control since 1984. In the Mieh Plain where areas close to the sea are used mainly for agriculture, the electric conductivity of groundwater is in the range 3, 000–8, 000 mmhos cm^{-1}. The annual recharge to groundwater in the Sharquiya Province and the coastal strips is estimated as 70.8×106 m^3 and to Al-Sur and Quryat as 36.8×10^6 m^3.

ii- *Oman Interior Region*: The highly elevated parts of this region receive most of the rainfall on the mountains. It helps to recharge the sedimentary plains, as is the case in Al-Batinah Region. The fertility of the sedimentary plains in both regions can be attributed to the active weathering factors and subsequent deposition of erosion products in the outlets of wadis leaving the mountainous parts to form the fan-like deltas in the lower wadi reaches. It may be of interest to mention here that the mountainous parts have several natural springs.

Groundwater can be found in fractured rocks in the mountainous parts, and moves under gravitational effect towards the sedimentary layers where it is abstracted by means of wells and falages. The yield of these wells and falages depends on the hydrogeological properties of the said rocks. The sedimentary plains underneath the mountainous wadis are thicker than in the vicinity of the rocky flanks. Additionally, the depth to groundwater ranges from 2 to 25 m depending on the geology and drainage pattern of the layer in question. The total discharge of the springs at the foothills, springing from limestone aquifers, is estimated as 0.1 m^3 s^{-1}.

From the water balance of the southern basins it appears that the annual recharge is about 75×10^6 m^3. The storage reservoir constructed on the Wadi Nezwa for water management prevents the surface water from percolating deeply in the gravel and conglomerate layers instead it raises the water level in it. Besides, the deposits that accumulated up to the dam level as a result of successive floods have added to the storage capacity of the conglomerate-gravel layer. Groundwater is abstracted at a slow rate either through the falages or the baseflow of Wadi Nezwa. The quality of groundwater in the sedimentary layers is generally good. The quality is likely to change adversely upon its movement underneath the heavily populated wadi basins where water is reused for several purposes a number of times. The electric conductivity of such water can easily exceed 2, 000 mmhos cm^{-1}.

iii- Ad-Dhahira Region: This region comprises the western foothills extending from the north extremity of Oman Mountains to the western international frontier between Oman and the United Arab Emirates near Al-Buraimi Oasis. It also comprises the southwestern foothills of Al-Asswad and Al-Abyadh Mountains and the extension of the flat plains in the southwestern direction up to the frontiers. The major basins in this region are Ibri and Al-Buraimi.

In the Ibri Basin there are two falages springing from the gravelly layer, with a combined discharge of 0.05 m^3 s^{-1}. The remaining demand on water is supplied by

40 dug wells abstracting water from the conglomerate-gravel layer. The depth to groundwater varies from 2 to 8 m. In addition to the falages and wells there are some springs with a total flow of 7.5 l s^{-1}, which is too little indeed. Studies have shown that the groundwater that can be abstracted from the wadi basins in Al-Buraimi Oasis, are basically available in the sedimentary deposits. These deposits are fed from the runoff generated by the rainfall on the rocky upper reach of each wadi and recharges the underlying water-bearing layer. Approximate estimates of the water balance of Al-Dhahira Region show that the annual recharge to groundwater is 58.8×10^6 m^3.

The groundwater quality in the areas crowded with wadis and sedimentary layers along the southern part of the region can be described as good and suitable for irrigation. When the flow of groundwater in the sedimentary layers is obstructed by impermeable formations or rocks buried under the surface, the quality of groundwater deteriorates to the extent that its use for agriculture is no longer acceptable.

iv- The Southern Region: This region is bounded on the east and south by the Arabian Sea, on the west by the frontier with the Republic of Yemen, on the north by the Central Plateau of Oman, and on the west by Ar-Rub el-Khali Desert. The region comprises a total of 51 wadis, 18 wadis in the northern part and 33 wadis in the southern part of the region. The northern wadis originate from Dhofar Mountains and discharge their water in Najd while as the southern wadis can be found in Salalah and surroundings and discharge their water into the Arabian Sea.

From the water balances of these two subregions it appears that the total recharge to groundwater bearing formations is 27×10^6 m^3 y^{-1} and 52.8×10^6 m^3 y^{-1} for the northern and southern basins respectively. In view of the presence of intensive green canopy, it has been found that the annual rain depth should exceed 200 mm before any recharge goes to the water-bearing formations in the region. As a result of the fast-moving surface runoff generated by heavy torrential rains, most of the water carried by the wadis is discharged into the sea. During this process certain quantities of the flow find their way to the coastal plains and thereupon recharge the underlying sedimentary layers. For example 60% of the surface runoff in the Salala Plain goes to the sea and the remaining 40% to the groundwater body. These percentages correspond to 10×10^6 m^3 y^{-1} and 7×10^6 m^3 y^{-1} respectively. It has been claimed that 50% of the runoff discharged into the sea can be saved through groundwater recharging reservoirs.

The discharge of four major springs was observed in the period 1977–83 and their combined average discharge was found to be 9.1×10^6 m^3 y^{-1}. The water balance of the Salala Plain shows that 10% of the average precipitation, 282×10^6 m^3 y^{-1}, moves annually towards the groundwater body in the coastal plain. This amount, i.e. 28×10^6 m^3 y^{-1}, is divided between the discharge of springs, 9×10^6 m^3 y^{-1} and the remaining 19×10^6 m^3 y^{-1} flows under the surface. When this figure is added to the 7×10^6 m^3 y^{-1} already mentioned earlier, the total annual flow under the ground surface becomes 26×10^6 m^3.

The groundwater body in the Carbonate rocks and the silt-pebbly formations in the Salala Plain has an average thickness of 50 m. Excessive amounts of groundwater are abstracted from this water-bearing formation to supply the demands for agricultural and domestic purposes. The quality of water in the sedimentary plains is at its best in the middle and deteriorates as it approaches the seacoast. The electric conductivity of water in the middle of the plain ranges between 800 and 1,200 mmhos cm^{-1}, increasing to 2,000 mmhos cm^{-1} in the vicinity of the sea. Recently, evidences have accumulated to show that seawater is intruding in the groundwater bearing formation. The length of the intrusion wedge has already exceeded 3.0 km. The best quality of water can be found in the recharge areas of the mountain range. It gets worse as water moves underground to Najd area in the north and Salala Plain in the south. The water gains more mineral salts upon its movement.

From the available hydrogeological information the availability of old groundwater is not ascertained except in the Dammam, Rus and Lower and Upper Umm er-Radhoma formations in the southern sub-region. Even in these formations within the territory covered by the Sultanate of Oman no more information of substance is yet available.

Brief review of two cases studies of groundwater recharge in Oman

i- Groundwater recharge Processes in Eastern Batinah: The Eastern Batinah (Figure 30) is the most populated area in Oman outside the Muscat area. It includes major wadis like W.Bani Ghafir, W. Al-Fara, W. el Taww, etc. The groundwater levels along much of the coastline are currently below mean sea level and

Figure 30. Location map of two study areas of groundwater recharge

saline water intrusion is emerging as a serious management issue. Quantification of recharge is essential in order to develop a realistic numerical model of the performance of the aquifer system.

Whenever rainfall records over a sufficiently long span of time are available, hydrograph analysis is considered to provide a reliable method for determining the spatial and temporal distribution of recharge (Lakey et al., 1995). The available rainfall data covering the period from October 1983 up to September 1992 were regarded adequate for the purpose of the study. It was observed that the dominating characteristic of the high altitude gauges is a significantly higher frequency of small events rather than a greater number of larger events. Besides, long-term rainfall records reveal a pattern of wet and dry periods, with a periodicity of 5 to 6 years. The pattern of wadi flow predictability resembles the rainfall pattern, with discrete events often lasting only a matter of hours. There is a rainfall threshold of about 10 mm below which runoff does not generally occur. As such, rain events of 15 mm or less, which comprise a substantial part of the total rainfall, may produce significant groundwater recharge both on the coastal plain and within the highlands.

The aquifer underlying the coastal zone is unconfined, with a relatively shallow water table within 20 m below the surface and water level fluctuations are dominated by the pattern of groundwater pumping. Coastal wells display relatively small amplitude fluctuations of a meter or less. As the response is relatively consistent throughout the zone, it is fair to conclude that the hydrogeological conditions near the coast are relatively uniform. Another important feature of the water level hydrographs is the short-term and localized nature of the response to wadi flow.

The total rainfall was determined for each event or group of events and the corresponding water level rise was determined from the well hydrograph. A linear regression of the data yields a gradient of 3.7, i.e. a rainfall of 10 mm will produce a water table rise of 37 mm. The lower boundary of the data scatter has a gradient of 1.5, which implies that the minimum hydrograph response to a rainfall of 10 mm is a water table rise of 15 mm.

Beneath the central and southern plain, the groundwater body occurs at a greater depth and the hydrographs are different from those in the coastal zone. The hydrographs in this zone show a long wavelength cyclic response. The rapid response to rainfall events observed in the coastal zone is generally absent in the plain zone. The amplitude of water level changes (up to 10 m) in the plain zone is small compared to aquifer thickness (200–300 m), but quite large compared to fluctuations in the coastal zone.

ii- Simulation model with application to Wadi Ghulaji, Ash Sharquiya Region: The modeling approach described briefly here is one part of detailed hydrotechnical recharge studies in the Wadi Ghulaji (Wheater et al., 1995). The wadi basin (Figure 30) has already been described in Chapter 7. A marked change in wadi slope occurs from steep, bare rock jebel to alluvial plain, which is characteristic of most arid areas.

The data employed in this study were daily rainfall within the catchment of Wadi Ghulaji, the stage hydrographs and stage-discharge rating curves. The Integrated water resources model consisted of three sub-models. The first is the rainfall sub-model whereby the probability of occurrence of rain on a given day is conditioned on rainfall occurrence on the previous day. The second component or sub-model is the water balance model. It comprises rainfall, runoff, evaporation losses and groundwater recharge. The curve number, CN, method was used to establish the rainfall-runoff relationship. The value of CN for the steep jebel plane ranged from 90 to 95 and the alluvial plain from 77 to 85. Additionally, the parameter α, which defines the transmission loss in the system of wadi channels, was assumed to vary from 0.01 to 0.10 depending on the type of territory. These values were found to produce runoff flows very close to the observed ones. The groundwater recharge simulated by the water balance model was used as input to third sub-model, the groundwater model.

The aquifer was conceptualized as a system of three interconnected layers of uncemented gravels, partly cemented gravels and cemented fissured gravel/ bedrock. The water levels and flow were modeled using the MODFLOW software package. Representative values of permeability and transmissivity values of each cell in the model were estimated from the analysis of pumping test data for six wells in the catchment.

Application of the above-described sub-models to the case of Wadi Ghulaji considering three different scenarios; wet (87.7 mm rainfall), average (87.0 mm) and dry (52.6 mm), resulted in recharges equal to 14.6, 12.9 and 10.5% of the corresponding rainfall depths.

9.4.22 Yemen

Based on the results accumulated from a number of hydrogeologic investigations, the principal aquifers in northern Yemen (formerly Arab Republic of Yemen, YAR) are: Tihama Plain alluvial aquifer, Kholan series, Tawilah and Mejd-Zer Series, Yemen volcanics and Quaternary alluvium and Quaternary volcanics. A summary of the lithology and hydrogeology of these aquifers is given in Table 26, Appendix II. A brief description of the mentioned aquifer systems is as follows:

– *Tihama Plain alluvial aquifer*: The aquifer formed by the alluvial deposits can be regarded as one single aquifer 500 km long and 10–15 m thick. The aquifer is composed mainly of cemented to uncemented boulders, pebbles, gravels and sands. It is expected that the aquifer is underlain below 200 m depth by a formation of Tertiary or pre-Tertiary sedimentary volcanic rocks. The average regional transmissivity of this aquifer was estimated as between 1,700 and 2,500 $m^2\ d^{-1}$ and the storativity between 0.01 and 0.02. Additionally, the estimated groundwater reserves are $1.9 \times 10^6\ m^3\ y^{-1}\ km^{-1}$ width of aquifer.

In the Tihama Plain Water Resources Study in 1985, 425 vertical electrical soundings were carried out throughout the plain to determine the geometrical properties of the aquifer system. The model used in that study identified six different layers, each having its own characteristics. The Tihama 'effective' aquifer is found to consist of two lithologically distinct units of Quaternary age (layers IV and V) underlain by two older formations saturated with saline groundwater (layers II and III) and bedrock (layer I), which forms the impermeable base of the aquifer system in the eastern part of the plain. In the western part, the bedrock is downthrown to great depth by several northsouth to northwest-southeast step faults (Smith & Al-Mooji, 1987). Layer V contains coarse wadi sediments containing freshwater, and layer IV is of slightly finer composition containing less fresh water.

The water quality is fair to good. However, it deteriorates in the lateral direction as one goes away from the mountain front, and with depth in the vertical direction.

Good quality water is generally associated with wadi flows and flood irrigation of areas within the plain. Large-scale withdrawal of water from the Tihama aquifer within the zone 20–40 km from the seacoast might result in serious quality deterioration by seawater encroachment.

- *Kholan Series aquifer*: It is composed of a series of cross-bedded sandstones with several thin layers of conglomerate. The aquifer thickness ranges from 100 to more than 500 m. Despite neither the aquifer parameters nor the quality of water are known for certain, it is believed to be an important aquifer in the interior areas of the country with high potential.
- *Tawilah and Mejd-Zir aquifer*: It is composed of coarse-grained sandstones with layers of conglomerate bringing the total aquifer thickness to about 400 m. The aquifer potential is good, particularly the lowest 300 m of its thickness, and is characterised by a well-developed fracture system, which results in good permeability.

The aquifer areal extent coincides with the outcrops of the Tawilah formation in the north and northeast, on the east bounded by the escarpment facing the Eastern Desert and on the west by the Red Sea Escarpment. The vertical boundary of the aquifer is marked by the Cretaceous-Jurassic shales at the bottom while the upper limit is marked by the overlying Tertiary basalts or the Quaternary basalts and/or sedimentaries.

The depth to water table below surface generally varies between 25 and 100 m. There are hardly any data known about the aquifer parameters. Several low yielding springs of the water table type are fed by the aquifer. Larger springs are known to be fed by this aquifer along the Red Sea Rift escarpment.

- *Yemen Volcanics aquifers*: A large part of the northern Yemen is covered by an extensive, thick layer of volcanic rock as can be observed from Figure 31 (UN Report Series on Water No.9, 1982). These volcanic rocks are predominantly lavas with alternating agglomerates. The series is intruded by basaltic rocks in the form of dykes and sills. The maximum thickness is 1,200 or more. The hydrogeology of the aquifer is complex and highly variable. The water-bearing characteristics are governed by the extent of fracturing of the rock. It expected that the values of the aquifer parameters are generally high (Country report of YAR, 1986).
- *Quaternary Volcanics aquifer*: Its outcrops are mostly basaltic flows in the extreme northeast corner of the country. As such, its areal extent is limited to a small area of about $1,000\ km^2$. The aquifer is composed predominantly of lava

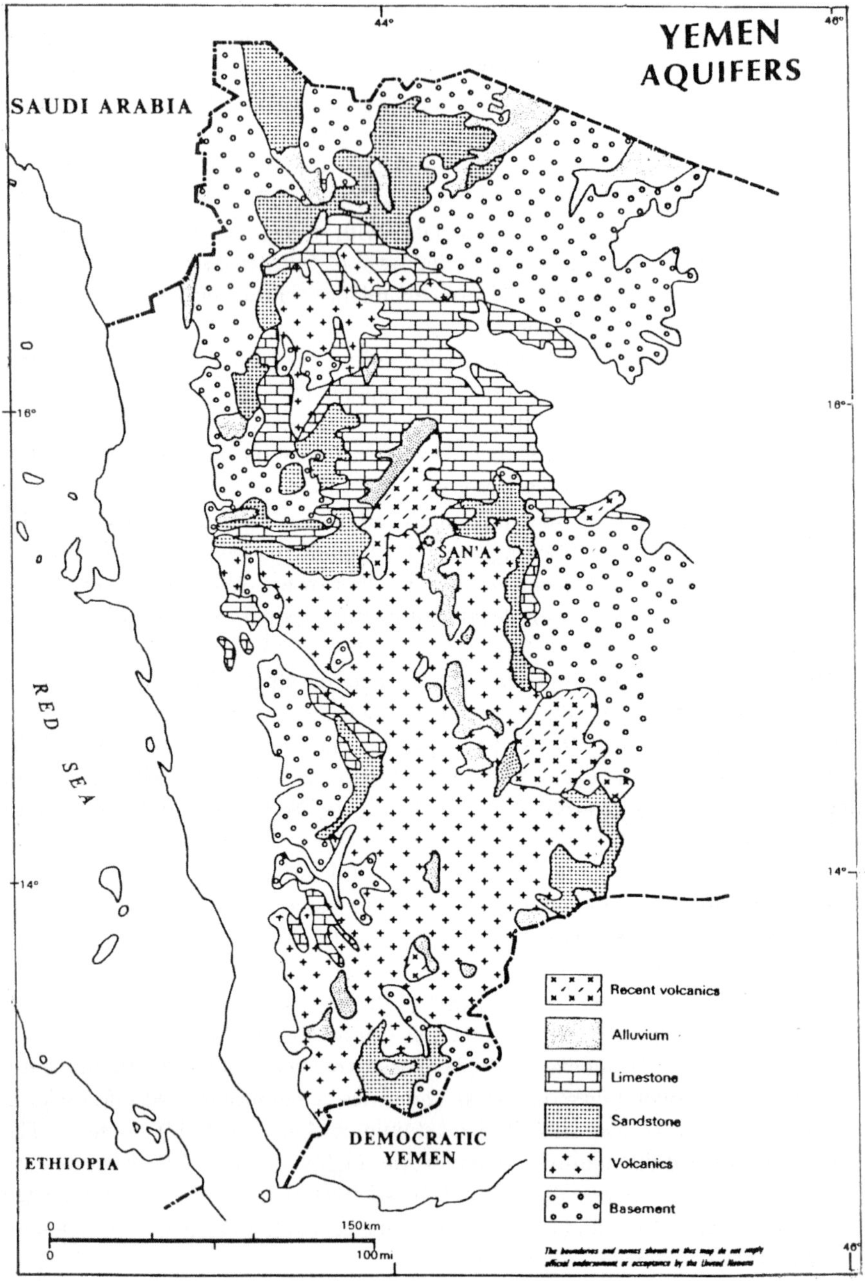

Figure 31. Groundwater aquifers in North Yemen (UNDP Report Series on Water No. 9, 1982)

flows reaching several hundred meters in thickness. The groundwater potential of this aquifer is still unknown.

- *Quaternary Valley Fill*: The aquifer is composed of alluvium varying from coarse boulder conglomerates through gravel, sand, silt and clay. This material is mainly deposited in the floors of most of the upper catchment wadis in the mountain regions. The aquifer thickness is variable and ranges from a few meters to over 300 m. Water table conditions predominate and semi-artesian conditions may occasionally occur. Despite the fact that the aquifer parameters are not well defined, the aquifer potential and quality of its water are likely to be good. This explains the high degree of aquifer exploitation through tapping it by a large number of wells scattered allover the Yemen Plateau. The permeability is a function of the percentage of fine material within the alluvial sediment.

Groundwater over-exploitation in North Yemen has been experienced during the last three decades, particularly in the urban areas and where intensive irrigation is being practiced. Recent investigations indicated a general water level decline in the Sana'a Basin. The annual decline ranged from 0.52 m to 3.75 m with an average rate of 1.85 m. Other areas like Taiz, Mabar, Dhamar and others are currently suffering from over-draft conditions. As a consequence, the quality of water is showing strong signs of deterioration, at least in certain locations.

South Yemen (formerly People's Democratic Republic of Yemen, PDRY) can be divided into four groundwater provinces; Wadi Hadramout, Wadi Tuban Watershed and Wadi Meafa'h Province. Each of these provinces has its aquifers which can be briefly described as follows:

– Wadi Hadramout Province:

i- Valley fill deposits: These deposits are almost the same as described in the case of northern Yemen. The thickness of the aquifer is widely variable and the transmissivity ranges from 140 to 1, 300 $m^2\ d^{-1}$ depending on the percentage of the fine material fraction. The thickness of the deposits in the northern areas of the country is very thin. The depth to water table ranges from 10 to 20 m, with a hydraulic gradient of about 1.1 m km^{-1}. Water wells tapping this aquifer have low to moderate yielding capacity of less than 10 $m^3\ h^{-1}$. One of the factors contributing to this situation is the poor recharge to several parts of the aquifer, especially in the northern areas.

ii- Rus-Jeza deposits (Eocene): These deposits are composed of limestone, chalk, marl and gypsum intercalations with dolomitic and silicified limestone. With the exception of the areas north of Hadramout Basin, the deposits are impervious and contain no water. The aquifer in the northern areas is composed of 2–15 m thickness of semi-impervious shales, marls and clays sandwiched between beds of limestone-dolomite deposits at fairly regular intervals.

Exploratory drilling in these deposits have indicated that the depth to the deeper regional groundwater level ranges from 60 to 120 m depending on the ground level. Besides, the water yielding capacity of wells in the northern areas was reported to fall in the range of 0.25 to 0.5 l s^{-1}. In general, the zone below the regional deeper groundwater level offers better prospects of striking groundwater than the top patchy zone. The expected direct groundwater recharge in the deeper zone is claimed to be very poor and the well yielding capacity from 0.38 to 0.76 l s^{-1} (Country Report of the PDRY, 1986).

iii- Umm er-Radhuma deposits (Paleocene): These deposits are composed of massive beds of limestone and marly limestone, occasionally fractured and solution-channeled. The part of this formation occurring within the Hadramout Province has proven to be of inadequate for economic extraction of groundwater through wells. In the areas located north of the Hadramout Province the hydrogeologic properties of the Umm er-Radhuma deposits are not well known. Exploratory drilling in the formation pointed out to the existence of a meager quantity of groundwater. Recharge by leakage from the overlying Rus-Jeza deposits is hardly possible due to the existence of shale beds in between.

iv- Mukalla Sandstone (Cretaceous): The aquifer is made up of a top layer of fine-grained uniform sandstone of moderate permeability, with thick beds of shales, siltstone and occasionally marls and clays occurring within these deposits. The lower layer of this aquifer has a low permeability.

Artesian conditions prevail particularly in the areas to the north of Hadramout Basin where depth to water table ranges from 50 to 100 m and is generally deeper in the northern areas. Wells tapping groundwater from Mukalla sandstones are generally deep (200–300 m) and their yielding capacity ranges from 30 to 60 m^3 h^{-1}. Groundwater salinity is in the order of 440–1,000 ppm of total dissolved solids (T.D.S.).

A preliminary estimate of the recharge to Hadramout aquifer was as 190×10^6 m^3 y^{-1}. One quarter of the rainfall (387×10^6 m^3) was assumed to go as a direct recharge, 75×10^6 m^3 and 20×10^6 m^3 as surface runoff from the southern plateau and the northern plateau respectively, 1.3×10^6 m^3 as surface inflow from tributaries and 1.0×10^6 m^3 as subsurface inflow from the Sabatayan district. About 4×10^6 m^3 were assumed to represent the outflow to Wadi Masila, Masila district and downward leakage to Mukalla aquifer.

The quality of aquifer water can be subdivided into three classes. The first class comprises groundwater of fair quality (1,000–3,000 ppm), occurs mainly along the flood plain of Wadi Hadramout and at the confluence of its tributaries. The second class is poor quality water (3,000–5,000) and can be found in the central part of the wadi. The third class has the most saline water (5,000–14,000 ppm). Groundwater encountered in the Mukalla Sandstone is of good quality (440–1,000 ppm), but deteriorates when mixed with waters from valley fill deposits. The groundwater tapped within the Rus-Jeza deposits in the northern areas is of moderate quality, with T.D.S. ranging from 1,400 to 3,700 ppm. Groundwater from the deeper zone is generally of better quality than that encountered in the shallow zone.

– *Wadi Tuban Watershed Province*: The Wadi Tuban watershed is about 5,600 km^2. Rock units ranging from Pre-Cambrian Complex, Mesozoic and Paleocene sedimentary formations, and Quaternary volcanics and sedimentary formations can be found in the Alluvial Plain of the wadi (1,800 km^2). The major aquifer in this province occurs in the alluvial deposits, which consist of an alternating series of highly transmissive coarse clastics and less permeable silt and clay. The aquifer, which is mostly unconfined with local confined or semi-confined condition in the lower part of the delta, varies in thickness from 150 m in the upper reach of the wadi to about 500 m in the delta area. According to the PDRY Report (1986), approximate values of the hydrogeologic parameters of the aquifer are as follows:

Area	*Depth to water, m*	*Average depth of wells, m*	*Hydraulic gradient, %*	*Transmissivity, $m^2\ d^{-1}$*	*Storativity %*
Upper Delta	*20–30*	*Dug wells, 35*	*8–12*	*950*	*0.12*
Central Delta	*15–20*	*Tube wells*			
Lower Delta	5–10	60–70	2.5–1.0	6,050	0.07

Nothing so far is known about the deeper zones of the aquifer of the Wadi Tuban Province. However, there are strong indications that their water-bearing characteristics are poor.

Hydrological investigations carried out more than three decades ago have shown that most of the aquifer recharge comes from the main wadi and its tributaries as well as from spate irrigation water diverted to those areas put under cultivation. The estimated recharge is between the limits of 25 and $40 \times 10^6\ m^3\ y^{-1}$. Model investigation has indicated that $40–60 \times 10^6\ m^3$ represents the upper limit of groundwater extraction so as to avoid any harmful consequences of the aquifer water.

Groundwater quality varies widely throughout the Wadi Tuban Delta. Three main water quality zones are differentiated. These are: 1- along a narrow strip in the upper course of Wadi Tuban with T.D.S. ranging between 650 and 1,000 ppm, 2- the central part of the delta where T.D.S. increases from 1,200 to 1,600 ppm and 3- the lower reaches of the Delta where the salinity ranges from 3,000 to 9,000 ppm. The salinity of water in one of the wells in zone-3 went up from 1,200 ppm in 1947 to between 2,000 and 3,200 ppm in the period 1965–70, and further up to between 3,200 and 4,500 ppm in 1973.

– *Wadi Meafa'h Province*: The aquifer system of Wadi Meafa'h is predominantly alluvium consisting of clastic deposits of boulders, cobbles, gravel and silt in varying proportions. It is generally coarse clastic in the upper reaches of the wadi and gets finer at the lower delta. The aquifer thickness is variable and not properly known. As usual, the permeability varies depending on the fraction of fine-textured grains within the complex of material forming the aquifer. The average depth to water level is 12–38 m, and the average depth of wells tapping the aquifer is about 60 m. The water yielding capacity of the wells can be described as fair. The potential utilisation of the aquifer was estimated as $22 \times 10^6\ m^3\ y^{-1}$. The annual runoff increases from 80 through 100 and further up to $165 \times 10^6\ m^3\ y^{-1}$ corresponding to probability of 90% dry year, 50% average year and 2% wet year respectively. The corresponding annual recharge to the aquifer in the given order of probability of occurrence is 40, 45 and $61 \times 10^6\ m^3$.

Hydrogeological studies, which began some decades ago in both parts of Yemen, have indicated that the Tihama Plain together with Hadramout, Tuban and Abian wadi basins are by far the most important groundwater basins in Yemen. The results available from the late 1980s suggest that the recharge to and the discharge from the aquifers of the Mawr, Surdud, Siham, Rima, Zabid and Rasyan Wadis (Chapter 7) are 40, 52, 18, 30, 82 and 25, all times 10^6 m^3 y^{-1} for the recharge, and 80, 26, 70, 100, 120 and 5, all times 10^6 m^3 y^{-1} for the discharge. Would these figures be correct, the total recharge for the said wadi aquifers must have been 247×10^6 m^3 y^{-1} and the total discharge 401×10^6 m^3 y^{-1}. As such, the discharge from the aquifers is 162% of the annual recharge. Figure 32 shows the rise in the number of pumped wells from 1960 to 1982.basin-wise. According to van der Gun et al. (1992) the number of mumped wells in the early 1990s reached 10,000. The yield of these wells support the groundwater-based irrigation much larger than the surface irrigated under surface water. The upcoming of saline or brackish water in some wells and the urgent need to deepen those wells that dried off added to the increase in the cost of pumping are among the adverse effects caused by intensive aquifer exploitation.

Regional model simulations were carried out to predict the long-term development of the aquifer conditions under alternative pumping schemes. The simulation period of 200 years starting from 1989 has been represented by two average abstraction rates; 159×10^6 m^3 for the first 25 years and 166×10^6 m^3 y^{-1} for the next 175 years. Different controlled regimes were also investigated. All model simulations clearly showed that the increasing depth to groundwater, up to 120 m in the eastern part of the basin, is more severe than the intrusion of saline water. Declining groundwater level leads to rising costs of water abstraction, thus adversely affects the profitability and sustainability of groundwater for irrigated agriculture.

Another example of uncontrolled abstraction is the Sana'a basin. Estimated aquifer recharge ranges from 28×10^6 m^3 to 50×10^6 m^3 and the annual abstraction

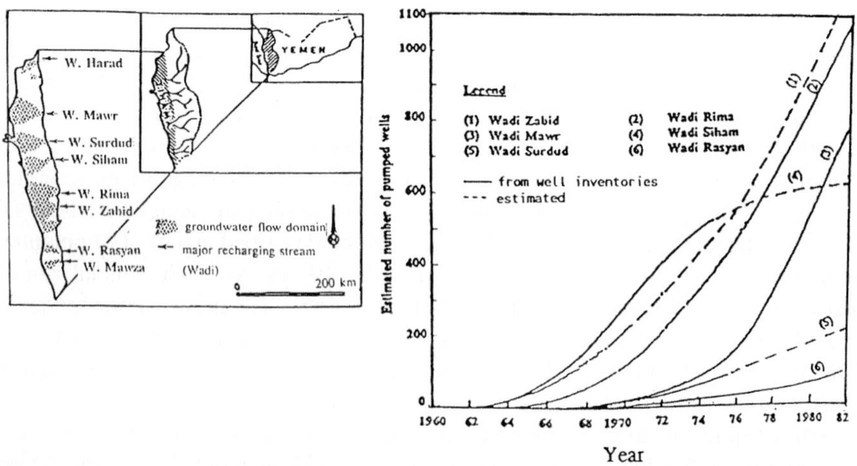

Figure 32. Increase in groundwater wells used for pumping water from the aquifer underneath the Tihama Plain (van der Gun et al., 1992)

as 30×10^6 m^3. The total decline of water level in the 1980s reached 6 m around the well fields. The model showed that the aquifer response to annual abstraction of 12.5×10^6 m^3 is 25 m decline from 1990 to 2050, 70 m decline from 1990 to 2050 for abstraction of 25×10^6 m^3 and about 100 m decline from 1990 to 2020 corresponding to withdrawal of 50×10^6 m^3 y^{-1}.

CHAPTER 10

NON-CONVENTIONAL (NEW) WATER RESOURCES

10.1. GENERAL INTRODUCTION

Surface runoff carried by streams, whether perennial, seasonal or intermittent, and groundwater abstracted from wells and boreholes, springs and aflags are often termed in hydrologic nomenclature as traditional or conventional water resources. One should not forget that stream and groundwater flows are essentially generated by precipitation and are closely related to it. So, in simple words, one cannot imagine a region scarce in precipitation, such as the Arab Region, will ever enjoy abundance of surface and groundwater resources.

Next to its natural scarcity of precipitation, the Arab Region is characterised by a high rate of population growth that exceeds by far the rate of development and conservation of natural water resources. As a result to the booming oil industry in the region, the last six decades or so have witnessed a dramatic increase in the number of skilled and non-skilled labourers coming from outside the region searching for work and better living conditions than those in their countries of origin. The immediate consequence to the coupling between these factors is the steady decline in the share of water per capita. Additionally, the abrupt wealth that struck a considerable part of the region is undoubtedly responsible for the increasing need of more water per capita compared to the situation in the pre-oil wealth episode. Another important factor is that a better conservation and management of the already existing water resources seem, at least in some countries of the region, to be lagging far behind the search for new sources of water.

The possible new or non-conventional sources can be grouped under the following six headings:

1- Reuse of wastewater: This group comprises agricultural drainage water, urban drainage water and industrial wastewater.
2- Desalination of saline and brackish water: This group comprises ocean, sea and gulf water (saline) and brackish groundwater.
3- Deep, fossil or non-rejuvenated groundwater. This refers to withdrawal of deeply situated (hundreds of meters below ground surface) groundwater.
4- Transport of water: This group includes inter-basin (trans-boundary) and intra basin transport.

5- Rain harvesting and cloud seeding.
6- Virtual water: This group includes import of goods, e.g. meat, rice and cane sugar, which need substantial amounts of water to produce them

One of the conclusions based on the statistics obtained from arid and semi-arid regions all over the world is that the major sources of non-conventional water are the reuse of wastewater and desalination of seawater or brackish water. Evidently, the technical, social economic and environmental feasibilities of any of the above-listed possibilities differ from one location to another and from time to time. Each of these possibilities has its advantages and disadvantages depending on the time, location, market conditions and many more factors. In some cases the best solution can be obtained by having a basket containing more than just one solution. One should not forget that with the everyday progress in science and engineering more efficient technologies are constantly becoming available for obtaining new water resources.

The next section of the present chapter will be devoted to brief reviews of the principles underlying the above-listed methods and techniques. The third section will be concerned with the practical applications as exercised in a number of the Arab countries and the experiences gained there from. The last section will be left to the discussion of some of the environmental and social impacts of the new water.

10.2. PRINCIPLES UNDERLYING THE SEARCH FOR NEW WATER RESOURCES

10.2.1 Reuse of wastewater

According to the Water Pollution Control Federation (Water Reuse Manual of Practice SM-3, 1983), factors that determine the quantity of wastewater that can be reused include:

- Geographic location of discharge points and potential users.
- Timing of wastewater discharges and changing requirements of potential users (for example, irrigation requirements vary with the season, but municipal discharges are relatively constant throughout the year); and
- Availability and cost of alternative supplies.

Reuse of wastewater comprises irrigation of agricultural land, landscape irrigation, recharging groundwater aquifers, domestic and other purposes.

10.2.1.1 Reuse of urban drainage water

Shuval (1995) emphasised the need for constructing a sewerage system infrastructure able to collect the wastewater from domestic, commercial and industrial sources for all major urban areas in the Middle East before achieving any level of recycling and reuse of wastewater. He also suggested a minimum water requirement per capita for survival in an urban/industrial community in the severely water-short areas of the Middle East at 125 $m^3\ y^{-1}$. Shuval (1992,1995) assumed further that between 65% and 80% of this amount could be recycled and reused. The corresponding amount, at least in his opinion, is enough to grow all of the fresh food crops required by the urban population, which is the source of the water. The area that can be irrigated with the suggested amount, i.e. 80-100 m^3, ranges between 100 and 200 m^2 depending on the kind of crop, and length and climate of the growing season. The combined cost of water recycling and treatment has been estimated as US $ 1.05–1.35 per m^3 (1995 prices).

10.2.1.2 Reuse of agricultural drainage water

Drainage water left from land irrigation can be a potential source for further irrigation when collected and not disposed of as wastewater. The extent to which the drainage water needs to be treated or rehabilitated depends largely on the kind of crop to be irrigated, type of soil raising the crop, climate and efficiency of the land drainage system.

Water with total dissolved solids (T.D.S.) below 450 mg l^{-1} may be suitable for irrigating any crop. Under extremely effective drainage conditions water with T.D.S. up to 2,000 mg l^{-1} can be used for irrigating nearly all crops except those that are highly sensitive to salts. Water with salinity above 2,000 mg l^{-1} can be used for irrigating salt-tolerant crops. It goes without saying that salt-leaching requirement increases with increasing water salinity. Relative salt-tolerance of agricultural crops can be found in literature dealing with irrigation using water with different levels of salinity (e.g. FAO, 1985). Another example is Table 16, Appendix III, which was prepared by Asano (1994). Despite its limited size compared to other tables, the information it contains is useful.

10.2.2 Water desalination

By distillation is meant here the process by which the salt content of saline (sea or ocean water) or brackish water (generally less saline than seawater, mostly groundwater) is reduced to a level that renders its use for a certain purpose acceptable. The terms distillation and desalination are used synonymously in this context. Desalination is often an attractive solution as desalination plants can be installed in a relatively short time. Theoretically, there is no limit to the amount of water to be desalinated as long as the country seeking a new or an additional freshwater resource is not land-locked.

According to Gleick (2000), desalination of sea or brackish water remains hindered by high economic costs in part because of the large amounts of energy required to strip salt ions from water. In addition, the high cost of moving water from one place to another further constrains desalination developments to areas within a limited distance from the coasts. Despite the constraints imposed by the price of energy and the transfer distance of water, a look at Figure 1 (Wangnick, 1998) shows that total capacity of desalination plants worldwide has grown from almost zero in 1960 to reach more than 20×10^6 m^3 d^{-1}.

Obviously, the amount of energy that has to be spent in such an operation is proportional to the range of water salinity between the initial and final conditions of the water. The availability of energy resources constitutes the backbone the

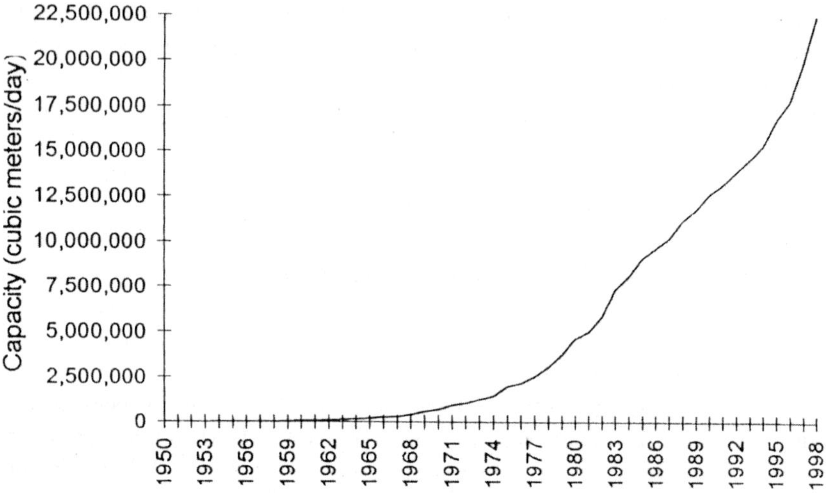

Figure 1. Cumulative desalination capacity worldwide, 1950–1998 (Source: Wangnick, 1998)

desalination process needs. Algeria, Libya, Iraq, Kuwait, Bahrain, Qatar, Saudi Arabia, the United Arab Emirates and Oman are all rich in oil and gas, and the production of these energy sources in each of Egypt, the Sudan, Syria and Yemen is on the increase. This shows that the majority of the countries in the Arab Region is secured as far as the energy source for saline/brackish water distillation is concerned. Besides, the region possesses abundance of solar energy during a considerable part of the year as indicated in Chapter 3, which is an effective and cheap source. Unfortunately, till now, the use of such a source does not go beyond the experimentation phase.

Three major distillation processes are currently in use and economically acceptable. These are; the multiple-effect distillation (MED), multi-stage flash (MSF) distillation and vapour compression (VC) distillation. Other than these processes there are two major membrane desalination processes in current use as they are commercially acceptable. These are; the reverse osmosis (RO) and the electro-dialysis (ED) processes. The next subsections briefly discuss the characteristics of each of the said processes.

Desalination costs are generally high due to the large amounts of energy needed to convert water into vapour. The location of the source of water, production capacity, distillation process and quality of feed water are major cost-determining factors. Desalting seawater can be from three to seven times more expensive than brackish water using the same method in both cases.

10.2.2.1 Multi-effect distillation (MED)

Historically speaking, this process can be ranked as the earliest process to be used in relatively large-scale water desalination projects. The percentage of MED plants to the total number of plants in service has declined rapidly in the last 50 years or so. Operation and maintenance problems are behind this decline. The MED desalination plants usually recover almost one-half (40–65%) of the supplied water as desalinated product water.

In the MED process, the inflowing saline water after being heated is forced to run through a series of evaporation units. In the first effect, water evaporates by condensing live steam. Water vapour released from the hot brine undergoes the second effect where it condenses. The heat produced in the second effect causes further evaporation of the brine. The pressure in each succeeding effect is lowered to permit boiling of water and further transform into vapour at lower temperatures.

10.2.2.2 Multi-stage flash (MSF) distillation

This process succeeded the MED process and shortly afterwards became the most widely used distillation method. It dominated the commercial distillation market from 1960 up to 1970. The introduction of the reverse osmosis in the early 1970s has led to the decline of the percentage of MSF plants to the total distillation plants from 69% in 1969 to 50% in 1991. Currently it represents about 48% of the total number of distillation plants having a capacity greater than 4,000 $m^3\ d^{-1}$.

The inflowing water after being heated in a so-called brine heater is led into the first chamber or stage. When the pressure in that chamber is reduced, the hot brine boils violently and a small portion of it instantaneously flashes into vapour. As the brine passes through successive stages (20–50) operated at continually lower pressures and temperatures, more and more of the brine flashes into steam. The water vapour thus produced is condensed on the outside walls of the tubes, which supplies the inflow to the brine heater. Increasing the number of stages means more capital cost while maintenance cost becomes less as the overall efficiency of the process increases. A sketch of the components forming a MSF distillation plant is shown in Figure 2.

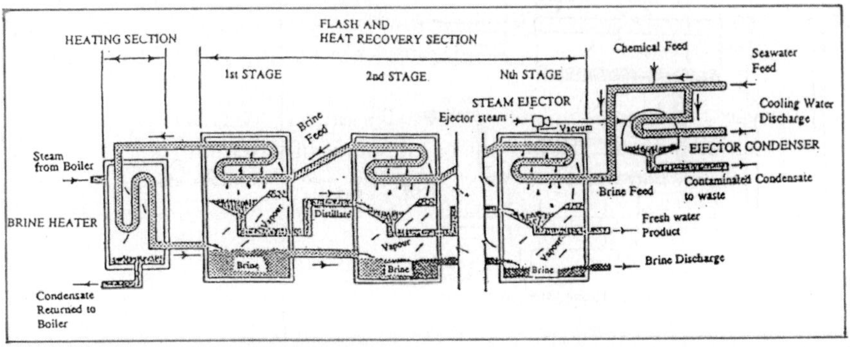

Figure 2. Simplified sketch of the components included in a once-through multistage flash (MSF) distillation plant (Source: Bushnak, 1995)

10.2.2.3 Vapour compression (VC)

In the MED and MSF distillation processes reduced pressure over the brine is used to enhance vapourisation. The resulting water vapour is directed to vapour compression units where it is collected and compressed. In this way the enthalpy (amount of heat in a unit of mass) as well as the vapour temperature are considerably augmented. As such that they can serve as a heat source throughout the condensation of vapour in a stage operating at a temperature higher than that of the vapour before having it compressed. The necessary energy to increase the enthalpy can be introduced either in a mechanical vapour compressor (MVC) driven by an electric motor or through thermal compression. Figure 3 illustrates the mechanical vapour compression process.

10.2.2.4 Reverse osmosis (RO)

This technique is used for both small and large plants, amounting to 22% of the world's larger plants having a capacity of more than 4,000 $m^3 d^{-1}$. Whenever a salty feed on one side of a semi-permeable membrane is exposed to a pressure higher than its natural osmotic pressure, the direction of flow is reversed and the pure water will diffuse through the membrane leaving behind a more salty water (brine). Figure 4 illustrates a schematic presentation of a reverse osmosis desalination plan

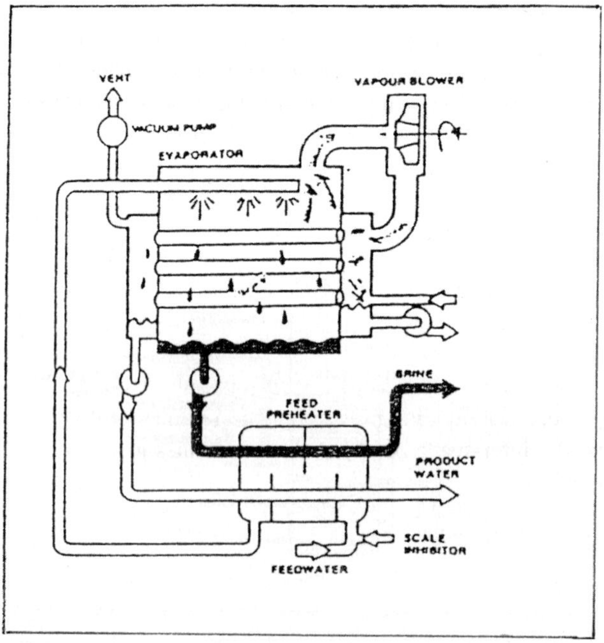

Figure 3. Simplified sketch of a mechanical vapour compression unit (Source: Bushnak, 1995)

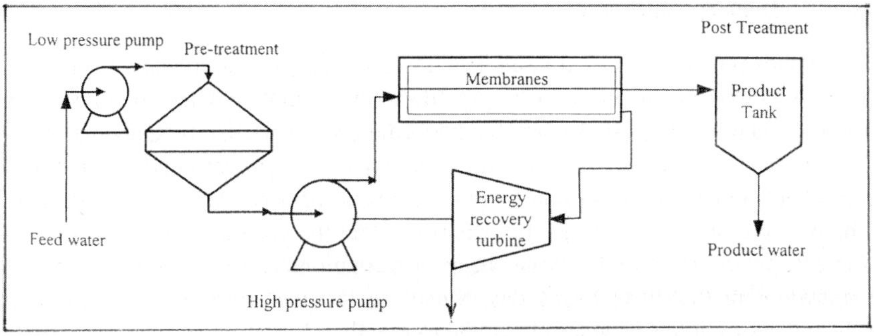

Figure 4. Schematic presentation of a reverse osmosis desalination plant (from Semiat, 2000)

(Semiat, 2000). Typical pressures applied to brackish water are 14–35 kg cm^{-2} (1,380–3,450 k Pa) and for sea or ocean water 56–85kg cm^{-2} (5,520–8,275 k Pa).

In reverse osmosis process the supply, or feed, to the distillation plant is pumped into a pressure vessel containing the membrane. Membranes have different forms and types. Each module type has its technical and economic advantages and disadvantages. The type of plants operating by the RO process, which are used in the Gulf States, is known as seawater reverse osmosis plants (SWRO). These plants are usually operated with 20–50% of the feed recovered as freshwater (Bushnak, 1995).

10.2.2.5 Electro-dialysis (ED)

This process has proven to be adequate for desalinating brackish water but not seawater. The process relies on the separation of the components of an ionic solution through placing certain membranes across the path of a direct electric current. The membranes are permeable to a certain type of ion but not to another. The desalination cell thus contains two different types of membranes; cation-permeable and anion-permeable. If a direct current flows across a stream of saline water passing between a pair of those membranes, ions acting as carriers of electricity will migrate across the stream. The cation-permeable membrane will permit the positive ions (e.g. Na^+) to pass through while repelling negative ions (e.g. Cl^-) and the other membrane allows negative ions to pass through but not the positive ions. The membranes act as one-way check valves thus preventing the re-entry of the ions they let through. In this way, the inter-space between the membranes gets desalted while the streams on the electrode sides become concentrated with the penetrating ions. In ED desalting devices several pairs of membranes are used between a single pair of electrodes forming an ED stock.

Before we leave this subsection, it is worthwhile to mention that it is possible to use a hybrid system in which two or more desalination processes are combined together. Examples of hybrid systems used in Arab countries are included in Sec. 10.3.

10.2.3 Deep groundwater

This groundwater is kept stored in ancient aquifers, deeply seated below the earth's surface. The volume of water in storage depends on the thickness of the water-bearing formations, their areal extent, and thus their storage capacity. One should not forget that the thickness of such aquifers is usually in the order of hundreds, if not in thousands of meters. Well drilling by petroleum companies while searching for oil has helped to collect some useful information about the geology and hydrogeology of these water-bearing strata. Water-bearing formations were filled with some of the rainwater in the pluvial episodes. Since the last wet episode there has been hardly any addition to the water in storage to mention. Some countries have already begun extracting some of this water to cover the deficit between demand and supply from other sources.

10.2.4 Import of water

Water trade, like any other trade between countries, is based on import and export. Whether the trade is between different countries or between two parts of the same one country, water has to be transferred from place to place. Mankind has diverted water across river basins for its cities in many civilisations along the centuries. The Roman system of aqueducts developed two millennia ago is a widely known example. Construction of storage reservoirs on the principal tributaries of the Ganga and the Brahmaputra in India and Nepal along with inter-liking with canal systems helps to transfer surplus flows of the eastern tributaries of the Ganga to the west. Besides, it helps linking the Main Brahmaputra and its tributaries with the Ganga, and the Ganga with Mahanadi. The project of transferring some of the discharge of the River Rhône in France to Barcelona, Spain, via the 'Languedoc-Rousillon Aqueduct' is currently under way. Upon its completion, the project will help to augment the available water in the Catalonia Region, Spain, in addition to development of hydroelectric power. It goes without saying that there is a wide diversity of water trade schemes in many parts of the world. Next to the technical and financial feasibilities, legal and environmental issues and repercussions, there is the issue of transport of water. This can be done either by sea or overland.

10.2.4.1 Water transport by sea

This operation involves the transportation of freshwater from rich sources to areas where the available supply is less than the demand. The transport can be accomplished by water-carrying ships, tankers, or by means of flexible bags made of durable material such as vinyl.

Due to factors related to sales price and maintenance it is advisable to buy oil tankers not older than 20 years of age, coat them with a new lining and convert them to water tankers. Newly manufactured tankers provide an alternative to old, used oil tankers. There are many examples of the latter such as transport of water from Spain to Mallorca Island in the Mediterranean Sea. The advantage of tankers is their speed, consequently the length of the travel time between the point of export and the point of import. This advantage is reduced by the long time a tanker needs for loading and unloading. The major concern in using oil tankers for transporting water is having them clean to secure the quality of exported water.

The second option, especially when the bags are large in size, is cheaper compared to import of water by tankers. The Norwegian Nordic Water Company, according to Haddad & Mizyed (2004), has been using since the year 2000 bags of 19,000 m^3 capacity to transport water from Turkey to northern Cyprus. Currently, the same company is busy constructing a commercial bag of 80,000 m^3 capacity. It is claimed that with such a large size, the bag is able to transport water for a distance up to about 600 km. The size of Medusa bags is relatively small, but the manufacturers, Medusa Corporation, have the ambition to produce commercial bags up to 1.75×10^6 m^3.

10.2.4.2 Transport over land through pipelines

There is a possibility to convey water from a water-rich region to a water-scarce region partly or completely through aqueducts or pipelines similar to oil transportation.

10.2.5 Rain harvesting and cloud seeding

Rainwater harvesting is an old technology that has been used for more than four centuries to supply freshwater for different uses. The idea is based on collecting rainwater and conveying it with the least possible losses to a collector. The annual rainfall depth can be as little as 50–80 mm. The conveyance network can be formed of conduits or open channels while cisterns, ponds and tanks are the usual types of collectors used.

Natural catchment surfaces are pervious to some extent. It is desirable to bring this porosity to the least possible extent. In Western Australia, catchment areas are graded and rolled; catchments are cambered so that the runoff produced by rainfall flows quickly to the sides of the roads, highways and airports. Catchment treatment can be achieved to some extent by compacting the earth after wetting the surface. A better result can be obtained by adding a dispersing agent such as sodium chloride, sodium silicate or sodium polyphosphate to the upper portion of the land surface. Spraying low-cost chemicals on the natural surfaces can improve the runoff coefficient. Paraffin wax when spread on the ground surface melts under the heat of the sun and seals the pores to form an impervious surface. Wax-treated plots were used successfully to establish trees on mine spoils and were found to provide a substantially greater amount of soil moisture to growing plants than did untreated plots (Aldon et al., 1974). Asphalt coated surfaces or the catchment treatment using asphalt-fabric membranes can raise the runoff coefficient. The asphalt coat can last up to 5 years while the asphalt-fabric membrane is more durable and lasts up to 20 years. A few data about the runoff coefficient and the duration for a number of catchment treatment methods:

Treatment	*Average runoff coefficient, %*	*Estimated life, years*
Compacted earth	*45*	*Indefinite*
Sodium salts	*75*	*10*
Concrete	*60–85*	*25*
Paraffin wax	*80–95*	*7*
Gravel covered sheeting	*85*	*15*
Asphalt-fabric membranes	*95*	*20*

Fok & Chu (1995) highlighted the Thai experience policy regarding the construction of $2-m^3$ reinforced concrete jars, each for a 6-person home in the rural areas to provide safe drinking water to the rural population. The policy adopted by the Thai Government made it possible to construct a large number (no less than 200,000) of $2-m^3$ jars by the end of 1990, thus conserving a considerable amount of rainwater from being lost.

Most of the attempts at modifying the amount and location of precipitation involve the addition of condensation nuclei at the appropriate levels in a cloud mass. Besides, one should bear in mind that it is impracticable to modify the humidity or stability conditions of an air mass. However, it is possible to identify the type, size and concentration of the nuclei that will produce a maximum amount of precipitation from a cloud containing precipitable water. Too few or too small nuclei will not result in precipitation. Likewise, too many nuclei may also have the same effect, as the result will be a large number of droplets too small to result in coalescence. Procedures used to provide condensation nuclei depend on temperature of the cloud. For warm clouds ($T > 0$ °C) the usual technique is to spray the clouds with Na Cl condensation nuclei from above. For cold clouds the addition of dry ice (dry CO_2) will result in the formation of ice nuclei within the cloud mass. Iodide salts such as Ag I (Silver iodine) and PB I_2 (lead iodine) also may be used, however the level of placement may be more important. The Ag I should be placed above the −6 °C level while Pb I_2 is more effective above the −10 °C level. Accurate placement of condensation nuclei requires special techniques to be used. The effectiveness of cloud seeding operations is difficult to evaluate, as long time is necessary to allow assessment of the possibility of natural occurrences.

Readers who are interested to learn more of the nuclei of cloudy condensation, growth of cloud droplets and some of the early rainmaking experiments, can consult any of the specialised material on this subject. The author recommends, as an example, a rather old book, yet a simple and well-written one by Mason (1962).

10.2.6 Virtual water

By virtual water is meant water mostly found in solid rather than in liquid form. This water is embedded in a wide variety of commodities people use daily, or at least frequently, like grains, cotton, fruit and conserved foodstuff. Despite the fact that virtual water is a convenient solution for the water scarcity in many parts of the world including the Arab Region it does not necessarily mean that every country receiving virtual water is a water-short country. When a country like Finland or Great Britain imports rice and cane sugar, which consume plenty of water before having them available to the consumer, it does not mean that any of these two countries is short of water. Instead, they are short of climate that helps these crops to grow and be produced in such lands.

10.3. NEW (NON-CONVENTIONAL) SOURCES OF WATER IN THE ARAB REGION

To enlighten the reader, especially the one from outside the Arab Region, of the amounts of water needed to produce common food items in the region, the figures listed below might serve as a guide:

Cereals	*1,500* $m^3\ t^{-1}$	*Meat bovine fresh*	*20,000* $m^3\ t^{-1}$
Roots and tubers	*1,000* $m^3\ t^{-1}$	*Meat sheep fresh*	*10,000* $m^3\ t^{-1}$
Citrus fruits	*1,000* $m^3\ t^{-1}$	*Meat poultry fresh*	*6,000* $m^3\ t^{-1}$
Palm oil	*2,000* $m^3\ t^{-1}$		

The rapid increase in the population of the Arab Region (average 3% per year) coupled with the ever-growing need of water and food per capita lead to a growing demand of water that keeps exceeding the supply from the conventional resources. The obvious consequence is the need to have (new) water from unconventional resources in order to fill in the gap between the demand and supply. The next subsections describe the experiences of some of the countries of the region to get water from one or more of the new resources. However, before embarking upon these experiences it should be well remembered that, at least, some of them might appear elsewhere under the title of water conservation techniques. As such, they will not be repeated in Chapter 11, which deals with water conservation in the Arab region. Instead, it will be confined to storage works of surface and groundwater only.

10.3.1 Wastewater Reuse

The search and exploiting new water of acceptable quality is of growing interest to meet agricultural, domestic and industrial demands in several parts of the Arab Region. Municipal wastewater is the most widespread and generally available source of reusable wastewater. The report prepared by ECWA (1984) highlights the potential of wastewater reuse as the most attractive future alternative to supplement water supplies in the Middle East Region. It is expected by the end of 2005 that the annual volume of wastewater reuse in the Gulf Council Countries, including Yemen, will exceed 0.5×10^9 m^3. The sources of this amount are the domestic and industrial wastewater.

The major factors affecting the process of wastewater reuse are the unit cost of reclaimed and the environmental repercussions. The unit cost depends upon the size of treatment plant, the composition of the wastewater and the required quality of the product water. As a rough cost estimate is SR (Saudi Riyals) 1.0–3.0 per m^3. Landscape irrigation such as that of parks and public gardens, playgrounds and cleaning public facilities with partly treated water in warm climates, though not unhealthy, can generate unpleasant odours and other discomforts. A summary of the national (USA) wastewater discharges available for reuse showed a decline from 10,710 m^3 s^{-1} in 1975 to 9,752 m^3 s^{-1} in 1985 and a predicted rate of 7,995 m^3 s^{-1} for the year 2000. The unavailability of wastewater measurements for the year 2000 makes it impossible to assess the credibility of the predicted wastewater figures.

Guidelines for wastewater effluents to be reused for crop irrigation generally consists of explicit standards, such as coliform concentrations, or the minimum treatment required, e.g. primary, secondary, etc, or both. The guidelines depend on the crop to be irrigated whether it is consumable or non-consumable, cooked or uncooked. There is a wide variety of guidelines mainly depending on the health risk to be anticipated. Sometimes the guidelines reflect, as already mentioned, the economic situation of the country in which treated wastewater is used. Table 1 includes the current microbiological standards for wastewater reuse in crop irrigation in some Arab countries (WHO, 1989).

The next subsections review practices followed in seven Arab countries for water reuse. In six of them; Saudi Arabia, Tunisia, Jordan, Oman, Yemen and Syria the main resources are domestic and industrial wastewater, and in the seventh country, Egypt, the source is agricultural drainage water.

10.3.1.1 Saudi Arabia

The concept of wastewater reuse was given impetus through a legal ruling issued by a Committee of eminent Muslim Scholars of Saudi Arabia in April 1979. The

Table 1. Current microbiological standards for wastewater reuse in crop irrigation in some Arab countries (WHO, 1989)

Country	Restricted Irrigation	Unrestricted Irrigation
Kuwait	< 1000TC*/100 ml	< 100 TC/100 ml not salade crops or strawberry
Saudi Arabia	Use of secondary effluent (vegetables, landscapes)	< 2.2 TC/100 ml
Oman	2.2 TC/100 ml	crops irrigation not permitted
	Greenbelt irrigation	
Tunisia	Secondary treatment (fruits and vegetables eatenvegetables cooked)	no irrigation of raw eaten

TC = Total coliforms*

Committee, after having taken evidence from experts that treated water is not only safe for human consumption but is free of anything that might be regarded as ritual impurity, reached its decision that treated wastewater can be used for all religious rituals. It was decided that water returns to its natural form after undergoing a thorough treatment. This ruling was contained in a 'fatwa', a legal decision based on Islamic laws and standards, and having the power of Law in the Kingdom of Saudi Arabia. After this legal ruling several projects for wastewater reuse have been envisioned and by now some of them have already been implemented, e.g. a certain amount of treated water from the urban drainage of the city of Mecca is used for irrigating an area of originally undeveloped land to produce fresh vegetables.

Farooq & Al-Lyla (1985) and later Mohorjy (1988) stated that evaluation of the potential for water reclamation and reuse is based on the following sequence:

i- Screening, analysing and evaluating-in terms of quantity and quality-the available data on existing water supply in the country

ii- Municipal, industrial and irrigation water requirements of the country with future projections are examined.

iii- Comparison between available supplies and demands should be carried out to serve as background for evaluating future water reuse potential.

iv- Impact of national socio-economic trends on the available ware supply is evaluated.

v- Potential of water reuse for meeting future demands on water for municipal, industrial and agricultural needs is evaluated.

The Arab-American Petroleum Company 'ARAMCO' developed reuse criteria for various uses in Saudi Arabia (Aramco, 1980). The criteria and related information are given in Table 17, Appendix III.

The total water requirement in Saudi Arabia for 1985 was estimated as about 9×10^9 m^3 y^{-1}, of which 1.85×10^9 m^3 y^{-1} were supplied by surface and shallow groundwater resources. The difference was supplied by non-renewable groundwater resources, reclaimed water (0.1×10^9 m^3 y^{-1}) and desalinated water (0.4×10^9 m^3 y^{-1}) (Ukayli & Husain, 1988). As a consequence of the heavy exploitation of groundwater, the depth to groundwater level dropped rapidly during the last decades and in some regions the quality deteriorated (Mohorjy, 1988). To reduce the ever-increasing exploitation of the aquifers, it became apparent that augmentation of water reuse and desalination of saline/brackish waters might be the most appropriate solutions. Currently all major cities such as Er-Riyadh, Jeddah, Mecca, Medina Dhahran, Al-Khobar and Dammam are already connected to the sewage systems and their drainage waters treated. The system of dual pipelines is adopted in some cities. One network supplies high-quality water mainly for drinking and the other purposes. The other line is installed in order to take care of the lower quality water obtained from reclaimed wastewater for watering the parks, car washing, fire fighting and certain industrial purposes. It seems, however, that the idea of using water already used before, even after having it treated, did not gain a wide acceptance among the users. Recent figures of the items constituting the water balance of Saudi Arabia (Hussein & Ahmed, 1997 and Al-Turbak, 1997) indicate a dramatic increase, more than 200% in about 10 years, in the abstraction of non-renewable groundwater resources while the annual reuse of wastewater was 0.110×10^9 m^3, i.e. just 10% more than the reuse the 1980s.

Generally, the quality of treated wastewater used in industrial processes ranges from raw sewage to deionised water. However, there are problems inherent in the use of reclaimed wastewater in industrial processes such as scaling in boilers and hot-water pipes and corrosion of metal surfaces.

10.3.1.2 Tunisia

There is a growing tendency to use treated wastewater for irrigating more land in Tunisia. The acreage currently under irrigation with treated wastewater is estimated as 25,000 ha.

Irrigation of citrus trees from coastal groundwater aquifers recharged by treated wastewater is being experienced in northeastern Tunisia since 1986. A wastewater treatment plant was built on the seacoast close to Nabeul with a potential capacity 8,000–9,000 m^3 d^{-1}. This created a problem of bad odour to the beach area there. Since groundwater resources in the area are limited, a program to reuse treated wastewater was established. This implied the construction of an artificial recharge station in an area where the aquifer hydrology and hydrogeology are known. The hydraulic parameters resulting from the recharge, the hydrodynamic impact of the recharge, the artificial recharge mechanism and the chemical aspects of artificial recharge were observed and the results reported by Rekaya & Bedmar (1995). It was concluded that the area (300 ha) raising citrus needs 8–10 recharge stations. Additionally, the aquifer put under experimentation in view of the limited thickness of its saturation zone, does not constitute an ideal case. The artificial radioactive tracers (e.g.Tritium) have proven to be useful tools for investigating recharge and chemical evolution of the injected water. Dilution tests in boreholes provide information on the characteristics of groundwater flow prior to and during the recharge operation.

The same experimental set up was used to investigate the microbiological aspects of artificial recharge with treated wastewater. The combined area of the four injection basins in the recharge station is 1,365 m^3. In that investigation, water from the purification plant was diverted to two basins up to 1.0 m height then diverted to the other two basins. The filling operation lasted 45 day during which a total volume of 43,000 m^3 was recharged to the aquifer. The resulting hydrologic and microbiological changes were observed through 18 piezometer tubes. The mean counts of total coliforms were 4.3×10^5 in 100 ml treated water and 5.4×10^2 in the groundwater of nearby wells. The corresponding faecal coliforms were 3.6×10^5 and 31, both in 100 ml.

The response of the groundwater to injection of artificial recharge brought by treated wastewater, as appeared from the piezometers, differed from one piezometer to another. The various responses can be classified into three main groups. In the first group, the quality of groundwater in certain locations did not show any substantial response to the recharge from the beginning of the operation to its end.

The piezometer representing the second class showed a considerable rise of the counts of bacteria during the injection operation, after which the pollution of water fell gradually till the end of the year. The third class indicates a quick response within a limited number of days then the aquifer loses its purification ability for a few weeks time. After the number of coltiforms reached its peak value, about 4 weeks from the beginning of the recharge operation, the quality of groundwater improved gradually and reached a stable, low value till the end of the year. Figure 5 shows the behaviour of the three groups in their respective order as observed from piezometers No Z_7, PC_1 and Z_4 respectively.

Those findings show that the bacteria carried by the injected recharge does not distribute itself uniformly all over the groundwater body. It is necessary to interpret the distribution pattern in the light of the flow direction of groundwater. It has also been concluded that the period in which the pollution lasts is relatively short. Almost three weeks after the cessation of the recharge operation the groundwater quality becomes comparable to the quality before recharge is injected. It is possible to improve the purification ability of the soil by introducing more appropriate recharge scenarios (Tarad, 1995).

10.3.1.3 Jordan

The use of treated wastewater in Jordan for irrigating agricultural crops began in the 1980s. The annual volume of recycled wastewater grew since 21×10^6 m^3 in 1986 to reach 46×10^6 m^3 in 1991 and further to 50×10^6 m^3 in 1993. The expected volumes to be treated in 2010 and 2020 are 144×10^6 m^3 and 219×10^6 m^3 respectively, with over 95% to be used for irrigation (Shantanawi & Jayousi, 1995).

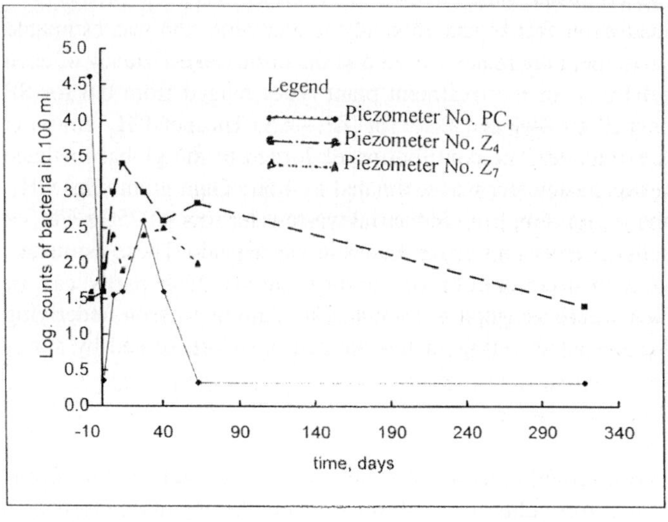

Figure 5. Development of number of bacteria in groundwater as a result of injecting the aquifer with treated wastewater (from Tarad, 1995)

The wastewater treatment plants can be divided into three major types: 1-activated sludge (AS) system and its modifications, 2-biological trickling filters (TF), and 3-stabilization pond (SP) system. The available data up to 1996 of the flow to the treatment plants, and the BOD_5 and TSS in the influent and effluent water are given in Table 27, Appendix II (Al-Sheriadah, 1997).

The compliance of the effluent BOD_5 and TSS from the 14 treatment plants to the design criteria (i.e. BOD^5 is 50 mg l^{-1} for S.P., 30 mg l^{-1} for other types, and TSS is 50 mg l^{-1} for all plants) was examined statistically. Based on the limited number of results the conclusion that has been drawn is as follow:

Type of wastewater Treatment plant	*Probability of failure, %, for*	
	BOD_5	*TSS*
Stabilization Pond	*94*	≈ 100
Trickling Filter	*77*	*66*
Activated Sludge	*15*	*20*

The above listed results, among others, show that activated sludge performs, well even at times of loading, above the design criteria. They also indicate that the activated sludge system shows the least variation in the effluent concentrations thus making it more stable compared to other types of treatment.

It is known that disposal and treatment of industrial and municipal wastewater can produce emissions of the most important 'Greenhouse Gases', which constitute a threat to the world's environment. In addition to the 14 domestic wastewater treatment plants already mentioned above, 22 Jordanian industries were investigated (Abdulla & Al-Ghazzawi, 1997).

The fraction of wastewater that is anaerobically treated (FWTA) was estimated based on the type of treatment technology used and the observed efficiency of each treatment plant. The FTWA for the treatment plant types ranged from 0.6 for SP type to 0.25 for TF, 0.20 for AS and 0.15 for TF + AS. The net CH_4 emission from each plant was estimated. The contribution of Jordan to the global methane emissions from domestic wastewater was estimated as 4.663 Giga gram (Gg) CH_4. The total annual methane emission from industrial wastewater was 0.075Gg CH_4 or just 1.61% of the methane emissions from domestic wastewater. These estimates, however, are subjected to a certain degree of uncertainty, which partly can be attributed to the FTWA figure assigned to each station. Another reason underlying the uncertainty is that not all dwellings and inhabited areas are served by sewer systems.

10.3.1.4 Oman

Traditionally, Oman is an agricultural country where most people rely on farming as a major source of income. Agricultural activities in Oman, like any place elsewhere, are mainly governed by the extent of water availability. The coastal plain of the Al-Batinah area accounts for more than 50% of the cropped area and consumes about 54% of the net water demand of Oman. The 1993 water balance of the country shows a total deficit of 378×10^6 m^3 shared by Muscat (27); Batinah (180); Musandam (4); Dhahirah (65); Dakhilah (65); Sharquiya (7); Wusta (0) and Dhofar (30), all figures between parentheses are in 10^6 m^3. In attempt to fill in part of this gap, the authority in Oman constructed about 250 sewage treatment plants (stabilisation pond system) with a total volume of treated wastewater of 28.6×10^6 m^3 y^{-1}, of which almost 75% is reused for agricultural purposes (Al-Ajimi & Abdel-Rahman, 2001). The cost price some 5 years ago used to be US $ 0.20–0.25 for 1 m^3 and the revenue US $ 0.25–1.0.

Besides, in some of the oil well fields in Oman, like the Nimr, Rima and Marmul fields the water produced with oil is moderately brackish. With the appropriate technology this water can be treated and reused for irrigating a variety of crops especially in the areas adjacent to the oil fields. Raw water from several sources in the Nimr area has salinity with concentration ranging from 1,390–4,400 mg l^{-1} and total dissolved solids (TDS) from 5,400 to 6,700 mg l^{-1}. The treatment strategy involves oil/water separation to remove free phase hydrocarbons. Certain processes might follow to reduce dissolved hydrocarbons and other organics. In some cases it might also be necessary to remove heavy metals. After completing these steps, the salinity may need to be reduced before water becomes suitable for irrigation purpose.

Water quality standards have been established in Oman for various irrigation purposes. A few of these standards is as follows (after Khan & Fuggle, 1995):

A- *Irrigation of vegetables likely to be eaten uncooked, and so are fruits eaten raw and within two weeks of any irrigation, and*

B- *Irrigation of vegetables to be cooked or processed; fruit if no irrigation takes place within 2-weeks of cropping; fodder, cereal and seed crops:*

Parameter	*A*	*B*
	less than, mg l^{-1}	*up to, mg l^{-1}*
Chemical Oxygen Demand (COD)	*150*	*200*
Phenols (total)	*0.001*	*0.200*
Oil and grease (total extractable)	*0.50*	——
Sulphate (as SO_4)	*400*	——
Sodium (as Na)	*200*	*300*
*Total Dissolved Salts (TDS)**	*1500*	*2000*

Below 500 mg l^{-1}, no detrimental effects are usually noticed. Between 500 and 1,000 mg l^{-1} TDS in irrigation water can affect sensitive plants. Between 1,000 and 2,000 mg l^{-1}, TDS levels can affect many plants and careful management practice is required. Above 2,000 mg l-1, water can only be used regularly for salt-tolerant plants and permeable soils.

10.3.1.5 Yemen

The country is heading towards a serious water crisis because its groundwater resources as already explained in Chapter 9 are depleting very rapidly. In two decades the number of deep wells has increased tenfold to over 25,000 well.

The authorities in Yemen thought of wastewater reuse as a solution to overcome this crisis. Wastewater provides an extra reliable water source, provides valuable fertilizers for improving the crop production and minimising environmental hazards caused by pollution.

The most suitable source of wastewater for plant growth is treated wastewater of the rapidly expanding urban sewerage systems in cities such as Sana'a, Taiz, Aden, Hodeidah, Ibb, Dhamar and even some of the less populated townships. The current wastewater standards of the National Water and Sanitation Authority are suitable to provide effluent useable for a large variety of crops. As a matter of fact, the farmers staying in the neighbourhood of urban wastewater disposal sites are already practising the reuse of non-treated or partially treated wastewater. In Sana'a, farmers near the stabilisation ponds treating the sewage of the old Sana'a city do use the wastewater for irrigation around the airport quarter. The effluent is transported via a system of open channels to the agricultural plots where maize, wheat, barley, grapes and qat are grown (Zaemey, 1992). The study of long-term effects of irrigation by treated wastewater is underway. No definite conclusions have so far been reached.

The combined wastewater flow to the six major cities of Yemen; Sana'a, Aden, Taiz, Hodeidah, Ibb and Dhamar was 45,000 m^3 d^{-1} in 1990 and expected to reach 112,060 m^3 d^{-1} by the end of 2005. Stabilisation ponds are used for wastewater treatment in Sana'a, Taiz and many more cities. The BOD of the effluent in Sana'a is about 10% of the BOD of the influent in the winter season and 20% in the summer season while the percentage of COD is 22–23% in the whole year round. The wastewater of the city of Aden is treated in a series of five ponds consisting of anaerobic and facultative ponds before it is reused. Most of the treated wastewater is used for the 'greenbelt' project where trees and crops are grown alternately in stripes of 500 m long by 50 m width.

A somewhat detailed study of the reuse of wastewater in Sana'a was carried out in the season from October 1991-till March 1992. In that study emphasis was laid on the characteristics of the raw domestic water, performance of the then existing stabilisation ponds and the suitability of the effluent for irrigation purposes (Nozaily, 1992).

The raw water (8,000 m^3 daily discharge) has a total BOD of 800 mg l^{-1}, total COD of 1,607 mg l^{-1}. The anaerobic ponds showed a high performance while the performance of the facultative ponds was poor. The sulphate dropped considerably from 128 mg l^{-1} at the inlet site to just 20 mg l^{-1} after the anaerobic ponds and rose again to 88 mg l^{-1} at the outlet. Some typical characteristics of the Sana'a effluents are: Sodium, 145; Potassium, 22; Calcium, 200; Magnesium, 79; and TDS 1,032, all in mg l^{-1}.

The quality of the treated wastewater in Sana'a is not acceptable for uncontrolled use in agriculture. This can be mainly attributed to the poor performance of the stabilisation ponds, which result in an effluent with high concentration of BOD and probably also high concentrations of faecal coliforms. A more profound and comprehensive study of the effluents is strongly recommended.

10.3.1.6 Syria

The annual water requirements for agriculture in the Damascus Basin was estimated in the base year, 1980, as 523×10^6 m^3, and as 857×10^6 m^3 from 2010 up to 2030. These amounts represent 60% of the total requirements for the basin. Similar estimates were developed for the Sahel (coastal) Basin and the figures obtained for agricultural requirements (50% of the total requirements) were 232×10^6 m^3 and 370×10^6 m^3 for the same years respectively. To meet the growing demand a study was done on the reuse of sewage effluent for irrigation in Syria (Aigho & Bournas, 1995).

Estimates of the annual sewage effluent for five major cities, Damascus, Homs, Hama, Lattakia and Tartus were 71.5, 20.0, 10.75, 17.65 and 6.0 for the year 2000, all in 10^6 m^3. The estimates for 2030 of these cities in their respective order are 197.5, 104.5, 75.5, 77.35 and 42.15×10^6 m^3. The growth in the available water leads to an increase in the irrigated surface in Damascus basin from about 45,000 ha in 1980 to 70,000 ha in 2000 and further up to 82,320 ha as from 2010 up to 2030. The corresponding figures for the Sahel Basin are 42,300 ha in 1980, 59,100 in 2000 and 67,500 ha in 2010 up to 2030 respectively.

Assuming that the median concentration of BOD_5 in the raw sewage water to be 300 mg l^{-1}, the reduction in the organic pollutants was found to be 30% for mechanical treatment, 95% for biological treatment without oxidation of the nitrogen and 95% for biological treatment with oxidation. As such it was concluded that the best ways for improving the quality of wastewater before using it for irrigating agricultural land are mechanical-biological treatment followed by treatment using Ozonation.

10.3.1.7 Egypt

The quality of agricultural drainage water in Egypt and its suitability for irrigation has been investigated and the results reported more than four decades ago (Ghaith et al, 1960; text in Arabic). Samples of water were taken monthly from May 1958 to April 1959 from several major drains in Upper, Middle and Lower Egypt. The results obtained have indicated that water from most of the investigated drains were suitable for irrigation, especially when the irrigated soil is light in texture such as silty loam, silt, sandy loam, etc. Unfortunately, the value of that investigation reduced very much after the construction of the High Aswan Dam in 1964, and the steady application of water conservancy measures.

The large size of agricultural land under irrigation in Egypt coupled with the availability of river water for irrigation and the moderate efficiency of irrigation brings the estimate of the drainage/irrigation ratio to no less than 40%. A review of the irrigation and drainage quantities of water given by Hashem et al. (1967) indicates that the ratio of drainage water to the irrigation quantity for the Nile Delta area expressed as percent was 45, 48, 48, 51, 52, 48 and 46 for the years 1960, 61, 62, 63, 64, 65 and 66 respectively. Since 1966, the average for the period 1960–66, 48.6%, has become less with time as a result of reducing the losses inherent in conveyance and distribution of irrigation water.

It might be of interest to mention here that in the said reference it has been reported that an amount of 0.2×10^9 m^3 of drainage water was used in 1964 for land irrigation in the Eastern part of the Nile Delta. Recently, the predicted reuse of agricultural drainage water will reach 4.6×10^9 m^3 y^{-1} by the year 2007 while the potential figure is 7.0×10^9 m^3 y^{-1} (Attia, 1996). Unfortunately, the criteria set for estimating the potential amount of reuse are not given rendering the annual potential as 7×10^9 m^3 questionable. Table 2 gives detailed figures of river water supply, drainage water quantities and salinities, and amounts of reuse for the eastern part of the Nile Delta during the period 1984–1988 (Roest et al., 1993). From thiat Table it is clear that for the period of study, 1984–88, the ratio of drainage to the Nile water supply ranges between 35.1% and 37.8%, with an average of 36.3%. The reuse as percentage of the drainage water for the same period falls in the range of 17.9–21.6, with an average of 19.8%.

Visser et al. (1993) investigated the effect of reuse on irrigation efficiencies in the Eastern part of the Nile Delta. The investigation was carried out using a Simulation Water Management in the Arab Republic of Egypt (SIWARE) model. The model package has been calibrated using data from the Eastern Nile Delta in 19886. Subsequently, the model was validated based on data from 1984 till 1988. Figure 6 is a schematic illustration of the model, sub-models and in-and output.

The results obtained show that with reuse (A) the average overall efficiency is slightly better than without reuse (B). The average improvement, i.e. (A-B)/B, was found to be about 5.7%. Most of the improvement was from the field application efficiency while conveyance and distribution efficiencies hardly showed any improvement. Oppositely, the average soil salinity for the period 1984–88 increased from 2.91 mmhos cm^{-1} (B) to 3.04 mmhos cm^{-1} (A), i.e. an average increase in soil salinity of about 4.5%. Generally speaking the overall irrigation efficiency in the Eastern part of the Delta is high compared to the efficiencies in the Western and Central parts.

In addition to the extensive, open drains network, a considerable surface in Lower Egypt is equipped with subsurface drainage, basically as a means of saving the land, which otherwise will be consumed by excavating open field drains. The subsurface

Table 2. Nile water supply to the Eastern Nile Delta, and annual drains discharge and salinity for the period 1984–88 (from Roest et al., 1993)

Year	Nile water supply, 10^6 m^3 y^{-1}	total 10^6 m^3 y^{-1}	Drainage water, reused 10^6 m^3 y^{-1}	salinity mol m^{-3} Cl	reuse as % of total	Drainage/ supply, %
1984	12.243	4.633	0.801	12.4	17.3	37.8
1985	11.969	4.355	0.804	14.3	18.5	36.4
1986	11.645	4.281	0.925	13.5	21.6	36.8
1987	11.249	3.948	0.814	13.2	20.6	35.1
1988	10.322	3.652	0.652	16.1	17.9	35.4
Average	11.486	4.173	0.799	13.9	19.2	36.3

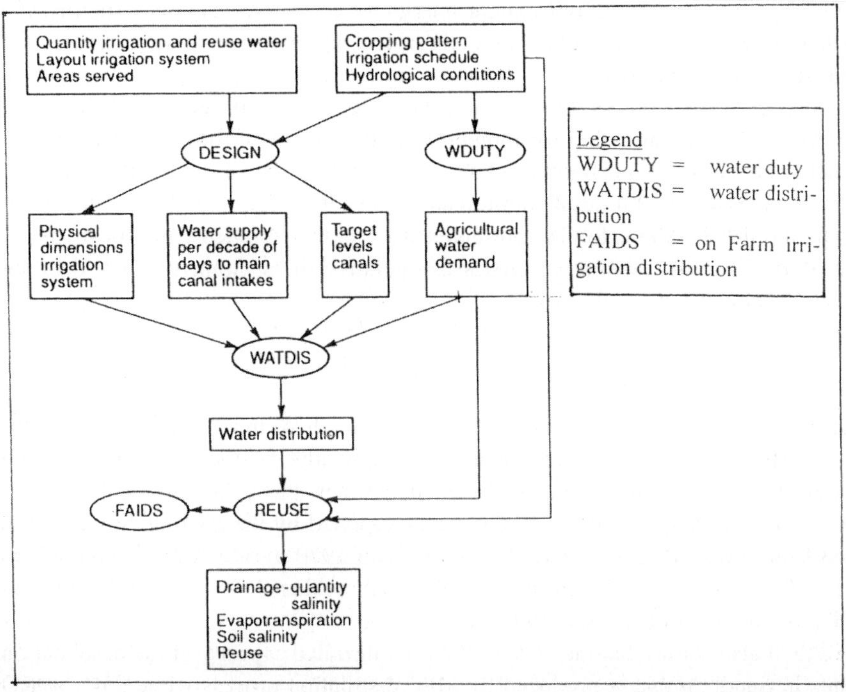

Figure 6. Schematisation of the SIWARE model, its submodels and in-and output (from Visser et al., 1993)

drainage rates are generally higher in summer than in winter at the north and centre of the Nile Delta area due to rice irrigation. Additionally, the salt concentration in the drainage water is higher in winter than in summer. Farther south, the drainage rate and water salinity remain almost the same during winter and summer. In summer time, the subsurface drainage rates corresponding to 10% non-exceedance probability of occurrence is 3.1, 2.0, 0.7 and 0.5 mm d^{-1} for northern Delta, middle Delta, southern Delta and upper Egypt respectively. The winter rates for the same areas in their respective order are 2.2, 0.7, NA (not available) and 0.5 mm d^{-1} (Abdel-Dayem et al., 1993). The figures obtained from that study as well as similar studies could be used as a guide to a more efficient design of subsurface drainage systems.

10.3.2 Desalinated water

The rapid growth of population and expansion of urban areas in most of the Arab countries make desalination of seawater and brackish water an attractive option for obtaining additional supplies of freshwater. This is especially true for water-short countries with substantial energy or economic resources such as the Arabian Gulf and North Africa where six of the top ten desalination countries are located.

The Gulf States Co-operative Council (GCC) Countries have always been world leaders in using desalination for municipal water supply. Two-thirds of the world installed capacity of desalination plants exist in these countries. Bushnak (1993, 1995) reviewed the total desalination plant's capacity sold to the Arab countries in the period 1963–1993. He also presented data of the installed capacity of plants in the GCC and the distillation processes operating them. Some of these data are given in Table 3.

10.3.2.1 Saudi Arabia

The installed capacity of desalination plants in Saudi Arabia equals to about 24% of the capacity of plants world-wide and about 50% of the GCC. Despite the increase of the cost of producing desalinated water, the volume annually desalinated is rising sharply with time. The water supplied by the distillation plants has risen from 4.6×10^6 gal d^{-1}(6×10^6 m^3 y^{-1}) in 1970 to 50×10^6 m^3 in 1980, to 331×10^6 gal d^{-1}(430×10^6 m^3 y^{-1}) in 1985 to more than 400×10^6 gal d^{-1} (around 520×10^6 m^3 y^{-1}) in the mid 1990s.

Table 3 also shows that the 70% of the total installed capacity of the desalination plants in Saudi Arabia is produced by MSF distillation process while 25% by RO process. The total desalinated water produced by these two processes represents 95% of the total amount of desalinated water, which is similar to water distillation in the other Gulf Countries. Hybrid systems of desalination are experienced in Saudi Arabia. The objective is to combine the different characteristics of each process productively. An example of such systems is the system used in Jeddah's desalination facility combining the MSF and RO processes together, aiming at reducing the overall unit cost of water. Some time later than the introduction of that system in Jeddah, the MSF facilities in Yanbu and Al-Khobar have been expanded by adding to them RO units. An additional advantage of combining the MSF and RO in dual purpose plants (plants producing desalinated water and electricity) is the relative ease of starting up or stopping an RO plant for certain operational aspects. The cost price of desalinated water as obtained for some plants is included in Table 28, Appendix II.

10.3.2.2 Kuwait

Distilled water is the major source of freshwater used for drinking and domestic purposes in Kuwait. The installed capacity of the plants operating by MSF process represents 95.5% of the total installed capacity. Together with those plants working by the RO process, they represent some 99.2% of the total installed capacity in Kuwait. The heavily growing demand for water has led to the steep rise of the installed capacity of desalination plants from 0.0075×10^6 m^3 y^{-1} in the year

Table 3. Total and installed capacities of desalination units in the Arab Region with emphasis on desalination plants in the GCC ((Gulf Council Countries), Sources: Bushnak, 1993 and 1995, and Wangnick, 1994))

Country	Capacity, $m^3\ d^{-1}$ Total*	Installed**	% ° of WW⁺	GCC⁺⁺	MSF°°, $m^3\ d^{-1}$	% of WW	GCC	RO°°, $m^3\ d^{-1}$	% of WW	GCC	MED°°, $m^3\ d^{-1}$	% of WW	GCC	Others°°° $m^3\ d^{-1}$	% of WW	GCC
Mauritania	4,654															
Morocco	15,325															
Algeria	204,312															
Tunisia	50,914															
Libya	766,750															
Egypt	87,044															
Sudan	1,776															
Djibouti	404															
Somalia	408															
Syria	7,703															
Jordan	8,445															
Iraq	333,093															
Kuwait	1,523,210	1,413,610	8.9	18.3	1,350,514	16.9	21.7	51,783	1.0	4.2	4,904	0.56	12.8	6,408	0.42	2.8
Bahrain	315,197	297,841	2.5	3.9	160,820	2.0	2.6	119,343	2.3	9.6	1,175	0.13	3.0	16,503	1.08	7.2
Qatar	562,074	398,189	2.6	5.2	386,025	4.8	6.2	4,715	0.1	0.3	3,642	0.41	9.4	3,787	0.24	1.6
S. Arabia	5,020,324	3,800,029	24.0	49.1	2,678,135	33.5	43.1	954,009	18.8	76.8	16,456	1.87	42.6	151,429	9.87	65.8
UAE	2,081,091	1,655,157	10.5	21.4	1,503,745	18,8	24.2	93,692	1.8	7.5	8,266	0.94	21.4	49,454	3.23	21.5
Oman	162,096	160,581	1.0	2.1	134,645	1.7	2.2	19,336	0.4	1.6	4,200	0.47	10.8	2,421	0.16	1.1
Yemen	37,188															
Total (GCC)		7,725,407	49.5	100.0	6,213,884	77.7	100.0	1,242,878	24.4	100.0	38,643	4.38	100.0	230,002	15.0	100
Total worldwide		15,582,443	100.0		7,994,244	100.0		5,080,207	100.0		882,254	100.0		1,532,744	100.0	

* *Total = potential (1994 IDA, Wangnick),* ** *Installed = actual (Bushnak, 1993),* ° *= related to installed capacity,* °° *MSF = multi-stage flash,* °° *RO = reverse osmosis,* °° *MED = multi-effect distillation,* °°° *Others = electrodialysis, vapour compression and other technologies,* ⁺ *WW = worldwide*

1957 to 0.225×10^6 m^3 y^{-1} in 1975, 0.79×10^6 m^3 y^{-1} in 1985 and further to 1.413×10^6 m^3 y^{-1} in 1995. Table 3 indicates that Kuwait ranks number 3 among the GCC in production of desalinated water.

The distillation plants in Kuwait are located at Al Shuwaikh, Shuaiba, southern Zour and Doha. As a general rule, lightly saline groundwater, 8% by volume, is added to the desalinated water to bring the mixture to the international standards of drinking water. Lightly saline groundwater, as shown later, contains T.D.S. in the range 1,000–10,000 ppm.

10.3.2.3 Bahrain

In 1976 two seawater distillation plants with a total capacity of 22,730 m^3 d^{-1} were put into operation. The production of the plants was basically supplied to those areas that were badly affected by the deterioration in groundwater salinity, e.g. the middle region of the country. The pressure caused by the lack of suitable groundwater forced the State of Qatar to increase the number of distillation plants to produce a total of about 114,000 m^3 d^{-1} around 1985 and further to increase the production to about 300,000 m^3 d^{-1}(110×10^6 m^3) by 1995. The MSF operated plants produce 54.0% of the total output while the RO plants produce 40.1%. The remaining 5.9% is obtained from MED and other types of plants. According to Zubari et al (1994) the three desalination plants constructed in the period 1975–93 at the eastern coast of Bahrain have a total output capacity of 65×10^6 m^3 y^{-1}. The annual volume of desalinated water by the time of writing these lines is expected to exceed 146×10^6 m^3.

10.3.2.4 Qatar

The Country Report of the State of Qatar to the Workshop on Water Resources in the Arab Region and their Utilisation (1986) stated that groundwater resources in the country have been fully exhausted and there is hardly any way to develop them any further. Efforts have to be directed towards treatment of wastewater and desalination of seawater to meet the demands of water for domestic, industrial and agricultural purposes.

Desalination of seawater in Qatar began in 1953 when a small plant with a capacity of 680 m^3 d^{-1} was installed at Doha to supply water for its domestic purposes. Since then, the annual capacity of desalination units increased gradually to reach 264,000 m^3 d^{-1} around mid 1980s and about 400,000 m^3 d^{-1} around mid 1990s. The consumption of the desalinated water constitutes more than 70% of the total capacity of the plants. All major plants, which are operated by the Ministry of Electricity and Water, work with the multi-stage flash (MSF) distillation process. Table 3 shows that 97% out of the total installed capacity of 398,189 m^3 d^{-1} belong to MSF plants and 1.2% only to reverse osmosis (RO) plants. It is planned that future expansion of desalination plants in Qatar shall take place in the north, northwest and northeast of the country.

10.3.2.5 United Arab Emirates (UAE)

Groundwater used to be and still is considered to be the main source of water supply in the Emirates. Due to the thrust of the increasing demand of water, the country began to install seawater distillation plants so as to fill in the gap between demand and supply. The UAE is for quite sometime occupying the second rank among the GCC in the annual amount of desalinated water. According to the Country Report on the water resources of the UAE (1986), there are 5 plants in Abu-Dhabi, 2 in Dubbai and 1 in Esh-Sharja, with a combined, annual production capacity of about 230×10^6 m^3 y^{-1}. This figure is about 15% larger than the total capacity listed in Table 3. From the data in the said Table it is clear that 90.85% and 5.65% of the installed capacity belongs to plants run by the MSF process and RO processes respectively. The survey given by Al-Sajwani & Lawrence 1995 gives the annual capacity of all plants as 1.678×10^9 m^3, a close figure to the installed capacity listed in Table 3. Table 29, Appendix II, gives detailed information of these plants while Figure 7 shows their locations.

10.3.2.6 Oman

The capacity of the MSF distillation plants in Oman produces from less than 100 m^3 d^{-1} to 132,000 m^3 d^{-1} at Ghubrah. Distillation plants are distributed all

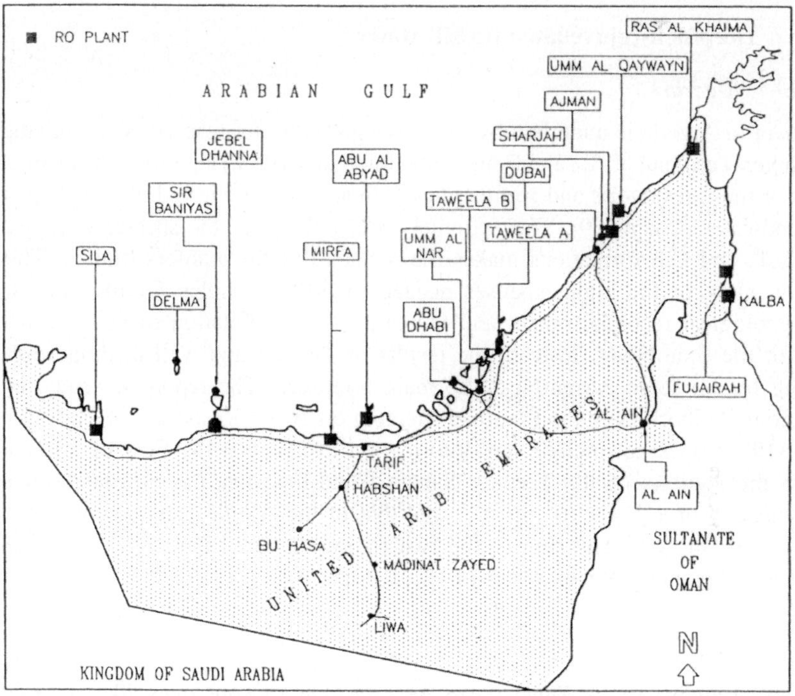

Figure 7. Locations of the major state-owned distillation plants in U.A.E. (Al-Sajwani & Lawrence, 1995)

over the country in 15 locations. All locations, except location No. 12, are situated along the coasts of the Gulf and the Arabian Sea in Oman. The exception, No. 12, is located inland and desalinates brackish water using the RO process. The Ghubrah units provide Muscat and certain parts of northern Oman with power and water. The initial capacity of 22,700 m^3 d^{-1} in 1975–76 rose to 77,300 in 1985–86 and further to 131,900 in 1992–93 and 159,200 m^3 d^{-1} in 1995–96 (Al-Sajwani et al., 1995).

The Ghubrah plants, the largest in the country with capacity equalling to 82.5% of the capacity of all distillation plants in Oman, are installed at five locations all along the coast as can be seen from the map in Figure 8. Besides Table 30, Appendix II, contains some detailed information about the distillation plants in Oman.

10.3.2.7 Palestine

Desalination has been for quite sometime considered as a serious potential source for improving the quality and increasing the quantity of the country's limited water resources. Since Gaza and the West Bank do not have, at least for the time being, oil or natural gas to serve as a source of energy, it is possible to exploit the difference in head between either the Red Sea or the Mediterranean Sea and the Dead Sea to generate power that can be used in desalinating seawater.

10.3.3 Deep/non-rejuvenated (fossil) water

10.3.3.1 Algeria

The Sahara in Algeria is underlain by extensive groundwater reservoirs comprising the Complex Terminal (C.T.) and Continental Intercalary (C.I.) aquifers. The characteristics of these reservoirs and aquifers have already been reviewed in the Chapter on groundwater (Chapter 9). The extremely arid climate of the surface occupied by the C.T. and C.I. formations makes the recharge of the aquifers limited. This is also in view of the high expenses needed to drill boreholes for the purpose of hydrgeological reconnaissance, and thereafter the exploitation of aquifers not urgent for the time being. Instead, the results of logging and well drilling while searching for oil serve currently as the main reference. The report presented by the Algerian Authorities to the Regional Meeting on Water Resources in The Arab Region (1983) does not include any information except that the flow of groundwater in the aquifers of the Tertiary Sahara (sand-limestone) is artesian/normal with a capacity of 1.474×10^9 m^3 y^{-1}, and that of the Albine Sahara (sandstone) is 0.984×10^9 m^3 y^{-1}. The sum of the two figures, 2.458 10^9 m^3 y^{-1}, has been reduced in the Country Report (1986) to 2.210^9 m^3 y^{-1}.

10.3.3.2 Tunisia

The deep groundwater in Tunisia can be found in certain areas like west of Gabes, Chott el-Fijaj, Nefzawa and Jerid. The depth to water exceeds 1,000 m and reaches 2,000 m in some parts of Tunisia. A small part of the flow out of the reservoir underlying the southern part of the Western Sahara (surface occupied by reservoir is 600,000 km^2) goes to the southwest and the larger part flows to the northeast.

The salinity of water varies from place to place; 2,500 mg l^{-1} at Chott el-Fijaj, Nefzawa and Jerid and 5,000 mg l^{-1} in the district of Burma. Groundwater is confined and well production is 300 l s^{-1}, with pressure around 10 bars and temperature of 70 °C.

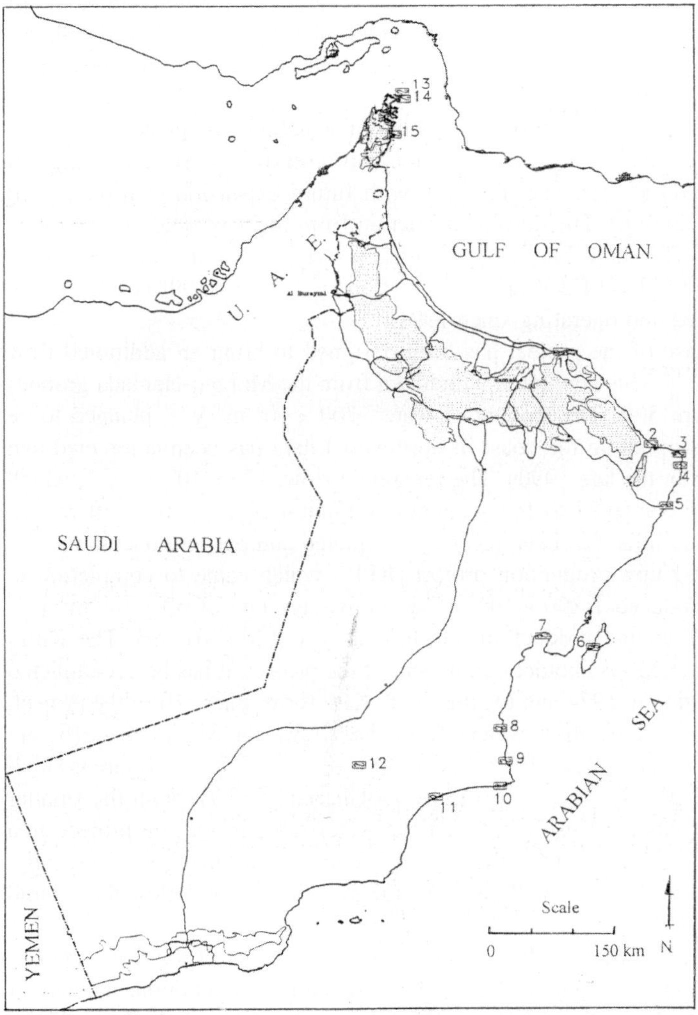

Figure 8. Locations of the desalination plants in Oman (Al-Sajwani & Lawrence, 1995)

10.3.3.3 Libya

In the 1960s as oil exploration moved further down south into the Libyan Desert in a search for new oil fields, the drilling rigs revealed the presence of immense groundwater reserves. The rainfall together with the production of the then existing water wells were not enough to cope with the growing demand for water by the inhabitants, most of whom are living along the fertile coastal strip. These conditions motivated the Libyan authorities to allocate part of the oil revenue to dig deep wells in the southern part of the country and convey the outflow through appropriate conduits to the coastal strip in the north. The project is called the Great Man-made River (GMmR).

The GMmR project is entrusted with conveying considerable amounts of good-quality, deep groundwater across thousands of kilometres of desert plains, mountains, hills, sabkhas and wadis to the inhabitants for their domestic use, agriculture and industries.

The implementation of the project consists of a number of phases. The first phase has been designed to carry water to the end reservoirs at Sirt and Benghazi from the well fields at Sarir and Tazerbo, with future expansion to a huge well field at Kufrah (Figure 6). The design abstraction from the well field at Tazerbo is 1.0×10^6 m^3 d^{-1}. The same daily design abstraction applies to the Sarir well field, i.e. the total yield of phase I is nearly 700×10^6 m^3 y^{-1}. Almost 99% of this phase has been completed and operating since 1991.

The second phase of the project has been designed to bring an additional flow of 875×10^6 m^3 y^{-1}. This amount is withdrawn from the Murzuq-Hamada groundwater basin (Figure 5(a). The amount of water, 700×10^6 m^3 y^{-1}, planned to be conveyed to the strip along the coast in northwest Libya has been completed and put into operation in the late 1990s. The remaining flow, 175×10^6 m^3 y^{-1}, which is supposed to be conveyed to the scattered communities along the nortwestern mountain range and hills, has been recently completed and put into operation.

Since 1960 the Kufra production project (KPP), which came to completion in 1968, till 1990 produced $1,800 \times 10^6$ m^3, at an average rate of 60×10^6 m^3 y^{-1}. As from 1987 the annual production rate has risen to 120×10^6 m^3. The Kufra settlement project (KSP) is another component of the project. It has been completed and put into operation in 1974 and together with KPP forms phase III of the project. The two well fields draw their waters from the Kufra aquifer (Figure 5(b) are supposed to yield together 0.6×10^6 m^3 d^{-1}(220×10^6 m^3 y^{-1}). This figure is small compared to the 560×10^6 m^3 y^{-1} given by Al-Ghariani (1997). With the smaller abstraction rate it has been estimated that the water table inside the project area will fall down by 5-13 m y^{-1} after the project becomes operational. Consequently, the pumping level in the wells will be in the range of 70–150 m below the ground surface.

Phase IV aims at connecting the western branch of phase I of the Sirte pipeline with the eastern pipeline branch of phase II and pump an additional amount of 350×10^6 m^3 y^{-1} to the north western region of the coastal strip and the Jefara Plain. To the best of the author's knowledge, there are as yet no measures undertaken concerning the implementation of this phase. It is likely that the economic sanctions imposed on the country during the last twenty-five years are behind the delay encountering this phase. Phase V comprises the extension of the eastern branch of phase I to the city of Tubruq in the extremity of the eastern region of the coastal strip. The aim of this phase is to provide about 200×10^6 m^3 y^{-1} for the municipal and industrial water uses by the dwellings along the route. Upon the completion of the project, the five phases together will transfer and redistribute a total volume of 2×10^9 m^3 y^{-1}.

Before closing this sub-section it should be borne in mind that the GMmR project comes under the titles of withdrawing deep groundwater and over-land transfer of huge quantities of water from place to place within the same one country. Beside the above-mentioned description, more information about the geology and hydrogeology, including number, depths and yields of wells, can be found in Chapters 2 and 9 respectively. Some information related to the GMmR project as a water-transfer project are given in sec. 10.4.

10.3.3.4 The Nubian sandstone in the western desert of Egypt and in the Sudan

The water levels in the various wells scattered in the oases far south of the coastal strip of the Mediterranean Sea suggest that they belong to one huge body of water. The geology and hydrogeology of this body, which is often referred to as the Nubian Sandstone reservoir, were investigated by a number of several research authorities as early as the beginning of the 20th century till present.

In order to check the performance of the Sandstone body under different climatic conditions, a model (AQUIFEM) was designed to represent a section extending from the Ennedi Highlands of north western Sudan to the to the Nile Delta, Egypt, passing through the Farafra and Bahariya Oases. The model thickness of sand under the desert surface in northern Sudan is constant. In Egypt, the basement dips from the southwest to the northeast. The thickness of sand increases from 500 m near the Kufra Oasis in Libya to 2,100 m in Farafra and Bahariya Oases and thins to 900 m at Wadi el-Natrun close to the Nile Delta. The coefficient of permeability adopted in the model was 6 m d^{-1}, the storativity 1.0×10^{-5} and the specific yield almost 0.10.

Three major mechanisms have been hypothesised and each of them tested. The backbones of each hypothesis were the depth and areal extent of recharge. The conclusions drawn from that investigation are as follows:

i- Recharge from the southern highlands can easily maintain the present piezometric head.

ii- Recharge from past fluvial periods is a 'plausible mechanism'.

iii- Current precipitation as a mechanism is unlikely to maintain groundwater levels by itself. It is probable that the groundwater in the Nubian sandstone is a combination of these mechanisms and others not taken into consideration (Chow & Wilson, 1981).

In the period 1984–87, a joint Egyptian-German project aiming at investigating the origin of groundwater in the Nubian Sandstone was carried out. The principal conclusions drawn from that investigation can be summarised as follows (Brinkmann et al., 1987):

i- Groundwater was recharged over large surface in the whole Nubian Sandstone system up to the end of the last pluvial episode some 8,000 years ago. Since then, infiltration that used to support the equilibrium condition stopped.

ii- Paleo recharge also occurred in the northern confined part of the aquifer. The old age of groundwater in the intermediate water-bearing formations can be explained by the low velocity of flow.

iii- The water balance for steady and transient states show a deficit in the confined part of the aquifer. The deficit is made up by the regional flow from north to south.

iv- While the aquifer system can be kept filled by just a few millimetres recharge every year, some millimetres (5–8) are needed each day for the purpose of land irrigation. So, if the latter amount has to be supplied for irrigating large surfaces of desert land, the groundwater level would drop then by 15–20 m annually. As such, the only possibility is to limit irrigation to selected areas of small size such as the Kharga, Dakhla Oases in Egypt and Kufra Oasis in Libya.

The Nubian sandstone in the Sudan covers a total surface ranging from 600,000 to 700,000 km^2. This area comprises six basins; the desert basin of the Nile, Nubian basin, Central Darfur, An-Nahoud, Saq el-Na'ām and River Atbara basins. The estimated storage capacity is 16.3×10^9 m^3 and the annual recharge between 42×10^6 and 257×10^6 m^3. The total withdrawal from the entire basin is in the order of magnitude of 19×10^6 m^3 y^{-1}, i.e. 7–45% of the annual recharge. The majority of groundwater extracted from the water-bearing formations in the Sudan comes from sedimentary basins where the depth to groundwater is rather shallow. One should not forget that the basic rocks and the volcanic rocks together cover almost 52% of the surface area of the country. With the exception of fissured and fractured rocks, it is hard to extract water of good quality from these rock groups. The total dissolved salts in the volcanic group have been estimated as 1,500–3,000 ppm (Country Report of the Sudan, 1986). Water with as high TDS as such is of limited usefulness.

Figure 9 is a map showing the groundwater heads in the Nubian Sandstone in the Sudan with small areas of Egypt and Libya.

10.3.3.5 Jordan

Groundwater is the main source of water in Jordan. It accounts for 70% of the total water budget of the country. Groundwater resources can be divided into two types; renewable and non-renewable. Studies have indicated that the safe yield of the renewable resources is 275×10^6 m^3 y^{-1}. In reality, Jordan extracts almost 150% of the safe yield. In attempt to bring the demand and supply in balance, Jordan is currently withdrawing groundwater from non-renewable (fossil) sources at a rate of 170×10^6 m^3 y^{-1}, mostly for agricultural purposes. Since the groundwater of most

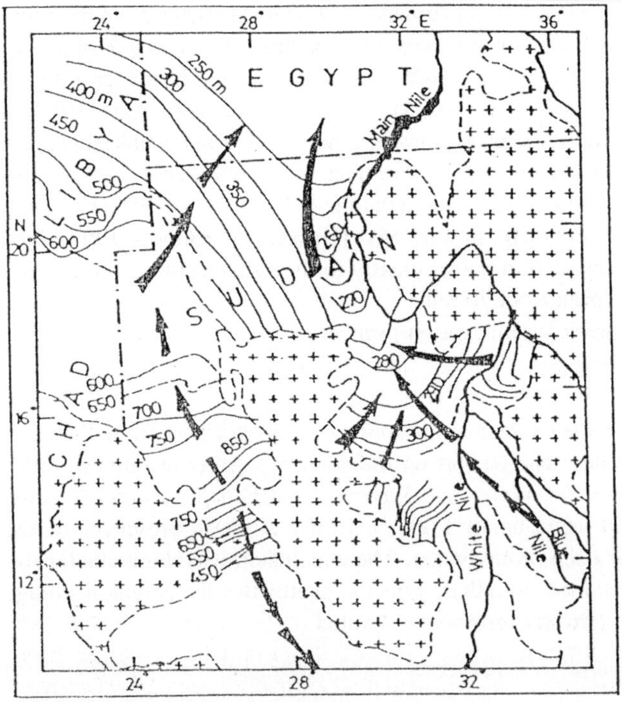

Figure 9. Groundwater heads in the Nubian Sandstone in the Sudan and small parts of Egypt and Libya (Country Report of the Sudan/Hedayet, 1986)

of the shallow aquifers is polluted, there is a growing tendency to withdraw more water from the deep aquifers. However, the cost price of water withdrawn from the fossil sources remains a limiting factor (Qaisi, 1995).

10.3.3.6 Saudi Arabia

The Saq and Tabuk formations in northern Saudi Arabia are the major groundwater systems. Both are deep aquifers. The major deep aquifers in Central Saudi Arabia are the Minjur and Biyadh-Wassia formations. In the eastern region of the country, Umm er-Radhuma and the Dammam are the most important formations. The thickness of the Umm er-Radhuma varies between 240 m and 700 m.

The water in the water-bearing formations can be subdivided into renewable and non-renewable. The renewable formations are post-Jurassic, characterised by the usual rejuvenation, downward flow to the underlying layers as well as drainage systems. The non-renewable formations belong to the pre-Cretaceous époque. The fossil water that filled the non-rejuvenated formations date back to the Pleistocene interglacial periods. These formations have neither drainage systems nor regular patterns of downward flow. As such they do not enjoy any rejuvenation. In the renewable formations it is possible to withdraw groundwater from the downward leakage, from the stored water or replenishment. The water abstracted from the deep, non-renewable formations is drawn from the stored water, e.g. the water supply to the city of Riyadh which is drawn from Al-Menjur water-bearing formation (UNDP Report Series # 9, Groundwater in Eastern Mediterranean and West Asia, 1982).

The 1982 volume of groundwater abstraction from the non-renewable resources reached 3.45×10^9 m^3, almost three times as much as the renewable groundwater (Farooq & Al-Lyla, 1985). This volume jumped to 6.48×10^9 m^3 in 1985 (Ukayli & Hussein, 1988) and to 13.48×10^9 m^3 in 1990 (Hussein & Ahmed, 1997). The sharp rise of the annual abstraction points clearly to the satisfaction of the Saudi's with groundwater and its preference to reuse of treated wastewater and desalinated water, especially for domestic and irrigation purposes.

10.3.3.7 United Arab Emirates

There is hardly any published information about water extraction from the deep groundwater in the Emirates. The Report on the country's water resources (1986) mentions that the Ministry for Water and Electricity carried out a large scheme entailing drilling of deep wells up to 1,000 m in the northern provinces of the country to collect data of deep groundwater. The conclusion was unsatisfactory as the preliminary results obtained from that project indicated that the potential storage capacity of the deep groundwater reservoir is limited.

10.3.3.8 Oman

It has been recently ascertained that old or non-rejuvenated groundwater in Oman does not exist except in the southern part of the country in the upper and lower Dammam, Rus and Umm er-Radhuma. The potentials of these formations, however, are not yet known with high accuracy. Some areas such as Nejd are currently under investigation. It is hoped that the final results of the investigation are satisfactory. One should not forget that the potentials of these formations in neighbouring countries like the United Arab Emirates and Saudi Arabia are substantial, and both countries abstract enormous amounts of water from them.

Besides, the preliminary results so far available indicate that the Umm er-Radhuma formation is the most promising one. The lower sub-formation appears to yield the best quality of water. The upper Umm er-Radhuma and the overlying Rus formations have about the same specific yield but the water quality is inferior, with salinity above 3,000 mmhos cm^{-1} which can be used for irrigating salt-tolerant crops.

It has been confirmed that strongly brackish groundwater exists in certain areas of Oman. However, due to lack of appropriate technology needed for using this water, the necessary investigation is currently out of the question.

10.3.4 Water import/transport

Problems associated with water scarcity in the Arab Region is a major factor underlying the conflicts among the countries of the region as well as between these countries and some of the surrounding countries from outside the region. An example of this can be found in the conflicts between Syria and Iraq on one side and Turkey on the other side and between Syria and Iraq, all connected with the Euphrates water. A much worse example is the conflict between Syria, Lebanon, Jordan and Palestine and Israel.

In sub-section 10.2.4 we have discussed the issue of transporting water from one location to another, both within the same one country, and from a water-exporting country to a water importing country. The former is done over-land while the latter is often done by sea.

10.3.4.1 Libya (The Great Man-made River)

The hydrogeological aspects of the GMmR project, Libya, have already been highlighted in sub-section 10.3.3.3. In the present sub-section we shall briefly review the water transport aspects.

It has already been mentioned that the objective of the project is to carry water to Sirte and Benghazi along the seacoast in the north from well fields located at Sarir, Tazerboo and Kufra in the south. The ground surface along the route of the pipeline falls steadily from 270 m a.m.s.l. to a low-lying point, just about the sea level, at Jalu and Maradah. From this low point the conduit rises gently across a limestone plain till the elevation of 95 m a.m.s.l. south of Ajdabiya, where a holding reservoir is sited. From Ajdabiya one branch of the pipeline continues north, over undulating, rocky ground crossed by wadis, to Benghazi; and a second branch turns west to Sirte over old beach deposits and numerous wadis draining into the sea.

In the western part of Libya, water from beneath Fezzan Desert is to be conveyed to Jefara Plain. Well-fields are located and developed in the Sarir-Qattush area as well as in other areas northeast of Jebel Fazzan. From Sarir Qattush the pipeline rises steeply to the Hamada Al-Hamra Plateau. After pumping water to a high point near Jebel Al-Sawda, the pipeline descends as it continues its route to the north. Beyond that it enters the gently rising Hamada Al-Hamra gravel plain, where wadis flowing northeast and draining into the sea frequently dissect the underlying limestone deposits and volcanic sheets of lava. The pipeline runs through the eastern flank of the Hamada Al-Hamra Plateau for about 300 km. The surface changes gradually from a gravel plain in the south to a thin sand cover in the north where the network of wadis at Suf al-Jin, Zamzam and Bani-Walid becomes denser. North of Suf al-Jin, the land rises more steeply to Jabal Nefussa with increasing vegetation cover and improved grazing land. As the pipeline proceeds northwards its route becomes torturous, crossing ever more deeply incised wadis. The crossings have been specially designed to protect the pipeline from the force of flash floods, which are known to sweep down the wadis following intense rain in the hills. Jabal Nefussa rises to 400 m above the Jefara Plain to the north. The crossing here has been designed on the basis of constructing a tunnel, and a hydropower plant incorporated at the foot of the north escarpment. The plant generates electricity as water flowing in the tunnel descends the escarpment to the Jefara Plain (the Management and Implementation Authority of the Great Man-made River, 1989).

10.3.4.2 Suggestions for Palestine and its neighbouring countries

In 1987, Late President T. Özal of Turkey suggested that a twin 'Peace pipeline' might be built from Seyhan and Ceyhan Rivers in southern Turkey to Ash-Sharja (east) in the United Arab Emirates and to Jeddah (west) on the Red Sea in Saudi Arabia. Such pipelines would carry 1.28×10^9 m^3 y^{-1} in the west and 0.9×10^9 m^3 y^{-1} in the east at one-third the cost of the same amounts of desalinated seawater. Recent cost estimates have indicated some change in the balance. Due to historical reasons added to the fears that the pipelines could easily be damaged by anyone from the Arab or non-Arab States or groups of individuals, similar to what is happening currently to the oil pipelines in Iraq, there was no public appreciation to that suggestion.

Kolars (1991) suggested that a Mini Peace Pipeline be built as far as Jordan, and that such a line draws its water from the Goksu or Manavgat Rivers West of the Seyhan and Ceyhan Rivers. This suggestion was debated between water resources experts from Jordan, Palestine and Israel. (Shuval (1992) suggested an even shorter pipeline, which is meant to supply water from Turkey as far south as southern Syria. An advantage of this suggestion is that the number of parties involved, Turkey and Syria, is less than any other suggestion thus rendering the necessary negotiations shorter and less cumbersome. Secondly, the length of the pipeline being short compared to those in other suggestions makes the cost less expensive. It can also improve the withdrawal of the waters of the Yarmouk River by Jordan and the West Bank.

Under the current conditions, a water tanker needs a total time for a single trip between Turkey and Israel or Gaza of about six days. Using gigantic, commercial bags with a capacity of 1.75×10^6 m^3 each, water can be transported from Turkey to Israel or Palestine in 6.3 days and return back to the same port after unloading its contents in 2.1 days. This means a total travel time of 8.4 days not including at least 2 days for unloading. As such, one bag needs almost 11 days to get loaded, transport the load, unload its consignment (average rate of 15,910 m^3 d^{-1}) and sail back to its port of embarkation. It is very unfortunate that the current circumstances in the Arab Middle East Region are not favourable to implement any of the above-mentioned suggestions.

Haddad & Mizyed (2004) carried out an evaluation non-conventional (import and desalination of water) water development options under prevailing conditions in the Middle East Region. The relevant results obtained from that evaluation are included in Table 4. The important conclusion to be drawn from this Table is that desalination of seawater and brackish water presents the best option. The second best is the transfer of water by sea using new tankers while import of water by land presents the worst option.

10.3.5 Water harvesting

The next paragraphs highlight the experiences gained by some Arab countries in the field of water harvesting through a variety of technologies.

Table 4. Evaluation of non-conconventional (import and desalination of water) water development options under prevailing conditions in the Middle East Region (from Haddad & Mizyed, 2004)

Option	Net cost, US$/ m^3	Evaluation criteria° Environ-mental	Political/ Legal	Economic/ Financial	Technical/ Technological	Total
Import						
by land						
Land	1.50–2.16	3	1	1	3	8
by sea						
N* tankers	0.73–1.36	4	4	3	4	15
O** tankers	0.73–1.36	2	4	2	3	11
Bags	1.00–6.00	4	4	1	2	11
Desalination						
seawater	0.84–1.70	4	5	3	5	17
brackish Water	0.30–1.00	4	5	3	5	17

Explanation
** N = new and** O= old.*
° Criteria: 5 = Excellent / highly feasible; 4 = good / feasible; 3 = Average feasibility; 2 = poor /questionable feasibility; and 1 = very poor / infeasible

10.3.5.1 Water harvesting in Jordan

This sub-section reviews briefly the Jordanian experience in rainfall-runoff water harvesting (RRWH) through efficient collection of runoff generated by rainfall on building roofs in residential, industrial and commercial areas in Jordan in general and the Greater Amman area in particular. The survey carried out in 1990 has shown that the number of households per donum (1, 000 m^2) in Jordan ranges from 1.8–2.7 for zone A, 2.7–3.6 for zone B, 4.3–4.8 for Zone C and 6.5–7.2 for Zone D. The corresponding roof area in percent of the total area of the building is taken as 30, 35, 40 and 44 for zones A, B, C and D respectively.

Assuming that the annual rainfall in the Greater Amman area to be 400 mm y^{-1}, the estimated gross yields from residential building roofs were given by Preul (1994) as:

		Yield	
Year	*m^3 y^{-1}*	*m^3 y^{-1}/HH*	*l d^{-1}/cap*
1985	*5,018,700*	*34.682*	*15*
1990	*6,365,000*	*35.850*	*16*
2005	*13,068,800*	*38.052*	*18*

To get an insight of the gain achieved through runoff collection before it enters the normal storm water runoff process, thereby avoiding certain losses, consider the estimated runoff coefficient for roofs = 0.8 and the estimated runoff coefficient for general urban runoff = 0.5, the gain is $(0.8 - 0.5)/0.5 = 60\%$. The estimate for the entire surface of Jordan is around 70% and for the Greater Amman area 60%. Further gains can be realised through RRWH in commercial and industrial areas, including roofs and parking lots. A preliminary estimate of the gross RRWH yields from commercial and industrial roofs and impervious lots in Greater Amman is about 4×10^6 m^3 y^{-1} for the year 1990, expected to rise by about 10% to reach 4.35×10^6 m^3 y^{-1} by 2005 (Preul, 1994).

Figure 10(a) represents the cross section of a runoff harvesting facility for residential and commercial buildings while Figure 10(b) is a facility for commercial and industrial buildings, both in Jordan.

It is noteworthy that Al-Kloub et al. (1995) have briefly reported on the pilot experiments carried out in Jordan using a cloud-seeding aircraft. The results obtained have shown an increase of almost 20% in the annual rainfall in an area of 8,000 km^2 in the northern part of the country. The corresponding gain is about 400×10^6 m^3 y^{-1} reaching the ground in that particular area. A certain portion of this gain flows to the neighbouring streams and wadis, another portion evaporates and the rest infiltrates the soil. Regrettably, no further information was ever disclosed about this experiment.

10.3.5.2 Fog collection in Oman

Weather modification was first thought of in 1946 and much has been learnt since then. The background is that fog particles, which by their nature, are so light that their fall velocities are negligible. Fog particles that come in contact with vegetation may adhere and coalesce with other droplets, and eventually form large drops

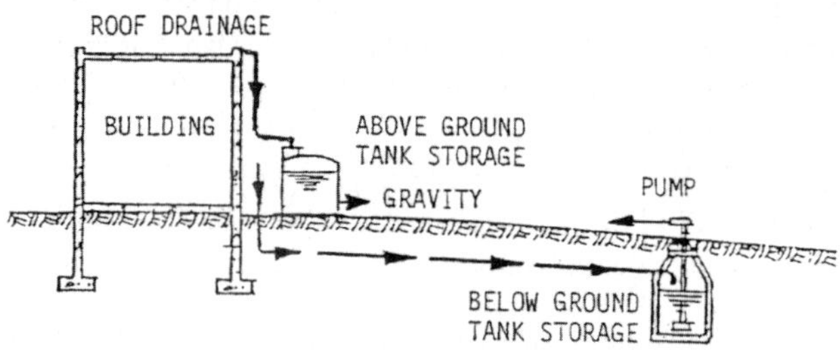

Figure 10(a). Roof water harvesting facility for residential or commercial buildings (from Preul, 1994)

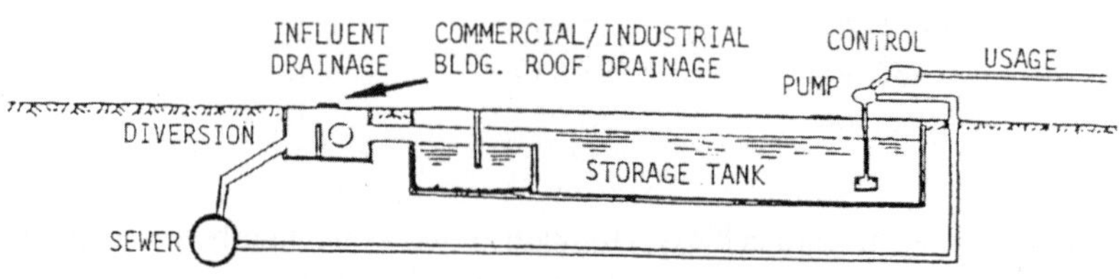

Figure 10(b). Roof drainage harvesting facility for commercial and industrial buildings (Preul, 1994)

enough to fall to the ground. Rich, water-bearing fogs and mists occur in a few places in the world. The area occupied by the Dhofar Mountains in the southern region of Oman is one of those places. The monsoon precipitation in Oman, different from any other precipitation, is composed of a vertical component in the form of light shower (drizzle) and a horizontal component in the form of fog moisture that is intercepted by vegetation such as shrubs and trees, native grass, etc. The latter represents an input to the catchment, which cannot be measured by the conventional rain gauges. This moisture does not help only to reduce evapotranspiration losses from soil and vegetation, but also may contribute to aquifer recharge by infiltration through the soil.

In studies conducted during the monsoon season at a certain location 865 m a.m.s.l. the amount of water collected was 2.86 $l\ m^{-2}\ h^{-1}$ at 4.2 m height, with a daily total of about 34.5 $l\ m^{-2}$ of mesh. Fog was collected using wire mesh collectors. According to Abdel-Rahman & Magid (1993), the amount of water collected is not substantial and can only provide an insignificant amount to a localised area in Oman for just a part of the year (locally known as Khareef, i.e. autumn, in which the number of rainy days with fog during July–October is 20–25). It would probably only benefit grazing cattle by supplying pools of drinking water.

Chebaane & Alesh (1995) contradicted that assessment on the grounds that fog water collected under a certain type of trees during the 1989 and 1990 seasons was in the range of 200-700 $l\ d^{-1}$.

An experiment was conducted at Jabal Qara, north of Salalah. It started in 1992 at a high elevation site (880 m a.m.s.l.) and in 1993 one foothill site (160 m a.m.s.l.) was added. Measurements were taken using three types of gauges; a conventional daily rain gauge, a Monsoon precipitation gauge with natural grass (savannah-type vegetation) and a Monsoon precipitation gauge with artificial type grass. The 1993 results of the high site showed that the ratio of fog water to the 'total precipitation' (vertical rainfall + horizontal fog water) varied between 48% and 70%, with an average of 67%, for natural grass and between 64% and 76%, with an average of 73%, for artificial grass. The results of the low site were much less, the ratio varied from a minimum of 6% to a maximum of 22%, with an average of 16% for the Khareef season.

The experiment of July 1993 extended to comprise the effect of wind speed on total precipitation as well as rainfall. The wind speed indicated a stronger effect on the total precipitation than the rainfall did. The wind speed generally increases the flow of fog moisture, but does not affect the horizontal precipitation in bare soils.

10.3.6 Virtual water

According to Allan (1999), countries of the MENA Region (most of the countries in this region also belong to the Arab Region) rely heavily on virtual water as a convenient solution to their water-scarcity problem. To produce one metric tonne of wheat, say, the farmer needs 1,000 m^3 of water. Importing a million tonnes of wheat is equivalent to importing $1 \times 10^9\ m^3$ of water. Allan (1999) added, "publicly available statistics clearly indicate the region's wheat imports are rising, and there is reason to believe that this trade will continue as countries seek seemingly inexpensive and readily available sources for dwindling water supplies". What is said about wheat can also be said about other commodities. The statistics gathered by the Food and Agriculture Organisation for 1994 (FAO, 1999) gives the estimates of virtual water contained in food imports in 1994 by some countries in the Arab Region as follows:

Country	*Water equivalent, $10^6\ m^3$*	*Country*	*Water equivalent, $10^6\ m^3$*
Algeria	*12.4*	*Mauritania*	−0.002
Bahrain	*0.68*	*Morocco*	2.4
Egypt	*18.2*	*Saudi Arabia*	*13.9*
Iraq	2.2	*Syria*	*1.0*
Jordan	*3.5*	*United Arab Emirates*	*3.4*
Kuwait	*2.8*	*Yemen*	*3.4*
Libya	3.3		

The above-listed figures show that Egypt followed by Saudi Arabia and Algeria are intensive virtual water importers. Contrarily, Mauritania, though to a very limited extent, is a virtual water exporter. In his discussion to the paper of Allan (1999), Shahin (1999) expressed his disagreement with the suggestion that the Arab countries should depend totally on virtual water to feed their inhabitants. It is more advisable to have a basket of solutions, not only one solution to depend on. It is true that complete agricultural self-sufficiency may well be impossible at least for some countries, yet water conservation and search for new resources should always receive the attention they deserve. Besides, one of the dangers of virtual water is that importing nations would be vulnerable in the event that grain-producing countries cut subsidies to their farmers, potentially leading to significantly higher prices. Furthermore, delaying or preventing export of virtual water commodities such as wheat can be used as a weapon in the event of political crises between exporting and importing countries similar to what used to happen between the United States and the former Soviet Union during the time of cold war (1960s up to 1980s).

10.4. ENVIRONMENTAL AND SOCIAL IMPACTS, AND SOME REMEDIAL MEASURES

The foregoing sections of this chapter address some of the technological aspects underlying the search for new water sources and their applications in the Arab region.

Next to the high unit price of having new or non-conventional water compared to the unit price of conventional water, at least in most parts of the region, the processes underlying those technologies might result in some detrimental impacts to the environment and adversely affect the interests of certain groups of the society. The major undesirable impacts will be briefly reviewed in this section.

10.4.1 Quality of treated wastewater

The quality of treated wastewater must be acceptable for the purpose of its use. High quality waters should be used, on a priority basis, for those uses requiring high-quality water. Appropriately lower quality waters should be used where high quality is not a requirement. As such, before embarking on wastewater reuse it is essential to specify the use(s) of recycled water and to judge its acceptability by the people, based on their cultural and religious values. A frequent complaint by the people from using not highly treated wastewater for watering public parking areas and sports facilities, and landscape irrigation is the discomfort they feel from the unpleasant odour left after using such water. This can be really disturbing in warm, arid climates.

It has been observed that the quality of reused water in Sana'a, Yemen, is not accepted for uncontrolled use in agriculture. The poor performance of the stabilisation ponds results in an effluent with still high concentrations of BOD and probably also high concentrations of faecal coliforms (Nozaily, 1992). Agricultural reuse of wastewater commonly results in an 'Actual Risk' to public health, especially in developing countries, if all of the following stages occur:

i- Either an infective dose of an excreted pathogen reaches an irrigation field or the pathogen multiplies in the field to form an infective dose.
ii- The infective dose reaches a human host.
iii- The human host becomes infected.
iv- The infection causes a disease or further transmission.

Certain survey research works in the areas of environmental toxicology and epidemiology can provide the interested reader with more information concerning health problems and wastewater use.

Despite the rapid growth of the water supply in Saudi Arabia to meet the demands for water, increase in wastewater reuse is rather limited and its percentage of total supply remains at about 1%. The Council of Leading Islamic Scholars (CLIS) strongly advised that *if there are negative impacts from its direct use on health, then it is better to avoid its use, not because it is impure but to avoid harming the human beings*. The CLIS also *prefers to avoid using it for drinking (as possible) to protect the health and not to contradict with human habits* (qutod from Abdel-Rahman, 2000)

In addition to the above-described impact there are many social, legal and institutional complexities that must be dealt with when initiating a wastewater use project. A three-stage approach of option evaluation and selection is recommended. This approach can be summarised as follows:

i- A large number of possible uses (options) for reclaimed water in the area to be served have to be identified. Identification should be followed by estimates of the initial feasibility, thus yielding a reduced set of options that would be worthy of further and more intensive study.
ii- For each option of the reduced set a thorough assessment of the technical, economic, social, legal environmental, political and institutional should be

carried out to enable the construction of a rating matrix so as to help in choosing two or three options for the final assessment;

iii- A great effort should spent to reach full and complete evaluation of this set of 2–3 options. Once the most preferred option has been selected; financing, construction and final implementation should proceed well (Water Polution Control Federation, 1983).

10.4.2 Availability of energy

As water distillation basically requires energy, the availability of energy sources constitutes the backbone water desalination needs. All Arab countries depending on desalinated water as a component of their freshwater supplies use oil and/or gas to operate desalination plants. There is hardly any attention to other sources of energy like hydropower and solar energy.

Using fossil fuels to run distillation plants increases the emission of Carbon dioxide gas thus intensifying the greenhouse process. This is not the case with solar energy or with hydropower, both are clean sources of energy. From Chapter 3, which deals with climate of the Arab region, one can easily observe that the region is generally privileged with clear skies and considerable amounts of solar radiation. Regrettably, there is hardly any comprehensive and long-lasting research carried out concerning these two sources. According to Hamid (cited in Gischler, 1979), the solar experimental set up installed on the roof of the Solar Energy Institute of the University of Khartoum, The Sudan, supplied 3.56 $l\ d^{-1}\ m^{-2}$. The costs when worked out on the basis of 10-years amortisation period were found negligible. This result was encouraging, and plans were laid down for constructing two-1000 m^2 pilot plants, one to be sited in Khartoum Province and the other in Kurdufan Province. Since then, no information has ever been disclosed about those plans leading to the conclusion that they were probably never implemented.

Experiments on desalination were conducted in the 1990s using the so-called distiller with film in capillary motion (DIFICAP) at Tripoli, Libya, on the Mediterranean Coast. A specific design of a multi-effect solar distiller was used. The distiller is equipped with a number of cells made up by spacing stainless steel plates (about 1 m^2 surface area, and 1.0 cm interspacing). It can be set up with vertical or inclined orientation, with or without reflectors, and with or without tracking. Obviously any combination of these elements will affect the yield of desalinated water. Figure 11 shows the DIFICAP system at vertical position. Misellati et al. (1997) gave the experimental results for a DIFICAP 3-cell unit using different combinations of the mentioned elements. The minimum productivity, 3.12 $m^3\ d^{-1}$, was obtained from a vertical system without any element added to it, and the maximum productivity, 5.5 $m^3\ d^{-1}$, was obtained from an inclined system with tracking and no reflectors.

With the exception of drawing some plans to use hydropower as a source of energy to operate any future desalination plants in Palestine, there is no real work done in this area in any Arab country. There are two major plans to implement such a scheme. The first is the Red-Dead Intersea scheme where water will be

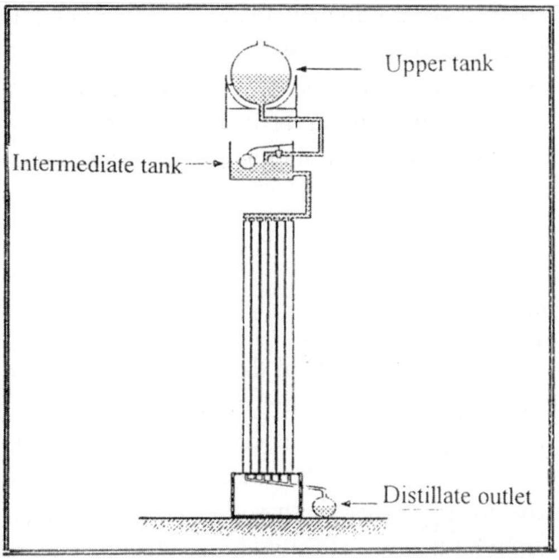

Figure 11. DIFICAP system placed in a vertical position (Misellati et al., 1997)

delivered through a canal taking its water from the Gulf of Aqaba to the Dead Sea. A desalination plant will be installed at the head race upstream the drop in elevation. Part of the diverted water will be desalinated while the rest will be dropped through turbines into the Dead Sea to generate hydropower needed to run the desalination plant. The alignment of the connecting canal will be selected in a way to minimise detrimental impacts on the environment such as pollution of groundwater. The second scheme is the Med-Dead Intersea scheme where water will be drawn from the Mediterranean Sea through an open canal for the first 20 km and a tunnel for the remaining 80 km. This scheme seems to affect adversely the groundwater aquifers, and thus needs to be constructed in a way to minimise such effect. Both schemes have high initial costs and require agreement, arrangement and coordination between the three parties involved (Israel, Jordan and Palestine).

In addition to the impacts already specified above, desalination of water might be detrimental to the environment depending on the possibility of disposal of the waste. This is not really a serious issue in the event of having the desalination plants located at or close to the seacoast. In this case desalination plants can easily discharge their waste into the sea or ocean, provided that every care should be undertaken to secure their mixing with the recipient water. Inland distillation plants do not enjoy this favourable situation, and the disposal of their waste remains an issue difficult to resolve. Collecting the waste from the location of the distillation plant and transporting it to a waste disposal site increases the unit price of the desalinated water. Obviously, the increase depends on the distance between the location of the desalination plant and the waste disposal site. The same conclusion applies to the transport of input water to the desalination plant.

The experiments already described in this sub-section and the few results obtained there from seem promising and deserve every attention, support and continuation. The time has come that Arab countries should really learn more and gain solid experiences using other sources of energy before thinking of reducing the fossil fuels they use for running, among others, water desalination plants. One should take into account the sharply rising cost of oil.

10.4.3 Groundwater abstraction

Abstraction of groundwater from any aquifer, regardless of the depth to water and the thickness of the formation, at rates exceeding the safe yield will lead in the course of time to a considerable drawdown. As a consequence, the lift head of the pumps withdrawing water will grow larger in time in the event of a non-artesian aquifer. Equivalently, the state of flow from an (originally) artesian aquifer after some time will change to pumped flow. In both cases, consumption of fuel needed to pump water will keep increasing, and thus will have some effect on the environment and on the cost price of extracted water.

In 1988, Mohorjy commented on the excessive abstraction of groundwater in Saudi Arabia saying "The groundwater level has dropped 30 ft (10 m) during the last three decades around the city of Er-Riyadh. The water table in Wadi Fatima and Wadi Khulys (around Jeddah) has dropped rapidly in the past few years with the Jeddah's growth". He added, "As a result deeper wells and more powerful pumps are needed to support agriculture in the wadis". In the Oasis of Al-Kharj, southeast of Riyadh, the recharge rate is low and water is extracted heavily for agriculture. Surface rock strata have collapsed, and the water level can be seen to have dropped 17–20 ft (4–5 m) in the past 10 years. Besides, it was observed that withdrawing excessive amounts of ground-water from the lower Wadi Fatima area in the western region of Saudi Arabia reduced the water table rapidly, causing all springs to dry up. Of 360 springs producing at one time, only four remain. Besides, the water level sank below the main channel of the wadi, causing more saline water in the surrounding areas to move laterally and down to the lower level.

The likelihood of increasing the salinity of water is a major difficulty associated with intensive withdrawal of groundwater. This is especially true when the site of the production wells is close to the sea or ocean. Excessive withdrawal of water from the coastal aquifer in the Gaza strip, Palestine, is causing the deterioration of water quality with time and expected to continue worsening unless drastic measures are undertaken. The installation of a water distillation plant on the seacoast near the city of Gaza seems to offer one of the most practical options for supplying the strip with additional water.

In sub-section 10.3.3.3 mention is given to the expected drawdown, 70–150 m, of groundwater in the Kufra Oasis, Libya. Similar to the situations of Saudi Arabia and Libya, the abstraction from the Nubian Sandstone in the Western Desert of Egypt to supply agriculture with adequate amounts of irrigation water and develop the living conditions in the Dakhla and Kharga Oases used to be excessive. The increase of abstraction rate and subsequent decline of groundwater head are plotted together versus time in years as shown in Figure 12 (Margat & Saad, 1984). From this Figure one can easily observe the drop of groundwater head reached about 35 m in the 20 years period, 1956–75. As a result, the free flowing artesian condition, which the aquifers used to enjoy, changed to non-artesian. The newly developed condition necessitated installation of pumps, fuel has to be regularly supplied to operate the pumps, every now and then pumps have to be changed, repaired and/or maintained, etc. The initial and running costs of these items have become a financial burden on the project, bringing the unit price of water much higher than was ever expected before project implementation. In attempt to ease the situation somewhat, the abstraction rate had been lowered since 1977, and drilling new wells became restricted to just replacing old ones. The change of water supply policy has reasonably helped to stop any further lowering of groundwater head though some parts of the land that were prepared for agriculture had to be deprived from being irrigated.

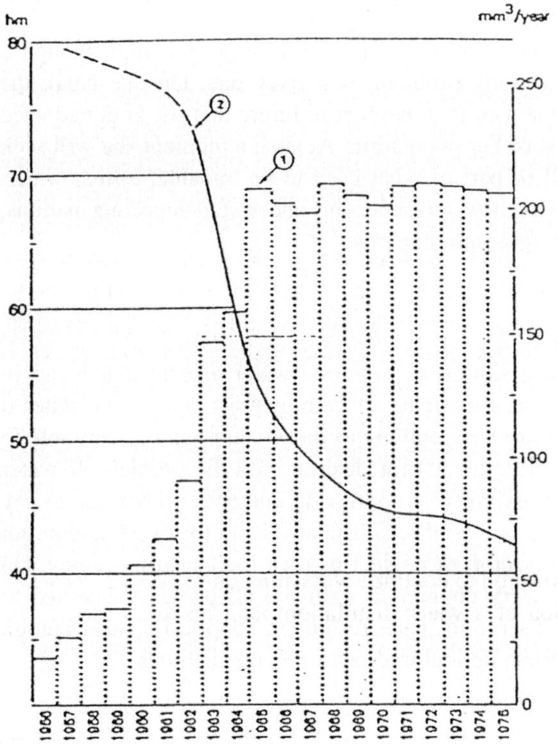

Figure 12. Change of abstraction rate and decline of groundwater head in the artesian aquifer underlying the Kharga and Dakhla Oases, Egypt (from Margat & Saad, 1984)

10.4.4 Water transport

Transport of water from a water-rich place to a water-poor place, both in the same country, is likely to improve the living condition of certain groups of people at the expense of other groups leading to social injustice. It is an established fact that the southern parts of many countries of the Arab Region are less developed compared to the northern parts of the respective countries. This may be because the northern parts are often situated along the seacoast, thus characterised by more agreeable climate and relatively easy communication than the south. As such, the northern parts generally enjoy a higher level of development. All these circumstances encourage the migration of many inhabitants from south to north.

A typical example is the Great Man-made River project in Libya. Water abstracted from the aquifers in the south of the country is conveyed hundreds of kilometres to the north, east and west, thus bringing an additional source for the development of the northern part. It also means that some areas in the southern part of the country are sentenced to life deprivation from means of development. It will be more appropriate to allocate fair portions of the project yield to develop the areas surrounding the well fields, such as Sarir and Fezzan, similar to the Kufra settlement project. This will improve to some extent the balance of social justice between the communities in the north and those in the south.

Transfer of water from a water-rich nation to a water-scarce nation, though a convenient solution to water scarcity problem, is a risky one. On one hand, the water-rich nation might realise at a certain moment in future that she is in bad need of its own water for the interest of her population. At such a moment she will seek everyway to stop exporting all or part of what used to be tradable, excess water. Besides, at times of crises between water-exporting and water-importing nations, water trade could easily be brought into jeopardy.

10.4.5 Virtual water

This aspect is in a certain sense similar to imported water, except that water in the former option is embedded in a diversity of commodities, and in the latter it is still in the liquid state. Both are easy options to those countries that are able to pay the price and provide effective means for dealing with the problem of water scarcity. Not withstanding the possibility of not having one or the other due to any reason, such as political disputes and conflicts, it is advisable for every nation not to abandon its own agricultural sector, or neglect its own food security policy. On the contrary, it is the task of every country to motivate all parties concerned to exploit the agricultural land, conserve its water resources, adopt efficient irrigation technologies and apply appropriate agricultural and drainage methods.

CHAPTER 11

WATER STORAGE

11.1. INTRODUCTION

Water resources in many Arab countries are equipped with more or less adequate water storage facilities for two major objectives. Water can be stored above the ground surface (surface storage) or below the ground surface (underground storage). Surface water is stored in reservoirs formed by damming off streams, mostly of perennial nature. The streambed forms the lower boundary of the reservoir. The capacity of a surface reservoir can be in the range from less than one 1×10^6 m^3 (small reservoir) to more than 1×10^{11} m^3 (large reservoir). The corresponding surface area at flood level varies from a few hectares to thousands of square kilometers. Impermeable beds form the lower boundary of underground reservoirs, and their sizes, depending on thickness and areal extent, vary in volume from a limited size to thousands of cubic kilometers.

Storage of surface water reduces that portion of runoff flowing to an outlet body such as sea, ocean or an inland lake. As storage helps to reduce the amount of water lost or wasted, storage of water in this sense can be viewed as a water conservation measure. This does not eliminate the fact that evaporation losses from reservoirs in arid and semi-arid areas can be so high as to minimise the gain achieved by saving water. Certain portions of surface runoff can be stored in underground reservoirs thereby increasing the volume of groundwater in storage. Generally, with this type of storage, the evaporation losses compared to the case of uncovered reservoirs is quite small. It is very likely, however, that the change in hydraulic conditions will cause some of the stored water to be lost by leakage to the neighbouring aquifers and/or find its way to a nearby outlet. Storage in underground reservoirs improves or, at least, preserves the quality of water. This is not the case with storage in surface reservoirs. One should not forget that sediment runoff and deposition in surface reservoirs often presents a difficult issue. Apparently, maintenance and repair costs are cheaper in storage of underground reservoirs than in the case of surface reservoirs.

Groundwater storage through recharge reservoirs requires the construction of retention dams. Water in such a reservoir is left to infiltrate the soil surface and percolate downward through the soil moisture profile till it joins the groundwater body. Recharge dams are usually constructed in wadis with adequate sealing elements to guarantee the safety of the structures. Depending on the local geology and existing depth to water table below ground surface, the so-called 'underground groundwater recharge dams'can be considered. In this particular type of storage reservoirs, only an impermeable underground barrier across the permeable part of the alluvial aquifers needs to be built. Additional works such as embankments and spillways are unnecessary (Strobl, 1995). Generally, wadi-related alluvial aquifers do not have large reserves. They are recharged directly by rainfall and to some extent by resurgence from the bedrock.

The volume of a storage reservoir is partly or completely filled with water during the high-flow season and released to the downstream during the low-flow season. The high-flow season lags a short period after the onset and offset of the rainy season. As a consequence, part or all of the excess water carried by a river or wadi is saved and the natural supply is augmented so as to the meet the demand in the period(s) of deficit. Storage aiming at stream flow regulation is in fact a component of water resources management, which is the second major aim of storage.

11.2. SOME HYDROLOGICAL ASPECTS OF RESERVOIR STORAGE

It is not the purpose of this section to review all hydrological aspects of storage reservoirs. Furthermore, those aspects presented here are not described in length. For completion and greater details, the reader is advised to consult with specialised textbooks, publications and research papers written about reservoir hydrology. As such, we shall briefly review in this section the water balance of each of the surface and subsurface water. Additionally, some of the methods used for determining the reservoir capacity are highlighted.

11.2.1 Water balance

Considering a time interval Δt, the notations used in this sub-section including Figure 1 can be defined as follows:

I = inflow to the reservoir,

P = direct precipitation over the average surface area of the reservoir during Δt,

E = evaporation from the average surface area of the reservoir during Δt,

V_s = outflow over the spillway,

V_g = release to the downstream through the dam sluices,

F = infiltration depth over the interval Δt, and

ΔS = change in surface water and groundwater storage over the interval Δt.

With these notations, the water balance can be expressed as:

$$(1) \qquad (I+P)-(F+E+V_s+V_g)\pm\Delta S$$

Eq. (1) is used for estimating one variable only provided that all other variables in the equation are known. It is customary to measure I, P, V_{sp}, V_g, and ΔS with a high

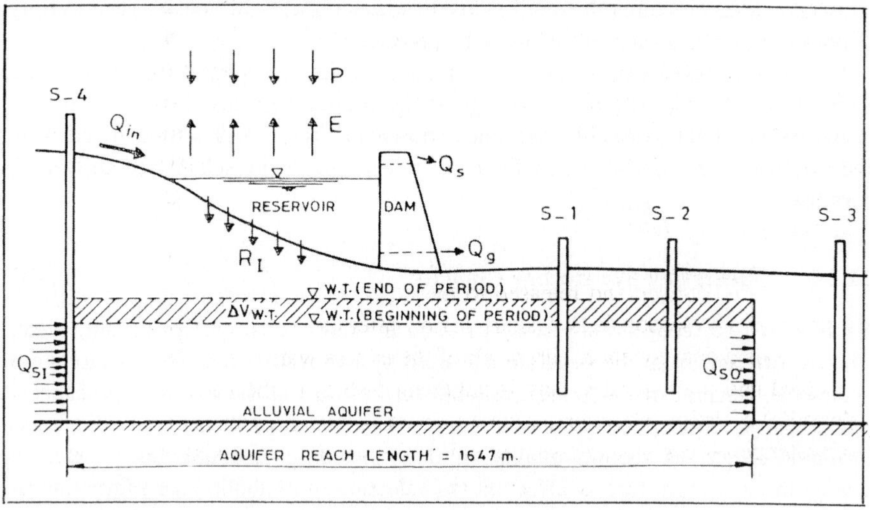

Figure 1. Cross section of a recharge dam and aquifer showing the notations used in the water balance equation

degree of accuracy. This leaves any of E or F to be estimated from the water balance equation. Experience has shown that after putting the reservoir into operation, the bed and walls become gradually sealed with a natural blanket of sediments that remains entrapped in it. If this is truly the case, and having chosen a balance interval that is not too short, e.g. one year, the amount F becomes infinitesimal and therefore can be neglected. This leaves E as the variable to be estimated from the balance equation without substantial error.

Before applying the water balance equation to arid zones, Eq. (1) needs some adjustment. The occurrence of precipitation and inflow is infrequent and lasts for only a short period of time each year. Flows over the spillway and through the dam sluices are rare, and if they occur they can be measured. Therefore, for most of the year, Eq. (1) reduces to:

(2) $$F = -E \pm \Delta\, S$$

If the infiltration F is the variable whose magnitude has to be estimated from Eq. (2), the evaporation E has to be calculated using any of the methods described earlier in Chapter 5

Denote the difference between subsurface inflow and subsurface outflow by ΔG, and the change in subsurface storage due to rise or fall of the groundwater table by ΔS_g, the water balance of the subsurface part of the flow system can be written as:

(3) $$\Delta S_g = F \pm \Delta G$$

Accurate knowledge of the inflow and outflow of groundwater and the change of storage due to water table fluctuation is unavoidable for estimating F using Eq. (3).

This knowledge requires the availability of measurements of aquifer permeability, depth of saturated zone, groundwater levels, etc.

Substituting these data in Darcy's equation yields the volumes of groundwater flow needed for Eq. (3). The readings of the piezometers installed upstream and downstream of the dam, like the ones shown in Figure 1, are used to calculate the average value for the hydraulic gradient that has to be substituted in Darcy's formula.

11.2.2 Reservoir live storage capacity

By live storage (conservancy storage) is meant here the storage of useable water. This is different from the dead storage used by contrast to describe the storage of sediments. Several methods are available for calculating the live storage capacity of a reservoir. Time description of storage (seasonal/annual, long- term or century), available length of record, quality of data including reservoir losses, etc. are among the important factors affecting the selection of method(s) for estimating the reservoir capacity. Missing inflow data can be supplemented using techniques such as regression models (simple and multiple), auto regressive models, unit hydrograph, etc.

In some cases the capacity of a reservoir is dictated by the topography of the storage site. The relationship between the inflow (Q_i), outflow including losses (Q_o) and the required storage capacity, assuming the reservoir to be empty at the beginning and end of the storage period, can be given by the expression:

$$R_n = \max.\left[\int_0^{\tau} Q_i dt - \int_0^{\tau} Q_o dt\right], n = 1, 2, \ldots\ldots, N \tag{4}$$

where,

$\tau =$ any arbitrary time during the chosen storage period,

$n =$ observation number, e.g. $n = 3$ considering annual storage means the third year of discharge observation in the record and,

$N =$ number of periods in record, also number of last period.

The storage volumes $R_1, R_2, , .R_N$ can be computed for the periods $n = 1, 2, \ldots$ *and* N using Eq. (4) and each of them compared to the available storage capacity, C. Besides, the graphical presentation of the cumulative inflow and outflow $\int_0^{\tau} Q_i dt$ and $\int_0^{\tau} Q_o dt$ versus time is known as Rippl diagram and the method itself is referred to as Rippl diagram method. Eq. (4) is written specifically for the case where the draft ratio, α, is equal to unity, i.e.

$$\alpha = \int_0^{\tau} Q_i dt / \int_0^{\tau} Q_o dt = 1 \tag{5}$$

It is quite possible that the ratio α in Eq. (5) equals 1 as the storage losses have been included in Q_o. Without including these losses the usual value of α falls in the range of 0.6-0.8.

The Rippl diagram method is characterised by its simplicity, and when the number, N, of observations is sufficiently large (50 observations or more), the result is not far from the capacity obtained from more sophisticated methods. Pieces of literature specialised in reservoir hydrology describe in details the positive as well as the negative aspects of the Rippl diagram method for estimating the design capacity of a reservoir.

Suppose that the available capacity, C, for a given situation happens to be m-times less than $R_n(n = 1, 2, \ldots, N)$ obtained from Eq. (4). The probability of failure, q, of the given capacity to store all excess water in the high-flow season and release it to the downstream during the low-flow season can be roughly taken as the relative frequency m/N. The value of q can be improved by adjusting the release during the periods of failure. This is possible in case the observations repeat themselves, in magnitude and sequential order, every N-periods.

Subsurface storage and outflow depend to a large extent on the depth of infiltration, F. The rate of infiltration, f, for the same soil conditions is at its maximum when the reservoir gets full and becomes smaller as the volume of water in the reservoir is reduced. Obviously f reaches zero when the reservoir gets empty. According to Al-Hassoun et al. (1995), infiltration from the reservoir is extremely high immediately after a flood. This is especially true for large floods occurring after a long dry period. The first few days following a flood may account for a large part of the head in the reservoir. Following these few days, the volume of water impounded in the reservoir becomes small and infiltration decreases rapidly resulting in a much lower values of F.

The groundwater outflow hydrograph consists of a slowly rising limb followed by a rapid rise till the peak outflow is reached. The recession limb, from the peak point of the outflow hydrograph falls steadily till the end of the season. The peak of the outflow curves lags some time behind the time of occurrence of the peak infiltration. The time lag depends on the magnitude of the flood and the history of the moisture conditions in the soil column underneath the bottom of the reservoir.

11.2.3 Design flood for the spillway of the dam

The general practice for designing a spillway till present is to rely on the concept of probable maximum flood (PMF) or the estimate of flood value corresponding to a chosen average recurrence interval. Many hydrologists follow an old rule of thumb that gives the PMF as the flood estimated from the general frequency function (Eq. 5) assuming a return period, T_r, of 10,000 years. According to Gumbel EV Type-1 distribution the variable y, known as reduced variate, can be obtained from the following equation:

$$(6) \qquad y = -\ln\left[-\ln\left(\frac{T_r - 1}{T_r}\right)\right]$$

The discharge corresponding to a return period T_r can be estimated from the equation:

$$(7) \qquad Q_{Tr} = Q_{av} + s(0.78y - 0.45)$$

According to UN-ECAFE/WMO (1967), the probable maximum flood, Q in $m^3 s^{-1}$, for a catchment area A in km^2, can be estimated from the following equation:

$$(8) \qquad Q = 1.3C(0.385A)^{0.935A^{-0.048}}$$

A formula, due to Fuller and widely used in the USA, gives the average design flood, Q_{av}(ft^{3}s^{-1}), which upon substitution in Eq. (7) yields the 'most probable' annual maximum flood, Q_m, as:

$$(9) \qquad Q_m = Q_{av}(1 + 0.8 \log T_r) = CA^{0.8}(1 + 0.8 \log T_r)$$

where

$A =$ catchment area in square miles and C is a coefficient often taken as 75.

Example: A catchment area, A, of 1,000 sq mi (2,590 km^2), according to Eq (9), is likely to receive an average flood discharge, Q_{av}, of 18,840 ft^3 s^{-1} or 533.5 m^3 s^{-1}. The corresponding Q_m, also from Eq. (9), is 2240 m^3 s^{-1}. Eq. (8) yields probable maximum flood, Q, corresponding to $A = 1{,}000$ mi^2(2,590 km^2) equal to Q = 108.84 C. Comparing this result to Q_m obtained from Eq (9), the coefficient C can be found as 20.6, a value that is usual for many arid zones including the Arabian Peninsula.

As Eq. (6) yields for $T_r = 10{,}000$ years a y-value of 9.21 the flood discharge can be obtained from Eq. (7) as $Q_{av} + 6.734s$, where s is the standard deviation from the mean. Knowledge of the values of Q_{av} and s from a sufficiently long series of Q will help to estimate the $Q_{Tr} = 10{,}000$.

11.3. SURFACE STORAGE RESERVOIRS IN THE ARAB COUNTRIES

The main objectives behind the construction of the surface storage reservoirs are to eliminate or at least reduce the loss of surface runoff water by storing all or part of the flood flow and to regulate the stream flow for the purposes of irrigation, flood control, hydropower development and navigation. Besides, in certain parts of the Arab region surface reservoirs are used for recharging the depleted groundwater aquifers. These are known as recharge dams.

The dimensions of reservoirs including their storage capacities vary in a wide range depending on the topography of the reservoir's catchment area and the flow conditions of the streams on which the dams forming the reservoirs are installed. Reservoirs on the Nile and its tributaries, the Euphrates and the Sénégal Rivers are of huge dimensions compared to those constructed on wadis in Morocco, Algeria, Tunisia, Jordan and other countries.

The next paragraphs have been prepared with the aim of reviewing the pertinent hydrologic features of the reservoirs in different parts of the Arab Region. While dwelling upon reservoir hydrology no discussion will be given to reservoir sedimentation as it has already been presented at large in Chapter 8.

11.3.1 Morocco

Moroccan wadi basins are equipped with a sufficiently large number of dams for water conservation and hydropower development. Besides, there are several dams under construction or planned for construction.

The Oum er-Rebia is the principal river of Central Morocco. The river basin has its first dam, Sidi Saïd Maachou, in 1929. This was soon followed by Kasb Tadla Dam in 1931, Im-Fout Dam in 1950, etc. Since the construction of Ayet Said Chouarit and Dashr el-Wadi Dams in the second half of the 1980s, the basin has more than 80% of its waters under control. Details of the storage capacities of these dams are included in Table 31, Appendix II, whereas the sites of existing, under construction and the projected ones are shown in the map, Figure 2. When the construction of the projected reservoirs is implemented, they will add 390×10^9 Watt to the power generated from the Oum er-Rebia Basin.

With its tributaries, the Sebou River accounts for almost half of Morocco's surface water resources. Despite the availability of water in the Sebou basin, the extent of flow regulation is smaller compared to Oum er-Rebia. The storage of water in the Kansara and Idriss Al-Awal (I) reservoirs has proven not to be fully adequate for the purposes of land irrigation and flood protection. The pre-1980 storage capacity was $1{,}542 \times 10^6$ m^3 expected to reach $6{,}095 \times 10^6$ m^3 after the completion of all projected works. This will lead to increasing the volume of regulated flow from $1{,}200 \times 10^6$ m^3 to $3{,}986 \times 10^6$ m^3 annually.

Moulouya is the principal river in northeastern Morocco. The Mechra Hamadi (1956) and Mohammed V (1967) have helped to some extent in providing water for land irrigation, mining and electricity. The Moh. Abdel-Karim Khattabi reservoir on the Nekor Wadi provides water for land irrigation and drinking purposes. Soil erosion is a major problem causing a considerable obstruction to the development of the water resources of the basin. Upon removal of sediments accumulated in the reservoir formed by the Khattabi Dam it will be able to provide a total volume of 62.3×10^6 m^3 annually; 9.3×10^6 m^3 for supplying Al-Huceima with drinking water, 41×10^6 m^3 for irrigation and 12×10^6 m^3 for recharging groundwater aquifers.

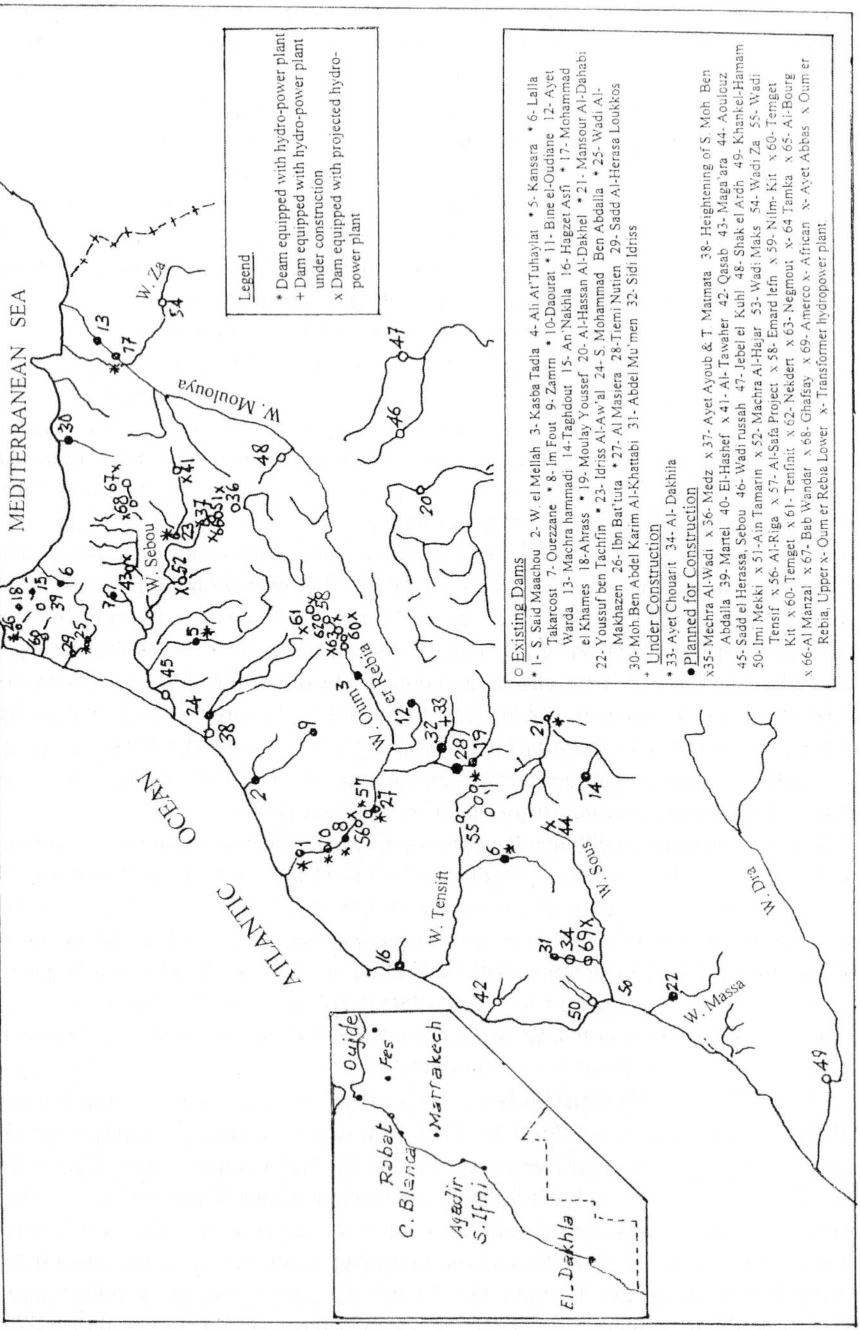

Figure 2. Location map of existing dams, dams under construction and dams planned for construction as from 1986 (Country Report of Morocco, 1986)

Dams have been built and building of more dams has long been planned to control as much as possible of the widely fluctuating flows in other basins. The dam on the Graw wadi helped to cover the water demands of the city of Rabat in the period 1969–74. This task has been shared since 1974 with Sidi Mohammed Ben Abdalla, (Bou Regreg) which extended some of its supply to Casa Blanca and its surroundings. Upon the heightening of Sidi Moh. Ben Abdalla Dam, the percentage control of the water in the Atlantic Basins will reach 50%. Two storage reservoirs (Wadi Al-Makhazen and Al-Lokkous) have been built in the Lokkous Basin with the objectives of flood protection, irrigation and hydropower development. These two reservoirs together regulate 480×10^6 m^3 y^{-1}. Wadi Ibn Battuta on the Wadi Al-Thulatha helps to supply the City of Tangier with water for its domestic needs.

Hydrology of the reservoirs built in the South Atlantic Basins is characterised by wide fluctuation of their inflows and intensive seepage of groundwater in the areas commanded by these reservoirs. Among the dams in these basins are; Al-Hassan Al-Dakhla on Wadi Ziz, Mansour El-Dahabi on the Drâa (sometimes written as Dra), Youssuf Ben Teshfin on Wadi Massa, Abdel Mu'men and Al-Dekhaila on the Wadi Eissen. Available particulars about these dams are listed in Table 31, Appendix II.

Since the late 1970s, Morocco has experienced a number of extremely dry winter seasons. This in itself has been an urgent reason calling for the development of sustainable water resources management. Consider the case of Wadi Drâa as an example. This wadi has two tributaries, Wadi Dades and Wadi Ouarzazat, which drain the southeastern and the south-western parts of the Atlas and join each other at Ouarzazate, thereby forming the Wadi Drâa. The construction of Mansour El-Dahbi Dam at the confluence site of these two wadis was completed in 1972 with an original storage capacity of 560×10^6 m^3.

Due to thick accumulation of deposited material in the reservoir, its storage capacity fell to 440×10^6 m^3 by the year 2002 (Speth & Christoph, 2004). The volume of water released for land irrigation in average years is 250×10^6 m^3. Figure 3 shows clearly that this condition was not met in 13 years out of 30 consecutive years, i.e. the chance of failure to meet this requirement in the period 1973–2002 was 43.3%. Figure 3 also shows that the April filling level of the reservoir in the said period was below the critical level. The limited amounts of snowmelt in the spring seasons of those below average years were one of the factors that contributed to the small inflow to the reservoir.

11.3.2 Algeria

Till 1961 the number of storage reservoirs in Algeria was 18 with a total storage capacity of 60×10^6 m^3. Seven more dams, with a total capacity of 360×10^6 m^3, were built in the period 1962–79. The five-year plan, 1979–84, witnessed the construction of 19 dams with a total storage capacity of 800×10^6 m^3. The next 5-year plan, 1984–89, included the projection and design of 16 more dams with a total capacity of $1,200 \times 10^6$ m^3 (Country Report on Water Resources in Algeria

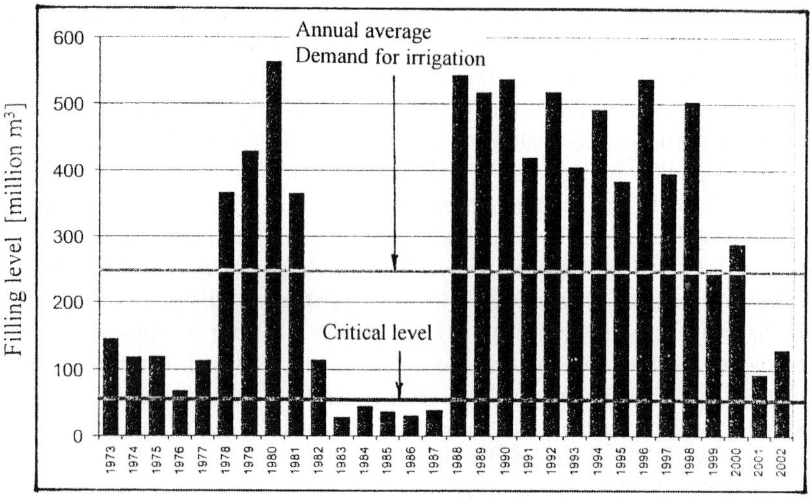

Figure 3. April filling levels of the Mansour El-Dahbi reservoir near Ouarzazate, Morocco for the period 1973–2002 (Speth & Christoph, 2004)

and their Utilisations, 1986). It was assumed then that the construction of those dams would have been completed before the year 2000. Some of the data obtained from the country survey of reservoirs up to 1986 in Algeria are listed in Table 32, Appendix II.

The Authorities in Algeria estimated that more storage works will be needed to meet the demands on water after the year 2000. In that estimate the mount needed for domestic purposes has been assumed to be shared by surface water resources (55%) and ground water resources (45%).

11.3.3 Tunisia

The total amount of surface runoff in Tunisia has been estimated as $2{,}630 \times 10^6$ m^3 y^{-1}. According to the Country Report of Tunisia (1986) the number of dams in 1982 was 12 with a total storage capacity of $1{,}047 \times 10^6$ m^3, the oldest of which is Wadi Al-Kabir, already built in 1923. The annual storage used to be 790×10^6 m^3. By the end of 1987, the number of dams reached 17 with a total storage capacity of 2.021×10^9 m^3. In addition to the storage reservoirs there are some inundation reservoirs in the middle and south of Tunisia with a total storage capacity of 21×10^6 m^3.

Two large wadis-the Zeroud and the Marguellil-cross Central Tunisia and converge in the Qairouan Plain. Water control of these two wadis formed the basis of a flourishing water management civilisation in the region. For 30 years ten dam sites have existed in the Foussana Basin alone and two dams in Kasserine, one dating back to the 18th century. All these dams have helped to irrigate 30,000 ha of land. The locations of the water management works are indicated on the map in Figure 4 (El-Amami, 1983).

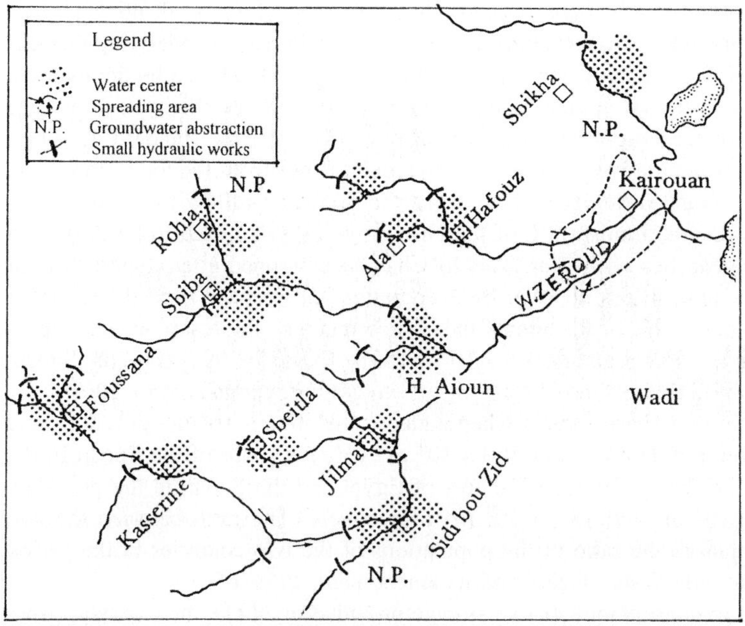

Figure 4. Management of Wadi Zeroud, Tunisia (El-Amami, 1983)

11.3.4 Egypt and the Sudan

Hydrology of the storage reservoirs in the Nile Basin has been described in many pieces of literature (e.g. Shahin, 1985, 2002) including Chapter 6 of this book. What interest us here are the reservoirs in Egypt and the Sudan without taking into account the reservoirs in the other riparian countries. These reservoirs listed from south to north are; Er-Roseires on the Blue Nile, Sennar on the Blue Nile, Khashm el Girba on the Atbara, Jabal Awlia on the White Nile and the High Aswan Dam (HAD) on the Main Nile. All of these reservoirs except the HAD are situated in the Sudan (Figure 6.9). The following tabulation gives the year when the reservoir was first put in operation and its design live storage capacity:

Reservoir	*River*	*Year of operation*	*Design live storage capacity*, 10^9 m^3
Er-Roseires	*Blue Nile*	*1966–67*	3.0 (1st stage) increasing
Sennar	*Blue Nile*	*1925–26*	*0.8*
Khashm el-Girba	*Atbara*	*1964–65*	*1.3*
Jabal el-Awlia	*White Nile*	*1937–38*	*3.5*
High Aswan Dam	*Main Nile*	*1965–68*	*90.0*

The above tabulation shows that over 90% of the storage capacity of reservoirs in Egypt and the Sudan combined is provided by the HAD in Egypt. In fact the capacity of this reservoir alone exceeds by far the capacities of all other surface water reservoirs in the Arab Region combined. The main reason underlying this fact is that the design capacity of the reservoir formed by HAD is based on the theorem of long-term or century storage. The reservoir capacity, S, for this type of storage is a function of the long-term mean annual river flow, M; the standard deviation of flow, σ; the cumulative departure, R, of annual flows from the long-term mean flow and the annual difference, B, between the draft and inflow to the reservoir. In the pre-HAD era, almost 35% of the Nile flow at Aswan used to be thrown in the Mediterranean Sea each year. This loss has been stopped after construction of the HAD. The annual loss of water by evaporation and seepage from the reservoir is, on the average, 12% of the annual mean flow reaching the reservoir, leaving an annual saving of 25% or about 21.5×10^9 m^3 to be shared by Egypt and the Sudan. The gain has been divided into 7×10^9 m^3 y^{-1} to go to Egypt and 14.5×10^9 m^3 y^{-1} to go to the Sudan. These figures when added to the 48.5×10^9 m^3 y^{-1}, the share of Egypt in the pre-dam era, and 4.0×10^9 m^3 y^{-1}, the share of the Sudan in the pre-dam era too, bring the annual shares of Egypt and the Sudan in the post-dam era to 55.5×10^9 m^3 and 18.5×10^9 m^3 respectively. The ratio between the two shares was equal to the ratio of the populations of the two countries in the period of drafting the Nile Water Treaty and its amendment, 1959–65.

In addition to conservancy or live storage provided by HAD, the reservoir gross capacity includes a certain volume kept for storage of sediments (dead storage). This volume is capable of storing all sediment inflow to the reservoir for a period longer than 300 years. This is not the case with the four reservoirs existing in the Sudan. With the exception of Sennar, which lost 20% of its capacity due to storage of sediments in the past 80 years, the remaining reservoirs have lost in 4 decades more than 50% of their original storage capacities. This situation is affecting adversely the interests of irrigation and hydropower development in the Sudan. Maintenance of these dams and cleaning their reservoirs is neither an easy job nor a cheap one, especially when the currently uncomfortable economic situation is taken into consideration.

11.3.5 Lebanon

In a country like Lebanon, where a considerable volume of water is lost to the sea, small dams are necessary for impounding and efficiently using water from rivers and other streams, and springs. Small reservoirs built for irrigation often has other advantages such as fish breeding, supplying water for stock and providing an aesthetic touch to landscape. A large number of dams have been recommended for construction after their sites have been geologically investigated. The capacities of these dams range from a minimum of 5×10^6 m^3 to a maximum of 120×10^6 m^3.

Water resources development in the basin of the Litani River, which lies entirely within the national borders of Lebanon, requires the use of storage reservoirs to regulate the cyclic fluctuations of the river flow (700×10^6 m^3 y^{-1} long-term average). The Qairouan reservoir has a gross capacity of 220×10^6 m^3 of which 150×10^6 m^3 is destined for live storage. Substantial diversion of waters of the

Litani was made to Awali River by a tunnel conduit, making the Awali the largest river in Lebanon (645×10^6 m^3 y^{-1}). This diversion for the hydropower scheme leaves only 125×10^6 m^3 y^{-1} of water for the Lower Litani.

The diversion of water from the Litani River to the Awali is accomplished via the Marakaba tunnel as can be seen from the map in Figure 5. Not all the diverted flow is for the purpose of hydroelectric power development. A certain amount of the diverted discharge flows in an irrigation canal and reaches the Nabatiya Plain (Murakami & Musiake, 1994).

11.3.6 Syria

The authorities concerned with water conservation and management in Syria have begun since the 1950s storing floodwater carried by rivers and other streams instead of leaving it to flow towards the sea or be wasted in the desert. The storage reservoirs have been arbitrarily classified into large, medium and small reservoirs depending on their storage capacities. A medium size reservoir is a reservoir with a storage capacity exceeding 15×10^6 m^3 up to 225×10^6 m^3. Reservoirs with smaller capacities are termed small reservoirs, and with larger capacities large reservoirs. Dams constructed up to 1986 are listed in Table 33, Appendix II.

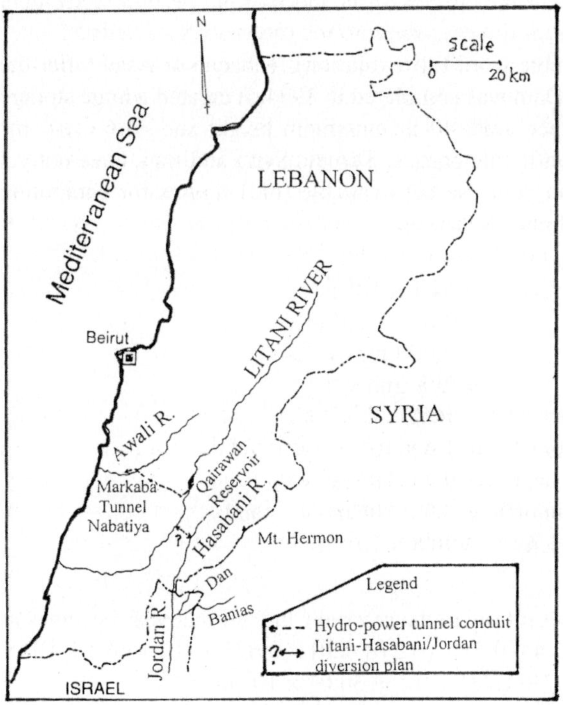

Figure 5. Water management in the Lower Litani (Murakami & Musiake, 1994)

According to the survey presented in the Country Report of Syria (1986), the sum of the capacities of the three groups of reservoirs is about 12.07×10^9 m^3, of which the Euphrates alone provides 11.6×10^9 m^3, i.e. 96.1% of the total capacity. The same survey lists 16 dams, some of which are under construction and the rest are planned for construction, aiming at adding an extra storage capacity of 147×10^6 m^3, or 1.2% of the pre-1986 capacity.

The first treaty controlling the utilisation of the Euphrates water was signed by Great Britain and France in 1920 soon after the decline of the Ottoman dominance on Syria and Iraq. Later, in 1946, the Governments of Turkey and Iraq signed a friendship treaty between the two nations. A special protocol regarding the control and sharing of waters of the Tigris and Euphrates Rivers and their tributaries was annexed to that Treaty. The protocol also included recommendations related to strengthening and upgrading the hydrometric networks for observing the river levels and discharges. Since 1962 two-parties and tripartite negotiations are taking place between; Syria and Turkey, Turkey and Iraq, Syria, Turkey and Iraq, with the aim of setting up rules for sharing the waters of the Tigris and Euphrates, and their tributaries. Negotiations also included the establishment of appropriate operation rules for filling and emptying the reservoirs of Keban in Turkey, Ath-Thawrah or Al-Asad (Syria) and Habbbnnyah (Iraq). Despite the frequent intermediation of the former Soviet Union, World Bank and other agencies between the three parties, and the Arab League and Saudi Arabia between Syria and Iraq, no definite agreement has so far been reached.

In Syria, Ath-Thawrah (Arabic word for revolution), Tabaqah or Asad (after the late President Asad of Syria) Dam was completed in 1974. It created a huge storage reservoir, with 625 km^2 surface area, 40 m maximum height and 11.6×10^9 m^3 storage capacity. Since the conflicting parties; Turkey, Syria and Iraq, have not yet settled their conflicts, one can only use the available (old) figures for computing the water balance of the Euphrates Reservoir.

Catchment area	$= 64.1 \times 10^9$ m^2,
Surface area of reservoir	$= 0.625 \times 10^9$ m^2,
Evaporation	$= 2{,}000$ mm y^{-1},
Precipitation	$= 278$ mm y^{-1}
Runoff coefficient (assumed)	$= 10\%$,
Inflow to reservoir (from springs)	$= 1.6 \times 10^9$ m^3 y^{-1},
Inflow to reservoir (groundwater)	$= 0.4 \times 10^9$ m^3 y^{-1},
Inflow to reservoir (surface runoff)	$= 0.5 \times 10^9$ m^3 y^{-1}, and
Inflow to reservoir (Upper reach)	$= 26.8 \times 10^9$ m^3 y^{-1}.

From the above figures, the net inflow to the reservoir in a normal year becomes as follows: $26.8 \times 10^9 + (1.6 + 0.4 + 0.5) + (0.278–2.0) \times 0.625 \times 10^9 + 64.1 \times 10^9 \times 0.278 \times 0.1 = [26.8 + 2.5–1.07 + 1.78] \times 10^9 = 30.01 \times 10^9$ m^3 y^{-1}.

It should be remembered that no account has been taken in the foregoing balance of any losses of water and/or abstraction from the river reach between the

Turkish-Syrian border and the location of the dam. Old data gives the mean river flow for the period 1974*-84 at the gauging station of Husseibah on the Syrian-Iraqi border as 23.61×10^9 m^3 y^{-1}. Should this figure stand for a long-term average, the volume of water representing the difference between inflow and outflow must be about 6.4×10^9 m^3 y^{-1}

11.3.7 Iraq

The period 1911–40 witnessed the construction of three major hydraulic works on rivers within Iraq. These were: a grand barrage (open-type dam) in 1913 on the Euphrates at Hindiyah, a grand barrage in 1939 on the Tigris at Al-Kut and a dam in 1940 on the River Diyala. These three works were built with the aim of raising the upstream water level to supply the main canals with water for irrigation. In the 1950s, following the end of the second world war, a barrage was built on the River Tigris at Samarrā, and the Tharthar regulator and a barrage on the Euphrates at Ramadi were also built for the same purpose as the earlier ones.

The Country Report of Iraq (1986) stated that the expected annual natural supply from river flows, with 95% probability of non-exceedance, is in the order of 41.5×10^9 m^3. The Iraqi authorities, which were entrusted with the development and management of water resources in the mid 1950s, considered this figure to be too low to meet the future demand of water. Storage of river flow during the wet season (March-May) to use it in the dry season appeared to provide an attractive solution. Among the recommendations reached by those authorities was the construction of new long-term storage reservoirs and regulators. Besides, they recommended the improvement of operation of the Dukan (on the lesser Zab) and the Darbandi-Khan (on the Diyala river) reservoirs. The locations of the major storage and control works are shown on the map in Figure 6.

Table 34, Appendix II, include some hydrological data relevant to the existing, under construction and planned for construction dams up to 1986. Since the early 1980s till the time of drafting these lines, the political, military, economic and security situations are imposing unprecedented downfall in the development of natural resources in Iraq, resulting in a halt in the construction of any new, major hydraulic structures. Discussion of this matter is totally outside the scope of this text.

11.3.8 Jordan

Malkawi & Abdulla (1997) gave a survey of the dams in Jordan. They classified the dams into four categories. These comprise the dams constructed in the Jordan Valley, and the hilly and arid regions of the country; dams under construction; dams about to be constructed; and dams under planning and design. Table 35, Appendix II, gives the location of each dam, height material of construction of the dam body, working or effective capacity and total design capacity. The locations of some of these dams are already indicated on the map in Figure 8.12.

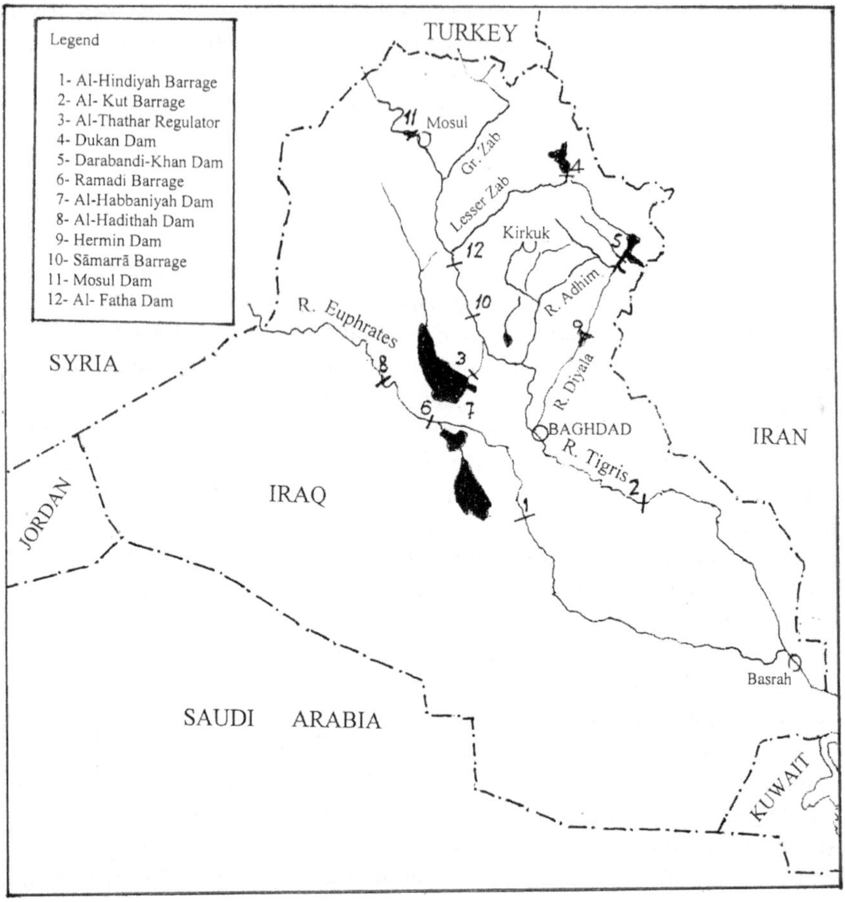

Figure 6. Location map of major hydraulic structures on the rivers in Iraq

The increase in the storage content of King Talal Reservoir from the original figure of 56×10^6 m^3 to 80×10^6 m^3, which took place in 1987, has already been highlighted in 8.5.3.1. An increasing proportion of the volume in storage comprises sewage effluents, of which the amount of treated water is expected to rise from 29×10^6 m^3 in 1985 to reach 116×10^6 m^3 in 2005 and further to 165×10^6 m^3 in 2015. Such a rise can be attributed to the ongoing projects as well as the expected execution of planned sewage collection and treatment works in urban and rural areas.

The survey published in 1997 lists Al-Wehda Dam under the group of dams yet to be constructed. Actually, this dam which was first conceived in 1956 to store the waters of the Yarmouk River, could have soon been built to the east of Maqarin about 20 km north of the town of Irbid. The average annual river flow at the

Maqarin measuring site is 273×10^6 m^3. Based on an agreement between Jordan and Syria signed in 1988, preliminary work on opening an 800 m long diversion tunnel was completed by the end of 1989. The reservoir to be formed by the dam would have a gross capacity of 225×10^6 m^3, with an effective storage volume of 195×10^6 m^3 annually. Syria too would have used part of the stored water for land irrigation and as much as 75% of the total hydroelectric power generated by the dam for industrial as well as other purposes. Strong opposition from Israel, which sought more water from the downstream reach of the Yarmouk River so as to meet the increasing need for water in order to cope with its expanding- settlement policy, however, brought the project to a halt (Murakami & Musiake, 1994).

Small embankment-type dams up to 80 m high have been constructed since 1968 in several rift-side wadis. Among these embankments are: Ziglab (sometimes written Ziqlab) Dam with 4.3×10^6 m^3 gross capacity; Shu'heib Dam with 2.3×10^6 m^3 gross capacity; and Kafrein Dam with a total storage capacity of 8.5×10^6 m^3. The reservoir formed by the Wadi Arab Dam, which was completed in 1987, has a total storage capacity of 20×10^6 m^3. The last scheme, i.e. the Wadi Arab Dam was originally designed to store 30×10^6 m^3 of spring flow in the wadi; however, the feeding spring suddenly ceased to flow as a consequence of groundwater development works in its neighbourhood.

11.3.9 Yemen

Since the historical event of destruction of the Ma'rib Dam, which was built centuries ago for water storage in Yemen, no other storage dams have ever been built. Instead, construction of water storage tanks on stages has become a tradition since 2,000 years ago. Such storage tanks can still be found in the Tawilah Basin, Hadramaut, in southern Yemen. At present there are 18 of such tanks collecting seasonal rainwater and supplying the township of Aden by some 40×10^6gals y^{-1} (about 180,000 m^3 y^{-1}) mainly for domestic purposes. These storage tanks and the associated channels and other appurtenances are regarded as places and items tourists travelling to southern Yemen are keen to visit.

The construction of flow deflectors and earthern bunds (ogmas) has often been associated with traditional irrigation systems in Yemen for some time. The availability of modern construction equipment is making the construction of such works faster and easier under difficult conditions. Diversion systems have been designed and constructed in Wadi Zabid (1979), Wadi Rima (1983) and Wadi Mawr (1987). Other diversion works have been designed at feasibility level for Wadi Rasyan (1983) and Wadi Siham (1985). In about the same time the design of diversion structures and bank protection works of the Wadis Jizan, Dhamad and Surdud were carried out.

In the 1970s, the southern part of Yemen (formerly the People's Democratic Republic of Yemen) paid attention to benefit from the floodwaters flowing in the major wadis by constructing a number of small earth dams, mainly diversion dams, to regulate floods and thus provide adequate water supply for irrigation. As such,

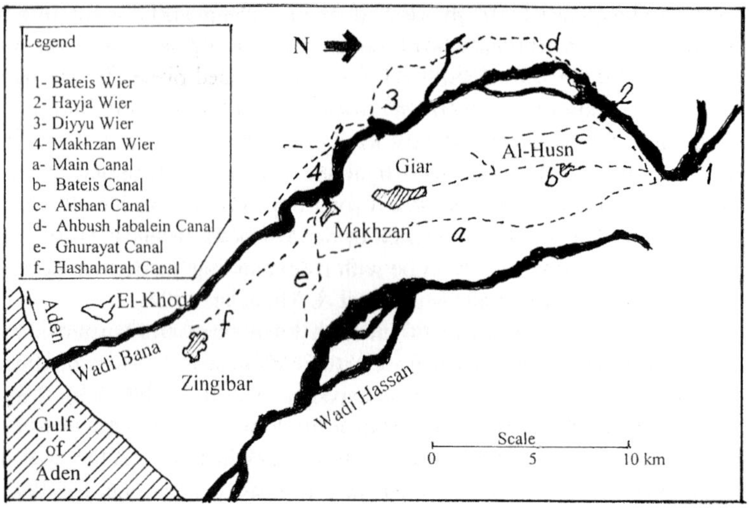

Figure 7. Map of the Abyan Delta project showing the locations of diversion structures and irrigation canals' systems (Macdonald, 1987)

these hydraulic structures aim at conserving surface runoff. Through cooperation with the former USSR technical aid, plans to construct eight small dams have been implemented since 1977. Different forms and sizes of ogmas (temporary diversion embankments), gabions and permanent weirs were built or planned for the purpose of water diversion to irrigation canals as well as flood protection. Figure 7 shows the locations of diversion works and irrigation canals in the Abyan Delta (Macdonald, 1987).

11.4. RECHARGE DAMS

11.4.1 General

Artificial recharge, despite its being a traditional experience, is steadily gaining ground in several countries in the Arab region. Examples can be found in the western sub-region and in the Arabian Peninsula. The advantages of artificial recharge are two fold: reduction of evaporation and other losses, and mitigation of the impacts of rainfall variability on agriculture. It could deal with a variety of problems such as aquifer depletion, salt water intrusion and contamination control. There are several ways for artificial recharging; open pits, recharge wells, water spreading and retention dams.

The major source of water for recharging groundwater aquifers in arid and semi-arid zones is wadi runoff. Groundwater dams may be used for temporary storage of water in near-surface aquifers. As mentioned earlier, there are two types of dams depending on whether the dam is built below or above the ground surface.

A recharging well is designed to absorb water coming from perforated pipes in the dam and its surface or subsurface reservoir. The recharging well is a radial collector well. It consists of a vertical shaft and radial perforated pipes. When the well is in operation, water passing through the back- pressure valves fitted at the end of the delivery pipes builds up in the well shaft. The resulting hydraulic head forces water to flow from the well shaft into the radial recharging perforated pipes ensuring a large contact area with the aquifer. At the end of the recharging process the sediments accumulated in the well are pumped. Traditional wells have been used in this way in Qatar. However, as the aquifer there is highly fractured, no radial screens are needed (UNESCO/ROSTAS, 1995).

Contrary to common belief, recharge mainly occurs downstream of the retention dam, not in the reservoir itself. The sediments accumulated in the reservoir, unless periodically cleaned, seal the bed and leads to a decline of the infiltration rate with time. Oppositely, water released from the reservoir contains small amounts of sediments, and easily infiltrates into the gravel aquifers downstream of the dam. Accordingly, the retention dams should be designed in such a way that the maximum discharge of the outlets ensures that the total wetted contact area in the downstream channel is sufficient to infiltrate the total release volume. Since most of the aflags or canats have been traditionally constructed in wadi fills and alluvial fans, their hydrogeologic set up is suitable for artificial recharge. An appropriate recharge method could therefore be used effectively to sustain the flow of foggaras.

11.4.2 Experiences of some countries with recharge dams

Some of the Arab countries such as Saudi Arabia, Yemen, Oman and the United Arab Emirates have been for quite some time embarking on dam construction programs to utilise as much as possible the surface water runoff and enhance a better management of their limited water resources. The number of dams constructed from the mid 1980s to the mid of 1990s has reached 256 distributed as follows:

Country	*Number of constructed dams, 1985–95*	*Storage capacity, 10^6 m^3*	*Number of planned dams*
Saudi Arabia	*190*	*0.01–325*	*9*
United Arab Emirates	*35*	*0.25–18.5*	*5*
Yemen	*17*	*0.50–369*	*12*
Oman	*14*	*0.01–12.5*	*32*
Total	*256*	*0.01–369*	*58*

Due to major differences in topography and relief of the country, one finds that the majority of the recharge dams in Saudi Arabia and Yemen have been built in mountainous regions because of the availability of surface runoff generated by the frequently occurring rainfall events. Differently, most of the recharge dams in Oman and the United Arab Emirates are built on wadi channels in flat areas, especially those close to the seacoast. Because of their flat topography, low land level and limited runoff due to scanty rainfall in the remaining countries of the Arabian Peninsula; Bahrain, Qatar and Kuwait, small diversion structures are built instead of retention dams to create the necessary detention basins for recharging the aquifers (UNESCO/ROSTAS, 1995).

The majority of small recharge dams are earth fill dams constructed of sands and gravels. They are commonly used in flat areas of low elevation and thickness of alluvial material exceeding 15 m. Besides, it is preferable, for the ease of transportation and thereupon reduced construction costs, that the dam site should be selected in a location with abundance of earth material for construction. Rock fill and concrete types are more suitable than earth fill for dams located in mountainous areas with a possibility of having extremely heavy torrential floods every now and then.

After this general introduction, experiences gained by some of the Arab countries using dams to recharge their aquifers are briefly described in the next paragraphs. Emphasis will be underlain on the experiences gained by Saudi Arabia not only because of the large number of retention dams it has but also because of the extensive and comprehensive research works carried out on them.

11.4.2.1 Tunisia

In addition to the traditional water systems, which contribute to the recharge of unconfined aquifers, Tunisia has developed a national plan for a wider-scale application of artificial recharge techniques. The main problems addressed are aquifer depletion, salt-water intrusion in coastal areas and increasing shortage of fresh-water resources. Some 30 wells were used in Tabalbah coastal area for injecting an alluvial aquifer. About 14×10^6 m^3 was recharged underground during a period of 757 days. The basin method was applied to an area in southeast Tunisia for underground storage of 100,000 m^3 (Rekaya, 1992).

A study based on comparison of different methods has indicated that recharge through wadi fill is the most viable method in arid zones. Spreading basins could be used if wadi fill and underlying aquifers do not furnish adequate media for artificial recharge. Basins need periodic cleaning from the silt deposited in them and so do recharge wells.

According to UNESCO/ROSTAS (1995), the amount of water stored underground in Tunisia reached 40×10^6 m^3 in 1994 and was expected to reach 100×10^6 m^3 in the year 2000. Sources of water to recharge the aquifers are flood flow, treated wastewater and untreated wastewater (Chapter 9).

11.4.2.2 Oman

Apart from a number of small storage tanks and birkats (birkat is an Arabic word meaning a pond or pool of water) in mountainous regions, Oman has no storage reservoirs, but only flood control and recharge dams. The common situation is so that wadi flows for most of the country's mountains meet up with large alluvial fans, either in the interior or along the coast, where flows are largely left to infiltrate the alluvial material. The infiltrated water recharges the underlying aquifer and becomes accessible for downstream use through abstraction wells.

According to Schmid & Al-Batashi (1995), extensive surveys and studies have indicated that the most efficient means to augment water resources in Oman is by construction of retention dams. Compared to other means of aquifer recharge, only storage of water in a surface reservoir does not affect adversely the existing water supply systems. Further, it is a traditional and well-known method in Al-Jabal Al-Akhdar area of Oman. It is therefore fully accepted by the villagers without any undesirable social effects on the communities occupying these areas.

In accordance with the above findings and comments, geotechnical investigations were carried out and new dam designs were developed to cope with the requirements of a project involving dam construction in 11 villages for domestic and irrigation purposes. The reservoir capacity ranged between 1,250 m^3 and 8750 m^3. The new dam design is similar to the existing dams in Al-Jabal Al-Akhdar, except that it meets the safety standards of modern dam engineering and therefore, the new dams appear to be voluminous compared to the previously built dams in the same locations.

Figures 8(a) and 8(b) show a plan view and cross-section of the newly designed retention dam (Schmid & Al-Batashi, 1995). The dam body consists of rubble masonry with a vertical concrete panel, which provides an impermeable sealing element for the dam. This panel is located upstream and connected to the cutoff wall, which counteracts under-seepage. The depth of the cutoff is adjusted to local underground conditions with a minimum depth of 1 m and anchored to the bedrock. At the wadi sites, the cutoff wall has been interlocked into the wadi slopes.

Surface water stored behind dams in Oman usually remains for periods not exceeding 15 days each, after which it is diverted to scattered areas adjacent to the natural channels of wadi courses. These areas usually exist in the downstream reaches of wadis, and water reaches them through channels especially dug for this purpose.

Most of the recharge dams, whether existing or projected are located in the coastal plains of Oman. The relief of the Batinah is uniform with typical wide spreading wadis towards the coast. The topography of the coastal wadis renders the reservoir width quite large (several kilometers) while its length might be limited to just a few hundred meters. The topography is more variable for recharge dams in the interior. Some sites correspond well with the uniform topography of the Batinah coast while others are typical mountainous sites with steep rock abutments, like in the narrow gorge of Wadi Tanuf, forming elongated reservoirs. The main features of recharge reservoirs in Oman are listed in Table 36, Appendix II (Sources: Abdel Rahman & Abdel-Magid, 1993; Al Muqbali & Schmid, 1995; and Muqabali & Kotwicki, 2000).

Other than the dams listed in Table 36, Appendix II, there is a 1.63 km long sea defense dam constructed at khor Al-Rusaq in Sur to prevent the intrusion of saline

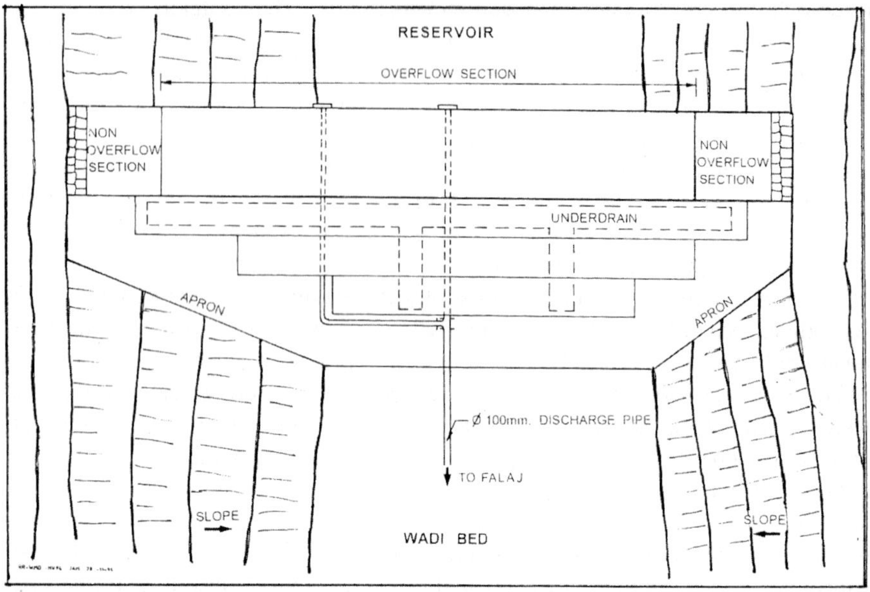

Figure 8(a). Plan view of a new retention dam (Schmid & Al-Batashi, 1995)

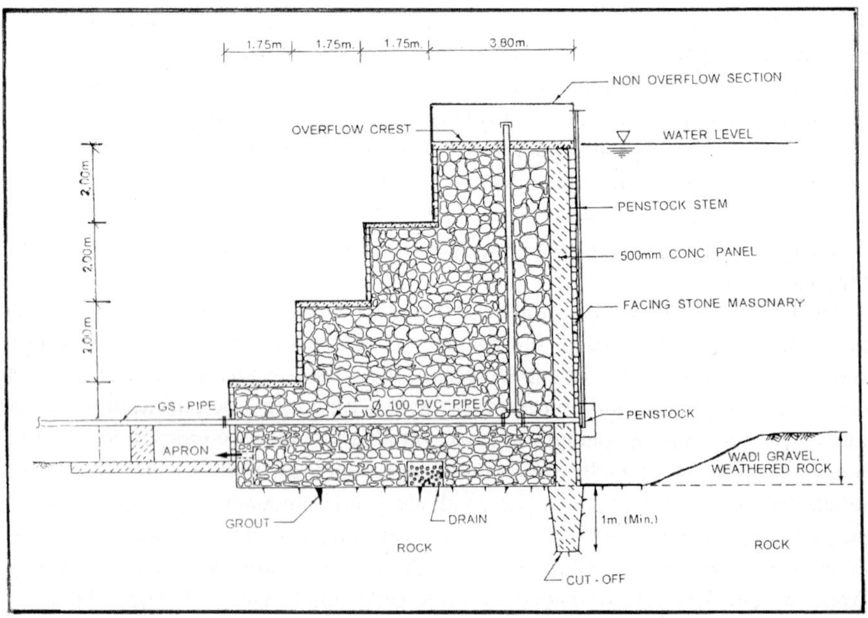

Figure 8(b). Typical cross section of a new retention dam (Schmid & Al-Batashi, 1995)

seawater and withhold part of wadi waters for recharge and leaching. Recharge dams in Oman aim at collecting part of the 120×10^6 m^3 of rainwater lost annually to the sea and/or desert. Lack of monitoring systems in most of the recharge dams made the assessment of their efficiency rather difficult. Visual surveys, however, have suggested that, on the average, dams have caught a sizeable amount of runoff. One of the major disadvantages is that the storage dams by having the sediments, carried by the surface runoff and being entrapped in their reservoirs, have deprived the coastal plains of the fertile silt the streams used to carry down the wadi valleys (Abdel Rahman & Abdel-Magid, 1993).

For all recharge dams, the spillway is the most dominant structural element. In general, the spillway design flood is often taken as 0.5 the probable maximum flood. Volumes of moderate floods, which are essential to determine the reservoir storage, are obtained from frequency analysis of reported flood volumes. Extreme floods are evaluated with unit hydrograph methods and compared with the flood frequency curves for Oman, which give the instantaneous flood peaks against the catchment area.

11.4.2.3 United Arab Emirates (U.A.E.)

The situation in the United Arab Emirates is not much different from that in Oman. The majority of the dams in these two countries as well as in other countries in the Arabian Peninsula are built for recharging the depleted aquifer systems. Dams are mainly built near the wadi outlets on the sea to prevent or at least to reduce the loss of some of the surface runoff. Instead, the retained water is left to infiltrate the ground surface and thereafter percolate the soil column to join the groundwater in the underlying aquifer. Based on the results obtained from geophysical investigation, a plan has been laid down for building retention dams in the following areas:

- Ras el-Kheima Area: Wadi Shahm (catchment area = 34.5 km^2), Ghalilah (70 km^2), Bieh (474.4 km^2) and Naqb (93.3 km^2),
- The Gravely Plain Area: Wadi Siji (86.6 km^2), Umm el-Nuqoul and al-Zayd,
- Al-Batinah Plain Area: Wadi Anina, Wadi Ham (190 km^2) and Kalba.

In the U.A.E. and Oman most of the dams built in the lowlands are claimed to be achieving their multiple objectives; aquifer recharge, flood protection of farmlands and reduction of saltwater intrusion. The storage of water in the reservoirs of al-Hod, Khilts and Quryat, all located in the coastal plain of U.A.E., during the severe floods of 1986–88 resulted in a rise of the water table in the range from 0.3 to 2.0 m. As expected, high rises of water level were reported in the main wadi channels, and salinity of water in some of the wells fell from 25,000 ppm to 10,000 ppm and in other wells from 10,000 ppm to 6,000 ppm. In cases where retention reservoirs do not serve their intended purposes, issuing problems can be attributed to poor maintenance and inappropriate rules of reservoir operation. Prolonged impoundment of water results in considerable loss of water by evaporation and increased accumulation of sediments in the reservoirs. Lack of regular release of water, especially during periods of low water level in the reservoir, contributes significantly to water loss through evaporation (UNESCO/ROSTAS, 1995). In the summer season when the rate of evaporation is in the range of 12 – 15 mm d^{-1} the corresponding volume of evaporatied water is 120- 150 m^3 ha^{-1} d^{-1}.

An example of the contribution of surface water runoff to groundwater recharge through some of the relatively large dams in the U.A.E., like those in the Wadi Ham, can be observed from the graphic plots shown in Figure 9. The rise in water table, as can be seen from the performance of some of the boreholes, depends on the amount of surface runoff brought by the flood and the location of the borehole. The water table level soon after reaching its peak level declines rapidly, especially in the first part of the dry season. This is followed by a slower rate of decline till the next flood event occurs and the cycle of water table rise and fall repeats itself, though at a different amplitude.

11.4.2.4 Saudi Arabia

Recharge dams are used in several parts of Saudi Arabia. The idea behind such dams is to hold the flowing surface water in the particular wadi at the site where recharge to the underlying aquifer is desired. Aquifer recharge in this method is

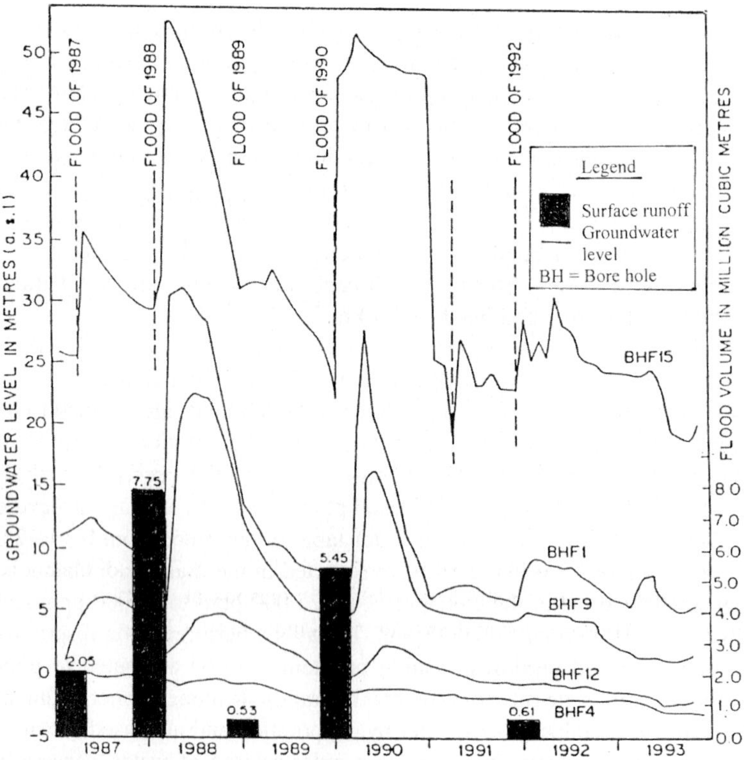

Figure 9. Runoff contribution to groundwater recharge through Wadi Al-Ham reservoir in the U.A.E. (redrawn from UNESCO/ROSTAS, 1995)

accomplished by supplying water to the aquifer from a surface source of relatively large area. This method is often criticised on the grounds that the evaporation loss from a storage reservoir in an arid area is certainly high and secondly, the rate of infiltration into the aquifer decreases with accumulation of silt and other deposits in the reservoir. The latter disadvantage can be eliminated or reduced by periodic removal of these deposits, e.g. dredging, which adds to the expenses of the recharging operation.

The 1982 statistics available at the Saudi Ministries of Planning, and Agriculture and Water list 66 dams which were completed or under construction as of September 1981. Of those dams 45 were built for aquifer recharge, 18 for recharge and flood control, and only 3 for purposes excluding recharge. Table 37, Appendix II, lists some of the design features of 45 dams; 25 constructed and 24 under construction or planned to be constructed. According to Al-Turbak & Al-Muttair (1989), the central part of Saudi Arabia comprised 35 dams, all within a 250 km radius from Riyadh. These dams differ in watershed area, year of completion of construction, dimensions such as average and maximum heights, length, and storage capacity, and in the construction material and type of construction.

Al-Turbak & Al-Muttair (1989) chose the Malham and Al-Maliha dams for a detailed and more objective evaluation of the effectiveness of dams as a means of artificial recharge. The Malham reservoir drains a catchment area of 289 km^2 and has a storage capacity of 0.5×10^6 m^3. Its construction was completed in 1970 and the silt deposits in the reservoir by the time it was investigated, 1985–86, ranged between 0.3 m and 1.2 m. The second dam, i.e. Al-Maliha, has a catchment area of 21.6 km^2 and a storage capacity of 1.0×10^6 m^3. The dam construction was completed in 1982 and by the time the study was carried out, the thickness of deposits in the reservoir ranged between 0 and 0.3 m.

The investigation, which began in 1985, comprised data collection; installation of new observation wells; providing each well with a water level recorder; etc. It also included topographic surveys of the two reservoirs, geophysical investigation of the reservoir in the U.A.E. (redrawn from UNESCO/ROSTAS, 1995) relevant aquifers and measuring rates of infiltration of the reservoir beds and wadi channels. A water budget model of a reservoir was developed to calculate the daily volumes and rates of infiltration from it for recharging the aquifer over a given period of time. The model was based on the set of equations (1)-(3), already described in sec. 11.2, together with the reservoir depth-volume and depth-surface area relationships.

The results obtained for the 1985/86 and 1986/87 seasons at the two recharge dams; Al-Malih and Malham, located in Central Saudi Arabia have shown that the dams are efficient in recharging their stored water. Evaporation amounts were relatively small when compared to the recharge volumes. The evaporation expressed as percent of the total water leaving the reservoir by evaporation and infiltration, was significantly high, up to almost 60% during summer months and the periods when water in storage was small. However, the effect of those high percentages on the overall efficiencies of the reservoirs was limited (Al-Turbak & Al-Muttair,

1989). Figure 10 shows the variation of recharge rate with depth to water in the Al-Malih reservoir below the spillway crest (Al-Muttair et al., 1986).

The most interesting finding related to the above study is that not all water infiltrated from the reservoir bed could be used downstream. A true measure of the effectiveness of the reservoir is to determine that proportion of the infiltrating water which flows downstream of the dam as groundwater. The previous study was extended to comprise a third season, 1987/88, using the available data of Al-Malih reservoir to determine the groundwater flow downstream of the dam. The second objective of the more recent study was to compare the groundwater flow downstream of the dam with the volume that infiltrated from the reservoir, using the water budget method as a tool for the required comparison. Last, but not least, was to study the time-variation of the results obtained over the three seasons 1985/86, 1986/87 and 1987/88.

The volume of water evaporated, E, was calculated by multiplying the evaporation from class A pan by an adjustment coefficient times the average surface area of the reservoir over the period chosen for the water balance; in this case one day.

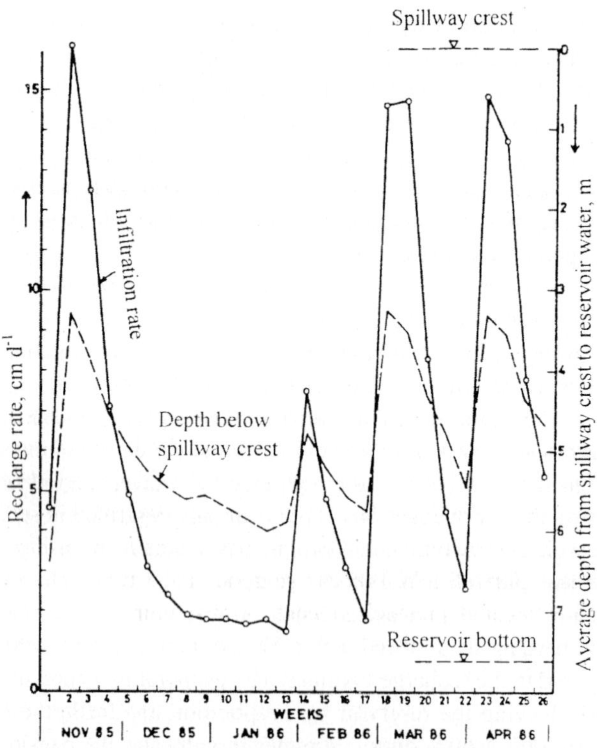

Figure 10. Weekly recharge rate and average depth to reservoir water from spillway crest of Al-Malih Reservoir, Central Saudi Arabia (Al-Muttair et al., 1986)

The relations depth-volume and depth surface area of the reservoir were applied for the calculation of the daily change in storage, ΔS. These two parameters were substituted in Eq. (2) on daily basis to determine the daily volume of water infiltrated from the reservoir. Table 1 gives the monthly volumes of water infiltrating from Al-Malih reservoir, F, for the three consecutive seasons 1985/86, 1986/87 and 1987/88.

The said Table includes also the volumes of infiltrated water as percentages of the corresponding volumes of water stored in the reservoir. These percentages, termed %*, is a measure of the reservoir's efficiency as a recharging means.

Daily subsurface inflows were calculated using Darcy's formula, which states that subsurface flow equals the product of the average gradient times the hydraulic conductivity of the aquifer multiplied by the cross sectional area of flow. The subsurface outflows were calculated in the same way as the inflows. The change in groundwater storage, ΔS_g, for each of the three seasons was estimated from the water levels at the beginning and end of the year, distance between observation wells, width of the aquifer measured from the available geophysical cross-sections, and estimates of porosities obtained at the dam site. Al-Hassoun & Al-Turbak (1995) described the calculations in details.

The net subsurface outflow, i.e. total subsurface outflow minus the subsurface inflow, divided by the volume of infiltrated water yields the percentage %**, which is a second measure of the efficiency of the recharge dam. This measure is more accurate than %*; the average %** for the three seasons is 75.1 compared to 94.5, the three seasons average for %*. The average reduction is in the order of 20%. Last, but not least, the third percentage, %***, represents the volume of net subsurface outflow (caused by the reservoir alone) divided by the volume of water stored in the reservoir in the same month. By virtue of this definition, the third measure of reservoir efficiency, %*** = %* multiplied by %**. Table 1 gives the annual values of %*** as 62.1, 71.7 and 90.4 for the years 1985/86, 1986/87 and 1987/88, with a 3-y average of 71.0, i.e. about 94.5% of %** and 75.1% of %*.

The above presentation shows that %*** is the most discriminating measure for assessing the efficiency of the reservoir as a means of artificial recharge of an aquifer. The average for the three years of study being 71% is still a reasonably high figure for the efficiency of a recharging dam.

Al-Muttair et al. (1994) examined the effect of management of the Malham and Al-Malih Recharge Dams in Central Saudi Arabia on their recharging efficiencies. Five alternative management plans were applied to the said dams. The management plans included the system that prevailed before carrying out the study, release of reservoir water to the downstream channel, release of reservoir water to a downstream basin, removal of the accumulated silt from the reservoir bed, and scratching of the reservoir bed. The results of field applications of the five management plans were not easily comparable due to changes of the configuration of reservoir bed, and the different magnitudes and times of occurrence of flood events throughout the seasons covered by the study period. This can be seen from the figures listed in Table 2. A rational comparison of recharge efficiencies was

Table 1. Results obtained from recharge dam efficiency study in Saudi Arabia based on subsurface flow analysis (Al-Hassoun & Al-Turbak, 1995)

Month & year	F, m^3	%*	Subsurface inflow, m^3	Subsurface outflow, m^3	Outflow due to reservoir, m^3	%**	%***
Nov. 1985	268660	95.8	4315	72016	67881		
Dec.	25113	88.3	6045	61919	55874		
Jan. 1986	36798	88.6	3978	43362	39382		
Feb.	36798	88.6	3082	37876	34794		
Mar.	355139	96.1	9505	98225	88720		
Apr.	305325	95.7	12743	107915	95172		
May	20874	66.9	12654	102892	90238		
Jun.	4329	36.4	9162	75978	66816		
Jul.	1719	31.8	6179	57364	51185		
Aug.	181	27.0	3478	39687	36209		
Sep.	83	27.0	1490	29937	28447		
Oct.	0	(-)	1108	23993	22885		
Total	1016844	93.3	73560	751164	677605	66.6	62.1
Nov.	408	84.5	1018	16228	15210		
Dec.	1864	78.0	1047	18417	17370		
Jan. 1987	393	72.2	1043	20709	19666		
Feb.	116	48.2	939	20253	19314		
Mar.	577383	97.1	9560	106138	96578		
Apr.	95147	88.6	10700	112247	101547		
May	7462	59.1	10253	88225	77972		
Jun.	2105	41.3	6849	62654	55805		
Jul.	570	30.0	4369	45936	41567		
Aug.	0	(-)	2089	33868	31779		
Sep.	0	(-)	1040	27070	26030		
Oct.	0	(-)	1055	18412	17357		
Total	685340	94.5	49962	570157	520195	75.9	71.7
Nov.	0	(-)	1019	18777	17758		
Dec.	0	0.0	1042	24273	23231		
Jan. 1988	123662	98.8	2512	54247	51735		
Feb.	17984	94.1	2456	35992	33536		
Mar.	1237	64.7	1792	32440	30648		
Apr.	261469	96.1	3174	55199	52025		
May	10994	76.6	6660	69877	63217		
Jun.	427	30.0	3779	42186	38407		
Jul.	0	(-)	1724	30536	28812		
Aug.	0	(-)	1134	23045	21911		
Sep.	0	(-)	786	17716	16930		
Oct.	0	(-)	561	14662	14101		
Total	415773	95.8	26639	418950	392311	94.4	90.4
Grand total	2117957	94.5	150161	1740271	1590111	75.1	71.0

%* = 100 *ltrated water/volume of water stored in the reservoir), %** = 100 (volume of sub-surface outflow due to reservoir alone/volume of infiltrated water) and %*** = %** x %*.

Table 2. Results obtained from a simulation model using five alternative management plans for two recharge dams in Central Saudi Arabia (Al-Muttair et al., 1994)

Parameter		% efficiency for reservoir and alternative			
		Malham reservoir		Al-Amalih reservoir	
		1	4	1	5
Flood volume for hypothetical	0.25	68.48	82.95	89.35	95.62
flood season, 10^6 m^3	0.50	80.96	88.62	92.36	96.88
	1.00	87.60	92.14	94.55	97.76
Alternative		2-1*	2–4**	3A-1^x	3B-1^{xx}
Total inflow, 10^6 m^3		0.500	0.500	0.500	0.500
Released volume, 10^6 m^3		0.147	0.094	0.029	0.044
Efficiency in downstream channel, %		95.72	95.84	91.45	91.59
Efficiency of reservoir alone, %		77.65	87.68	92.56	92.54
Efficiency of alternative, %		82.97	89.22	92.50	92.46

Explanation
2-1* = *Water release before silt removal,* 2-4** = *Water release after silt removal,* 3*A*-1^x = *Water release whenever the daily drop in reservoir water level is less than that in the basin, and* 3*B*-1^{xx} = *Release of water from the reservoir during periods of low evaporation provided that there is sufficient head to make the release possible.*

made through simulations. The results obtained have shown that for the Malham Reservoir (alternatives 1 and 4) and Al-Amalih Reservoir (alternatives 1 and 5) the recharge efficiency of each reservoir increased with increasing inflow volume. This can be attributed to the increase of the ponded head, and hence increased infiltration rate, extending the stored water further upstream from the dam, where the infiltration rate is greater than that near the dam site. Table 2 shows that for the three inflow levels considered, the recharge efficiency has improved as a consequence of the removal of the sediments in the Malham reservoir. The improvement is largest for the smallest inflow and smallest for the largest inflow, with an average of bout 12%.

As the Amalih reservoir was new it was very efficient in the recharging process, even before applying any of the suggested alternatives, as be seen from the figures in Table 2.

Recharging efficiencies with Alternative 5 when compared to those with Alternative 1 showed an average improvement for the three inflow levels of 5.1%. Additionally, the results obtained from simulation of Alternative 3 regarding release to a recharge basin, using two release policies (Alternatives 3 A-1 and 3 B-1) showed an insignificant improvement in the recharge efficiency over that of Alternative 1.

In general, It can be stated that field experiments have shown that the two alternatives involving silt removal and scratching of the reservoir bed silt produced reasonable amount of improvement in the recharge efficiency.

CHAPTER 12

WATER SCARCITY IN THE ARAB REGION-MAJOR PROBLEMS AND ATTEMPTS TO ALLEVIATE THEIR IMPACTS

12.1. WATER SCARCITY

12.1.1 Addressing water shortage

Scant rainfall and expansive desert surfaces attest to the limited availability of water to the land area in the Arab Region. Existing water resources are also in short supply when measured against population. The average rate of population growth in the region being one of the highest in the world is causing a considerable demand for water. The projected demand for water will more than double between 1985 and 2000 in Jordan and Oman; in Morocco, Tunisia, Egypt, Sudan and Yemen it will increase by half (Shahin, 1989).

Water resources have only a limited capacity for matching the burgeoning population growth. Surface runoff can be captured and stored for a short-term net gain, as can redirecting and damming surface water and reusing or reallocating water supplies to more critical or higher value uses. Exploiting deep groundwater reserves, however, consumes a nonrenewable resource and the currently available desalinisation technology is prohibitively expensive for all but the wealthiest states in the region.

None of the above mentioned remedial measures will adequately expand the resource base enough to meet expected future demands. Given the unlikelihood of dramatically expanding water resources in the Arab region, there are primarily six ways to address the water shortage problem: (a)- inter-basin transfer, (b) reallocation, (c) water conservation, (d)- wastewater reuse, (e)- institutional innovations, and (f)- technological breakthroughs (United States Agency for International Development, 1993). Since items (a), (b), (c), and (d) have already been covered in earlier chapters, mainly 10 and 11, we shall deal here only with the last two items, i.e. (e) and (f).

12.1.2 Demography and share in water resources per capita in the region

It is a known fact that there is hardly any economic activity whether in production of commodities or provision of services that can be accomplished or sustained without availability of water. The same condition applies to social well-being. In addition to meeting the basic requirements of human, animal, and plant life, water is needed to preserve the environment and to maintain acceptable personal and household hygienic standards (Haddadin, 2001). Consumption of water in a country tends to increase with its economic development. The available figures for the year 1990 show that the gross domestic production (GDP) for Tanzania was US $ 110 while the sum of the annual withdrawals for domestic, industrial and agricultural sectors was 36 m^3 cap^{-1}. This pair of figures rises to US $ 2,530 and 404 m^3 cap^{-1} for South Africa, and further to US $ 21,790 and 2,162 m^3 cap^{-1} respectively for the United States.

Besides, with adequate supply of energy inputs to distillation plants, man can turn saline or brackish water into fresh-water. The resulting brine, as mentioned earlier, is mostly returned back into the sea, but not without adverse impacts on local marine environment.

From the foregoing chapters, the reader should have realised the extent of scarcity of water in most parts of the Arab Region. Exceptions can be found in certain areas located in the north west of the region in Morocco, the thin coastal strip in Algeria and Tunisia along the Mediterranean, the coastal and mountainous strips of Lebanon and Syria, the southern part of the Sudan, the Comoros and the extreme south west of the Arabian Peninsula.

It is customary to express the extent of water scarcity by an index relating the quantity of water withdrawn in a unit of time to the number of consumers in a certain area, e.g. X-m^3 y^{-1} cap^{-1}. It goes without saying that while the withdrawals of water in a certain region do not change substantially over short periods of time, the number of consumers can show a more substantial change depending primarily on the growth rate of population.

Table 1 gives population estimates of the countries of the region in the period from 1950 up to 2000 and the U.N. ‘medium’ projections for the years 2025 and 2050 (United Nations Population Division, 1994, and World Resources Institute, 1998). The population estimate of about 90.8×10^6 for the entire region in 1955 reached 276×10^6 inhabitants for the year 2000. This means a three-fold increase in the population number in less than 50 years. The predicted figures for the years 2025 and 2050 are expected to reach about 482×10^6 and 665×10^6 consumers respectively, i.e. an increase of about 240% in the first half of the 21st century.

The last decades have witnessed an increasing amount of attention dedicated to water resources development and management in the Arab Region. Trusting that the data given in Table 1 do not depart sensibly from the actual figures, one needs then an equally reasonable estimate of the available quantity of water per country

Table 1. Population and population growth in the Arab Countries (sources: United Nations Population Division, 1994, and The World Resources Institute, 1998)

Country	Population estimates, 10^3 inhabitants, for the years								Growth
	1950	1955	1982	1990	1995	2000	2025	2050	% ge
Mauritania	825	901	1730	2003	2274	2446	4443	6077	2.02
Morocco	8953	10132	21670	24334	27028	28774	40650	47885	1.69
Algeria	8753	9715	20294	24935	27939	30010	47322	58991	1.93
Tunisia	3530	3860	6670	8080	8896	9377	13290	15607	1.50
Libya	1029	1126	3220	4545	5407	6084	12285	19109	2.96
Subtotal	23090	25734	53584	63897	71544	76691	117990	147669	1.87
Egypt	21834	24692	44670	56312	62931	67046	97301	117398	1.76
Sudan	9190	10150	19450	24585	28098	30712	46850	59947	1.89
Djibouti	N.A.	69	N.A.	517	N.A.	611	1055	1403	1.07
Somalia	3072	3401	5120	8677	9250	10967	23669	36408	2.50
Comoros	N.A.	194	N.A.	543	N.A.	710	1646	2484	2.72
Subtotal	N.A.	38506	N.A.	90634	N.A.	110046	170521	217640	1.84
Syria	3495	3967	9660	12348	14661	16023	26903	34463	2.31
Lebanon	1443	1613	2740	2555	3009	3208	4424	5189	1.29
Jordan	1237	1447	3490	4259	5439	6199	12039	16874	2.65
Palestine	1560	4360	3440	4360	5060	6175	12270	24413	2.79
Iraq*	5158	5911	14000	18078	20449	23006	42656	57691	2.44
Kuwait*	152	199	1560	2143	1574	1733	2805	3384	3.15
Bahrain	N.A.	134	330	490	N.A.	609	922	1046	2.19
Qatar	N.A.	35	N.A.	485	N.A.	682	799	889	3.46
S. Arabia	3201	3593	9680	16048	17880	21546	42651	60897	2.99
U. Arab Emirates	70	79	790	1671	1904	2466	2958	3423	3.97
Oman	456	503	950	1751	2163	2305	6094	10005	3.14
Yemen	4316	4734	7710	11311	12779	14437	39589	61129	2.69
Subtotal	N.A.	26575	N.A.	75499	N.A.	89389	194120	279403	2.51
Total	N.A.	90815	N.A.	230030	N.A.	276126	482631	664712	2.08

Explanation
The last column gives the percentage rate of growth based on the assumption that the growth rate over the period from 1950 or 1955 up to 2050 is constant.
The growth rate ranks 1 for Djibouti, 2 for Lebanon, 3 for Tunisia, 4 for Morocco, 5 for Egypt, 6 for the Sudan, 7 for Algeria, 8 for Mauritania, 9 for Bahrain, 10 for Syria, 11 for Iraq, 12 for Somalia, 13 for Jordan, 14 for Yemen, 15 for Comoros, 16 for Palestine, 17 for Libya, 18 for S. Arabia, 19 for Oman, 20 for Kuwait, 21 for Qatar, and 22 for United Arab Emirates. The central sub-region ranks as 1, the western sub-region as 2 and the eastern sub-region as 3.

so as to obtain an estimate of the water availability-index. Table 2 provides the reader with the salient statistics of water availability and withdrawal as produced by different authorities (United Nations Population Division, 1994, World Resources Institute, 1998 and Seckler et al., 1998). The figures listed in Table 2 for available water serve as a basis for computing the share of water per capita. The results are listed in column B for the year 2000.

Table 2. Salient statistics of water availability and withdrawal in the Arab Countries (Council on Environmental Quality, 1982, United Nations Population Division, 1994, World Resources Institute, 1998 and Seckler et al., 1998)

Country	Annual water resources (AWR), 10^9 m^3										Available water, 10^3 m^3 cap^{-1} for the years				
	Precipitation	Stream flow	Ground-Water abs-Traction	Desalinated water	Waste-water reuse	IRWR^x	WWD^{xx}	UN Popultion Division	IWMI^o	WWD^{xx}, % of AWR	2000	2000	2000	2025	2050
						1992	1970/87	1994	1998	1998	A	B	C	B*	B**
Mauritania	157.2	1.0	0.9	0.002	0.000	0.40	0.73	8.88	N.A.	N.A.	N.A.	3.630	N.A.	1.999	1.461
Morocco	82.4	22.5	3.5	0.006	0.100	30.00	10.85	28.00	30.0	36.0	0.900	0.973	0.427	0.689	0.585
Algeria	192.5	13.0	2.0	0.075	0.050	18.90	3.00	17.20	14.3	30.0	1.000	0.573	0.180	0.363	0.242
Tunisia	39.8	2.7	1.5	0.019	0.000	3.75	2.30	4.36	N.A.	N.A.	0.400	0.465	N.A.	0.328	0.279
Libya	49.0	0.0	1.7	0.248	0.000	0.70	2.62	4.87	N.A.	N.A.	1.200	0.800	N.A.	0.396	0.255
Subtotal	520.9	39.2	9.6	0.350	0.150	53.75	19.50	63.31				0.826		0.537	0.429
Egypt	15.3	55.5	3.8	0.032	4.700	2.60	56.40	64.03	68.5	97.0	0.050	0.990	0.956	0.658	0.545
Sudan	1094.4	20.5	0.9	0.001	0.000	30.00	18.60	120.78	154.0	12.0	1.900	3.930	0.633	2.578	2.015
Djibouti	190.6	0.0	0.02	0.000	0.000	0.30	0.01	0.05	N.A.	N.A.	N.A.	0.080	N.A.	0.047	0.036
Somalia	4.0	8.2	0.3	0.000	0.000	11.50	0.81	8.50	N.A.	N.A.	1.800	0.775	N.A.	0.359	0.233
Comoros	6.0^o	0.9^{oo}	0.0	0.000	0.000	*1.02*	*1.02*	1.02	N.A.	N.A.	N.A.	1.411	N.A.	0.620	0.454
Subtotal	1310.3	85.1	5.0	0.033	4.700	45.42	76.84	194.26				1.765		1.139	0.893
Syria	52.7	22.1	2.7	0.003	0.126	7.60	5.60	25.79	26.3	9.0	1.000	1.610	0.435	0.956	0.748
Lebanon	6.8	3.9	0.2	0.000	0.000	4.80	0.75	4.98	N.A.	N.A.	N.A.	1.552	N.A.	1.126	0.960
Jordan	6.7	0.5	2.0	0.003	0.077	1.02	0.45	3.05	N.A.	N.A.	N.A.	0.666	N.A.	0.253	0.181
Palestine	8.0	0.1	0.2	0.000	0.050	*0.80*	*0.60*	0.43	N.A.	N.A.	N.A.	0.070	N.A.	*0.035*	*0.018*

Iraq	99.9	43.2	1.5	0.121	0.000	34.00	42.80	42.66	75.4	43.0	1.300	1.948	4.575	1.000	0.739
Kuwait	2.4	0.0	0.1	0.516	0.430	0.00	0.50	0.16	N.A.	N.A.	0.300	0.606	N.A.	0.057	0.048
Bahrain	0.05	0.0	0.1	0.109	0.045	0.00	0.00	0.09	N.A.	N.A.	0.300	0.411	N.A.	0.098	0.086
Qatar	0.8	0.0	0.1	0.145	0.075	0.00	0.00	0.05	N.A.	N.A.	0.300	0.337	N.A.	0.063	0.056
S. Arabia	126.2	2.0	2.3	1.307	0.170	2.20	3.60	5.78	2.4	164.0	N.A.	0.268	0.497	0.136	0.095
U.A.E.	7.4	0.2	0.3	0.604	0.050	0.15	0.90	0.49	N.A.	N.A.	0.300	0.462	N.A.	0.166	0.123
Oman	15.0	0.9	0.4	0.059	0.075	0.99	0.48	1.93	N.A.	N.A.	0.300	0.837	N.A.	0.317	0.193
Yemen	67.2	3.8	1.3	0.014	0.100	4.10	3.40	5.20	4.1	136.0	0.300	0.360	0.335	0.131	0.085
Subtotal	393.1	76.7	11.2	2.881	1.198	55.66	59.05	50.26				0.562		0.259	0.180
Total	2224.3	201.0	25.8	3.264	6.048	154.83	155.39	307.83				1.115		0.638	0.463

Explanation

IRWR[x] *= Internally renewed water resources, WWD*[xx] *= water withdrawal (World Resources Institute, 1998), IWMI*[o] *= Irrig. Wat.Mgmt. Institute,*

A = Council on Environmental Quality & Dept. of State, 1982,

B = United Nations Population Division, 1994, C = World Resources Institute, 1998, B = same as 1 assuming a medium projection for population growth in the period 2000–2025, B** = same as B assuming a medium projection for population growth period 2000–2050, and the underlined figures in Italics are obtained from other sources.*

12.1.3 Projection of future shares per capita of the available water resources

The share per capita for the year 2000 listed in Table 2 ranges from a maximum of more than 3,600 m^3 y^{-1} for each of the Sudan and Mauritania to close to 1,950 m^3 y^{-1} for Lebanon and further to a minimum of 70 m^3 y^{-1} for Palestine. The share is largest (1,765 m^3 y^{-1}) for the middle section of the region. This is followed by 826 m^3 y^{-1} for the western sub-region and finally 562 m^3 y^{-1} for the eastern sub-region including the Arabian Peninsula. The average rate for the whole Arab Region in the year 2000 has been estimated as 1,115 m^3 y^{-1}.

Assume that the annually available water resources remain unchanged for the various States and sections of the region. The population data in Table 1 for the years 2025 and 2050 with the available water resources figures have been used for estimating the shares per capita for these two years. The results obtained are listed in columns B* and B** of Table 2 for 2025 and 2050 respectively. The global share for the region drops from 1,115 in 2000 to 638 and 463 m^3 y^{-1} for 2025 and 2050, respectively. The figures for 2025 and 2050 represent 57.2 and 41.5% respectively of the share of the available resources for the year 2000. The continuous decline of the share per capita is catastrophic and very likely will result in social and economic unrest in many parts of the Arab Region. It is therefore essential that the concerned authorities in such areas meet this possible calamity with the seriousness it deserves. National as well as international money granting agencies should not hesitate in providing the necessary funds for water resources development projects. It is equally important that all possible efforts, social, medical etc) should be spent in order to reduce the rates of population growth to the required level.

It is worthwhile to remind the reader here that the above figures express the quantities and shares per capita of the total annually renewable freshwater available resources (ARAR) per country, sub-region and region. One should not forget that these quantities are different from the annually internally renewed water resources (IRWR). The former does not take into account geographic location of the source of available water. A striking example of this situation is the Nile water in Egypt. Table 2 shows that the IRWR of Egypt is 2.6×10^9 m^3 y^{-1} whereas ARAR for the same State has been estimated as 68.5×10^9 m^3 y^{-1} (IWMI, 1998). Postel (1989) commented on this situation stating that: "Egypt, where rain is sparse, may be the region's most extreme case. It's (55 million people around 1990) depend almost entirely on waters of the Nile, none of which originates within the desert nation's borders." A more or less similar situation occurs in the Sudan where IRWR and ARAR are about 30×10^9 m^3 y^{-1} and 121×10^9 m^3 y^{-1} respectively. More examples can be found in Syria and Iraq and in all Arab countries traversed by international rivers, which originate in countries located outside the region.

The substantial difference between ARAR and IRWR for some countries brings the total annual amount of IRWR for the region to 154.39×10^9 m^3 compared to 307.83×10^9 m^3, which is the annual regional amount of ARAR. This means that the estimate of the internally renewed water resources in the region counts to almost one half of the annually available water resources. This difference, no doubt, is alarming as it constitutes one of the major reasons behind the conflict between some of the States inside the region like Syria and Iraq and between the countries within the region and source countries outside the region like Syria and Turkey; Iraq, and Iran and Turkey; and Egypt, Sudan and/or Ethiopia. Keeping this fragile situation unsettled can easily be disturbed by changes in the balance of political power or a stronger determination by the source countries to reserve for themselves bigger shares of water to meet the demands of their future populations. The calculations presented above are based on the assumption that the volumes of available fresh water resources remain unchanged in the period 2000–2050. The same assumption holds for those countries, which for one reason or another, do not seem to have a fair chance for developing their water resources in the coming decades. On the other hand, some of are Arab countries working according to preset plans to build storage reservoirs, import water, desalinate sea and brackish water, extract deep groundwater as discussed in detail in previous chapters. Despite the questionability of the issue of share per capita of the annual internally renewed water, at least for the time being, we find it necessary to bring it here to the fore in attempt to present the situation of the region as complete as possible. The figures for the year 2000 and the projections for 2025 and 2050 have been estimated using the data in Tables 1 and 2, and the results obtained are listed in Table 3.

12.1.4 Distribution of water resources among the three major sectors of water use

These sectors are domestic, industrial and irrigation water use. The second half of Table 3 gives the sectoral water withdrawal, in percent, for each country in the region. With the exception of Lebanon and Kuwait, where the percentage water use in the irrigation sector comprises 60% of the total withdrawal of water, and Jordan and Palestine, where the same percentage reaches 75, the remaining countries of the region use at least 80% of their withdrawn waters for irrigation. The average percentages for the whole region are 85% for irrigation followed by 15% for the domestic sector and 3% for the industrial sector.

The above-mentioned percentages clearly indicate that most of the withdrawn water goes to land irrigation. In several cases, like Jordan for example, despite the stability of the withdrawal, in percent, for irrigation, the quantity withdrawn increases substantially as a result of the rise in total withdrawn quantity of water. This is evident from the graphs in Figure 1. The increase of water withdrawn for irrigation is necessary for producing more agricultural crops through the increase of surface area under irrigation and/or covering the irrigation requirements for inadequately watered areas. According to Hani (1995) the population of Jordan increased by almost 50% in the period 1986–1993. The corresponding increase in water withdrawal for irrigation in the same period was about 60%.

Table 3. Per capita share of IRWR and percent of sectoral withdrawal of water in the Arab Region (source: World Resources 1998–99)

Country	Per capita IRWR, $m^3\ y^{-1}$ for the years			Sectoral withdrawal of water, %		
	2000	2025	2050	domestic	industrial	irrigation
Mauritania	164	90	66	6	2	92
Morocco	1043	738	627	5	3	92
Algeria	630	400	211	25	15	60
Tunisia	400	282	240	9	3	88
Libya	115	57	37	11	2	87
Subtotal	701	456	364			
Egypt	39	27	19	6	8	86
Sudan	976	640	500	4	1	94
Djibouti	49	28	21	6	0	94
Somalia	1008	486	316	3	0	97
Comoros	1437	620	410	*20**	*0*	*80*
Subtotal	413	266	209			
Syria	815	485	379	4	2	94
Lebanon	1496	1085	925	28	4	68
Jordan	165	85	60	22	3	75
Palestine	130	65	33	15	10	75
Iraq	1477	797	589	3	5	92
Kuwait	11	7	6	37	2	60
Bahrain	49	33	29	12	2	86
Qatar	44	38	34	15	5	80
Saudi-Arabia	102	52	36	9	1	90
U.A.E.	61	51	44	7	1	92
Oman	430	162	99	5	2	93
Yemen	284	103	67	7	1	92
Subtotal	623	287	199			
Total/average*	561	321	233	12*	3*	85*

Explanation
**Underlined figures are pure estimates*

12.1.5 Land use

Agricultural activities form the backbone of economy in the developing countries, especially those that do not possess any mineral wealth of substance. In several Arab States agriculture is the sector that requires and consumes most of the withdrawn water.

In arid areas, as a rule, regular dry-land farming is impossible without irrigation; in semi-arid areas a cereal monoculture is practiced according to a fallow/cereals or pasture land/ cereal rotation. The risk of a failure of the harvest caused by insufficient rainfall is in the order of 2 years in 10 but diversification of agriculture and intensive rotation are limited by lack of water (Wallén, 1968). According to these definitions, stock-raising is the predominant agricultural activity in the arid

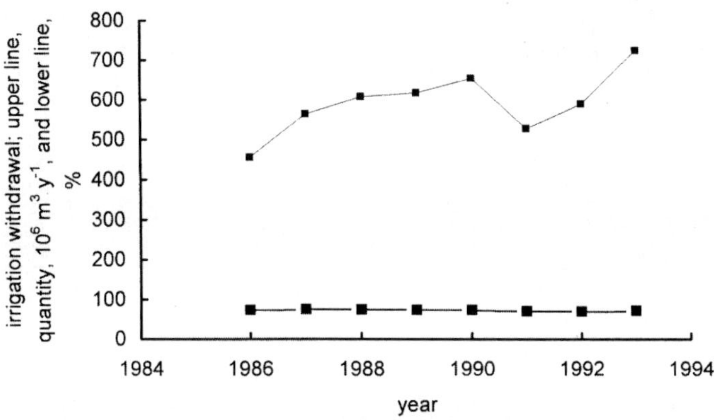

Figure 1. Irrigation withdrawal and percentage of withdrawal in the period 1987/93, Jordan (Hani, 1995)

areas apart from the irrigated parts. Cereal production and stock-raising are the main agricultural activities. Table 4 gives some basic statistics of land use in the Arab Region, average for the period 1992/94. From this Table one can easily observe that the area suitable for agriculture in the whole region accounts for just 13.8% of the total land area. The total cropland represents about 4.16% of the total land area or about 30% of the area suitable for agriculture. It is important to mention here that the cropland under irrigation represents 24% and the rain-fed area is 76% of the total surface area of the cropland. This means that the land surface under rain-fed agriculture in the Arab Region is larger than three times the surface area put under irrigated agriculture.

According to Wallén (1968), the annual precipitation in the Mediterranean climate of the region is approximately the same as the amount during the growing season, and this is related to the dry-land farming possibilities. The possibility of dry-land farming is affected by the interannual variability of rainfall $(IaV)_{rel}$, which can be expressed as:

$$(1) \qquad (IaV)_{rel} = \left[100 \sum_{r=2}^{r=n} (P_{r-1} - P_r) \right] / P_m(n-1)$$

where,

$P =$ annual rainfall, $P_m =$ annual average rainfall over the whole period of record, $r =$ year number and $n =$ number of years in the period of record

The studies carried out in the Levant have shown that an annual average rainfall 240 mm with an interannual variability of 37% are the most probable values for the precipitation conditions at the limit for dry-land farming. An approximate linear

Table 4. Surface areas occupied by land under cultivation, forests and woodland, pastures and other land (Source: World Resources Institute, 1998)

Country	Land area, 1000 ha	Area, 10^3 ha, 1992/94					
		suitable for agriculture.	total rainfed	cropland irrigated	forest and woodland	permanent pastures	other land
Mauritania	102,522	1,100	158	50	4,410	39,250	58,654
Morocco	44,630	35,250	8,083	1,208	8,613	20,933	5,397
Algeria	238,174	39,536	7,480	563	3,949	31,024	195,197
Tunisia	15,536	11,000	4,428	385	666	3,416	6,602
Libya	175,954	3,800	1,693	477	840	13,300	159,644
Subtotal	576,816	90,686	21,842	2,683	18,478	107,923	425,494
Egypt	99,545	4,452	0	3,265	34	0	96,374
Sudan	237,600	58,900	11,029	1,946	42,367	110,000	72,258
Djibouti	23,000	10	4	3	2	2,200	N.A.
Somalia	62,734	8,850	826	194	16,000	43,000	2,708
Comoros	2,235	10	235	0	2,000	0	N.A.
Subtotal	425,114	72,222	12,094	5,408	60,403	155,200	171,340
Syria	18,378	5,864	4,956	1,015	484	8,191	3,718
Lebanon	1,023	350	217	89	80	14	623
Jordan	8,893	1,465	336	69	70	791	7,627
Palestine	6,000	3,000	N.A.	N.A.	N.A.	N.A.	N.A.
Iraq	43,737	11,500	2,070	3,680	192	4,000	33,995
Kuwait	1,782	163	1	3	2	137	1,638
Bahrain	62	7	0	4	0	4	N.A.
Qatar	11,000	6	0	3	0	5	N.A.
S. Arabia	214,969	2,508	1,292	201	1,800	120,000	89,392
U.A.E.	8,360	15	9	73	3	280	8,002
Oman	21,246	56	3	60	0	1,000	20,183
Yemen	52,797	3,708	1,093	447	2,000	16,065	33,212
Subtotal	388,247	28,642	9,977	5,644	4,631	150,487	198,390
Total	1,390,177	191,550	43,913	13,735	83,512	413,610	795,224
%	100	13.779	3.159	0.998	6.007	29.752	57.203

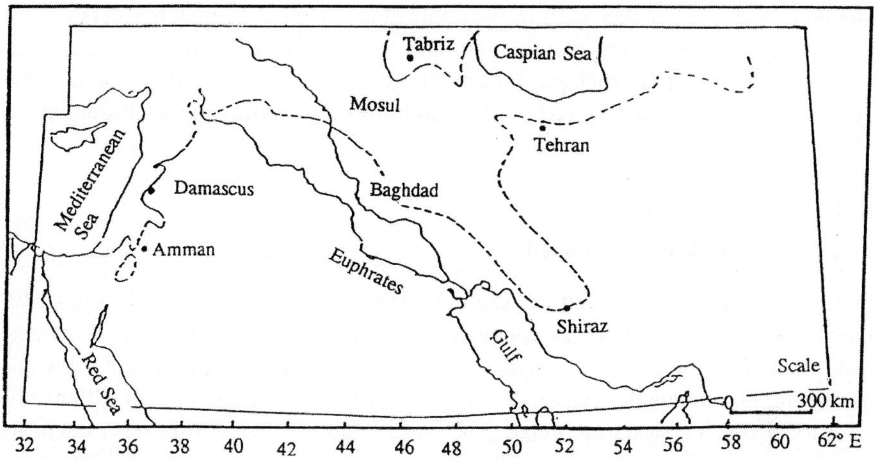

Figure 2. Map showing the limits of dry-land farming in a certain part of the Arab region and its surroundings (Wallén, 1968)

relation between the average annual amount of precipitation and the variability could be established at the limit of dry-land farming:

(2) $(IaV)_{rel} = 0.07P_m + 21.7$

Allowing for failure of the yield due to insufficient rain in 2 years out of 10, it has been found that the minimum amount possible in an individual year to permit dry-land

Farming is different from one region to another. The amount was as low as 180 mm in the area of regular rainfall in Jordan and southwestern Syria and as high as 210 mm in the Zagros Mountains in Iraq. The map indicating the limits for regular dry-land farming in the north of the Eastern Sub-region is shown in Figure 2 (Brichambaut & Wallén, 1963).

The same source described a wet month as a month with rainfall, *p*, not less than 25 mm and $(IaV)_{rel}$ not exceeding 100%. The parameter q links the monthly rain, p, to the $(IaV)_{rel}$ by the expression:

(3) $q = p/(IaV)_{rel} \geq 0.25$

By plotting *q* versus the time in months for any climate station it becomes possible to indicate the month where $q = 0.25$, thereby establishing the beginning and end of the wet season.

12.2. MEASURES OF WATER POVERTY

A number of approaches has been developed for the purpose of assessing the extent of water poverty or scarcity experienced in a certain country or region. The simplicity/complexity met with in the application of these approaches depends on the number of parameters they include and the ease or difficulty encountered in measuring or evaluating each parameter. The next sub-sections highlight some of the approaches that are widely used for the said purpose according to the chronological order of their development

12.2.1 Water stress concept

Falkenmark (1992) suggested the use of 'water stress index' as a tool to describe the availability of water for an adequate quality of life in a moderately developed country in an arid zone. She stated that a country whose renewable freshwater availability, on an annual per capita basis, exceeds about 1,700 m^3 will suffer only occasional or local water shortage problems. Countries below this level begin to experience periodic or regular 'water stress'. When fresh water availability falls below 1,000 $m^3\ y^{-1}\ cap^{-1}$, countries experience 'chronic water scarcity'. At this stage the lack of water begins to hamper economic development and human health well being. When renewable freshwater supplies drop to below 500 $m^3\ y^{-1}\ cap^{-1}$ countries suffer from 'absolute water scarcity'.

Engelman & LeRoy (1993) referred to the above-mentioned rates as rough benchmarks rather than precise thresholds. The exact level at which water stress sets in varies from region to region; it is a function of climate, level of economic development and other factors. Despite the weaknesses in the water stress concept, the 1,000 $m^3\ y^{-1}\ cap^{-1}$ benchmark has been accepted as a general indicator to water scarcity by many individuals and agencies. According to the same reference the oil rich Arab States-Kuwait, Qatar, Bahrain, Saudi Arabia and the United Arab Emirates-make up five of the nine countries with the least water per capita. The same source lists Djibouti, Bahrain, Kuwait and Jordan as water-scarce countries in 1955, and Qatar, Saudi Arabia, United Arab Emirates, Yemen, Tunisia, Algeria and Somalia as water-scarce in 1990. It adds Libya, Oman, Morocco, Egypt, Comoros and Syria to the list of water-scarce countries by 2025 even under the UN low-scenario of population growth projection.

The difference between the data sources used in preparing this list and those used in preparing Table 2 is the main reason behind the discrepancies in the results appearing in them. Despite these discrepancies the two sets of results show a strong agreement with regard to two basic conclusions; the big majority of the States in the Arab Region suffer from absolute or chronic water stress already for quite sometime, and the extent of water stress is getting worse by the day reflecting the low pace of water resources development compared to the rate of population growth.

The fact that population growth in most of the Arab countries is unparalleled by the rate of increase of the renewable water resources results in a decline in the annual share per capita of available water. Even those countries, which do not suffer currently from water stress, will soon reach it unless urgent, remedial measures to alleviate its detrimental consequences are implemented.

Some countries are using more freshwater than they have, or at least what they have in the form of renewable resources. The ratio of water withdrawal to renewable water supplies in all countries of the Arabian Peninsula exceeds 100%. This ratio is referred to as index of unsustainability. The values of this measure, according to Engelman & LeRoy (1993), ranged from 374% for Libya in late 1980s to 140% for the United Arab Emirates to about 110% in Jordan and Saudi Arabia, all in the same time. Despite Libya's increasing use of desalination and treated wastewater, rapid growth in the country's demand for water has led it to rely on mining of non-renewable groundwater. In other countries the imbalance between the withdrawals and renewable resources is caused by migratory flows which originate in other countries.

12.2.2 Social water stress index (SWSI)

In 1999, Yoffe & Ward presented a paper in which they proposed a methodology for defining potential indicators of international water conflict and spatially analysing and portraying these indicators within a Geographic Information System. Indicators are defined across multiple scales in a parallel analysis of global and basin attributes.

The river basin was chosen in this method as the key unit of research since it fits well the framework of Trans-boundary Freshwater Dispute Database (TFDD), which is designed around the concept of international river basins. The approach combines global level analysis with basin-specific information. The per capita water availability, border stretches under dispute, land use (agriculture versus industry) and hydrology of river flow are examples of country attributes, country boundary attributes, river basin attributes, and river attributes, respectively.

A database of water-related treaties is available through the TFDD. The treaties deal with one or more of the following issues: water rights, water allocations, water pollution, principles for equitably addressing water needs, hydropower/reservoir storage/flood control development, and environmental issues and the rights of riverine ecological systems. The gross national production (GNP), disputed border stretches, land use (agriculture versus industry), and water quantity are four examples of the indicators representing a country, country boundary, river basin and river hydrology respectively. Indicators used to simplify and amalgamate huge amounts of data. They represent a compromise between accuracy and precise information, but provide a simplification of complex systems that facilitates evaluation of those systems. In general good indicators should satisfy a number of criteria such as corresponding to the selected application. In addition, there should also be a means for testing an indicator's accuracy and relevance.

The SWSI has two components; the Human Development Index (HDI) and the Water Stress Index (WSI). The latter is a water-specific, global-level, which measures freshwater availability per capita within a country. The WSI divided by twice the HDI gives the WSI. While the SWSI is an improvement, neither of its key components, the HDI nor the WSI incorporates measures of percentage of population with access to freshwater supply or sanitation facilities; water-related diseases, water quantity and efficiency of existing water delivery systems. These measures provide more accurate representations of water stress and become increasingly important as the analysis moves from large scale to small scale. There is a number of other potential indicators, each of which requires more detailed research in order to evaluate their validity.

A number of the indicators data like the examples given above can effectively be represented within a Geographic Information System. Examples of key water-related indicators include: studying areas that are prone to flooding or drought, performing spatial analysis on road networks within watersheds and countries to determine areas of high industrialization. Indicators that provide an indication of potential water conflict, even when tested against historical evidence, should be viewed with some skepticism. This skepticism, however, does not suggest that water-related indicators of potential instability or conflict are without value. Through research, access to potentially useful information, and contact with various regional experts, reasonable indicators can be developed.

12.2.3 Water poverty index

According to Sullivan (2001), establishing a water poverty index (*WPI*) will contribute a more equitable allocation of water resources, by considering water issues from the perspective of both the supply of water, and the demand for it in order to identify who needs water, when and where. The index is essentially a statistical concept that measures a quantity relative to a specified period. As such, the *WPI* fits this concept of an index that measures something indirectly and that is made up of well-defined components. The *WPI* is a composite index, comprising various elements such as water availability, access to safe water and clean sanitation, and time and effort required while collecting domestic water.

The same source adds: "Understanding of the factors that influence the relationship between ecosystems, water and poverty and the unsteady nature of this relationship, will enable decision-makers to make better informed decisions about how financial resources can be used most effectively and equitably deal with water allocation problems." The development of *WPI* according to this approach requires a database of information relating the index to water availability and to water demand. A computer-based geographical information system (GIS) can provide one way in which such an integrated database can be developed. Additionally, the use of an object-orientated approach to database construction can provide a foundation for future change.

The formula expressing the index constructed by Sullivan (2001) can be written as follows:

$$WPI = 1/3\,[w_a A + w_s S + w_t (100 - T)] \quad (4)$$

where,

$A =$ adjusted water availability assessment (AWA), in percent, calculated on the basis of surface and groundwater availability,

$S =$ population with access to safe water and clean sanitation, in percent,

$T =$ sub-index ranging between 0 and 100, representing time and effort spent in collecting water for household, and

$w_a, w_s,$ *and* $w_t =$ the weights given to the A, S and T components of the WPI. The sum of the three weights $w_a + w_s + w_t = 1.0$

Numerical example: The values *A, S and T* (time expressed as a percentage of say per capita labour-time) for a certain region are 80, 50 and 30 and the weights w_a, w_s and w_t are 0.50, 0.25 and 0.25, respectively. Substituting the given values in Eq. (1), the *WPI* in points can be found as 23.3. It should always be remembered that the smaller the index value is the greater the extent of water poverty would be.

Another possible way to construct a WPI is a time-analysis approach, in which the time needed by an average person to collect a certain quantity say 1,000 m^3 is considered. This can be expressed as follows:

$$(5) \qquad WPI = T/1000m^3$$

The method expressed by Eq. (5) is no doubt simple but not without weaknesses. It depends solely on domestic issues without taking ecosystems needs and commercial interests into account. Besides, it is difficult to evaluate the means by which water may be provided or transported.

12.2.4 Expanded water poverty index (EWPI)

The expanded index, which was constructed by Lawrence et al. (2003), differs from the previous indexes in that it contains a lot more components and parameters. The make up of this index as summarised by Al-Hamoud & Edwards (2005) is listed in Table 5. The following six regression models were run:

$$(6) \qquad \ln(EWPI_i) = \alpha + \alpha_1 \ln(PrivateInvestment_i) + e_i$$

$$(7) \qquad \ln(Resources_i) = \beta_{0r} + \beta_1 \ln(PrivateInvestment_i) + u_i$$

$$(8) \qquad \ln(Access_i) = \chi_{0r} + \chi_1 \ln(PrivateInvestment_i) + \partial_i$$

$$(9) \qquad \ln(Capacity_i) = \gamma_{0r} + \gamma_1 \ln(PrivateInvestment_i) + r_i$$

$$(10) \qquad \ln(Use_i) = \varepsilon_{0r} + \varepsilon_1 \ln(PrivateInvestment_i) + t_i$$

$$(11) \qquad \ln(Environment_i) = \phi_{0r} + \phi_1 \ln(PrivateInvestment_i) + q_i$$

The variables used in Eqs. (3) thru.' (8) are as explained in Table 5 whereas e, u, ∂, r, t *and* q are normal, independent, and identically distributed errors. Results of the regression tests has shown that private investment does increase access

Table 5. The make-up of the expanded water poverty index (Al-Hamoud & Edwards, 2005)

Resources	i- Internal freshwater flows ii- External inflows iii- Population
Access	i- % of population with access to clean water ii- % of population with access to sanitation iii- % of population with ccess to irrigation adjusted by per capita water resources
Capacity	i- Income per capita ii- Under age of five mortality rates iii- Education enrollment rates iv- Gini coefficient
Use	i- Domestic water use, litres per day ii- Share of use adjusted by the sector's share of GDP
Environment	i- Water quality ii- water pollution iii- Environmental regulation and management iv- Informational capacity v- Biodiversity based on threatened species

to water and sanitation and its use in a statistically-significant fashion. On the other hand, private investment does not impact the environment, the extraction of resources or the capacity as measured by accessibility through wealth. In addition to these conclusions, one can fairly add that a country should go through significant institutional management, regulatory and policy changes before undergoing any change in its management doctrine. More importantly, a government should educate its citizens through public awareness campaigns on the need for, the benefit of, privatising a vital sector such as water sanitation sector (Al-Hamoud & Edwards, 2005).

12.2.5 Comparative global scene

The percent increase in water withdrawals over the 1990 to 2025, and the withdrawals in 2025 as a percentage of the annual water resources of a country have been suggested to be used as criteria for classifying the globe into a number of groups. The grouping, generally, also reflects climatic-hydrologic and socio-economic homogeneity.

Most of the countries comprising the Arab Region are encountered in groups I, II and III as follows:

Group I: This group includes the largest number of countries forming the region; Tunisia, Libya, Egypt, Djibouti, Syria, Jordan, Palestine, Iraq, Kuwait, Bahrain, Qatar, Saudi Arabia, Oman, United Arab Emirates and Yemen. These countries are scarce by the above-mentioned criteria. Water scarcity is already a constraint on food production and will grow to become a major constraint in the near future.

Group II: The Sudan and Somalia are the two Arab countries included in this group. According to Chaturvedi (2000), "The countries in this group must develop more than twice the amount of water they currently use to meet reasonable future requirements. They include some of the least developed countries, socio-politically and techno-economically. As water withdrawal figures reveal they have undertaken very little development of water resources. Irrigated cropland and yields are generally low."

Group III: This group includes from the Arab Region three countries situated in northwestern and northern Africa; Mauritania, Morocco and Algeria. The countries in this group need to increase their withdrawals by between 24 and 100%, with an average of 48% to meet the future requirements (Chaturvedi, 2000).

12.2.6 Water crisis

According to Biswas (1999), water abstraction, which is widely used as a proxy for water use, is methodologically wrong. Water is a reusable source, which can be used and reused many times. The same source adds: "no reasonable estimates exist for reuse of water, even at the national levels, let alone for the world as a whole." So far, water professionals have not yet considered reuse of water an important factor in global water availability and use considerations, rendering the predictions for future highly suspect.

There are strong indications that the world in 2025 will be different from what it was in 2000, in the same way the world of the year 2000 is significantly different form what it was in 1975. This change can be attributed to, among others, changes in the demographic and human living conditions. Advances in technology, improvement of national and inter-governmental policies and the growing tendency to put price on water consumption also play a certain role in the availability of-demand for water balance. The changing world implies a need to change the existing water management practices.

12.3. ADVERSE IMPACTS OF WATER SCARCITY AND REMEDIAL MEASURES

12.3.1 Priority water resources issues in the Arab Region

Three priority issues dominate water resources in the Arab Region. The three issues stem from the region's central problem: the mismanagement of water resources. These issues can be briefly summarised as follows (United States Agency for International Development, 1993):

– *Water shortage*: As mentioned in several sub-sections in this chapter, water deficits already exist in many parts of the Arab Region. Water transfers and reuse positively affect supply by directing water to areas that are drastically short of water, but this merely redistributes existing and to some extent new resources. Transfer and reuse will never satisfactorily augment supply. Such measures are predominantly marginal, short-term corrections that fail to alter the basic problem of limited water resources.

– *Deterioration of water quality*: Under conditions of scarcity as manifested by the intensity of population pressure on the water resources, the quality of the resource is threatened by environmental hazards and water pollution becomes a real menace. It needs only a small amount of pollutants to degrade the quality of a modest water body. The degradation and depletion of water resources have exacerbated the limited supply of water and are constraints to economic development in several Arab States. This situation has stemmed from many factors, including lack of an environmental ethic throughout the region, a general lack of public concern with water resources issues, inadequate regulatory and enforcement capability, and restrictions on economic forces that value water in relationship to demand.

– *Inadequate management performance of water resources by public and private sectors:* Arab States' governments traditionally focused on specific water sectors for investment, usually agriculture. Subsidies that discourage the efficient use of water, e.g. no or too little pricing of water, are common in the agricultural sector.

It is essential that bodies concerned with water resources in the region must experiment with policy and institutional options to improve the management of their limited resources. Morocco, for example, has recently adopted a policy of disengagement from its large-scale irrigation projects. Management responsibility has been shifted from close governmental control to private sector. Tunisia is actively supporting the formation of water users associations for water supply, and Egypt is creating similar associations for irrigation systems.

All three countries are also developing programs to recover costs for operation and maintenance in order to reduce the government's financial burden and encourage users participation. Integrated planning of water resources remains an issue as long as responsibility is divided among a number of ministries. Some countries, like Algeria and Libya, have made good progress by establishing water authorities that can plan and coordinate supply and use.

12.3.2 Adverse impacts

In his paper on water scarcity impacts and potential conflicts in the Middle east and North Africa (MENA) Region, Haddadin (2001) focused his discussion on the impacts on domestic and regional stability, and on the social value of water.

12.3.2.1 Impacts on domestic stability

Scarcity of water helps to create an increase in competition among the different water sectors. Such a competition can be at the expense of water essentially needed for irrigated agriculture. This in turn can lead to reduced food production and probably to less available clean water for drinking and other domestic purposes. The tendency to compensate all or part of the deficit in water resources through pumping from the groundwater aquifers more than their sustainable yields has two conflicting results. On one hand it helps to ease the deficit for some time, and on the other hand it has its perilous consequences. Mining deep, non-renewable groundwater reservoirs means leaving a sourceless resource for future generations. Should aquifers be located along the seacoast, they could lead to the deterioration of groundwater quality. These consequences can be listed under the category of environmental impacts.

Despite the demerits of water scarcity, irrigation has a potential for adverse environmental impacts which include: water logging and salinisation of soils, increased incidence of water borne and water related diseases, resettlement of changes in lifestyle of local populations, and increases of agricultural pests and diseases resulting from the elimination of dry season-die-back and the creation of a more humid microclimate. The expansion and intensification of agriculture has the potential to cause: increased erosion; pollution of surface and groundwater from agricultural biocides; deterioration of water quality; and increased nutrient levels in the irrigation and drainage water resulting in algal blooms, proliferation of aquatic weeds and eutrification in irrigation canals and downstream waterways (Chaturvedi, 2000). Some of these factors have a delayed response to water scarcity. They pertain to the integrity of the environment, and the deterrence it imparts on the development and economy of the country. Such a situation could adversely affect the chances for creating new jobs and increase unemployment thus causing social difficulties.

12.3.2.2 Impacts on regional stability

There is a serious, and steadily growing challenge on surface as well as subsurface water resources in the Arab region. This is the challenge of resolving international water conflicts peacefully and concluding all-inclusive and comprehensive riparian agreements on international water basins. Areas of potential conflict are classified and reviewed in the next section. It is feared that unless such conflicts, added to the other and already existing disputes, may trigger more violence between the conflicting parties and to regional instability.

12.3.2.3 Social impacts

Adverse impacts of water scarcity are more pronounced in economically developing countries than in developed countries. The comparative advantage of most developing countries lies in the skills the inhabitants have acquired over the years and the low cost of the added value they contribute to any product. At present these countries have a long-standing heritage in agriculture coupled with a recent, limited exposure to modern technology, refined industries or sophisticated levels of services.

A stringent marketing requirement is the high quality of the product, whether it is agricultural, industrial or service. The attention, effort and money added in agriculture are higher than the values developing countries are able to add to industrial products, and quality control is easy to achieve in their agricultural production than industries or services. Another competitive advantage for agriculture is the low capital needed to be invested to create a job in the agriculture sector as compared to the cost of creating a job in the industrial or service sector.

Many other social benefits are accrued by the society from the use of water. In addition to the factors of survival, domestic hygiene, public health, enhancement of the environment, and job creation, water use in irrigation has created monolithic societies in agricultural settlements.

12.4. POTENTIAL AREAS OF CONFLICT ON WATER RESOURCES IN THE ARAB REGION

12.4.1 Surface Water Resources

12.4.1.1 Areas of conflict in the Western Sub-region

The Sénégal River in Mauritania is the only river of international character in the western sub-region. Surface runoff of other areas in the western sub-region is carried by watercourses of the wadi type.

Water management works of the 'highly irregular' of the Sénégal River include, among others, the Diama Barrage near St. Louis on the Atlantic Ocean, the Manantali Dam on the Bafing River, a tributary of the Sénégal, in Mali. The barrage that was completed in 1990s has been designed essentially to reduce the penetration of the saline seawater into the river. The Manantali has been operating since 1986/87, and since then helps to control almost 50% of the flow down the Sénégal Valley. The Sénégal River Commission (OMVS) that takes care of the management of the river water is composed of representatives of the four riparian states; Mauritania, Sénégal, Mali and Guinea.

The environmental and health changes seen in the Sénégal River Basin have not happened in isolation; the change in water management and the expansion of irrigated agriculture have brought broad-based social changes, including tensions between pastoralists and farmers and among ethnic groups as well. Additionally, it has been reported that the net impact of the Sénégal development projects on the health of the riparian people has clearly been negative. Schistosomiasis, especially intestinal schistosomiasis, has been widely observed at high infection rates along the Mauritanian shore of the river. There is an increased risk of cholera and other water borne diseases in the lower basin of the river, principally Mauritania. The same can be said about diarrhea while the news is conflicting about the spread of malaria and other kinds of fevers (World Resources, 1998).

12.4.1.2 Areas of conflict in the Central Sub-region

– *The Nile River*: Egypt, the Sudan, Ethiopia, Rwanda, Burundi, Kenya, Uganda, Tanzania, Central African Republic and Congo (Kinshasa) are the riparian countries sharing the Nile Basin. The river water is almost exclusively used for the interests, mainly irrigation and power development, by Egypt and the Sudan. By the power of the bilateral agreement between these two countries, 55.5×10^9 m^3 y^{-1} is consumed by Egypt and 18.5×10^9 m^3 y^{-1} by the Sudan.

From time to time complaints are raised by source countries like Ethiopia and Uganda voicing up the need for their respective countries for a fair share of the river water. There are calls for a multi lateral treaty on the Nile water encompassing the other riparian States in addition to Egypt and the Sudan.

– *The Shebelle and Juba Rivers*: The riparian countries of the Shebelle River are Ethiopia and Somalia while those of the Juba River are Ethiopia, Kenya and Somalia. So far there are no disputes or conflicts to mention between them about the flow of any of these rivers.

12.4.1.3 Areas of conflict in the Eastern Sub-region

There are two groups of countries constituting potential conflict areas in this sub-region. The first group (a) comprises: Turkey, Iraq, Iran and Syria, and the second group (b) comprises Lebanon, Syria, Israel, Jordan and West Bank (Palestine).

Group (a)

– *The Tigris River*: Turkey, Iraq and Syria are the riparian countries sharing the basin of the Tigris River. However, half of the river flow originates in Turkey and the other half in Iraq. As yet, there is no agreement between the riparian countries regarding division and full utilisation of the river water.

– *The Euphrates River*: Turkey, Syria and Iraq are the countries sharing the basin and flow of the Euphrates river. Since 1990 there is an agreement between Syria and Iraq. This agreement stipulates that 42% of the flow (500 m^3 s^{-1}) entering Syria after leaving Turkey goes to Syria and the remaining 58% is utilized by Iraq. In view of the frequent complaints of the riparian countries, an all-inclusive multilateral agreement might provide the optimum solution to this issue.

– *Shatt el-Arab*: This waterway is formed by the confluence of the Tigris and the Euphrates Rivers in the southern part of Iraq, and discharges its water into the Gulf. The agreement that was reached in 1974 between Iraq and Iran, the two riparian countries, was revoked on the onset of the war between them in 1981. The dispute between the two parties over their international borders affects navigation. The high salinity of water of the Shatt makes it unsuitable for irrigation.

Group (b)

– *Orontes (ʻĀsi) River*: This river springs in Lebanon and flows northwards through Iskenderon, a disputed region between Turkey and Syria. A bilateral agreement between Lebanon and Syria was worked out in 1996, but Turkey so far is not included. Syria's territorial claims to the disputed province, which annexed by France to Turkey when the former was the mandatory power over Syria, has to be settled before Syria approves of the riparian status of Turkey on the river.

– *The Hasabani and Wazzani Tributaries*: These tributaries spring in Lebanese territories. The Hasabani forms the border between Lebanon and Syria for a few kilometers before entering Israel. With the exception of a meager volume of water used for domestic purposes in the village of Wazzani, Lebanon, and the Ghajar in the occupied Syrian Golan, Israel is currently utilizing the entire flow of the Hasabani and the Wazzani Rivers. Lebanon has legitimate rights to use the waters of both rivers. Syria can claim rights on the Hasbani as it forms part of the Syrian-Lebanese border.

– *The Banyas River*: The Banyas, as described in an earlier chapter, rises in Syria and flows entirely within Syrian territory before joining the Hasbani and the Dan Rivers to form the Jordan River. The Banyas falls within the area of the Golan occupied by Israel since 1967. The legitimate rights of Syria and Israel in the waters of the Banyas River should be negotiated bilaterally between them.

– *The Upper Jordan River*: As mentioned earlier, the Upper Jordan River is formed by junction of the three tributaries: Hasabni, Dan, and Banyas. It flows through Israel and flows southwards towards Lake Tiberias, which had been converted by Israel into a storage reservoir for the purpose of regulating the flow of the Jordan River. As a consequence, the floodwater level is about 4 m than used to be in the pre reservoir era, and there upon the lake shoreline is moving eastwards through the Syrian territories. Syria is legitimately entitled to recover its international border that gets submerged by the lake on the northeast shore. The West Bank too has legitimate rights on the Jordan. These rights are supposed to be negotiated in the final status negotiations between Palestine and Israel.

– *The Yarmouk River*: The Yarmouk is the largest tributary of the Jordan River south of Lake Tiberias. Syria, Jordan, Israel and Palestine are the users of the Yarmouk River. Currently, Syria uses close to one half of the river water. Israel is entitled to 25×10^6 m^3 y^{-1}. The bilateral agreement of 1987 between Jordan and Syria gives Jordan the right to the river flow and to the springs rising in Syrian territories at an elevation below 250 m a.m.s.l. and the flood flow in excess of what some 26 storage reservoirs could impound. These flows when summed up give Jordan between 140 and 200×10^6 m^3 y^{-1}. The remaining river flow, if there is any, goes to Palestine. Jordan is linked with Syria by a bilateral agreement and with Israel by another bilateral agreement. According to Haddadin (2001) it would be to the benefit of all parties if they were linked together by one multilateral agreement.

– *The Lower Jordan River*: This reach of the Jordan River serves as a drain collecting agricultural drainage water, flood waters of the Yarmouk and Jordan Rivers in excess of the available storage capacities, and the outflow of saline springs, naturally draining in the Tiberias Lake. Israel, and Jordan to a much smaller scale, are using the Lower Jordan water for fish breeding. The bilateral agreement between Israel and Jordan obliges the riparian parties to clean up the Lower Jordan water and share equally the reclaimed water where the river forms their common boundary between the Yarmouk-Jordan confluence of the Wadi Yabis on the Lower Jordan River.

12.4.2 Groundwater resources

12.4.2.1 Western and central sub-regions

There is hardly any mention of conflicts between the States in the western sub-region or the central sub-region regarding the exploitation of the aquifers underlying their territories.

12.4.2.2 Areas of conflict in the eastern sub-region (Haddadin, 2001)

– *South Lebanon-North Israel Coastal Aquifer*: Israel and Lebanon share this groundwater aquifer. Israel has been exploiting it, and Lebanon needs to exploit it for the development of the poor south. Sharing the water of this aquifer has to be negotiated between these two countries.
– *The Mountain Aquifer (Yarqon-Taninim)*: This aquifer underlies the territories of Israel and the West Bank and drains westwards. The use of this aquifer is only available to Israel at present.
– *The Eastern Aquifer*: This aquifer lies entirely in the West Bank. The complication in this aquifer originates from the Israeli settlements built in the West Bank since 1967. Some of the settlers there depend entirely on water from this aquifer. Negotiations on this aquifer are strongly linked to the future of the Israeli settlements.
– *The Northern Aquifer*: This aquifer underlies the territories in the north of the West Bank and Israel. The Palestinians have been using a modest portion of this aquifer to irrigate limited areas north of the West Bank, but Israel uses most of the aquifer's water inside the Israeli territories.
– *The Gaza Aquifer*: The recent removal of the Israeli settlements from the Gaza Strip is giving the Palestinians a full access to the aquifer in the strip.
– *Wadi Araba Aquifer*: This aquifer lies entirely within Jordan south of the Dead Sea. The agreement between Jordan and Israel allows Israel to continue farming an area within Jordan in Wadi Araba. Besides, Israel was allowed to increase its withdrawal from the aquifer by 10×10^6 m^3 y^{-1} provided that an equal amount is pumped to Jordan in the north where Jordan most needs such water.
– *The Golan Aquifer*: This aquifer underlies the Golan Heights in Syria and drains into Lake Tiberias. The aquifer's water is fully utilised by Israel since the 1967 war and is adversely affecting the withdrawal of water by Syria. This issue can possibly be negotiated bilaterally between the two countries in future.
– *The Desi Aquifer*: This aquifer, which consists mainly of sandstones and quartizites with an average thickness of about 350 m, belongs to the Paleozoic Era. The main outcrops of the aquifer occur in the southwestern part of Jordan with an extension in Saudi Arabia. It forms the most important freshwater producing aquifer in south Jordan. The range for the depth to water surface is 60–80 m, for the permeability 0.34–1.0 m d^{-1} and for water quality (total dissolved solids) 170–1,020 ppm. Long-term pumping test data have indicated that the mean transmissivity of the aquifer system is close to 720 m^3 d^{-1} m^{-1} and the range for the storage coefficient 0.01–0.03. The same aquifer appears

Table 6. International water disputes in the Arab Region (Shahin, 1989, Renner, 1989 and USAID, 1993)

Source of Water	Countries in dispute	Disputed issues
Nile	Egypt, Ethiopia and Sudan	Water flow/diversion, flooding and siltation
Euphrates and Tigris	Iraq, Syria, Turkey and Iran	Water flow and salination
Jordan, Yarmouk, Litani and West Bank aquifer	Lebanon, Jordan, Israel and Palestine	Water flow/diversion
Shatt el-Arab	Iraq, Kuwait and Iran	Water flow and salination
Desi aquifer	Jordan and Saudi Arabia	Shared aquifer and lack of groundwater management

in Saudi Arabia under the group of the Saq formation. It is regarded as one of the most productive sandstone aquifers in Saudi Arabia, with flowing and non-flowing artesian conditions.

The Desi aquifer is exploited by both Jordan and Saudi Arabia. Unfortunately, there is hardly any cooperation between these two countries for a better management of this valuable resource of water.

A brief summary of the areas and issues of dispute in the Arab Region is given in Table 6 (Shahin, 1989, Renner, 1989, and USAID, 1993)

12.4.3 Water resources development, Case studies from the Arab region

12.4.3.1 Water resources development in Jordan

This is an example of freshwater resources development scheme in an Arab country suffering from an absolute state of water stress. The scheme is shown diagrammatically in Figure 3. "To meet the peak demand during the summer, which coincides with a surface water shortage, a considerable hydraulic development program has been carried out over the last 40 years. This development comprises storage, transport and water distribution structures. The Jordan Valley Hydraulic Scheme consists of the following components:

- *Storage Works*: King Talal (KT.D) Dam on the Zarqa River is the main storage structures. Karameh Dam (KD) on the Lower section of the Valley to the south of the Zarqa River and is supplied only by the King Abdalla Canal (KAC). Wadi Arab (WAD) Dam is the third largest structure located on the Wadi Arab in the northern part of the Jordan Valley. The Ziqlab, Wadi Shueib, and Wadi Kafrein Reservoirs contribute reasonable water quantities for irrigation uses. Both the KTD and WAD are connected to KAC, and can be integrated into overall management of resources in the valley.
- *Water Transport Structures*: The KAC is shown in Figure 3 forms the backbone of the scheme along the River Jordan for a length of 110 km and constitutes the

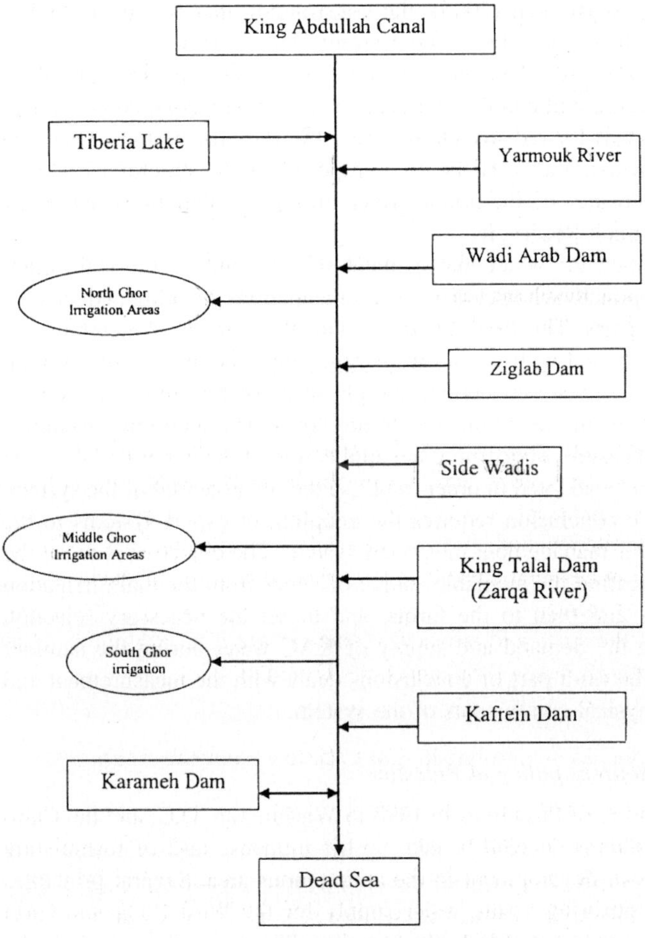

Figure 3. Schematic diagram for King Abdullah Canal; inputs to and outputs of the system (from Hussein, 2005)

main valley water conveyance carrier. The canal is supplied from the north by the Yarmouk River and Mukheibeh wells. The reach of the canal between the Yarmouk River and the Zarqa syphon is referred to as the KAC North, and south of the Zarqa syphon as KAC South. The system is operated by 36 crosscheck gates and a number of conveyance appurtenances. These comprise the Zarqa Carriers, the connection between the canal and WAD, and piping and pumping systems to supply water from the KAC to Amman.

- *Future Projects*: Several plans are laid down with the aim of increasing the currently available storage capacities. Any development on the Yarmouk River will provide a reasonable annual volume of the Yarmouk River for domestic and irrigation in the Jordan Valley. The main project here is the Unity (Wehda in

Arabic). According to Hussein (2005), the construction that began in 2003 is expected to be completed by 2008. The Kafrein Dam should be connected to the KAC and its banks raised in order to increase its capacity. In view of its location, the connection will enable the unused water to be recovered and stored in the Kafrein reservoir for redistribution during the shortage periods. Dams on side wadis like Kufrinja, Yabis, Hisban are proposed for construction. It is also suggested that floodwater of the Jordan river during its high flow season be diverted to the Karameh Reservoir.

Hussein (2005) described the water management strategy and operational organisation of the development scheme. The recommendations he obtained can be subdivided into three parts. The first part deals with the software concerning the human resources; the second with the software concerning the operational system; and the third with the hardware concerning the physical component of the system. As far as the first part of the recommendation are concerned, additional technical and administrational measures concerning the application of the integrated decision support system need to be adopted in order to utilise the full potential of the system. The second part of the conclusion requires the coupling of expert systems to the existing system of water management integrated system. The objective is to set the operational rules of locating the available source of water from the main irrigation system to the laterals and then to the farms, and to set the necessary selection criteria for optimising the demand and supply of KAC water during the summer and winter seasons. The third part of conclusions deals with the measurement and maintenance of the physical components of the system.

12.4.3.2 Water resources policy of Palestine

Following the Declaration of Principles in 1993 in Washington, D.C. and the Cairo Agreement, the Palestinians decided to take up the immense task of formulating policies for all sectors of development in the autonomous area. Several principles establish the basis of ensuring a safe water supply for the West Bank and Gaza Strip and guide the development of Palestinian policy. These principles include the observation of aquifers' safe yields to avoid over pumping; adequate provision of secure domestic water supply; efficient and productive agricultural and industrial sectors; fair water pricing rules; water exchange between the West Bank and Gaza Strip; and in the long-run water exchange between Palestine and Israel. Each item is necessarily interlinked with the other items, as no one option alone can solve Palestinian water needs.

An estimated flow of $180–200 \times 10^6\ m^3\ y^{-1}$ can be supplied by surface runoff and from the Jordan River to the Palestinians in the occupied territories. Israel and Jordan have signed a peace treaty that included an agreement on water allocation from The Yarmouk and Jordan Rivers. This requires construction of two dams in addition to a number of development projects that will adversely affect the flow to the Lower Jordan. Palestinians should have a full partnership in the development of the Jordan Rift Valley and their rights to get its share from the Jordan River should be preserved.

The West Bank Mountain Aquifer includes the area that used to be controlled by the Jordanian Administration before the 1967 war and since then became under Israeli occupation. The annual renewable freshwater of this aquifer ranges between 600 and 650×10^6 m^3 y^{-1}. The groundwater basins are recharged directly from rainfall on the outcropping geologic formations forming the phreatic portion, while the greatest part of the storage areas is located in the confined portions. The phreatic portions constitute the subsurface area under the West Bank Mountains where the Palestinians dug their groundwater wells to tap the shallow unconfined aquifers. The Israelis, however, dug their wells to tap the confined aquifers whose quantity and quality are better. Issues mainly related to the West Bank Mountain aquifer and the adjacent aquifers, and which have to be addressed through bilateral tracks between Lebanon, Israel, Syria and Palestine.

12.4.4 Water resources remedial measures

Since the United Nations Water Conference in March 1977 at Mar del Plata until present, many pieces of literature on water resources development and management have been published. Items of importance in this respect are highlighted in the next subsections.

12.4.4.1 Controlled (reduced) population growth

Any reduction in the rate of population growth generally reacts positively on water shortage. Using Egypt and Saudi as two important examples from the Arab Region, the updated version (1994) of the report issued by the United Nations Population Division gives the following predictions:

Country	*Annual share per capita, m³, for the year and projection level*					
	2025			*2050*		
	Low	*Medium*	*High*	*Low*	*Medium*	*High*
Egypt	*676*	*605*	*546*	*644*	*502*	*398*
Saudi Arabia	*110*	*107*	*103*	*84*	*75*	*67*

Comparing the low-projection value of 2025 to the high-projection value of 2050 shows a decline in the share per capita from 607 m^3 to 398 m^3, i.e. 41%, for Egypt, and from 110 m^3 to 67 m^3, i.e. 39%, for Saudi Arabia. Oppositely, the change from the high-projection level value of 2025 to the low-projection level value of 2050 shows a rise of about 15% for Egypt and a reduction of just 18%. Social and decision makers should undertake strong efforts to improve the public awareness of the consequences reduction in population growth rate can bring to their lives.

More elaborate review of this point can be found in sub-section 12.1.3.

12.4.4.2 Technological breakthroughs

These events may increase water supply in the future. Experiments include seeding clouds in Morocco and Jordan, towing icebergs from Antarctica to the Gulf, and capturing dew condensation from olive trees in Oman. Such efforts are promising, but none is yet either economically viable or practical on a large scale. Desalination of sea or brackish water could be expanded if the cost can be reduced. At current cost levels, only Saudi Arabia and the Gulf States can afford to expand their water supply by developing seawater desalinisation. More than 20% of the water used at present in Saudi Arabia is desalinised water, making waste disposal an increasing problem. As an example the pipeline has been constructed to transport groundwater from the Sahara Desert to the Mediterranean Coast. Plans are already laid down to convey water from the Mediterranean Sea inland to desalinisation plants in Jordan and Israel. The cost, among other factors, remains a serious constraint.

12.4.4.3 Water conservation schemes

One of the feasible remedial measures, though neither easy nor cheap to implement, is based on augmentation of accessible water by reducing the non-productive losses. All measures for water conservation already reviewed in earlier chapters and sections apply here.

In the Arab region most of the annual withdrawal, reaching some 80% of the annual total withdrawal, is consumed in the agricultural sector. Application of improved measures of water use can contribute substantially to save the water that is unnecessarily lost by letting it flow to the sea or go as a waste. This is the case of many wadis in the Arabian Peninsula and North Africa. Maintenance of irrigation structures, remodeling or replacement of the old structures, such as irrigation inlet and outlet structures, and storage and other hydraulic works, though in some cases could be expensive yet they are useful.

12.4.4.4 Reduced demand on Water

Closely related to sub-section 12.4.3.3, application of irrigation methods and techniques aiming at improving the efficiency of irrigation, and putting price on water, might help to reduce the waste and make the demand on water less. Suppose that 80% of the annual withdrawal of water is consumed in irrigated agriculture, and the overall irrigation efficiency is 50%. If this efficiency, upon using recent methods and techniques of irrigation application, rises to 65% the gain in water will be as much as 12% of the total water withdrawal. The cost of changing the old irrigation methods and introducing new ones can be gained from improving the quantity and quality of crop yield of crop, growing high-valued market cash crops, and putting price on water used. The same principles apply to other sectors such as industries and service activities.

12.4.4.5 Institutional innovations

Scarcity of water resources, threats of water resources development and the conflicting interests of water users imply that water resources planning and management should be an integral part of the overall national economic planning. Capacity building activities should be undertaken both within and among water sub-sectors, like irrigation, water supply and sanitation, and hydropower, thereby providing improved coordination. The proposed strategy included in the Delft Declaration, 1991, involves, inter alia, the following approaches (Dumitrescu, 1992):

- Developing improved policy and legal frameworks, institutional development and a commitment to development of human resources and marginal systems for this sector.
- Having the external support agencies adopt capacity building as an essential element of their assistance efforts.
- Involving, wherever appropriate, the private sector in managing or providing water related services.
- Encouraging national and foreign universities, research institutes, consulting organisations, professional associations and other agencies to participate in the process of capacity building.
- Encouraging countries to conduct water sector assessment, which include the need for capacity building and
- Creating awareness of the vital role and finiteness of water on the part of decision-makers and the public at large.

The author wishes to seize this opportunity and calls for a balanced approach to water management policy that emphasises more efficient use of available supplies, water conservation, and demand management. He also refers to pollution control and environment protection measures. Special attention should be focused on problems typical of arid countries, and the urgent need to carry out the necessary research to identify the problems, define project objectives and evaluation criteria, formulation and assessment of alternatives, implementation of selected alternative and monitoring ex-post analysis.

REFERENCES

Abdel-Dayem S, El-Atfy H, El-Gammal H (1993) Regional and seasonal changes of subsurface drainage water in Egypt. Transactions of the 15th Congress of the International Commission on Irrigation and Drainage (ICID), vol 1-A, Q. 44 R. 13, The Hague, pp 145–156

Abdel-Rahman WA (2000) Application of Islamic legal principles for advanced water management. Wat. Int. 25(4):513–518

Abdel-Rahman HA, Abdel-Magid IM (1993) Water conservation in Oman. Wat. Int. 18(2):95–102

Aboukhaled A (1972) Crop water requirements. Proc. of water use Sem., Damascus, FAO Irr & Dr. Paper No.13: 42–60, Food and Agriculture Organisation of the United Nations, Rome

Abdulai BI, Stigter CJ, Ibrahim AA, Adeeb AM, Adam HS (1990) Evaporation calculations for Lake Sennar (Sudan): a search for a meteorological minimum input approach for shallow lakes. The Netherlands J Agr. Sci. 38:725–730, Wageningen, The Netherlands

Abdulla FA, Al-Ghazzawi ZD (1997) Assessment of methane emissions from industrial and municipal wastewater treatment plants in Jordan. In: Proc. of Int. Conf. on Wat. Problems in the Mediterranean Countries, vol I. Near East University, Lefkoşa, Turkish Republic of Northern Cyprus, pp 477–484

Ackers P, White WR (1973) Sediment transport: new approach and analysis. Proc. ASCE 99(HY 11):2041–2060

Agib el NA (1995) Study of hydrological variations, drought and desertification in the Sudan. M.Sc. Thesis, University of Bahrain

Agricultural Development Organization for the Arab States (1980) Report on food security programme in the Arab region. In: Natural Resources, vol II. Khartoum

Aguado E (1987) A time-series analysis of the Nile River low flows. Ann. Assoc. Amer. Geogr. 72(1):109–119

Aigo F, Bouras C (1995) Reuse of sewage effluent for irrigation in Syria. In: Proc. of Int. Conf. on Wat. Resourc. Mgmt. in Arid Countries, vol 2. Muscat, Oman, pp 683–688 (text in Arabic)

Al-Ajimi HA, Abdel-Rahman HA (2001) Water management intricacies in the Sultanate of Oman: the augmentation-conservation conundrum. Wat. Int. 26(1):68–79

Alatas SH (1963) Reflections on the theories of religion. Thesis, University of Amsterdam for the Ph. D. Degree in Political and Social Sciences. Pasmans Publishers, The Hague, The Netherlands

Aldon EF, Springfield HW (1974) Using paraffin and polyethylene to harvest water for growing shrubs. In: Proc. of Water Harvesting Symp, Phoenix, Arizona Agricultural Research Station, USDA, pp 251–257

Al-Ghariani SA (1997) Man-made rivers: a new approach to water resources development in dry areas. Wat. Int. 22(2):113–117

Al-Garoo AS (1987) Historic review of spate irrigation and its effect on agricultural development. In: Spate Irrigation, Proc. Subregional Expert Consult. organized by UNDP and FAO, Aden (formerly People's Democratic Republic) Yemen, pp 34–40

Algeria (1983) Rapport national de la République Algérienne Démocratique et Populaire. Réunion Régionale sur les Resources en Eau dans les Pays Arabes. Report, UNESCO Regional Office for Science and Technology in the Arab States (ROSTAS)

Algeria, National Institute for Hydraulic Research (I.N.R.H.) of Algeria (1986) Water resources and their utilisations in Algeria. Document, the panel on water resources and their utilization in the Arab region. ACSAD-AFESD-KFAED, Kuwait

Al-Hamoud R, Edwards J (2005) Water poverty and private investment in the water and sanitation sector. Wat. Int. 30(3):350–355

Al-Hassoun SA, Al-Turbak AS (1995) Recharge dam efficiency based on subsurface flow analysis. Wat. Int. 20(1):40–45

Ali H, Madramootoo CA, Abdel-Dayem S, Amer MH (2000) Water and salt balance of model of lake Qaroun, Egypt. J ICID 49(3):25–39

Ali W, Glaser J, Thiel M, Hőtzel H, Werz H (2004) Hydrogeological investigations in the North-Eastern Dead Sea area, Sumeimeh, Jordan. In: Zereini F, Jaescheke W (eds) Water in the Middle East and North Africa. Springer, Berlin New York London Paris Tokyo, pp 22–30

Al-Jabbari MH, Al-Ansari NA (1986) Dissolved load and ionic balance in the upper reaches of the Euphrates River. J Wat. Resourc. 5(1):98–116, Baghdad, Iraq

Al-Jabbari MH, Mansour NB (1986) Sediment discharge in Adhaim River within drought conditions. J Wat. Resourc. 5(1):117–134, Baghdad, Iraq

Al-Kharabsheh A, Ta'Any R (2005) Challenges of water demand management in Jordan. Wat. Int. 30(2):210–219

Al-Klouub B, Al-Shemmeri TT (1995) Sustainable development of water resources and possible enhancement technologies and application of water supply in Jordan. Wat. Int. 20(2):106–109

Allam MN, Al-Wagdany AS (1989) A physically-based flood volume distribution model. Wat. Resourc. Mgmt. 3:205–230, Kluwer Academic Publishers, Dordrecht, The Netherlands

Allam MN (1990) Geomorphologic rainfall-runoff model: incorporating Philip's infiltration expression. J Wat. Resourc. Plan. And Mgmt./ASCE, 116(2), Paper No. 24470:262–281

Allan JA (1999) Now you see it, now you don't....... A convenient solution. The UNESCO Courier 52(2):29–32, UNESCO, Paris

Allen RG, Pruitt WO (1991) FAO-24 reference evapotranspiration factors. J Irr & Dr Div ASCE 117(IR 5):758–773

Allen RG, Smith M, Perrier A, Pereira LS (1994) An update for the definition of reference evapotranspiration. ICID Bull., 43:1–92

Allen RG, Pereira LS, Raes D, Smith M (1998) Crop evapotranspiration guidelines for computing crop water requirements. FAO Irr & Dr Paper 56, Food and Agriculture Organisation of the United Nations, Rome

Al-Madina Daily Newspaper (1987) Countries of Islamic World: basic facts and indicators. Issue No. 7219 (date: 27.01.87), p 2 of the special extra issue, Jeddah, Saudi Arabia

Al-Muqbali N, Schmid R (1995). Design features of recharge dams in the Sultanate of Oman. In: Proc of Int Conf on Wat Resourc Mgmt. in Arid Countr, vol. 1. Muscat, Oman, pp 160–165

Al-Muqbali N, Kotwicki V (2000) Design criteria and effectiveness of recharge dams in arid climates. In: Goosen M, Shayya WH (eds) Water management, purification & conservation in arid climates, vol 3. Water Conservation, Technomic Publ. Co., Lancaster Basel, pp 287–302

Al-Muttair FF, Sendil U, Al-Turbak AS (1994) Management of recharge dams in Saudi Arabia. J Wat. Resourc. Plan. and Mgmt. 120(6):749–763

Al-Nozaily FA (1992) Reuse of wastewater in Sana'a. In: Proc. of the Natl. Sem. on Wastewater Reuse, WHO/Sana'a University, Yemen, pp 103–109

Al-Ruwaih F (1995) Assessment and management of a desert basin in Kuwait. Wat. Int. 20(4):213–224

Al-Ruwaih F, Shehata M (1998) The chemical evolution and hydrogeology of Al-Shagaya Field-B, Kuwait. Wat. Int. 23(2):75–83

Al-Sajwani T, Lawrence RJ (1995) Small desalination plants in rural communities in Oman. In: Proc. of Int. Conf. on Wat. Resourc. Mgmt. in Arid Countries, vol 2. Muscat, Oman, pp 617–626

Al-Sawaf FD (1973) Hammam Al-Alil spa-the relationship between temperature, chemical composition and origin of the water. In: Proc. Sem. on Groundwater, organised by Institute for Applied Research on Natural Resources and UNESCO, Baghdad, p 19

Al-Sheriadeh MS (1997) A comparative study of different types of wastewater treatment plants, Jordan case. In: Proc. Int. Conf. on Wat. Problems in the Mediterranean Countries, vol I. Near East University, Lefkoşa, Northern Cyprus, pp 395–401

Al-Turbak AS, Al-Muttair FF (1989) Evaluation of dams as a recharge method. Wat. Resourc. Dev. 5(2):119–124

Al-Turbak AS (1997) Future water supply and demand predictions in Saudi Arabia. In: Proc. Int. Conf. on Wat. Problems in the Mediterranean Countries, vol II. Near East University, Lefkoşa, North Cyprus, pp 701–713

Al-Washah RAM (1992) Jordan's water resources: technical perspective. Wat. Int. 17(3):124–132

Amerongen van A, Hijmans MJ (1993) Algerije. Koninklijk Instituut van de Tropen (KIT), Amsterdam & Novib, Den Haag

Ansari Al N, Ali JL (1986) Suspended load and solute discharge in River Tigris within Baghdad. J Wat. Resourc. 5(1):35–50, Baghdad

Ansari Al N, Al-Jabbari MH, Al-Sinawi G (1986) Suspended sediment in the upper parts of River Euphrates. J Wat. Resourc. 5(1):67–83, Baghdad

Arab Organisation for Agricultural Development/Arab League (1980) Document on Arab Food Security Programme, Part 2: Natural Resources (Arabic text), Kharoum, Sudan

Arab Science Establishment for Research and Transfer of Technology (ASER) (1986) The situation of water resources in the West Bank and Gaza Strip. Document, the panel on water resources and their uses in the Arab region, ACSAD-AFESD-KFAED, Kuwait

Aramco (1980). Wastewater effluent reuse and disposal. Tech. Services Dep. AER-5139, TSI, ARAMCO, Dhahran, Saudi Arabia, pp 57–104

Asano T (1994) Reusing urban wastewater-an alternative and a reliable water resource. Wat. Int. 19(1):36–42

Attia F, Fekry A, Smidt E (1995) Artificial recharge of groundwater: a tool for Egypt's future water management. In: Proc. of the Intl. Conf. on Wat. Resourc. Mgmt. in Arid Countries, vol 1. Muscat, Oman, pp 183–189

Attia F (1996) Agricultural threats to groundwater, the Egyptian situation. In: Candela L, Aureli A (eds) Agricultural threats to groundwater quality, workshop proceedings. Zaragoza, Spain, pp 191–201

Baban RT, Gaish AL (1987). Statistical analysis of flood hydrographs of Wadi Beihan. In: Spate Irrigation, Proc. Subregional Expert Consult. organized by UNDPand FAO. Aden (formerly People's Democratic Republic) Yemen, pp 58–59

Baghdadi AI (1973) Water springs in Iraq, their geological characteristics and utilization. In: Proc. Sem. on groundwater, organised by Institute for Applied Research on Natural Resources and UNESCO, Baghdad, p 14

Bahrain (1986) Country report on water resources and their utilizations in the state of Bahrain (prepared by Al-Mansour K, Al-Aradi A). Document, the Panel on Water Resources and their Utilisations in the Arab Region. ACSAD-AFESD-KFAED, Kuwait

Bajjali W, Clark ID, Fritz P (1997) The artesian thermal groundwaters of northern Jordan: insights into their recharge history and age. J Hydrol. 192:355–382

Balek J (1977) Hydrology and Water Resources in Tropical Africa. Develop. Wat. Sci. 8, Elsevier Sci. Publ. Co., Amsterdam Oxford New York

Ba'Momen AM (1995) Water resources and its applications in agricultural development in Yemen: a special study on irrigation in Wadi Hadramout (Arabic text with an English summary). In: Proc. Int. Conf. on Wat. Resourc. Mgmt. In Arid Countries, vol 1. Muscat, Oman, pp 303–320

Barbar W, Carr DP (1976) Preliminary appraisal of water resources of the United Arab Emirates. Progress Report, Phase I, United Arab Emirates, p 31

Bartholomew JC (editorial director) (1980) The times concise atlas of the world (revised edition). Times Books Ltd, London

Beater AB (1989) The applicability of two single event rainfall-runoff models to catchments with different climate and physiography. M.Sc. Thesis, Department of Geography, Rhodes University, Grahamstown, Republic of South Africa, p 158

Beaumont P, Blake G, Wagstaff JM (1976) The Middle East: a geographical study. Wiley, London New York Sydney

Belaid MN (1995) Seawater intrusion into a quaternary aquifer west of Tripoli, Libya. In: Proc. Intl. Conf. on Wat. Resourc. Mgmt. in Arid Countries, vol 2. Muscat, Oman, pp 547–552

Bell FC (1969) Generalised rainfall-duration-frequency relationships. J Hydr. Div. Am. Soc. Civ. Engrs. HY, 1:311–327

Belloum A (1993) Hydrologie agricole en Algérie-une double problématique. J Hydrol. Sci. 38(6): 479–495

Bennett T, Aller L, Lehr JH, Petty RJ (1985) DRASTIC: a standardized system for evaluating groundwater pollution using hydrogeologic settings. US Environment Protection Agency (EPA), EPA/600/2-85/018, Ada (Oklahoma), p 163

Bienert HD (2000) Water shortage in Jordan since prehistoric times-lessons from history. J ICID 49(4):17–37

Biswas AK (1967). Hydrologic engineering prior to 600 B.C. Proc. ASCE 93(HY 5):115–135

Biswas AK (1972) History of hydrology. North-Holland Publ. Co., Amsterdam/London

—— (ed) (1994) International waters of the Middle East from Euphrates-Tigris to the Nile. Oxford University Press, Mombay Delhi Calcutta Madras, p 221

—— (1999) Water crisis: current perceptions and future studies. Wat. Int. 24(4):363–367

Blaney HF, Criddle WD (1950) (improved 1952). Determining water requirements in irrigated areas from climatological and irrigation data. USDA, SCS TPs: 96p, 48p

Blaney HF (1957) Monthly consumptive use of water by irrigated crops and natural vegetation. IAHS Gen Assem of Toronto, vol II. IAHS Publ 44:431–439

Blaney HF, Criddle WD (1966) Determining consumptive use for planning water developments. Proc. of the Irr. and Dr. Spec. Conf. on 'Methods for Estimating Evapotranspiration', by ASCE, New York, pp 1–34

Bouguerra K (1986) Quantification of different erosion from micro-basin studies in semi-arid zones in Algeria. J Wat. Resourc. 5(1):223–250, Baghdad

Boumans JH, Hulsbos WC, Lindenbergh HLJ, van der Sluis PM (1963) Reclamation of salt-affected soils in Iraq. Dieleman PJ (ed). Int Inst for Land Reclam and Improv (ILRI), Publ. No 11, H. Veenman & Zonen N.V., Wageningen, The Netherlands

Bouzaiane S, Lafforgue A (1986) Interpretation des données hydrologiques du Bassin de l'Oued Zeroud. Mono. Hydrol. des Oueds Zeroud et Merguillil III, Dir. Centr. Des Resourc. en Eau/Inst. Français de Rech. Sci. pour le Develop. en Coopération, Ministère de l'Agriculture, Tunisie, pp 474–604

Box GEP, Jenkins GM (1976) Time series analysis, forecasting and control. Holden-Day, USA

Brichambaut P, Wallén CG (1963) A study of agrometeorology in semi-arid and arid zones of the Middle East. WMO-No. 141, TP. 66, Technical Note No. 56, World Meteorological Organisation, Geneva

Brinkmann PJ, Hein M, Holländer L, Reich G (1987) Retrospective simulation of groundwater flow and transport in the Nubian aquifer system. Berliner geowiss. Abb., Reihe A/Band 75.2, Dietrich Reimer, Berlin, pp 465–516

Brune GM (1951) Sediment record in Midwestern U.S. IASH 3:29–38

Bureau de Rechèrches Géologiques et Miniéres (BRGM) (1986) Water, agriculture and soil studies of the Saq and overlying aquifers. Tech. Rep. (unpublished), Ministry of Agriculture and Water. Riyadh, Saudi Arabia

Burman R, Pochop IO (1994) Evaporation, evapotranspiration and climatic data. Dev. Atm. Sci. 22, Elsevier, Amsterdam New York Tokyo

Bushnak AA (1993) Water desalination, the experience of GCC countries. Paper presented to the regional symposium on water use and conservation, p 63, ESCWA (UN)/CEHA (WHO), p 63, Amman, Jordan

Bushnak AA (1995) Desalination technology and economics. In: Proc. of Middle East Consultants Conference "Putting Price on Water", Bahrain, p 50

Butler MA (1933) Irrigation in Persia by qanāts: an ancient method of collecting and conducting water in long underground galleries in use today. Civil Engineering, ASCE 3: 69–73

Caponera AD (1973) Water laws in Moslem countries. FAO Irr. & Dr. Paper No.20. Food and Agriculture Organisation of the United Nations, Rome

Chaturvedi MC (2000) Water for food and rural development-developing countries. Wat. Int. 25(1):40–53

Chebaane M, Alesh SA (1995) An experiment on Monsoon precipitation measurement in Dhofar Mountains. In: Proc. of Int. Conf. on Wat. Resourc. Mgmt. in Arid Countries, vol 2. Muscat, Oman, pp 392–400

Chow VT (1964) Handbook of applied hydrology. McGraw-Hill Book Co., New York Toronto London

Chow JS, Wilson JL (1981) A qualitative review of Nubian Sandstone regional aquifer behaviour. In: Proc. of Int. Conf. on Wat. Resourc. in Egypt, Cairo University, Cairo, Egypt, pp 363–382

Christiansen JE (1966) Estimating pan evaporation and evapotranspiration from climatic data. In: Proc ASCE Irr. & Dr. Speciality Conf. on 'Methods for estimating Evapotranspiration', Las Vegas, Nevada, pp 193–231

Colombani J, Olivry JC, Kallel R (1984) Phénomènes exceptionnels d'érosion et de transport solide en Afrique aride et semi-aride. IAHS Publ. No. 144, pp 295–300

Combrémont R (1972) Crop water requirements in Tunisia (summary). In: Proc. Wat. Use Sem. Damascus, FAO Irr. & Dr. Paper No.13, Food and Agriculture Organisation of the United Nations, Rome, pp 138–141

Cooper RH, Peterson AW (1970) Discussion of Paper 6884 by Tom Blench of Coordination in Mobile-bed Hydraulics. ASCE, vol. 96 (HY 9), Proc. Paper 9747, 1970

Daniel H (1980) Man and climatic variability. Publ of the World Clim Prog, world Meteorological Organization, Genéva

Dawdy RD, O'Donnel T (1965) Mathematical models of catchment behaviour. J Hydr. (HY 4), ASCE, pp 123–137

Diaz-Grandos M, Bras RL, Valdes JB (1983) A derived flood frequency distribution based on geomorphologic IUH and the density function of rainfall excess. TR No. 292, Dept. of Civ. Engg., Massachusetts Inst. Of Tech. (MIT), Massachusettes, USA

Doorenbos J (1976) Agrometeorological field stations. FAO Irr. & Dr. Paper No. 27, (Prepared in consultation with the World Meteorological Organisation), Food and Agriculture Organisation of the United Nations, Rome

Doorenbos J, Kassam AH (1986) Yield response to water. FAO Irr. & Dr.Paper No 33, Food and Agriculture Organisation of the United Nations (FAO), Rome

Doorenbos J, Pruitt WO (1977) Guidelines for predicting crop water requirements. FAO Irr. & Dr. Paper No. 24, Food and Agriculture Organisation of the United Nations (FAO), Rome

Droogers P, Allen RG (2002) Estimating reference evapotranspiration under inaccurate data conditions. J Irr & Dr Systems 16:33–45, Kluwer Ac. Publ., Dordrecht, The Netherlands

Dumitrescu S (1992) Summary report of the conference papers. International Conference on Water and the Environment, Dublin, with annexes

Dunne T (1979) Sediment yield and land use in tropical climates. J Hydrol. 42:281–300

Eckstein Y, Eckstein GE (2003) Groundwater resources and international law in the Middle East peace process. Wat. Int. 28(2):154–161

Economic Commission for Western Asia (ECWA) (1981) Assessment of water resources in the ECWA Region. United Nations Economic and Social Council, Commission forWestern Asia, Natural Resources, Science and Technology Division, Water Programme, Doc. E/ECWA/NR/L/1, New York, p 290

—— (1984) Wastewater reuse and its application in ECWA Region. Tech. Publ., Natural Resources and Technology Div., ECWA, New York, p 166

Egypt (1986) Country report on water resources and their utilisations in Egypt. Document, the Panel on Water Resources and their Utilisations in the Arab Region. ACSAD-AFESD-KFAED, Kuwait

Einstein HA (1950) The bed load function for sediment transportation in open channel flows. USDA, SCS TB No.1026, USA

El-Amami S (1983) Changing concepts of water management in Tunisia. Impact of Science on Society (1):57–64

El-Ghandour MFM, Hefny K, Khalil JB, Atta SA (1983) Suitability of groundwater in the Delta Region, Egypt, for irrigation. Middle East Water & Sewage, **7**(2):146–152, Fuel & Metallurgical Journals Ltd, Surrey, United Kingdom

El-Sheikh S, Kaikai A, Andah K (1991) Intensive sediment transport from the Upper Nile and water resources management in Sudan. IAHS Publ. No. 201:291–300

El-Tahir EAB (1996) El Niño and the natural variability of the flow in the Nile River. Wat. Resourc. Res. 32(1):131–137

Elwan MY (1995) Synthetic generation of missing runoff data of the White Nile, River Sobat and Bahr el-Jebel. Thesis, the Int. Inst. Infrastruct., Hydr. and Env. Engg. (IHE) for the Degree of Masters of Science in Hydrological Engineering, Delft, The Netherlands

Elwan MY, Shahin M, Hall MJ (1996) Synthetic generation of missing runoff data for the White Nile, Sobat River and Bahr el-Jebel. In: Proc. 17th IAHR Int. Symp. on Stochastic Hydraulics, Mackay, Australia, pp 333–340

El-Zaemey AK (1992) Wastewater reuse practices in Yemen. In: Proc. of the Natl. Sem. on Wastewater Reuse, WHO/ Sana'a University, Yemen, pp 35–43

Engelman R, Le Roy P (1993) Sustaining water, population and the future of renewable water supplies. Population and Environment Program, Population Action international, Washington DC 20036, USA

Englund F, Hansen E (1967) A Monograph on Sediment Transport in Alluvial Streams. Teknisk Forlag, Copenhagen

Fahmy A, Panattoni L, Toddini E (1982) A mathematical model of the River Nile. In: Abbott MB, Cunge JA (eds) Engineering Applications of Computational Hydraulics, vol I. Pitman Advanced Publ. Progr., UK, pp 111–130

Farooq S, Al-Lyla R (1985) Potential of water development in Saudi Arabia. Wat. Int. 10(4):151–155

Farquharson FAK, Plintson DT, Sutcliffe JV (1996) Rainfall and runoff in Yemen. J Hydrol. Sci. 41(5):797–811

Fathi AM (1995) Water consumptive use and crop coefficient for some field crops in desert sandy soils using simple field volumetric lysimeters. In: Proc. Int. Conf. on 'Water Resources Management in Arid Countries', vol 1. Muscat, Oman, pp 106–113

Fok Yu-Si, Chu SC (1995) Rain harvesting; varieties of practical applications for arid countries. In: Proc. of Int. Conf. on Wat. Resourc. Mgmt. in Arid Countries, vol 1. Muscat, Oman, pp 190–195

Food and Agriculture Organisation of the United Nations (FAO) (1981) Arid zone hydrology for agricultural development (based on the work of K.R. Jones, with contributions from O. Berney, D.P. Carr, E.C. Barrett and FAO Staff). FAO Irr. & Dr. Paper No. 37, FAO, Rome

—— (1986) African agriculture, the next 25 years. In: Atlas of Agriculture. FAO, Rome

—— (1989) Manual for CROPWAT Computer Programme. FAO Land and Wat. Develop. Div., FAO, Rome

—— (1995) Irrigation in Africa in figures/ L'irrigation en Afrique en chiffres. Water Report **7**, FAO, Rome

Fournièr F (1960) Climat et Erosion. Presse Université Francaise, Paris, France

Garbrecht G (1987) Irrigation throughout history-problems and solutions. In: Wunderlich W, Prins JE (eds) Proc. of Int. Sym. on Water for the Future. Rome, pp 3–17

Gentili J (1972) Australian Climate Patterns. T. Nelson, Melbourne, Australia

Ghaith AM, Ra'fat I, Helmy AA, Baghdady FA, Mikhael MT (1960) Investigation of water quality of the main drains in the United Arab Republic of Egypt (text in Arabic/8181 1959–3000). The General Organisation for Government Publishing Offices, Cairo, Egypt, p 57 (with 1 map)

Gibali el A, Shenouda F, Tawdros S, Gamal M (1966) Irrigation requirements and frequency of late corn. Agr Res. Rev. 44(1):139–157, Cairo, Egypt

Gibbons JD (1971) Nonparametric statistical inference. McGraw Hill, London New York

Girgirah AA, Maktari MS, Mohammed MF, Abbas HH, Shoubihi HM (1987) Wadi development of agriculture in PDR (People's Democratic Republic of Yemen). In: Spate Irrigation, Proc. Sub-regional Expert Consult. organised by UNDP and FAO, Aden, Aden, PDR (formerly People's Democratic Republic), Yemen, pp 9–26

Gischler CE (1979) Water resources in the Arab Middle East and North Africa. Middle East and North African Studies Press Ltd, Cambridge

Gleick PH (2000) The changing water paradigm. A look at twenty first century water resources development. Wat. Int. 25(1):127–138

Griffiths JF (ed) (1972) Climates of Africa. In: World Survey of Climatology, vol 10. Elsevier Publ. Co. Amsterdam London New York

Griffiths JF, Soliman KH (1972) The northern Sahara (Chapter 3). In: climates of Africa. World survey of climatology, vol 10. Elsevier Publishing Co. Amsterdam London New York

Grove AT (1972) The dissolved and solid load carried by some West African rivers: Sénégal, Niger, Benue and Shari. J Hydrol. 16(4):277–300

Gulick RL (1941) Muhammad the educator. Thesis, University of California, Berkeley, in partial satisfaction of the requirements for the Degree of Master of Arts in Education, California, USA

Gun van der JA, Elderhorst WI, Kruseman GP (1992) Predicting long-term impacts of groundwater abstraction from an intensively exploited coastal aquifer. In: Simmers I, Villaroya F, Rebollo LF (eds) Proc. 23rd Congress of Int. Assoc. of Hydrogeologists, 1991, 3:303–318, Tenerife

Haddad M, Mizyed N (2004) Non-conventional options for water supply augmentation in the Middle East: a case study. Wat. Int. 29(2):232–242

Haddad RH, Jawad SB, Haddad WI (1970) Preliminary studies on groundwater in Samara-Tikrit area. Tech, Rep. No. 13, p 16 (with maps and tables) In: Dougrameji J, Ubel K (eds) Institute for Applied Research on Natural Resources, Baghdad

Haddad RH, Jawad SB, Haddad WI, Younan AI, Sadov AV (1971) Hydrogeological investigations in Jolak Basin of the Altun-Kupri project area. Tech. Rep. No.25, p 38 (with maps and tables), Institute for Applied Research on Natural Resources, Iraq

Haddadin MJ (2001) Water scarcity impacts and potential conflicts in the MENA Region. Wat. Int. 26(4):460–470

Hadley RF (1986) Fluvial transport of sediment in arid and semi-arid regions. Iraqi J Wat. Resourc. 5(1):335–345, Baghdad

Hamad OET, Mohamed TES (1986) Some aspects of the recent drought on the operation of the Gezira and Managil main canals. In: Proc. of Int. Conf. on Wat. Resourc. Needs and Planning in Drought Prone Areas, Khartoum, Islamic African Centre Press, Khartoum, Sudan, pp 933–946

Hani B (1995) Jordan's strategies for freshwater resources development. In: Proceedings of the general seminar on options and strategies for freshwater development and utilisation in selected Arab countries, Centre for Environment & Development for the Arab Region and Europe (CEDARE), Amman, Jordan, pp 183–206

Hargreaves GH (1956) Irrigation requirements based on climatic data. J Irr & Dr Div 82 (IR 3):1–10, Proc. ASCE Paper 1105

Hargreaves GH, Samani ZA (1982) Estimating potential evapotranspiration. J Irr & Dr Div ASCE 108 (IR 3): 225-230

Hargreaves GH, Samani ZA (1985) Reference crop evapotranspiration from temperature. Applied Engineering in Agriculture, Publ. by ASAE I(2):96–99

Harrold LL (1966) Measuring evapotranspiration by lysimetry. In: Proc. Conf. on Evapotranspiration and its Role in Water Resources Management. ASCE, Michigan, pp 28–33

Hashem AS, Said NF, Abu Zeid M (1967) Drainage of agricultural lands (text in Arabic). The Survey Department (560/67), Ministry of Irrigation, United Arab Republic, Cairo, Egypt

Hassan AH, van der Sluijs AA (1970) Compilation of hydrogeological data from the area between Baghdad and Hillah anticipating more detailed investigations in the Greater Mussayeb Project. Tech. Rep. No.10, p 8 (with maps and tables), Institute for Applied research on Natural Resources, Iraq

Hedayat A (1986) See Country Report of the Sudan

Hefny K (1991) Planning for groundwater development of Nubian sandstone aquifer for sustainable agriculture. Proc. Roundtable meeting (RTM-91) on Planning for Groundwater Development in arid and Semi-arid regions, Cairo, pp 113–124 (RIGW/IWACO eds.), Cairo Rotterdam

Herschel C (1899) The two books on water supply of the City of Rome of Sextus Julius Frontinus. Dana Gstos & Co., Boston

Hesse K, Hissine A, Kheir O, Schnäker E, Schneider M, Thorweihe U (1987) Hydrogeological investigations in the Nubian Aquifer System, Eastern Sahara. Berliner Geowissen Abhandlung A. 75.2, pp: 397–464, Technische Universität Berlin, Germany

Heusch B, Cayla O (1986) Assessment of sediment discharge measurements in the Maghreb countries. J Wat. Resourc. 5(1):349–366, Baghdad, Iraq

Himmelblau DM (1969) Process analysis by statistical methods. Wiley, Inc., New York London
Hipel KW, McLoed AI (1978) Preservation of the rescaled adjusted range: 2: simulation studies using the Box-Jenkins models. Wat. Resourc. Res. 14(3): 509–518
Hobler M, Rajab R (2002) Groundwater vulnerability and hazards to groundwater in the Damascus Ghouta. Report prepared by ACSAD & BGR, Tech. Coop. Proj. 'Management, Protection and Sustainable Use of Groundwater and Soil Resources', vol 2(2), Damascus
Hoff R (1995) Jemen (mensen, politiek, economie, cultuur). Royal Institute for Tropics (KIT), Amsterdam/Novib, The Hague, The Netherlands
Hofmann G, Rambow J (1995) Spatial and temporal rainfall characteristics of the Wadi al Jizi Basin in northern Oman. In: Proc. Int. Conf. Wat. Resourc. Mgmt. in Arid Countries, vol 2. Oman, pp 411–418
Husary S, Najjar T, Aliewi A (1995) Analysis of secondary source rainfall data from the Northern West bank. Report on Water resources Management: West Bank and Gaza Strip (WARMP/TEC/J/07), University of Newcastle upon Tyne/UK Overseas Development Administration
Hussein MT (1986) Groundwater potentials of the Eastern region of the Sudan. In: Proc. Int. Conf. on Wat. Resourc. Needs and planning in drought prone areas, vol 2. The Islamic African Centre Press, Khartoum, Sudan, pp 1141–1150
Hussein ASA, El-Daw AK (1989) Evapotranspiration in Sudan Gezira irrigation scheme. Proc. ASCE Paper No. 24187, J Irr & Dr Div 115(IR 6):1018–1033
Hussein T, Ahmed AH (1997) Environmental and economic aspects of wastewater reuse in Saudi Arabia. Wat. Int. 22(2):108–112
Hussein MT (2004) Hydrochemical evaluation of groundwater in the Blue Nile Basin. Hydrogeol. J 12:144–158
Hussein I (2005) Application of expert and decision support systems for optimizing water supply in the Jordan valley: the case of King Abdullah Canal. Wat. Int. 30(3):304–313
Ibrahim GAM (1993) Application and comparison of rainfall-runoff models on the Blue Nile Basin. M.Sc. Thesis in Hydrological Engineering, the Int. Inst. for Infrastruc, Hydr. and Env. Engg, Delft, The Netherlands
Iraq (1986) Country report on water resources and their utilisations in the republic of Iraq, Document (in Arabic), the Panel on Water Resources and their Utilisations in the Arab Region. ACSAD-AFSED-KFAED, Kuwait
Israelsen OW (1959) Irrigation principles and practices, 2nd edn. John Wiley & Sons Inc., New York
Jacobi K (1991) Water atlas of Yemen. Certain chapters from the report on Yemen Water Master Plan. Agrar und Hydrotechnik GMBH, Essen, Germany
Jadid El AG, Şen Z (1997a) Regional wet and dry period statistics of Libyan rainfall. Proc. Med. Conf. On Water Problems in the Mediterranean Countries, vol I: 23–29, Lefkoşa, Turkish Republic of Northern Cyprus
——1997b. Gamma distribution of Libyan rainfall records. Proc. Med. Conf. on water problems in the mediterranean countries, vol I: 53–60, Lefkoşa, Turkish Republic of Northern Cyprus
Jansen JM, Painter RB (1974) Predicting sediment yield from climate and topography. J Hydrol. 21:371–380
Jensen ME (1966) Empirical methods for estimating or predicting evapotranspiration using radiation. Proc. ASCE Conf. on "Evapotranspiration and its Role in Water Rresources Management", ASAE, pp 49–53
Jensen ME, Haise HR (1963) Estimating evapotranspiration from solar radiation. J Irr & Dr Div ASCE 89(IR 4):15–41
Jensen ME, Burman RD, Allen RG (eds) (1990) Evapotranspiration and irrigation water requirements. ASCE Manuals and Reports on Engineering Practice, 70:332
Jibrael N (1973) An evaluation of groundwater resources in Iraq. In: Proc. Groundwater Seminar, Institute for Applied Research and UNESCO, Baghdad, p 14
Jones PD, Wigley TML, Wright PB (1986) Global temperature variations between 1851–1984. J Clim. and Appl. Meteorol. 25:1213–1230
Jordan (1986) Water resources in Jordan and their utilisationss. Country Report (in Arabic), symposium on water resources in the Arab countries and their utilisations. ACSAD-AFESD-KFAED, Kuwait

Kamal ed-Din IAR (1981) Stochastic analysis of streamflow data in Iraq. M. Sc. Thesis, Faculty of Engineering, University of Baghdad, Iraq

Kashef AZI (1981) The Nile – one river and nine countries. J Hydrol. 53:53–71

Kashef AZI (1983) Salt-water intrusion in the Nile Delta. J Groundwater (2):160–167

Kaul FJ (1995) Investigation of potential for storage dams in Wadi Dayqah, Oman. In: Proc. Int. Conf. Wat. Resourc. Mgmt. In Arid Countries, vol 1. Oman, pp 166–175

Khadam MAA, Salih MK (1986) Assessment of water resources systems for a semi-arid savannah urban region (Reference: Nyala Town). In: Proc. Intl. Conf. on Wat. Resourc. Needs and planning in drought prone areas, Part II: 1237–1247, Khartoum, Sudan

Khan MO, Fuggle RW (1995) Beneficial reuse of contaminated water in Oman. In: Proc. of Int. Conf. on Wat. Resourc. Mgmt. in Arid Countries, vol 2. Muscat, Oman, pp 666–674

Khoury J, Rasoul'allah WA, Drouby A (1986) Water resources in the Arab world and their future perspectives. Paper, Symp. Wat. Resourc. and their Utilisations in the Arab Region, ACSAD-AFESD-KFAED, Kuwait

KNMI (Royal Dutch Meteorological Institute) (2002) World climate in the period 1971–2000. Elmar CD-Rom, Elmar B.V. Publishers, Rijswijk, The Netherlands

Kolars J (1991) The future of the Euphrates River. Prepared for the World Bank Conference: Int. Workshop on Comprehensive Wat. Resourc. Mgmt. Policy, Washington, D.C., p 33

Köppen W (1931) Die Klimat der Erde. W. de Gruyter Publishing Co., The Netherlands

Krishna R (1986) The Nile River basin. Paper prepared for the Near East studies program and its conference. "U.S. Foreign Policy on Water Resources in the Middle East for Peace and Development", Washington, D.C.

Kurzon Y (ed) (1979) Water balance and water resources of the Earth (translated from Russian, original printing in 1974). UNESCO Press, Paris

Kuwait (1986) Country report on water resources and their utilizations in Kuwait. Document, the panel on the water resources and their utilisations in the Arab region. ACSAD-AFESD-KFAED, Kuwait

Lakey R, Easton P, Al-Hinai HA (1995) Groundwater recharge processes, Eastern Batinah, Oman. In: Proc. Int. Conf. on Wat. Resourc. Mgmt. in Arid Countries, vol 2. Muscat, Oman, pp 511–520

Lamb C (1988) Climate in the last thousand years: natural climatic fluctuation and change. In: Flohn H, Fantechi R (eds) The climate of Europe: past, present and future. D. Reidel Publishing Co., The Netherlands

Laurenson EM (1964) A catchment storage model for runoff routing. J Hydrol. 2:141–163

Lawrence P (1987) Sediment control in wadi irrigation systems. In: Proc. Sub-regional Expert Consult. on Wadi Develop. for Agric. (UNDP/FAO), Aden, Yemen, pp 91–101

Linden W van der (1986) Groundwater modeling and the optimum operation of the wadi Nyala aquifer. In: Proc. Intl. Conf. on Wat. Resourc. Needs and planning in drought prone areas, Part I: 455–466, Khartoum, Sudan

Linsley RK, Kohler MA, Paulhus JL (1958) Hydrology for Engineers. McGraw Book Co., Inc., New York Toronto London

Lloyd JW (1990) Groundwater conditions and development of the Eastern Sahara. J Hydrogeol. 4:35–48

Lloyd JW (1992) Protective and corrective measures with respect to the over-exploitation of groundwater, in selected papers on aquifer exploitation. International Association of Hydrogeologists 3:167–182

Lloyd JW (1994) Groundwater management problems in the developing world. J Appl. Hydrogeol. 4:35–48

Lockwood JG (1974) World climatology: an environmental approach. E. Arnold Publishers, London New York

Macdonald AK (1987) Aspects of spate irrigation in People's Democratic Republic (PDR) Yemen. In: Proc. of sub-regional expert consultation on wadi development for agriculture in the natural Yemen, Aden (PDR), UNDP/FAO, pp 80–90

Mahdi El AE (1996) The impact of rainfall reduction on arid and semi-arid regions-an example: Sudan. M.Sc. Thesis in water and environmental resources management, the Int. Inst. for Infrastruc. Hydr. and Env. Engg, Delft, The Netherlands

Maklawi AH, Abdulla FA (1997) Sedimentation problems in Jordanian reservoirs. In: Proc. Int. Conf. on Wat. Probl. in the Mediter. Countr., vol I. Lefkoşa, North Cyprus, pp 133–140

Management & Implementation Authority of the Great Man-made River Project (1989) The great man-made river project. G. Canale & C.S. p. A. Turin, Italy

Mandelbrot BB, Wallis JR (1968) Noah, Joseph and operational hydrology. Wat. Resourc. Res. 12(6):909–918

Mankarious WF (1979) Hydrology of Lake Qaroun, Egypt. M.Sc. Thesis, Faculty of Engineering, Cairo University, Cairo, Egypt

Margane A, Hobler M, Droubi A, Rajab R, Subah A, Khater AR (2004) Groundwater vulnerability mapping in the Arab countries. In: Zereini F, Jaeshke W (eds) Water in the Middle East and in North Africa, Springer, Berlin London Tokyo, pp 145–155

Margat J (1979) Aridité et resources en eau. CIEH/CEFIGRE. In: Sémin. Intl. Politique de l'Eau pour l'Agr.en Zones Arides et Semi-arides. Niamey, Niger. Also Doc. BRGM 79 SGN 255 HYD, Orléans

Margat J, Saad KF (1984) Deep-lying aquifers: water mines under the desert? Nature and Rresources XX(2):7–13, UNESCO, Paris

Martins O, Probst JL (1991) Biogeochemistry of major African rivers: carbon and mineral transport (Chapter 6). In: Degens ET, Kempe S, Richey JE (eds) Biogeochemistry of major world rivers. SCOPE. John Wiley & Sons Ltd., New York

Martonne E. de (1926) Aréisme et indice d'aridité. C. R. Acad Sci. (Paris) 182:1395–1398

Martyn D, (1992) Climates of the World. Develop. in Atm Sci. 18, Elsevier Publ. Co., Amsterdam New York and PWN, Polish Sc. Publishers

Mason BJ (1962) Clouds, rain and rainmaking. Cambridge at the University Press, Cambridge, UK

Mauritania (1986) Country report on water resources and their utilisations in the Islamic Republic of Mauritania. Document, the Panel on Water Resources and their Utilisations in the Arab Region. ACSAD-AFESD-KFAED, Kuwait

McCabe GJ Jr, Hay LE (1995) Hydrological effects of hypothetical climate change in the East River Basin, Colorado, USA. Hydrol. Sci. J 40(3):303–318

Mehanna AMM (1976) Estimation of potential evapotranspiration over Egypt. Meteo Res. Bull. 8(1):69–93, Cairo, Egypt

Meigs P (1951) La repartition mondiale des zones arides ou semi-arides. UNESCO, NSAZ 37, UNESCO, Paris

——1966. Geography of coastal deserts, Arid Zone Research, UNESCO, Paris

Meijer R (1997) Jordanië (mensen, politiek, ecomomie, cultuur). Royal Institute for the Tropics (KIT), Amsterdam/Novib, The Hague, The Netherlands

Mein RG, Laurenson EM, McMahon RA (1974) Simple non-linear model for flood estimation. Proc. ASCE (HY 11):1507–1518

Meinzer OE (ed) (1942) Hydrology (Physics of the Earth IX). Dover Publications, Inc., New York

Messahel M, Boudjadja A, Hadi-Kaddour B, Khouli RM (1997) Groundwater quality degredation in coastal aquifers, Case study: Nador and Oued Mazafran Aquifers, Algiers Region. Proc. of the Intl. Conf. on Water Problems in the Mediterranean Countries, vol II:1071–1080, Lefkoşa, Turkish Republic of Northern Cyprus

Meteorological Service of Algeria (2002) Meteorological data of Algeria (unpublished reports, obtained through personal communication), Algiers, Algeria

Ministry of Agriculture and Land Reclamation of Morocco (1986) Water resources and their utilisations in the Kingdom of Morocco. Document, the Panel on Water Resources and their Utilisations in the Arab Region, ACSAD-AFESD-KFAED, Kuwait

Ministry of Hajj and Endowments, The Kingdom of Saudi Arabia (1990/1411 H.) The Holy Qur-ān (English translation of the meanings and commentary). King Fahd Holy Qur-ān Printing Complex, Al-Madinah Al-Munawarah, Saudi Arabia

Ministry of irrigation and water resources of the Arab Republic of Egypt, Mott MacDonald International Limited & W.S. Atkins International Limited (1991) Review of crop water requirements. In: Nile water resources management. Cairo, Egypt/Cambridge, England

Ministries of Irrigation, Planning and Housing of Syria (1986) Water resources and their utilisations in the Syrian Arab Republic. Document, the Panel on Water Resources and their Utilisations in the Arab Region, ACSAD-AFESD-KFAED, Kuwait

Ministry of Oil and Mineral Resources of Yemen and The Dutch Institute of Applied Geoscience, TNO, The Netherlands (1995) Water resources assessment in the Yemen Arab Republic (WRAY). Series of reports prepared in the period 1987–91 by the Yemen Ministry of Oil and Mineral Resources, San'a, and the Dutch Institute of Applied Geoscience, Delft, The Netherlands

Ministry of War and Marine (formerly) of Egypt (1950) Climatological normals for Egypt. C. Tsoumas & Co. Press, Cairo, Egypt

Ministry of Water Resources of Oman (1995) Water resources of the Sultanate of Oman: an introductory guide. Muscat, Oman

Misellati MM, Al-Garyani SA, Howas AA (1997) Multi-effect solar desalination. In: Proc. Int. Conf. on Wat. Problems in the Mediterranean countries, vol II. Near East University, Lefkoşa, Turkish Republic of Northern Cyprus, pp 1095–1102

Mogheir Y (1997) Saltwater intrusion modeling and monitoring in the Gaza Strip Aquifer, Palestine. M.Sc. Thesis, International Institute for Infrastructural, Hydraulic and Environmental Engineering (IHE) for the degree of Masters in Hydrological Engineering, Delft, The Netherlands

Mohorjy AM (1988) Water resources management in Saudi Arabia and water use. Wat. Int. 13(3): 161–171

Mohsin A, Obrien W, Bradford M (1995) Hydrotechnical feasibility of recharge dams in eight interior catchments. In: Proc. Int. Conf. on Wat. Resourc. Mgmt. in Arid Countries, vol 1. Muscat, Oman, pp 196–203

Morocco (1986) Country report on water resources and their utilisations in Morocco. Document, the Panel on Water Resources and their Utilisations in the Arab Region. ACSAD-AFESD-KFAED, Kuwait

Morton FI (1979) Climatological estimates of lake evaporation. Wat. Resourc Res. 15(1):64–76

—— (1983) Operational estimates of lake evaporation. J. Hydrol. 66:76–100

—— (1986) Practical estimate of lake evaporation. J Theor. Appl. Climatol. 25:371–387

Mu'Allem AS (1987) Crop production under spate irrigation in coastal areas of PDRY (People's Democratic Republic of Yemen). In: Spate Irrigation, Proc. Sub-regional Expert Consult. organized by UNDP and FAO, Aden (formerly People's Democratic Republic) Yemen, pp 51–57

Müller MJ, Baltes K, Werle D (1980) Handbuch Ausgewählter Klimastationen der Erde. G. Richter Publisher, Trier, Germany

—— (1996) Handbuch Ausgewählter Klimastationen der Erde (expanded and improved print). G. Richter Publisher, Trier, Germany

Murad AA, Krishnamurthy RV (2004) Factors controlling groundwater quality in Eastern United Arab Emirates: a chemical and isotopic approach. J Hydrol. 286:227–235

Murakami M, Musiake K (1994) The Jordan River and the Litani. In: Biswas AK (ed) International waters of the Middle East from Euphrates-Tigris to Nile, Wat. Resourc. Mgmt. Ser. 2:117–155, Oxford University Press, Bombay Delhi Calcutta Madras

Murray GW (1955) Water from the desert: some ancient Egyptian achievements. Geogr. J 121:171–181

Naji F (2004) Water resources management in Palestine: political, technical and financial obstacles. In: Zereini F, Jaescheke W (eds) Water in the Middle East and North Africa. Springer, Berlin New York London Paris Tokyo, pp 239–249

Naqash A, Marrei SH (1983) The occurrence of groundwater resources in arid zone, with emphasis on United Arab Emirates. In: Proc. Int. Sym. on Groundwater in Water Resources Planning, UNESCO-IAH-IAHS, Koblenz, Germany, pp 159–170

Nash I (1957) The form of the Instantaneous unit hydrograph. IAHS Publ No. 45(3):114–121

NEDECO (1975) Flood control study. Final report for mission 2: Development of the Sebou River. Netherlands Engineering Consultants, The Hague, The Netherlands

Neumann J (1953) Energy balance and evaporation from sweet water lakes of the Jordan Rift. Bibl of the Res. Council of Israel II(4):337

Niemcyzyncowicz J (1989) Some examples of important problems connected to rainfall-runoff modeling in semi-arid zones. In: Stout GE, Demissie M (eds) Proc. Sahel Forum "State-of-the-Art of Hydrology and Hydrogeology in the Arid and Semi-arid Areas of Africa, UNESCO/IWRA, pp 255–266

Nouh MA (1988) On the prediction of flood frequency in Saudi Arabia. Proc. Instn. of Civ. Engrs. Part 2, 85:121–144

Nyrop RF (1976) Area handbook for Egypt. DAPam 550–43, U.S. Government Printing Office, Washington D.C.

Okabe T, Amou S, Ishigaki M (1993) A simulation model for sedimentation process in gorge-type reservoirs. IAHS Pub. No. 217:119–126

Oman (1986) Country report on water resources and their utilisations in the Sultanate of Oman. Document (in Arabic), the Panel on Water Resources and their Utilisations in the Arab Region. ACSAD-AFSED-KFAED, Kuwait

Omar MH (1960) Evapotranspiration at Giza. In: Publ. UNESCO Reg. Training Course in Microclimat. for Ecol. and Soil Sc., Cairo, Egypt

Omar MH, el-Bakry MM (1981) Estimation of evaporation from the lake of the Aswan dam (Lake Nasser) based on measurements over the lake. Agr Meteo 23:293–308, Elsevier Publ. Co., Amsterdam

Palestine (1986) Country report on the water situation in the West Bank and Gaza Strip. Document, the Panel on Water Resources and their Utilisations in the Arab Region. ACSAD-AFESD-KFAED, Kuwait

Pallas P (1978) Water resources of the socialist people's Lybian Arab Jamahiriya. Tech. Rep. (TF 9184/79/23), the Food and Agriculture Organisation of the United Nations, Tripoli/Rome

Pallas P (1980) Water resources of the socialist people's Libyan Arab Jamahiriya. 2nd Sym. on the Geology of Libya, vol II. Tripoli, Libya, pp 539–594

Penman HL (1948) Natural evaporation from open water, bare soil and grass. Proc Roy Soc (London) 193:120–145, UK

Pike JG (1964) The estimation of annual runoff from meteorological data in a tropical climate. J Hydrol. 2(2):116–123

Pike JG (1983) Groundwater resources development and the environment in the central region of the Arab Gulf. Water resources Development, 1(2):115–132, Tycooly International Publishing Ltd

Pitt JD, Thompson G (1984) The impact of sediment on reservoir life. IAHS Pub No. 144:541–548

Popper W (1951) The Cairo Nilometer. Studies in Ibn Taghri Birdi's Chronicles of Egypt: I. University of California Publications in Semitic Philology, vol 12. Berkeley and Los Angeles, University of California Press

Postel S (1989) Water for agriculture: facing the limits. Worldwatch Paper 93, Worldwatch Institute, USA

Poulter S (1995) Regional monitoring of unconfined aquifers. In: Proc. Int. Conf. on Water Resources Management in Arid Zones, vol 2. Muscat, Oman, pp 457–465

Preul HC (1994) Rainfall-runoff water harvesting prospects for Greater Amman and Jordan. Wat. Int. 19(2):82–85

Qaisi K (1995) Limited water resources in Jordan. In: Proc. of Int. Conf. on Wat. Resourc. Mgmt. in arid Countries, vol 1. Muscat, Oman, pp 352–359

Qatar (1986) Country report on water resources and their utilisations in Qatar. Document (in Arabic), the panel on water resources in the Arab region and their utilisations. ACSAD-AFESD-KFAED, Kuwait

Quackenbush TH, Phelan JT (1965) Irrigation water requirements of lawns. Paper 4350, Proc. ASCE, J Irr & Dr Div 91(IR 2):11–19

Rákóczi L (1986) Measurement techniques and networks 1986. Iraqi J Wat. Resourc. 5(1):570–598, Baghdad, Iraq

Rendel JC (1992) A dwindling natural resource: Israel relies heavily on water from Arab territories. In: World News, The Washington Post (Wednesday, May 13, 1992) A25–A29

Rekaya ME (1992) La recharge artificielle des nappes du Cap Bon. Actes de la 10ème Journée des resources en eau, DGRE, Tunis

Rekaya ME, Plata Bedmar A (1995) The groundwater artificial recharge by treated wastewater in Tunisia. In: Proc. of Int. Conf. on Wat. Resourc. Mgmt. in Arid Countries, vol 2. Muscat, Oman, pp 657–665

Renner M (1989) National security: the economic and environmental dimensions. Worldwatch Paper 89, Washington: Worldwatch Institute

Revelle R, Waggoner PE (1983) Effects of a carbon dioxide-induced climate change on water supplies in the Western United States. In: Changing climate. Report of the carbon dioxide assessment committee, National Academy Press, Washington, D.C.

Reusing G (1994) Contribution to the risk analysis of hydrologic droughts based on hydrologic time-series of the River Nile. Berliner Geowissenschaftliche Abhandlungen, Reihe D, Band 7, Technical University of Berlin, Berlin

Ritchie JC, Murphy JB, Grissinger EH, Garbrecht JD (1993) Monitoring streambank and gully erosion by airborne laser. IAHS Pub No. 217:161–166

Rochette C (1974) Le Bassin du Fleuve Sénégal. Monograhies Hydrologiques de l'O.R.S.T.O.M. 1, O.R.S.T.O.M., Paris

Rodier J (1964) Régimes hydrologiques de l'Afrique Noir àl'Ouest du Congo. Mém.O.R.S.T.O.M. 6, O.R.S.T.O.M., Paris

—— (1989) General characteristics of surface hydrology in arid and semi-arid areas of Africa and their consequences for development. In: Proc. Sahel Forum: state-of-the-art of hydrology and hydrogeology in the Arid and Semi-arid areas of Africa, Ouagadougou, Burkina Faso, pp 15–24

Rodriguez-Iturbe I, Valdes JB (1979) The geomorphologic structure of hydrologic response. Wat. Resourc Res. 15:1409–1420

Roest CWJ, Abdel-Gawad ST, Abdel-Khalek MA (1993) Water management in the Eastern Nile delta of Egypt, validation of the SIWARE model. Trans. 15th Congress of Int. Com. on Irr & Dr (ICID), vol 1-C, Q 44, R 99, pp 1263–1283, The Hague, The Netherlands

Sadek MF (1992) Evaporation from the reservoir of the High Aswan Dam, Egypt: a comparison of relevant methods with limited data. M.Sc. Thesis, Int. Inst. Infrasruct., Hydr. and Env. Eng. (IHE), Delft, The Netherlands

Sadek MF, Shahin M, Stigter CJ (1997) Evaporation from the reservoir of the High Aswan Dam, Egypt: a new comparison of relevant methods with limited data. J Theor. Appl. Clim. 56:57–66

Sadler IC (1968) Average cloudiness in the tropics from satellite observations. International Indian Ocean Expedition, Meteorological Monograph No. 2, Honolulu: East-West center Press, USA

Salaam A (1966) The groundwater geology of the Gezirah. Thesis, the University of Khartoum, The Sudan, for the Degree of M.Sc., Khartoum, Sudan

Salas JD, Obeysekera JB, Boes DC (1981) Modeling of equatorial lakes outflows. In: Singh VP (ed) Proc. Int. Sym. on Rainfall-Runoff Modeling, Mississippi, Wat. Resourc. Publ., Littleton, Colorado, pp 431–440

Saleh Al MA (1983) Statistical characteristics of rainfall in the Syrian Badyat (Desert) (text in Arabic). Bull. of the Arab Engineer, 68:9–14

Salem OM (1991) The great man-made river project: a partial solution to Libya's water supply. Proc. round table meeting (RTM-91) on Planning for Groundwater Development in Arid and Semi-arid Regions, Cairo: 221–238 (RIGW/IWACO eds.), Cairo/Rotterdam

Salih AM, Sendil U (1984) Evapotranspiration under extremely arid climates. J Irr & Dr Div, ASCE, 110(IR 3):289–303

Saudi Arabia (1986) Country report on water resources and their utilisations in Saudi Arabia. Document (in Arabic), the panel on water resources in the Arab region and their utilisations. ACSAD-AFESD-KFAED, Kuwait

Sayed SA, Al-Ruwaih FM (1995) Relationships among hydraulic characteristics of the Dammam Aquifer and wells in Kuwait. J Hydrogeol. 3(1):57–70

Schmid R, Al-Batashi MB (1995) Water harvesting in the Mountains of Al-Jabal Al-Akhdar by small retention dams. In: Proc. Int. Conf. on Wat. Resourc. Mgmt. in Arid Countries, vol I. Muscat, Oman, pp 152–159

Seckler D, Amarsinghe U, Molden D, De Silva R, Barker R, (1998) World water demand and supply, 1990 to 2025: scenarios and issues. Int. Wat. Mgmt. Inst. (IWMI), Colombo

Semiat R (2000) Desalination: present and future. Wat. Int. 25(1):54–65

Shahin M (1985) Hydrology of the Nile basin. In Developments in Water Science No. 21. Elsevier Publ., Amsterdam Oxford New York Tokyo

Shahin M (1986) Prediction of the concentration of suspended matter in the main Nile between Halfa and Aswan. J Wat. Resourc. 5(1):683–704. Baghdad, Iraq

Shahin M, (1986) Stochastic structure of the annual discharge series of some African rivers. In: Proc. Int. Conf. on Water Resources Needs and Planning in Drought Prone Areas, Khartoum, The Sudan, Part II, pp 189–210

Shahin M (1989) Review and assessment of water resources in the Arab region. Wat. Int. 14(4):206–219

Shahin M (1993) An overview of reservoir sedimentation in some African river basins. In: Proc. Int. Symp. on Sediment Problems: Strategies for monitoring, prediction, and control. Yokohama, Japan, IAHS Pub. No. 217:93–100

Shahin M (1998) Estimation of reference evapotranspiration and crop water use in the Arabian Peninsula. Arab J Sci. and Engg. 23(1c):27–42. King Fahd University for Petroleum and Minerals, Dhahran 31261, Saudi Arabia

Shahin M (1999) Discussion of the article 'Now you see it, now you don't…. A convenient solution/An economic mirage? By Allan JA. The UNESCO Courier, 52(2):29–32, UNESCO, Paris

Shahin M (2002) Hydrology and water resources of Africa. Water Science and Technology Library, vol 41. Kluwer Academic Publishers, Dordrecht Boston London

Shahin M, El-Shal MI (1969) An investigation of the consumptive use of water for crops and the frequency of irrigation in the United Arab Republic (Egypt). Trans. 7th ICID Congress, New Mexico Q 23 (R 2):27–51

Shalabei Al NM, Khalil AD, Al Ek F, Awaidah Y, Eido M, Nahawi S, Nseir R (1996) Implications of expected changes for the Syrian coast. In: Jeftić L, Kečkeš S Pernetta JC (eds) Climatic change and the Mediterranean, vol 2. Arnold, London New York Sydney Auckland

Shalash S, Makary A (1986) Mathematical modeling for sedimentation process in the High Aswan Dam Reservoir. Iraqi J Wat. Resourc. 5(1):654–682, Baghdad

Shantanawi MR, Al-Jayousi O (1995) Evaluating market-oriented water policies in Jordan: a comparative case study. Wat. Int. 20(2):88–97

Shata MA, El-Fayoumy I (1969) Remarks on the hydrogeology of the Nile Delta, U.A.R. In: Proc. Intl. Symp. on Hydrology of Deltas. IAHS/UNESCO, vol. II: 385–396

Shaw EM (1983) Hydrology in practice. Van Nostrand Reinhold (UK) Co. Ltd

Sherif MM, Singh VP, Amer AM (1990) A two dimensional finite element model for dispersion (2D-FED), a model for seawater encroachment in leaky coastal aquifers, J Hydrol. 118:343–356

Shih SF (1984) Data requirement for evapotranspiration estimation. J Irr Dr Div 110 (IR 3), ASCE Paper 19142: 263–274

Shuval H (1992) Approaches to resolving the water conflicts between Israel and its neighbours – a regional water-for-peace plan. Wat. Int. 17(3):133–143

Shuval H (1995) Optimising wastewater recycling and reuse as a sustainable water resource and environmental control strategy in arid countries. In: Proc. of Int. Conf. on Wat. Resourc. Mgmt. in Arid Countries, vol 2. Muscat, Oman, pp 649–656

SIDA (1978) Etude qualitative et quantitative de l'érosion dans le nord et le center de la Tunisie. FAO/SIDA Tunis, Report SIDA/Tun 5/13 No. 83, p 72

Simons M (1967) Deserts: the problem of water in arid lands. Oxford University Press, London

Singh M, Cherchali SA (1995) Vegetation change detection and GIS input into groundwater modeling – an integrated approach. In: Proc. Int. Conf. on Wat. Resourc. Mgmt. in Arid Countries, vol 2. Oman, pp 592–598

Singh VP, Banijklewicz A, Ram RS (1981) Some empirical methods of determining the unit hydrograph. In: Singh VP (ed) Rainfall–Runoff relationship, Proc. Int. Sym. on Rainfall-Runoff Modelling, Mississippi. Wat. Resourc. Pub., Colorado, USA, pp 67–90

Smith GV, Al-Mooji YA (1987) Groundwater development in the Tihama coastal plain. In: Proc. of Sub-regional Expert Consultation on Wadi development for Agriculture in Natural Yemen 'Spate', organised by UNDP and FAO, Aden, (former) People's Democratic Republic of Yemen (PDRY), pp 151–161

Soil Conservation Service, USDA, (1964) A method of estimating volume and rate of runoff in small watersheds. Tech. Pub. 149, Washington, DC, USA

Solaiman AA, Salih MA (1987) Evapotranspiration estimates in extremely arid areas. J Irr & Dr Div ASCE 113(IR 4):565–574

Sorman AU, Abdulrazzak MJ (1993) Infiltration-recharge through wadi beds in arid regions. J Hydrol. Sci. 38(3):173–186

Sorman AU, Qari MY, Hassani MM (1995) Flood peak estimation using satellite information, Case study: Wadi Itwad, southwestern Saudi Arabia. In: Proc. Int. Conf. on Wat. Resourc. Mgmt. in Arid Countries, vol 2. Oman, pp 742–746

Speth P, Christoph M (2004) IMPETUS West Africa: an integrated approach to the efficient management of scarce water resources in West Africa – Case studies from selected river catchments in different climatic zones. In: Zereini F, Jaeschke W (eds) Water in the Middle East and in North Africa, Springer Publishers, Berlin Heidelberg New York London Tokyo, pp 275–286

Stamp LD, Morgan WT (1972) Africa: a study in tropical development. Wiley Inc., New York London Sydney

Stigter CJ, Kisamo EAC (1978) Evaporation determinations by Penman's method: a rational and up-to-date approach for Tanzanian conditions. Univ. Sci. J (4):53–72, Dar es Salam, Tanzania

Starosoloszky Ö (ed) (1987) Applied surface hydrology. Wat. Resourc. Pub., Colorado, USA

Strobl TH (1995) Advanced developments in hydraulic structures for groundwater recharging. In: Proc. Int. Conf. on Wat. Resourc. Mgmt. in Arid Countries, vol 1. Muscat, Oman, pp 137–143

Sudan The (see Hedayat) (1986) Country report on water resources and their utilisations in the Sudan. Document, the Panel on Water Resources and their Utilisations in the Arab Region. ACSAD-AFESD-KFAED, Kuwait

Sullivan C (2001) The potential for calculating a meaningful water poverty index. Wat. Int. 26(4): 471–480

Sutcliffe JV, Parks YP (1999) The hydrology of the Nile. IAHS Pub 5:179

Sundborg Å, Nilsson B (eds) (1985) Qattara hydrosolar power project, environmental assessment. UNGI Report No. 62, Uppsala University, Department of Physical Geography, Lindbergs Grafiska Hb Uppsala, Sweden

Syria (1986) Country report on water resources and their utilisations in the Syrian Arab Republic. Document, the Panel on Water Resources and their Utilisations in the Arab Region, ACSAD-AFESD-KFAED, Kuwait

Takahashi K, Arakawa H (eds) (1981) Climates of Southern and Western Asia, world survey of climatology, vol 9. Elsevier Scientific Publishing Co. Amsterdam Oxford New York

Tarad M (1995) Microbiological aspects of artificial groundwater recharge with treated wastewater (text in Arabic). In: Proc. Int. Conf. on Wat. Resourc. Mgmt. in Arid Countries, vol 2. Muscat, Oman, pp 689–695

Taylor K (1971) The living Bible (paraphrased). Tyndale House Publishers, Great Britain

Thompson CG, Gardner EW (1932) The Prehistoric geography of the Kharga Oasis. Geogr. J LXXX(5):369–409

Thompson G (1986) Reservoir sedimentation. J Wat. Resourc. 5(1):755–774, Baghdad, Iraq

Thornthwaite CW (1948) An approach toward a rational classification of climate. Geogr. Rev. 38(1): 55–94

Toribio IS, Bermúdez FL, del Amor F, León A (1996) Assessment of reference evapotranspiration (E_{to}) in semi-arid Mediterranean climate conditions. J ICID 45(1):1–20

Touaïbia B, Dautrebande S, Dieter G, Aidaoui A (1999) Approche quantitative de l'érosion hydrique à difféntes échelles spatiales: basin versant de l'Oued Mina. J Hydrol. Sci. 44(6):973–986

Toupet C, Pitte JR (1977) La Mauritanie. Presses Universitaires de France (puf), Paris

Toussoun O (1925) Mémoirs sur l'histoire du Nil (trios parties). Mémoirs présentés à l'Institut d'Egypte. vol 8–10. Le Caire, Impremeie de l'Institut Français d'Archéologie Orientale

Trewartha GT (1962) The earth's problem climates. University of Wisconsin Press, Madison

Tunisia (1986) Country report on water resources and their utilisations in Tunisia. Document, the Panel on Water Resources and their Utilisations in the Arab Region. ACSAD-AFESD-KFAED, Kuwait

Turc L (1954) Calcul du bilan de l'eau, évaluation en fonction des précipitations et des temperatures. IAHS Rome Sym, Pub. No. 38:188–202

Ukayli MT, Hussein T (1988) Comparative evaluation of surface water availability, wastewater reuse and desalination in Saudi Arabia. Wat. Int. 13(4):218–225

UNDP (1982) Groundwater in eastern Mediterranean and western Asia. Natural resources/report series of studies on water, No. 9, United Nations Publication ST/ESA/112, New York

UN-ECAFE/WMO (1969) Proceedings of the second symposium on the development of deltaic areas. Water Resources Series No. 39, Tokyo, Japan

UNEP (1993) Poverty and Environment. Information Document for the International Environment Day for the Year 1993 (in Arabic). United Nations Environmental Programme, Regional Office for West Asia, Bahrain, pp 6–7

UNESCO/ROSTAS (1995) Rainfall water management. State of the art report [(Khouri J, Amer A,. Salih A (eds)], ROSTAS, Cairo, p 147

UNESCO (1993) Discharges of selected rivers of the world, vol. II (part II) Monthly and annual discharges recorded at various selected stations, 20-Year Catalogue (1965–84), UNESCO. Paris

UNESCO (1995) Discharges of selected rivers of Africa. Studies and reports in hydrology 52, UNESCO, Paris

UNESCO (1996) Global river discharge data base (RivDIS v1.0), vol I. Africa. Technical Documents in Hydrology, International Hydrological Programme (IHP), UNESCO, Paris

United Arab Emirates (1986) Country report on water resources and their utilisations in The United Arab Emirates. Document, the Panel on Water Resources and their Utilisations in the Arab Region. ACSAD-AFESD-KFAED, Kuwait

United Nations Administration of Technical Cooperation for Development (1982) Groundwater in the Eastern Mediterranean and Western Asia. Natural resources/series of studies on water, Report No. 9 (in Arabic)

U.N. Population Division (1994) Sustaining water: an update. World population prospects, the 1994 revision. The United Nations, New York

United States Agency for International Development (USAID) (1993) Water Resources Action Plan for the Middle East, Bureau for the Near East. USA

USDA/SCS (1967) Irrigation water requirement. Tech. Release 21, Eng. Div. Soil. Cons. Serv., Dept. of Agr., USA

Verhagen BT, Geyh MA, Frőhlich K, With K (1991) Isotope hydrological methods for the quantitative evaluation of groundwater resources in arid and semi-arid areas. Research Reports of the Federal Ministry for Economic Cooperation of the Federal Republic of Germany, vol 81. Bonn/Federal Institute of Geosciences and Mineral Resources (BGR), Hanover, Germany, p 164

Verlinden P (2000) Oman en de Eemiraten aan de Golf. Konninklijk Instituut voor Tropen (KIT), Amsterdam and Novib, Den Haag, The Netherlands

Visser TNM, Wolters W, Smit MFR, Abdel-Khalek M (1993) Effect of reuse on irrigation efficiencies in the eastern Nile Delta, Egypt, using the SIWARE model. Trans. 15th Congress of the Int. Com. on Irr. & Dr. (ICID), vol 1-D, Q 44, R 118, pp 1485–1500, The Hague, The Netherlands

Wakil M, Bonneli R (1996) Salt tolerance of wheat in the semi-arid Khabur Basin Plains, Syria. ICID J 45(1):11–20

Wallén CC (1967) Aridity definitions and their applicability. Geografiska Annale 49:367

Wallén CC (1968) Agricultural studies in the Levant. Proc. of the Reading Sym., UK. UNESCO in Natural Resources Research No. VII, pp 225–233

Walling DE (1984) The sediment yields of African rivers. In: Proc. Harare Sym. on Challenges in Afr. Hydrol. and Wat. Resourc. IAHS Publ. No. 144:265–283

Walling DE (1986) Sediment yields and sediment delivery dynamics in Arab countries: some problems and research needs. Iraqi J Wat. Resourc. 5(1):775–799, Baghdad

Walling DE, Kleo AH (1979) Sediment yields of rivers in areas of low precipitation: IAHS Publ. No. 128:479–493

Wan P (1976) Point rainfall characteristics of Saudi Arabia. Proc. Instn Civ. Engrs, Part 2(61):179–187

Wangnick K (1998) Worldwide desalting plants inventory. No. 15. Report prepared by Wanganick Consulting Co. to the International Desalination Association (IDA), Gnarrenburg, Germany

Wartena L (1959) Het klimaat en de verdamping van een meer in Centraal Irak (The Climate and the evaporation from a lake in Central Iraq; Dutch text). Thesis, the Agricultural University of Wageningen for the degree of Doctorate in Agricultural Sciences, Wageningen, The Netherlands

Water Pollution Control Federation (WPCF) (1983) Water reuse, manual of practice SM-3 prepared by the Technical Practice Committee, Task Force on Water Reuse (Avendt RJ Chairman), Washington D.C., USA, p 118

Water Resources Council (WRC) (1967) A uniform technique for determining flood flow frequencies. Washington DC, USA

Wheater HS, Bell NC (1983) Northern Oman flood studies. Paper 8659, Proc. Instn of Civ. Engrs, Part 2, 75:453–473

Wheater HS, Brown RPC (1989) Limitations of design hydrographs in arid areas – an illustration from southwest Saudi Arabia. Proc. Instn Civ. Engrs, Part 2, 87, Pap. 9461:517–538

Wheater HS, Butler AP, Stewart EJ, Hamilton GS (1991a) A multivariate spatial-temporal model of rainfall in southwest Saudi Arabia. I. Spatial-rainfall characteristics and model formulation. J Hydrol. 125:175–199

Wheater HS, Onof C, Butler AP, Hamilton GS (1991b) A multivariate spatial-temporal model of rainfall in southwest Saudi Arabia. II.Regional analysis and long-term performance. J. Hydrol. 125:201–220

Wheater HS, Jolley TJ, Peach D (1995) A water resources simulation model for groundwater recharge studies: an application to Wadi Ghulaji, Sultanate of Oman. In: Proc. Int. Conf. Wat. Resourc. in Arid countries, vol 2. Oman, pp 502–510

Wigley TML (1992) Future climate of the Mediterranean basin with particular emphasis on changes in precipitation. In: Jeftif L, Milliman JD, Sastini G (eds) Climatic change and the mediterranean. Arnold, London New York Melbourne Auckland

Wilson EM (1983) Engineering hydrology, 3rd edn. MacMillan Education Ltd., London

Wischmeier WH, Smith DD (1965) Predicting rainfall-erosion losses from cropland east of the rocky mountains. USDA, Handbook 282, Agr. Res. Service, USDA, Washington DC

—— (1978) Predicting rainfall-erosion losses: a guide to conservation planning. USDA-SEA, Agric. Handbook 537, Agric. Res. Service, USDA, Washington D.C.

Wischmeier WH, Smith DD (1978) Predicting rainfall-erosion losses. Agriculture Handbook No. 573, US Dept. of Agr., Washington DC, USA

World Health Organisation (WHO) (1989) Guidelines for safe use of wastewater and excreta in agriculture and aquaculture (prepared by Mara D, Carincross S). WHO, Geneva

World Health Organisation Regional Office for Africa (1996) Comoros. In: Water, sanitation and people: partners in development, case studies. First Regional Consultation on Africa 2000 Initiative for Water Supply and Sanitation. Brazzaville, Republic of the Congo

World Guide (1997) A View from the South (1997–1998). New Internationalist Publications Ltd, Oxford OA4 IBW, United Kingdom

World Guide (1999) A View from the South (1999–2000). New Internationalist Publications Ltd., Oxford United Kingdom, 1999

World Meteorological Organisation (1981) Guide to Hydrological Practices, vol. I: Data Acquisition and Processing, WMO-No. 168, WMO Geneva

—— (1966) Measurement and estimation of evaporation and evapotranspiration. WMO-No. 201. TP. 105, WMO, Geneva

—— (1971) Climatological normals (CLINO) for climate and climate ship stations for the period 1931–1960. WMO-No. 117 (reprint 1982), WMO, Geneva

—— (1983) Guide to Hydrological Practices. WMO Tech. Note 126, World Meteorological Organization, Geneva

Wright JL (1982) New evapotranspiration crop coefficients. J Irr Dr Div ASCE 108(IR 2):57–74

World Resources Institute, the United nations Environment Programme, The United Nations development Programme, and the World Bank (1998) World Resources 1998/99. Oxford University Press, New York Oxford

Yemen (YAR) (1986) Country report on water resources and their utilisations in the (Arab Republic of Yemen). Document (in Arabic), the Panel on Water Resources and their Utilisations in the Arab Region, ACSAD-AFSED-KFAED, Kuwait

Yemen (PDRY) (1986) Country report on water resources and their utilisations in the (former) People's Democratic Republic of Yemen. Document (in Arabic), the Panel on Water Resources and their Utilisations in the Arab Region, ASAD-AFSED-KFAED, Kuwait

Yevjevich V (1992) Water and civilization. Wat. Int. 17(4):163–171

Yoffe S, Ward B (1999) Water resources and indicators of conflict-A proposed spatial analysis. Wat. Int. 24(4):377–384

Young PC (1984) Time series Methods and Recursive Estimation, UK

Yousef AS (2004) Water resources, protection and management in Palestine. In: Zereini F, Jaescheke W (eds) Water in the Middle East and North Africa, Springer, Berlin New York London Paris-Tokyo, pp 87–99

Zaki NA, Al-Hadithy AA, Sazonov BI (1970) A study of precipitation in Iraq. Technical Report no.6, Institute of Applied Research on Natural Resources, Abu-Ghraib, Iraq, Government of Iraq/United Nations Special Fund (UNESCO)

Zebidi H (1980) Bilan des sources en eau. Progress Report, Department of Land and Water Resources of Tunisia, Ministry of Agriculture, Tunisia

Zein el-Abedine A, Abdalla MM, Abdel-Al SI (1967) Evapotranspiration studies on maize, UAR (Egypt). In: Proc. Symp. on 'Isotope and Radiation Techniques in Soil Physics and Irrigation Studies'. IAEA/FAO, Istanbul

Zubari WK, Mubarak AM, Madany IM (1993) Development impacts on groundwater resources in Bahrain. Wat. Resourc. Dev. 9(3):263–279

Zubari WK, Madany IM, Al-Junaid SS, Al-Manali S (1994) Trends in the quality of groundwater in Bahrain with respect to salinity, 1941–1992. Env. Int. 20(6):739–746

REFERENCES INDEX

GEOGRAPHICAL INDEX

Sultanate (Sl)/Monarchy (M)/Republic (R)/ State (St)/Emirate (E)

NOTE TO THE READER/USER OF THE BOOK

The present book comprises 12 Chapters dealing with different hydrometeorological and water resources aspects of the Arab Region. These Chapters include maps, graphs and tables containing results obtained from data analyses. The 'raw data' used for producing the 'processed data' have been classified into meteorological, hydrologic and water quality data. The raw data in their respective order form Appendixes I, II and III respectively, all three together occupying 146 pages.

In order to avoid having this heavy bulk of pages containing the raw data placed in one location or dispersed over the 12 Chapters as an integrated part of the book, it has been placed as a separate item on a compact disc (CD Rom) at the back cover of the book.

In order to browse the CD files containing the raw data, the user needs to have a MS EXCEL software installed on his P.C.

It is sincerely hoped that, with the said arrangement, the user will be able to achieve the utmost benefit of the book with a minimum effort.

The author

APPENDIX I: METEOROLOGICAL DATA

CONTENTS OF APPENDIX I

APPENDIX II: HYDROLOGICAL DATA

CONTENTS OF APPENDIX II

APPENDIX III: WATER QUALITY DATA

CONTENTS OF APPENDIX III

Water Science and Technology Library

1. A.S. Eikum and R.W. Seabloom (eds.): *Alternative Wastewater Treatment.* Low-Cost Small Systems, Research and Development. Proceedings of the Conference held in Oslo, Norway (7–10 September 1981). 1982 ISBN 90-277-1430-4
2. W. Brutsaert and G.H. Jirka (eds.): *Gas Transfer at Water Surfaces.* 1984 ISBN 90-277-1697-8
3. D.A. Kraijenhoff and J.R. Moll (eds.): *River Flow Modelling and Forecasting.* 1986 ISBN 90-277-2082-7
4. World Meteorological Organization (ed.): *Microprocessors in Operational Hydrology.* Proceedings of a Conference held in Geneva (4–5 September 1984). 1986 ISBN 90-277-2156-4
5. J. N.ĕmec: *Hydrological Forecasting.* Design and Operation of Hydrological Forecasting Systems. 1986 ISBN 90-277-2259-5
6. V.K. Gupta, I. Rodríguez-Iturbe and E.F. Wood (eds.): *Scale Problems in Hydrology.* Runoff Generation and Basin Response. 1986 ISBN 90-277-2258-7
7. D.C. Major and H.E. Schwarz: *Large-Scale Regional Water Resources Planning.* The North Atlantic Regional Study. 1990 ISBN 0-7923-0711-9
8. W.H. Hager: *Energy Dissipators and Hydraulic Jump.* 1992 ISBN 0-7923-1508-1
9. V.P. Singh and M. Fiorentino (eds.): *Entropy and Energy Dissipation in Water Resources.* 1992 ISBN 0-7923-1696-7
10. K.W. Hipel (ed.): *Stochastic and Statistical Methods in Hydrology and Environmental Engineering.* A Four Volume Work Resulting from the International Conference in Honour of Professor T. E. Unny (21–23 June 1993). 1994
    10/1: Extreme values: floods and droughts ISBN 0-7923-2756-X
    10/2: Stochastic and statistical modelling with groundwater and surface water applications ISBN 0-7923-2757-8
    10/3: Time series analysis in hydrology and environmental engineering ISBN 0-7923-2758-6
    10/4: Effective environmental management for sustainable development ISBN 0-7923-2759-4
    Set 10/1–10/4: ISBN 0-7923-2760-8
11. S.N. Rodionov: *Global and Regional Climate Interaction: The Caspian Sea Experience.* 1994 ISBN 0-7923-2784-5
12. A. Peters, G. Wittum, B. Herrling, U. Meissner, C.A. Brebbia, W.G. Gray and G.F. Pinder (eds.): *Computational Methods in Water Resources X.* 1994 Set 12/1–12/2: ISBN 0-7923-2937-6
13. C.B. Vreugdenhil: *Numerical Methods for Shallow-Water Flow.* 1994 ISBN 0-7923-3164-8
14. E. Cabrera and A.F. Vela (eds.): *Improving Efficiency and Reliability in Water Distribution Systems.* 1995 ISBN 0-7923-3536-8
15. V.P. Singh (ed.): *Environmental Hydrology.* 1995 ISBN 0-7923-3549-X
16. V.P. Singh and B. Kumar (eds.): *Proceedings of the International Conference on Hydrology and Water Resources* (New Delhi, 1993). 1996
    16/1: Surface-water hydrology ISBN 0-7923-3650-X
    16/2: Subsurface-water hydrology ISBN 0-7923-3651-8

Water Science and Technology Library

16/3: Water-quality hydrology ISBN 0-7923-3652-6
16/4: Water resources planning and management ISBN 0-7923-3653-4
Set 16/1–16/4 ISBN 0-7923-3654-2

17. V.P. Singh: *Dam Breach Modeling Technology*. 1996 ISBN 0-7923-3925-8
18. Z. Kaczmarek, K.M. Strzepek, L. Somlyódy and V. Priazhinskaya (eds.): *Water Resources Management in the Face of Climatic/Hydrologic Uncertainties*. 1996 ISBN 0-7923-3927-4
19. V.P. Singh and W.H. Hager (eds.): *Environmental Hydraulics*. 1996 ISBN 0-7923-3983-5
20. G.B. Engelen and F.H. Kloosterman: *Hydrological Systems Analysis*. Methods and Applications. 1996 ISBN 0-7923-3986-X
21. A.S. Issar and S.D. Resnick (eds.): *Runoff, Infiltration and Subsurface Flow of Water in Arid and Semi-Arid Regions*. 1996 ISBN 0-7923-4034-5
22. M.B. Abbott and J.C. Refsgaard (eds.): *Distributed Hydrological Modelling*. 1996 ISBN 0-7923-4042-6
23. J. Gottlieb and P. DuChateau (eds.): *Parameter Identification and Inverse Problems in Hydrology, Geology and Ecology*. 1996 ISBN 0-7923-4089-2
24. V.P. Singh (ed.): *Hydrology of Disasters*. 1996 ISBN 0-7923-4092-2
25. A. Gianguzza, E. Pelizzetti and S. Sammartano (eds.): *Marine Chemistry*. An Environmental Analytical Chemistry Approach. 1997 ISBN 0-7923-4622-X
26. V.P. Singh and M. Fiorentino (eds.): *Geographical Information Systems in Hydrology*. 1996 ISBN 0-7923-4226-7
27. N.B. Harmancioglu, V.P. Singh and M.N. Alpaslan (eds.): *Environmental Data Management*. 1998 ISBN 0-7923-4857-5
28. G. Gambolati (ed.): *CENAS. Coastline Evolution of the Upper Adriatic Sea Due to Sea Level Rise and Natural and Anthropogenic Land Subsidence*. 1998 ISBN 0-7923-5119-3
29. D. Stephenson: *Water Supply Management*. 1998 ISBN 0-7923-5136-3
30. V.P. Singh: *Entropy-Based Parameter Estimation in Hydrology*. 1998 ISBN 0-7923-5224-6
31. A.S. Issar and N. Brown (eds.): *Water, Environment and Society in Times of Climatic Change*. 1998 ISBN 0-7923-5282-3
32. E. Cabrera and J. García-Serra (eds.):*Drought Management Planning in Water Supply Systems*. 1999 ISBN 0-7923-5294-7
33. N.B. Harmancioglu, O. Fistikoglu, S.D. Ozkul, V.P. Singh and M.N. Alpaslan: *Water Quality Monitoring Network Design*. 1999 ISBN 0-7923-5506-7
34. I. Stober and K. Bucher (eds): *Hydrogeology of Crystalline Rocks*. 2000 ISBN 0-7923-6082-6
35. J.S. Whitmore: *Drought Management on Farmland*. 2000 ISBN 0-7923-5998-4
36. R.S. Govindaraju and A. Ramachandra Rao (eds.): *Artificial Neural Networks in Hydrology*. 2000 ISBN 0-7923-6226-8
37. P. Singh and V.P. Singh: *Snow and Glacier Hydrology*. 2001 ISBN 0-7923-6767-7
38. B.E. Vieux: *Distributed Hydrologic Modeling Using GIS*. 2001 ISBN 0-7923-7002-3

Water Science and Technology Library

39. I.V. Nagy, K. Asante-Duah and I. Zsuffa: *Hydrological Dimensioning and Operation of Reservoirs*. Practical Design Concepts and Principles. 2002 ISBN 1-4020-0438-9
40. I. Stober and K. Bucher (eds.): *Water-Rock Interaction*. 2002 ISBN 1-4020-0497-4
41. M. Shahin: *Hydrology and Water Resources of Africa*. 2002 ISBN 1-4020-0866-X
42. S.K. Mishra and V.P. Singh: *Soil Conservation Service Curve Number (SCS-CN) Methodology*. 2003 ISBN 1-4020-1132-6
43. C. Ray, G. Melin and R.B. Linsky (eds.): *Riverbank Filtration*. Improving Source-Water Quality. 2003 ISBN 1-4020-1133-4
44. G. Rossi, A. Cancelliere, L.S. Pereira, T. Oweis, M. Shatanawi and A. Zairi (eds.): *Tools for Drought Mitigation in Mediterranean Regions*. 2003 ISBN 1-4020-1140-7
45. A. Ramachandra Rao, K.H. Hamed and H.-L. Chen: *Nonstationarities in Hydrologic and Environmental Time Series*. 2003 ISBN 1-4020-1297-7
46. D.E. Agthe, R.B. Billings and N. Buras (eds.): *Managing Urban Water Supply*. 2003 ISBN 1-4020-1720-0
47. V.P. Singh, N. Sharma and C.S.P. Ojha (eds.): *The Brahmaputra Basin Water Resources*. 2004 ISBN 1-4020-1737-5
48. B.E. Vieux: *Distributed Hydrologic Modeling Using GIS. Second Edition*. 2004 ISBN 1-4020-2459-2
49. M. Monirul Qader Mirza (ed.): *The Ganges Water Diversion: Environmental Effects and Implications*. 2004 ISBN 1-4020-2479-7
50. Y. Rubin and S.S. Hubbard (eds.): *Hydrogeophysics*. 2005 ISBN 1-4020-3101-7
51. K.H. Johannesson (ed.): *Rare Earth Elements in Groundwater Flow Systems*. 2005 ISBN 1-4020-3233-1
52. R.S. Harmon (ed.): *The Río Chagres, Panama*. A Multidisciplinary Profile of a Tropical Watershed. 2005 ISBN 1-4020-3298-6
53. L.S. Kuchment: *Generation of River-Runoff*. Planned 2007 ISBN 1-4020-3965-4
54. V. Badescu, R.B. Cathcart and R.D. Schuiling (eds): *Macro-Engineering: A Challenge for the Future,* 2006 ISBN 1-4020-3739-2
55. K.-P. Seiler and J.R. Gat: *Groundwater Rechange from Run-off, Infiltration and Percolation*. Planned 2007 ISBN 1-4020-5305-3
56. G. Salvadori, C. De Michele, N.T. Kottegoda and R. Renzo: *Extremes in Nature*. An Approach to Using Copulas. Planned 2007 ISBN 1-4020-4414-3
57. S.K. Jain, R.K. Agarwal and V.P. Singh: *Hydrology and Water Resources of India*. 2006 ISBN 1-4020-5179-4
59. M. Shahin: *Water Resources and Hydrometerology of the Arab Region*. 2007 ISBN 1-4020-4577-8